TROPICAL

FOREST PLANT

ECOPHYSIOLOGY

TROPICAL

FOREST PLANT

ECOPHYSIOLOGY

EDITED BY

STEPHEN S. MULKEY

ROBIN L. CHAZDON

ALAN P. SMITH

CBS PUBLISHERS & DISTRIBUTORS

4596/1 A, 11-DARYAGANJ, NEW DELHI-110002

CBS Pub. ISBN : 81-239-0437-1
Chapman ISBN : 0-412-03571-5

First Indian Edition : 1997

Published by S.K. Jain for CBS Publishers & Distributors,
4596/1A, 11 Darya Ganj, New Delhi - 110 002 (India).

Printed at : Asia Printograph,
Shahdara, Delhi - 110 032

Alan Paul Smith
(1945 – 1993)

On August 26, 1993 Alan Paul Smith died, bringing to a close a remarkable career in plant ecology. Alan's contributions began in high school when he and a companion carried out a biological inventory of wetlands near Madison, New Jersey, in a successful effort to acquire funding to set aside the area as a nature preserve. Over the next thirty years, Alan's research programs expanded to be truly global in scope. His dissertation on tropical alpine plants was completed at Duke under W. D. Billings and included study sites in Central and South America, Hawaii, New Guinea, and Mount Kenya. Alan's later research included African savannas, Panamanian forests, and mountain ranges on two continents. Along the way, he was an inspiration and a consistently positive force for all of the students and field assistants lucky enough to cross his path. His important contributions to the literature are many, but he will perhaps be remembered most for his insightful views of tree biomechanics and the unique long-term studies of his beloved tropical alpine rosette plants. Alan's research on rainforest plants over the last decade at the Smithsonian Tropical Research Institute in Panama was among the most ambitious and multifaceted programs in tropical plant ecology. His visionary approach included the development of a new canopy photo analysis system, experimental treefalls, and most recently, a construction crane for access to the upper canopy. From the demography of the smallest herb in a Smoky Mountain cove forest to the physiology of a rainforest giant, Alan's curiosity and enthusiasm were boundless. This book is dedicated to the memory of Alan Smith, who conceived of this volume but was unable to complete it.

STEPHEN S. MULKEY
TRUMAN P. YOUNG
January 1995

Contents

Contributors

David D. Ackerly
Department of Organismic and
    Evolutionary Biology
Harvard University
Cambridge, MA, USA

Marilyn C. Ball
Research School of Biological
    Sciences
Australian National University
Canberra, Australia

Fakhri A. Bazzaz
Department of Organismic and
    Evolutionary Biology
Harvard University
Cambridge, MA, USA

Jaime Cavelier
Departamento de Ciencias
    Biologicas
Universidad del los Andes
Santa Fe de Bogota, Colombia

Robin L. Chazdon
Department of Ecology
    and Evolutionary Biology
University of Connecticut
Storrs, CT, USA

Phyllis D. Coley
Biology Department
University of Utah
Salt Lake City, UT, USA

Elvira Cuevas
Centro de Ecología
Instituto Venezolano de
    Investigaciones Científicas
Caracas, Venezuela

Frank Ewers, Associate Professor
Department of Botany and
    Plant Pathology
Michigan State University
East Lansing, MI, USA

Ned Fetcher
Department of Biology
University of Puerto Rico
San Juan, Puerto Rico

Christopher B. Field
Carnegie Institution of
    Washington
Stanford, CA, USA

Arthur L. Fredeen
Carnegie Institution of
    Washington
Stanford, CA, USA

Guillermo Goldstein
Department of Botany
University of Hawaii
Honolulu, HI, USA

Thomas Herbert
Department of Biology
University of Miami
Coral Gables, FL, USA

Kevin Hogan
Manaaki Whenua Landcare
  Research
Lincoln, New Zealand

N. Michele Holbrook
Department of Organismic and
  Evolutionary Biology
Harvard University
Cambridge, MA, USA

Paula Jackson
Laboratory of Biomedical and
  Environmental Sciences
University of California Los
  Angeles
Los Angeles, CA, USA

Kaoru Kitajima
Department of Biology
University of Missouri, St. Louis
St. Louis, MO, USA

Thomas Kursar
Biology Department
University of Utah
Salt Lake City, UT, USA

David W. Lee
Department of Biological Sciences
Florida International University
Miami, FL, USA

Ernesto Medina
Centro de Ecología
Instituto Venezolano de
  Investigaciones Científicas
Caracas, Venezuela

Frederick C. Meinzer
Hawaiian Sugar Planter's
  Association
Aiea, HI, USA

Stephen S. Mulkey
Department of Biology
University of Missouri, St. Louis
St. Louis, MO, USA

A. Orozco-Segovia
Centro de Ecología
UNAM
Ciudad Universitaria, Mexico

Robert W. Pearcy
Plant Sciences
University of California, Davis
Davis, CA, USA

Francis E. Putz
Department of Botany
University of Florida
Gainsville, FL, USA

Robert L. Sanford, Jr.
Department of Biological Sciences
University of Denver
Denver, CO, USA

Leonel da Silveira Lobo
  Sternberg
Department of Biology
University of Miami
Coral Gables, FL, USA

Sylvia Strauss-Debenedetti
Sigma Xi, Princeton Chapter
Princeton, NJ, USA

Melvyn T. Tyree
Aiken Forest Science Laboratory
Burlington, VT, USA

C. Vázquez-Yanes
Centro de Ecología
UNAM
Ciudad Universitaria, Mexico

Klaus Winter
Smithsonian Tropical Research
  Institute
Balboa, Republic of Panama

S. J. Wright
Smithsonian Tropical Research
  Institute
Balboa, Republic of Panama

Truman P. Young
Calder Center
Fordham University
Armonk, NY, USA

Gerhard Zotz
Smithsonian Tropical Research
  Institute
Balboa, Republic of Panama

Acknowledgements

Much of the research presented in this book has been facilitated by two premier research institutions: the Smithsonian Tropical Research Institute and the Organization for Tropical Studies. We thank these institutions and the U. S. National Science Foundation for providing the facilities, a stimulating research climate, and financial support that has enabled progress in tropical plant ecophysiology, particularly in the Neotropics. We gratefully acknowledge the anonymous reviewers of these chapters.

Preface

Tropical forests pose a daunting but exceedingly stimulating challenge to plant physiological ecologists. In 100 m² of lowland tropical forest in Costa Rica live 233 vascular plant species, including 5 canopy tree species, 102 species of woody seedlings, 30 species of vines and lianas, and 59 species of epiphytes (Whitmore, Peralta & Brown, 1985). This overwhelming taxonomic, morphological, and functional diversity begs the fundamental question: how do all of these species manage to coexist? How do they partition or compete for resources? How do species differ in their responses to resource heterogeneity, herbivory, or vagaries of the environment? Does functional diversity reflect taxonomic or morphological diversity?

In 1977, faced with the imminent threat of massive conversion of the world's tropical forests to agricultural development, the National Research Council of the U. S. National Academy of Sciences appointed a committee to develop research priorities in tropical biology. The committee's recommendations broadly included increased emphasis on studies of plant physiological ecology, particularly in relation to ecosystem properties and forest dynamics (NRC, 1980). In their review of the state of tropical plant ecophysiology, Mooney et al. (1980) called attention to the general lack of information on resource availability and distribution in tropical forests and emphasized the need to study how tropical species respond to their unique environmental conditions. At the same time, Bazzaz and Pickett (1980) provided a critical working framework for studies of tropical forest plant ecophysiology within the context of forest regeneration and gap dynamics. These key developments ushered in a new era of tropical ecophysiological research.

As fledgling graduate students during 1980, we witnessed the tremendous growth of ecophysiological studies in the tropics. Increased funding opportunities for research in tropical plant ecophysiology coupled with the development of tropical field stations provided major impetus for this growth. During this period, two intern; tional symposia fostered interaction among researchers from different backgrounds, perspectives, and geographic regions, leading to new syntheses of pioneering studies (Medina, Mooney & Vázquez-Yanes, 1984; Clark, Dirzo & Fetcher, 1987). Through these developments, we have acquired a more detailed understanding of microenvironmental variability within and among tropical forests, allowing for more sophisticated field-based studies of plant responses to environmental heterogeneity (Chapter 1). Improved canopy access now makes virtually any tropical leaf a target for gas exchange measurements (Chapter 3). The time is approaching when computer storage capacity, rather than field time, will pose the major limitation to data acquisition in the field.

Tropical plant ecophysiology has achieved the stature of a mature scientific discipline, and with this maturity comes a new sense of purpose. The 21 chapters of this book synthesize the major developments in the field during the last ten years. Most of these developments have been directly linked to technological advances that have brought the study of physiological ecology out of the laboratory and into the hot, humid, buggy forest. Applications of pressure-volume techniques initially developed for desert species are giving us new insights into the importance of seasonal drought stress for species of evergreen tropical forests (Chapter 7). Scaling up carbon gain and water flux from individual leaves to whole trees and stands, an unthinkable exercise 15 years ago, is now an attainable goal (Chapter 4).

The development of our field has also brought about an important paradigm shift. During the 1980's, gap dynamics was the backdrop for virtually all ecophysiological studies in tropical as well as temperate forests (Bazzaz & Pickett, 1980; Bazzaz, 1984). This focus led to important advances in our understanding of specific physiological traits in relation to regeneration requirements of herbs, shrubs, and tree species (Chapter 6). But, as Fetcher, Oberbauer and Chazdon (1994) point out in a recent review, no single physiological trait fits this theoretical construct in a satisfactory way. Functional traits are subject to structural and developmental constraints that vary widely among life forms and life history stages. Most species of tropical woody plants are neither obligate shade species nor pioneers, but occupy an intermediate position along the shade-tolerant/light-demanding continuum. A new paradigm has emerged that focuses on functional links among traits in relation to ecological patterns of distribution and abundance of species, rather than on a simple (often unsubstantiated) categorization of regeneration mode. The leaf-level approach has given way to a broader approach focusing on the integration of traits at the organismal level. Chapters 19 and 21 of this volume illustrate the power and broad applicability of this synthetic approach. For large and long-lived species, such as canopy trees, a whole-plant approach poses enormous methodological and theoretical challenges.

The chapters of this book signal many exciting future developments in tropical forest ecophysiology. We emphasize four frontiers of research. First, we envision a rapid vertical expansion of studies: upwards into the canopy and downwards beneath the soil surface. Within the forest canopy, we need to know more about environmental heterogeneity within individual tree crowns, physiological integration within and among tree branches, and the role of epiphytes, hemiepiphytes, vines, and lianas in carbon, water, and nutrient fluxes within the forest ecosystem. Active research in the tropical forest canopy is needed to provide critical data linking canopy chemistry to leaf physiology, spectral reflectance, nutrient cycling, and remote sensing. As discussed in Chapter 10, belowground processes in tropical forests remain poorly studied. Particular areas of priority for study are mycorrhizal relations, allocation of carbon and nutrients to roots, and root architecture in relation to water and nutrient acquisition. These new areas of investigation will enable more detailed approaches to scaling up leaf-level processes to the whole-plant and stand level.

Second, we need to explore in far greater detail the nature of functional links among traits across the entire spectrum of tropical species. This effort will require large-scale synthetic studies focused on particular life forms, life-history stages, or ecological groups of plants. The approach of Reich, Walters and Ellsworth (1992) focusing on leaf longevity is noteworthy and should be expanded to include other sets of functionally-linked traits. Ontogenetic changes in functional links as well as developmental constraints on trait expression during different life-history stages deserve more detailed, comparative study.

The linkage between physiology and demography of tropical plants constitutes a third focus for future work. This linkage is particularly challenging for tropical trees lacking growth rings. Long-term studies are shedding light on complex life-history patterns of tropical trees in relation to light availability and carbon metabolism (Clark & Clark, 1992). Nevertheless, we know remarkably little about ecophysiological determinants of mortality, growth, and reproductive effort for most tropical species. New approaches are needed to integrate short-term ecophysiological measurements with the significantly larger spatial and temporal scale of critical demographic processes. Demographic studies based on the use of Geographic Information Systems (GIS) offer much promise in linking underlying physiological responses to species distributions and demographic processes.

A fourth frontier lies in studies of intraspecific genetic diversity in quantitative characters and of ecotypic differentiation in tropical species. Although we are gaining some understanding of genetic structure and mating systems of tropical tree populations (Hamrick & Loveless, 1986; Bawa & O'Malley, 1987; Murawski et al., 1990), to our knowledge, there has been only a single published study documenting ecotypic differentiation in ecophysiological traits of a tropical forest plant species (Hogan et al., 1994; Chapter 17). Ecophysiologists and foresters share a common goal in assessing the genetic basis for functional traits in tree species appropriate for management and reforestation. Studies of species with distributions spanning wet and dry tropical forests or broad elevational ranges may reveal significant ecotypic differentiation in phenology, growth patterns, and underlying physiological traits that potentially influence processes at the community, ecosystem, and landscape level.

As we confront the reality of deforestation and land degradation in tropical regions, we must seriously consider how advancements in tropical plant ecophysiology can contribute to the design of ecologically sound forest management, agroforestry, conservation, and restoration projects. Research decisions, such as the choice of species to investigate, study area and habitat, collaborative arrangements, and access to project results, should consider the basic research needs of applied fields such as forestry, wildlife management, and watershed management. If, through our research, we can provide essential ecophysiological background for local resource managers and conservation practitioners, we can maximize the practical application of our science.

Plant ecophysiology is by nature a synthetic field. The studies described in this book emphasize the linkage of ecophysiological patterns with processes affecting populations, communities, ecosystems, and landscapes in the tropics. Although we are proud of the progress that this book illustrates, each chapter points out the gaps in our knowledge and suggests new directions for research. We encourage all readers to use the work described here as a point of embarkation for new research voyages.

REFERENCES

BAWA K. & O'MALLEY, D. M. (1987) Estudios genéticos de sistemas de cruzamiento en algunas especies arboreas de bosques tropicales. *Revista de Biologia Tropical*, **35 (Suppl. 1)**, 177–188.

BAZZAZ, F. A. (1984) Dynamics of wet tropical forests and their species strategies. *Physiological Ecology of Plants of the Wet Tropics* (eds. E. MEDINA, H. A. MOONEY & C. VÁZQUEZ-YANES) Dr. Junk Publishers, The Hague, pp 233–244.

BAZZAZ, F. A. & PICKETT, S. T. A. (1980) Physiological ecology of tropical succession: A comparative review. *Annual Review of Ecology and Systematics*, **11**, 287–310.

CLARK, D. A. & CLARK, D. B. (1992) Life history diversity of canopy and emergent trees in a Neotropical rain forest. *Ecological Monographs*, **62**, 315–344.

CLARK, D. A., DIRZO, R. & FETCHER, N. (eds). (1987) Ecología y Ecofisiología de Plantas en los Bosques Mesoamericanos. *Revista de Biología Tropical* **35**, (Suppl. 1).

FETCHER, N., OBERBAUER, S. F. & CHAZDON, R. L. (1994) Physiological ecology of plants at La Selva. *La Selva: Ecology and Natural History of a Neotropical Rainforest* (eds. L. McDADE, K. S. BAWA, H. HESPENHEIDE & G. HARTSHORN) University of Chicago Press, Chicago, pp 128–141.

HAMRICK, J. L. & LOVELESS, M. D. (1986) Isozyme variation in tropical trees: Procedures and preliminary results. *Biotropica*, **18**, 201–207.

HOGAN, K. P., SMITH, A. P., ARAUS, J. L. & SAAVEDRA, A. (1994) Ecotypic differentiation of gas exchange responses and leaf anatomy in a tropical forest understory shrub from areas of contrasting rainfall regimes. *Tree Physiology*, **14**, 819–831.

MEDINA, E., MOONEY, H. A. & VÁZQUEZ-YÁNES, C. (eds.) (1984) *Physiological Ecology of Plants of the Wet Tropics*. Dr. W. Junk Publishers. The Hague.

MOONEY, H. A., BJÖRKMAN, O., HALL, A. E., MEDINA, E. & TOMLINSON, P. B. (1980) The study of the physiological ecology of tropical plants – Current status and needs. *Bioscience*, **30**, 22–26.

MURAWSKI, D. A., HAMRICK, J. L., HUBBELL, S. P. & FOSTER, R. B. (1990) Mating systems of two Bombacaceous trees of a Neotropical moist forest. *Oecologia*, **82**, 501–506.

NRC (1980) *Research Priorities in Tropical Biology*. Washington, D. C., National Academy of Sciences.

REICH, P. B., WALTERS, M. B. & ELLSWORTH, D. S. (1992) Leaf lifespan in relation to leaf, plant, and stand characteristics among diverse ecosystems. *Ecological Monographs*, **62**, 365–392.

WHITMORE, T. C., PERALTA, R. BROWN K. (1985) Total species count in a Costa Rican tropical rainforest. *Journal of Tropical Ecology*, **1**, 375–378.

I

Resource Acquisition

INTRODUCTION

Tropical forests are unique among terrestrial forest ecosystems because of the way in which resources are distributed within them. In this section we present reviews of the influence of variation in spatial and temporal distributions of resources on the physiology and, where there are data, the growth and reproduction of tropical species. Three broad classes of resources are considered: light, water, and soil. Of these, light shows the strongest variation and influence on the characteristics of tropical plants. Tropical forests contain some of the darkest and brightest terrestrial habitats as one views the continuum of irradiance from the forest floor to the top of the canopy. Near the equator, irradiance levels during the clear dry season often considerably exceed 2200 μmol m^{-2} s^{-1} at the top of the forest canopy. Values near the forest floor are routinely less than 1 percent of the available sunlight, which may be particularly low during the cloudy rainy season. Similarly, most tropical forests show some seasonal variation in rainfall. Despite high annual totals, the dry season for many tropical forests is a time with extended periods of little or no rain. High annual rainfall, warm temperatures, and the year-round activity of organisms results in soil nutrient regimes that differ markedly from other regions. In general, primary productivity and nutrient cycling rates are high on an annual basis.

Over the past decade, photosynthesis of tropical forest species has received considerable attention because of the extreme variation in light in these habitats and the importance of this process for understanding plant, stand, and ecosystem function. In Chapter 1, Chazdon et al. review spatial and temporal variation of light, photosynthetic acclimation, and photosynthetic variability in contrasting light environments. Despite the central role of leaves, these studies show that leaf-level variation alone does not adequately describe the distribution and abundance of tropical species with respect to irradiance regime. C_3 plants, which dominate much of the tropics, are the focus of most of this work, although there are habitats where other photosynthetic pathways are highly functional. Medina (Chapter 2) reviews the photosynthetic characteristics of CAM and C_4 species and their respective advantages as epiphytes and as herbs in seasonal tropical habitats. Recent advances in canopy access and instrumentation have resulted in some of the first comprehensive data on diurnal and seasonal variation in photosynthesis of canopy leaves (Zotz & Winter, Chapter 3).

Meinzer and Goldstein (Chapter 4) focus on the constraints to scaling gas exchange from single leaves to the whole plant and canopy levels. Although this chapter focuses on stomatal conductance to water vapor, we include it in the section on photosynthesis because of the implications of this approach for understanding limitations to carbon gain. Meinzer and Goldstein argue that success in linking leaf gas exchange to individual, stand, and regional levels depends on the ability to collect compatible data concurrently at more than one spatial reference point in the canopy. A possible solution to acquiring such data is to model photosynthetic gain of contrasting canopy sections with leaves of known (or estimated) orientation and position. In Chapter 5, Herbert provides examples of the effect of leaf position on carbon gain. The great diversity of leaf form and function in tropical plants has always required explanation. Strauss-Debenedetti and Bazzaz (Chapter 6) evaluate the hypothesis that species such as gap colonists, which experience great environmental heterogeneity, have evolved greater physiological flexibility than, for example, shade tolerant plants that spend their lives in the relatively constant understory.

Plant growth and reproduction in tropical forests are usually coordinated in some fashion with seasonal variation in rainfall. Despite the high annual rainfall, limitations to photosynthesis during drought can constrain the whole-plant carbon budget, resulting in reduced growth and reproduction, and not infrequently, death (Mulkey & Wright, Chapter 7). In general, plants with higher annual biomass turnover are more sensitive to carbon limitations during drought, especially in the light-limited understory. Limits to continued carbon gain in the face of drought are partially determined by rooting depth and the resistance to water movement, an area that until recently has received very little attention. Tyree and Ewers (Chapter 8) show how hydraulic architecture of tropical trees and lianas influences the flow of water from roots to leaves. Many tropical species are deeply-rooted and evade the negative effects of drought on gas exchange by maintaining high hydraulic conductances during drought. Those with shallow roots frequently experience water limitations to metabolism and control water loss through stomatal closure. Goldstein et al. (Chapter 9) explore how stable isotopes can be used to determine the depth from which water is being drawn, and to obtain an integrated, long-term index of the frequency and duration of stomatal closure.

Root function and acquisition of soil nutrients are the least-studied areas of tropical forest ecophysiology. It is apparent that our understanding of plant carbon budgets cannot be complete without knowledge of the seasonal and spatial distribution of carbon in roots. Roots of tropical trees are the most diverse in morphology and habitat, and thus their potential effect on the whole-plant carbon budget is highly variable from species to species. Sanford and Cuevas (Chapter 10) examine the possible controls on allocation of carbon to roots and mechanisms of root function in tropical systems. Studies of root turnover, respiration and carbon budget responses to environmental variation in tropical forests are just beginning.

1

Photosynthetic Responses of Tropical Forest Plants to Contrasting Light Environments

Robin L. Chazdon, Robert W. Pearcy,
David W. Lee and Ned Fetcher

Across the complex matrix of microsites that compose tropical forests, light availability varies more dramatically than any other single plant resource. On a sunny day, instantaneous measurements of photosynthetically active radiation range over 3 orders of magnitude, from less than 10 μmol m^{-2} s^{-1} in closed-canopy understory of mature forests to well over 1000 μmol m^{-2} s^{-1} in exposed microsites of gaps and large clearings, or at the top of the forest canopy (Chazdon & Fetcher, 1984b; Figure 1.1). Among the environmental factors that influence plant growth and survival in tropical forests, light availability is likely to be the resource most frequently limiting growth, survival, and reproduction (Chazdon, 1988; Fetcher, Oberbauer & Chazdon, 1994). Photosynthetic utilization of light is therefore a major component of the regeneration responses of forest species within the larger context of forest dynamics and succession.

In this chapter, we explore relationships within and among tropical forest species in photosynthetic responses to these contrasting light

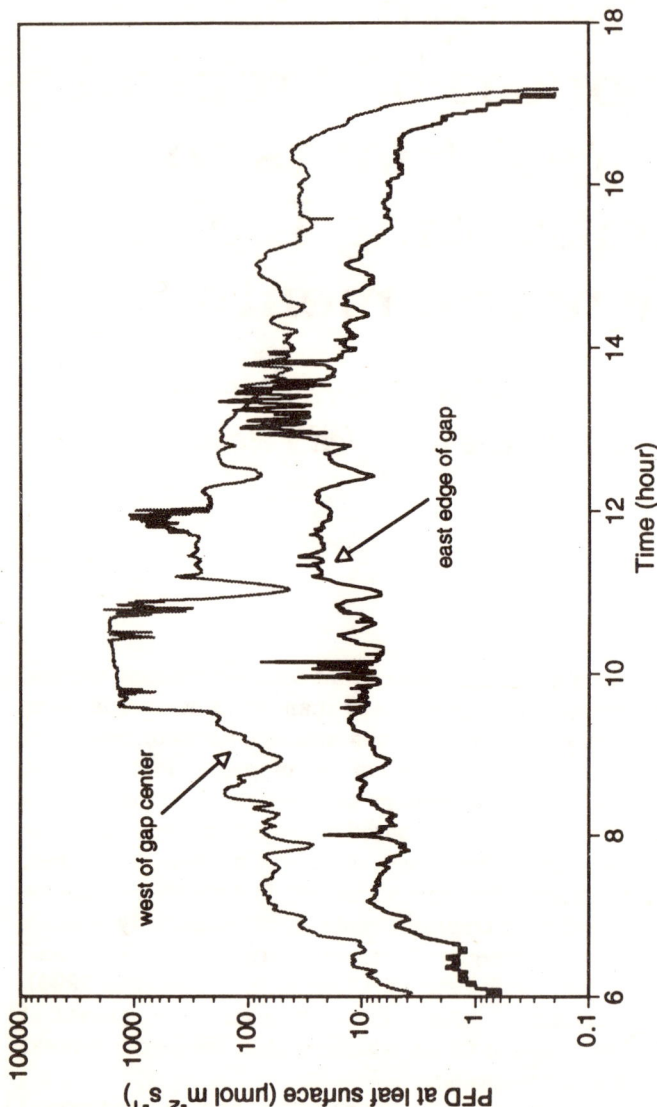

Figure 1.1. Time-course of instantaneous photon flux density (PFD) incident on leaves at two positions along a transect through a 350 m² canopy gap in a Costa Rican rainforest for November 2, 1990. Daily PFD for the sensor on the east gap edge was 0.55 mol m⁻² d⁻¹ (1.55% of full sun), whereas the sensor west of center received 10.51 mol m⁻² d⁻¹ (29.61% of full sun). Sensor readings were taken every 5 s.

environments. We first describe spatial and temporal patterns of light variation in tropical forest habitats in relation to forest dynamics. We then discuss photosynthetic differences among species that differ in their ecological distribution. Finally, we examine the effects of light availability on the expression of photosynthetic variability within a species. We conclude with a discussion of photosynthetic light responses in relation to regeneration patterns of tropical plants. Although tropical forest species illustrate a wide range of photosynthetic light responses and acclimation capacities, our synthesis suggests that a consideration of these leaf-level responses alone does not clearly segregate tropical plants according to their ecological distribution, growth requirements, or successional status. Species differences in photosynthetic light responses are linked with variation in a suite of related plant traits, including maximum growth potential, leaf turnover and longevity, and leaf and canopy structure. For a broader discussion of these linkages, we refer the reader to chapters by Strauss-Debenedetti and Bazzaz (Chapter 6), Mulkey and Wright (Chapter 7), Fredeen and Field (Chapter 20), and Ackerly (Chapter 21).

1.1 LIGHT VARIATION IN TROPICAL FOREST HABITATS

Our understanding of light environments in tropical forests has progressed significantly during the past two decades, as theory and techniques initially developed in agricultural meteorology have been increasingly applied to forest ecosystems (Lemeur & Blad, 1974; Norman & Welles, 1983; Myneni, Ross & Asrar, 1989; Pearcy, 1990). Additionally, improvements in instrumentation have facilitated accurate field measurements at ecologically significant sample sizes and locations. These innovations include the development of sensors calibrated for photosynthetically active radiation (Biggs et al., 1971; Gutschick et al., 1985), the commercialization of field spectroradiometers for the characterization of the spectral distribution of radiation, and the use of microprocessors for the collection of large data sets (Gutschick et al., 1985).

For the purposes of photobiology, solar radiation is measured as a photon flux density (PFD), the numbers of quanta (moles or Einsteins of photons) arriving at a unit surface (1 m^2) per unit time (1 s or 1 day) (Pearcy, 1989). The range of biological processes affected by solar radiation depends upon the bandwidth considered. For example, UV-A

(320–380 nm) and UV-B (280–320) nm may damage molecules essential to photosynthesis, or may contribute to photoinhibition. The bandwidth 400–700 nm has been defined as photosynthetically active radiation (PAR). Roughly 50% of the energy of global solar radiation (300–3000 nm) lies within this photosynthetically active waveband (Lee & Downum, 1991). Blue (400–500 nm), red (600–700 nm), and far-red (700–800 nm) wavelengths may indirectly influence photosynthesis through effects on photomorphogenesis. The most important shift in spectral quality, because of its effect on phytochrome equilibrium, is undoubtedly the red:far-red ratio (R:FR), expressed as the quantum ratio in the 655-665 nm waveband to that in the 725–735 nm waveband (Smith, 1982). Here we describe patterns of variation of photon flux density (PFD) and R:FR ratio among understory, gap, and clearing habitats in tropical forests throughout the world.

1.1.1   Variation above the forest canopy

Sunlight arriving at the forest canopy or in a large clearing is influenced by atmospheric conditions, season (solar declination), and time of day (solar elevation). At tropical latitudes, the northernmost and southernmost positions of the sun at mid-day are within zenith angles of 23.5°. The high solar elevation angles produce greater average and maximum solar flux than at higher latitudes. Typical PFD values at midday frequently reach instantaneous values of 1800–2200 $\mu mol\ m^{-2}\ s^{-1}$ (Figure 1.1). Irradiances up to 2600 umol $m^{-2}\ s^{-1}$ have been recorded from the canopy (S. Mulkey, personal communication). Diurnal and seasonal patterns of cloud formation profoundly affect the radiation environment; cloud formation typically reduces PFD by 75% or more (Lee & Downum, 1991). In the absence of cloud cover, the daily photons received may exceed 47 mol $m^{-2}\ d^{-1}$ (Lee and Downum, 1991; Lee, 1989; Raich, 1989).

Annual variation in photoperiod in the tropics varies from 1-3 hr, depending on latitude. The extent of seasonal variation in daily PFD depends on the interaction between solar angles with seasonal changes in atmospheric conditions. In Miami, Florida, just north of the tropic of Cancer, winter daily total (10.5 hr photoperiod) reached 21.5 $\mu mol\ m^{-2}\ day^{-1}$, with a peak PFD of 1520 $\mu mol\ m^{-2}\ s^{-1}$. The highest daily summer totals (13.5 hr photoperiod) were 43.1 mol m^{-2} d^{-1}, with peak PFD of 2224 $\mu mol\ m^{-2}\ s^{-1}$ (Lee & Downum, 1991). Measurements in an open site in Malaysia (5° 28'N) over a nine-month period, however, showed no apparent seasonal variation, with a mean daily PFD of 31.3 mol m^{-2} d^{-1} (Raich, 1989). In a large clearing of a Costa Rican rainforest, mean daily PFD varied significantly across seasons, with a 24% higher daily total during the dry season (Chazdon & Fetcher, 1984b). Over a 65-day period from November to March, daily PFD in another clearing at this location ranged from 6.9 to 46.1 mol m^{-2} d^{-1}, with an average of 27.9 (Oberbauer et al., 1989). Over an entire year, mean monthly values of daily PFD in this clearing reached a minimum of 23-24 mol m^{-2} d^{-1} in July and August, and a maximum of 32 mol m^{-2} d^{-1} in April (Rich et al., 1993).

Atmospheric conditions can influence the spectral distribution of radiation in open habitats. A water vapor-absorbing band centered on 728 nm significantly alters R:FR ratios (Gorski, 1976). Red:far-red ratios of unfiltered sunlight generally exceed 1.0 (Lee & Downum, 1991; Turnbull & Yates, 1993). The red:far-red ratio increased from 1.05 to 1.48 with increasing relative humidity and decreasing solar elevations in Miami, FL (Lee & Downum, 1991).

1.1.2 Understory light regimes

Solar radiation is spectrally altered by passage through the canopy, affecting both the quantity and the quality of photosynthetically active radiation in the understory (Figure 1.2). Leaves reflect and transmit little of the visible wavelengths (400-700 nm), but most of the far-red wavelengths (> 700 nm). The greatest spectral change after canopy filtering, therefore, is the reduction in R:FR (Stoutjestijk, 1972, Lee 1987; Turnbull & Yates, 1993; see Endler, 1993 for a broader discussion of spectral quality). Mean R:FR of background radiation (PFD = 11 µmol m^{-2} s^{-1}) was 0.40 at La Selva, Costa Rica and 0.35 (PFD = 19 µmol m^{-2} s^{-1}) at Barro Colorado Island, Panama, whereas mean canopy level ranged from 1.22 to 1.33 (Lee, 1987). Over the course of an entire year, the mean R:FR ratio in an Australian subtropical forest was 0.37, although daily mean values ranged from 0.2 to 0.5 (Turnbull & Yates, 1993). Lee (1987, 1989) demonstrated a log-linear relationship between percent of full sun PFD and R:FR, although this relationship is considerably more complex when more extensive sampling is undertaken (Turnbull & Yates, 1993). The lowest R:FR values (< 0.20) occur immediately underneath leaves, areas unaffected by the spectrally neutral contributions of diffuse sky and penumbral radiation. Crowns of three canopy tree species showed similar shifts in spectral quality (Lee, 1989). Reflection from objects in the understory, particularly tree trunks, may further reduce the R:FR of nearby plants (Lee & Richards, 1991). Within the understory, the most remarkable shifts in spectral quality occur during the transition from background radiation to sunflecks, when R:FR ratios may approach that of direct sunlight (Chazdon & Fetcher, 1984a; Lee, 1987; Turnbull & Yates, 1993; Figure 1.2).

Within closed-canopy understory sites, small holes in the canopy contribute to penumbral effects from the solar disk (Miller & Norman, 1971; Smith, Knapp & Reiners, 1989). At typical canopy heights of 30 m, holes less than 0.5° wide do not allow exposure of the entire solar disk, reducing the PFD of solar radiation penetrating to the forest floor. Larger holes allow increased penetration of diffuse skylight and, potentially, the incidence of sunflecks of relatively increased duration and peak PFD (Chazdon, 1988). The resulting spatial and temporal heterogeneity creates a challenge for measuring light conditions beneath tropical forest canopies (Reifsnyder, Furnival & Horovitz, 1971; Chazdon, 1988; Rich et al., 1993; Baldocchi & Collineau, 1994).

One useful approach to quantifying understory light environments involves estimates of diffuse and direct light penetration, based on computerized analyses of digitized hemispherical photographs (Anderson, 1964; Chazdon & Field, 1987b; Becker, Erhart & Smith, 1989; Rich, 1989; Rich et al., 1993; Whitmore, 1993). The earlier history of this technique was reviewed by Chazdon (1988). This technique facilitates assessment of long-term light conditions within understory and gap sites, and is most valuable when used to compare different sites within a single geographic area. Hemispherical photographs can yield good predictions of monthly PFD, particularly when separate weighting factors are used for diffuse and direct site factors (Rich et al., 1993).

For purposes of examining leaf carbon gain, however, direct measurement using photon-sensitive sensors is a more suitable technique. Although this task is magnified by the large samples needed for adequate documentation of a site, the availability of inexpensive PAR sensors and microprocessor-based data storage and reduction have made such research possible. Chazdon and Fetcher (1984a) reviewed the earlier research in tropical forests. More recent and thorough studies based on PAR sensors include Costa Rica (Chazdon & Fetcher, 1984a,b; Chazdon, 1986; Oberbauer et al., 1988, 1989; Rich et al., 1993), Mexico (Chazdon, Williams & Field, 1988), Panama (Smith, Hogan & Idol, 1992), Australia (Pearcy, 1987; Turton, 1988, 1990; Yates, Unwin & Doley, 1988; Turnbull & Yates, 1993), Sumatra (Torquebiau, 1988), Sri Lanka (Ashton, 1992), Malaysia (Raich, 1989; Whitmore, 1993), Puerto Rico (Fernandez & Fetcher, 1991a), Hawaii

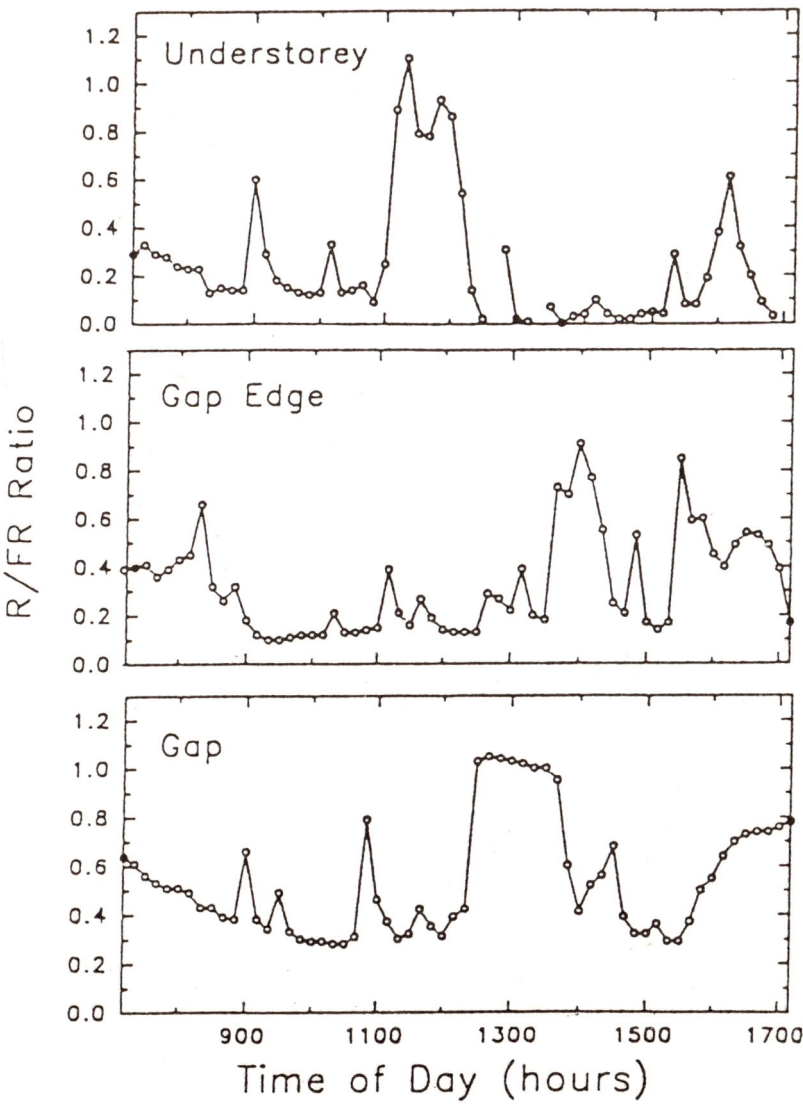

Figure 1.2.   Time course of variation in 10-min average R:FR ratio for three rainforest sites on December 22, 1990. Each 10-min average is based on 10 instantaneous sensor readings at 1-min intervals. (From Turnbull & Yates, 1993).

(Pearcy, 1983), India (Lee, 1989; Lee & Paliwal, 1988); and the Ivory Coast (Alexander, 1982).

Understory light environments consist of low levels of diffuse irradiance, punctuated by brief sunfleck events (Figures 1.1, 1.2 and 1.3). Diffuse radiation in the understory results from transmission of light through leaves, reflection from leaves and woody surfaces, diffuse skylight, and penumbral light. Within closed-canopy and gap-edge microsites, diffuse radiation predominates during the day; more than 90% of PFD readings are below 25 μmol m^{-2} s^{-1} (Pearcy, 1987; Chazdon, 1988; Oberbauer et al., 1989; Turnbull & Yates, 1993). Typical values of diffuse PFD in mature forest understory range from 5 to 25 μmol m^{-2} s^{-1}, depending on weather conditions and forest structure (Torquebiau, 1988). Under cloudy skies, a greater portion of photons is contributed by diffuse sky radiation, and the fraction of canopy radiation penetrating to the understory increases (Chazdon & Fetcher, 1984b). Although sunflecks may occur very briefly, their contribution to the total quantum flux can be substantial. On days with some sunny periods, 10-85% of the daily photons received may be contributed by sunflecks exceeding 50 μmol m^{-2} s^{-1} (Pearcy, 1987; Chazdon, 1988). Peak PFD during brief sunflecks, particularly those less than 1 min duration, is often much less than full-sun PFD because of penumbral effects of small canopy openings (Chazdon, 1986; Chazdon & Pearcy, 1991). Sunflecks of longer duration, although much less frequent, may exhibit peak PFD approaching full-sun irradiances. The frequency of sunflecks is influenced by patterns of cloud cover, which is typically greater in the afternoons at most tropical sites and greater during the wet season. At lower solar elevations, such as during early morning and late afternoon, or during winter at the latitudinal limits of the tropics, the probability of sunfleck occurrence is reduced (Smith, Hogan & Idol, 1992). Sunfleck frequency is strongly correlated with daily PFD in forest understory microsites (Chazdon, Williams & Field, 1988; Chazdon & Pearcy, 1991).

Within the understory, sunfleck activity occurs on a spatial scale smaller than most plant crowns; different leaves within the same shrub or sapling often receive dramatically different light environments (Chazdon, Williams & Field, 1988; Oberbauer et al., 1988, 1989). Analysis of spatial autocorrelation has revealed that there is little probability of sensors (as close as 15 cm apart) receiving the same sunfleck and the same diurnal pattern of PFD (Chazdon, Williams & Field, 1988). Microsite variation in daily total PFD within the understory is largely due to differences in sunfleck incidence, rather than differences in background diffuse PFD (Chazdon & Pearcy,

1991). More conservative estimates of light environments based on analysis of hemispherical photographs have shown similar spatial patterns of variation (Becker & Smith, 1990). Mature forest understories, however, exhibit high spatial autocorrelation of light environments as assessed by hemispherical photographs at distance intervals of 2.5 m or less (Becker & Smith, 1990).

Few measurements of understory light regimes have been described for tropical secondary forests. Mean PFD decreased from 815 μmol m^{-2} s^{-1} in one-year-old second growth to less than 400 μmol m^{-2} s^{-1} four years after abandonment in sites near San Carlos de Rio Negro, Venezuela (Ellsworth & Reich, 1995). In a nine-year-old second-growth stand, mean PFD further decreased to 57 μmol m^{-2} s^{-1}, similar to readings in a 200 m^2 gap in nearby mature forest. In secondary tropical wet forests 7–16 years old, diffuse PFD levels in the understory may exceed those measured in closed-canopy microsites of old-growth stands by as much as 25%, although daily PFD levels are similar (Chazdon, unpublished data).

1.1.3 Gap light regimes

Openings in the forest canopy dramatically affect the light environment (Fernandez & Fetcher, 1991a). These openings may be relatively restricted (branch or tree fall, selective logging) or extensive (landslide, hurricane damage, or clearcutting). As the size of the canopy opening increases, the incidence of higher overall and peak irradiances at ground level increases accordingly (Brown, 1993). Mean daily PFD increased from 3.4 to 10.6 mol m^{-2} d^{-1} in the center of gaps in a Costa Rican forest as gap size increased from 71 m^2 to 615 m^2 (Barton, Fetcher & Redhead, 1989). The relationship between PFD and gap size is not linear; microclimates are more sensitive to changes in gap size among small gaps than among large gaps (Brown, 1993). Dramatic changes in PFD within gaps can occur over relatively small spatial scales (Chazdon & Fetcher, 1984b; Chazdon, 1986; Pearcy, 1987; Chazdon, Williams & Field, 1988; Raich, 1989; Chazdon, 1992). Mean PFD in a Malaysian forest varied from less than 30 to over 350 μmol m^{-2} s^{-1} across a distance less than 25 m along a transect through a gap (Raich, 1989). Within a large gap in a Costa Rican forest, mean daily PFD ranged from less than 1 mol m^{-2} d^{-1} at the east gap edge to over 10 mol m^{-2} d^{-1} only 14 m to the west in the gap center (Fig. 1.1). Although instantaneous PFD may reach full-sun irradiances for periods of up to several hours, integrated daily PFD in the center of relatively large gaps rarely exceeds 50% of full-sun values (Chazdon, 1992; Figures 1.1 and 1.4). The influence of a gap may extend far beyond the immediate gap area, as shown by elevated PFD values in understory locations up to 20 m from a gap edge (Canham et al., 1990; Raich, 1989), although this effect has not always been observed (Rich et al., 1993).

Variation in photosynthetic light regimes within and among gaps is clearly seen by examining the daily frequency distribution of different PFD intervals (Chazdon & Fetcher, 1984b; Chazdon, 1986; Chazdon, Williams & Field, 1988). Within a Costa Rican forest, less than 0.5% of daily PFD measurements in the center of a small gap (approx. 50 m^2) exceeded 500 μmol m^{-2} s^{-1}. In contrast, more than 15% of the daily PFD measurements exceeded this threshold in the center of a large gap (approx. 350 m^2) (Chazdon, 1992). Gap light regimes exhibit substantial yearly variation due to changes in both solar angle and weather conditions (Rich et al., 1993). Mean monthly PFD in a small gap within a Costa Rican forest (94 m^2) reached a major peak in April, a minor peak in September, and a minimum in December and January (Rich et al., 1993). Dramatic annual variation in light availability in the center of two Malaysian forest gaps is also described by Raich (1989). Compared to seasonal variation in daily mean PFD, spectral quality remained fairly constant across the year in gap and gap-edge microsites in a Queensland rain forest (Turnbull & Yates, 1993). Over an entire year, the mean R:FR ratio in gap and gap-edge microsites was 0.512 and 0.540, respectively (Figure 1.2; Turnbull & Yates, 1993).

Distinctions between gap and understory environments exaggerate more subtle differences in light environments that may occur among these sites (Lieberman, Lieberman & Peralta, 1989; Smith, Hogan & Idol, 1992). Processes of gap regeneration, recovery from large-scale forest damage by landslides and hurricanes, and forest regrowth from abandonment of plantations or pastures have yet to be examined in detail with respect to long-term changes in photosynthetic light environments. Because of the dynamic nature of tropical forests, light environments within a microsite are almost certain to change dramatically at least once during the lifespan of most species (Smith, 1987). Smith, Hogan and Idol (1992) document this long-term variation at Barro Colorado Island, Panama, based on analyses of canopy photographs taken over a two-year period along a 2.5 km transect. Their analysis suggests that changes in light environments within microsites may occur at a far more rapid time scale than changes in plant growth and reproduction, creating substantial time lags in whole-plant response to light availability.

1.1.4 Vertical light gradients

Depending on the structure and distribution of foliage in the forest canopy, pronounced vertical gradients of PFD and spectral quality occur within the forest. Yoda (1974) documented such a light gradient in Malaysia, showing an exponential decrease of PAR from the primary canopy at 30 m to the forest floor. In an Australian rainforest, the PFD incident on leaves decreased from 40 mol m^{-2} d^{-1} at the top of an *Argyrodendron peralatum* crown to 5 mol m^{-2} d^{-1} at the bottom of the crown (Doley, Unwin & Yates, 1988). In terms of integrated and peak PFD, leaves at the bottom of the tree's crown experienced irradiances similar to those measured in the center of small to medium canopy gaps. The vertical distribution of PFD as measured from a tower may differ considerably from measurements made from within tree crowns using a crane (Parker, Smith & Hogan, 1992). Although poorly documented, knowledge of these vertical gradients will help in understanding the physiology and ecology of species that encounter different light environments at different stages in their life histories (Chapters 6 and 21). Theoretical effects of canopy structure on vertical light gradients have been modeled by Pukkala et al. (1991). Torquebiau (1988) described effects of canopy structure on vertical gradients in PAR at two Sumatran sites.

1.2 PHOTOSYNTHESIS IN THE RAINFOREST UNDERSTORY

The shaded understory of a tropical forest presents a particular challenge for photosynthetic acquisition of sufficient energy and carbon to support growth and survival. The strong limitation by light is illustrated in the CO_2 assimilation (A) of seedlings of the Australian tropical forest tree, *Argyrodendron peralatum*, in the understory (Figure 1.3 B & C). This type of diurnal response is broadly representative of tropical tree seedlings (Pearcy & Calkin, 1983) as well as understory herbs (Björkman, Ludlow & Morrow, 1972). During most of the day when only diffuse light (10-20 μmol m^{-2} s^{-1}) was incident, A was very low (<1 μmol CO_2 m^{-2} s^{-1}). Under these conditions, A was less than 10% of the maximum light saturated value, well within the linear initial portion of the light response curve (Figure 1.3 A). Carbon assimilation was strongly dependent on sunflecks, which were present for only 16 minutes of the day, but provided 38% of the daily PFD.

Utilization of these sunflecks accounted for 32% of the carbon gain (Pearcy, 1987). In microsites where sunflecks are more abundant, sunflecks could be expected to account for an even larger fraction of the daily carbon gain of understory plants (Chazdon, 1988). Stomatal conductances (g_s), although low, were high enough that they provided little limitation to assimilation except during the brightest sunflecks (Figure 1.3 D).

Maximizing photosynthesis under shade light requires maximizing the amount of light absorbed and the quantum yield for CO_2 uptake, while minimizing respiratory carbon losses. Indeed, understory plants have been shown to have extremely low light-compensation points, ranging from 1–5 µmol photons m^{-2} s^{-1}, primarily because of their very low rates of dark respiration (Björkman, 1981; Chazdon, 1986; Sims & Pearcy, 1991; Fredeen & Field, 1991; Ellsworth, 1991). The requirement for low respiratory losses in deeply shaded environments, however, restricts maximum photosynthetic capacity to relatively low levels (Sims & Pearcy, 1991; Chazdon, 1992). Constraints on photosynthetic capacity and associated low PFD required for light saturation may further restrict assimilation during bright sunflecks, particularly in gap-edge or gap microsites (Chazdon & Pearcy, 1986b).

1.2.1 Maximizing light absorption and quantum yield in the shade

Maximizing light absorption depends on both leaf and whole plant properties. Increases in chlorophyll content per unit area cause a semilogarithmic increase in absorptance of PAR. The localization of the chlorophyll in chloroplasts causes greater transmissivity than found in a chlorophyll solution, but this is offset by the high internal scattering which increases the effective pathlength of the light (Björkman, 1981). Despite the greater priority expected for maximizing light absorption in the shade as compared to the sun, chlorophyll concentrations per unit area do not differ significantly. Chlorophyll content per chloroplast is much higher in leaves of shade plants relative to sun plants (Anderson, Chow & Goodchild, 1988). However, the number of chloroplasts per unit area of leaf is often lower in the shade leaves (Chow et al., 1988). While this at first seems surprising in view of the priority for maximizing light capture, there is a diminishing return for increasing chlorophyll concentrations per unit area. Most leaves already absorb 80 to 85% of the available light; a further doubling of chlorophyll concentration only increases absorption by 3 to 6% (Björkman, 1981; Evans, 1988).

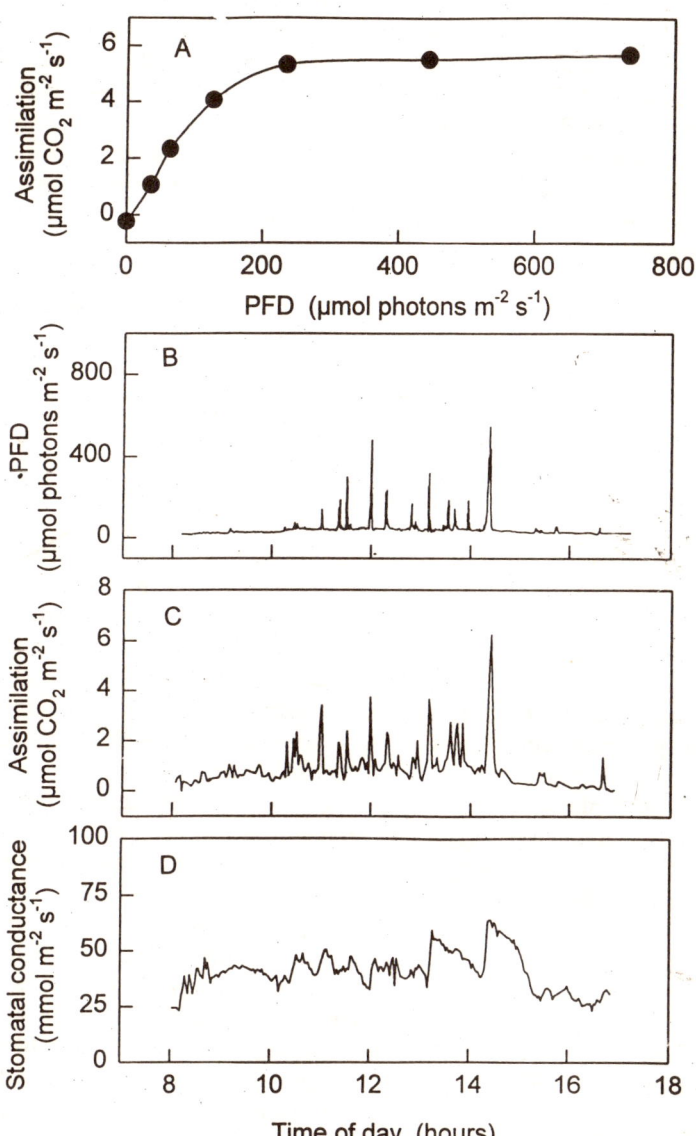

Figure 1.3.   The dependence of steady-state assimilation rate on PFD for leaves of Agyrodendron peralatum *seedlings in the understory of a tropical forest in Northeastern Australia (panel A). Panels B, C, and D show the natural course of PFD, net CO_2 assimilation, and stomatal conductance, respectively, for a comparable leaf measured on September 13, 1983. (Based on Pearcy, 1987).*

Shade leaves are relatively enriched in chlorophyll b, as shown by the low chlorophyll a/b ratio that is characteristic of understory plants (Björkman, 1981; Chow et al., 1988; Chazdon, 1992; see also Chow, Adamson & Anderson, 1991). This enrichment is due to the extensive grana stacking present in chloroplasts of these leaves. These stacks are where the light harvesting complex II, containing most of the chlorophyll b, is localized (Anderson, Chow & Goodchild, 1988). Extensive grana stacking may function to increase chlorophyll and hence light harvesting per chloroplast. Decreased chlorophyll a/b has been suggested to be a chromatic adaptation to help balance the absorption of light between the two photosystems (Anderson, 1986), but its overall consequences may be small (Evans, 1988). Photosystem stoichiometry is known to be shifted to higher photosystem I/photosystem II ratios in rain-forest plants (Chow, Anderson & Melis, 1990), but the overall consequences of these shifts in quantum yield for photosynthesis in the understory have not been examined.

There is probably relatively little scope for an adaptive enhancement of quantum yield in understory plants via modification of photosynthetic reaction efficiency. Substantial differences in quantum yields have sometimes been reported between sun and shade plants, especially when curve-fitting procedures are used, but these may be biased by curvature in the response at relatively lower PFD in shade as compared to sun plants. Moreover, stresses such as photoinhibition usually reduce the quantum yield, preventing a realistic comparison in sun and shade plants. When careful comparisons are made of unstressed plants in white light and CO_2-saturating conditions, the quantum yields of both sun and shade plants approach the theoretical maximum set by the stoichiometry of photosynthesis (Björkman & Demmig, 1987).

The quantum yield of shade plants can, however, be enhanced by the environmental conditions of the understory. Quantum yield in C_3 plants is strongly CO_2 dependent because oxygen and CO_2 compete with each other as substrates for the primary CO_2-fixing enzyme, ribulose-1,5-bisphosphate carboxylase/oxygenase, Rubisco (Ehleringer & Björkman, 1987). When the oxygenase reaction occurs, leading to photorespiration, quantum yield is reduced because of the energy loss. High CO_2 concentrations near the forest floor (Sternberg, Mulkey & Wright, 1989) partially offset this reduction. Moreover, the high values of g_s relative to A yields high intercellular CO_2 pressures, further enhancing the quantum yield (Pfitsch & Pearcy, 1989). Pearcy (1987) calculated that this effect could enhance quantum yields by about 14%. At a PFD of 15 μmol m^{-2} s^{-1} and with a respiration rate of 0.2 µmol CO_2 m^{-2} s^{-1}, values that are typical for understory environment and plants, this enhancement will increase A by about 10%. High stomatal conductances of understory plants in the shade are, in part, due to stomatal responses to high humidity characteristic of plants in the rainforest understory (Mooney et al., 1983).

1.2.2 Photosynthetic capacity of understory plants

The light-saturated photosynthetic capacity of a leaf (A$_{max}$) primarily comes into play in the understory when sunflecks occur. Even during sunflecks, however, maximum PFD is often below 500 µmol m^{-2} s^{-1} (Chazdon & Pearcy, 1991). Light saturation in understory plants is often attained at PFD from 200–500 µmol m^{-2} s^{-1} (Chazdon, 1986; Fetcher, Oberbauer & Chazdon, 1994).

Studies of both tree seedlings and understory plants show that A at light saturation ranges between 1 and 5 µmol CO_2 m^{-2} s^{-1} (Fetcher, Oberbauer & Chazdon, 1994). These rates are from 5 to 50 percent of the typical values measured for open-habitat tropical plants or canopy leaves on trees. The reason for the wide variation among shade species is unknown, although different life forms (tree seedlings, shrubs, herbs, and vines) do not seem to differ significantly in photosynthetic capacity in the shade. Within and among species growing in the understory of wet tropical lowland forests, A$_{max}$ varies linearly with leaf nitrogen content on both an area and mass basis (Chazdon & Field, 1987a), although statistical relationships among A$_{max}$, N, and leaf specific mass vary across species (Reich et al. 1994; Reich & Walters, 1994).

Variation in light spectral quality, although influencing the morphology of high-light species grown in the shade, has no consistent effect on photosynthetic rates of tropical plants. One study (Kwesiga & Grace, 1986) reported higher photosynthetic rates for tree seedlings grown in simulated shadelight enriched in far-red radiation as compared to neutral shade, while Turnbull (1991) reported an opposite effect. In a third study, seedlings of three tropical tree species showed no effect of enriched far-red radiation on either photosynthetic rates, apparent quantum yields, or sunfleck utilization (Tinoco-Ojanguren & Pearcy, 1995). There is now abundant evidence that the low A$_{max}$ of shade plants results from low concentrations per unit area of photosynthetic enzymes and electron transport carriers (Anderson & Osmond, 1987). A primary reason for the low concentrations is the construction of thin leaves with low cell volumes per unit area and low chloroplast numbers per unit area (Chow et al., 1988; Sims & Pearcy, 1989;

Chazdon & Kaufmann, 1993). These area-based measures may well be more important determinants of photosynthetic capacity than variation in the concentration of enzymes per cell or per chloroplast, although the latter also varies among species. A key advantage of reduced concentrations of enzyme concentrations and electron transport components is the low cost of construction and maintenance respiration (Williams, Field & Mooney, 1989; Sims & Pearcy, 1991).

1.2.3 Photosynthetic utilization of sunflecks

As discussed above, sunflecks provide increased PFD for brief periods lasting from a few seconds to 10 minutes or longer. In extreme shade plants, longer sunflecks may cause photoinhibition (Le Gouallec & Cornic, 1988; Le Gouallec, Cornic & Blanc, 1990). Photoinhibition during sunflecks, however, may not ultimately cause much reduction in carbon gain unless it is so severe that repair overnight is incomplete, which is likely only in gaps (Mulkey & Pearcy, 1992). Photosynthetic utilization of sunflecks depends not only on the steady-state characteristics that determine the equilibrium photosynthetic rates at any PFD, but also on the dynamic properties of the photosynthetic apparatus. These dynamic characteristics determine the induction requirement of photosynthesis that limits the capacity to utilize sunflecks, as well as the capacity for post-lightfleck CO_2 assimilation that can enhance their utilization under some circumstances (Chazdon & Pearcy, 1991).

Leaves can potentially respond rapidly to a sunfleck and reach their light-saturated photosynthetic rate within a few seconds. However, the realization of this potential is influenced by the induction requirement that causes photosynthesis to increase much more slowly in response to a sunfleck following a long period of deep shade. The induction requirement is a consequence of slow stomatal opening as well as light activation requirements of photosynthetic enzymes. In particular, light-regulation of Rubisco, which increases activity over a 5–10 min period after a light increase, has been shown to be important in limiting photosynthesis during induction (Seemann et al., 1988). Shading causes the induction requirement to slowly develop again because of stomatal closure and deactivation of enzymes. The loss of induction is, however, much slower than the gain because both stomatal closure and the decrease in Rubisco activity are slower than the corresponding increases (Chazdon & Pearcy, 1986a; Pearcy, 1988; Woodrow & Mott, 1988; Poorter & Oberbauer, 1993). Induction can be carried over and actually increase between sunflecks in a series because of this hysteresis (Chazdon & Pearcy, 1986a; Poorter & Oberbauer, 1993).

Rates of induction are fairly rapid for leaves of shade-tolerant species measured in the forest understory (Küppers & Schneider, 1993; Kursar & Coley, 1993; Poorter & Oberbauer, 1993). Initial stages of photosynthetic induction (induction to 50% of maximum photosynthetic rates) occurred within 1–3 min for 8 species from Barro Colorado Island (Kursar & Coley, 1993). The final stage of induction varied greatly among species, however, ranging from 3-5 min for species with short-lived leaves, to 10–36 min for species with long-lived leaves (Kursar & Coley, 1993). Rates of induction were significantly longer in potted plants compared to field-grown individuals of the same species (Kursar & Coley, 1993). Rates of induction (time to 90% A_{max}) in sapling leaves of the late-successional tree *Dipteryx panamensis* were faster in the morning (16 min) than in the afternoon (25 min; Poorter & Oberbauer, 1993). Induction rates of the pioneer species *Cecropia obtusifolia* also varied diurnally, ranging from 9 min in the morning to 43 min in the afternoon. Following exposure to low light, induction loss followed a logarithmic decline, but at a faster rate for *Cecropia*. For both species, induction was nearly completely lost within one hour of low light (Poorter & Oberbauer, 1993). Induction state following a sequence of five 30-s lightflecks was negatively correlated with indirect site factor in *Dipteryx* saplings. These findings suggest that the capacity for carbon gain during sunflecks may be more highly developed for species adapted to deep shade (Poorter & Oberbauer, 1993).

Stomatal responses to light and humidity have been shown to enhance sunfleck utilization, as well as increase quantum yield in low PFD because of resulting higher intracellular CO_2 pressures. The hysteretic response of stomatal conductance to light increases causes a delay of 15–20 min in stomatal opening in response to a short sunfleck (Kirschbaum, Gross & Pearcy, 1988). In the understory shrub *Piper aequale*, opening in response to a sunfleck caused little increase in carbon gain during that sunfleck, but substantially improved carbon gain during subsequent sunflecks (Tinoco-Ojanguren & Pearcy, 1992). The high relative humidity characteristic of the understory of tropical forests is important in this regard because of decreased rates of stomatal closure (Mooney et al., 1983; Tinoco-Ojanguren & Pearcy, 1992).

Post-illumination CO_2 fixation (PICF) following a sunfleck results from the buildup of metabolite pools in the chloroplast during the sunfleck proper (Pearcy, 1990; Sharkey, Seemann & Pearcy, 1986). These high energy pools are used to support continued CO_2 fixation at a gradually diminishing rate as they are depleted. In uninduced leaves, where the rate of utilization of the pools is low chiefly because of low Rubisco activity, PICF may continue for a minute (Chazdon & Pearcy, 1986b). In contrast, PICF lasts for less than 10 seconds in induced leaves. Nevertheless, for short sunflecks (< 10 s), PICF may constitute a large fraction of the total photosynthesis and may therefore significantly enhance the utilization of sunflecks both under induced and uninduced conditions (Chazdon & Pearcy, 1986b). Since short sunflecks make only a small contribution to the available PFD in tropical forest understories, however, the absolute contribution of PICF to daily carbon gain is also small (Pearcy et al., 1994). PCIF may be more important in short canopies with many rapid sunflecks due to windy conditions, as is characteristic of some Hawaiian forests (Pearcy, 1983), or in microsites of gap-edges and small gaps within high forests.

Simulation studies indicate that under natural sunfleck regimes, induction may reduce daily carbon gain of *Alocasia* in the understory by up to 25% over that expected if there were no induction requirement (Pearcy et al., 1994). Thus, there should be strong selection pressure for minimizing induction limitations. Induction appeared to be less limiting for sunfleck use in the understory shrub *Piper aequale* than in the pioneer species *Piper auritum,* when both were grown in the shade (Tinoco-Ojanguren & Pearcy, 1992).

Little information is available beyond the scale of a day with respect to the overall importance of sunfleck utilization to the carbon balance of tropical understory plants. Because weather conditions greatly influence the frequency of sunflecks, their contribution on a daily basis measured on clear days is an overestimate for an annual basis. Moreover, photosynthetic responses to light are most sensitive to variations at low PFD than at high (near-saturating) PFD. A strong correlation has been shown, however, between the estimated potential duration of sunfleck activity and growth of tree seedlings in forest understories, indicating that their photosynthetic contribution may be substantial on long time scales (Pearcy, 1983; Oberbauer et al., 1988). Other studies suggest that both direct and diffuse components of light availability are key determinants of sapling growth (Oberbauer et al., 1993).

1.3 PHOTOSYNTHESIS IN GAPS, CLEARINGS, AND TREE CANOPIES

Since 1980, much progress has been made in understanding photosynthetic responses of tropical plants under naturally variable conditions within gap, clearing, and canopy habitats. These advances are significant in view of the logistical problems that plague field measurements of photosynthetic responses, such as rainy weather, transporting equipment, canopy access, and variable light conditions. Although it may be possible to simulate gap conditions in laboratory or controlled-environment experiments (Wayne & Bazzaz, 1993), these studies are limited in their potential for revealing the full complexity of photosynthetic responses to varying spatial and temporal light availability within and among clearings in tropical forests.

1.3.1 Photosynthetic capacity and dark respiration

Although few tree species have been studied in detail, photosynthetic capacity as measured by light-saturated rates of photosynthesis at optimal leaf temperature and ambient CO_2 concentration (A_{max}) ranges from 5 µmol CO_2 m^{-2} s^{-1} in the case of canopy leaves of *Castanospermum australe* and 6.5 µmol CO_2 m^{-2} s^{-1} for *Pentaclethra macroloba* to 11.5 µmol CO_2 m^{-2} s^{-1} for *Argyrodendron peralatum* (Oberbauer & Strain, 1986; Pearcy, 1987; Doley, Unwin & Yates, 1988). For shrubs in the genera *Miconia* and *Piper*, A_{max} ranged from 4.0 to 14.5 µmol CO_2 m^{-2} s^{-1} (Walters & Field, 1987; Denslow et al., 1990; Chazdon, 1992).

A four-fold range in A_{max}/area was observed among seedlings of six tropical tree species grown under growth-chamber or greenhouse conditions at a constant PFD of 500 µmol m^{-2} s^{-1} (Oberbauer & Strain, 1984). Seedlings of the early-successional pioneer *Ochroma lagopus* attained maximum rates of 27.8 µmol CO_2 m^{-2} s^{-1}; CO_2 uptake was not light-saturated until nearly 1500 µmol m^{-2} s^{-1}. Seedlings of early-successional trees, such as *Bursera simaruba* and *Hampea appendiculata*, reached A_{max} of 11.8 and 13.9 µmol CO_2 m^{-2} s^{-1}, respectively. Species that typically regenerate beneath mature forest canopies, such as *Pentaclethra macroloba*, achieved low A_{max} of 5–6 µmol CO_2 m^{-2} s^{-1}; CO_2 uptake was nearly light-saturated at a PFD of 500 µmol m^{-2} s^{-1} (Oberbauer & Strain, 1984). Photosynthetic capacity (area basis) varied three-fold among five tree species in sites spanning 1 to 10 years following secondary succession on abandoned slash-and-burn sites near San Carlos, Venezuela (Ellsworth & Reich, 1995). Light-saturated rates ranged from less than 5 μmol CO_2 m^{-2} s^{-1} in *Licania heteromorpha* 10 years after abandonment to over 20 μmol CO_2 m^{-2} s^{-1} for *Cecropia ficifolia* during the first year after abandonment. Differences among species were even more pronounced when comparing A_{max} on a leaf dry mass basis (Reich, Ellsworth & Uhl, 1995). It is not clear, however, whether these differences reflect species differences or differences in light availability across the successional chronosequence examined. Decreases in A_{max} during the first 10 years of succession reflect changes in species composition as well as plastic responses to decreasing resource availability (Reich et al., 1994; Ellsworth & Reich, 1995; Reich, Ellsworth & Uhl, 1995).

Few data are available on variation in rates of photosynthesis within and among tree canopies (Chapter 3). Leaves of some canopy species exhibit mid-day depression of photosynthesis (Roy & Salager, 1992). Net photosynthesis of leaves at the top of a lowland rain forest canopy in Cameroon reached a peak of 10–12 μmol CO_2 m^{-2} s^{-1} during mid-morning and commonly declined through midday, only occasionally recovering late in the day (Koch, Amthor & Goulden, 1994). Leaves within different positions of the canopy of *Argyrodendron peralatum* in an Australian rainforest did not appear to vary in photosynthetic light responses (Doley, Unwin & Yates, 1988), although Pearcy (1987) reports that canopy leaves of this species attained nearly two-fold higher A_{max} than understory leaves. Increases in A_{max} from the understory to the canopy were correlated with increases in leaf conductance to water vapor, leaf nitrogen content, and leaf mass per area. Similar differences were observed between canopy and understory leaves of the shade-tolerant species *Castanospermum australe* and the gap-requiring species *Toona australis* (Pearcy, 1987).

Because of cloudiness and shading, however, PFD within canopy and gap environments is often below the maximum required to saturate photosynthesis (Figure 1.1). Thus, Oberbauer and Strain (1986) observed that most values of PFD measured in conjunction with photosynthesis in the canopy of a lowland forest in Costa Rica were less than 400 μmol m^{-2} s^{-1}. On the other hand, Doley, Unwin and Yates (1988) observed that leaves in the upper crown of an *Argyrodendron peralatum* tree received approximately 70% of the daily PFD received by a horizontal sensor above the canopy. On a cloudless day, most of the foliage in the upper portion of this crown received PFD exceeding saturating irradiances (above 500 μmol m^{-2} s^{-1}) for most of the day.

The effect of water deficits in the soil or atmosphere on photosynthesis in high-light environments needs further study (Mulkey & Wright, Chapter 7). Pearcy (1987) found little evidence to suggest that atmospheric humidity deficits limited photosynthesis within canopies of *Agryrodendron peralatum*. As evidence for limitation by soil moisture deficits, Doley, Unwin and Yates (1988) reported reductions in the photosynthetic rate of canopy leaves of *Argyrodendron peralatum* on the third day without rain. Their study was performed at the end of the dry season, when the soil may not have been completely recharged. Photosynthesis in gaps was mildly affected by water availability in herbaceous perennials and shrubs in a seasonally dry tropical forest in Panama (Mulkey, Smith & Wright, 1991; Mulkey, Wright & Smith, 1993). In addition to restricting carbon gain as a consequence of stomatal closure, soil water deficits may further reduce photosynthesis by increasing the susceptibility of leaves to damage from high PFD (Osmond, 1983; Demmig-Adams & Adams, 1992).

Photosynthetic capacity in high PFD has been correlated with increased leaf nitrogen content, although the nature of the A_{max}/N relationship varies considerably among groups of species. In a comparison of 23 Amazonian species varying in leaf lifespan, Reich et al. (1991) found that A_{max} per unit weight was positively correlated with N per unit weight, but correlations between area-based measures of A_{max} and N were not statistically significant (see also Reich & Walters, 1994). Similar findings in a survey of leaves from different origins suggest that there is a general relationship between photosynthetic capacity and leaf N per unit weight (Field & Mooney, 1986; Reich, Walters & Ellsworth, 1992; Reich et al., 1994). Within species, however, these relationships do not always strictly apply, due to plasticity in leaf thickness and mass per unit area. Walters and Field (1987) found positive correlations between photosynthetic capacity and leaf N expressed on an area basis for *Piper hispidum* and *Piper auritum* growing in a broad range of light environments in lowland forest in Mexico, but they found a significant correlation between photosynthetic capacity and leaf N expressed on a weight basis for only one species, *Piper auritum*. Similarly, area-based regressions of A_{max} on leaf nitrogen were highly significant for two *Piper* species growing across gap transects in a Costa Rican forest, but weight-based regressions were not significant (Chazdon, 1992). Of 23 species sampled across four Amazonian communities, 17 showed statistically significant mass-based A_{max}/N relationships, and 21 species showed significant area-based A_{max}/N relationships (Reich et al., 1994).

Area-based dark respiration rates are higher in gap environments than in the forest understory (Fredeen & Field 1991; Chazdon & Kaufmann, 1993; Mulkey, Wright, & Smith, 1983; Fredeen & Field, Chapter 20). Variation in daytime dark respiration rates across gap transects within two *Piper* species was correlated with thickness of the mesophyll layer (Chazdon & Kaufmann, 1993). *Piper* species normally found in large gaps and clearings had approximately twice the dark respiration per unit of leaf area or dry mass as species found predominantly in shaded understory sites, although relationships between daily PFD and respiration did not vary among species (Fredeen & Field, 1991; Chapter 20). Interspecific differences in the sensitivity of respiration to daily PFD have been shown between *Piper* species in other studies, however (Walters & Field, 1987). Among eight species in the Venezuelan Amazon, leaf dark respiration rates were significantly correlated with peak photosynthesis rates, suggesting that early morning dark respiration was associated with maintenance costs of the photosynthetic apparatus (Ellsworth, 1991). In this field study, crop and pioneer tree species had twofold higher rates of leaf dark respiration than did tree species characteristic of later successional stages (Ellsworth, 1991).

Tropical plants typically encounter high levels of PFD in clearings and gaps, although total daily fluxes are often less than those experienced by many plants growing in open environments in the temperate zone. Consequently, leaves are often likely to experience PFD in excess of the requirements of photosynthesis, which can lead to a reduction of the quantum yield of photosynthesis and possible damage to the photosynthetic apparatus, particularly photosystem II. This phenomenon, termed photoinhibition, was first proposed as a factor affecting photosynthesis of tropical trees by Langenheim et al. (1984). The reduction of quantum yield measured under saturating CO_2 conditions is strongly correlated with characteristics of chlorophyll fluorescence, in particular the ratio of variable to maximum fluorescence (F_v/F_m) (Björkman & Demmig 1987; see Lovelock, Jebb & Osmond, 1994).

Plants accustomed to high light are able to cope with high PFD through non-radiative dissipation of the absorbed energy without experiencing permanent damage (summarized in Demmig-Adams & Adams, 1992). The principal mechanism for non-radiative dissipation is thought to be the xanthophyll cycle whereby violaxanthin is converted to zeaxanthin via antheraxanthin (Demmig-Adams & Adams, 1992). This conversion is associated with an increase in reversible quenching of fluorescence, with consequent reduction in F_v/F_m. Upon exposure to darkness or low PFD, the zeaxanthin is reconverted to violaxanthin, and F_v/F_m increases to levels found originally. The proportion of carotenoid pigments associated with the xanthophyll cycle appears to be a good indicator of the capacity of plants to tolerate high PFD environments. In a survey of 20 species, including rainforest understory species and crop species, Demmig-Adams and Adams (1992) found that for the rainforest species, an average 12% of the total carotenoids were associated with the xanthophyll cycle, whereas for the crop species, the average was 30%.

Field studies of photoinhibition in the tropics indicate little evidence for permanent damage to the photosynthetic apparatus resulting from exposure to high PFD when other factors, such as water or nutrients, are not limiting. Fernandez and Fetcher (1991b) found little evidence for photoinhibition as measured by reduction in F_v/F_m for saplings of eight native species growing in a plantation exposed to full sun in lowland Costa Rica. Castro, Fernandez and Fetcher (1995) grew seedlings of five species from lowland forest in Costa Rica on benches under full shade, partial shade, and full sun and measured F_v/F_m on leaves that were allowed to recover overnight. When compared to plants grown in partial or full shade, only one species, the emergent tree *Dipteryx panamensis*, exhibited significant reduction in F_v/F_m. The reduction in F_v/F_m was highly correlated with a reduction in quantum yield. The persistence of reduced F_v/F_m and quantum yield following a long period of darkness suggests that there might have been permanent damage to the photosynthetic apparatus, although this is by no means certain. Leaf nitrogen (weight basis) was not significantly greater for *Dipteryx panamensis* when grown in full sun compared to partial shade, whereas it was significantly greater for *Ochroma lagopus* and *Inga edulis*, two species that did not show photoinhibition.

Several comparative studies confirm that early successional species exhibit less photoinibition in gaps than shade-tolerant species. In an experimental study of *Cecropia schreberiana*, a pioneer tree, *Manilkara bidentata*, and *Palicourea riparia*, a shrub, growing on a landslide in Puerto Rico, neither fertilization nor light treatment significantly affected F_v/F_m measured in the morning or the afternoon for the two tree species (Fetcher, unpublished data). *Palicouria* growing in the open had significantly lower F_v/F_m values than two tree species growing in the open or *Palicouria* growing in the edge sites. This reduction in F_v/F_m is consistent with the normal habitat of *Palicouria*, which is normally confined to the understory and small gaps (Lebron, 1979). Reduction in F_v/F_m was not severe, however, and may only reflect increased levels of photoprotection.

In a field experiment simulating rainforest disturbance following logging, Lovelock, Jebb & Osmond (1994) documented abrupt decreases in F_v/F_m within a few hours of exposure to high PFD in seedlings of four tree species. The decline in F_v/F_m was greatest for tree species common in mature forest understory sites; for these species, F_v/F_m did not recover pre-exposure values within 21 days (Lovelock, Jebb and Osmond, 1994). A field survey of F_v/F_m ratios in rainforest gaps revealed that shade-tolerant species have lower midday ratios compared to gap-opportunists and species commonly found in gaps (Figure 1.4). A statistical analysis of these data shows that F_v/F_m of shade-species in full sunlight differed signficantly from F_v/F_m of the other two groups of species (P = 0.045), but the two types of gap species did not differ significantly (Figure 1.4). Thus, although all species undergo some degree of photoinhibition when exposed to full sunlight, species commonly found in gaps recover quickly and show

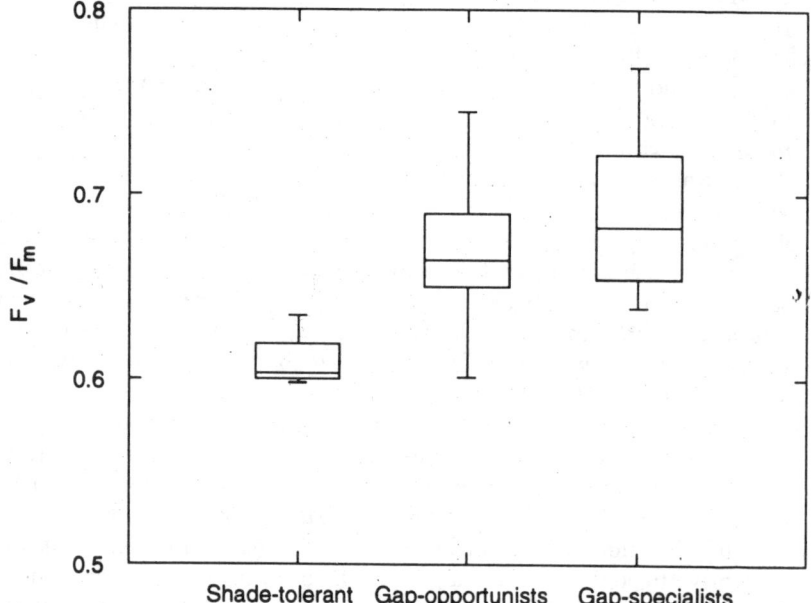

Figure 1.4.   The ratio of variable to maximum fluorescence of chlorophyll (F_v/F_m), a measure of photoinhibition, for leaves of 13 tropical forest species growing in full sun conditions in disturbed forest clearings. Fluorescence measurements were taken between noon and 2 pm for 4 leaves per plant. Shade-tolerant species growing in forest clearings exhibited a greater degree of photoinhibition than gap specialists or gap opportunists (p < 0.45). (Based on published data [Table 2] from Lovelock, Jebb & Osmond 1994).

less long-term depression of F_v/F_m ratios (Lovelock, Jebb & Osmond, 1994). Avoidance of severe photoinhibition in these species results from near vertical leaf orientation in gaps as well as greater concentrations of xanthophyll cycle pigments and increased capacity for photosynthetic carbon reduction.

Two species of neotropical palms exhibited photoinhibition under natural field conditions in a large forest clearing during the dry season in a Panamian forest (Araus & Hogan, 1994). Leaves of high-light grown plants of *Socratea exorrhiza* exhibited a greater reduction in F_v/F_m than the more xeromorphic leaves of *Scheelea zonensis*. Unlike *Scheelea*, *Socratea* leaves did not completely recover F_v/F_m overnight. Photoinhibitory damage of *Socratea* leaves probably results in the destruction of photosynthetic machinery, whereas *Scheelea* leaves showed little evidence of disruption of chloroplast structure under high-light conditions. Sensitivity of species to drought-enhanced photoinhibition is likely to restrict their ability to colonize large clearings.

1.3.2 Daily carbon gain in tropical forests

Few direct measurements of daily carbon gain have been made for tropical forest species in natural environments. These limited studies demonstrate that the frequency and duration of sunflecks are major determinants of leaf carbon gain for understory plants (Chazdon, 1988; Figure 1.3). For individuals of *Alocasia macrorrhiza* in the understory of a Queensland rain forest, daily carbon gain on a clear and an overcast day were 1.24 and 1.04 μmol cm^{-2} d^{-1}, respectively (Björkman, Ludlow & Morrow, 1972). On an overcast day, the daily quantum efficiency was calculated to be 0.07 mol CO_2 per mol of incident photons, close to the maximum efficiency obtained for short-term photosynthetic measurements (Björkman, Ludlow & Morrow, 1972). In the understory of a Hawaiian forest, total daily carbon gain for *Euphorbia forbesii* and *Claoxylon sandwicense* was 2.92 and 4.18 μmol cm^{-2} d^{-1}, respectively, for two days that varied in cloudiness (Pearcy & Calkin, 1983). As was the case for *Alocasia*, the lower carbon gain for *Euphorbia* reflects the lower number of sunflecks received. Pearcy and Calkin (1983) estimated that 60% of the carbon gain for *Claoxylon* and 40% in *Euphorbia* on these days was attributable to sunfleck activity.

Across contrasting rainforest microsites, variation in carbon gain is considerably more restricted than variation in light availability. Pearcy (1987) measured daily carbon gain of *Argyrodendron peralatum* in understory, gap, and canopy environments of a Queensland rainforest. Whereas daily PFD varied nearly 30-fold, from 1.17 to 34.5 mol m^{-2} d^{-1} from the understory to the canopy, daily carbon gain ranged only 10-fold, from 2.4 to 25.0 µmol cm^{-2} d^{-1}. In this study, the contribution of sunflecks to daily carbon gain was considerably lower (32–38%) than in the Hawaiian understory described by Pearcy and Calkin (1983) because sunflecks were much less frequent beneath the tall canopy.

In the absence of measured daily courses of CO_2 uptake, computer simulations permit estimation of carbon gain under different measured light regimes. Modeling efforts have generally assumed that net photosynthesis responds instantly to a change in PFD (Chazdon, 1986; Field, 1988). Chazdon (1986) estimated daily (24-hr) carbon gain for leaves of understory palms in Costa Rica and found that values ranged between slightly greater than 0 to 4.8 µmol CO_2 cm^{-2} d^{-1} in habitats ranging from understory to the center of small gaps. When sunflecks contributed less than 50% to total daily PFD, simulated carbon gain was highly correlated with total daily PFD, whereas when sunflecks contributed more than 50%, the correlation was weaker.

Field (1988) simulated carbon gain for *Piper hispidum* and *Piper auritum* growing in a range of light environments in Mexico. He found that daily carbon gain would be expected to increase with increasing leaf nitrogen in nearly all cases except for leaves of *P. hispidum* under simulated low light levels. Fetcher et al. (unpublished data) simulated carbon gain for seedlings of pioneer and non-pioneer canopy tree species in Costa Rica. Carbon gain of both groups was similar under light conditions characteristic of understory, gaps, and gap edges, but pioneer species had greater carbon gain in the clearing environment. In all of the studies mentioned, carbon gain increased in a curvilinear fashion with increasing total daily PFD, but the location of the curve and the degree of curvature varied considerably. Palms had considerably higher carbon gain under low light levels than shrubs or tree seedlings. Nighttime respiration is such a significant component of the daily carbon budget of a leaf growing at low light that many carbon gain simulations for tree seedlings showed a negative carbon budget in the understory.

Steady-state models most likely overestimate daily carbon gain because induction limitations to photosynthesis during sunflecks are ignored (Chazdon & Pearcy, 1986a,b). More accurate projections of daily carbon gain have been achieved using a simulation model based on the dynamic responses of photosynthesis to light fluctuations (Gross, Kirschbaum & Pearcy, 1991). Using this approach, Pearcy et al. (1994) compared steady-state and dynamic simulations of carbon gain in *Alocasia macrorrhiza* using PFD data measured in *Alocasia* sites in a Queensland rainforest. Steady-state and dynamic simulations differed by less than 2.5% for PFD measurements from open (canopy) or gap environments. In understory light environments, however, the dynamic model predictions of daily carbon gain were 1.2 to 25.5% lower than steady-state predictions (Pearcy et al., 1994). For the understory simulations, steady state and dynamic predictions differed most when sunflecks contributed more than 50% of total daily PFD. These simulations strongly suggest that induction limitations are indeed an important constraint on leaf carbon gain in understory microsites. Moreover, a negligible proportion of daily carbon gain could be attributed to post-illumination CO_2 fixation for these particular understory measurements because brief sunflecks were so infrequent (Pearcy et al., 1994).

Despite the overall importance of sunflecks for carbon gain in the understory, small microsite variations in diffuse light levels can strongly affect daily carbon gain (Chazdon, 1986). For two microsites with similar daily total PFD but different background diffuse PFD, daily carbon gain will always be greater for the microsite with enhanced diffuse PFD. Rates of carbon gain per unit leaf area are therefore likely to be considerably higher in building-phase microsites of gaps and in young secondary forests compared to mature-phase microsites in old-growth forests.

1.4   ACCLIMATION AND PLASTICITY IN PHOTOSYNTHETIC RESPONSES

Tropical forest species often experience substantial daily and seasonal variation in light availability. These variations may be relatively subtle, such as daily changes in sunfleck incidence within understory sites, or they may be dramatic, such as sustained increases in light availability following a large-scale canopy disturbance. Tropical forest species illustrate a continuum of photosynthetic responses to changing light availability (Anderson & Osmond, 1987). These responses involve environmentally-induced changes in steady-state and dynamic photosynthetic parameters that influence daily carbon gain: photosynthetic capacity, induction rates, quantum yield, or dark respiration.

In this review, we distinguish between the process of light acclimation of photosynthesis in response to a sudden change in light conditions and plasticity of photosynthesis, which is a broader concept. In our view, plasticity describes the range of potential phenotypes that can be expressed within a single genotype in response to variation in environmental conditions. Acclimation is therefore viewed as one process by which plasticity is expressed. This view differs somewhat from that described by Strauss-Debenedetti and Bazzaz (Chapter 6).

Light acclimation is the process that allows for environmentally-induced changes in photosynthetic utilization of light, depending upon the light regime under which leaves develop. For species adapted to the low end of the light-availability spectrum, light acclimation involves a complex set of physiological, biochemical, and structural responses that increase the capacity for utilization of light levels higher than those routinely encountered under natural growing conditions (Björkman, 1981). For species adapted to high-light environments, however, a different concept of light acclimation applies. Here, we are examining responses to decreased light availability relative to optimal growth conditions. In this case, it is difficult to determine to what extent acclimation is an active physiological process or simply a passive (resource deprivation) response. Comparisons between low- and high-light specialists suggest that these two groups of species generally exhibit different capacities for light acclimation (Björkman, 1981; Strauss-Debenedetti & Bazzaz, Chapter 6). Most species of plants, however, occupy intermediate light environments, and naturally experience varying degrees of temporal and spatial variation in growth conditions. For these species, acclimation is of critical ecological importance. Among these intermediate species with relatively wide ecological amplitude, light acclimation responses appear to be far more similar than they are different.

1.4.1   Time course of acclimation to changes in light availability

Unfortunately, virtually all of what we know about light acclimation responses of tropical species is based on greenhouse studies of small plants (usually first-year seedlings) grown in controlled or artificial light environments. To date, only one field study has followed individual plants or leaves in natural gaps over time to examine the time-course of light acclimation (Newell et al., 1993). Photosynthetic responses of three *Miconia* species to canopy opening were very similar, despite differences in growth form and ecological distributions within a Costa Rican rainforest. Although no changes were observed in A_{max} two weeks following treefa l, by four months, A_{max} in all three species nearly doubled compared to closed-canopy levels (Newell et al., 1993). Overall, no significant changes were observed in rates of dark respiration and apparent quantum yield before and after canopy opening. Acclimation in these species required the production of new leaves following treefall.

In leaves transferred from low to high light, successful light acclimation appears to be clearly related to coordinated increases in both carboxylation and electron transport activity (Boardman, 1977; Langenheim et al., 1984; Thompson, Stocker & Kriedemann, 1988; Thompson, Huang & Kriedemann, 1992; Mulkey, Smith & Wright, 1991). High-light acclimation is also associated with changes in chloroplast ultrastructure and chlorophyll a:b ratios (Lichtenthaler et al., 1981; Pearcy & Franceschi, 1986; Anderson, Chow & Goodchild, 1988; Evans, 1988), although these changes are not strictly prerequisite for light acclimation in the shade species *Tradescantia albiflora* (Adamson et al., 1991; Chow, Adamson & Anderson, 1991).

Acclimation of A_{max} following transfer from low to high light may be strongly influenced by nutrient supply (Thompson, Stocker & Kriedemann, 1988; Riddoch, Lehto & Grace, 1991). Seedlings of the Australian rainforest tree *Flindersia brayleyana* grown at high light but low nutrient supply showed decreased quantum yield and A_{max} compared to high-light grown seedlings provided with high levels of nutrients (Thompson, Stocker & Kriedemann, 1988). High nutrient availability is further associated with protection from photooxidative damage in plants transferred from low to high light (Ferrar & Osmond, 1986; Demmig-Adams & Adams, 1992). These nutrient effects suggest a physiological linkage between acclimation potential and susceptibility to photoinhibition (Langenheim et al., 1984; Anderson & Osmond, 1987).

Effects of changing light availability on photosynthetic traits are further mediated by the developmental stage of leaves at the time of the change (Sims & Pearcy, 1992). When shade-grown plants are transferred to a substantially higher PFD, mature leaves often show rapid photoinhibition, leaf chlorosis, and early leaf abscission (Björkman, 1981; Langenheim et al., 1984; Oberbauer & Strain, 1985; Strauss-Debenedetti & Bazzaz, 1991; Turnbull, Doley & Yates, 1993). Newly produced leaves, however, exhibit a range of acclimation responses ranging from little or no change in photosynthetic properties to pronounced physiological changes. For example, fully-developed leaves of seedlings of the canopy dominant *Pentaclethra macroloba* in Costa Rica showed little capacity for acclimation to a change in light regime (Oberbauer & Strain, 1985). Longer-term adjustments of photosynthetic capacity in response to growth light regime were observed in *Pentaclethra* leaves, however. Maximum photosynthetic rates increased from 5.4 μmol CO_2 m^{-2} s^{-1} in full shade to 7.0 and 6.7 μmol CO_2 m^{-2} s^{-1} in partial shade and full sun, respectively. Leaves developed in full shade reached light saturation at less than half the PFD of leaves developed in full sun, and apparent quantum yields of full shade plants were close to double those developed in full sun (Oberbauer & Strain, 1985).

Strauss-Debenedetti and Bazzaz (1991) studied photosynthetic acclimation to light of seedlings of five co-occurring tree species in the Moraceae (Strauss-Debenedetti & Bazzaz, Chapter 6). Although all species exhibited some degree of leaf injury when initially transferred from low to high light, three of the species, *Cecropia*, *Ficus*, and *Brosimum*, showed complete acclimation of photosynthetic capacity within 3 to 5 months following transfer. In contrast, *Poulsenia* and *Pseudolmelia* exhibited little or no increase in A$_{max}$ following transfer from low to high light, suggesting morphological or biochemical constraints on acclimation in these species (Strauss-Debenedetti & Bazzaz, 1991).

In low- to high-light transfers of three Australian tree species, significant levels of photosynthetic acclimation in expanded leaves were observed in all species, with significant increases in A$_{max}$ within 28 days of transfer, although one species (*Omalanthus populifolius*) showed foliar injury and localized photobleaching (Turnbull, Doley & Yates, 1993). For all species, acclimation was incomplete, indicating that mature leaves were constrained in their ability to respond to increased light availability compared to plants accustomed to growth under high light conditions. The lack of significant increases in stomatal conductances following low- to high-light transfer may partially explain constraints on photosynthetic acclimation in these species (Turnbull, Doley & Yates, 1993).

Following transfer from low to high light, respiration rates changed far more rapidly than did photosynthetic capacity in leaves of *Alocasia macrorrhiza* (Sims & Pearcy, 1991) and three Australian tree species (Turnbull, Doley & Yates, 1993). Following reciprocal transfers to high and low light, respiration rates of *Alocasia* adjusted within one week to those measured in plants grown in high and low light environments, respectively (Sims & Pearcy, 1991). In this case, increases in dark respiration were more closely associated with accumulation of photosynthate rather than maintenance of high photosynthetic capacity. For *Alocasia*, photosynthetic acclimation to high light environments appears to involve greater respiratory costs in leaf construction than for maintenance respiration (Sims & Pearcy, 1991).

Leaves of *Alocasia macrorrhiza* transferred while at intermediate stages of development acclimated only partially to the high-light conditions (Sims & Pearcy, 1992). Like the temperate understory herb *Fragaria virginiana* (Jurik, Chabot & Chabot, 1979), photosynthetic acclimation to light in *Alocasia* occurs through changes in leaf thickness rather than changes in photosynthetic capacity per unit mesophyll volume (Sims & Pearcy, 1992). The primary response of this species to increased PFD is abscission of mature leaves and their replacement with new leaves, which have substantially higher photosynthetic capacity in the new light regime (Sims & Pearcy, 1992). Nevertheless, mature leaves of *Alocasia* do show significant capacity for recovery from photoinhibition following a sudden exposure to simulated gap conditions (Mulkey & Pearcy, 1992). Complete acclimation to light conditions in a simulated gap was observed only in newly produced leaves (Mulkey & Pearcy, 1992).

The ability for mature leaves to acclimate to an increase in PFD may depend, in part, on the potential for anatomical changes, such as increased mesophyll tissue layers. Fully developed leaves of low-light-grown *Bischofia javanica* increased A_{max} when transferred to high light, but only after recovering from initial photoinhibition (Kamaluddin & Grace, 1992a,b). Palisade thickness increased by 132% over that of leaves before exposure to high light, but was still 32% lower than high-light-grown leaves. Thus, acclimation of mature leaves was substantial, but was incomplete with respect to leaf anatomy and photosynthetic capacity (Kamaluddin & Grace, 1992a,b). In contrast, new leaves developed following a switch to contrasting light conditions showed nearly complete anatomical and photosynthetic adjustment compared to high- and low-light controls (Kamaluddin & Grace, 1993).

Within a species, acclimatory responses to light increases appear to be more restricted than responses to light decreases. Following transfer from high to low PFD, photosynthetic capacity of mature leaves of three Australian rainforest tree species remained at pre-transfer levels for 7-14 days and then decreased in all species, with most of the response complete within 28 days for *Omalanthus populifolius* (an early-successional tree) and *Duboisea myoporoides* (a mid-successional tree), and within 56 days for *Acmena ingens* (a late-successional tree) (Turnbull, Doley & Yates, 1993). In the former two species, leaves achieved complete acclimation in A_{max} following transfer to deep shade (Turnbull, Doley & Yates, 1993). Decreases in A_{max} were associated with lower quantum yields, decreased rates of dark respiration, and decreased stomatal conductance.

1.4.2   Plasticity of photosynthesis in response to light availability

Early research on both temperate and tropical forest species led to the widely held notion that photosynthetic rates of shade-adapted (late-successional) species are less plastic in response to growth light environments compared to shade-intolerant (early-successional) species (Bazzaz, 1979; Bazzaz & Pickett, 1980; Bazzaz & Carlson, 1982; Strauss-Debenedetti & Bazzaz, Chapter 6). Indeed, many species of the tropical forest understory show little or no potential to increase A_{max} in response to increasing light availability (Chazdon, 1986; Mulkey, 1986; Mulkey, Smith & Wright, 1991; Fetcher et al., 1987; Ramos & Grace, 1990; Riddoch et al., 1991). Recent studies show, however, that shade-adapted herbs, tree seedlings, and shrubs can exhibit substantial photosynthetic plasticity across a wide range of light levels (Chow et al., 1988; Sims & Pearcy, 1989; Adamson et al., 1991; Chow, Adamson & Anderson, 1991; Chazdon, 1992). One consistent problem in comparing photosynthetic responses to contrasting light conditions has been the application of high-light treatments that far exceed measured PFD in large gaps within tropical forests (Figure 1.1). In many greenhouse studies, the high-light treatment is often equivalent to full-sun irradiance. Field data show, however, that even in large forest gaps, daily PFD rarely exceeds 50% of full sun PFD (Chazdon, 1992; Brown, 1993; Whitmore, 1993). Comparisons of the degree of plasticity in photosynthetic responses clearly depend on the range of light conditions examined.

Early-successional pioneer species generally exhibit a high degree of plasticity in photosynthetic capacity compared to species characteristic of later stages of forest succession (Chapters 6, 20, 21). In the pioneer species *Cecropia obtusifolia*, A_{max} ranged from 1.7 μmol CO_2 m^{-2} s^{-1} in low light to 9.9 under full sun conditions in a greenhouse (Strauss-Debenedetti & Bazzaz, 1991). Not all pioneer species, however, exhibit statistically significant changes in A_{max} under different light regimes. Seedlings of the Australian pioneer species *Solanum aviculare* did not show significant plasticity from 1% to 60% of full sun, regardless of whether shade light was neutral or enhanced in far-red wavelengths (Turnbull, 1991).

Canopy trees that regenerate in mature forests often show restricted variation in A_{max} over a wide range of light conditions (Fetcher et al., 1987; Strauss-Debenedetti & Bazzaz, 1991; Thompson, Huang & Kriedemann, 1992; Poorter & Oberbauer, 1993). Yet, wide photosynthetic plasticity was observed in the late-successional Australian species *Acmena ingens* (Turnbull, 1991) and the neotropical *Carapa guianensis* (Fetcher et al., 1987). The extent of photosynthetic plasticity exhibited by a species is not always a good predictor of its acclimation behavior, although high plasticity is often associated with high acclimation potential (Strauss-Debenedetti & Bazzaz, 1991).

Several studies have examined variation in photosynthetic capacity within tropical species growing naturally under a range of environmental conditions (Walters & Field, 1987; Chazdon & Field, 1987a, Pearcy, 1987; Mulkey, Smith & Wright, 1991; Chazdon, 1992; Chazdon & Kaufmann, 1993; Oberbauer et al., 1993; Poorter & Oberbauer, 1993). Measurements of A_{max} across natural gap transects illustrate differences in photosynthetic plasticity of shade-tolerant and shade-intolerant shrub species (Fig. 1.5). Photosynthetic capacity of *Piper sancti-felicis*, an early-successional shrub, is closely associated with microsite variation in light availability across gap transects in a Costa Rican rainforest (Chazdon, 1992; Figure 1.5), whereas A_{max} of *P. arieianum*, a shade-tolerant shrub, showed a weak but statistically significant association. Nitrogen-use efficiency increased significantly with light availability among leaves of *P. sancti-felicis*, but not of *P. arieianum* (Chazdon, 1992).

Species differences in photosynthetic plasticity are most pronounced when photosynthetic capacity is compared on a leaf dry mass basis rather than the usual area basis. Increased responsiveness of the photosynthetic system to microsite variation in light availability may be due, in part, to plasticity in the allocation of leaf nitrogen to photosynthetic vs. non-photosynthetic functions, resulting in increased carboxylation capacity per unit leaf biomass (Field & Mooney, 1986). Constraints on increasing carboxylation capacity are also associated with the increased susceptibility of shade species to photoinhibition (Demmig-Adams & Adams, 1992). In the center of a large gap, photosynthetic capacity decreased in leaves of *P. arieianum*, but not in *P. sancti-felicis,* suggesting differential susceptibility to photoinhibition (Chazdon, 1992; Fig. 1.5). In parallel to the two *Piper* species described above, *Pleiostachya*, common along forest edges and large clearings, showed significant differences in A_{max} in gaps compared to forest understory, whereas *Calathea inocephala*, common in the forest understory, did not vary significantly in A_{max} per unit leaf mass

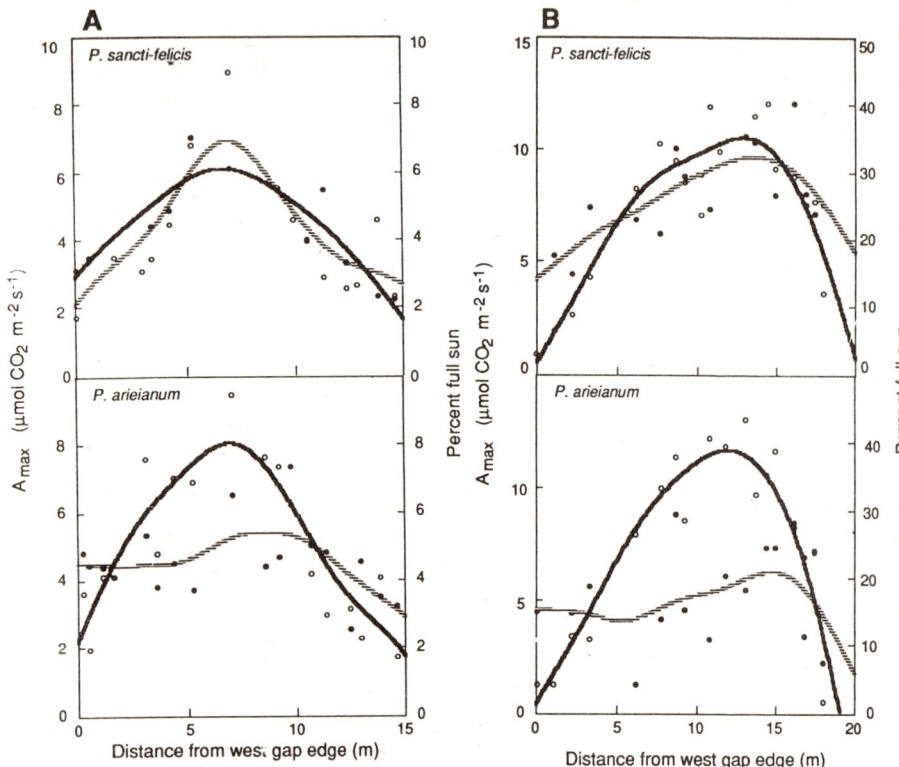

Figure 1.5.   Variation in daily photon flux density (open circles; % full sun for a 3-day period) and photosynthetic capacity (closed circles; $\mu mol\ CO_2\ m^{-2}\ s^{-1}$) for 6-month old rooted cuttings of two Piper *species growing in pots across east-west transects through A) a small gap (projected area 54 m^2) and B) a large gap (projected area 300 m^2). Lines indicate least squares smoothed curves for daily photon flux density (solid) and photosynthetic capacity (hatched).* Piper sancti-felicis *shows a greater capacity to adjust A_{max} to light variability across the gap transects compared to* P. arieianum. *Based on Chazdon (1992).*

(Mulkey, Smith & Wright, 1991). Carboxylation capacity, as measured by the initial slope of the relationship between A_{max} and intracellular CO_2 pressure, increased more dramatically in *Pleiostachya* than *Calathea* in gaps (Mulkey, Smith & Wright, 1991). Moreover, electron transport capacity increased significantly in gaps only in *Pleiostachya*.

Several studies have documented environmentally-induced changes in dynamic photosynthetic responses in both laboratory and field conditions (Chazdon & Pearcy, 1986b; Tinoco-Ojanguren & Pearcy, 1992; Poorter & Oberbauer, 1993; Kursar & Coley, 1993). Growth of

Alocasia macrorrhiza under different regimes of sunfleck frequency and duration does not appear to affect the efficiency of sunfleck utilization (Sims & Pearcy, 1993), although sunfleck utilization efficiency is greater for plants grown under low vs. high light (Chazdon & Pearcy, 1986b). Rates of induction of *Dipteryx panamensis* saplings were positively correlated with sapling light availability as measured with hemispherical photographs (Poorter & Oberbauer, 1993). Kursar and Coley (1993), however, found no evidence for differences in induction times for leaves from gap and understory plants of the same species.

1.5 SYNTHESIS: PHOTOSYNTHETIC SPECIALISTS AND GENERALISTS IN THE CONTEXT OF RAINFOREST DYNAMICS

Photosynthetic responses to contrasting light environments within tropical forests provide clear examples of adaptive specialization on the one hand and opportunistic plasticity on the other. The photosynthetic apparatus of species in the deeply shaded understory reflects selection pressures to maximize quantum yield, light absorption, and sunfleck utilization, while minimizing respiratory costs associated with high photosynthetic capacity. At the other extreme, species growing in open environments receiving more than 50–70% of full sun PFD maximize carbon gain through a high photosynthetic capacity, high stomatal conductances, increased light saturation points, and increased capacity for photoprotection involving the xanthophyll cycle. One price paid for these high metabolic capacities is increased respiration rates, a condition that is incompatible with positive carbon balance in low-light environments. Associated with high rates of photosynthesis are high rates of transpiration, requiring a high hydraulic conductivity (Chapters 7 and 8).

Plastic or acclimatory responses of species within these two extreme categories are highly asymmetrical. When grown under the same high-light conditions, shade-adapted species never attain photosynthetic capacities exhibited by light-demanding species (Strauss-Debenedetti & Bazzaz, Chapter 6). Many light-demanding species, on the other hand, illustrate a high capacity to modulate or down-regulate photosynthetic capacity and dark respiration in response to decreased light availability (Strauss-Debenedetti & Bazzaz, 1991; Chazdon, 1992; Turnbull, Doley & Yates, 1993). Although some shade-tolerant species do show a significant capacity to increase photosynthetic capacity in response to increased light availability (Chow et al., 1988; Turnbull, 1991; Thompson, Huang & Kriedemann, 1992), these adjustments are generally observed within a restricted range of light conditions (Chazdon, 1992), and are often incomplete compared to leaves that have developed in the higher light regimes (Sims & Pearcy, 1989, 1991, 1992; Adamson et al., 1991; Strauss-Debenedetti & Bazzaz, 1991; Turnbull, Doley & Yates, 1993). Across rainforest habitats, leaf light availability is a far better predictor of photosynthetic capacity for leaves in light-demanding species than for shade-tolerant species (Chazdon & Field, 1987a; Chazdon, 1992).

The differential capacity for photosynthesis in light-demanding and shade-tolerant species to respond opportunistically to changes in leaf light availability becomes more explicable when low-light acclimation is viewed as a distinct process from high-light acclimation. According to this view, low-light acclimation involves a down-regulation of electron transport and carboxylation capacity in response to reduced light availability. The intrinsically high photosynthetic capacity of light-demanding species provides a "built-in" capacity for plasticity (Chazdon, 1992). In contrast, plasticity of photosynthesis for shade-tolerant species depends on their ability to respond to increased light availability through increased electron transport, carboxylation capacity, and stomatal conductance. These types of responses are significantly constrained in mature leaves (Sims & Pearcy, 1991), and depend strongly upon increased tissue nutrient concentration and rates of water supply (Field & Mooney, 1986; Chazdon & Field, 1987a; Walters & Field, 1987). Acclimation responses in shade species often involve anatomical changes, and do not appear to involve increasing photosynthetic capacity on a leaf dry mass basis (Sims & Pearcy, 1991; Chazdon & Kaufmann, 1993; see Mulkey, Wright & Smith, 1993). Light-demanding species, in contrast, tend to show light-dependent variation in photosynthetic capacity per unit leaf biomass, as well as anatomical changes (Chazdon & Kaufmann, 1993)

Given that it is theoretically advantageous for all forest species to respond to increases in light availability through increasing leaf carbon gain, why does acclimation potential vary so widely across species? To examine how different species accomplish variation in carbon gain, we must consider the nature of light variation among forest microsites, as well as the metabolic costs of acclimation (Sims & Pearcy, 1991). The recent focus on gap dynamics in tropical forests has led to a somewhat exaggerated view of the frequency distribution of gap environments within the forest understory. Recent surveys of wet tropical forest light conditions indicate that microsites with diffuse canopy transmittance or global site factors above 5% occur relatively rarely in the forest landscape in secondary and primary stands (D. Clark et al., 1996; Chazdon & Nicotra, unpublished data). Within the exception of pioneer species, virtually all forest species must be able to tolerate shaded conditions, at least during early stages of their life history (Clark & Clark, 1992). For these species, microsite variation in daily PFD is largely determined by cummulative duration and peak PFD of sunflecks, and by relatively small variations in background diffuse PFD (Chazdon, 1986; Oberbauer et al., 1988, 1989). Changes in photosynthetic capacity are unlikely to yield substantial benefits in terms of daily carbon gain under these conditions. Rather, maximizing carbon gain depends on minimizing carbon losses through dark respiration and maintaining high leaf induction states. Metabolic adaptations that increase the efficiency of sunfleck utilization are not likely to require increases in leaf construction costs or maintenance respiration. Carbon-balance simulations during high- to low-light acclimation suggest that carbon gain is closely correlated with rates of dark respiration, but not with A_{max} (Turnbull, Doley & Yates, 1993).

Within clearings and large canopy gaps, saturating or near-saturating irradiances are sufficiently frequent that increasing photosynthetic capacity is likely to provide significant carbon gain benefits (Figure 1.1). For leaves of three Australian tree species transferred from low to high light, simulated daily carbon gain was highly correlated with A_{max} during light acclimation (Turnbull, Doley & Yates, 1993). A strong, predictive relationship between A_{max} and daily carbon gain has been demonstrated for leaves within the forest canopy (Zotz & Winter, 1993, Chapter 3). Moreover, photoinhibition in full sunlight is far more likely in leaves with low carboxylation capacities (Lovelock, Jebb & Osmond, 1994). Differences in species potential to acclimate to light conditions within large gaps and clearings are therefore likely to limit carbon gain, growth rates, and competitive ability. Across a gradient of light availability, photosynthetic capacity and plant growth should therefore be highly correlated in light-demanding species, but unrelated or weakly correlated in shade-adapted species lacking significant acclimation potential. Among the three species of saplings studied by Oberbauer et al. (1993), only *Simarouba amara*, a light-demanding species, showed significant, positive correlations between A_{max} and growth.

Because small gaps are frequent within most tropical forests, species typical of understory and gap habitats do not appear to differ significantly in their photosynthetic responses under these conditions

(Newell et al., 1993). Many shade-tolerant species exhibit significant plasticity of photosynthetic capacity within a range of approximately 1 to 20% of full sun PFD (Sims & Pearcy, 1989; Chazdon, 1992). In microsites ranging in daily PFD from 0.5 to 4 mol m^{-2} d^{-1}, photosynthetic capacity did not differ significantly between the shade-tolerant shrub *Piper arieianum* and the pioneer shrub *P. sancti-felicis* (Chazdon, 1992; Chazdon & Kaufmann, 1993). It is therefore not surprising that many, if not most, species of tropical forests appear to be photosynthetic generalists, with limited acclimation potential to full-sun conditions, but ample acclimation potential over the range of most forest light conditions.

Species differences in photosynthetic responses to light availability, regardless of their importance for carbon gain, do not adequately explain contrasting light requirements for regeneration among tropical tree seedlings (Strauss-Debenedetti & Bazzaz, Chapter 6). Variation in photosynthetic responses of seedlings was far more restricted than variation in growth rates and mortality rates across a range of light conditions (Ramos & Grace, 1990; Kitajima, 1994). Among 15 tree species on Barro Colorado Island, Panama, shade tolerance of young seedlings did not depend on higher rates of photosynthesis or biomass accumulation in shade (Kitajima, 1994; & Chapter 19). Rather, shade tolerance depended more on morphological characteristics and carbon allocation patterns that enabled survivorship under low light conditions (Kitajima, 1994). The high degree of photosynthetic plasticity exhibited by seedlings of shade-intolerant species grown in low light may enable these seedlings to achieve a positive carbon gain for a limited time, but if light conditions do not improve, these seedlings will die.

Photosynthetic responses of rainforest species are best viewed in the context of whole-plant capacities for growth and survival under a range of light conditions (Ackerly, Chapter 21). Although differences in maximum photosynthetic capacity and acclimation potential do not appear to directly determine life or death, these responses mediate relationships between light availability and carbon gain across a highly dynamic matrix of light environments.

1.6 DIRECTIONS FOR FUTURE RESEARCH

Presently, our understanding of the ecological significance of photosynthetic responses to contrasting light environments is hampered by a lack of thorough, field-based investigations of photosynthesis in species shown to represent a wide range of regeneration patterns. These studies should not be limited to seedlings, but should include larger saplings and mature plants whenever possible. In particular, measurements of daily and seasonal fluctuations in photosynthesis and detailed time-courses of light acclimation in mature and newly-developed leaves are needed to illustrate the range of responses observed within tropical species. Studies focused on different species in similar environments are as insightful as studies of species' responses to different environments (Mulkey, Wright & Smith, 1993; Ackerly, Chapter 21). Ideally, future studies should also provide detailed analyses of leaf light environments so that the carbon-gain consequences of acclimatory behavior can be evaluated quantitatively as well as qualitatively. Ultimately, these data can be used, in concert with detailed studies of photosynthetic dynamics, to derive accurate carbon-balance simulations based on a mechanistic, dynamic model of photosynthesis (Gross, Kirschbaum & Pearcy, 1991). Moreover, leaf carbon-gain needs to be evaluated over periods of time significantly longer than a single day. The scaling between A_{max} and daily carbon gain exhibited by canopy leaves (Zotz & Winter, Chapter 3) should be examined in more detail across the full range of light conditions throughout the forest.

Differences in the nature of light acclimation across tropical forest species bears further scrutiny, particularly under field conditions. To what extent are anatomical adjustments required for changes in photosynthetic capacity? Are adjustments in all photosynthetic components achieved during light acclimation? How are acclimation responses mediated by nutrient and water availability? Why does quantum yield vary during light acclimation in some species, but not in others? To some extent, the different pathways and constraints shown among species may become more clear if low- and high-light acclimation are viewed as distinct physiological processes with different sets of constraints.

For some species, high light availability functions as a stress factor rather than as a resource that can be utilized by leaves. Protection from photoinhibition may involve production of new leaves (Mulkey & Pearcy, 1992), or may occur in leaves produced before high-light exposure (Demmig-Adams & Adams, 1992). The extent to which leaves can recover from photoinhibition, and the synergistic effects of leaf temperature, humidity, soil moisture, and nutrient availability are poorly understood for most tropical species, and particularly within the forest canopy, where irradiance can vary two orders of magnitude within a single tree crown (Zotz & Winter, Chapter 3). We need to understand better how these stress factors interact to influence seedling carbon gain following large-scale human, as well as natural, disturbances.

Traditionally, field-based studies of growth and survivorship and laboratory-based studies of photosynthetic responses have been conducted by different sets of investigators with distinct research objectives. Consequently, the lack of an integrated ecophysiological approach has impeded ecologically-relevant studies of leaf carbon gain. Photosynthetic responses to light availability and its variation within tropical forests must be assessed critically in terms of whole-plant growth and survivorship across forest habitats. Ecologically-based studies of photosynthesis and growth responses to light availability within primary and secondary tropical forests will greatly enhance the potential utilization of many tropical tree species in forest management or reforestation projects. These analyses should include cost-benefit studies of photosynthetic acclimation (Sims & Pearcy, 1991), analyses of leaf light environment and carbon gain in the context of plant canopy structure (Chazdon, Williams & Field, 1988; Field, 1988; Ackerly, Chapter 21), studies of plant growth as a function of A_{max} and daily carbon gain, and demographically-oriented studies of survivorship and reproduction as influenced by daily carbon gain as well as carbon allocation patterns.

REFERENCES

ADAMSON, H. Y., CHOW, W. S., ANDERSON, J. M., VESK, M. & SUTHERLAND, M. W. (1991) Photosynthetic acclimation of *Tradescantia albiflora* to growth irradiance: Morphological, ultrastructural, and growth responses. Physiologia Plantarum, **82**, 353–359.

ALEXANDER, D. Y. (1982) Etude de l'eclarement du sous-bois d'une foret dense humid sempervirente (Tai, Cote-d'Ivoire). *Acta Oecologica*, **3**, 407–447.

ANDERSON, J. M. (1986) Photoregulation of the composition, function, and structure of thylakoid membranes. *Annual Review of Plant Physiology*, **37**, 93–136.

ANDERSON, J. M. & OSMOND, C. B. (1987) Shade-sun responses: Compromises between acclimation and photoinhibition. *Topics in Photosynthesis, Vol. 9, Photoinhibition* (eds. D. J. KULE, C. B. OSMOND, & C. J. ARNTZEN,) Elsevier, Amsterdam, pp 1–38.

ANDERSON, J. M., CHOW, W. S. & GOODCHILD, D. J. (1988) Thylakoid membrane organization in sun/shade acclimation. *Australian Journal of Plant Physiology*, **15**, 11–26.

ANDERSON, M. C. (1964) Studies of the woodland light climate I. The photographic computation of light condition. *Journal of Ecology,* **52,** 27–41.

ARAUS, J. L. & HOGAN, K. P. (1994) Leaf structure and patterns of photoinhibition in two neotropical palms in clearings and forest understory during the dry season. *American Journal of Botany,* **81,** 726–738.

ASHTON, P. M. S. (1992) Some measurements of the microclimate within a Sri Lankan tropical rainforest. *Agricultural and Forest Meteorology,* **59,** 217–235.

BALDOCCHI, D. & COLLINEAU, S. (1994) The physical nature of solar radiation in heterogeneous canopies: Spatial and temporal attributes. *Exploitation of Environmental Heterogeneity by Plants* (eds. M. M. CALDWELL & R. W. PEARCY), Academic Press, Inc., San Diego, pp 21–71.

BARTON, A. M., FETCHER, N. & REDHEAD, S. (1989) The relationship between treefall gap size and light flux in a neotropical rain forest in Costa Rica. *Journal of Tropical Ecology,* **5,** 437–439.

BAZZAZ, F. A. (1979) The physiological ecology of plant succession. *Annual Review of Ecology and Systematics,* **10,** 351–371.

BAZZAZ, F. A. & CARLSON, R. W. (1982) Photosynthetic acclimation to variability in the light environment of early and late successional plants. *Oecologia,* **54,** 313–316.

BAZZAZ, F. A. & PICKETT, S. T. A. (1980) Physiological ecology of tropical succession: A comparative review. *Annual Review of Ecology and Systematics,* **11,** 287–310.

BECKER, P. & Smith, A. P. (1990) Spatial autocorrelation of solar radiation in a tropical moist forest understory. *Agricultural and Forest Meteorology,* **52,** 373–379.

BECKER, P., ERHART, D. W. & SMITH, A. P. (1989) Analysis of forest light environments. Part I. Computerized estimation of solar radiation from hemispherical canopy photographs. *Agricultural and Forest Meteorology,* **44,** 217–232.

BIGGS, W. W., EDISON, A. R., EASTIN, J. W., BROWN, J. W., MARANVILLE, J. W. & CLEGG, M. D. (1971) Photosynthesis light sensor and meter. *Ecology,* **52,** 126–131.

BJÖRKMAN, O. (1981) Responses to different quantum flux densities. *Physiological Plant Ecology I. Encyclopedia of Plant Physiology* (eds. O. L., LANGE, P. S. NOBEL, C. B. OSMOND, & H. ZIEGLER) Springer-Verlag, New York, pp 57–107.

BJÖRKMAN, O. & DEMMIG, B. (1987) Photon yield of O_2 evolution and chlorophyll fluorescence characteristics at 77K among vascular plants of diverse origins. *Planta,* **170,** 489–504.

BJÖRKMAN, O., LUDLOW, M. M. & MORROW, P. A. (1972) Photosynthetic performance of two rainforest species in their native habitat and analysis of their gas exchange. *Carnegie Institute of Washington Yearbook,* **71,** 94–102.

BOARDMAN, N. K. (1977) Comparative photosynthesis of sun and shade plants. *Annual Review of Plant Physiology*, **28**, 355–377.

BROWN, N. (1993) The implications of climate and gap microclimate for seedling growth conditions in a Bornean lowland rainforest. *Journal of Tropical Ecology*, **9**, 153–168.

CANHAM, C. D., DENSLOW, J. S., PLATT, W. J., RUNKLE, J. R., SPIES, T. A. & WHITE, P. S. (1990) Light regimes beneath closed canopies and tree-fall gaps in temperate and tropical forests. *Canadian Journal of Forest Research*, **20**, 620–631.

CASTRO, Y., FERNANDEZ, D. & FETCHER, N. (1995) Chronic photoinhibition in attached leaves of tropical trees. *Plant Physiology*, in press.

CHAZDON, R. L. (1986) Light variation and carbon gain in rainforest understory palms. *Journal of Ecology*, **74**, 995–1012.

CHAZDON, R. L. (1988) Sunflecks and their importance to forest understory plants. *Advances in Ecological Research*, **18**, 1–63.

CHAZDON, R. L. (1992) Photosynthetic plasticity of two rainforest shrubs across natural gap transects. *Oecologia*, **92**, 586–595.

CHAZDON, R. L. & FETCHER, N. (1984a) Light environments of tropical forests. *Physiological Ecology of Plants of the Wet Tropics* (eds. E. MEDINA, H.A. MOONEY, & C. VAZQUEZ-YANES) Dr. W. Junk Publishers, The Hague, pp 553–564.

CHAZDON, R. L. & FETCHER, N. (1984b) Photosynthetic light environments in a lowland tropical rainforest in Costa Rica. *Journal of Ecology*, **72**, 553–564.

CHAZDON, R. L. & FIELD, C. B. (1987a) Determinants of photosynthetic capacity in six rainforest *Piper* species. *Oecologia*, **73**, 222–230.

CHAZDON, R. L. & FIELD, C. B. (1987b) Photographic estimation of photosynthetically active radiation: Evaluation of a computerized technique. *Oecologia*, **73**, 525–532.

CHAZDON, R. L. & KAUFMANN, S. (1993) Plasticity of leaf anatomy of two rainforest shrubs in relation to photosynthetic light acclimation. *Functional Ecology*, **7**, 385–394.

CHAZDON, R. L. & PEARCY, R. W. (1986a) Photosynthetic responses to light variation in rain forest species. I. Induction under constant and fluctuating light conditions. *Oecologia*, **69**, 517–523.

CHAZDON, R. L. & PEARCY, R. W. (1986b) Photosynthetic responses to light variation in rainforest species. II. Carbon gain and photosynthetic efficiency during lightflecks. *Oecologia*, **69**, 524–531.

CHAZDON, R. L. & PEARCY, R. W. (1991) The importance of sunflecks for forest understory plants. *BioScience*, **41**, 760–766.

CHAZDON, R. L., WILLIAMS, K. & FIELD, C. B. (1988) Interactions between crown structure and light environment in five rain forest *Piper* species. *American Journal of Botany*, **75**, 1459–1471.

CHOW, W. S., ADAMSON, H. Y. & ANDERSON, J. M. (1991) Photosynthetic acclimation of *Tradescantia albiflora* to growth irradiance: Lack of adjustment of light-harvesting components and its consequences. *Physiologia Plantarum*, **81**, 175–182.

CHOW, W. S., ANDERSON, J. M. & MELIS, A. (1990) The photosystem stoichiometry in thylakoids of some Australian shade-adapted plant species. *Australian Journal of Plant Physiology*, **17**, 665–674.

CHOW, W. S., LUPING, Q., GOODCHILD, D. J. & ANDERSON, J. M. (1988) Photosynthetic acclimation of *Alocasia macrorrhiza* (L.) G. Don to growth irradiance: Structure, function and composition of chloroplasts. *Australian Journal of Plant Physiology*, **15**, 107–122.

CLARK, D. A. & CLARK, D. B. (1992) Life history diversity of canopy and emergent trees in a Neotropical rain forest. *Ecological Monographs*, **62**, 315–344.

CLARK, D. B., CLARK, D. A., RICH, P., WEISS, S., & OBERBAUER, S. F. 1996. Landscape-level analyses of forest structure and understory light environments in a neotropical rain forest. *Canadian Journal of Forest Research* (in press).

DEMMIG-ADAMS, B. & ADAMS, W. W. I. (1992) Photoprotection and other responses of plants to high light stress. *Annual Review of Plant Physiology and Plant Molecular Biology*, **43**, 599–626.

DENSLOW, J. S., SCHULTZ, J. C., VITOUSEK, P. M. & STRAIN, B. R. (1990) Growth responses of tropical shrubs to treefall gap environments. *Ecology*, **71**, 165–179.

DOLEY, D., UNWIN, G. L. & YATES, D. J. (1988) Spatial and temporal distribution of photosynthesis and transpiration by single leaves in a rainforest tree. *Australian Journal of Plant Physiology*, **15**, 317–326.

EHLERINGER, J. & BJÖRKMAN, O. (1987) Quantum yields for CO_2 uptake in C_3 and C_4 plants. Dependence on temperature, CO_2 and O_2 concentrations. *Plant Physiology*, **73**, 555–559.

ELLSWORTH, D. S. (1991) Variation in leaf structure, nitrogen and photosynthesis across light gradients in a temperate and tropical forest. *Ph.D. dissertation*, University of Wisconsin. Madison, Wisconsin.

ELLSWORTH, D. S. & REICH, P. B. (1995) Photosynthesis and leaf nitrogen in five Amazonian tree species during early secondary succession. *Ecology*, in press.

ENDLER, J. A. (1993) The color of light in forests and its implications. *Ecological Monographs*, **63**, 1–27.

EVANS, J. R. (1988) Acclimation by the thylakoid membranes to growth irradiance and the partitioning of nitrogen between soluble and thylakoid proteins. *Australian Journal of Plant Physiology*, **15**, 93–106.

FERNANDEZ, D. & FETCHER, N. (1991a) Changes in light availability following hurricane Hugo in a subtropical montane forest in Puerto Rico. *Biotropica*, **23**, 393–399.

FERNANDEZ, D. & FETCHER, N. (1991b) Chlorophyll fluorescence and quantum yield in seedlings and saplings of tropical forest. *Primer Congreso Venezolano de Ecologia. Resumenes, Sociedad Venezolana de Ecologia, Caracas*, pp 13.

FERRAR, P. J. & OSMOND, C. B. (1986) Nitrogen supply as a factor influencing photoinhibition and photosynthetic acclimation after transfer of shade grown *Solanum dulcamara* to bright light. *Planta*, **168**, 563–570.

FETCHER, N., OBERBAUER, S. F. & CHAZDON, R. L. (1994) Physiological ecology of plants at La Selva. *La Selva: Ecology and Natural History of a Neotropical Rainforest* (eds. L. MCDADE, K. S. BAWA, H. HESPENHEIDE & G. HARTSHORN) University of Chicago Press, Chicago, pp 128–141.

FETCHER, N., OBERBAUER, S. F., ROJAS, G. & STRAIN, B. R. (1987) Efectos del regimen de luz sobre la fotosıntesis y el crecimiento en plántulas de árboles de un bosque lluvioso tropical de Costa Rica. *Revista de Biologia Tropical*, **35 (Suppl.)**, 97–110.

FIELD, C. B. (1988) On the role of photosynthetic responses in constraining the habitat distribution of rainforest plants. *Australian Journal of Plant Physiology*, **15**, 343–358.

FIELD, C. & MOONEY, H. A. (1986) The photosynthesis-nitrogen relationship in wild plants. *On the Economy of Plant Form and Function* (ed. T. J. GIVNISH,) Cambridge University Press, Cambridge, pp 25–55

FREDEEN, A. L. & FIELD, C. B. (1991) Leaf respiration in *Piper* species native to a Mexican rainforest. *Physiologia Plantarum*, **82**, 85–92.

GORSKI, T. (1976) Annual cycle of red and far-red radiation. *International Journal of Biometeorology*, **24**, 361–365.

GROSS, L. J., KIRSCHBAUM, M. U. F. & PEARCY, R. W. (1991) A dynamic model of photosynthesis in varying light taking account of stomatal conductance, C^{13}-cycle intermediates, photorespiration and Rubisco activation. *Plant, Cell and Environment*, **14**, 881–893.

GUTSCHICK, V. P., BARRON, M. H., WAECHTER, D. A. & WOLF, M. A. (1985) Portable mcnitor for solar radiation that accumulates irradiance histograms for 32 leaf-mounted sensors. *Agricultural and Forest Meteorology*, **33**, 281–290.

JURIK, T. W., CHABOT, J. F. & CHABOT, B. F. (1979) Ontogeny of photosynthetic performance in *Fragaria virginiana* under changing light regimes. *Plant Physiology*, **63**, 542–547.

KAMALUDDIN, M. & GRACE, J. (1992a) Acclimation in seedlings of a tropical tree, *Bischofia javanica*, following a stepwise reduction in light. *Annals of Botany*, **69**, 557–562.

KAMALUDDIN, M. & GRACE, J. (1992b) Photoinhibition and light acclimation in seedlings of *Bischofia javanica*, a tropical forest tree from Asia. *Annals of Botany*, **69**, 47–52.

KAMALUDDIN, M. & GRACE, J. (1993) Growth and photosynthesis of tropical forest tree seedlings (*Bischofia javanica* Blume) as influenced by a change in light availability. *Tree Physiology*, **13**, 189–201.

KIRSCHBAUM, M. U. F., GROSS, L. J. & PEARCY, R. W. (1988) Observed and modelled stomatal responses to dynamic light environments in the shade plant *Alocasia macrorrhiza*. *Planta*, **174**, 527–533.

KITAJIMA, K. (1994) F ᵃtive importance of photosynthetic traits and allocation pattern as corrᵉ ates of seedling shade tolerance of 13 tropical trees. *Oecologia*, **98**, 419–428.

KOCH, G. W., AMTHOR, J. S. & GOULDEN, M. L. (1994) Diurnal patterns of leaf photosynthesis, conductance, and water potential at the top of a lowland rainforest canopy in Cameroon: Measurements from the Radeau des Cimes. *Tree Physiology*, **14**, 347–360.

KÜPPERS, M. & SCHNEIDER, H. (1993) Leaf gas exchange of beech (*Fagus sylvatica L.*) seedlings in lightflecks: Effects of fleck length and leaf temperature in leaves grown in deep and partial shade. *Trees*, **7**, 160–168.

KURSAR, T. A. & COLEY, P. D. (1993) Photosynthetic induction times in shade-tolerant species with long-and short-lived leaves. *Oecologia*, **93**, 165–170.

KWESIGA, F. R. & GRACE, J. (1986) The role of the red/far-red ratio in the response of tropical tree seedlings to shade. *Annals of Botany*, **57**, 283–290.

LANGENHEIM, J. H., OSMOND, C. B., BROOKS, A. & FERRAR, P. J. (1984) Photosynthetic responses to light in seedlings of selected Amazonian and Australian rainforest tree species. *Oecologia*, **63**, 215–224.

LE GOUALLEC, J.-L. & CORNIC, G. (1988) Photoinhibition of photosynthesis in *Elatostema repens*. *Plant Physiology and Biochemistry*, **26**, 705–712.

LE GOUALLEC, J.-L., CORNIC, G. & BLANC, P. (1990) Relations between sunfleck sequences and photoinhibition of photosynthesis in a tropical rain forest understory herb. *American Journal of Botany*, **77**, 999–1006.

LEBRON, M. L. (1979) An autecological study of *Palicourea riparia* Bentham as related to rain forest disturbance in Puerto Rico. *Oecologia*, **42**, 31–46.

LEE, D. W. (1987) The spectral distribution of radiation of two neotropical rainforests. *Biotropica*, **19**, 161–166.

LEE, D. W. (1989) Canopy dynamics and light climates in a tropical moist deciduous forest in India. *Journal of Tropical Ecology*, **5**, 65–79.

LEE, D. W. & DOWNUM, K. R. (1991) The spectral distribution of biologically active solar radiation at Miami, Florida, USA. *International Journal of Biometeorology*, **35**, 48–54.

LEE, D. W. & PALIWAL, K. (1988) The light climate of a South Indian tropical evergreen forest. *GeoBios*, **15**, 3–6.

LEE, D. W. & RICHARDS, J. H. (1991) Heteroblastic development in vines. *The Biology of Vines* (eds. F. E. PUTZ, & H. A. MOONEY) Cambridge University Press, New York, pp 205–244.

LEMEUR, D. R. & BLAD, B. L. (1974) A critical review of light models for estimating the shortwave radiation regime of plant canopies. *Agricultural Meteorology*, **14**, 255–286.

LICHTENTHALER, H. K., BUSCHMANN, C., DOLL, M., FIETZ, H.-J., BACH, T., KOZEL, U., MEIER, D. & RAHMSDORF, U. (1981) Photosynthetic activity, chloroplast ultrastructure, and leaf characteristics of high-light and low-light plants and of sun and shade leaves. *Photosynthesis Research*, **2,** 115–141.

LIEBERMAN, M., LIEBERMAN, D. & PERALTA, R. (1989) Forest canopies are not just Swiss cheese: Canopy stereogeometry of non-gaps in tropical forest. *Ecology*, **70,** 550–552.

LOVELOCK, C. E., JEBB, M. & OSMOND, C. B. (1994) Photoinhibition and recovery in tropical plant species: Response to disturbance. *Oecologia*, **97,** 297–307.

MILLER, E. E. & NORMAN, J. M. (1971) A sunfleck theory for plant canopies. I. Lengths of sunlit segments along a transect. *Agronomy Journal*, **63,** 735–738.

MOONEY, H. A., FIELD, C. B., VÁZQUEZ-YÁNES, C. & CHU, C. (1983) Environmental controls on stomatal conductance in a shrub of the humid tropics. *Proceedings of the National Academy of Science*, **80,** 1295–1297.

MULKEY, S. S. (1986) Photosynthetic acclimation and water-use efficiency of three species of understory herbaceous bamboo (Gramineae) in Panama. *Oecologia*, **70,** 514–519.

MULKEY, S. S. & PEARCY, R. W. (1992) Interactions between acclimation and photoinhibition of photosynthesis of a tropical forest understory herb, *Alocasia macrorrhiza*, during simulated canopy gap formation. *Functional Ecology*, **6,** 719–729.

MULKEY, S. S., SMITH, A. P. & WRIGHT, S. J. (1991) Comparative life history and physiology of two understory Neotropical herbs. *Oecologia*, **58,** 26–32.

MULKEY, S. S., WRIGHT, S. J. & SMITH, A. P. (1993) Comparative physiology and demography of three Neotropical forest shrubs: Alternative shade-adaptive character syndromes. *Oecologia*, **96,** 526–536.

MYNENI, R. B., ROSS, J. & ASRAR, G. (1989) A review on the theory of photon transport in leaf canopies. *Agricultural and Forest Meteorology*, **45,** 1–153.

NEWELL, E. A., McDONALD, E. P., STRAIN B. R. & DENSLOW, J. S. (1993) Photosynthetic responses of *Miconia* species to canopy openings in a lowland tropical rainforest. *Oecologia*, **94,** 49–56

NORMAN, J. M. & WELLES, J. M. (1983) Radiative transfer in an array of canopies. *Agronomy Journal*, **75,** 481–488.

OBERBAUER, S. F. & STRAIN, B. R. (1984) Photosynthesis and successional status of Costa Rican rain forest trees. *Photosynthesis Research*, **5,** 227–232.

OBERBAUER, S. F., CLARK, D. A., CLARK, D. B. & QUESADA, M. (1989) Comparative analysis of photosynthetic light environments within the crowns of juvenile rainforest trees. *Tree Physiology*, **5,** 13–23.

OBERBAUER, S. F., CLARK, D. B., CLARK, D. A. & QUESADA, M. (1988) Crown light environments of saplings of two species of rainforest emergent trees. *Oecologia*, **75,** 207–212.

OBERBAUER, S. F., CLARK, D. B., CLARK, D. A., RICH, P. M. & VEGA, G. (1993) Light environment, gas exchange, and annual growth of saplings of three species of rainforest trees in Costa Rica. *Journal of Tropical Ecology* **9**, 511–523.

OBERBAUER, S. F. & STRAIN, B. R. (1985) Effects of light regimes on the growth and physiology of *Pentaclethra macroloba* (Mimosaceeae) in Costa Rica. *Journal of Tropical Ecology*, **1**, 303–320.

OBERBAUER, S. F. & STRAIN, B. R. (1986) Effects of canopy position and irradiance on the leaf physiology and morphology of *Pentaclethra macroloba* (Mimosaceae). *American Journal of Botany*, **73**, 409–416.

OSMOND, C. B. (1983) Interactions between irradiance, nitrogen nutrition, and water stress in the sun-shade responses of *Solanum dulcamara. Oecologia*, **57**, 316–321.

PARKER, G. C., SMITH, A. P. & HOGAN, K. P. (1992) Access to the upper forest canopy with a large tower crane. *BioScience*, **42**, 664–670.

PEARCY, R. W. (1983) The light environment and growth of C_3 and C_4 species in the understory of a Hawaiian forest. *Oecologia*, **58**, 26–32.

PEARCY, R. W. (1987) Photosynthetic gas exchange responses of Australian tropical forest trees in canopy, gap, and understory microenvironments. *Functional Ecology*, **1**, 169–178.

PEARCY, R. W. (1988) Photosynthetic utilization of lightflecks by understory plants. *Australian Journal of Plant Physiology*, **15**, 223–238.

PEARCY, R. W. (1989) Radiation and light measurements. *Plant Physiological Ecology: Field Methods and Instrumentation* (eds. R. W. PEARCY, J. R. EHLERINGER, H. A. MOONEY, & P. W. RUNDEL) Chapman and Hall, New York, pp 97–135.

PEARCY, R. W. (1990) Sunflecks and photosynthesis in plant canopies. *Annual Review of Plant Physiology and Molecular Biology*, **41**, 421–453.

PEARCY, R. W. & CALKIN, H. (1983) Carbon dioxide exchange of C_3 and C_4 tree species in the understory of a Hawaiian forest. *Oecologia*, **5**, 26–32.

PEARCY, R. W. & FRANCESCHI, V. R. (1986) Photosynthetic characteristics and chloroplast ultrastructure of C_3 and C_4 tree species grown in high- and low-light environments. *Photosynthesis Research*, **9**, 317–331.

PEARCY, R. W., CHAZDON, R. L., GROSS, L. J. & MOTT, K. A. (1994) Photosynthetic utilization of sunflecks, a temporally patchy resource on a time scale of seconds to minutes. *Exploitation of Environmental Heterogeneity by Plants: Ecophysiological Processes Above and Below Ground* (eds. M. M. CALDWELL, & R. W. PEARCY) Academic Press, New York, pp 175-208.

PFITSCH, W. A. & PEARCY, R. W. (1989) Daily carbon gain by *Adenocaulon bicolor*, a redwood forest understory herb, in relation to its light environment. *Oecologia*, **80**, 465–470.

POORTER, L. & OBERBAUER, S. F. (1993) Photosynthetic induction responses of two rainforest tree species in relation to light environment. *Oecologia*, **96**, 193–199.

PUKKALA, T., BECKER, P., KUULUVAINEN, T. & OKER-BLOM, P. (1991) Predicting spatial distribution of direct radiation below forest canopies. *Agricultural and Forest Meteorology*, **5**, 295–307.

RAICH, J. W. (1989) Seasonal and spatial variation in the light environment in a tropical dipterocarp forest and gaps. *Biotropica*, **21**, 299–302.

RAMOS, J. & GRAECE J. (1990) The effects of shade on the gas exchange of four tropical trees from Mexico. *Functional Ecology*, **4**, 667–677.

REICH, P. B., ELLSWORTH, D. S., & UHL, C. (1995) Leaf carbon and nutrient assimilation and conservation in species of differing successional states in an oligotrophic Amazonian forest. *Functional Ecology, 9*, 65-76.

REICH, P. B., UHL, C., WALTERS, M. B. & ELLSWORTH, D. S. (1991) Leaf lifespan as a determinant of leaf structure and function among 23 Amazonian tree species. *Oecologia*, **86**, 16–24.

REICH, P. B., WALTERS, M. B. & ELLSWORTH, D. S. (1992) Leaf life-span in relation to leaf, plant, and stand characteristics among diverse ecosystems. *Ecological Monographs*, **62**, 365–392.

REICH, P. B. & WALTERS, M. B. (1994) Photosynthesis-nitrogen relations in Amazonian tree species II. Variation in nitrogen vis-a-vis specific leaf influences mass- and area-based expressions. *Oecologia*, **97**, 73–81.

REICH, P. B., WALTERS, M. B., ELLSWORTH, D. S. & UHL, C. (1994) Photosynthesis-nitrogen relations in Amazonian tree species I. Patterns among species and communities. **97**, 62–72.

REIFSNYDER, W. E., FURNIVAL, G. M. & HOROVITZ, J. L. (1971) Spatial and temporal distribution of solar radiation beneath forest canopies. *Agricultural Meteorology*, **9**, 21–37.

RICH, P. M. (1989) A manual for analysis of hemispherical canopy photography. *Los Alamos National Laboratory, Los Alamos, New Mexico LA-11733-M.*

RICH, P. M., Clark, D. B., CLARK, D. A. & OBERBAUER, S. F. (1993) Long-term study of solar radiation regimes in a tropical wet forest using quantum sensors and hemispherical photography. *Agricultural and Forest Meteorology*, **65**, 107–127.

RIDDOCH, I., GRACE, J., FASEHUN, F. E., RIDDOCH, B. & LADIPO, D. O. (1991), Photosynthesis and successional status of seedlings in a tropical semi-deciduous rainforest in Nigeria. *Journal of Ecology*, **79**, 491–503.

RIDDOCH, I., LEHTO, T. & GRACE, J. (1991) Photosynthesis of tropical tree seedlings in relation to light and nutrient supply. *New Phytologist*, **119**, 137–147.

ROY, J. & SALAGER, J. L. (1992) Midday depression of net CO_2 exchange of leaves of an emergent rainforest tree in French Guiana. *Journal of Tropical Ecology*, **8**, 499–504.

SEEMANN, J. R., KIRSCHBAUM, M. U. F., SHARKEY, T. D. & PEARCY, R. W. (1988) Regulation of ribulose 1,5-bisphosphate carboxylase activity in *Alocasia macrorrhiza* in response to step changes in irradiance. *Plant Physiology,* **88,** 148–152.

SHARKEY, T. D., SEEMANN, J. R. & PEARCY, R. W. (1986) Contribution of metabolites of photosynthesis to post-illumination CO_2 assimilation in response to lightflecks. *Plant Physiology,* **82,** 1063–1068.

SIMS, D. A. & PEARCY, R. W. (1989) Photosynthetic characteristics of a tropical forest understory herb, *Alocasia macrorrhiza,* and a related crop species, *Colocasia esculenta* grown in contrasting light environments. *Oecologia,* **79,** 53–59.

SIMS, D. A. & PEARCY, R. W. (1991) Photosynthesis and respiration in *Alocasia macrorrhiza* following transfers to high and low light. *Oecologia,* **86,** 447–453.

SIMS, D. A. & PEARCY, R. W. (1992) Response of leaf anatomy and photosynthetic capacity in *Alocasia macrorrhiza* (Araceae) to a transfer from low to high light. *American Journal of Botany,* **79,** 449–455.

SIMS, D. A. & PEARCY, R. W. (1993) Sunfleck frequency and duration affects growth rate of the understory plant *Alocasia macrorrhiza. Functional Ecology,* **7,** 683–689.

SMITH, A. P. (1987) Respuestas de hierbas del sotobosque tropical a claros ocasionados por la caída de árboles. Ecología y ecofisiología de plantas en los bosques mesoamericanos. *Revista de Biología Tropical,* **35 (Suppl.),** 111–118.

SMITH, A. P., HOGAN, K. P. & IDOL, J. R. (1992) Spatial and temporal patterns of light and canopy structure in a lowland tropical moist forest. *Biotropica,* **24,** 503–511.

SMITH, H. (1982) Light quality, photoreception and plant strategy. *Annual Review of Plant Physiology,* **33,** 481–518.

SMITH, W. K., KNAPP, A. K. & REINERS, W. A. (1989) Penumbral effects on sunlight penetration in plant communities. *Ecology,* **70,** 1603–1609.

STERNBERG, L. d. S. L., MULKEY, S. S. & WRIGHT, S. J. (1989) Ecological interpretation of leaf carbon isotope ratios: Influence of respired carbon dioxide. *Ecology,* **70,** 1317–1324.

STOUTJESTIJK, P. (1972) A note on the spectral transmission of light by tropical rainforest. *Acta Botanica Neerlandica,* **21,** 346–350.

STRAUSS-DEBENEDETTI, S. & BAZZAZ, F. A. (1991) Plasticity and acclimation to light in tropical Moraceae of different sucessional positions. *Oecologia,* **87,** 377–387.

THOMPSON, W. A., HUANG, L.-K. & KRIEDEMANN, P. E. (1992) Photosynthetic response to light and nutrients in sun-tolerant and shade-tolerant rainforest trees. II. Leaf gas exchange and component processes of photosynthesis. *Australian Journal of Plant Physiology,* **19,** 19–42.

THOMPSON, W. A., STOCKER, G. C. & KRIEDEMANN, P. E. (1988) Growth and photosynthetic response to light and nutrients of *Flindersia brayleyana.* F.

Muell. A rainforest tree with broad tolerance to sun and shade. *Australian Journal of Plant Physiology*, **15**, 299–315.

Tinoco-Ojanguren, C. & Pearcy, R. W. (1992) Dynamic stomatal behavior and its role in carbon gain during lightflecks of a gap phase and an understory *Piper* species acclimated to high and low light. *Oecologia*, **92**, 222–228.

Tinoco-Ojanguren, C. & Pearcy, R. W. (1995) A comparison of light quality and quantity effects on the growth and steady-state and dynamic photosynthetic characteristics of three tropical tree species. *Functional Ecology*, **9**, 222–230.

Torquebiau, E. F. (1988) Photosynthetically active radiation environment, patch dynamics, and architecture in a tropical rainforest in Sumatra. *Australian Journal of Plant Physiology*, **15**, 327–342.

Turnbull, M. H. (1991) The effect of light quantity and quality during development on the photosynthetic characteristics of six Australian rainforest tree species. *Oecologia*, **87**, 110–117.

Turnbull, M. H. & Yates, D. J. (1993) Seasonal variation in the red/far-red ratio and photon flux density in an Australian sub-tropical rainforest. *Agricultural and Forest Meteorology*, **64**, 111–127.

Turnbull, M. H., Doley, D. & Yates, D. (1993) The dynamics of photosynthetic acclimation to changes in light quantity and quality in three Australian rainforest tree species. *Oecologia*, **94**, 218–228.

Turton, S. M. (1988) Solar radiation regimes in a north Queensland rainforest. *Proceedings of the Ecological Society of Australia*, **15**, 101–105.

Turton, S. M. (1990) Light environments within montane tropical rainforest, Mt. Bellenden Ker, North Queensland. *Australian Journal of Ecology*, *Australian Journal of Ecology* **15**, 35–42.

Walters, M. B. & Field, C. B. (1987) Photosynthetic light acclimation in two rainforest *Piper* species with different ecological amplitudes. *Oecologia*, **72**, 449–456.

Wayne, P. M. & Bazzaz, F. A. (1993) Birch seedling responses to daily time courses of light in experimental forest gaps and shadehouses. *Ecology*, **74**, 1500–1515.

Whitmore, T. C. (1993) Use of hemispherical photographs in forest ecology: Measurement of gap size and radiation totals in a Bornean tropical rainforest. *Journal of Tropical Ecology*, **9**, 131–151.

Williams, K., Field, C. B. & Mooney, H. A. (1989) Relationships among leaf construction cost, leaf longevity, and light environments in rainforest plants of the genus *Piper*. *American Naturalist*, **133**, 198–211.

Woodrow, I. E. & Mott, K. A. (1988) Quantitative assessment of the degree to which ribulose bisphosphate carboxylase/oxygenase determines the steady-state rate of photosynthesis during sun-shade acclimation in

Helianthus annuus L. *Australian Journal of Plant Physiology,* **15,** 253–262.

WRIGHT, S. J., MACHADO, J. L., MULKEY, S. S. & SMITH, A. P. (1992) Drought acclimation among tropical forest shrubs (*Psychotria,* Rubiaceae). *Oecologia,* **89,** 457–463.

YATES, D. J., UNWIN, G. L. & DOLEY, D. (1988) Rainforest environment and physiology. *Proceedings of the Ecological Society of Australia,* **15,** 31–37.

YODA, K. (1974) Three-dimensional distribution of light intensity in a tropical rainforest of West Malaysia. *Japanese Journal of Ecology,* **24,** 247–254.

ZOTZ, G. & WINTER, K. (1993) Short-term photosynthesis measurements predict leaf carbon balance in tropical rainforest canopy plants. *Planta,* **191,** 409–412.

2

CAM and C₄ Plants in the Humid Tropics

Ernesto Medina

2.1 INTRODUCTION

Climatic diversity between the Tropics of Cancer and Capricorn results from variation in the amount, seasonality, and distribution of rainfall, and from changes in evapotranspiration determined by altitudinal location in tropical mountains. This paper encompasses plant communities in the tropics growing in areas where rainfall is much higher than potential evapotranspiration during most of the year. A convenient separation of the wet and dry realms of tropical environments is provided by Bailey's index, calculated from rainfall and average temperature data (Bailey, 1979). According to Bailey's model, the lowland humid tropics are characterized by average annual temperatures around 26°C and average annual rainfall above 1500 mm. As the temperature decreases with increasing elevation in tropical mountains, the amount of rainfall required for humid climatic conditions also decreases. The sequence of lowland tropical forests that occurs along rainfall gradients, including dry woodlands at the lower end and rainforests at the higher end of rainfall, can also be observed along altitudinal gradients (Beard, 1944).

The plants with Crassulacean Acid Metabolism (CAM) and C_4-photosynthesis (C_4) considered in this paper are those occurring in tropical rain forests, semi-evergreen lowland forests, and cloud forests. In addition, I include lowlands in high rainfall areas that are prone to inundation by rainfall or changes in river levels, where grasslands communities with variable proportions of grasses and sedges gain dominance.

In what follows I will discuss the ecological significance of CAM and C_4 plants in humid tropical ecosystems, based on their distribution and pertinent ecophysiological studies. Although I will attempt to provide a pan-tropical perspective, most of the examples will derive from ecosystems in which I have the most experience—those located in northern South America and the Caribbean. I think, however, that the generalizations apply to the whole of the humid tropical belt of the world.

2.2   PHYSIOLOGICAL PROPERTIES OF CAM AND C₄ PLANTS

The biochemical characterization of the photosynthetic pathways in higher plants was completed many years ago (Hatch, 1976), but research on biochemical and environmental regulation is still a very fertile field (Geiger & Servaitis, 1994; Schulze & Caldwell, 1994). To discuss the field performance of plants with different photosynthetic pathways, a brief description of the most relevant ecophysiological properties is given. In all higher plants, CO_2 is incorporated into carbohydrates in a reaction with ribulose bisphosphate (RuBP) mediated by the enzyme ribulose bisphosphate carboxylase-oxygenase (Rubisco). In C_3 plants, CO_2 diffuses through the stomata from the surrounding air to Rubisco. Under normal O_2 partial pressures (ca. 200 mbar), the carboxylation efficiency of Rubisco is reduced because of its oxygenase activity. The enzyme is capable of oxidizing RuBP leading to a complex biochemical pathway known as photorespiration. Increasing CO_2 partial pressure above normal (up to 1000 µbar) or reducing O_2 partial pressure increases carboxylation efficiency by as much as 40%.

Crassulacean acid metabolism (CAM) plants are characterized by an inverted stomatal rhythm; stomata open at night and remain closed at least during the first half of the day. This rhythm is associated with the activity of PEP-carboxylase in the cytoplasm of chloroplast-bearing cells. This enzyme fixes CO_2 during the night, leading to an accumulation of organic acids, mainly malate, in the large vacuoles characteristic of the photosynthetic tissues of these plants (Phase 1 of Osmond, 1978). Malate is transported back to the cytoplasm at the beginning of the next light period and is rapidly decarboxylated. The resulting high CO_2 partial pressures induce stomatal closure and inhibit photorespiration in the chloroplasts of the same cell (Phase III). PEP-carboxylase activity, separated in time from the Rubisco activity of chloroplasts, generates large internal CO_2 partial pressures increasing the efficiency of carboxylation by Rubisco. Only when the malate accumulated during the previous night has been completely consumed will the CO_2 partial pressure decrease, causing the stomata to reopen. Then atmospheric CO_2 can diffuse directly into the chloroplasts as in C_3 plants (Phase IV). Crassulacean acid metabolism results in comparatively high water-use efficiencies, but lower growth rates than those of C_3 and C_4 plants.

In C_4 plants, the oxygenase activity of Rubisco has been reduced through the development of a highly compartmentalized leaf anatomy and biochemistry. This anatomical specialization allows for CO_2 fixation in mesophyll cells mediated by PEP-carboxylase, an enzyme with high affinity for CO_2 and an insensitivity to O_2. The result of this fixation is malate and/or aspartate, which are transported to the bundle sheath cells where Rubisco-containing chloroplasts are active. Decarboxylation of organic acids in the bundle sheath cells results in high internal CO_2 partial pressure, inhibiting the oxygenase activity of Rubisco. The PEP-carboxylase system located in the mesophyll cells acts then as a CO_2 pump maintaining high CO_2 partial pressures in the bundle sheath. As a consequence, C_4 plants frequently have higher photosynthetic rates than C_3 plants (as photosynthesis is insensitive to O_2 partial pressure), and saturate at lower CO_2 partial pressures than those required by C_3 plants. Other consequences of great ecological significance include a higher water-use efficiency and a higher nitrogen-use efficiency. The latter results from the smaller amounts of nitrogen required to maintain a given carboxylation capacity in the bundle sheath cells (Pearcy & Ehleringer, 1983).

Crassulacean acid metabolism- and C_4-plants have higher natural abundances of ^{13}C, a stable isotope of carbon. This results from the nearly insignificant discrimination of PEP-carboxylase against ^{13}C in comparison with the strong discrimination of Rubisco (Farquhar et al., 1988). Values of $\delta^{13}C$ approaching $-12‰$, indicate that most carbon is entering the plant via PEP-carboxylase, while values approaching $-28‰$ indicate that primary fixation of CO_2 was performed by Rubisco (assuming a $\delta^{13}C$ of the air of $-8‰$ and a discrimination due to diffusion through stomata approaching $-4‰$). Slight variations are expected if the $\delta^{13}C$ of the source is more negative than indicated or if diffusion is restricted by water stress in C_3 plants (Farquhar et al., 1988).

2.3 ECOPHYSIOLOGY AND OCCURRENCE OF CAM PLANTS IN HUMID TROPICAL FORESTS

The criteria to identify a plant with CAM have been described above. It is convenient to distinguish between constitutive and facultative CAM plants (Winter, 1985). The former show CAM characteristics under a wide range of environmental conditions, while the latter only show CAM under conditions of water or salt stress. As will be seen below, the absolute majority of CAM plants in the tropics are constitutive, as the number of facultative species described are very limited. There are two variations in the nocturnal organic acid accumulation process that are considered to be related to CAM and are frequently found in epiphytes of humid forests: CAM-idling and CAM-cycling (Kluge & Ting, 1978; Benzing, 1990). In both processes, nocturnal increases of acid concentration are observed without net CO_2 uptake from the atmosphere. CAM-idling occurs in water-stressed constitutive CAM plants and results from the refixation of respiratory CO_2 (Szarek, Johnson & Ting, 1973; Ting, 1985), while CAM-cycling occurs in tissues without water stress (Ting, 1985). CAM-cycling has been hypothesized to be a precursor of full-CAM plants (Monson, 1989; Martin, 1994), but it is still not clear why lower internal CO_2 partial pressure resulting from fixation of respiratory CO_2 during the night does not influence stomatal behavior.

Several ecophysiological features are essential to understand the abundance and ecological significance of CAM in humid tropical forests:

(a) *High water-use efficiency.* Constitutive CAM plants take up CO_2 mainly during the night, when the leaf-air water vapor saturation deficit is lower. High water-use efficiency associated with the succulent morphology of photosynthetic organs allow the maintenance of a positive carbon balance in drought-prone habitats, and is therefore of paramount importance for the occupation of exposed epiphytic habitats in rainforests.

(b) *Respiratory CO$_2$ recycling.* In many constitutive CAM plants, simultaneous measurements of CO$_2$ exchange and organic acid accumulation during the night frequently reveal that the increase in acid concentration is larger than that expected from the amount of CO$_2$ taken up by PEP-carboxylase. Dark CO$_2$ fixation should yield 2 equivalents of acidity (malic acid) per CO$_2$ molecule taken up. In the Crassulaceae and Agavaceae, this 2:1 stoichiometry held for various temperature conditions (Nobel & Hartsock, 1978; Medina, 1982); however, this is not the case with several species of the Bromeliaceae and Orchidaceae, of the genus *Clusia*, and in CAM ferns. This difference is apparently due to refixation of respiratory CO$_2$ (Griffiths, 1988). Respiratory CO$_2$ recycling can be calculated in absolute or relative terms. Absolute recycling (AR) is the amount of CO$_2$ originated in mitochondrial respiration converted to malic acid during the night (= Δmalic acid – CO$_2$ taken up at night). The relative contribution of respiratory CO$_2$ recycling to total acid accumulation is calculated as: (Absolute recycling / Δmalate) × 100.

Recycling of respiratory CO$_2$ may be important because it helps to maintain a positive carbon balance under stressful environmental conditions by reducing carbon losses due to nocturnal respiration. It is not yet clear if the CO$_2$ recycled in many succulent CAM plants originates from the respiration of the CAM cells themselves or from the respiration of non-photosynthetic tissues (see Loeschnen et al., 1993; Martin, 1994). Besides, there is no proper account of the competition for substrate between CAM resulting in malic acid synthesis and the consumption of PEP in the respiratory pathway. CAM may regulate the activity of the respiratory pathway by modifying the availability of pyruvate for the tricarboxylic acid cycle (Morel, 1979). On the other hand, nocturnal acidification is inhibited in absence of oxygen (Watanabe et al., 1992), probably because of the ATP requirement for malate transport into the vacuole (Lüttge et al., 1981). The question is important to understand the carbon balance of CAM plants, and to explain the highly variable recycling values reported for different constitutive CAM plants (see Griffiths, 1989; Medina, 1990; Loeschen et al., 1993; Martin, 1994).

(c) *Adaptation to shady environments.* The numerous taxa with CAM in humid tropical forests that can grow and compete with C$_3$ plants in shady environments indicate that CAM by itself does not appear to be a liability in low-light environments. In humid forest environments shade-adapted CAM plants have growth rates and net carbon gains similar to those of C$_3$ plants that are limited by low light (Martin, Eades & Pither, 1986; Medina, 1990). It seems that CO$_2$ fixation during the night by CAM plants might be favored by the high CO$_2$ concentrations of the air under the canopy of humid tropical forests (Martin, 1994).

2.3.1   The epiphytic synusia in humid tropical forests

One of the most vivid scientific accounts on the epiphytic flora of tropical latitudes was first published by Schimper (1884), who described the epiphytic vegetation of the West Indies. This paper was followed by a classical monograph of the epiphytic vegetation of the American continents (Schimper, 1888), where taxonomy, ecology, and physiology of epiphytes was analyzed with insight that continues to inspire research on these plants. The ecophysiological interest in the epiphytic synusia in the humid tropics is exemplified by a large body of literature that has recently appeared and is summarized in reports of specialized workshops (e.g., Lüttge, 1989) and monographs (Benzing, 1990).

Schimper (1888) listed 22 families and 199 genera in the epiphytic communities of northern South America. The current numbers are much higher, amounting in the neotropics to 36 angiosperm families, 10 pteridophyte families, and several thousand species (Madison, 1977; Kress, 1986). In these families, the following contain CAM or CAM-related species: Asclepiadaceae, Bromeliaceae, Cactaceae, Clusiaceae, Crassulaceae, Gesneriaceae, Orchidaceae, Piperaceae, Polypodiaceae, and Rubiaceae (Table 2.1). The orchids and bromeliads contain the largest number of CAM genera, but each family has species that are of ecophysiological interest because of variability in their metabolism under natural conditions. This variability offers numerous insights into the evolution of this type of carbon acquisition in higher plants.

All epiphytic habitats potentially restrict plant growth as a result of low water availability. In addition, the supply of nutrients is restricted because of the exclusion of the plant roots from soil nutrient reserves. Many epiphytes are therefore dependent on their capability to accumulate organic debris and water by way of specialized morphology. This is the case of most tank bromeliads and nest-forming pteridophytes and orchids.

The availability of water and nutrients varies with the position of the epiphyte within the forest. Epiphytes located in the higher branches are more dependent on the accumulation of dust and nutrients in rainfall, and are exposed to higher light intensities and to more frequent dry spells. In lower branches, evapotranspiration demands are smaller, water availability is greater, and nutrient availability is improved due to the accumulation of organic debris and the nutrient-

Table 2.1.   *Families and genera in which epiphytic CAM species of humid tropical forests have been reported. (with information from Coutinho 1963, 1969; McWilliams, 1970; Medina 1974; Medina et al., 1977, 1989; Griffiths & Smith, 1983; Ting et al., 1985; Winter et al., 1983; 1985; Earnshaw et. al., 1987; Griffiths, 1989; Benzing, 1990; Lüttge, 1991; Martin, 1994; Franco et al., 1994).*

Family	Genera with CAM species		
Polypodiaceae	*Pyrrosia* *Microsorium*		
Bromeliaceae	*Acantostachys* *Aechmea* *Araeococcus* *Billbergia* *Bromelia* *Canistrum* *Hohenbergia*	*Neoregelia* *Nidularium* *Portea* *Quesnelia* *Streptocalyx* *Tillandsia* *Wittrockia*	
Orchidaceae ca. 500 genera and more than 20,000 epiphytic	*Bulbophyllum* *Cadetia* *Campylocentrum* *Cattleya* *Chilochista* *Cymbidium* *Dendrobium* *Epidendrum* *Eria* *Flickingeria*	*Luisia* *Micropera* *Mobilabium* *Oberonia* *Oncidium* *Phalenopsis* *Pholidota* *Plectorrhiza* *Pomatocalpa* *Rhinerrhiza*	*Robiquetia* *Saccolabiopsis* *Saccolabium* *Sarchochilus* *Schoenorchis* *Taeniophyllum* *Thrixspermum* *Trachoma* *Trichoglottis* *Vanda*
Asclepiadaceae	*Hoya* *Dischidia*		
Cactaceae 25 epiphytic genera, most probably all CAM	*Epiphyllum* *Hylocereus* *Rhipsalis* *Strophocactus* *Zygocactus*		
Clusiaceae	*Clusia* *Oedematopus*		
Crassulaceae	*Echeveria* *Sedum*		
Gesneriaceae	*Codonanthe*		
Piperaceae	*Peperomia*		
Rubiaceae	*Myrmecodia* *Hydnophytum*		

enriched throughfall. On the other hand, these plants must cope with lower light intensities.

2.3.2 Occurrence of CAM in specific plant groups

2.3.2.1 Pteridophytes

Only species of the genera *Pyrrosia* (Hew & Wong, 1974; Ong, Kluge & Friemert, 1986; Winter, Osmond & Hubick, 1986; Kluge et al., 1989b) and *Microsorium* (Earnshaw et al., 1987) have been reported as CAM plants among ferns. *Pyrrosia longifolia* can be found growing under fully exposed or shaded conditions in humid tropical forests. This plant performs CAM in both conditions, but both acidification and Phase IV CO_2 fixation are higher in the exposed plants. No differences in $\delta^{13}C$ could be detected, indicating that in both populations, the proportion of total carbon gain derived from night CO_2 fixation is similar (Winter, Osmond & Hubick, 1986). Photosynthesis of sun and shade fronds of *Pyrrosia* saturate at similar light intensities between 200–400 µmol quanta $m^{-2} s^{-1}$. Utilization of high light in sun fronds might have been precluded by the low nitrogen content of those fronds. *Pyrrosia* spp. appear to have a high degree of respiratory CO_2 recycling (Griffiths, 1988; Griffiths et al., 1989).

Shade plants transferred to high light and, to a lesser degree, exposed plants, exhibited photoinhibition (Winter, Osmond & Hubick, 1986). Similar observations were made by Griffiths et al. (1989) in sun- and shade-adapted fronds of *Pyrrosia piloselloides* under natural conditions. These authors suggested that the release of CO_2 derived from nocturnal CO_2 fixation could have alleviated photoinhibition during the day in the sun-adapted populations. True shade plants, such as *Pyrrosia confluens*, were not able to acclimate to high light intensity (Adams, 1988). For sun fronds of *P. confluens*, it was shown that their habitat light regime was not static—periods of low light during the day allowed for recovery from photoinhibition.

The question of how CAM may have evolved within the Polypodiaceae has been discussed by Kluge, Avadhani and Goh (1989a) and Benzing (1990). All relatives of the genus *Pyrrosia* grow in humid, shady environments, therefore, it appears improbable that the CAM species derived from xerophytic ancestors. In this particular case, it appears that high light and frequent low humidity, characteristic of epiphytic environments, may have been the selective factors leading to the development of CAM in this genus. In addition, ferns are a group of plants highly dependent on continuous water availability during their gametophytic stage. This makes more feasible the assumption that CAM ferns are derived from shade forms in humid forests, and that CAM evolved secondarily in the epiphytic environment (Kluge, Avadhani and Goh, 1989).

2.3.2.2   Epiphytic Gesneriaceae and Piperaceae

The Gesneriaceae and Piperaceae are families with a large number of epiphytic species in the humid tropics, many of which are succulent. In a few species of these families (*Codonanthe crassifolia* in the Gesneriaceae and about 15 species within the genus *Peperomia*), CAM activity has been detected only as acid accumulation during the night period. None of those species shows the characteristic nocturnal CO_2 uptake of CAM plants (Ting et al., 1985; Guralnick, Ting & Lord, 1986). These species also exhibit evidence of CAM in their isotopic composition. Although their $\delta^{13}C$ values are always identical to those expected for C_3 plants, indicating that all carbon used for growth derives from daytime CO_2 uptake, their δD values are more similar to those of CAM plants than to those of C_3 or C_4 plants (Figure 2.1) (Ting et al., 1985). CAM plants are comparatively rich in deuterium as a result of diffusive and biochemical fractionations (Ziegler et al., 1976; Sternberg, DeNiro & Johnson, 1984).

2.3.2.3   Epiphytic orchids and bromeliads of humid forests

Orchids and bromeliads contain the largest number of epiphytic species that are constitutive CAM, plants. The orchids occupy pantropical epiphytic habitats, while the bromeliads are restricted to the neotropics. Both families contain epiphytic and terrestrial species that exhibit typical CAM, expressed in gas exchange patterns, nocturnal malic acid accumulation, and reduction of non-structural carbohydrate content during the night (Coutinho 1963, 1969; McWilliams, 1970; Medina, 1974; Neales & Hew, 1975; Goh et al., 1977; Griffiths & Smith, 1983; Lüttge et al., 1985; Goh & Kluge, 1989). In both families, a considerable number of epiphytic species are found that perform CAM under partial or heavy shade conditions. A thorough account on the morphology, physiology, and ecology of epiphytes in these two families was published by Benzing (1990), while the ecophysiology of bromeliads was recently revised by Martin (1994).

Distribution of epiphytic species of orchids and bromeliads is correlated with atmospheric humidity. In Trinidad, the proportion of CAM

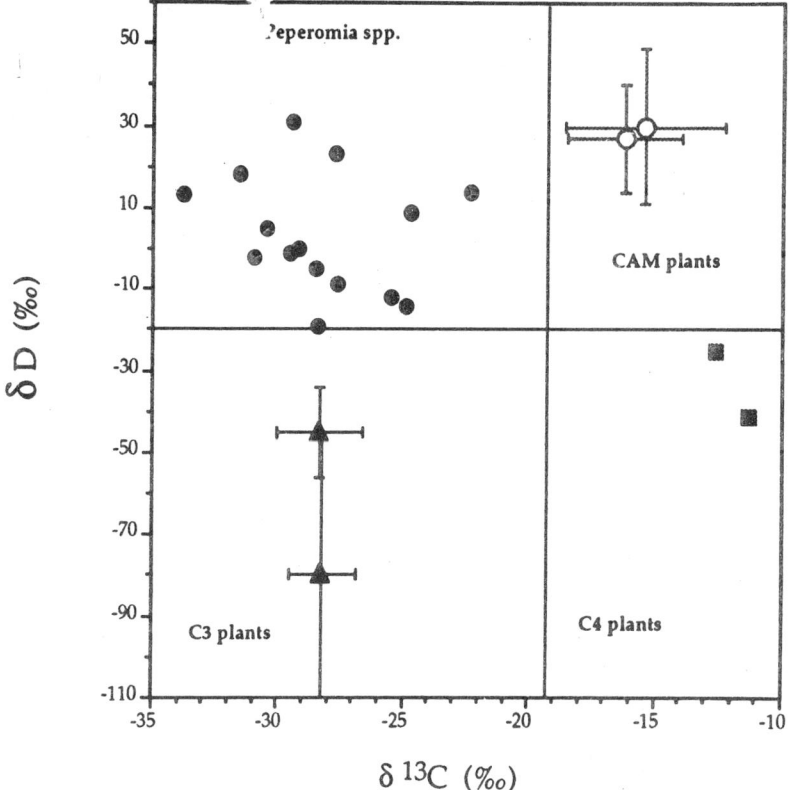

Figure 2.1.   *Distribution of δD and δ¹³C values of species of the genus Peperomia compared to C₃, C₄ and CAM species (with data from Ting et al. 1985)*

epiphytes decreases with altitude in montane forests (Smith, 1989); similar results were obtained with CAM orchids in Papua, New Guinea (Earnshaw et al., 1987).

The larger genera of bromeliads and orchids contain both C_3 and CAM species. The genus *Tillandsia* in the Bromeliaceae, for instance, has species ranging from dry to humid forests. CAM species predominate in the former, while C_3 species predominate in the latter. Another bromeliad genus, *Aechmea*, contains only CAM species, some of which are restricted to the floor of humid forests. In the Orchidaceae, similar photosynthetic diversity can be observed in the genera *Bulbophyllum* (1000 epiphytic species) and *Dendrobium* (900 epiphytic species). In these genera, there is an almost continuous variation in $\delta^{13}C$ values,

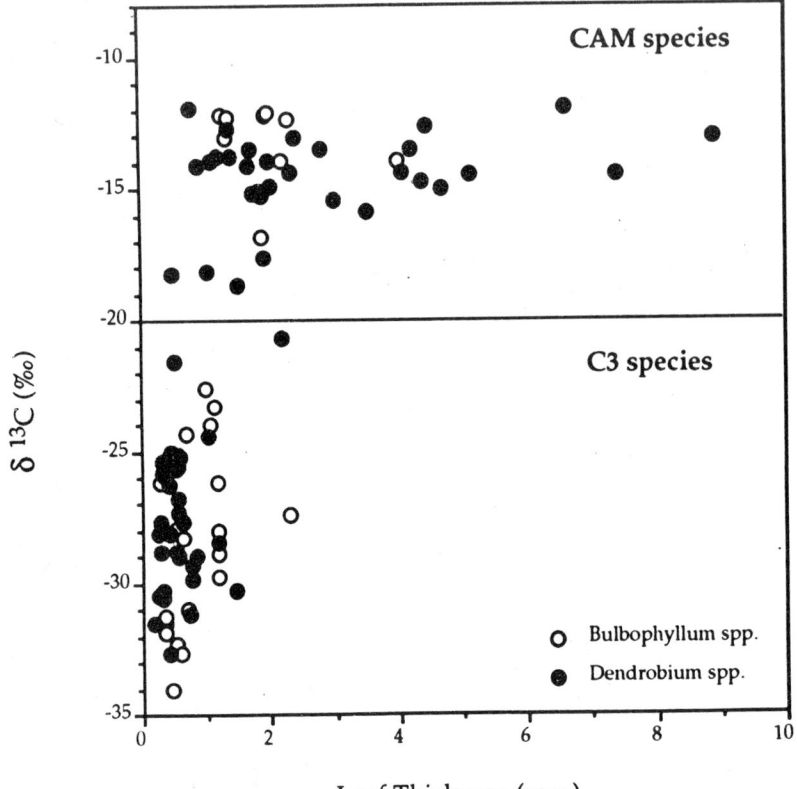

Figure 2.2.   Relationship between leaf thickness and $\delta^{13}C$ values of leaves of several orchid species belonging to the genera Bulbophyllum *and* Dendrobium *(with data from Winter, Osmond & Hubick, 1986).*

from less than $-30‰$, to $-12‰$. CAM activity in these orchid genera is correlated with the development of succulence (Figure 2.2). This morphological property is associated both with the presence of cells with big vacuoles appropriate for accumulation of large quantities of malic acid during the night, and with the development of water-storage parenchyma, which are often photosynthetically inactive. For example, the leaf thickness of high altitude orchids in upper montane forests of Papua, New Guinea is due to the development of a chlorophyll-free hypodermis (Earnshaw et al., 1987), very similar to that described for *Peperomia* spp. (Ting et al., 1985) and *Codonanthe* spp. (Guralnick, Ting & Lord, 1986; Medina, Delgado & García, 1989).

In many orchid species, a strong reduction in the development of the shoots is observed. In addition, roots of many orchid species contain chloroplasts that are photosynthetically active (Benzing et al., 1983). In some leafless species, chlorophyll-containing roots are capable of CAM (Cockburn, Goh & Avadhani, 1985; Winter et al., 1985). The nutritional relationships in these shootless orchids deserve investigation because chloroplast-bearing cells, specialized in photosynthesis, and mycorrhizal cells, specialized in nutrient uptake, coexist in the same root tissues (Winter et al., 1985).

The three subfamilies of Bromeliaceae (Pitcairnioideae, Tillandsioideae, and Bromelioideae) exhibit rather clear differentiation in habitat preferences and ecophysiology (Medina, 1990). Species of Pitcairnioideae are predominantly terrestrial, and in humid environments, regardless of habitat light level, they are exclusively C_3 (for example, species of Pitcairnia, Brocchinia, and Navia). In drier savanna environments in central Brazil or subtropical semi-arid areas in Argentina and Chile, CAM genera have evolved with a strong succulent habit and high resistance to dessication (for example, species of Abromeitiella, Dyckia, Encholirion and Deuterocohnia).

The Tillandsioideae contain mostly epiphytic species (only a few species of Vriesea and Tillandsia are lithophytes), which are represented by sun and shade C_3 species and sun CAM species. All these variations may be observed within the genus Tillandsia. Other Tillandsioid genera, such as Catopsis, Glomeropitcairnia, and Vriesea, appear to be exclusively C_3. In the genus Guzmania, an intermediate species has been described (G. monostachia, Medina et al., 1977; Griffiths and Smith, 1983), a condition which is rare in the family.

In the Bromelioideae, a more complicated picture is found. Practically all the species studied thus far, with the exception of those of the genus Greigia, are CAM plants. Greigia is a terrestrial genus found in northern South America and central and southern Chile, growing under high light intensity in relatively humid environments. The Bromelioideae contains a large number of both terrestrial and epiphytic CAM species, some of which grow well in both habitats (e.g., Aechmea aquilega). Both terrestrial and epiphytic species may be well adapted to grow in deep shade. For example, Aechmea, subgenus Chevaliera, has several shade-adapted species that perform full CAM (Pfitsch & Smith, 1988; Medina, 1990). Other remarkable examples are Ananas, Canistrum, Nidularium, Portea, and Quesnelia. The genus Ananas deserves further consideration. The wild species of the genus grow under partial or deep shade in humid forests (Ananas parguazensis, A. ananassoides), while primitive cultivars of Ananas comosus (cultivars Brecheche and Panare) are cultivated under partial shade along the Orinoco river. The latter can also grow well under high light intensity, provided that water availability is high in well-drained soils (Medina et al., 1991). Interestingly, the commercial cultivars of *Ananas comosus* are grown under full sun exposure, frequently in relatively dry environments. Perhaps CAM in this genus made it possible to convert drought-sensitive plants into drought-resistant cultivars.

The distribution of constitutive CAM species within the Bromeliaceae has been well documented using $\delta^{13}C$ values. The distribution of all the values published thus far by different authors reveals the predominance of C_3 types in the Pitcairnioideae, the predominance of CAM species in the Bromelioideae, and the similar abundance of C_3 and CAM types in the Tillandsoideae (Medina, 1990; Martin, 1994).

Analysis of distribution of a number of epiphytic CAM bromeliads showed that they can occur in a wide range of light conditions, although their abundance tends to increase under full sun exposure (Pittendrigh, 1948; Smith, Griffiths & Lüttge, 1986). Detailed physiological analyses of gas exchange and nocturnal acidification in the highly morphologically-modified *Tillandsia usneoides* indicate that daily carbon balance reaches a maximum at intermediate light intensities (Martin, Eades & Pither, 1986). These authors found that CAM in this species saturates at integrated photosynthetic active radiation fluxes of around 10 mol m^{-2} day^{-1}, approximately one third of the amount available in a clear day in the tropics. These results have been interpreted as an indication of shade adaptation of this species in spite of its frequent occurrence under full sun exposure. The process of shade adaptation of CAM plants in view of the relatively higher energetics requirement merits more research at the biochemical level.

2.3.2.4 Epiphytic cacti

In the Cactaceae, the differentiation between terrestrial and epiphytic species is more pronounced. There is no genus with both terrestrial and epiphytic species; however, several climbing and decumbent species are found within the genera *Acanthocereus* and *Heliocereus*. The epiphytic cacti are confined within the tribes Rhipsalidanae (8 genera), Epiphyllanae (9 genera), and Hylocereanae (9 genera) within the subfamily Cereoideae (Britton & Rose, 1963). The photosynthetic metabolism of epiphytic cacti has been studied in only a few genera (Table 2.1), but there is little doubt that all the species are most likely constitutive CAM plants. Species growing in deep shade, such as *Strophocactus witti* (Medina et. al., 1989), are damaged when exposed to high light intensity, yet they show typical CAM gas exchange and acid accumulation.

2.3.2.5 Occurrence of CAM in arborescent species

Arborescent CAM plants are well known in semi-arid areas within the Cactaceae in America, the Euphorbiaceae in Africa, and the Didiereaceae in Madagascar. In humid forests, no arborescent CAM plants were known until Tinoco-Oranjuren and Vázquez-Yánes (1983) reported strong diurnal acid oscillations in *Clusia lundelli* (Clusiaceae, previously included in the family Guttiferae) in the rainforest of Los Tuxtlas in southern Mexico. This species is a strangler, beginning its life cycle as an epiphyte, then developing roots that eventually anchor in the soil (Holbrook & Putz, Chapter 13). The subsequent development of the strangler may lead to the death of the host tree. This finding of CAM was thoroughly documented in the large tree strangler *Clusia rosea* in the US Virgin Islands (Ting et al., 1985; Ball et al., 1991). This species is widely distributed in the Caribbean region and grows on a range of soil types. Numerous papers have confirmed these observations in several species of the genus *Clusia* in natural conditions (Ting et al., 1987; Sternberg et al., 1987; Popp et al., 1987; Borland et al., 1992; Winter et al., 1992; Franco et al., 1994). The process has also been analyzed under controlled conditions where the effects of light and water stress, temperature, and nutrient supply on the pattern of acid accumulation and gas exchange have been measured (Schmitt, Lee & Lüttge, 1988; Lee, Schmitt & Lüttge, 1989; Franco, Ball & Lüttge 1990, 1991, 1992; Haag-Kerwer, Franco & Lüttge, 1992).

The gas exchange of *Clusia* spp. is of great ecophysiological interest because of a number of peculiarities that can be summarized as follows: a) several *Clusia* species present the largest acidity oscillations ever measured in CAM plants (Popp et al., 1987; Borland et al., 1992), b) a large fraction of the acidity observed in *Clusia* species is due to citric acid (Popp et al., 1987; Franco, Ball & Lüttge, 1990); c) in spite of nocturnal CO_2 fixation and large diurnal acidity oscillations, several *Clusia* species have values of $\delta^{13}C$ more typical of C_3 than of CAM plants (Figure 2.3). On a molar basis, the nocturnal oscillations of malate are more pronounced than those of citrate, as is usual in typical CAM plants (Figure 2.4A). On the other hand, to explain the increase in the proton concentration during the night, in all CAM *Clusia* species with nocturnal acidification investigated so far, it is necessary to take into consideration the amount of citrate accumulated. The addition of the two acidity equivalents per mol corresponding to malic acid with the three acidity equivalents per mol corresponding to citric acid correlates linearly with the amount of protons accumulated during the night period (Figure 2.4B). With species of this genus, no conclusion on CAM activity can be deduced by measuring acidity oscillations alone because citric acid accumulation is not related to net carbon gain by the plant (Franco, Ball & Lüttge, 1992; Haag-Kerwer, Franco & Lüttge, 1992).

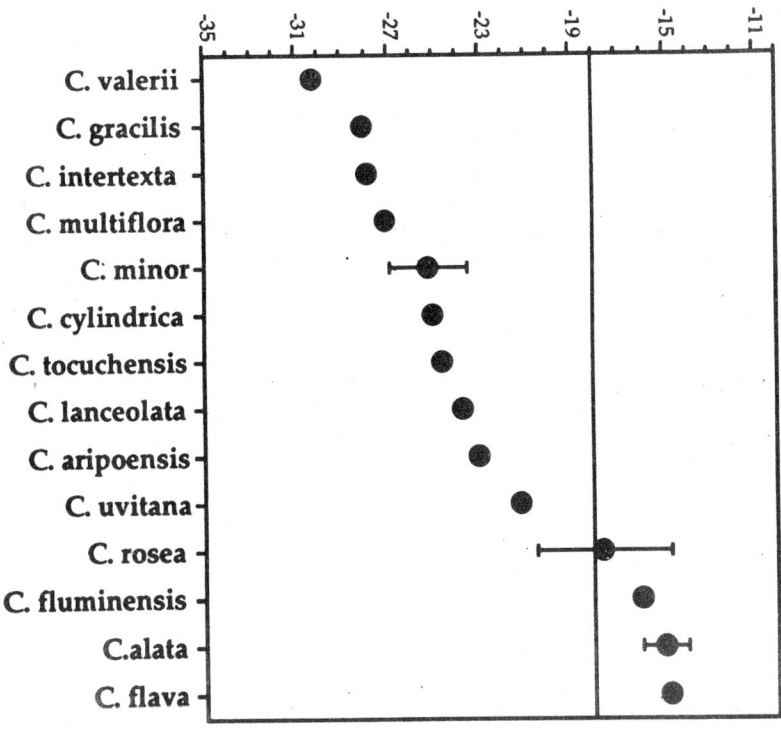

Average δ¹³C

Figure 2.3.   Ordination of δ¹³C values of Clusia spp. *All the species included in the figure have substantial diurnal acidity oscillations (data from Ting et al., 1985, 1987; Popp et al., 1987; Sternberg et al., 1987; Franco et al. 1994).*

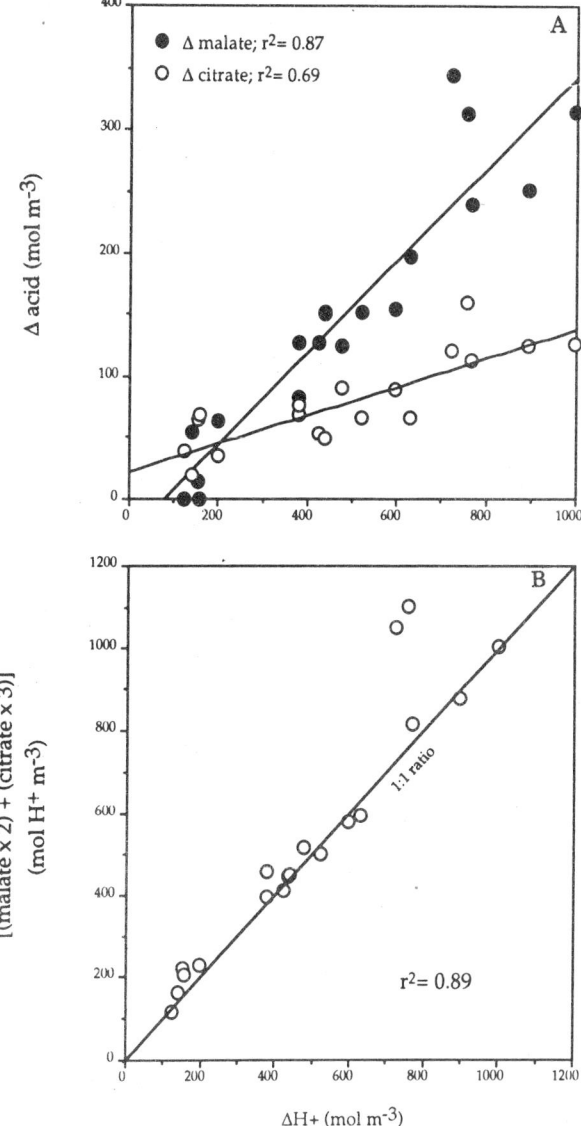

Figure 2.4.   A. *Relationship between Δnocturnal acidification and Δmalate and Δcitrate measured in the same leaf extracts.*
B. *Relationship between equivalents of acidity corresponding to the sum of malate and citrate accumulated during the night and the total amount of protons accumulated during the same period (data from Popp et al. 1987; Franco et al. 1990, 1991, 1992; Borland et al. 1992).*

Photosynthetic performance in *Clusia* species provided with sufficient water and high light range from the pure C$_3$ type (e.g., in *Clusia multiflora* and *C. venosa*) to typical CAM (e.g., in *C. minor* and *C. major*) (Franco, Ball & Lüttge, 1990). Furthermore, gas exchange of *Clusia* CAM species has been shown to be highly flexible, going from typical CAM to typical C$_3$, including continuous net fixation of CO$_2$ during a 24 hour cycle, depending on light intensity during growth, day-night temperature cycles, drought, and even nutrition (Schmitt, Lee & Lüttge, 1988; Lee, Schmitt & Lüttge, 1989; Franco, Ball & Lüttge, 1990; Zotz & Winter, 1993; Holbrook & Putz, Chapter 13).

Light requirement for saturation of photosynthesis varies considerably depending on the light intensity provided during growth. *Clusia minor* cultivated in growth chambers at comparatively low light saturate at low light intensity (120 µmol m^{-2} s^{-1}; Schmitt, Lee & Lüttge, 1988). In the field, shade plants of the same species saturated around 200 µmol m^{-2} s^{-1}, while exposed leaves saturated at nearly 600 µmol m^{-2} s^{-1} (Borland et al., 1992).

Drought induced contrasting effects in *Clusia* species that fix substantial amounts of CO$_2$ during the night. In *C. uvitana*, daytime uptake decreases rapidly without irrigation, while nocturnal fixation increases throughout the treatment (Winter et al., 1992). In *C. minor* and *C. rosea*, both photosynthesis and night CO$_2$ fixation decrease with water stress, while in *C. lanceolata*, nocturnal uptake is barely affected after 16 days without irrigation (Franço, Ball & Lüttge, 1992).

In most species where CO$_2$ exhange and acid accumulation have been measured simultaneously, large rates of respiratory CO$_2$ recycling have been reported. Interestingly, the actual level of recycling varies considerably among species and growth conditions. Under natural conditions in the Virgin Islands, Ball et al. (1991) reported that in *C. rosea*, 20% of malate accumulation derives from refixation of respiratory CO$_2$. In the laboratory under favorable growth conditions, for two species in which nocturnal CO$_2$ fixation provided more than 70% of the total leaf carbon acquisition, percent recycling varied from 65 in *C. alata* to 98 in *C. major* (Franco et al., 1990). Recycling of respiratory CO$_2$ does not appear to be significant under low light conditions (Franco, Ball & Lüttge, 1991; Haag-Kerwer, Franco & Lüttge, 1992). The reasons for these differences are not clear. The respiratory metabolism during the dark period and the fate of the resultant CO$_2$ certainly deserves special attention in the immediate future, not only in *Clusia* but in all CAM plants (Medina et al., 1993).

In conclusion, *Clusia* species present a highly diversified photosynthetic metabolism, perhaps explaining the wide range of habitats occupied by this genus in the tropics (Lüttge, 1991; Holbrook & Putz, Chapter 13). CAM operates as a mechanism to both conserve water and carbon under water stress, and to recycle respiratory CO_2 under high light conditions (Franco, Ball & Lüttge, 1992). Substantial contributions of nocturnal CO_2 fixation to long-term carbon gain have been documented for only a few species so far. In most species, nocturnal malate fluctuations appear to be related to the recycling of respiratory CO_2. The role of high concentrations and nocturnal increases of citrate in the leaves of most *Clusia* species examined so far may be indirectly related to CAM because it may increase the buffering capacity of the vacuole, maintaining favorable pH conditions for the activity of tonoplast ATPase transporting 2 H^+/ATP (Franco, Ball & Lüttge, 1992).

2.4 C_4 PLANTS IN THE HUMID TROPICS

In the tropics, grasslands are frequently the dominant vegetation type in flood-prone areas or swamps (Hueck and Seibert, 1972). The flooded grasslands and palm savannas in South America occupy extended areas in the lowlands of the Magdalena and Atrato rivers in Colombia, the lower Orinoco Llanos in southwestern Venezuela and Colombia (Sarmiento 1984), the flooded savannas of the Orinoco Delta (Vareschi, 1980), the Llanos de Mojos in Bolivia (Haase, 1989; Gunnemann, 1991), and the Gran Pantanal region in southern Brazil (Hueck, 1966). This vegetation type is also widespread in tropical Africa (Vesey-Fitzgerald, 1963). Along large rivers such as the Orinoco and the Amazonas with pronounced seasonal variations in river water levels, a vegetation dominated by floating grasslands and trees tolerating flooding periods of several months gains dominance (Junk, 1970). These areas have high economic significance because they can be cultivated during the low-water season and are rich in nutrients deposited by the river during the high-water season (Junk, 1989).

Besides tropical wet grasslands, a few arborescent C_4 species of the genus Euphorbia in Hawaii have been described (Pearcy & Troughton, 1975; Robichaux & Pearcy, 1980). Two species, *Euphorbia forbesii* and *E. rockii*, grow as understory trees in Hawaiian rainforests, and a few other species grow as shrubs in open wet environments. The study by Pearcy (1983) on *Euphorbia forbesii* raised considerable interest because of the unique adaptation of the C_4 pathway to low-light environments. In this species, the C_4 pathway did not confer any competitive advantage in the shaded cool environment, but it was not deleterious, as the species had similar growth rates as other understory trees in the same forest (Pearcy, 1983; Pearcy & Calkin, 1983). The studies of Hawaiian *Euphorbia*, a genus with a large number of C_4 species distributed in hot, high-light environments in the tropics, suggests that succesful establishment in a given site is highly dependent on the flexibility of photosynthate allocation during growth. The successful adjustment of energy requirements of a C_4 species to low light suggests a larger adaptability of this photosynthetic pathway than generally recognized.

2.4.1 Composition and productivity of flooded savannas

Flooded savannas have a characteristic zonation of vegetation according to the duration of flooding (Escobar & González-Jiménez, 1979; Gunnemann, 1991). The higher areas correspond to seasonal savannas and are always dominated by C_4 grasses. At lower levels, duration of the inundation period increases, promoting the establishment of C_3 grasses, particularly in areas where nutrient availability is high. In still lower areas, the inundation period lasts for more than 9 months, and floating C_3 grasses dominate in relatively nutrient-rich environments, while floating C_4 grasses occur in more stressful environments (lower nutrient availability or mechanical stress exerted by river water flow). This sequence has been described in detail in the flooded savannas of southwestern Venezuela (Escobar and González-Jiménez, 1979) and more recently in the savannas de Mojos in Bolivia (Haase, 1989; Haase, & Beck 1989; Gunnemann, 1991). Essentially the same conditions prevail in other flooded savannas in central Brazil. The production of organic matter is related to the duration of the flooding period. In Venezuela, the lower areas dominated by C_3 grasses are more productive (Escobar and González-Jiménez, 1979; Medina and Motta, 1990), while in Bolivia, the Tajibales (marshy grasslands), with a mixture of C_3 and C_4 grasses are the most productive community (Gunnemann, 1991) (Table 2.2).

2.4.2 Photosynthetic performance of C_3 and C_4 herbaceous species from tropical flooded areas

There are a number of dominant C_4 plants, mainly grasses and sedges, in tropical flooded grasslands and tropical swamps (Medina,

Table 2.2. Aboveground organic matter production in seasonally flooded grasslands in Llanos de Mojos, Bolivia (after Gunnemman; 1991). Dominant species are indicated by the annual average dominance figure determined following the method of Braun-Blanquet. Biomass production was estimated as the difference between maximum and minimum total aboveground biomass in burned plots. Photosynthetic types from Downton (1975), Raghavendra & Das (1978), and Medina & Motta (1990).

Topographical level	Dominant species	Dominance	Photosynthesis type	Production g m^{-2} year^{-1}
Semi-altura (seasonal grassland)				762
	Cynodon dactylon	(4.0)	C_4	
	Leersia hexandra	(1.8)	C_3	
	Cyperus surinamensis	(1.1)	C_4	
	Eragrostis acutiflora	(0.9)	C_4	
Tajibal (marshy grassland)				1022
	Paspalum plicatulum	(3.5)	C_4	
	Leersia hexandra	(1.3)	C_3	
	Panicum laxum	(0.9)	$C_3 - C_4$	
Bajío (flooded grassland)				758
	Hemarthria altissima	(1.9)	C_4	
	Leersia hexandra	(1.7)	C_3	
	Melochia arenosa	(1.1)	C_3 (Dicot)	
	Cyperus surinamensis	(1.0)	C_4	
	Diodia kuntzei	(1.0)	C_3 (Dicot)	
	Acroceras zizanoides	(0.7)	C_3	
	Panicum tricholaenoides	(0.7)	C_4 (?)	
	Andropogon bicornis	(0.5)	C_4	

de Bifano & Delgado, 1976; Hesla, Tieszen & Imbamba, 1982; Jones 1987, 1988; Junk 1991; Piedade, Junk & de Mello, 1992). They all have the typical characteristics of C_4 photosynthesis, such as higher water-use and nitrogen-use efficiencies compared to the C_3 species. As for C_3 plants, the development of aerenchyma or hollow stems in soil-rooted, flood-tolerant sedges and grasses is an absolute requirement for successful establishment and competition in wetlands.

There are exclusively C_4 genera in the Poaceae (e.g. *Andropogon, Axonopus, Heteropogon, Paspalum*) and the Cyperaceae (e.g., *Bulbos-*

tylis, Fimbristylis, Kyllinga, and *Mariscus*). These genera have species occurring both in dry and wet areas. Within genera including both C$_3$ and C$_4$ species, such as *Panicum* in the Poaceae and *Cyperus* and *Eleocharis* in the Cyperaceae, C$_4$ species tend to be more abundant in the drier and hotter areas, while the C$_3$ species dominate in more humid environments. Indeed, for grasslands in Australia, Africa, and South America, relative abundance of C$_4$ grasses and sedges is positively correlated with hot and wet summers, and they decrease in dominance in cool, moist environments (Meinzer, 1978; Tieszen et al., 1979; Rundel, 1980; Ellis, Vogel & Fuls, 1980; Hesla, Tieszen & Imbamba, 1982; Hattersley, 1983; Takeda et al., 1985; Ueno, Samejima & Koyama, 1989).

Only a few intermediate C$_3$-C$_4$ species have been reported for tropical Poaceae and Cyperaceae. Those are the species of the 'laxum' group in the genus *Panicum* (Brown & Brown, 1975; Medina & Motta, 1990) and species in the sedge genus *Eleocharis* (Ueno, Samejima & Koyama, 1989).

Photosynthetic rates of C$_4$ grasses and sedges under field conditions are not as high as those of cultivated species (Jones, 1987, 1988; Medina, 1986), possibly because they are limited by the supply of nitrogen (Jones, 1988; Piedade, Junk & de Mello, 1992). In tropical savanna grasses, photosynthetic rate is linearly related to the leaf nitrogen content (Baruch, Ludlow & Davis, 1985). In environments with high water availability, the greater water-use efficiency characteristic of C$_4$ grasses does not provide any competitive advantage against C$_3$ plants; however, many swampy environments in the tropics are reportedly limited by nitrogen. Under these conditions, C$_4$ plants may be better competitors compared to C$_3$ plants. Jones (1988) reported nitrogen-use efficiencies between 12 and 19 μmol CO$_2$ g N^{-1} s^{-1} in *Cyperus* spp., compared to 5 μmol CO$_2$ g N^{-1} s^{-1} in the C$_3$ *Typha latifolia* coexisting in the same swamp. For grasses from seasonal savannas grown under optimal conditions, Baruch, Ludlow and Davis (1985) reported values ranging from 30 to 40 μmol CO$_2$ g N^{-1} s^{-1}.

Most of the semi-aquatic C$_4$ plants in the tropics are characterized by very high production rates and are able to accumulate large amounts of biomass quickly (Table 2.3). Piedade, Junk and de Mello (1992) indicate that the relatively small areas covered by *Echinochloa polystachia* in the Amazon river may constitute important sinks for CO$_2$. These authors estimated that an area of about 5,000 km^2 dominated by *E. polystachia* can consume about 75 $\times$ 10^6 tons of carbon per year. Even considering the high rate of organic matter

Table 2.3.   Maximum aerial biomass recorded in C_3 and C_4 species in flooded areas of the Amazonas and Apure rivers and tropical swamps in Africa (Data from Escobar & González-Jiménez, 1979; Escobar, 1977; Jones, 1988; Junk, 1991; Piedade, Junk & de Mello, 1992).

Species	Biomass ton ha^{-1}
C_3 species	
Panicum laxum	3.2
Leersia hexandra	3.7
Luziola spruceana	5.5
Typha latifolia	3.5
Oryza perennis	17.2
C_4 species	
Paspalum repens	13.7–22.1
Cyperus papyrus	20.5
Cyperus latifolius	21.7
Paspalum fasciculatum	4.4–51.2
Echinochloa polystachia	70–80

decomposition in these hot and wet environments, substantial amounts of this carbon are transported down river to the oceans or are deposited in anoxic layers in the bottoms of lakes along the river.

2.4.3   Composition and growth rates of the floating meadows and the 'varzea' flooded vegetation

The growth patterns of the grasses in flooded grasslands and in 'Varzeas' are not homogeneous. Junk (1991) distinguishes between terrestrial and aquatic phases. C_4 grasses such as Paspalum fasciculatum and Paratheria prostrata reach their peak biomass before the highest inundation level, while others such as Paspalum repens and Echinochloa polystachia attain the largest biomass levels in the middle of the flooding period. An example from the Venezuelan savannas comparing the growth patterns of relatively pure communities of a C_4 grass Paspalum fasciculatum, a C_3-C_4 intermediate grass, Panicum laxum, and a C_3 grass, Leersia hexandra, illustrates the temporal displacement of productivity in these species (Figure 2.5).

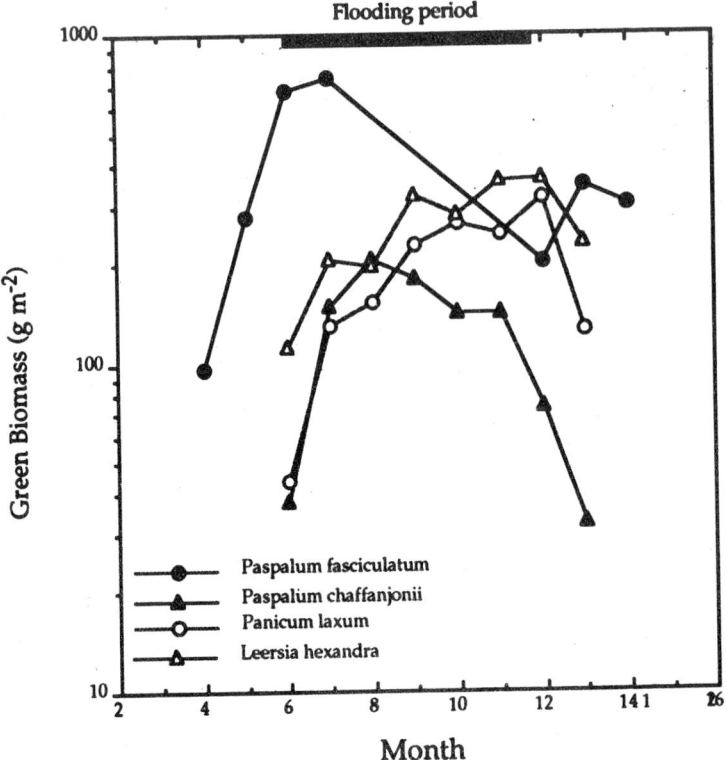

Figure 2.5.   Accumulation of living aboveground biomass in grasses from different flooded savannas in southeastern Venezuela (with data from Escobar & González-Jiménez, 1979; Escobar, 1977).

2.5   CONCLUDING REMARKS

CAM plants are a common component of the more exposed epiphytic synusia in humid forests. There, the largest number of CAM, epiphytic species belong to the Orchidaceae and Bromeliaceae. Apparently, relative high water-use efficiency, succulent habit, and cuticular protection against desiccation give a competitive advantage to constitutive CAM plants in epiphytic, exposed habitats. Moreover, a large number of constitutive CAM species have been able to successfully occupy shady forest environments, becoming adapted to low light. In this process the anatomical protection against cuticular water loss is reduced, increasing the sensitivity of those shade-adapted constitutive

CAM plants to dry air and photodamage when exposed to full sunlight. In some species, the process is reversible, as in several species of Bromeliaceae and Orchidaceae. In other species, apparently more frequently among epiphytic Cactaceae, drought sensitivity is permanent and those species are restricted to shaded habitats.

CAM taxa have evolved independently within several plant families (Monson, 1989). The environmental constraints determining a competitive advantage for constitutive CAM plants appear to be related to increased water-use efficiency and drought tolerance (Ehleringer & Monson, 1993). In epiphytic ferns, stimulus for development of CAM seems to be the occupation of drier epihytic environments. In the families Orchidaceae, Bromeliaceae, and Cactaceae, however, CAM evolved in terrestrial species in xeric habitats. The adaptation to epiphytic or humid, shady habitats is then a secondary phenomenon, and CAM remains as a non-deleterious physiological property.

A number of species from the families Gesneriaceae and Piperaceae accumulate significant amounts of organic acids during the night without net CO_2 fixation. This CAM-cycling, considered to be a precursor of full-CAM plants, is found in succulent epiphytic plants. In these species, acidification represents recycling of respiratory CO_2, and carbon used for growth derives exclusively from fixation during the day.

Respiratory CO_2 recycling is apparently widespread in epiphytic CAM plants. There are still a number of problems that require resolution for an integral interpretation of the ecological significance of recycling. For example, it is necessary to clarify the substrate competition for phosphoenolpyruvate between synthesis of malic acid in the CAM pathway and the tricarboxylic acid cycle.

A currently very active item in CAM research in the humid tropics is the peculiar acid metabolism and gas exchange of species in the arborescent genus *Clusia*. Several *Clusia* species show comparatively large accumulations of organic acids during the night. In many species, a portion of this acidification is due to refixation of respiratory CO_2 and the accumulation of citric acid, apparently without net CO_2 fixation. Further biochemical analysis of organic acid metabolism in *Clusia* species may help to clarify the regulation of organic acid accumulation and its physiological role.

Humid tropical forests are devoid of C$_4$ plants, with the significant exception of two *Euphorbia* species inhabiting rainforests in Hawaii. C$_4$ plants become predominant in flooded grasslands in humid climates. In this ecosystem, fast growth rates and efficient use of nitrogen for the production of photosynthates apparently provide a competitive advantage to C$_4$ plants. There are a number of species that can be classified as truly aquatic C$_4$ plants, some of them attaining their highest productivity rates as floating plants. It seems clear that for these plants, water-use efficiency provides no advantage relative to C$_3$ floating grasses.

Aquatic C$_4$ plants seem to be more competitive than C$_3$ plants in nutrient-poor waters. However, the nutrient-productivity relationships of tropical grasses still require much research to explain the distributional patterns observed in the field. A similar situation can be observed in the Cyperaceae. Some of the most productive species inhabiting wetlands belong to the genus *Cyperus* and are typical C$_4$ plants. In these habitats, tolerance to flooding, particularly regarding the development of physical mechanisms for oxygen transport to the root environment, are perhaps more crucial than photosynthesis for competitive success.

The dominance of C$_4$ herbaceous plants in highly seasonal lowland tropical environments has been explained on the basis of higher water-use efficiency and productivity. However, CO$_2$ starvation since the Late Tertiary, and not aridity, seems to be the environmental stress leading to the development of the C$_4$ pathway (Ehleringer & Monson 1993; Robinson, 1994). Therefore, the assessment of responses of tropical C$_4$ plants to an atmosphere with elevated CO$_2$ is of considerable ecological importance to predict rearrangements in biodiversity in these grass-dominated ecosystems in the tropics.

ACKNOWLEDGEMENTS

I thank Dr. Craig E. Martin, University of Kansas, for his thorough review and comments that contributed to improving an earlier version of this paper.

REFERENCES

ADAMS, W. W., III (1988) Photosynthetic acclimation and photoinhibition of terrestrial and epiphytic CAM tissues growing in full sun light and deep shade. *Australian Journal of Plant Physiology,* **15,** 123–134.

BAILEY, H. P. (1979) Semi-arid climates: Their definition and distribution. *Agriculture in Semi-Arid Environments* (eds. A. E. HALL, G. H. CANNELL, & H. W. LAWTON). Springer-Verlag, Berlin, pp 73–97.

BALL, E., HANN, J., KLUGE, M., LEE, H. S. J., LÜTTGE, U., ORTHEN, B., POPP, M., SHMITT, A. & TING, I. P. (1991) Ecophysiological comportment of the tropical CAM-tree C· *xia* in the field. II. Modes of photosynthesis in trees and seedlings. *New P. ?tologist*, **117**, 483–491.

BARUCH, Z., LUDLOW, M. M. & DAVIS, R. (1985) Photosynthetic responses of native and introduced C_4 grasses from Venezuelan savannas. *Oecologia*, **67**, 288–293.

BEARD, J. S. (1944) Climax vegetation in tropical America. *Ecology*, **25**, 127–158.

BENZING, D., FRIEDMAN, W. E., PETERSON, G. & RENFROW, A. (1983) Shootlessness, velamentous roots, and the pre-eminence of Orchidaceae in the epiphytic biotope. *American Journal of Botany*, **70**, 121–133.

BENZING, D. H. (1990), *Vascular Epiphytes*. Cambridge University Press, Cambridge.

BORLAND, A. M., GRIFFIYHS, H., MAXWELL, C., BROADMEADOW, M. S. J., GRIFFITHS, N. M. & BARNES, J. D. (1992) On the ecophysiology of the Clusiaceae in Trinidad: Expression of CAM in *Clusia minor* L. during the transition from wet to dry season and characterization of three endemic species. *New Phytologist*, **122**, 349–357.

BRITTON, N. L. & ROSE, J. N. (1963) *The Cactaceae*. Republication of the 2nd original edition of 1937, Carnegie Institution of Washington, Dover Publications Inc., New York.

BROWN, R. H. & BROWN, W. V. (1975) Photosynthetic characteristics of *Panicum milioides*, a species with reduced photorespiration. *Crop Science*, **15**, 681–685.

COCKBURN, W., GOH, C. J. & AVADHANI, P. N. (1985) Photosynthetic carbon assimulation in a shootless orchid *Chilochista usneoides* (Don.) Ldl. : A variant on crassulacean acid metabolism. *Plant Physiology*, **77**, 83–86.

COUTINHO, L. M. (1963) Algumas informações sôbre a ocorrência do "Efeito de DeSaussure" em epifitas e herbáceas terrestres da mata pluvial. *Boletim Faculdade de Filosofia, Ciencias e Letras Universidade de São Paulo, No.* **288**, 81–98.

COUTINHO, L. M. (1969) Novas observações sôbre a ocorrência do "Efeito de DeSaussure" e suas relaç—es com a suculencia, a temperatura folhear e os movimentos estomáticos. *Boletim Faculdade de Filosofia, Ciencias e Letras Universidade de São Paulo*, **331**, 79–102.

DOWNTON, W. J. S. (1975) The occurrence of C_4 photosynthesis among plants. *Photosynthetica*, **9**, 96–105.

EARNSHAW, M. J., WINTER, K., ZIEGLER, H., STICHLER, W., CRUTTWELL, N. E. G., KERENGA, K., CRIBB, P. J., WOOD, J., CROFT, J. R., CARVER, K. A., & GUNN, T. C. (1987), Altitudinal changes in the incidence of crassulacean acid metabolism in vascular epiphytes and related life forms in Papua New Guinea. *Oecologia*, **73**, 566–572.

EHLERINGER, J. R. & MONSON, R. K. (1993) Evolutionary and ecological aspects of photosynthetic pathway variation. *Annual Review of Ecology and Systematics*, **24**, 411–439.

ELLIS, R. P., VOGEL, J. C. & FULS, A. (1980) Photosynthetic pathways and the geographical distribution of grasses in South West Africa/Namibia. *South African Journal of Science,* **76**, 307–314.

ESCOBAR, A. (1977) *Estudio de las sabanas inundables de* Paspalum fasciculatum. Magister Scientiarum Thesis, Instituto Venezolano de Investigaciones Científicas, Caracas.

ESCOBAR, A., GONZÁLEZ-JIMÉNEZ, E. (1979) La production primaire de la savane inondable d'Apure (Venezuela). *Geo-Eco-Trop,* **3**, 53–70.

FARQUHAR, G. D., HUBICK, K. T., CONDON, A. G. & RICHARDS, R. A. (1988) Carbon isotope fractionation and plant water-use efficiency. *Stable Isotopes in Ecological Research* (P. W. RUNDEL, J. R. EHLERINGER, & K. A. NAGY, eds.). *Ecological Studies* **68**, Springer-Verlag, New York. pp. 21–40.

FRANCO, A. C., BALL, E. & LÜTTUGE U. (1990) Patterns of gas exchange and organic acid oscillations in tropical trees of the genus *Clusia. Oecologia,* **85**, 108–114.

FRANCO, A. C., BALL, E. & LÜTTGE, U. (1991) The influence of nitrogen, light, and water stress on CO_2 exchange and organic acid accumulation in the tropical C_3-CAM tree, *Clusia minor. Journal of Experimental Botany,* **42**, 597–603.

FRANCO, A. C., BALL, E. & LUTTGE, U. (1992) Differential effects of drought and light levels on accumulation of citric acids during CAM in *Clusia. Plant, Cell and Environment,* **15**, 820–929.

FRANCO, A. C., OLIVARES, E., BALL, E., LÜTTGE, U. & Haag-Kerwer, A. (1994) *In situ* studies of Crassulacean acid metabolism in several sympatric species of tropical trees of the genus *Clusia. New Phytologist,* **126**, 203–213.

GEIGER, D. R. & SERVAITIS, J. C. (1994) Diurnal regulation of photosynthetic carbon metabolism in C_3 plants. *Annual Review of Plant Physiology and Plant Molecular Biology,* **45**, 235–256.

GOH, C. J. & KLUGE, M. (1989) Gas exchange and water relations in epiphytic orchids. *Vascular Plants as Epiphytes* (U. LÜTTGE, ed.) *Ecological Studies,* Springer-Verlag Berlin, **76**, pp. 139–166.

GOH, C. J., AVADHANI, P. N., LOH, C. S., HANEGRAF, C. & ARDITTI, J. (1977) Diurnal stomata and acidity rhythms in orchid leaves. *New Phytologist,* **78**, 365–372.

GRIFFITHS, H. (1988) Crassulacean acid metabolism: A re-appraisal of physiological plasticity in form and function. *Advances in Botanical Research,* 15, 43–92.

GRIFFITHS, H. (1989), Carbon dioxide concentrating mechanisms and the evolution of CAM in vascular epiphytes. *Vascular Plants as Epiphytes* (U. LÜTTGE, ed.) *Ecological Studies* Springer-Verlag Berlin **76**, pp. 42–86.

GRIFFITHS, H. & SMITH, J. A. C. (1983) Photosynthetic pathways in the Bromeliaceae of Trinidad: Relations between life-forms, habitat preference, and the occurrence of CAM. *Oecologia,* **60**, 176–184.

GRIFFITHS, H., ONG, B. L., Avadhani, P. N., and GOH, C. J. (1989) Recycling of respiratory CO_2 during crassulacean acid metabolism: Alleviation of photoinhibition in *Pyrrosia piloselloides*. *Planta*, **179**, 115–122.

GUNNEMANN, H. (1991) Phytomasseproduktion und Nährstoffumsatz von Vegetationstypen bolivianischer Überschwemmungs-Savannen. *Göttinger Beiträge zur Land und Forstwirtschaft in den Tropen un Subtropen*, Heft, **67**, Göttingen.

GURALNICK, L. J., TING, I. P. & Lord, E. M. (1986) Crassulacean acid metabolism in the *Gesneriaceae*. *American Journal of Botany*, **73**, 336–345.

HAAG-KERWER, A., FRANCO, A. C. & LÜTTGE, U. (1992) The effect of temperature and light on gas exchange and acid accumulation in the C_3-CAM plant *Clusia minor L*. *Journal of Experimental Botany*, **43**, 345–352.

HAASE, R. (1989) Plant communities of a savanna in northern Bolivia I. Seasonally flooded grassland and gallery forest. *Phytocoenologia*, **18**, 55–81.

HAASE, R. & BECK, S. G. (1989) Structure and composition of savanna vegetation in northern Bolivia: A preliminary report. *Brittonia*, **41**, 80–100.

HATCH, M. D. (1976) Photosynthesis: The path of carbon. *Plant Biochemistry* (J. Bonner & J. E. Varner eds.). Academic Press, New York.

HATTERSLEY, P. W. (1983) The distribution of C_3 and C_4 grasses in Australia in relation to climate. *Oecologia*, **57**, 113–128.

HESLA, B. I., TIESZEN, L. L. & IMBAMBA, S. K. (1982) A systematic survey of C_3 and C_4 photosynthesis in the *Cyperaceae* of Kenya, East Africa. *Photosynthetica*, **16**, 196–205.

HEW, C. S. and WONG, Y. S. (1974) Photosynthesis and respiration of ferns in relation to their habitat. *American Fern Journal*, **64**, 40–48.

HUECK, K. (1966) *Die Wälder Südamerikas*. Gustav Fischer Verlag, Stuttgart.

HUECK, K. & SEIBERT, P. (1972) *Vegetationskarte von Südamerika*. Gustav Fischer Verlag, Stuttgart.

JONES, M. B. (1987) The photosynthetic characteristics of papyrus in a tropical swamp. *Oecologia* (Berlin), **71**, 355–359.

JONES, M. B. (1988) Photosynthetic responses of C_3 and C_4 wetland species in a tropical swamp. *Journal of Ecology*, **76**, 253–262.

JUNK, W. J. (1970) Investigations on the ecology and production-biology of the 'floating meadows' (*Paspalo-Echinocloetum*) of the middle Amazon. I. The floating vegetation and its ecology. *Amazoniana*, **2**, 449–495.

JUNK, W. J. (1989) The use of Amazonian floodplains under ecological perspective. *Interciencia*, **14**, 317–322.

JUNK, W. J. (1991) *Die Krautvegetation der Überschwemmungsgebiete des Amazonas (Várzea) bei Manaus und Ihre Bedeutung für das Ökosystem*. Max-Planck-Institut für Limnologie, Arbeitsgruppe Tropenökologie, Plön/Holstein, Germany.

KLUGE, M. & TING, I. P. (1978) Crassulacean Acid Metabolism. *Ecological Studies*, **30**. Springer-Verlag, New York.

KLUGE, M., AVADHANI, P. N. & GOH, C. J. (1989) Gas exchange and water relations in epiphytic tropical ferns. *Vascular Plants as Epiphytes* (U. LÜTTGE, ed.) *Ecological Studies* **76**, Springer-Verlag, Berlin, pp. 87–108.

KLUGE, M., FRIEMERT, V., ONG, B. C., BRULFERT, J. & GOH, C. J. (1989) *In situ* studies of the crassulacean acid metabolism in *Drymoglossum piloselloides*, an epiphytic fern of the humid tropics. *Journal of Experimental Botany*, **40**, 441–52.

KRESS, W. J. (1986) The systematic distribution of vascular epiphytes: An update. *Selbyana* **9**, 2–22.

LEE, H. S. J., SCHMITT, A. K. & LÜTTGE, U. (1989) The response of the C_3-CAM tree *Clusia rosea*, to light and water stress. II. Internal CO_2 concentration and water use efficiency. *Journal of Experimental Botany*, **40**, 171–179.

LOESCHEN, V. S., MARTIN, C. E., SMITH, M. & EDER, S. L. (1993) Leaf anatomy and CO_2 recycling during Crassulacean acid metabolism in twelve epiphytic species of *Tillandsia* (Bromeliaceae). *International Journal of Plant Sciences*, **154**, 100–106.

LÜTTGE, U. (ed.) (1989) *Vascular Plants as Epiphytes: Ecology and Evolution. Ecological Studies*, **76**, Springer-Verlag. Berlin.

LÜTTGE, U. (1991) *Clusia*: Morphogenetische, physiologische und biochemische Strategien von Baumwürgern im tropischen Regenwald. *Naturwissenschaften*, **78**, 49–58.

LÜTTGE, U., SMITH, J. A. C., MARIGO, G. & OSMOND, C. B. (1981) Energetics of malate accumulation in the vacuoles of *Kalanchoe tubiflora*. *FEBS Letters*, **126**, 81–84.

LÜTTGE, U., STIMMEL, K.-H., SMITH, J. A. C. & GRIFFITHS, H. (1985), Comparative ecophysiology of CAM and C_3 bromeliads. II. Field measurements of gas exchange of CAM bromeliads in the humid tropics. *Plant, Cell and Environment*, **9**, 377–384.

MADISON, M. (1977) Vascular epiphytes: Their systematic occurrence and salient features. *Selbyana*, **2**, 1–13.

MARTIN, C. E. (1994) Physiological ecology of Bromeliaceae. *Botanical Review*, **60**, 1–82.

MARTIN, C. E., EADES, C. A. & PITHER, R. E. (1986) Effects of irradiance on Crassulacean Acid Metabolism in the epiphyte *Tillandsia usneoides* L. (Bromeliaceae). *Plant Physiology*, **80**, 23–26.

McWILLIAMS, E. L. (1970) Comparative rates of dark CO_2 uptake and acidification in Bromeliaceae, Orchidaceae, and Euphorbiaceae. *Botanical Gazette*, **131**, 285–290.

MEDINA, E. (1974) Dark CO_2 fixation, habitat preference, and evolution within the Bromeliaceae. *Evolution*, **28**, 677–686.

MEDINA, E. (1982) Temperature and humidity effects on dark CO_2 fixation by *Kalanchoe pinnata*. *Zeitschrift für Pflanzenphysiologie*, **107**, 251–258.

MEDINA, E. (1986) Forests, savannas, and montane tropical environments. In: *Photosynthesis in Contrasting Environments* (N. R. Baker, & S. P. Long, editors). *Topics in Photosynthesis*, **7**, Elsevier Science Publisher B. V., Amsterdam, pp. 139–171.

MEDINA, E. (1990) Eco-fisiología y evolucion de las Bromeliaceae. *Boletin de la Academia Nacional de Ciencias, Córdoba*, **59**, 71–100.

MEDINA, E. & MOTTA, N. (1990) Metabolism and distribution of grasses in tropical flooded savannas in Venezuela. *Journal of Tropical Ecology*, **6**, 77–89.

MEDINA, E., DE BIFANO, T. & DELGADO, M. (1976) *Paspalum repens* Berg., a truly aquatic C_4 plant. *Acta Cientifica Venezolana*, **27**, 258–260.

MEDINA, E., DELGADO M. & GARCÍA, V. (1989) Cation accumulation and leaf succulence in *Codonanthe macradenia* J. D. Smith (Gesneriaceae) under field conditions. *Amazoniana*, **11**, 13–22.

MEDINA, E., DELGADO, M., TROUGHTON, J. H. & MEDINA, J. D. (1977) Physiological ecology of CO_2 fixation in Bromeliaceae. *Flora*, **166**, 137–152.

MEDINA, E., LÜTTGE, U., LEAL, F. & ZIEGLER, H. (1991) Carbon and hydrogen isotope ratios in bromeliads growing under different light environments in natural conditions. *Botanica Acta*, **104**, 47–52.

MEDINA, E., OLIVARES, E., DÍAZ, M. & van der Merwe, N. (1989), Metabolismo ácido de crassuláceas en bosques humedos tropicales. *Monographs in Systematic Botany, Missouri Botanical Garden* St. Louis, Missouri, pp 56–67.

MEDINA, E., POPP, M., OLIVARES, E., JANETT, H. -P. & LÜTTGE, U. (1993) Daily fluctuations and titratable acidity, content of organic acids (malate and citrate) and soluble sugars of varieties and wild relatives of *Ananas comosus* L. growing under natural tropical conditions. *Plant, Cell and Environment*, **16**, 55–63.

MEINZER, F. C. (1978) Observaciones sobre la distribución taxonomica y ecológica de la fotosíntesis C_4 en la vegetación del noroeste de Centroamérica. *Revista de Biología Tropical*, **26**, 359–369.

MONSON, R. K. (1989) On the evolutionary pathways resulting in C_4 photosynthesis and Crassulacean Acid Metabolism (CAM). *Advances in Ecological Research*, **19**, 57–110.

MOREL, C. (1979) Rôle coordinateur du CAM dans le métabolisme intermediaire. I. Fonction anaplérotique de l'enzyme malique et cycle des acides tricarboxyliques. *Physiologie Vegetale*, **17**, 697–712.

NEALES, T. F. & HEW, C. S. (1975) Two types of carbon fixation in tropical orchids. *Planta*, **123**, 303–306.

NOBEL, P. S. & HARTSOCK, T. L. (1978) Resistance analysis of nocturnal carbon dioxide uptake by a crassulacean acid metabolism succulent, *Agave desertii*. *Plant Physiology*, **61**, 510–514.

ONG, B. L., KLUGE, M. & FRIEMERT, V. (1986) Crassulacean acid metabolism in the epiphytic ferns *Drymoglossum piloselloides* and *Pyrrosia longifolia*: Studies on responses to environmental signals. *Plant, Cell and Environment*, **9**, 547–557.

OSMOND, C. B. (1978) Crassulacean acid metabolism. A curiosity in context. *Annual Review of Plant Physiology*, **29**, 379–414.

PEARCY, R. W. (1983), The light environment and growth of C$_3$ and C$_4$ tree species in the understory of a Hawaiian forest. *Oecologia*, **58**, 19–25.

PEARCY, R. W. & CALKIN H. C. (1983) Carbon dioxide exchange of C$_3$ and C$_4$ species in the understory of a Hawaiian forest. *Oecologia*, **58**, 26–32.

PEARCY, R. W. & EHLERINGER, J. (1983) Comparative ecophysiology of C$_3$ and C$_4$ plants. *Plant, Cell and Environment*, **7**, 1–13.

PEARCY, R. W. & TROUGHTON, J. H. (1975), C$_4$ photosynthesis in tree form *Euphorbia* species from Hawaiian rainforest sites. *Plant Physiology*, **55**, 1055–1056.

PFITSCH, W. A. & SMITH, A. P. (1988) Growth and photosynthesis of *Aechmea magdalenae*, a terrestrial CAM plant in tropical moist forests, Panama. *Journal of Tropical Ecology*, **4**, 199–207.

PIEDADE, M. T. F., JUNK, W. J. de MELLO, J. A. N. (1992) A floodplain grassland of the Central Amazon. *Primary Productivity of Grass Ecosystems of the Tropics and the Subtropics* (S. P. Long, M. B. Jones, M. B., M. J. Roberts, eds.), Chapman and Hall, London, pp 127–158.

PITTENDRIGH, C. S. (1948) The bromeliad-*Anopheles*-malaria complex in Trinidad. I. The bromeliad flora. *Evolution*, **2**, 58–89.

POPP, M., KRAMER, D., LEE, H., DIAZ, M., ZIEGLER, H. & LÜTTGE, U. (1987) Crassulacean acid metabolism in tropical dicotyledonous trees of the genus *Clusia. Trees*, **1**, 238–247.

RAGHAVENDRA, A. S. & DAS, V. S. R. (1978) The occurrence of C$_4$ photosynthesis: A supplementary list of C$_4$ plants reported during late 1974–mid-1977. *Photosynthetica*, **12**, 200–208.

ROBICHAUX, R. H. & PEARCY, R. W. (1980) Photosynthetic responses of C$_3$ and C$_4$ species from cool shaded habitats in Hawaii. *Oecologia*, **47**, 106–9.

ROBINSON, J. M. (1994) Speculations on carbon dioxide starvation, late Tertiary evolution of stomatal regulation and floristic modernization. *Plant, Cell and Environment*, **17**, 345–354.

RUNDEL, P. W. (1980) The ecological distribution of C$_3$ and C$_4$ grasses in the Hawaiian islands. *Oecologia*, **45**, 354–359.

SARMIENTO, D. (1984) *The Ecology of Neotropical Savannas.* Harvard University Press, Cambridge, Massachusetts.

SCHIMPER, A. F. W. (1884) Über Bau und Lebensweise der Epiphyten Westindiens. *Botanisches Zentralblatt*, **17**, 192–195; 223–227; 253–258; 284–294; 319–326; 351–359; 381–389.

SCHIMPER, A. F. W. (1888) Die *Epiphytische Vegetation Amerikas*. Botanische Mittleilunges aus den Tropen II. Jena. Fischer Verlag.

SCHMITT, A. K., Lee, H. S. J. & LÜTTGE, U. (1988) The response of the C_3-CAM tree *Clusia rosea*, to light and water stress I. Gas exchange characteristics. *Journal of Experimental Botany*, **39**, 1581–1590.

SCHULZE, E. -D. & CALDWELL, M. M. (eds.)(1994), *Ecophysiology of Photosynthesis*. *Ecological Studies*, **100**, Springer-Verlag, Berlin.

SMITH, J. A. C. (1989) Epiphytic bromeliads. *Vascular Plants as Epiphytes* (U. LÜTTGE, U., ed.) *Ecological Studies*, **76**, Springer-Verlag, Berlin, pp. 109–38.

SMITH, J. A. C., GRIFFITHS, H. & LÜTTGE, U. (1986) Comparative ecophysiology of CAM and C_3 bromeliads I. The ecology of Bromeliaceae in Trinidad. *Plant, Cell and Environment*, **9**, 359–376.

STERNBERG, L. S. L., DE NIRO, M. J. & JOHNSON, H. B. (1984), Isotope ratios of cellulose from plants having different photosynthetic pathways. *Plant Physiology*, **74**, 557–561.

STERNBERG, L. S. L., TING, I. P., PRICE, D. & HANN, J. (1987) Photosynthesis in epiphytic and rooted *Clusia rosea* Jacq. *Oecologia*, **72**, 457–60.

SZAREK, S. R., JOHNSON, H. B. & TING, I. P. (1973) Drought adaptation in *Opuntia basilaris*. *Plant Physiology*, **52**, 539–541.

TAKEDA, T., UENO, O., SAMEJINA, M. & OHTANI, T. (1985) An investigation for the occurrence of C_4 photosynthesis in the Cyperaceae from Australia. *Botanical Magazine*, **98**, 393–411.

TIESZEN, L. L., SENYIMBA, M. M., IMBAMBA, S. K. & TROUGHTON, J. H. (1979) The distribution of C_3 and C_4 grasses and carbon isotope discrimination along an altitudinal and moisture gradient in Kenya. *Oecologia*, **37**, 337–350.

TING, I. P. (1985) Crassulacean acid metabolism. *Annual Review of Plant Physiology*, **36**, 595–622.

TING, I. P., BATES, L., STERNBERG, L. O. & DENIRO, M. J. (1985) Physiological and isotopical aspects of photosynthesis in *Peperomia*. *Plant Physiology*, **78**, 246–249.

TING, I. P., HANN, J., HOLBROOK, N. M., PUTZ, F. E., STERNBERG, L. S. L., PRICE, D. & GOLDSTEIN, G. (1987) Photosynthesis in hemiepiphytic species of *Clusia and Ficus*. *Oecologia*, **74**, 339–346.

TING, I. P., LORD, E. M., STERNBERG, L. S. L. & DENIRO, M. J. (1985) Crassulacean acid metabolism in the strangler *Clusia rosea* Jacq. *Science*, **229**, 969–971.

TINOCO-OJANGUREN, C. & VÁZQUEZ-YANES, C. (1983) Especies CAM en la selva tropical húmeda de los Tuxtlas, Veracruz. *Boletin Sociedad Botánica Mexico*, **45**, 150–153.

UENO, O., SAMEJINA, M. & KOYAMA, T. (1989) Distribution and evolution of C_4 syndrome in *Eleocharis*, a sedge group inhabiting wet and aquatic environments, based on culm anatomy and carbon isotope ratios. *Annals of Botany,* **64,** 425–438.

VARESCHI, V. (1980) *Vegetationsökologie der Tropen.* Verlag Eugen Ulmer, Stuttgart.

VESEY-FITZGERALD, D. F. (1963) Central African grasslands. *Journal of Ecology,* **51,** 243–274.

WATANABE, S., TAMAI, N., HOSHINA, S., SANADA, Y., NISHIDA, K. & WADA, K. (1992) The respiration dependent malate accumulation in leaves of a Crassulacean Acid Metabolism plant, *Kalanchoe daigremontiana* Hamet et Perr. *Journal of Plant Physiology,* **140,** 129–133.

WINTER, K. (1985) Crassulacean acid metabolism. *Photosynthetic Mechanisms and the Environment* (J. BARBER, N. R. BAKER eds.) Elsevier Science Publishers B. V., Amsterdam, pp 328–387.

WINTER, K., MEDINA, E., GARCÍA, V., MAYORAL, M. L. & MUÑIZ, R. (1985) Crassulacean acid metabolism in roots of leafless orchids, *Campylocentrum tyrridion* Garay and Dunsterville. *Plant Physiology,* **118,** 73–78.

WINTER, K., OSMOND, C. B. & HUBICK, K. T. (1986) Crassulacean acid metabolism in the shade. Studies on an epiphytic fern, *Pyrrosia longifolia,* and other rainforest species from Australia. *Oecologia,* **68,** 224–230.

WINTER K., WALLACE B. J., STOCKER G. C., ROKSANDIC Z. (1983) Crassulacean acid metabolism in Australian vascular epiphytes and some related species. *Oecologia,* **57,** 129–141.

WINTER, K., ZOTZ, G., BAUR, B. & DIETZ, K. J. (1992) Light and dark CO₂ fixation in *Clusia uvitana* and the effects of plant water status and CO₂ availability. *Oecologia,* **91,** 47–51.

ZIEGLER, H., OSMOND, C. B., STICHLER, W. & TRIMBORN, P. (1976) Hydrogen isotope discrimination in higher plants: Correlations with photosynthetic pathway and environment. *Planta,* **128,** 85–92.

ZOTZ, G. & WINTER, K. (1993) Short-term regulation of crassulacean acid metabolism in a tropical hemiepiphyte, *Clusia uvitana. Plant Physiology,* **102,** 835–841.

3

Diel Patterns of CO_2 Exchange in Rainforest Canopy Plants

Gerhard Zotz and Klaus Winter

3.1 INTRODUCTION

Over the course of a day, plants in their natural environments are exposed to fluctuations in temperature, light, relative humidity, precipitation, and, to a lesser extent, atmospheric CO_2 concentration. These factors, combined with seasonal changes in soil water content, nutrient availability, and leaf developmental status, profoundly affect the plant's gas exchange with the surrounding atmosphere. Diel (24 h) measurements of CO_2 and water vapor exchange *in situ* provide information on how these environmental and leaf ontogenetic factors affect photosynthesis and respiration, leaf primary productivity, and the water and carbon economy of plants. Diel gas exchange measurements, in combination with other analytical tools (e.g., fluorescence analysis, enzyme assays, pressure bomb measurements, stem flow measurements, lysimetry) allow integrative studies from the leaf level to the entire plant.

Diel changes in gas exchange have been carefully documented in a variety of plants in different ecosystems, for example in temperate-zone trees, shrubs, and lianas (Schulze, 1970; Linder & Lohammar, 1981; Küppers, 1984a,b), in subtropical trees, shrubs, and epiphytes (Larcher, 1961; Martin, Christensen & Strain, 1981; Beyschlag, Lange & Tenhunen, 1986), and in arid-zone trees, shrubs, and succulents (Schulze, Lange & Koch, 1972; Pearcy et al., 1974). By contrast, many of the published data on net CO_2 exchange and transpiration in tropical species refer to studies with potted plants under managed conditions in greenhouses and growth chambers. We are only beginning to understand CO_2 and water vapor exchange characteristics of tropical trees, lianas, epiphytes, and hemiepiphytes *in situ*.

In this chapter, we focus on leaves of tropical canopy plants and provide an overview of the currently available information on diurnal and diel gas exchange from contrasting growth forms under natural conditions. After addressing certain aspects concerning instrumentation and measuring protocol, daily gas exchange patterns are presented for species of differing life form. A major outcome of these studies has been the finding that it is possible to extrapolate from spot measurements of leaf photosynthesis to diurnal and diel carbon gain of leaves.

3.2 METHODOLOGY AND MEASURING PROTOCOL

Technical problems with electronic measuring equipment have created obstacles to field studies in the humid tropics in the past (Field & Mooney, 1984). Modern instrumentation largely overcomes these problems, although the precise tracking of ambient conditions inside the leaf cuvette is still difficult. For example, it is necessary to lower the relative humidity (RH) of the air entering the leaf chamber below ambient levels in order to compensate for transpiration. Given the high RH in the wet tropics, the relative humidity of the air must often be lowered even further, especially at night and during rainstorms, to avoid condensation in the pneumatic system (Lüttge et al., 1986). Ventilation inside the leaf cuvette is necessary for controlling leaf temperature. By removing the leaf-boundary layer resistance, ventilation also allows for the calculation of stomatal conductance (g_w) and intercellular CO_2 partial pressure (p_i). However, ventilation may strongly overestimate *in* situ rates of transpiration (Fichtner & Schulze, 1990; Meinzer & Goldstein, Chapter 4), and may greatly underestimate naturally occurring leaf temperatures when ambient air movement is low and photon flux density (PFD) is high. Light-weight and readily portable gas-exchange measuring equipment is now available, facilitating studies in the rainforest canopy. If equipment is protected from rain and a weather-proof cuvette is used, even measurements during rainstorms are possible (Zotz & Winter, 1993). Access to the upper strata of tall forests via rope-climbing techniques remains cumbersome, while towers permit only limited access. The use of construction cranes holds more promise (Parker, Smith & Hogan, 1992) and will greatly increase sample sizes, which otherwise have been chronically small.

Depending on the scope of a particular study, the minimum frequency of individual measurements throughout a day may differ considerably. For example, if responses of net CO$_2$ uptake and stomatal conductance to rapidly changing light conditions are to be investigated, data collection every second may be necessary. For determining total daily carbon gain of leaves, surprisingly low frequencies are sufficient. Our studies on rainforest canopy leaves showed that estimates of daily carbon gain were essentially the same when measurements were performed at 5, 15, 30, or 60 min intervals throughout a day (Table 3.1). Only much longer intervals (one measurement every 120 min) led to unreliable results that deviated up to 43% from the measurements taken every 5 min.

Table 3.1. Dependence of the estimate of 24-h CO$_2$ gain on the frequency of gas-exchange measurements. 24-h CO$_2$ gain was integrated for 6 study days with sun leaves of four tropical species (Ceiba pentandra, Clusia uvitana, Polypodium crassifolium, Catasetum viridiflavum), with net CO$_2$-exchange measured every 5 minutes. Longer measurement intervals (15, 30, 60, and 120 min) were simulated by using only every third, sixth, twelvth, and twenty-fourth data point of the original data sets for the determination of the 24-h integrals. The differences between the integrals indicate that estimates of diel carbon gain based on hourly measurements of net CO$_2$ exchange are sufficiently accurate.

Measurement interval (min)	% Deviation	Deviation (Range)
5	—	—
15	1.1 ± 0.7	0.5-2.3
30	2.9 ± 2.7	0.9-6.9
60	3.2 ± 2.4	0.9-7.2
120	14.9 ± 15.1	5.6-43

3.2.1   Trees

Water status of leaves of evergreen trees in moist tropical forests may be little affected during prolonged periods of reduced rainfall (Wright & Cornejo, 1990; Zotz, 1992; Zotz & Winter, 1994a). Daily carbon gain of two evergreen tree species (*Anacardium excelsum* and *Virola surinamensis*) in moist forest on Barro Colorado Island was only slightly reduced during the 4-month dry period from January to April relative to the wet season (Zotz, 1992). In another species, *Ceiba pentandra*, which produces new leaves at the beginning of the dry season, carbon gain was higher during the dry than during the following wet season when leaves were older and daily light levels were lower (Zotz & Winter, 1994a). Seasonal changes in daily carbon gain of trees in moist tropical forests are thus mainly determined by the interplay between leaf phenology and light availability (Wright & Van Schaik, 1994; Wright, Chapter 15). In dry tropical forests, water availability is a much more important factor. In *Exostema caribaeum*, a dry forest species in Puerto Rico, gas exchange is severely restricted during the dry season and daily carbon gain is less than 20% of that during the rainy season (Lugo et al., 1978).

Midday reductions in stomatal conductance and net CO_2 uptake (A) are frequently observed in tropical trees on bright days when leaf temperatures increase to 35 °C or higher, and vapor pressure differences between the leaf and the air (Δw) increase above 40 mPa Pa^{-1} (Medina, Sobrado & Herrera, 1978; Aylett, 1985; Oberbauer, 1985; Roy & Salager, 1992; Zotz, Königer, Harris & Winter, unpublished data; Figure 3.1.). This phenomenon, which has been known for a long time for plants in semiarid and arid regions (Schulze, Lange & Koch, 1972; Tenhunen, Pearcy & Lange, 1987), occurs in both early and late successional species of moist tropical forests and is associated with decreased 24-h carbon gain. For example, in *Pseudobombax septenatum,* daily carbon gain is markedly higher on overcast than on clear days because on clear days, a pronounced midday-depression of net CO_2 uptake occurs (Winter & Virgo, unpublished data).

Maximum rates of net CO_2 uptake (A_{max}) of recently-matured outer-canopy leaves of tropical forest trees range, on average, from 10 to 20 $\mu mol \ m^{-2} \ s^{-1}$ (Table 3.2), and are similar to those of temperate-zone trees. As in temperate forests, tropical pioneer trees usually exhibit higher A_{max} than late successional species (Bazzaz & Pickett, 1980; Strauss-Debenedetti & Bazzaz, Chapter 6). Persisting pioneer trees such as *Ceiba pentandra* sustain high rates of net CO_2 uptake as large emergents (Zotz & Winter, 1994a).

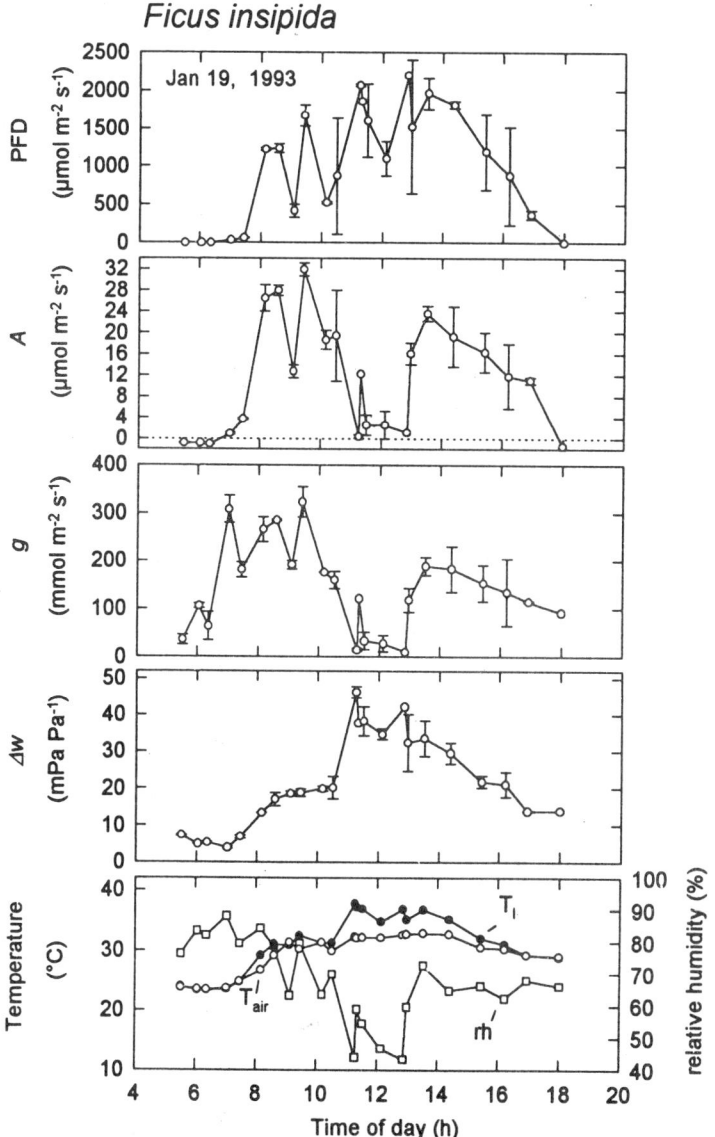

Figure 3.1.   Diurnal course of photon flux density (PFD, μmol m⁻² s⁻¹), net CO₂-exchange (A, μmol m⁻² s⁻¹), stomatal conductance (g_w), leaf-air vapor pressure gradient (Δw), leaf (T_L) and ambient temperature (T_air), and relative humidity (rh) for Ficus insipida Willd. Data are means ± SD of 3 − 5 measurements (modified after Zotz et al. 1995).

Table 3.2.   *Maximum in situ rates of net CO$_2$ uptake (A$_{max}$) of several tropical forest trees. Measurements were made on naturally exposed sun leaves of mature trees.*

Species	Study Site	A$_{max}$ (μmol m^{-2} s^{-1})	Source
Ficus insipida	Dry forest, Panama	33.1	G. Zotz, M. Königer, G. Harris, K. Winter, unpubl.
Didymopanax morototoni	Dry forest, Panama	23.3	G. Zotz, M. Königer, G. Harris, K. Winter, unpubl.
Ceiba pentandra	BCI, Panama	18.2	Zotz & Winter 1994a
Virola surinamensis	BCI, Panama	12.5	Zotz 1992
Dialium pachyphyllum	Cameroon	> 12	Koch et al. 1994
Argyrodendron peralatum	Queensland, Australia	> 10	Pearcy 1987
Argyrodendron peralatum	Queensland, Australia	c. 9	Doley et al., 1987
Penthaclethra macroloba	Costa Rica	> 9	Oberbauer & Strain 1986*
Toona australis	Queensland, Australia	c. 8	Pearcy 1987*
Anacardium excelsum	BCI, Panama	7.9	Zotz 1992

* only spot measurements. Other data from diel measurements.

Table 3.3.   Photosynthetic and leaf characteristics of Anacardium excelsum *(Bertero & Balb) Skeels (Anacardiaceae). All gas exchange data are from diel courses in situ. Measurements are from fully expanded leaves in ca. 30 m (canopy), 26 m (mid-canopy) and in the understory on approx. 0.4 m tall seedlings. Data are means ± SD (n). Measurements were made on Barro Colorado Island, Panama, in February, March, July, September 1990 and June 1992.*

Parameter	canopy	mid-canopy	understory
Max. net photosynthetic rate, μmol m^{-2} s^{-1}	6.8 ± 0.9 (5)	3.4 ± 0.4 (3)	3.8 (2)
Max. dark respiration rate, μmol m^{-2} s^{-1}	1.1 ± 0.07 (5)	0.42 ± 0.05 (3)	0.15 (2)
Light compensation point, μmol photons m^{-2} s^{-1}	28.2 ± 6.3 (5)	8 ± 2.3 (3)	6 (2)
Stomatal density, mm^{-2}	743 ± 57 (7)	672 ± 93 (7)	302 ± 45 (7)
Specific leaf mass, g m^{-2}	98 ± 6 (9)	70 ± 10 (4)	34 ± 4 (3)

Late successional tree species are exposed to contrasting light environments over time, since they normally spend the first years as seedlings in the understory but eventually reach the forest canopy. In laboratory studies, seedlings of such plants showed a limited ability to acclimate photosynthetically to different light regimes (Langenheim et al., 1984; Strauss-Debenedetti & Bazzaz, 1991; Chapter 6; Thompson, Huang & Kriedemann, 1992; Chazdon et al., Chapter 1). In field studies with *Argyrodendron peralatum* in Australia (Pearcy, 1987), *Penthaclethra macroloba* in Costa Rica (Oberbauer & Strain, 1986), and *Anacardium excelsum* in Panama (Zotz & Winter, unpublished data; Table 3.3; Figure 3.2), changes in photosynthetic capacity between understory seedlings and canopy leaves of emergent trees were surprisingly small (less than 2- fold difference), although rates of dark respiration showed a larger plasticity (5-to 7-fold difference).

3.3.2   Lianas

According to Putz and Windsor (1987), "lianas epitomize tropical forests." Croat (1978) called the presence of lianas the single most important physiognomic feature differentiating tropical from temperate forests. However, there are still very few comprehensive field studies of the ecology and physiology of these climbing plants (Gentry,

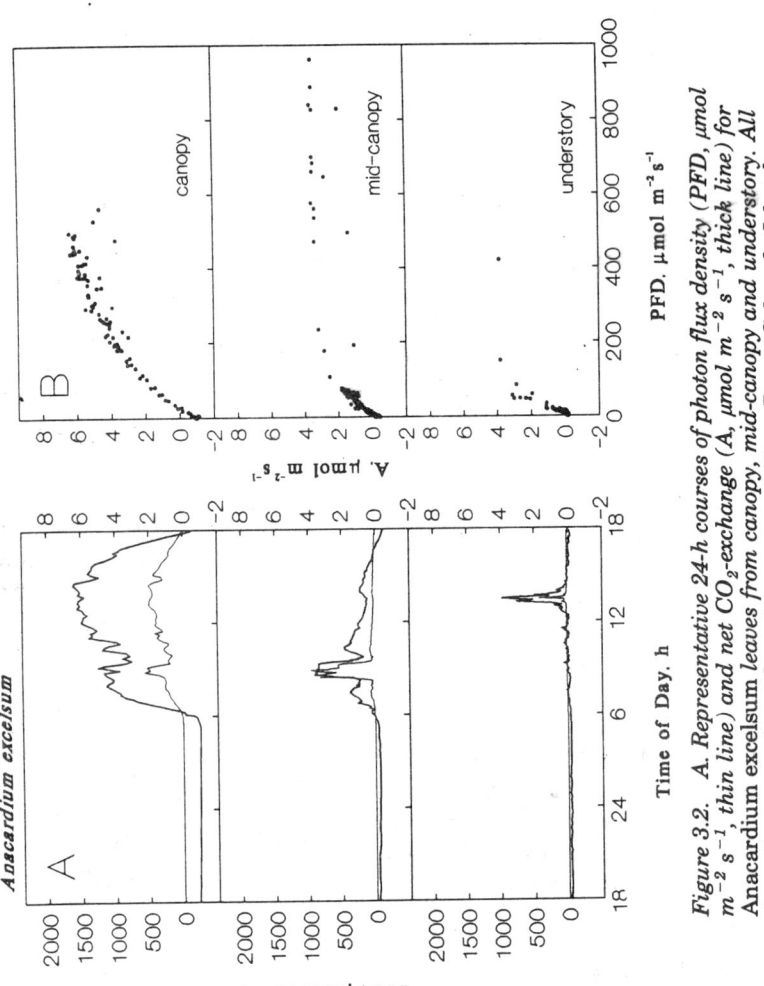

Figure 3.2.   A. Representative 24-h courses of photon flux density (PFD, μmol m⁻² s⁻¹, thin line) and net CO₂-exchange (A, μmol m⁻² s⁻¹, thick line) for Anacardium excelsum leaves from canopy, mid-canopy and understory. All days shown are from the rainy season 1990 on Barro Colorado Island. B. Light-response curves for the days shown in A.

1991; Holbrook & Putz, Chapter 13). The available information on photosynthesis and gas exchange has been reviewed recently by Castellanos (1991) and by Holbrook and Putz (Chapter 13). In general, lianas and vines allocate a larger proportion of their entire biomass to leaves compared to trees; however, net photosynthetic rates and stomatal conductances are not exceptionally high. In a survey of seven tropical lianas, A_{max} did not exceed 10 µmol CO_2 m^{-2} s^{-1} (Castellanos, 1991). Midday-reductions of stomatal conductance have been shown to occur in several species of vines in a Mexican dry forest (Fichtner & Schulze, 1990; Castellanos, 1991).

Available field data on diurnal CO_2 exchange of lianas from tropical moist forests are restricted to a single species, *Uncaria tomentosa* (Rubiaceae), a common liana on Barro Colorado Island (Zotz & Winter, unpublished data). This evergreen species reaches the upper forest canopy and often covers large parts of the canopy of its host trees. Measurements were performed on three days in the dry season and on three days during the following wet season, respectively (Table 3.4, Figure 3.3). Maximum rates of net CO_2 uptake reached 13.9 µmol m^{-2} s^{-1}. Stomatal conductance, g_w, rarely exceeded 300 mmol m^{-2} s^{-1}. Water use efficiency (WUE) was as low as 2.2 mmol CO_2 [mol H_2O]$^{-1}$ in the dry season and as high as 4.8 mmol CO_2 [mol H_2O]$^{-1}$ in the rainy season. Integrated diurnal carbon gains ranged from 181 to 286 mmol CO_2 m^{-2} d^{-1}, 10.2 ± 1.5% of which was lost by respiration at night. There were no significant differences in diurnal carbon gain, stomatal

Table 3.4. Photosynthetic and leaf characteristics of Uncaria tomentosa *(Willd.) DC. (Rubiaceae). Gas exchange was measured in situ on fully expanded sun leaves. The liana was growing in the outer canopy of a 30-m tall tree* (Virola surinamensis) *on Barro Colorado Island, Panama. Data are means ± SD (n).*

Parameter	Mean ± SD
Max. net photosynthetic rate, µmol m^{-2} s^{-1}	11.5 ± 2.4 (6)
Dark respiration rate, µmol m^{-2} s^{-1}	0.84 ± 0.08 (6)
Light compensation point, µmol photons m^{-2} s^{-1}	17 ± 3 (6)
Water use efficiency, mmol CO_2 mol H_2O^{-1}	3.2 ± 0.7 (6)
Quantum yield, mol CO_2 mol photons^{-1}	0.046 ± 0.005 (6)
Leaf longevity, months	4.9 ± 2.0 (56)
Specific leaf weight, g m^{-2}	78.5 ± 10.3 (6)
Leaf thickness, mm	0.21 ± 0.05 (6)
Leaf nitrogen content, % wt	2.2 ± 0.13 (6)
Leaf chlorophyll content, g m^{-2}	0.41 ± 0.14 (6)

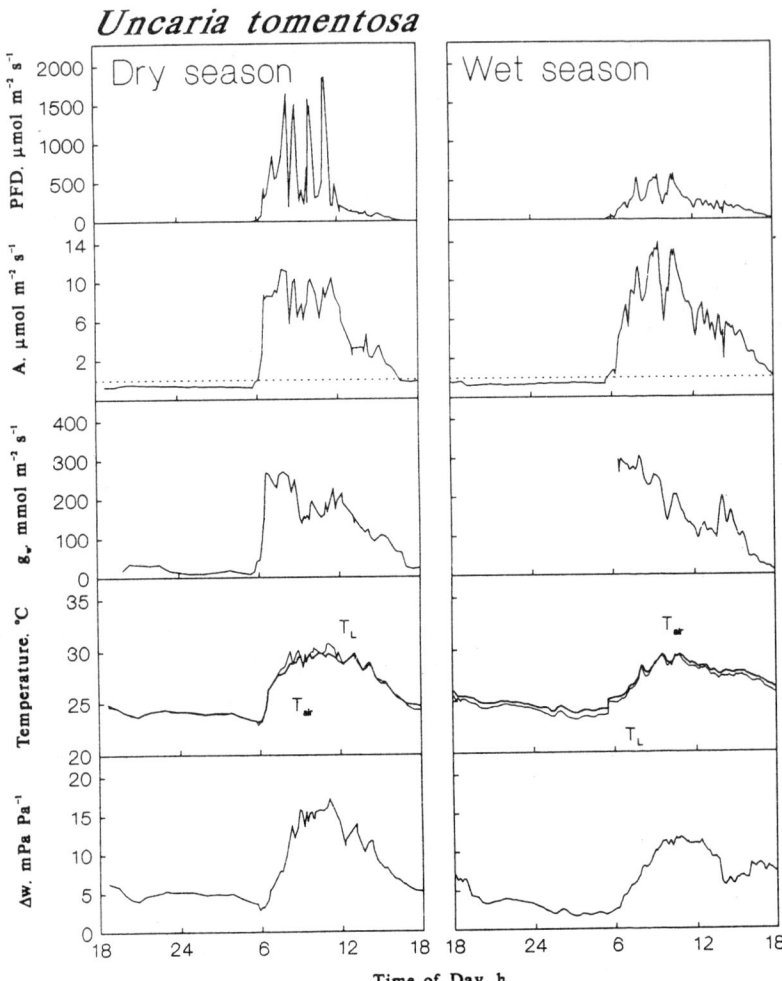

Figure 3.3. Diel courses of photon flux density (PFD), net CO₂-exchange (A), stomatal conductance (g_w), leaf (T_L) and ambient temperature (T_air), and leaf-air vapor pressure gradient (Δw) for Uncaria tomentosa *(Willd.) DC. in the dry (February 20, 1990) and wet season (July 14, 1990). On February 20, g_w and Δw are not shown at night and during the early morning because condensation occurred in the pneumatic system. Leaf water status was determined psychrometrically as described in Zotz & Winter (1994a). Leaf water potential (Y) before dawn and at noon was – 0.85 and – 1.05 MPa, respectively, in the dry season and – 0.6 and – 0.7 MPa, respectively, in the wet season. Turgor pressure, estimated from the difference of Y and osmotic potential, was always > 0.4 MPa, even in the dry season.*

conductance, and leaf water status (water potential, osmotic pressure, turgor pressure) between the wet and the dry season. However, on all three measuring days during the dry season, g_w and A slightly decreased at noon. As with tree species (*Anacardium excelsum, Ceiba pentandra, Virola surinamensis*), overall gas exchange of this liana was little affected during the dry season on BCI. Many lianas root very deeply (F. Ewers, personal communication) and possess an efficient hydraulic conduit system (Ewers, Fisher & Fichtner, 1991; Tyree & Ewers, Chapter 8), features that may aid photosynthetic performance during seasonal drought.

3.3.3. Epiphytes

Moisture is probably the single most important abiotic factor that influences photosynthetic gas exchange of epiphytes (Benzing, 1990). In contrast to trees and lianas, relatively short periods of drought can markedly affect their gas exchange. Epiphytes show various structural adaptations to drought (Benzing, 1990) and not surprisingly, a large proportion of vascular epiphytes use the water-conserving CAM pathway of photosynthesis (Winter, 1985; Winter & Smith, 1995; Medina, Chapter 2).

Much of our current knowledge of the physiological ecology of epiphytes has been published in a recent book (Lüttge, 1989), which contains detailed information on diel patterns of gas exchange of species from the most species-rich taxonomic groups of vascular epiphytes—the orchids, ferns, and bromeliads.

The first comprehensive field measurements of gas exchange and water relations of epiphytes in the humid tropics were reported by Lüttge and coworkers (Lüttge et al., 1986; Griffiths et al., 1986; Smith et al., 1986). Net CO_2 exchange rates of C_3 bromeliads were similar to those of CAM bromeliads. Remarkably, the C_3 bromeliads were very similar to the CAM bromeliads, in terms of their water use efficiency. Another important finding was that in CAM bromeliads, a high percentage of nocturnal accumulation of organic acids was derived from refixation of respiratory CO_2. Subsequent studies have documented substantial CO_2 recycling in many (Griffiths et al., 1989; Kluge et al., 1989; Zotz, 1992), but not all, tropical CAM epiphytes (Goh & Kluge, 1989; Ball et al., 1991).

Documented net CO_2 uptake rates of C_3 and CAM epiphytes in the field do not often exceed 3 μmol m^{-2} s^{-1}, but some species, like the fern *Polypodium crassifolium* or the orchid *Dimerandra emarginata*, reach 7 to 8 μmol m^{-2} s^{-1} (Zotz & Winter, 1994b; Zotz, unpublished data; Coley & Kursar, Chapter 12). The generally low photosynthetic rates and concomitantly low stomatal conductances reflect the high risks of dehydration, even in moist forests. With increasing drought, gas exchange is gradually reduced, and this can eventually lead to a complete cessation of CO_2 uptake during the day.

A previously neglected research area is the study of photosynthesis in tropical, non-vascular epiphytes. Lichens, liverworts, mosses, and algae are not a striking physiognomic feature in most lowland tropical forests, although their abundance increases strongly in montane cloud forests. Recent field studies with the epiphytic lichen *Leptogium azureum* on Barro Colorado Island have shown that net CO_2 uptake was generally restricted to the first daylight hours before photosynthetic activity was inhibited by decreasing water content. This pattern occurred even on moist days during the rainy season (Zotz & Winter, 1994c; Figure 3.4). During the warm nights (c. 23–25°C), high air humidities usually keep the thallus sufficiently hydrated for considerable respiratory CO_2 losses to occur. Respiratory carbon loss was so high on many days that integrated 24 h carbon balances were negative. Even in premontane rainforests where nighttime temperatures are considerably lower (c. 15–17°C), a large percentage (c. 70%) of diurnal carbon gain of the basidiolichen *Dictyonema glabratum* was lost at night (Lange et al., 1994). It is therefore conceivable that the combination of relatively high nocturnal temperatures and high humidities, resulting in high respiratory CO_2 losses, is the key factor in limiting lichen growth in lowland tropical forests.

3.2.4   Hemiepiphytes

The life cycle of hemiepiphytes includes an early epiphytic and a later rooted phase. Primary hemiepiphytes are found in at least 20 families of dicotyledonous plants, the majority of species belonging to the genera *Ficus* (500 species) and *Clusia* (150 species) (Putz & Holbrook 1986; Holbrook & Putz, Chapter 13). The genus *Clusia* has attracted the interest of many plant ecophysiologists over the last decade because several arborescent species exhibit crassulacean acid metabolism (Tinoco-Ojanguren & Vásquez-Yánes, 1983; Sternberg et al., 1987; Ting et al., 1987; Ball et al., 1991; Lüttge, 1991; Borland et al., 1992).

Hemiepiphytic species provide the unique opportunity to compare the same species in two distinct growth forms: as an epiphyte and as

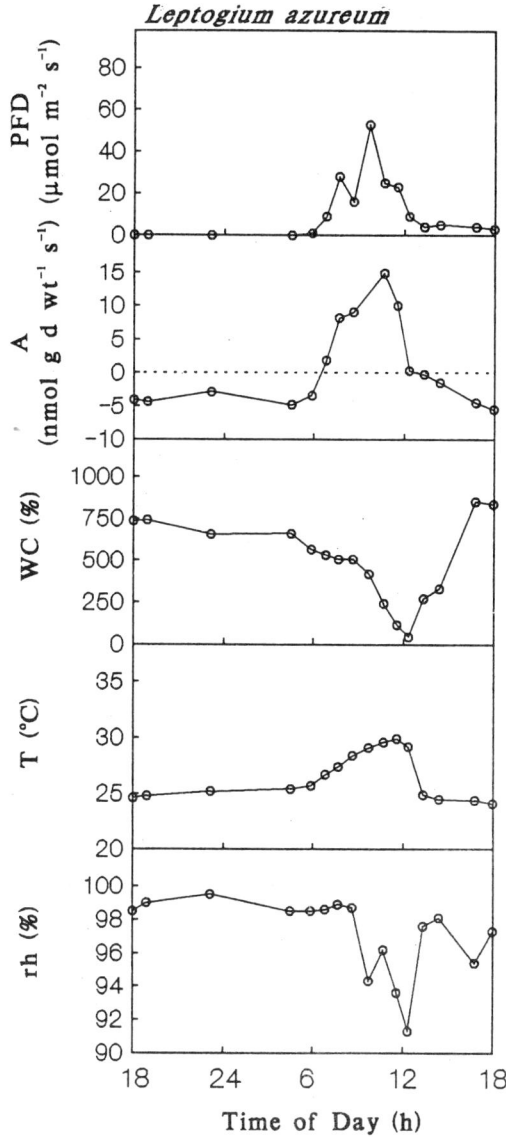

Figure. 3.4.   Diel pattern of photon flux density (PFD, $\mu mol\ m^{-2}\ s^{-1}$), net CO$_2$-exchange (A, $\mu mol\ m^{-2}\ s^{-1}$), relative water content (WC, % wt.), ambient temperature, and air relative humidity for Leptogium azureum *(Sw. ex Ach.) Mont. The lichen was naturally exposed in the inner canopy of an emergent tree (modified from Zotz & Winter, 1994c).*

a tree. A general expectation would be to encounter higher CAM activity in the drought-prone epiphytic stage than in the hemiepiphytic stage, when plants have access to soil water (Medina, Chapter 2; Holbrook & Putz, Chapter 13). Field studies have, however, yielded contrasting results. In *C. rosea* on St. John Island, freestanding trees operated mainly in the CAM mode, while most epiphytic seedlings exhibited gas exchange features typical of C_3 plants (Ball et al., 1991). In the same and other *Clusia* species studied in Costa Rica, $\delta^{13}C$ values of epiphytic and rooted plants showed no clear differences, indicating that the expression of CAM in *Clusia* is highly variable (Ting et al., 1987). In contrast, in epiphytic individuals of *C. uvitana* on Barro Colorado Island, nocturnal CO_2 fixation contributed much more to total carbon gain than in hemiepiphytic individuals, particularly during the dry season (Zotz & Winter, 1995; Figure, 3.5).

In all studies of *Clusia* species, 24 h carbon gain is lower in epiphytic than in hemiepiphytic plants (Table 3.5). Because this difference is not restricted to the dry season, variation in water supply is not a sufficient explanation for these differences. Lower photosynthetic carbon gain in epiphytes might also be the result of lower leaf nitrogen content (Zotz & Winter, 1994d).

3.3   THE A_{MAX}-A_{24H} RELATIONSHIP

Diel measurements of net CO_2 exchange are the most direct method to determine the primary photosynthetic productivity of leaves in the field. Although the relationship between photosynthetic rate per unit leaf area and growth is not always clear (Körner, 1991), knowledge of leaf primary productivity is essential for any analysis of carbon allocation or growth. Field measurements of CO_2 exchange are time-consuming and require sophisticated instrumentation in order to simulate ambient conditions inside the gas exchange cuvette. Thus, in most ecophysiological studies with tropical (and temperate) plants, the photosynthetic performance of leaves is typically characterized with light response curves or the short-term determination of the photosynthetic capacity under natural conditions (A_{max}). A common implicit assumption seems to be that a higher A_{max} coincides with a higher diel carbon gain (A_{24h}). This is surely the case under constant conditions in the laboratory (Königer & Winter, 1993), or if simple models are used to calculate carbon gain (Turnbull, Doley & Yates, 1993). However, the nature of such a relationship in situ had not been assessed until we (Zotz & Winter, 1993) showed that the 24-h carbon balance

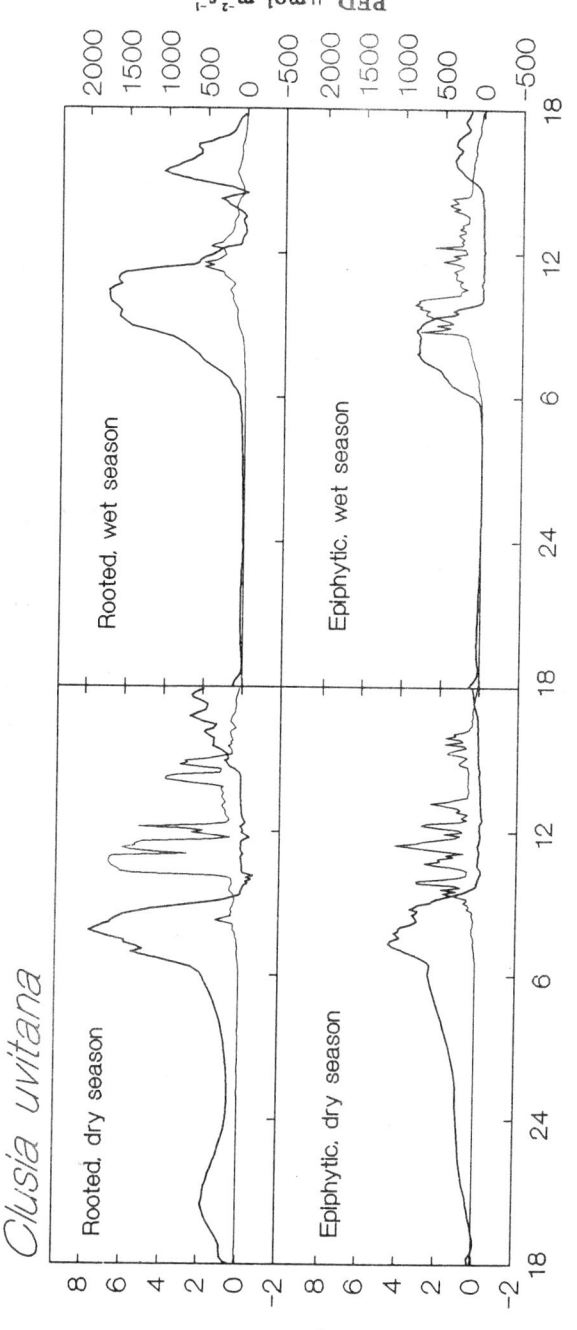

Figure 3.5.   Diel pattern of photon flux density (PFD, μmol m^{-2} s^{-1}) and net CO$_2$-exchange (A, μmol m^{-2} s^{-1}) in Clusia uvitana Pittier in the dry and wet season. Depicted are representative days for an epiphyte and a rooted specimen.

Table 3.5. *Diel net CO_2 gain ($mmol\ CO_2\ m^{-2}\ d^{-1}$) in epiphytic and rooted Clusia and Ficus species. Integrals were given in original publications or were estimated by planimetry from plots of net CO_2 exchange. Data are means ± SD (n) or represent single days. Diel carbon gains of epiphytes are also expressed as percentage of those of conspecific rooted plants.*

Species	Site and Season	Growth form		%	Source
		rooted	epiphyte		
C. rosea	St. John, July 85	84.3	43.9	52	Ting et al. 1987
	St. John, Bordeaux mountain, March 90	93 ± 30 (12)	24.2 (8)	26	Ball et al. 1991
	Miami, Florida, dry season	95.6	4.7	5	Sternberg et al. 1987
	Miami, Florida, wet season	138.6	39.1	28	
C. minor	Barinas, Venezuela, dry season	220.9	30.7	14	Ting et al. 1987
	Barinas, Venezuela, wet season	195.3	103.4	53	
C. uvitana	BCI, Panama, dry season	115.3 ± 31.2 (20)	50.2 ± 27.8 (12)	44	Zotz & Winter 1994b, d
	BCI, Panama, wet season	110.1 ± 40 (37)	79.9 ± 29.4 (21)	73	
F. aurea	Sarasota, Florida, wet season	202.5	116.7	58	Ting et al. 1987
F. nymphaeifolia	Barinas, Venezuela, dry season	168.6	57.8	34	Ting et al. 1987
	Barinas, Venezuela, wet season	214.5	78.8	37	

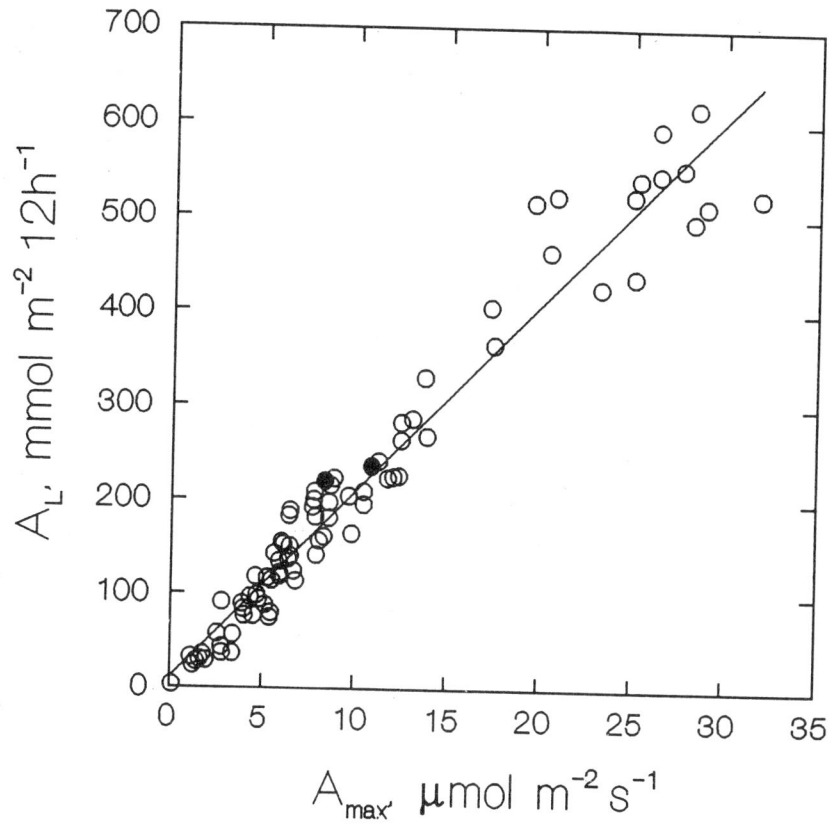

Figure 3.6.   Integrated diurnal net CO_2 gain (A_L) versus the maximum rate of net CO_2 assimilation (A_{max}). Data were obtained from 82 diel courses of net CO_2 exchange in nine rain forest species on Barro Colorado Island and Parque Natural Metropolitana, Panama, 1990 through 1993. The regression is $A_L = 19.8 A_{max} + 10.5$ ($R^2 = 0.95$; $p < 0.001$). Also included are two days of Argyrodendron peralatum (closed circles; estimates from figures in Pearcy, 1987 and Doley et al., 1987)

of exposed canopy leaves can be derived very reliably from spot measurements of net CO_2 exchange ($A_{24h} = 19.2 A_{max} - 7.7$, $R^2 = 0.92$; Zotz & Winter, 1993). Nocturnal CO_2 loss was approximately 10% of daytime carbon gain (A_L). Subsequent work has extended the list of species and the range of A_{max} for which the correlation of A_{max} and A_{24h} or A_L, respectively, holds (Zotz, Königer, Harris & Winter, unpublished data; Zotz & Winter, 1995). The few suitable data of other tropical canopy species also follow the same relationship (Figure 3.6).

Such a close relationship between A_{max} and A_{24h} seems surprising under natural, fluctuating environmental conditions. The weather during the days used to establish this relationship was highly variable, ranging from overcast or rainy, partly cloudy to sunny. The resulting variability of the time period at which leaves operate at light saturation, should lead to a rather loose correlation between A_{max} and A_{24h}. As shown in Figure 3.7, the same A_{max} could theoretically correspond with widely differing daytime carbon gains.

One of the reasons for the close A_{max} / A_{24h} relationship lies in the rather low light saturation point of photosynthesis of many canopy leaves. Even in sun leaves A normally does not track PFD at higher light levels (Figure 3.7A). Light saturation of photosynthesis generally occurs at less than 50% of full sun light (Figure 3.7B; Zotz & Winter, 1993). Thus, much of the daytime fluctuation in light does not result in any changes in A. Also, under water stress, plants will both decrease A_{max} and increase the length of the midday depression simultaneously, resulting in a concomitant decrease of A_{max} and A_L.

Over longer time periods, a changing biochemical capacity for CO_2 fixation due to changes in the leaf nitrogen content or changing plant water status are more important than fluctuating light conditions in determining A_{max} and 24 h carbon gain. The average length of the light period, however, should strongly influence the slope of the A_{max} / A_{24h} relationship. There is evidence that this is the case. In the temperate zone, where day length during the growing period is considerably longer compared to the tropics, similar A_{max} corresponds to a higher daily carbon gain (Beyschlag, Lange & Tenhunen, 1987).

The practical implications of this relationship are manifold. It can be used to estimate the long-term carbon balance of leaves by periodic spot measurements of light-saturated rates of net CO_2 uptake. To our knowledge, the only estimates of annual leaf carbon gain in tropical canopy species are from our work on Barro Colorado Island (Table 3.6.). In the future the long-term carbon gain of canopy leaves can be estimated much easier using the A_{max} / A_{24h} relationship. This seems particularly important in the context of scaling from the leaf level to canopy and ecosystem (Ehleringer & Field, 1993). It will also be a valuable tool in the study of the economy of resource use, such as the relationship between leaf nitrogen and long-term carbon gain (Zotz & Winter, 1994e).

The discovery of this relationship does not immediately render the measurements of diel courses of net CO_2 exchange obsolete, since it must be tested for more species in the tropics and other ecosystems. Of particular interest is the question of whether a similar A_{max} / A_{24h} rela-

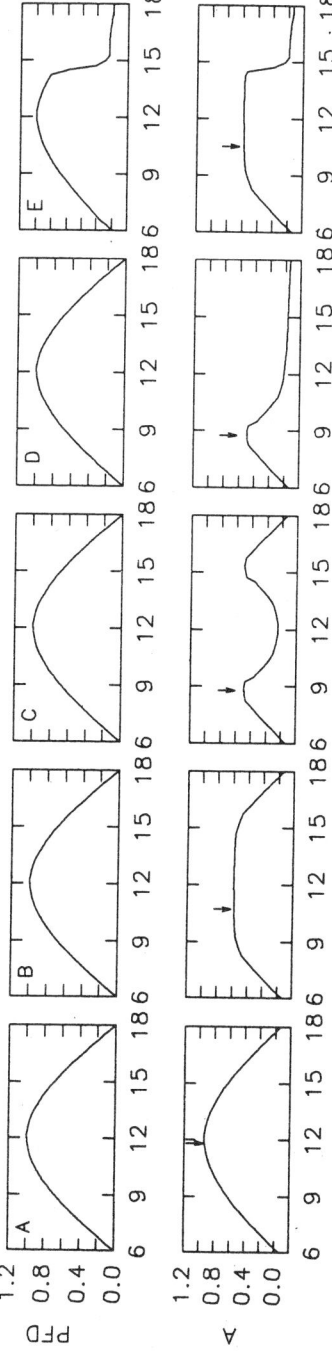

Figure 3.7.   Model diagrams of diurnal courses of PFD and net CO$_2$ exchange (A); units for A and PFD are relative. (A) A following PFD without light saturation. (B) A following PFD with light saturation at intermediate levels of PFD. (C) A with midday depression of net CO$_2$ uptake. (D) A restricted to the early morning hours. (E) As in B, but with A being strongly restricted by low light in the afternoon, e.g. because of a rainstorm, A$_{max}$ is indicated by a vertical arrow. Note that A$_{max}$ is identical in case B–E, but the integrated diurnal carbon gain differs strongly.

Table 3.6.   Leaf annual carbon budgets of four canopy species. Estimates are derived from integration of diel gas exchange measurements on fully mature leaves during the entire year of 1991 (Zotz and Winter 1994 a,b,d).

Species	Life form	Leaf primary production	
		g CO_2 m^{-2}	g CO_2 g wt^{-1}
Ceiba pentandra	Tree	2640	21.0
Clusia uvitana	Hemiepiphyte	1780	9.7
Clusia uvitana	Epiphyte	1060	6.1
Catasetum viridiflavum	Epiphyte	1090	26.3
Polypodium crassifolium	Epiphyte	840	7.4

tionship holds true for shade leaves. First evidence based on studies on shade leaves of *Clusia uvitana* (Zotz & Winter, 1995) suggests that this is indeed the case.

REFERENCES

AYLETT, G. P. (1985) Irradiance interception, leaf conductance, and photosynthesis in Jamaican upper montane rainforest trees. *Photosynthetica*, **19**, 323–337.

BALL, E., HANN, J., KLUGE, M., LEE, H. S. J., LÜTTGE, U., ORTHEN, B., POPP, M., SCHMITT, A., & TING, I. P. (1991) Ecophysiological comportment of the tropical CAM-tree *Clusia* in the field. II. Modes of photosynthesis in trees and seedlings. *New Phytologist*, **117**, 483–492.

BAZZAZ, F. A. & PICKETT, S. T. (1980) Physiological ecology of tropical succession: A comparative review. *Annual Review of Ecology and Systematics*, **11**, 287–310.

BENZING, D. H. (1990) *Vascular Epiphytes. General Biology and Related Biota.* Cambridge University Press, Cambridge.

BEYSCHLAG, W., LANGE, O. L., & TENHUNEN, J. D. (1986) Photosynthese und Wasserhaushalt der immergrünen mediterranen Hartlaubpflanze *Arbutus unedo* L. im Jahreslauf am Freilandstandort in Portugal. I. Tagesläufe von CO_2-Gaswechsel und Transpiration unter natürlichen Bedingungen. *Flora,* **178**, 409–444.

BEYSCHLAG, W., LANGE, O. L., & TENHUNEN, J. D. (1987) Photosynthese und Wasserhaushalt der immergrünen mediterranen Hartlaubpflanze *Arbutus unedo* L. im Jahreslauf am Freilandstandort in Portugal. II. Charakteristische Gaswechselparameter von CO_2-Aufnahme und Transpiration. *Flora,* **179**, 399–420.

BORLAND, A. M., GRIFFITHS, H., MAXWELL, C., BROADMEADOW, M. S. J., GRIFFITHS, N. M., & Barnes, J. D. (1992) On the ecophysiology of the Clusiaceae in Trinidad: Expression of CAM in *Clusia minor* L. during the transition from wet to dry season and characterization of three endemic species. *New Phytologist*, **122**, 349–357.

CASTELLANOS, A. E. (1991) Photosynthesis and gas exchange of vines. *The Biology of Vines*, (eds. F. E. PUTZ, & H. A. MOONEY) Cambridge University Press, Cambridge, pp 181–204.

CROAT, T. (1978) *Flora of Barro Colorado Island*. Stanford University Press, Stanford.

DOLEY, D., YATES, D. J., & UNWIN, G. L. (1987) Photosynthesis in an Australian rainforest tree, *Argyrodendron peralatum*, during the rapid development and relief of water deficits in the dry season. *Oecologia*, **74**, 441–449.

EHLERINGER, J. R. & FIELD, C. B. (eds.) (1993) *Scaling Physiological Processes*. Academic Press, San Diego.

EWERS, F. W., FISHER, J. B., & FICHTNER, K. (1991) Water flux and xylem structure in vines. *The Biology of Vines* (eds. F. E. PUTZ, & H. A. MOONEY) Cambridge University Press, Cambridge, pp 127–160.

FICHTNER, K. & SCHULZE, E.-D. (1990) Xylem water flow in tropical vines as measured by a steady state heating method. *Oecologia*, **82**, 355–361.

FIELD, C. & MOONEY, H. A. (1984) Measuring gas exchange of plants in the wet tropics. *Physiological Ecology of Plants of the Wet Tropics* (eds. E. MEDINA, H. A. MOONEY, & C. VÁSQUEZ-YÁNES). Junk Publishers, The Hague, pp 129–138.

GENTRY, A. H. (1991) The distribution and evolution of climbing plants. In *The Biology of Vines*, (eds. F. E. PUTZ, & H. A. MOONEY), Cambridge University Press, Cambridge.

GOH, C. J. & KLUGE, M. (1989) Gas exchange and water relations in epiphytic orchids. *Vascular Plants as Epiphytes: Evolution and Ecophysiology*, (ed. U. LÜTTGE), Springer-Verlag, Heidelberg.

GRIFFITHS, H., LÜTTGE, U., STIMMEL, K. H., CROOK, C. E., GRIFFITHS, N. M., & SMITH, J. A. C. (1986) Comparative ecophysiology of CAM and C$_3$ bromeliads. III. Environmental influences on CO$_2$ assimilation and transpiration. *Plant, Cell and Environment*, **9**, 385–393.

GRIFFITHS, H., SMITH, J. A. C., LÜTTGE, U., POPP, M., CRAM, W. J., DIAZ, M., LEE, H. S. L., MEDINA, E., SCHAFER, C., & STIMMEL, K. H. (1989) Ecophysiology of xerophytic and halophytic vegetation of a coastal alluvial plain in northern Venezuela. IV. *Tillandsia flexuosa* Sw. and *Schomburgkia humboldtiana* Reichb., epiphytic CAM plants. *New Phytologist*, **111**, 273–282.

KLUGE, M., FRIEMERT, V., BRULFERT, J., & GOH, J. (1989) *In situ* studies of crassulacean acid metabolism in *Drymoglossum piloselloides*, an epiphytic fern of the humid tropics. *Journal of Experimental Botany*, **40**, 441–452.

KOCH, G. W., AMTHOR, J. S., & GOULDEN, M. L. (1994) Diurnal patterns of leaf photosynthesis, conductance, and water potential at the top of a lowland forest canopy in Cameroon: Measurements from the Radeau des Cimes. Tree Physiology, 14, 347–360.

KÖNIGER M., WINTER, K. (1993) Reduction of photosynthesis in sun leaves of Gossypium hirsutum L. under conditions of high light intensities and suboptimal leaf temperatures. Agronomie, 13, 659–668.

KÖRNER, C. (1991) Some often overlooked plant characteristics as determinants of plant growth: A reconsideration. Functional Ecology, 5, 162–173.

KÜPPERS, M. (1984a) Carbon relations and competition between woody species in a Central European hedgerow. I. Photosynthetic characteristics. Oecologia, 64, 332–343.

KÜPPERS, M. (1984b) Carbon relations and competition between woody species in a Central European hedgerow. II. Stomatal responses, water use, and hydraulic conductivity in the root/leaf pathway. Oecologia, 64, 344–354.

LANGE, O. L., BÜDEL, B., ZELLNER, H., ZOTZ, G., & MEYER, A. (1994) Field measurements of water relations and CO_2 exchange of the tropical basidiolichen Dictyonema glabratum in a Panamanian rainforest. Botanica Acta, 107, 279–290.

LANGENHEIM, J. H., OSMOND, C. B., BROOKS, A., & FERRAR, P. J. (1984) Photosynthetic responses to light in seedlings of selected Amazonian and Australian rainforest tree species. Oecologia, 63, 215–224.

LARCHER, W. (1961) Jahresgang des Assimilations- und Respiratonsvermögens von Olea europaea L. ssp. sativa Hoff. et Link., Quercus ilex L. und Quercus pubescens Willd. aus dem nördlichen Gardaseegebiet. Planta, 56, 575–606.

LINDER, S. & LOHAMMAR, T. (1981) Amount and quality of information on CO_2-exchange required for estimating annual carbon balance of coniferous trees. Studia Forestalia Suecica, 160, 73–87.

LUGO, A. E., GONZALEZ-LIBOY, J. A., CINTRÓN, B., & DUGGER, K. (1978) Structure, productivity, and transpiration of a subtropical dry forest in Puerto Rico. Biotropica, 10, 278–291.

LÜTTGE, U. (ed.) (1989) Vascular Plants as Epiphytes: Evolution and Ecophysiology. Ecological Studies, Springer-Verlag, Berlin.

LÜTTGE, U. (1991) Clusia. Morphogenetische, physiologische und biochemische Strategien von Baumwürgern im tropischen Regenwald. Naturwissenschaften, 78, 49–58.

LÜTTGE, U., STIMMEL, K. H., SMITH, J. A. C.,& GRIFFITHS, H. (1986) Comparative ecophysiology of CAM and C_3 bromeliads. II. Field measurements of gas exchange of CAM bromeliads in the humid tropics. Plant, Cell and Environment, 9, 377–383.

MARTIN, C. E., CHRISTENSEN, N. L., & STRAIN, B. R. (1981) Seasonal patterns of growth, tissue acid fluctuations, and $^{14}CO_2$ uptake in the crassulacean acid metabolism epiphyte *Tillandsia usneoides* L. (Spanish moss). *Oecologia*, **49**, 322–328.

MEDINA, E., SOBRADO, M., & HERRERA, R. (1978) Significance of leaf orientation for leaf temperature in an Amazonian sclerophyll vegetation. *Radiation and Environmental Biophysics*, **15**, 131–140.

OBERBAUER, S. F. (1985) Plant water relations of selected species in wet and dry tropical lowland forest in Costa Rica. *Revista Biología Tropical*, **33**, 137–142.

OBERBAUER, S. F. & STRAIN, B. R. (1986) Effects of canopy position and irradiance on the leaf physiology and morphology of *Pentaclethra macroloba* (Mimosaceae). *American Journal of Botany*, **73**, 409–416.

PARKER, G. G., SMITH, A. P., & HOGAN, K. P. (1992) Access to the upper forest canopy with a large tower crane. *BioScience*, **42**, 664–670.

PEARCY, R. W. (1987) Photosynthetic gas exchange responses of Australian tropical forest trees in canopy, gap, and understory micro-environments. *Functional Ecology*, **1**, 169–178.

PEARCY, R. W., HARRISON, A. T., MOONEY, H. A., & BJÖRKMAN, O. (1974) Seasonal changes in net photosynthesis of *Atriplex hymenelytra* shrubs growing in Death Valley, California. *Oecologia*, **17**, 111–121.

PUTZ, F. E. & HOLBROOK, N. M. (1986) Notes on the natural history of hemiepiphytes. *Selbyana*, **9**, 61–69.

PUTZ F. E. & WINDSOR, D. M. (1987) Liana phenology on Bàrro Colorado Island, Panama. *Biotropica*, **19**, 334–341.

ROY, J. & SALAGER, J.-L. (1992) Midday depression of net CO_2 exchange of leaves of an emergent rain forest tree in French Guiana. *Journal of Tropical Ecology*, **8**, 499–504.

SCHULZE, E.-D. (1970) Der CO_2-Gaswechsel der Buche (*Fagus silvatica* L.) in Abhängigkeit von den Klimafaktoren im Freiland. *Flora*, **159**, 177–232.

SCHULZE, E.-D., LANGE, O. L., & KOCH, W. (1972) Ökophysiologische Untersuchungen an Wild- und Kulturpflanzen der Negev-Wüste. III. Tagesläufe von Nettophotosynthese und Transpiration der Trockenzeit. *Oecologia*, **9**, 317–340.

SMITH, J. A. C., GRIFFITHS, H., LLÜTTGE, U., CROOK, C. E., GRIFFITHS, N. M., & STIMMEL, K. H. (1986) Comparative ecophysiology of CAM and C_3 bromeliads. IV. Plant water relations. *Plant, Cell and Environment*, **9**, 395–410.

STERNBERG, L. d. S. L., Ting, I. P., Price, D., & HANN, J. (1987) Photosynthesis in epiphytic and rooted *Clusia rosea* Jacq. *Oecologia*, **72**, 457–460.

STRAUSS-DEBENEDETTI, S. & BAZZAZ, F. A. (1991) Plasticity and acclimation to light in tropical Moraceae of different successional positions. *Oecologia*, **87**, 377–387.

TENHUNEN, J. D., PEARCY, R. W., & LANGE, O. L. (1987) Diurnal variation in leaf conductance and gas exchange in natural environments *Stomatal function,* (eds. E. ZEIGER, G. D. FARQUHAR,& I. R. COWAN) Stanford University Press, Stanford, pp 323–352.

THOMPSON, W. A., HUANG, L.-K., & KRIEDEMANN, P. E. (1992) Photosynthetic response to light and nutrients in sun-tolerant and shade-tolerant rainforest trees. II. Leaf gas exchange and component processes of photosynthesis. *Australian Journal of Plant Physiology,* **19,** 19–42.

TING, I. P., HANN, J., HOLBROOK, N. M., PUTZ, F. E., STERNBERG, L. d. S. L., PRICE, D., & GOLDSTEIN, G. (1987) Photosynthesis in hemiepiphytic species of *Clusia* and *Ficus. Oecologia,* **74,** 339–346.

TINOCO-OJANGUREN, C. & VÁSQUEZ-YÁNES, C. (1983) Especies CAM en la selva húmeda tropical de los Tuxtlas, Veracruz. *Boletin de la Sociedad Botánica de México,* **45,** 150-153.

TURNBULL, M. H., DOLEY, D., & YATES, D. J. (1993) The dynamics of photosynthetic acclimation to changes in light quantity and quality in three Australian rainforest tree species. *Oecologia,* **94,** 218-228.

WINTER, K. (1985) Crassulacean acid metabolism. *Photosynthetic Mechanisms and the Environment* (eds. J. BARBER & BAKER), Elsevier, Amsterdam.

WINTER, K. & SMITH, J. A. C. (1995) An introduction to crassulacean acid metabolism: Biochemical principles and biological diversity. *Crassulacean Acid Metabolism. Biochemistry, Ecophysiology and evolution,* (eds. K. WINTER & J. A. C. SMITH), Springer-Verlag, Berlin, in press.

WRIGHT, S. J. & CORNEJO, F. D. (1990) Seasonal drought and leaf fall in a tropical forest. *Ecology,* **71,** 1165-1175.

WRIGHT, S. J. & VAN SCHAIK, C. P. (1994) Light and the phenology of tropical trees. *The American Naturalist,* **143,** 192-199.

ZOTZ, G. (1992) *Photosynthese und Wasserhaushalt von Pflanzen verschiedener Lebensformen des tropischen Regenwalds auf der Insel von Barro Colorado, Panama.* Doctoral thesis, Würzburg.

ZOTZ, G. & WINTER, K. (1993) Short-term photosynthesis measurements predict leaf carbon balance in tropical rainforest canopy plants. *Planta,* **191,** 409-412.

ZOTZ, G. & WINTER, K. (1994a) Photosynthesis of a tropical canopy tree, *Ceiba pentandra,* in a lowland tropical forest in Panama. *Tree Physiology,* **14,** 1231-1301.

ZOTZ, G. & WINTER, K. (1994b) Annual carbon balance and nitrogen use efficiency in tropical C₃ and CAM epiphytes. *New Phytologist,* **126,** 481-492.

ZOTZ, G. & WINTER, K. (1994c) Photosynthesis and carbon gain of the lichen *Leptogium azureum,* in a lowland tropical forest. *Flora,* **189,** 179-186.

ZOTZ, G. & WINTER, K. (1994d) A one-year study on carbon, water, and nutrient relationships in a tropical C₃-CAM hemiepiphyte, *Clusia uvitana. New Phytologist,* **127,** 45-60.

ZOTZ, G. & WINTER, K. (1994e) Predicting annual carbon balance from leaf nitrogen. *Naturwissenschaften,* **81,** 443.

ZOTZ, G. & WINTER, K. ᴠ ˀ5) Seasonal changes in daytime versus nighttime CO$_2$ fixation of *Clusi* *uvitana in situ. Crassulacean Acid Metabolism. Biochemistry, Ecophysiology and Evolution*, (eds. K. WINTER, & J. A. C. SMITH), Springer-Berlin, in press.

ZOTZ, G., KÖNIGER, M., HARRIS, G. & WINTER, K. (1995) High rates of photosynthesis in the tropical pioneer tree, *Ficus insipida*, Willd. *Flora*, **190**, 265–272.

4

Scaling up from Leaves to Whole Plants and Canopies for Photosynthetic Gas Exchange

Frederick C. Meinzer and Guillermo Goldstein

A gap in interpretation and understanding currently exists between two expanding databases on photosynthetic gas exchange behavior obtained from tropical forests. One of these databases is derived from observations of ecophysiological behavior at the single leaf scale using leaf chambers, and the other is derived from observations of micrometeorological properties of entire forest canopies. Bridging this gap is important for understanding the influence of species composition and individual species characteristics on ecosystem function. In agricultural crops and other vegetation characterized by one or few dominant species, scaling from the leaf directly to the canopy and higher levels may capture enough detail to link quantitatively variations in gaseous fluxes from vegetation to the behavior of individual leaves. In species-diverse tropical forests, however, direct scaling from leaf to canopy and larger scales may result in considerable loss of information concerning the role of a particular species in regulating fluxes of water vapor, CO_2, and heat between vegetation and the atmosphere.

The ability to assess the impact of individual leaf behavior on gas exchange from a particular portion of the canopy mosaic occupied by a given species or group of similar species has important implications for predicting the effects of deforestation and global climate change on the water and carbon balance of tropical regions. In this chapter, we focus on three of the principal constraints to scaling photosynthetic gas exchange from single leaves to the whole plant and canopy levels. The first is interactions and feedbacks at higher scales that influence the perception of environmental variables by individual leaves. The second is a tendency toward homeostasis of gas exchange rates on a unit leaf area basis but vastly different rates of gas exchange per individual in plants experiencing different levels of availability of resources such as water or nutrients. The third constraint is developmental adjustments in gas exchange per unit leaf area that reflect the changing balance between total leaf area and root system size. Ways in which these constraints may be partially overcome by making simultaneous observations at contrasting scales are discussed.

General approaches for scaling processes directly from the single leaf to the canopy and larger scales have been discussed in detail in a recent volume edited by Ehleringer and Field (1993). In attempting to provide a conceptual framework for moving between measurements at increasingly disparate scales, the contributors to the above volume have carefully considered constraints such as stratification of samples to adequately represent horizontal and vertical heterogeneity (e.g., Norman, 1993; Schimel, Davis & Kittel, 1993), errors in predictions resulting from transposition of temporal scales (e.g., Reynolds, Hilbert & Kemp, 1993), and feedbacks operating across scales (Caldwell et al., 1993). In these and other recent treatments of scaling, more emphasis is often placed on evaluating predictions across large gaps in scale than on reconciling measurements made at distinct but closely related scales. A mechanistic understanding of plant-environment interactions, the goal of much ecophysiological research, requires the ability to move back and forth between concurrent measurements at different but precisely defined scales. In many cases, conventional ecophysiological techniques relying on measurements at a single scale cannot distinguish between intrinsic differences in physiological responsiveness to a given environmental variable, or merely differences in the magnitude of the variable perceived at the leaf surface. As discussed below (4.2), this problem can be reduced to one of selecting the appropriate reference point for measurement of environmental variables. Here we argue that future progress in understanding plant-environment interactions at the individual, stand, and regional levels will require that compatible data be gathered concurrently at more than one reference point (spatial scale).

4.1   DEFINING THE PROBLEM

Before undertaking an exercise in scaling, it is necessary that individual scales be unamibiguously and consistently defined. Although the term "canopy" is sometimes used with reference to the foliage of a single individual, the entire complement of leaves belonging to a given plant is more appropriately referred to as the crown. In this chapter, we will adhere to the well established convention of using "canopy" in referring to vegetation and its properties on a land area basis. Nevertheless, gas exchange properties expressed on a unit leaf or crown area basis may be partially analogous to canopy properties if similar reference points are selected for measurements of environmental variables such as partial pressures of water vapor and CO_2.

Although concurrent measurements of leaf gas exchange and biochemistry may permit scaling down to make inferences concerning biochemical and biophysical control mechanisms (von Caemmerer & Farquhar, 1981; Woodrow & Berry, 1988), the reverse problem of scaling up from observations on single leaves to entire plants is larger and more complex than the relatively staightforward problem of selecting a representative sample of leaves in different positions and exposures. Perhaps the potentially most severe constraint encountered in scaling up gas exchange from single leaves to whole plants and canopies is the decoupling influence of the unstirred air boundary layers surrounding each leaf and the entire canopy. This results in a form of micrometeorological feedback in which the gaseous fluxes themselves modify the environment near the leaf surface, altering the driving forces for gas exchange and therefore ultimately rates of gas exchange (Jarvis & McNaughton, 1986). In making single-leaf gas exchange measurements, the unstirred air boundary layer that normally surrounds the leaf is disrupted. The total diffusive resistance is therefore invariably underestimated. Thus, even if gas exchange is measured for every leaf on a plant, the average or sum of these measurements may not give an accurate picture of absolute fluxes of CO_2 and water vapor or of the role of stomata in regulating these fluxes. In dense lowland tropical forests characterized by a high degree of spatial heterogeneity, the errors associated with scaling up from leaf level measurements can be expected to be particularly serious. For example, extreme vertical stratification in the magnitude of micrometeorological feedback associated with the boundary layer should occur from the upper canopy to the lower strata because wind is rapidly attenuated down through the canopy (Shuttleworth, 1989; Roberts, Cabral & De Aguiar, 1990). This could make comparative single leaf gas exchange measurements of upper canopy and gap and understory species difficult to interpret in terms of actual fluxes of water vapor and carbon dioxide.

Application of conventional ecophysiological techniques to measurements on single leaves has led to great progress in understanding the regulation of leaf gas exchange (Farquhar & Sharkey, 1982; Schulze & Hall, 1982; Schulze, 1986). Because these measurements are typically made on several replicate leaves of intact plants, they are often implicitly assumed to reflect accurately processes occurring at the whole-plant and canopy levels. However, it is important to recognize that in single leaf gas exchange measurements fluxes of carbon dioxide and water vapor are most often expressed on a unit leaf area basis because total leaf area per individual is rarely determined. Due to inevitable differences in plant size, the whole individual, rather than a unit area basis, may be the most appropriate scale for evaluating photosynthetic gas exchange and its response to the environment in most ecophysiological studies. Consideration of the whole plant assumes added importance in the context of evidence that gas exchange rates per unit leaf area may be linked to plant developmental stage and the total amount of leaf area present, independently of any external environmental stress factors (Meinzer & Grantz, 1990; Donovan & Ehleringer, 1992; Ovaska, Walls & Mutikainen, 1992).

4.2 MICROMETEOROLOGICAL FEEDBACK

The impact of micrometeorological feedback on scaling up from single leaf gas exchange is more easily assessed with respect to fluxes of water vapor than of carbon dioxide. Water vapor loss from leaves is limited by the stomatal conductance in series with the conductances of the boundary layers surrounding individual leaves and the entire canopy. The conductance of the combined leaf and canopy boundary layers may be low enough to promote local equilibration of transpired water vapor near the leaf, within the boundary layer, uncoupling the vapor pressure at the leaf surface from that in the bulk air (Jarvis & McNaughton, 1986). Vapor pressure will thus decline from the leaf surface, through the boundary layers surrounding each leaf and the entire canopy, into the global air mass above the canopy. The leaf surface must therefore be used as the reference point for determining the vapor pressure gradient driving transpiration from the leaf interior through the stomatal pores. Any measurement of vapor pressure except that at the leaf surface overestimates the vapor gradient across the stomatal pores and consequently both the magnitude of transpiration and the extent to which it is controlled by stomata.

The problem of scaling transpiration up from leaves to whole plants to entire canopies is thus essentially one of selecting appropriate reference points, with respect to the transpiring leaf surface, for measuring the driving force for transpiration. This is illustrated by the relatively common practice of estimating transpiration as the product of stomatal conductance measured with a porometer and the leaf-to-bulk air vapor pressure difference (e.g., Ludlow & Ibaraki, 1979; Meinzer, Seymour & Goldstein, 1983; Bennet et al., 1986; Koch & Rawlik, 1993). It is often not fully recognized that the resulting estimates of transpiration will be strongly dependent on the location of the external reference point at which ambient vapor pressure is determined. In a porometer chamber, the properties of the bulk air approximate those at the leaf surface because the boundary layer is disrupted with fans, permitting stomatal conductance to be calculated from the transpiration rate and the gradient of vapor pressure between the leaf interior and the bulk value within the chamber. The conductances determined by porometry or other ventilated leaf chamber techniques therefore do not predict actual fluxes of water vapor because the driving forces for water loss acting on unencumbered leaves are smaller than in ventilated chambers. This problem is not overcome by enclosing entire plants or groups of plants in large, ventilated chambers because these too will alter boundary layer conditions compared with those prevailing around leaves unencumbered by chambers. Reliance solely on porometric measurements under these circumstances can therefore seriously misrepresent transpirational fluxes and therefore the extent to which transpiration is controlled by stomatal movements.

Recent studies suggest that porometric measurements can lead to considerable overestimates of transpiration in tropical forest species. For example, in *Anacardium excelsum*, a canopy tree species, crown transpiration predicted from porometric measurements of stomatal conductance (as stomatal conductance x leaf-to-air VPD) was generally 10 to 100% higher than transpiration determined from concurrent measurements of sap flow using the heat balance technique (Figure 4.1A). Similarly, in 2 to 3 m-tall *Miconia argentea* plants growing in treefall gaps, porometric estimates of transpiration were up to 300% higher than rates determined from sap flow (Figure 4.1B). The heat balance method provides accurate estimates of transpirational water loss when precautions are taken to prevent environmentally induced stem temperature gradients (Gutiérrez et al., 1994) and to account for heat storage in the stem and water movement associated with growth when flow rates are low. Since mass flow of water through the stem is detected, no assumptions about the relative magnitudes of stomatal and boundary layer conductance are required. On the other hand, the magnitude of the error associated with leaf chamber-based estimates of transpiration will depend on the extent to which prevailing boundary layer conditions uncouple vapor pressure at the leaf surface from that in the bulk air. Both *A. excelsum* and *M. argentea* have relatively large leaves, which should promote uncoupling through reduced boundary layer conductance (Grace, Fasehun & Dixon, 1980; Meinzer et al., 1993). Lower wind speed near the forest floor in gaps than in the canopy would be expected to further reduce boundary layer conductance for species such as *M. argentea*. The relationship between predicted and actual transpiration obtained for *M. argentea* (Figure 4.1B) suggests that the magnitude of the error associated with porometry-based estimates of transpiration increased sharply with increasing actual transpiration and therefore with increasing stomatal conductance. The reasons for this are discussed in detail below.

As implied in the discussion above, the extent to which stomatal aperture and stomatal movements control transpiration is determined largely by the ratio of stomatal to boundary layer conductance. Stomatal control of transpiration is strong only when boundary layer conductance is high in relation to stomatal conductance. When boundary layer conductance is low in relation to stomatal conductance, local equilibration of transpired water vapor near the leaf within the boundary layer uncouples the evaporative demand at the leaf surface from that in the bulk air. Thus, as stomatal conductance increases in the presence of a fixed external boundary layer, the flux of vapor through the boundary layer becomes increasingly limiting and further increases in stomatal conductance have a diminishing effect on transpiration (Figure 4.2A).

The decoupling influence of the boundary layer has been described quantitatively in terms of a dimensionless decoupling coefficient, Ω, representing the sensitivity of transpiration to a marginal change in stomatal conductance (Jarvis & McNaughton, 1986; McNaughton & Jarvis, 1991). The decoupling coefficient is derived from a formulation of the Penman-Montieth equation that partitions control of transpira-

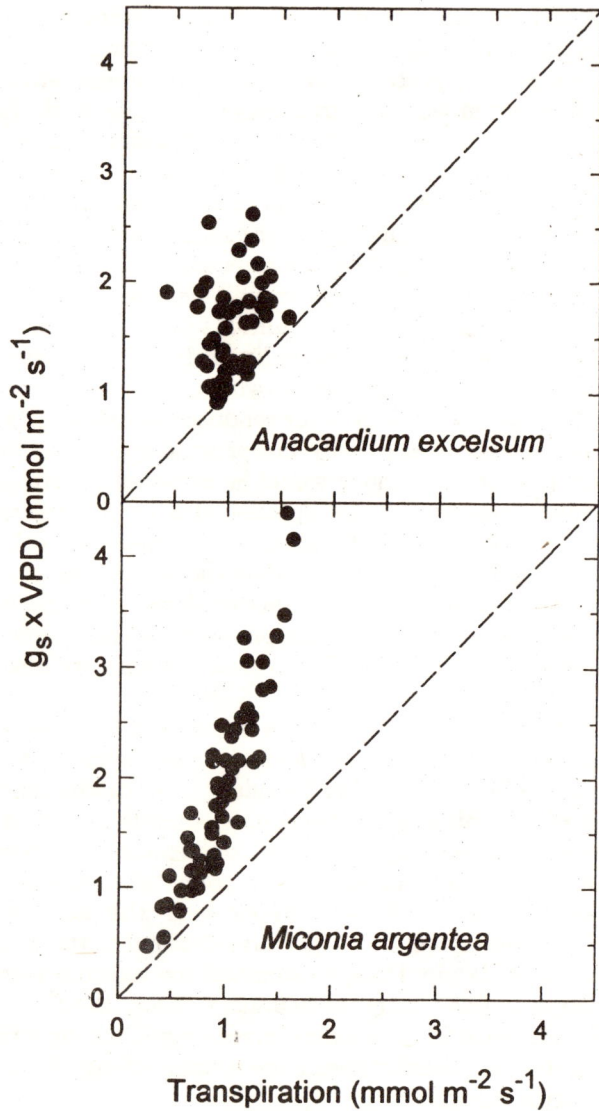

Figure 4.1.   *Predicted transpiration of two lowland tropical forest species in relation to transpiration determined from sap flow measurements using the stem heat balance technique. Predicted transpiration was obtained from the product of stomatal conductance (g_s) and the leaf-to-bulk air vapor pressure difference (VPD). The dashed lines represent 1:1 relationships between predicted and measured transpiration. Data for A. excelsum adapted from Meinzer et al. (1993).*

tion between an equilibrium component driven by net radiation, and a component driven by the extent to which the atmospheric saturation deficit is imposed at the leaf surface (Jarvis & McNaughton, 1986). In general, Ω, which ranges from zero to 1, approaches 1 with increasing stomatal conductance in a manner determined by the prevailing boundary layer conductance (Figure 4.2B). For a given species, Ω is thus characterized by a range of values rather than a specific value. Stomatal control of transpiration grows progressively weaker as Ω approaches 1, because the vapor pressure at the leaf surface becomes increasingly decoupled from that in the bulk air.

A survey of values of Ω estimated for different species and types of vegetation reveals a large range. In a number of broadleaved tropical forest tree species, estimates of Ω ranged from 0.6 to 0.9 (Table 4.1). On the other hand, in needle-leaved, aerodynamically rough coniferous trees with their relatively low stomatal conductance and high boundary layer conductance, Ω is only about 0.1 (Table 4.1). In small individuals (2–3 m tall) of several forest gap pioneer species growing in sites on Barro Colorado Island, Panama, the average value of Ω was about 0.62 during the dry season (Table 4.2), indicating that stomatal control of transpiration was relatively weak even though values of stomatal conductance were substantially lower then those measured during the wet season. This was chiefly a consequence of low boundary layer conductances, which strongly attenuated the influence of fluctuations in stomatal conductance on the overall vapor phase conductance. Average wind speeds of less than 0.5 m s^{-1} in this site and the relatively large leaf size of the species studied contributed to the low values of boundary layer conductance observed. Boundary layer conductance was calculated as the reciprocal of the difference between total vapor phase resistance and stomatal resistance. Comparison of stomatal conductance with the much lower total vapor phase conductance that actually limits water loss (Table 4.2) provides an idea of the extent to which transpiration would have been overestimated from porometric measurements of stomatal conductance alone. Increases in Ω following irrigation during the dry season (Table 4.2) suggest that stomatal control of transpiration in these species would be even weaker during the wet season, when stomatal conductance is higher in relation to boundary layer conductance. The values of Ω presented in Table 4.2 are consistent with those estimated by Roberts et al. (1990) for Amazonian rainforest understory species. Poor ventilation in understory sites might be expected to result in even lower values of boundary layer conductance and higher values of Ω in understory

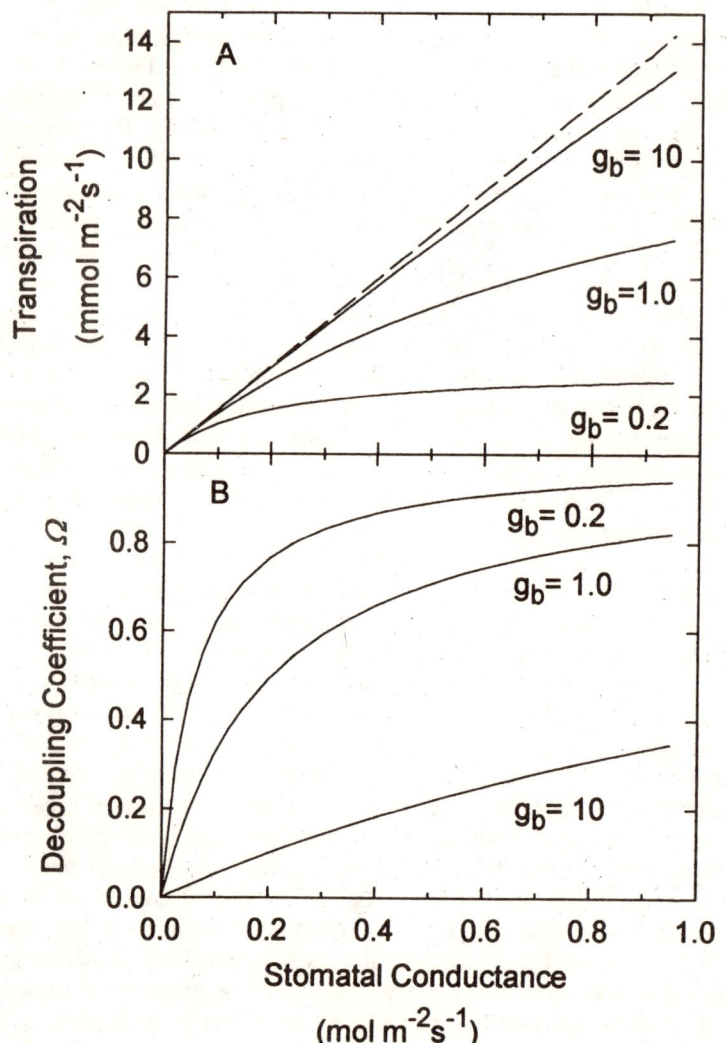

Figure 4.2.   The influence of boundary layer conductance (g_b, mol m^{-2} s^{-1}) on stomatal control of transpiration. In (A), decreasing g_b diminishes the impact of changes in stomatal conductance on transpiration. If g_b were infinite, the relationship between transpiration and stomatal conductance would be linear (dashed line). In (B), the stomatal decoupling coefficient, Ω, increases steeply with stomatal conductance when g_b is low. The slope of each curve in (A) at a given value of stomatal conductance is $1-\Omega$. Adapted from Meinzer, (1993).

species than for species growing in nearby gaps. Restriction of stomatal conductance by low light levels, however, may exert a compensating effect on Ω for species growing in the understory.

Derivation of a stomatal decoupling coefficent for water vapor represents a significant advance in the interpretation of stomatal behavior in the field because it provides a means of predicting the influence of stomatal movements in individual leaves on transpiration per unit leaf area and larger scales from whole individuals to entire stands. It would be highly desirable if a similar approach could be applied in characterizing coupling of leaves and the atmosphere with respect to stomatal control of CO_2 fluxes at the single leaf and larger scales. Unfortunately, the analysis of driving forces and conductances governing CO_2 assimilation is not entirely analogous to the situation with regard to water vapor. In addition to purely physical limitations on diffusion, CO_2 assimilation is regulated by biochemical reactions that do not respond linearly to changes in the CO_2 gradient between the leaf interior and the air. Furthermore, environmental variables such as light, which have no direct effect on transpiration, directly influence the biochemical capacity for CO_2 fixation. This makes it much more difficult to apply an analysis of coupling to predict the impact of a marginal change in stomatal conductance on CO_2 assimilation. For example, the stomatal control coefficient for photosynthesis can be negative if an increase in stomatal conductance enhances transpirational cooling, lowering leaf temperature sufficiently to reduce assimilation capacity despite the rise in internal CO_2 partial pressure. Another example is the covariation between boundary layer conductance and vapor pressure at the leaf surface in the field. Although an increase in boundary layer conductance should enhance CO_2 assimilation, the boundary layer control coefficient will be diminished by the accompanying decrease in vapor pressure at the leaf surface, which will limit assimilation by reducing stomatal conductance. Despite these potential interactions, control coefficients describing the sensitivity of photosynthetic CO_2 fixation to small changes in stomatal conductance, boundary layer conductance, and other variables have been derived recently for leaves held in gas exchange cuvettes (Woodrow, Ball & Berry, 1990) and for leaves under field conditions (Woodrow, Ball & Berry, 1986; Woodrow, 1989).

Table 4.1. Values of Ω in relation to leaf size and stomatal conductance for several forest tree species.

Species	Location	Leaf size (cm)	Stomatal conductance (mol m^{-2} s^{-1})	Ω
Tropical Forest				
[1] *Tectona grandis*	Nigeria	26	1.26	0.9
[1] *Gmelina arborea*	Nigeria	14	0.78	0.9
[2] *Anacardium excelsum*	Panama	12	0.20	0.6
[1] *Triplochiton scleroxylon*	Africa	20	0.28	0.6
[3] *Piptadenia suaveolens*	Brazil	–	0.23	0.4
[3] *Licania micrantha*	Brazil	–	0.14	0.4
Temperate Forest				
[1] *Picea sitchensis*	Scotland	0.2	0.05	0.1
[1] *Pinus sylvestris*	Scotland	–	0.16	0.1
[1] *Fagus sylvatica*	Scotland	4	0.10	0.2

Source of data: [1] Jarvis & McNaughton (1986), [2] Meinzer et al. (1993), [3] Roberts et al. (1990).

Table 4.2. Values of Ω in relation to stomatal, boundary layer and total vapor phase conductance for high light requiring species growing in treefall gaps on Barro Colorado Island, Panama during a severe dry season. Average values of stomatal conductance are shown. Maximum values were two to three times higher but occurred only briefly.

Species	Conductance			
	Stomatal	Boundary layer (mol m^{-2} s^{-1})	Total	Ω
Miconia argentea				
dry	0.149	0.237	0.088	0.65
irrigated[a]	0.218	0.190	0.099	0.80
Palicourea guianensis				
dry	0.027	0.046	0.016	0.56
irrigated	0.049	0.075	0.026	0.64
Cecropia insignis	0.166	0.326	0.110	0.62
Cecropia obtusifolia	0.097	0.192	0.057	0.62
Coccoloba spp.	0.156	0.286	0.093	0.65

[a] Measurements were made after two days of irrigation during February.

4.3 GAS EXCHANGE ON A UNIT AREA AND WHOLE PLANT BASIS

Physiological adjustment to resource limitation often confers a marked degree of homeostasis of certain properties such as photosynthesis and nutrient content on a unit leaf area basis (Pereira, 1990). Partial maintenance of photosynthesis per unit leaf area as total leaf area declines under limited resource supply is reasonably well documented. Nevertheless, it does not appear to be widely recognized that this behavior imposes a number of constraints on interpretation of unit leaf area-based gas exchange measurements. For example, it may be necessary to consider an individual's previous growth and environmental history. This is illustrated by the long-term response of the tropical forest understory species *Coffea arabica* (coffee) to different soil moisture regimes. After 120 days under different irrigation regimes, the total leaf area of coffee plants irrigated twice weekly was only about half that of plants irrigated twice daily, although initially their total leaf areas were similar and their assimilation rates on a unit area basis had been nearly equal throughout the experiment (Table 4.3). The maintenance of nearly identical assimilation rates in coffee plants irrigated twice daily and twice weekly despite markedly reduced stomatal conductance in the latter implies that photosynthetic activity increased in the plants irrigated twice weekly. This pattern suggests that a major mode of adjustment to reduced soil moisture availability in coffee is the maintenance of a nearly constant photosynthetic activity on a unit leaf area basis through a reduction in the rate of increase in total leaf area per plant (Meinzer, Saliendra & Crisosto, 1992).

A similar form of adjustment to reduced water availability apparently occurred in greenhouse-grown individuals of the desert shrub *Simmondsia chinensis* (jojoba) held under different water regimes for

Table 4.3. *Total final leaf area and gas exchange characteristics of coffee plants growing under three different soil moisture regimes for 120 days.*

Irrigation frequency	Leaf area (m^2)	CO_2 Assimilation ($\mu mol\ m^{-2}\ s^{-1}$)	Stomatal conductance ($mol\ m^{-2}\ s^{-1}$)
Twice daily	2.24	4.50	0.060
Twice weekly	1.25	4.22	0.045
Weekly	0.53	2.22	0.023

Table 4.4.   Leaf area and gas exchange characteristics of Simmondsia chinensis plants grown in a greenhouse under three different levels of
water supply for 7 months. From Goldstein et al. (unpublished).

Water Supply[a]	Leaf area (m^2)	CO$_2$ Assimilation (μmol m^{-2} s^{-1})	Stomatal conductance (mol m^{-2} s^{-1})
High	0.60	5.34	0.091
Intermediate	0.23	4.63	0.047
Low	0.13	4.10	0.041

[a] Plants were supplied with 6, 3 and 1.2 l of water per week for the high, intermediate and low
water availability regimes, respectively.

seven months (Table 4.4). The average CO$_2$ assimilation rate of plants
subjected to the intermediate water supply treatment was only about
13% lower than that of plants subjected to the high water supply
treatment, despite a 50% reduction in stomatal conductance of the
plants grown under intermediate water supply. The nearly complete
homeostasis of photosynthesis on a unit leaf area basis apparently
resulted from a markedly reduced rate of leaf expansion and maintenance of a lower total leaf area by plants subjected to the intermediate
water supply treatment. Even at the lowest level of water supply,
severe restriction of leaf area development to only one-fifth that of the
high water supply treatment caused the CO$_2$ assimilation rate per unit
leaf area to remain at 77% of that in the high water supply treatment.

Responses of single leaf gas exchange in the Hawaiian forest tree
species Metrosideros polymorpha to variations in the physiological
availability of soil moisture provide another illustration of partially
compensating adjustments in single leaf gas exchange and total leaf
area. Metrosideros polymorpha, as its species name suggests, is notable for extreme morphological and physiological variability
(Stemmermann, 1983). Several varieties of M. polymorpha are recognized in Hawaii and occur across a broad environmental range of
elevation, precipitation, and soil type (Vitousek, Field & Matson,
1990). Maximum stomatal conductance was lower in M. polymorpha
populations growing in a dry site and in a bog, but maximum assimilation rates at a given stomatal conductance were higher in these
populations than in populations growing in sites where soil moisture
was more freely available (Figure 4.3). It is likely that increased A$_{max}$
was achieved partially through reduced rates of leaf expansion. Restricted leaf area development was particularly apparent in the plants
growing in the bog, which were severely stunted.

Interpretation of single leaf gas exchange behavior may be further
confounded by developmental adjustments linked not directly to plant

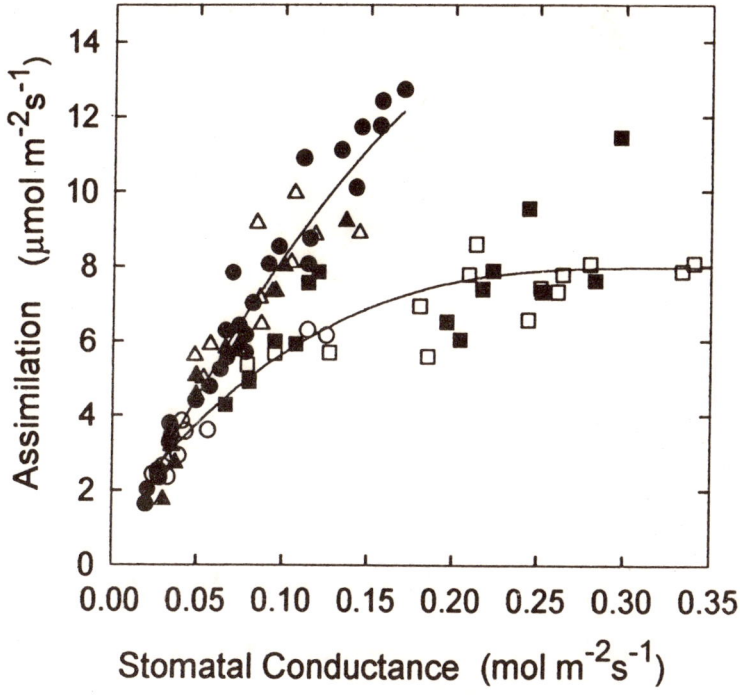

Figure 4.3 Relationship between CO_2 assimilation and stomatal conductance in six Metrosideros polymorpha *populations on the island of Kauai, Hawaii. Assimilation rates at a given stomatal conductance were higher in populations growing in a bog (▲ △) where the physiological availability of soil moisture was low, and in a dry site (●) where rainfall was low. The remaining populations (■ □ ○) experienced greater soil moisture availability. Adapted from Meinzer et al., (1992).*

age, but to the total amount of leaf area present at a given time (Black & Squire, 1979; Squire & Black, 1981; Meinzer & Grantz, 1990; Donovan & Ehleringer, 1992; Ovaska, Wallis & Mutikainen, 1992). For example, in the tropical grass sugarcane (*Saccharum* spp. hybrid), stomatal conductance per unit leaf area of well-irrigated plants increased with leaf area up to about 0.2 m^2 plant^{-1}, then declined with further increases in plant size (Figure 4.4A). These responses caused conductance (and transpiration) on a per plant basis to saturate rather than increase linearly with further increase in leaf area (Figure 4.4B). Within a few minutes following partial defoliation, stomatal conductance increased, re-establishing the original relationship with

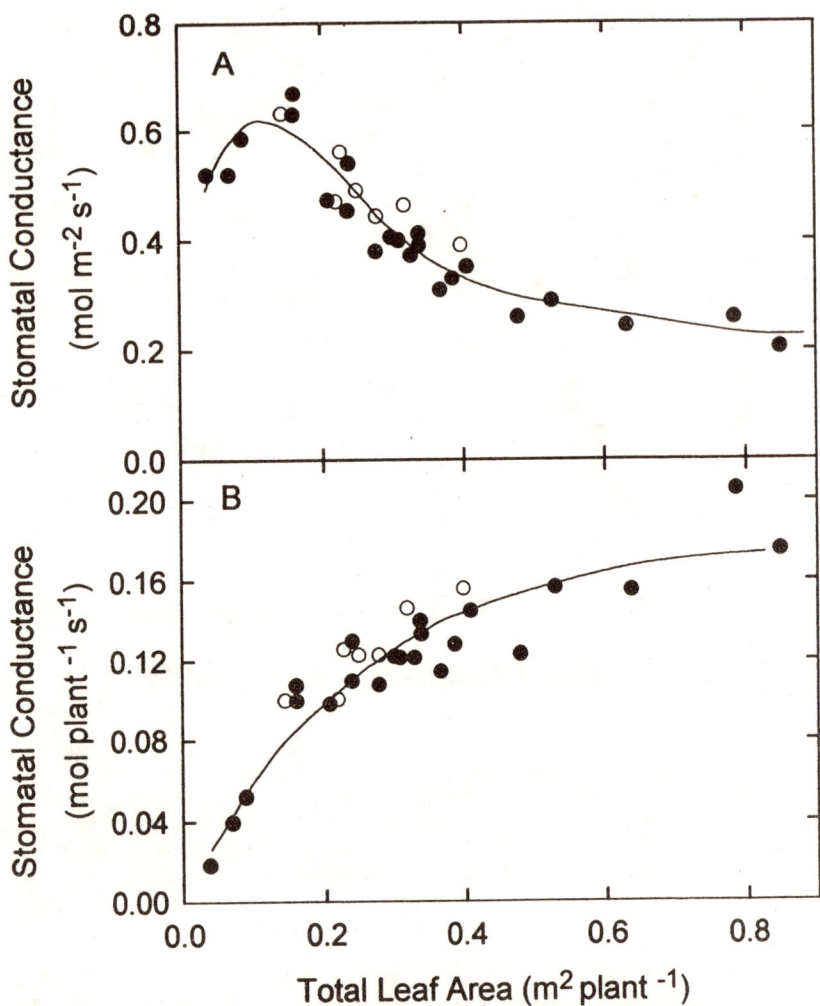

Figure 4.4 (A) Relationship between maximum stomatal conductance and total leaf area for sugarcane plants. Open circles represent plants subjected to removal or covering of 20–40% of their leaf area prior to measurement of stomatal conductance. (B) Relationship between maximum stomatal conductance per plant and total leaf area. Conductance values shown in (A) were multiplied by the total leaf area to yield total conductance per plant. Adapted from Meinzer and Grantz (1990).

remaining leaf area. Similarly, pruning of roots caused rapid reductions in stomatal conductance, which maintained or improved leaf water status. These rapid, dynamic responses indicate that the observed stomatal adjustments were not related to plant or leaf age per se, leaf water status, or to shading, but rather to changes in the ratio of transpiring leaf area to root system area (Meinzer & Grantz, 1990). These and other observations (Black & Squire, 1979; Squire & Black, 1981) suggest that the downward trend of stomatal conductance observed over the lifespan of individual leaves may often be reversible, although it is usually ascribed to the irreversible effects of leaf ageing (Solarova, 1980).

Adjustments in single leaf gas exchange following removal of leaf area are often interpreted as compensatory responses to herbivory (Detling, Deyer & Winn, 1979; Nowak & Caldwell, 1984; Wallace, McNaughton & Coughenour, 1984). Nevertheless, these responses may be a manifestation of a more general phenomenon (Tschaplinski & Blake, 1989; Ovaska, Wallis & Mutikainen, 1992). Maximum rates of leaf gas exchange may be limited by the hydraulic efficiency of the soil, roots and stem in supplying a given amount of foliage with water (Küppers, 1984; Meinzer et al., 1988; Reich & Hinckley, 1989; Meinzer, Goldstein & Grantz, 1990). This developmental coordination between liquid and vapor phase water transport properties apparently serves to regulate leaf water status within narrow limits (Meinzer et al., 1992) and may be mediated by fluxes of root-derived materials in the xylem sap (Meinzer, Grantz & Smit, 1991).

The examples given in this section suggest that even when micrometeorological feedback has been taken into account, it may not be possible to determine whether similar leaf gas exchange rates among individuals of contrasting size are attributable merely to differences in plant age or to resource limitations on leaf expansion unless comparative measurements of growth are made over a longer timescale than that required for gas exchange measurements.

4.4   SIMULTANEOUS OBSERVATIONS AT DIFFERENT SCALES

As discussed above, the problem of scaling consists of selecting the appropriate reference points, with respect to the leaf surface, for characterizing the driving forces for gas exchange. Thus, the principal constraint on extrapolating from conventional single-leaf measurements is that concurrent measurements of fluxes and driving forces referenced to the leaf surface and to a point in the bulk air away from the leaf surface are required. The simplest approach for moving up-scale from conventional single leaf measurements is to make porometric measurements of stomatal conductance and simultaneous measurements of leaf temperature and actual water vapor flux from unenclosed branches, entire individuals, or extensive canopies in which the leaf and canopy boundary layer elements have not been disturbed. This procedure allows the total vapor phase conductance, boundary layer conductance, and vapor pressure at the leaf surface to be determined by applying an Ohm's law analogy to the measurements obtained (Meinzer et al., 1993). Micrometeorological approaches such as the Bowen ratio and eddy correlation techniques have been widely employed for measuring transpiration at the canopy scale in tropical forests (Doley, 1981; Shuttleworth et al., 1984; Shuttleworth, 1988). However, these canopy and landscape level approaches have limited application to traditional ecophysiological studies in multispecific stands because they do not permit partitioning of transpiration among individual species and plants. Measurements of vapor fluxes at the canopy scale also do not distinguish between transpiration and evaporation from the soil and wet leaf surfaces. These limitations can be partially overcome by use of techniques such as the heat balance method for measuring transpiration of individual branches or entire plants (Cermák, Deml & Penka, 1973; Steinberg, van Bavel & McFarland, 1989). On the other hand, minimum sample sizes required to obtain representative data may become burdensome in heterogeneous environments when heat balance and other techniques for measuring sap flow are used to estimate transpiration.

The ability to make simultaneous observations at different scales and external reference points is particularly crucial for characterizing stomatal responses to environmental variables such as atmospheric humidity. In poorly coupled leaves and canopies with their low ratio of boundary layer conductance to stomatal conductance, stomatal responses to humidity in the field may not be apparent from conventional measurements of stomatal conductance, and humidity differences calculated from leaf temperature and ambient vapor pressure. For example, stomatal conductance in two large-leaved tropical forest heliophile understory species appeared to be unresponsive to variations in the leaf to bulk air vapor pressure difference (Fanjul & Barradas, 1985). Similarly, in the large-leaved canopy tree species *Anacardium excelsum*, no clear response of stomatal conductance to variation in the leaf to bulk air VPD was observed (Figure 4.5A). However, when the leaf surface was used as the reference point to calculate ambient vapor pressure, stomatal conductance of *A. excel-*

sum was observed to decline sharply with increasing evaporative demand (Figure 4.5B). In addition to large leaf size (ca. 12 x 25 cm), leaf arrangement and branch architecture contribute to low boundary layer conductance and poor coupling in *A. excelsum*. The leaves are supported by short, rigid petioles that restrict their movement, and short internodes cause the distal portions of older leaves to extend beyond the bases of younger leaves. The preceding examples show how the interpretation of experimental results can be strongly influenced by the location of the external reference point (spatial scale) for measurement of environmental variables such as humidity. If the boundary layer substantially decouples humidity at the leaf surface from that in the bulk air, use of a reference point in the bulk air to characterize stomatal response to humidity will give misleading results.

The physiological and ecological consequences of contrasting patterns of stomatal behavior in different species often cannot be determined from measurements of stomatal conductance alone because plant water balance is governed largely by the actual flux of water vapor rather than stomatal conductance. Nevertheless, traditional measurements of stomatal conductance should not be abandoned in favor of measuring actual transpiration rates at the whole plant and larger scales. An integrated approach, involving concurrent measurements of fluxes, driving forces and conductances across different spatial scales from individual leaves to entire canopies is likely to reveal patterns of stomatal regulation, and evidence concerning underlying mechanisms that would be undetectable at a single scale of observation.

For example, canopy conductance in sugarcane was observed to remain essentially constant over a wide range of leaf area index (LAI). Concurrent measurements at the single leaf scale revealed that stomatal conductance declined with increasing LAI, so that the product of stomatal conductance and LAI, a potential canopy conductance in the absence of boundary layer effects, remained constant as LAI increased (Meinzer & Grantz, 1989). The mechanism by which stomata "sensed" increasing LAI to maintain transpiration constant on a ground area basis during canopy development is unclear. Measurements of declining stomatal conductance with increasing leaf area in isolated, greenhouse-grown sugarcane plants (see section 4.3) suggest that the decline in stomatal conductance with increasing LAI in the field was not mediated by increasing self-shading. This regulatory phenomenon would not have been apparent from observations at either the canopy or single leaf scales alone.

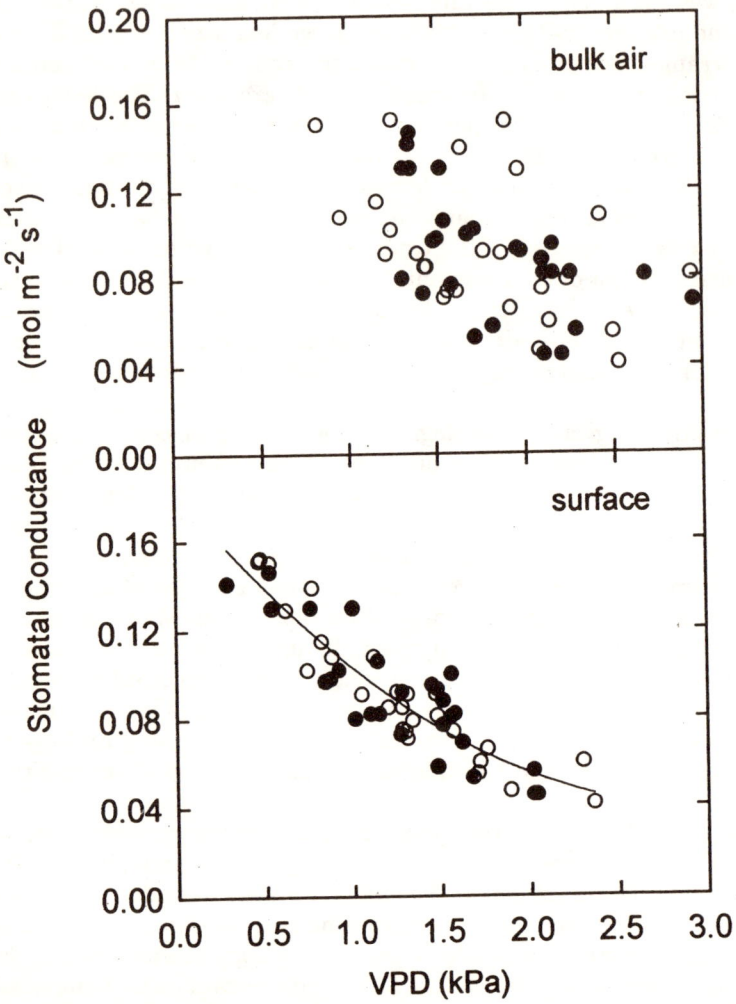

Figure 4.5 *Stomatal conductance in relation to the leaf-to-air vapor pressure difference (VPD) in the crown of an* Anacardium excelsum *tree during the dry season. Vapor pressure difference was determined using an external reference point in the bulk air ca 1 m away from the leaves and at the leaf surface. Adapted from Meinzer et al., (1993).*

4.5 CONCLUSIONS

Further research on patterns of stomatal behavior in relation to boundary layer conductance and the stomatal decoupling coefficient (Ω) under well-defined conditions in the field is needed in tropical forests with their high degree of vertical and horizontal heterogeneity and species diversity. Additional research will permit the relative roles of factors such as phylogeny, leaf size, and canopy microenvironment in determining stomatal behavior and transpiration to be assessed. Increased recognition of the decoupling influence of the external boundary layer surrounding each leaf and the entire canopy will doubtless lead to changed perceptions about the relative roles of intrinsic and external factors in determining patterns of stomatal behavior. This is illustrated by a recent model that treats boundary layer conductance as an independent variable in order to predict diurnal courses of stomatal conductance (Collatz, et al., 1991). The model suggests that the diurnal course of stomatal conductance is dramatically altered if boundary layer conductance undergoes a four-fold to tenfold change with all other variables held constant. As an example of how this would relate to actual field conditions, the boundary layer conductance for a needle-leaved conifer and a broadleaf tree having leaves with a mean length of 20 cm would differ by about an order of magnitude if both were exposed to the same wind speed. These results raise the possibility that contrasting stomatal behavior under superficially similar environmental conditions may be governed as much by the external boundary layer as by intrinsic genetic and physiological properties.

In contrast to typical temperate coniferous forests, many tropical forests contain a wide variety of emergent canopy species and associated variation in stomatal conductance, leaf size, canopy structure, and roughness. This suggests that in addition to exhibiting greater species diversity than temperate coniferous forests, tropical forests may be characterized by greater interspecific variation in coupling between leaves and the atmosphere, and therefore in the extent to which stomatal movements regulate photosynthetic gas exchange. This has important implications for predicting the effects of deforestation and global climate change on the water and carbon balance of tropical regions. Physiological responses governing exchange of CO_2 and water vapor that can be observed on a small scale may not control stand or regional level responses. In a poorly coupled canopy, for example, the air inside the thick boundary layer becomes enriched in water vapor and depleted in CO_2. When this occurs, changes in stomatal opening may have little or no effect on fluxes of water vapor and CO_2 from the stand as a whole. Thus, the effect on evapotranspiration of changes in species composition arising from deforestation will depend both on stomatal properties of the new group of species and leaf-atmosphere coupling. Conversely, poorly coupled canopies may exhibit considerable self-regulation or homeostasis in the face of changes in bulk atmospheric variables such as humidity and CO_2 levels expected to occur as a result of global climate change. Some of these relationships and their consequences for water balance are considered in the diagram shown in Figure 4.6. In this scheme, influences on evapotranspiration and regional water balance are partitioned between two components, represented by Ω and $1-\Omega$. The decoupling coefficient, Ω, represents the influence of variation in the energy supply driving evaporation (net radiation). The corresponding coupling coefficient $(1-\Omega)$ represents the influence of all other factors determining boundary layer conductance and stomatal aperture. If Ω is larger than $1-\Omega$, then evapotranspiration will be controlled primarily by variation in net radiation rather than bulk atmospheric evaporative demand. It thus seems clear that the degree of coupling of individual species and entire canopies to the atmosphere will have a

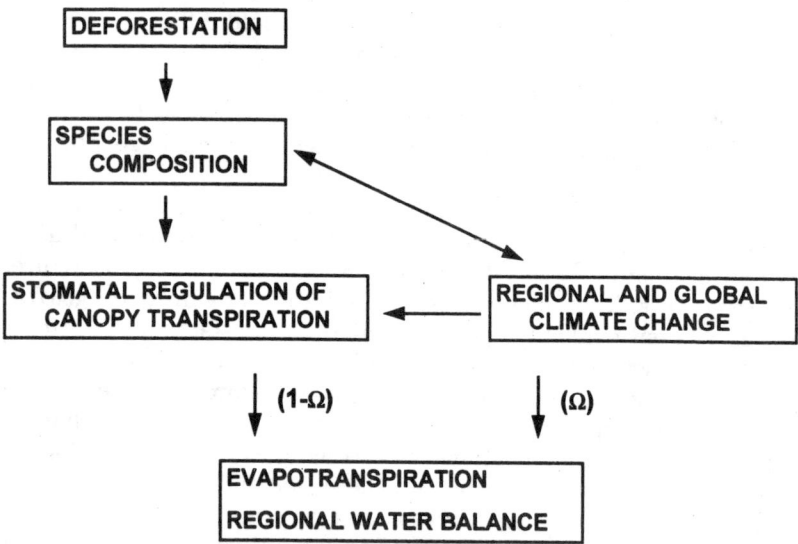

Figure 4.6   Conceptual scheme summarizing influences of regional and global climate change and changes in species composition on stomatal $(1-\Omega)$ and environmental (Ω) control of evapotranspiration and regional water balance.

substantial influence on their responses to changing environmental conditions.

REFERENCES

BENNET, J. M., JONES, J. W., ZUR, B. & HAMMOND, L. C. (1986) Interactive effects of nitrogen and water stresses on water relations of field-grown corn leaves. *Agronomy Journal*, **78**, 273–280.

BLACK, C.R. & SQUIRE, G. R. (1979) Effects of atmospheric saturation deficit on the stomatal conductance of pearl millet (*Pennisetum typhoides* S. and H.) and groundnut (*Arachis hypogaea* L.). *Journal of Experimental Botany*, **30**, 935–945.

CALDWELL, M. M., MATSON, P. A., WESSMAN, C. & GAMON, J. (1993) Prospects for scaling. *Scaling Physiological Processes Leaf to Globe* (eds. J.R. Ehleringer & C.B. Field) Academic Press, San Diego.

CERMÁK, J., DEML, M. & PENKA, M. (1973) A new method of sap flow rate determination in trees. *Biologia Plantarum*, **15**, 171–178.

COLLATZ, G. J., BALL, J. T., GRIVET, C. & BERRY, J. A. (1991) Physiological and environmental regulation of stomatal conductance, photosynthesis, and transpiration: A model that includes a laminar boundary layer. *Agricultural and Forest Meteorology*, **54**, 107–136.

DETLING, J. K., DEYER M. I. & WINN, D. T. (1979) Net photosynthesis, root respiration, and regrowth of *Bouteloua gracilis* following simulated grazing. *Oecologia*, **41**, 127–134.

DOLEY, D. (1981) Tropical and subtropical forests and woodlands. *Water Deficits and Plant Growth, Vol. VI* (ed. T.T. KOZLOWSKI) Academic Press, New York.

DONOVAN, L. A. & EHLERINGER, J. R. (1992) Contrasting water-use patterns among size and life-history classes of a semi-arid shrub. *Functional Ecology*, **6**, 482–488.

EHLERINGER, J. R. & FIELD, C. B. (eds.) (1993) *Scaling Physiological Processes Leaf to Globe*. Academic Press, San Diego.

FANJUL, L. & BARRADAS, V. L. (1985) Stomatal behavior of two heliophile understory species of a tropical deciduous forest in Mexico. *Journal of Applied Ecology*, **22**, 943–954.

FARQUHAR, G. D. & SHARKEY, T. D. (1982) Stomatal conductance and photosynthesis. *Annual Review of Plant Physiology*, **33**, 317–345.

GRACE, J., FASEHUN, F. E. & DIXON, M. (1980) Boundary layer conductance of the leaves of some tropical timber trees. *Plant, Cell and Environment*, **3**, 443–450.

GUTIÉRREZ, M. V., HARRINGTON, R. A., MEINZER, F. C., & FOWNES, J. H. (1994) The effect of environmentally induced stem temperature gradients on transpiration estimates from the heat balance method in two tropical woody species. *Tree Physiology,* **14,** 179–190.

JARVIS, P. G. & MCNAUGHTON, K. G. (1986) Stomatal control of transpiration: Scaling up from leaf to region. *Advances in Ecological Research,* **15,** 1–49.

KOCH, M.S & RAWLIK, P. (1993) Transpiration and stomatal conductance of two wetland mesophytes (*Cladium jamaicense* and *Typha domingensis*) in the subtropical Everglades. *American Journal of Botany,* **80,** 1146–1154.

KÜPPERS, M. (1984) Carbon relations and competition between woody species in a central European hedgerow II. Stomatal responses, water use, and hydraulic conductivity in the root/leaf pathway. *Oecologia,* **64,** 344–354.

LUDLOW, M. M. & IBARAKI, K. (1979) Stomatal control of water loss in siratro (*Macroptilium atropurpureum* (DC) Urb.), a tropical pasture legume. *Annals of Botany,* **43,** 639–647.

MCNAUGHTON, K. G. & JARVIS, P. G. (1991) Effects of spatial scale on stomatal control of transpiration. *Agricultural and Forest Meteorology,* **54,** 279–301.

MEINZER, F. C. (1993) Stomatal control of transpiration. *Trends in Ecology and Evolution,* **8,** 289–294.

MEINZER, F. C., GOLDSTEIN, G. & GRANTZ, D. A. (1990) Carbon isotope discrimination in coffee genotypes grown under limited water supply. *Plant Physiology,* **92,** 130–135.

MEINZER, F. C., GOLDSTEIN, G., HOLBROOK, N. M., JACKSON, P. & CAVELIER, J. (1993) Stomatal and environmental control of transpiration in a lowland tropical forest tree. *Plant, Cell and Environment,* **16,** 429–436.

MEINZER, F. C.. GOLDSTEIN, G., NEUFELD, H. S., GRANTZ, D. A. & CRISOSTO, G. M. (1992) Hydraulic architecture of sugarcane in relation to patterns of water use during plant development. *Plant, Cell and Environment,* **15,** 471–477.

MEINZER, F. C. & GRANTZ, D. A. (1989) Stomatal control of transpiration from a developing sugarcane canopy. *Plant, Cell and Environment,* **12,** 635–642.

MEINZER, F. C. & GRANTZ, D. A. (1990) Stomatal and hydraulic conductance in growing sugarcane: Stomatal adjustment to water transport capacity. *Plant, Cell and Environment,* **13,** 383–388.

MEINZER, F. C., GRANTZ, D. A. & SMIT, B. (1991) Root signals mediate coordination of stomatal and hydraulic conductance in growing sugarcane. *Australian Journal of Plant Physiology,* **18,** 329–338.

MEINZER, F. C., RUNDEL, P. W., GOLDSTEIN, G. & SHARIFI, M. R. (1992) Carbon isotope composition in relation to leaf gas exchange and environmental conditions in Hawaiian *Metrosideros polymorpha* populations. *Oecologia,* **91,** 305–311.

MEINZER, F. C., SALIENDRA, N. Z. & CRISOSTO, C. H. (1992) Carbon isotope discrimination and gas exchange in *Coffea arabica* during adjustment to different soil moisture regimes. *Australian Journal of Plant Physiology,* **19,** 171–184.

MEINZER, F. C., SEYMOUR, V. & GOLDSTEIN G. (1983) Water balance in developing leaves of four tropical savanna woody species. *Oecologia*, **60**, 237–243.

MEINZER, F. C., SHARIFI, M. R., NILSEN, E. T. & RUNDEL, P. W. (1988) Effects of manipulation of water and nitrogen regime on the water relations of the desert shrub *Larrea tridentata. Oecologia*, **77**, 480–486.

NORMAN, J. M. (1993) Scaling processes between leaf and canopy levels. *Scaling Physiological Processes Leaf to Globe* (eds. J.R. EHLERINGER & C.B. FIELD) Academic Press, San Diego.

NOWAK, R. S. & CALDWELL, M. M. (1984) A test of compensatory photosynthesis in the field: Implications for herbivore tolerance. *Oecologia*, **61**, 311–318.

OVASKA, J., WALLS, M. & MUTIKAINEN, P. (1992) Changes in leaf gas exchange properties of cloned *Betula pendula* saplings after partial defoliation. *Journal of Experimental Botany*, **43**, 1301–1307.

PEREIRA, J. S. (1990) Whole-plant regulation and productivity in forest trees. *Importance of Root to Shoot Communication in Responses to Environmental Stress* (eds. W. J. DAVIES & B. JEFFCOAT) British Society for Plant Growth Regulation, Bristol.

REICH, P. B. & HINCKLEY, T. M. (1989) Influence of pre-dawn water potential and soil-to-leaf hydraulic conductance on maximum daily leaf diffusive conductance in two oak species. *Functional Ecology*, **3**, 719–726.

REYNOLDS, J. F., HILBERT, D. W. & P. R. KEMP (1993) Scaling ecophysiology from the plant to the ecosystem: A conceptual framework. *Scaling Physiological Processes Leaf to Globe* (eds. J.R. EHLERINGER & C.B. FIELD) Academic Press, San Diego.

ROBERTS, J., CABRAL, O. M. R. & DE AGUIAR, L. F. (1990) Stomatal and boundary layer conductances in an Amazonian terra firme rainforest. *Journal of Applied Ecology*, **27**, 336–353.

SCHIMEL, D. S., DAVIS, F. W. & KITTEL, T. G. (1993) Spatial information for extrapolation of canopy processes: Examples from FIFE. *Scaling Physiological Processes Leaf to Globe* (eds. J. R. EHLERINGER & C. B. FIELD) Academic Press, San Diego.

SCHULZE, E.-D. (1986) Carbon dioxide and water vapor exchange in response to drought in the atmosphere and the soil. *Annual Review of Plant Physiology*, **37**, 247–274.

SCHULZE, E.-D. & HALL, A. E. (1982) Stomatal responses, water loss, and CO_2 assimilation rates of plants in contrasting environments. *Encyclopedia of Plant Physiology, New Series, Vol. 12B, Physiological Plant Ecology II* (eds. O. L. LANGE, P. S. NOBEL, C. B. OSMOND & H. ZIEGLER) Springer-Verlag, Berlin.

SHUTTLEWORTH, W. J. (1988) Evaporation from Amazonian rainforest. *Proceedings of the Royal Society, London Series B*, **233**, 321–46.

SHUTTLEWORTH, W.J. (1989) Micrometeorology of temperate and tropical forest. *Philosophical Transactions of the Royal Society, London Series B,* **324**, 299–334.

SHUTTLEWORTH, W. J., GASH, J.H., LLOYD, C. R., MOORE, C. J., ROBERTS, J. R., FILCHO A., FISCH, G., FILHO, V., RIBEIRO, M., MOLION, L. C. B., SA, L. D., NOBRE, C., CABRAL, O. M. R., PATEL, S. R. & MORAES, J. C. (1984) Eddy correlation measurements of energy partition for Amazonian forest. *Quarterly Journal of the Royal Meteorologica. Society,* **110**, 1143–1162.

SOLAROVA, J. (1980) Diffusive conductances of adaxial (upper) and abaxial (lower) epidermes: Response to quantum irradiance during development of primary *Phaseolus vulgaris* L. leaves. *Photosynthetica,* **14**, 523–31.

SQUIRE, G. R. & BLACK, C. R. (1981) Stomatal behavior in the field. *Stomatal Physiology* (eds. P.G. JARVIS & T. A. MANSFIELD) Cambridge University Press, Cambridge.

STEINBERG, S., VAN BAVEL, C. H. M. & MCFARLAND, M. J. (1989) A gauge to measure mass flow rate of sap in stems and trunks of woody plants. *Journal of the American Society of Horticultural Science,* **114**, 466–472.

STEMMERMANN, L. (1983) Ecological studies of Hawaiian *Metrosideros* in a successional context. *Pacific Science,* **37**, 361–373.

TSCHAPLINSKI, T. J. & BLAKE, T. J. (1989) Photosynthetic reinvigoration of leaves following shoot decapitation and accelerated growth of coppice shoots. *Physiologia Plantarum,* **75**, 157–165.

VITOUSEK, P. M., FIELD, C. B. & MATSON, P. A. (1990) Variation in foliar $\delta^{13}C$ in Hawaiian *Metrosideros polymorpha*: A case of internal resistance? *Oecologia,* **84**, 362–370.

VON CAEMMERER, S. & FARQUHAR, G. D. (1981) Some relationships between the biochemistry of photosynthesis and the gas exchange of leaves. *Planta,* **153**, 376–387.

WALLACE, L. L., MCNAUGHTON, S. J. & COUGHENOUR, M. B. (1984) Compensatory photosynthetic responses of three African graminoids to different fertilization, watering, and clipping regimes. *Botanical Gazette,* **145**, 151–156.

WOODROW, I. E. (1989) Limitation to carbon dioxide fixation by photosynthetic processes. *Photosynthesis* (ed. W. BRIGGS) Liss, New York.

WOODROW, I. E. & BERRY, J. A. (1988) Enzymatic regulation of photosynthetic CO_2 fixation in C_3 plants. *Annual Review of Plant Physiology and Plant Molecular Biology,* **39**, 533–594.

WOODROW, I. E., BALL, J. T. & BERRY, J. A. (1986) A general expression for control of the rate of photosynthetic CO_2 fixation by stomata, the boundary layer, and radiation exchange. *Progress in Photosynthesis Research Vol. 4* (ed. J. BIGGINS) Martinus Nijhoff, Dordrecht.

WOODROW, I. E., BALL, J. T. & BERRY, J. A. (1990) Control of photosynthetic carbon dioxide fixation by the boundary layer, stomata, and ribulose 1,5–bisphosphate carboxylase/oxygenase. *Plant, Cell and Environment,* **13**, 339–347.

5

On the Relationship of Plant Geometry to Photosynthetic Response

Thomas J. Herbert

The chapter by Chazdon et al. (Chapter 1) concerned relationships between the light environment and the photosynthetic response of tropical forest plants. This chapter builds upon these relationships by exploring the dependence of total plant photosynthesis upon orientation and position of photosynthetic surfaces and characteristics of the photosynthetic response. A simple two-dimensional model of an individual plant or plant stand will be used to explore the dependence of total plant photosynthetic rate on plant geometry, particularly with respect to two common characteristics of tropical forest plants and their environment, heliotropism and high solar elevation angles.

5.1 MODELING OF THE PHOTOSYNTHETIC RESPONSE

The solar radiation regime in the tropics is characterized by high elevation angles for the direct solar beam, resulting in high incident flux upon leaf lamellae and other photosynthetic elements such as stems and branches. High incident flux presents a complex challenge to tropical plants, which must balance the benefits of high incident flux as photosynthetic substrate with the problems of high temperature, resulting water loss, and photoinhibition. Additionally, tropical sun tracks place the sun at high solar elevation for longer periods of time than sun tracks in temperate and polar latitudes. Thus, one might expect that the spatial position and orientation of photosynthetic elements optimal for maximum photosynthesis and minimum effects of heat and photoinhibition would be quite different than for temperate and polar plants. Finally, heliotropism of tree leaves seems more common in tropical than in temperate environments and requires that we consider the dynamic relationship between the structure of tropical forest canopies and the solar radiation environment, involving both the movement of the sun and photosynthetic elements of the plant.

The photosynthetic response of a plant to the incident solar beam strongly depends upon the geometry of photosynthetic elements of the plant in relation to the spatial orientation of incident light flux. This dependence is principally the result of the complexity of dependence of photosynthetic rate upon the spatially and temporally varying incident flux. Instantaneous photosynthetic rate increases in a nonlinear fashion with incident flux, resulting in a maximization of photosynthetic rate, spatially integrated over all photosynthetic elements, when the array of elements is structured so that upper canopy elements are inclined at a higher angle (with respect to the horizontal) than lower canopy elements. Thus, the functional dependence of photosynthetic rate upon the magnitude of incident flux has direct influence upon the optimal geometry of a plant structure if optimal plant geometry is related to photosynthetic response. Similarly, the temporal dependence of photosynthetic rate affects the relationship between total plant photosynthesis and geometry. As the sun moves during the day and leaves move in response to phototropic response to direct and reflected light, wilting, and wind, light flux incident upon photosynthetic elements changes. These changes will produce a temporally integrated photosynthetic response which is also a function of plant geometry.

5.1.1   Application of photosynthetic and geometrical models

Mathematical models are essential for exploration and understanding of the relationships of photosynthetic response and incident flux to plant geometry. Two types of models are commonly used in biology; empirical models and mechanistic models. Empirical models are descriptions of observational data and thus have limited power to predict new phenomena. Mechanistic models are reductionist and can lead to a deeper understanding of the underlying processes and predict new phenomena. The usefulness of reductionist modeling as a predictive tool in plant physiology and ecophysiology can be shown in several ways: 1) *Reductionist models are successful. Even simple models often predict and help in understanding important phenomena.* Herbert and Nilson (1991) have shown that a pattern of decreasing leaf inclination from top to bottom of a plant is one of the most important geometrical characteristics that maximizes total plant photosynthetic rate and minimizes or nearly minimizes between-leaf variance of total plant photosynthetic rate. Honda and Fisher (1978) have shown that tree branch angles in *Terminalia* can be predicted with a simple model that maximizes total leaf surface area. 2) *Reduction of a problem and modeling in terms of essential elements is necessary in order to dissect the overall result in terms of understandable fragments.* In fractal geometry, a simple quadratic relationship results in spatial patterns of high complexity with infinitely recurring detail. Even model plants which consist of just a few vertically stacked leaves illuminated by the direct solar beam have a complex functional relationship between total plant photosynthetic rate and inclination angle of each of the leaves. 3) *The relationship of plant geometry and physiology is near to being understood at a physical level and so realistic models can be constructed.* Fairly complete models of plant physiology are within reach. The principal problem of using modeling may be existing limits on computational power. Farquhar (1989) and Laisk (Laisk & Walker, 1989; Laisk & Eichelmann, 1989) have successfully used modeling of photosynthetic kinetics to understand fundamental problems in plant physiology and to predict new phenomena. Barriers to more complex calculations are beginning to fall as computational power of modern computers increases dramatically. Much of the modeling of plant photosynthesis has centered on using complex mechanistic models incorporating a full range of physical and physiological phenomena to predict light interception or photosynthesis for a single set of parameters characterizing a model plant or for a limited set of such parameters (Geller & Nobel, 1984, 1986; Myneni & Impens, 1985a,b; Myneni et al., 1986a,b). In this paper and the studies that it describes, a different approach is used. Very simple models, reduced in scope to describing the dependence of total plant photosynthesis upon leaf inclination and position for two dimensional plant canopies, are used to explore the parametric 'landscape' around the set of leaf inclination angles or positions that maximize total plant photosynthetic rate. Simple models are required if optimization techniques that require the evaluation of total plant photosynthetic rate for thousands to millions of individual plant geometries are to be of use. But these methods permit a comprehensive investigation and resulting description of the dependence of photosynthesis upon plant geometry. Since there may be many plant geometries that result in nearly optimum photosynthetic response, an investigation of the parametric 'landscape' may be more important than determination of a single optimum geometry. Even if that determination may have high precision because of the incorporation of many details of geometry and physiology, it may not be an accurate representation of biological significance.

5.2   RELATIONSHIP OF LEAF INCLINATION TO PLANT PHOTOSYNTHETIC RESPONSE

5.2.1   A simple two-dimensional model

This paper considers a very simple two-dimensional model of plant structure, consisting of an array of flat photosynthetic leaf lamellae positioned in space (Figure 5.1). Non-photosynthetic elements, such as stems, branches, and trunks and curved photosynthetic elements including, stems, branches, trunks and curved leaves or petioles are not incorporated in this model. The simplified model permits isolation and study of the fundamental principles of leaf shading without secondary effects resulting from curvature of photosynthetic elements and shading by non-photosynthetic elements. The model depends only on the photosynthetic response function, the relationship between incident light flux and spatially integrated photosynthetic rate, and geometrical characteristics of position, orientation, and relative size of each of the flat photosynthetic lamellae. Thus, the model focuses on the basic characteristics of any array of flat photosynthetic surfaces and predicts the instantaneous photosynthetic response. This model is very general and independent of scale and so is applicable to any individual plant or plant stand, or forest in full sun that is not shaded by plants not included in the model.

One version of this model plant consists of N leaves, all of the same shape and size, vertically stacked above each other and irradiated

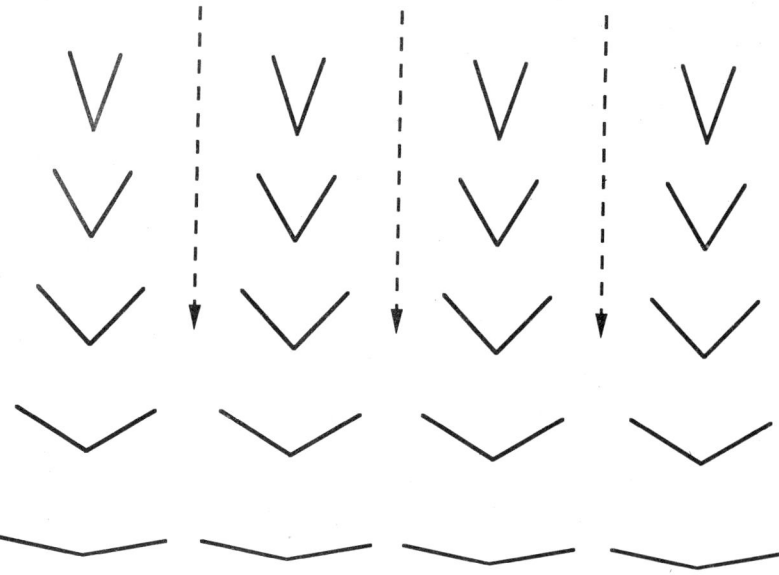

Figure 5.1.   Qualitative description of the array of plant leaves that maximizes total canopy photosynthetic rate when illuminated by solar radiation incident from directly above. Each line represents an individual flat photosynthetic surface. Petioles, stems, branches, and trunks are omitted both from the drawing and from the model and are not photosynthetic, nor do they shade the leaf surfaces in the model (viewed from a ground-based coordinate system in which the vertical axis is perpendicular to the gound).

from directly above by the solar beam. Then, total plant photosynthetic rate, ΣP, is:

$$\Sigma P = \frac{1}{N} \sum_{i=1}^{N} P_i \frac{\mu_i - \mu_{i-1}}{\mu_i}, \qquad (1)$$

where P_i is the relationship between photosynthetic rate and incident flux and μ_l is the cosine of the inclination angle β_l for a leaf in layer, i, where $i = 1$ for the top layer and $i = N$ for the bottom layer (Herbert & Nilson, 1991; Herbert, 1992). A useful relationship for P_i is given by the non-rectangular hyperbolic formula:

$$P_i = \frac{1}{2\Theta} \{ \alpha I_0 \mu_i + P_{max} - [(\alpha I_0 \mu_i)^2 + P_{max}^2 - \alpha I_0 \mu_i P_{max}(4\Theta - 2)]^{1/2} \}, \; (2)$$

where I_0 represents incident radiant flux and α, P_{max}, and Θ are parameters that characterize the photosynthetic response of the leaf. In particular, if $\Theta = 0$, the nonrectangular hyperbolic function becomes a familiar rectangular hyperbolic or Michaelis-Menten function, and if $\Theta = 1$, the non-rectangular hyperbolic function becomes a piecewise linear or Blackman function (Leith & Reynolds, 1987). Typically, the photosynthetic response may be characterized by values of $\Theta \approx 0.8$. P_{max} is the maximum value of P_i, and α is the slope of the P_i vs. I_0 relationship for $I_0 = 0$.

5.2.2 Maximization of total plant photosynthesis

Equation (2) can be used to explore relationships between leaf inclination and total plant photosynthesis resulting from illumination only by the direct solar beam. One such investigation has looked at the leaf inclination angles that result from maximization of total plant photosynthetic rate (Herbert & Nilson, 1991). In this study, several important results are apparent. First, the upper leaves are more steeply inclined than the lower leaves, as shown qualitatively in Figure 5.1. With realistic parameters for α, P_{max}, and Θ, maximization of total plant photosynthetic rate predicts that leaves in the top layer are very steeply inclined, with inclination angles in the range of 85°-89°. Second, the set of leaf inclination angles that maximizes total plant photosynthetic rate also nearly minimizes between-leaf variance of photosynthetic rate (Herbert & Nilson, 1991). If the photosynthetic response function is a rectangular hyperbola, maximization of total plant photosynthetic rate results in identical integrated photosynthetic rates for each leaf. For other photosynthetic response functions, the top one or two layers of leaves each contribute somewhat more to total plant photosynthetic rate than do the lower layers. But below the very top of the plant canopy, other layers contribute nearly equally to the total plant photosynthetic rate. Finally, if P_{max} decreases from the top to bottom of the canopy, a strategy that maximizes total plant photosynthetic rate may predict a set of leaf inclination angles that includes a layer, other than the bottom layer, that has zero inclination. In this model, a layer of horizontal leaves blocks all direct solar radiation from reaching lower layers of leaves and thus predicts no photosynthetic role for those leaves. Thus, the model described here suggests that a canopy illuminated only by direct solar radiation will have a limit to the number of leaf layers that receive illumination under a maximization strategy.

For an array of vc .ically stacked leaves illuminated from directly above, a procedure th..t maximizes total plant photosynthetic rate or minimizes variance in that rate always predicts that upper canopy leaves will be most steeply inclined, lower canopy leaves will be less steeply inclined, and there will be a monotonic decrease in inclination angle from canopy top to bottom. This pattern of leaf inclination is a direct result of the nonlinear form of the photosynthetic response, P_i. Consider an array of two horizontal leaves, one placed directly above the other and illuminated from above. If P_i is linear, then when the upper leaf is inclined, the decrease in integrated photosynthetic rate from that leaf is exactly offset by the increase in photosynthetic rate for the lower leaf. But if P_i is a nonlinear function of incident flux for which the slope decreases with increasing flux, the decrease in integrated photosynthetic rate for the upper leaf is smaller than the increase in integrated photosynthetic rate for the lower leaf, and total plant photosynthetic rate is increased by inclination of the top leaf. However, the results obtained by Herbert and Nilson (1991) show that the pattern of leaf inclination angles predicted by optimization procedures is not very sensitive to the exact shape of the photosynthetic curve.

The optimization procedure described by Herbert and Nilson (1991) finds only the optimum set of inclination angles, whether that is subject to the constraint of maximization of total plant photosynthetic rate or minimization of between-leaf variance of total photosynthetic rate. This procedure does not describe the shape of the multidimensional surface which represents total plant photosynthetic rate as a function of the inclination angles or positions of each of the individual leaves or photosynthetic elements. In one approach to this problem, Herbert and Nilson (1991) have calculated the total plant photosynthetic rate for 3601 vertically stacked arrays of 10 leaves for which inclination angles were chosen at random from a uniform distribution of angles distributed between 0° and 90°. For those arrays for which the top leaf was inclined at approximately 45° or less, total plant photosynthetic rate was nearly identical for any array that had any particular top leaf inclination. In these cases, all arrays with a particular top leaf inclination angle gave about the same total photosynthetic rate, irrespective of the inclination angles of the lower layers. But for higher values of top leaf inclination, none of the random arrays resulted in a total photosynthetic rate that was within 10% of the maximum rate and most arrays with highly inclined top leaves had an even lower photosynthetic rate.

5.2.3   Leaf orientation at angles that do not maximize photosynthesis

In another study (Herbert, 1992a,b), vertically stacked leaf arrays were characterized by their inclination angles and by a metric characteristic of the angular distance between any particular set of inclination angles, β_i', and the set, β_i, that maximizes total plant photosynthetic rate. The metric, which describes the radial distance in multidimensional space from the set of inclination angles that maximizes photosynthetic rate, is given as:

$$M\{i\} = \frac{1}{N\{i\}} \sqrt{\sum_{\{i\}} (|\beta_i'| - |\beta_i|)^2}, \tag{3}$$

where $N_{\{i\}}$ is the number of leaf layers in the set $\{i\}$. If the set of leaf angles $\{i\}$ contains just one leaf layer, e.g., $\{i\}$ for the top leaf layer, the metric $M\{i\}$ equals the difference between the two values for leaf inclination angle. If the set of leaf angles $\{i\}$ is multidimensional, e.g., $\{2–9\}$ for the next–to–top through the next–to–bottom leaf layer for a ten leaf–deep array, then the metric is a value between $0°$ and $90°$, representing a radial distance in multidimensional space, and can be directly compared with the one-dimensional measure of angular distance $|\beta_i' - |\beta_i||$.

Total plant photosynthetic rate will decrease from its maximum value as the metric $M\{i\}$ increases. Herbert (1992a,b) has described this dependence as a function of the metric for the top leaf layer, $\{1\}$, and the metric for all other layers except the lowest layer. The lowest layer in the leaf array is always fixed in a horizontal orientation and captures all the direct solar radiation not captured by the upper layers of leaves. If the lowest leaf layer were not fixed in a horizontal orientation, there would be an infinite number of leaf arrays that would maximize total plant photosynthesis, and the computer search procedure would not converge. Figure 5.2 shows the results of a simulation given by Herbert (1992a,b) for a ten-leaf deep array with leaves having a photosynthetic response for which $\Theta = 0$ and $\alpha I_0 / P_{max} = 3$. For $\alpha I_0 / P_{max} = 3$, photosynthetic rate P_i is 75% of the maximum photosynthetic rate at saturation, P_{max}, for a leaf that is oriented perpendicularly to the direct solar beam ($\beta = 0$ or $\mu_l = 1$ for a vertical solar beam).

This plot demonstrates several general results: *1) There is only one set of leaf inclination angles that results in a maximum in total plant photosynthetic rate.* This set of leaf angles gives a total plant photo-

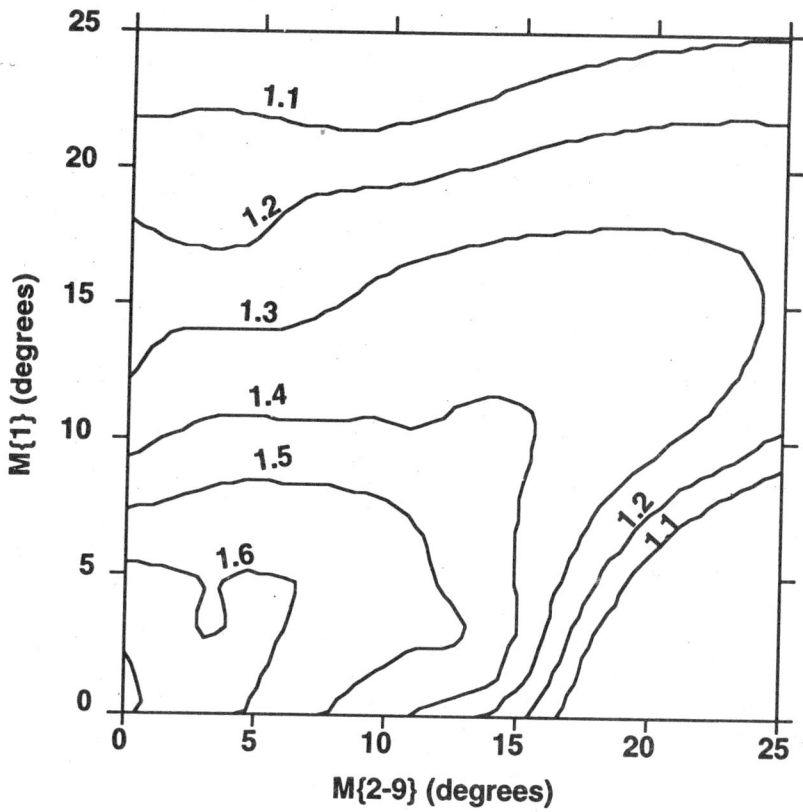

Figure 5.2. Total plant photosynthetic rate as a function of the metrics $M\{1\}$ and $M\{2-9\}$ for the top layer ($i = 1$) and lower layers ($i = 2, ..., 9$), respectively (in degrees). Total plant photosynthetic rate is given as a fraction of the total photosynthetic rate for a single leaf with $\beta = 0$.

synthetic rate that is more than 1.7 times the rate produced by a single unshaded leaf oriented perpendicular to the direct solar beam. In this two-dimensional representation of a nine-dimensional space, there are no other regions with high total photosynthetic rates. *2) There are many different sets of leaf inclination angles that result in nearly maximum total plant photosynthetic rates.* There are large regions in the contour plot that have similar values of total plant photosynthetic rate. It is most likely that variation in incident illumination, leaf or leaflet orientation or position would produce changes in total plant photosynthetic rate that would be much greater than those resulting from a less than optimum set of inclination angles for one particular geometry. *3) The dependence of total plant photosynthetic rate upon leaf angle is complex with considerable 'structure,' even for a simple array of ten vertically stacked leaves.* In Figure 5.2, there is a long ridge or plateau that extends out from the maximum value of total plant photosynthetic rate. This ridge represents a complex set of changes in leaf inclination angle. Clearly, total photosynthetic rate varies much less rapidly as one changes inclination angles in this complex way as compared with changing just the top or lower leaf layer inclination angles.

Niklas and Kerchner (1984) constructed a model of vascular sporophytes that maximized light interception and minimized bending moments. The multidimensional surface of light interception plotted against structural parameters was complex, but qualitatively very similar to the results shown by Herbert (1992a,b) and in this paper (Figure 5.1). In both types of simulations, the surface of light interception or total plant photosynthetic rate shows broad plateaus or ridges on which there is little dependence upon structural parameters and other regions for which light interception or photosynthetic rate decreases rapidly with changing structural parameters. Niklas and Kerchner (1984) note that structures that maximize light interception and minimize total bending moment have a main vertical trunk with photosynthetic elements oriented principally in horizontal planes. This geometry is very similar to the array of vertically stacked leaves assumed by Herbert (1992), except that the optimal geometry described by Niklas and Kerchner (1984) does not have the uppermost photosynthetic elements inclined because of developmental and growth constraints on the early Devonian plants. But decreasing inclination from top to bottom of a plant canopy is the result of a nonlinear relationship between photosynthetic rate and incident flux, if incident flux is vertical. Niklas and Kerchner (1984) implicitly assumed a linear relationship between incident flux and photosynthesis in maximizing interception of flux.

5.3   DEPENDENCE OF PHOTOSYNTHETIC RESPONSE UPON SOLAR POSITION

Nilson (1968) and Niklas and Kerchner (1984) have investigated light interception by plant canopies illuminated by a direct solar beam that is oriented in other than a vertical direction. Fundamentally, there is no difference between this geometry and one for which the solar beam is incident from directly above the canopy. In both cases, a strategy that maximizes total canopy photosynthesis will have the upper leaves oriented at an acute angle with respect to the solar beam, and the lower leaves oriented perpendicularly to that beam as shown previously in Figure 5.1. However, as the direct solar beam moves towards the horizon, the relative positions of the leaves in a plane perpendicular to the solar beam will change. Thus, shading of lower canopy leaves and the average canopy depth (as the number of layers of leaves in the direction of the solar beam) varies with solar position.

The photosynthetic consequences of solar movement can be inferred by careful examination of diagrams of the relative position and orientation of leaves in an extended canopy. Presently, a model similar to that used in studies by Herbert and Nilson (1991) and Herbert (1992a,b), but which permits arbitrary orientation of the direct solar beam, is being used to study the dependence of total plant photosynthetic rate upon leaf position and inclination. With this model, convergence times for calculation of leaf inclination angles that maximize total plant photosynthetic rate are long and do not permit exact calculation of leaf angles at the present time. However, results are qualitatively correct and show that maximization of total plant photosynthetic rate predicts that upper canopy leaves are inclined at a more acute angle with respect to the direct solar beam than lower canopy leaves, which are oriented nearly perpendicular to the beam. This general pattern of predicted leaf orientation for non-vertical solar flux incident upon an array of vertically stacked leaves is shown in Figure 5.3. Rotation of the coordinate system so that the solar beam is incident vertically, but the entire array of leaves is inclined clearly demonstrates that movement of the sun is equivalent to a change in canopy depth (number of leaf layers) and a horizontal shift in leaf position (Figure 5.4).

An increase in effective canopy depth and a horizontal shift in leaf position both have important consequences for total plant photosynthetic rate. *1) A change in apparent canopy depth affects the set of leaf inclination angles that maximize total plant photosynthesis.* Two scenarios are possible. First, for an isolated plant that is taller than it is wide, movement of the sun from the zenith towards the horizon would likely result in a decreased number of layers with respect to the canopy depth along the solar beam. The results of Herbert and Nilson (1991) indicate that if total plant photosynthetic rate is maximized for a canopy with fewer leaf layers, then the top canopy layers are inclined at a less steep angle with respect to the direct solar beam. Secondly, if the canopy depth is only a few layers deep, but largely in

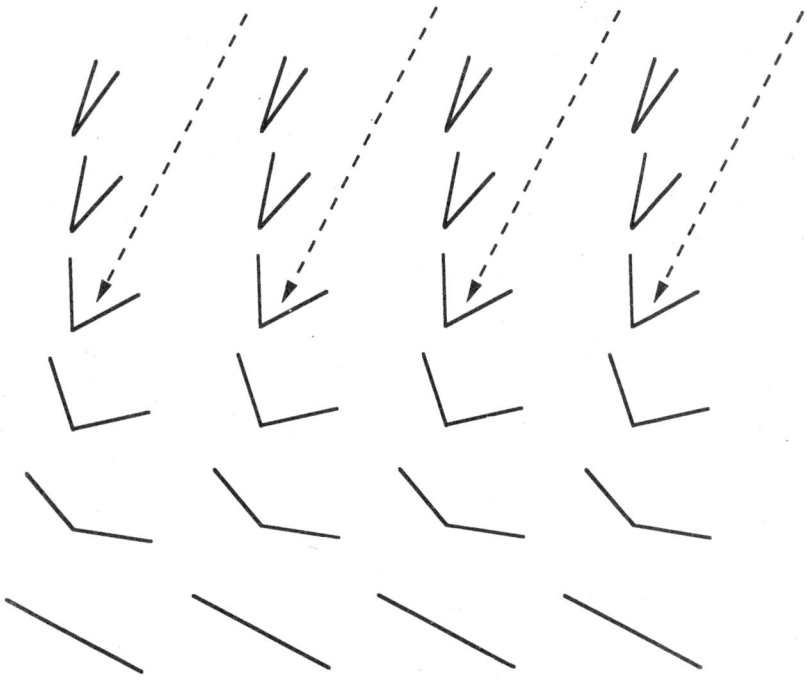

Figure 5.3 *First approximation to a qualitative description of the array of plant leaves that maximizes total canopy photosynthetic rate when illuminated by solar radiation incident from an arbitrary direction, (viewed from a ground-based coordinate system).*

horizontal extent, movement of the sun towards the horizon increases the apparent canopy depth and results in upper canopy leaves inclining to form a more acute angle with the solar beam. This result is shown in Figures 5.3 and 5.4, in which top canopy layer leaves are not only inclined as if to 'track' the sun, but also form a more acute angle with the direct beam than for the same canopy illuminated from directly above.

2) *A change in apparent horizontal packing of leaves affects the set of leaf inclination angles that maximize total plant photosynthesis.* Careful examination of Figure 5.4 shows that movement of the sun and a subsequent apparent horizontal shift in leaf position has created asymmetry in the leaf-sun geometry. Those layers just below the top canopy layer are now shaded asymmetrically such that one leaf of a leaf pair is still shaded by top canopy leaves, but the other leaf is now unshaded and has become a top canopy leaf. Maximization of total plant

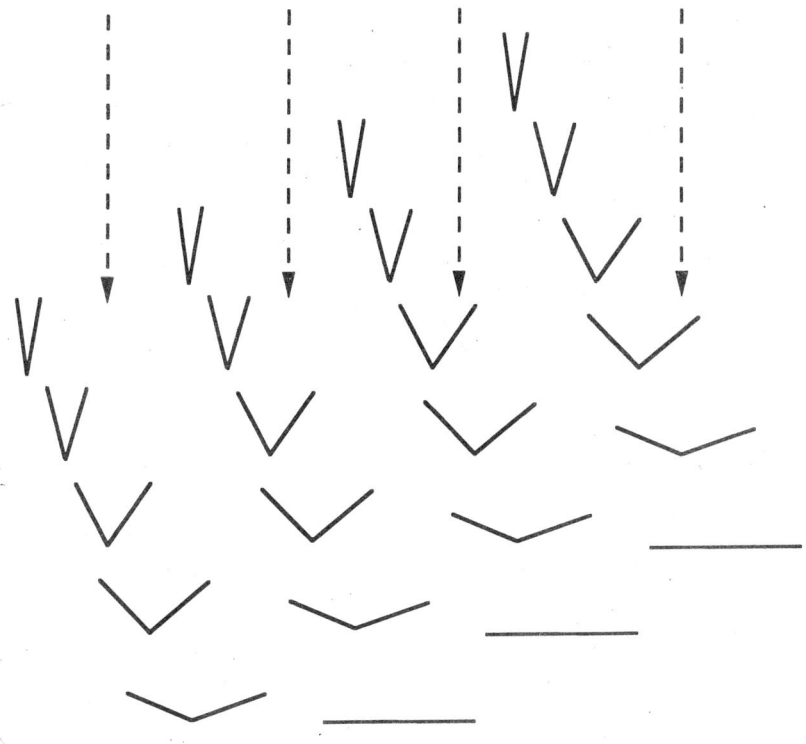

Figure 5.4.   First approximation to a qualitative description of the array of plant leaves that maximizes total canopy photosynthetic rate when illuminated by solar radiation incident from an arbitrary direction. (viewed from a sun-based coordinate system in which the vertical axis is parallel to the direct solar beam).

photosynthesis predicts that unshaded leaves are inclined at a more acute angle with respect to the solar beam than lower, shaded leaves. Therefore, the unshaded half of a leaf pair located below the top of the canopy should be inclined at a more acute angle with with respect to the solar beam than the shaded or partially shaded half of a leaf pair. The prediction of a model based upon maximization of total plant photosynthetic rate is that the orientation of some leaf pairs may not be symmetric with respect to the solar beam. (Figure 5.5). The mean of vectors perpendicular to the leaf or leaflet surface leads the sun position in the morning and lags in the afternoon. Additionally, predicted incli-

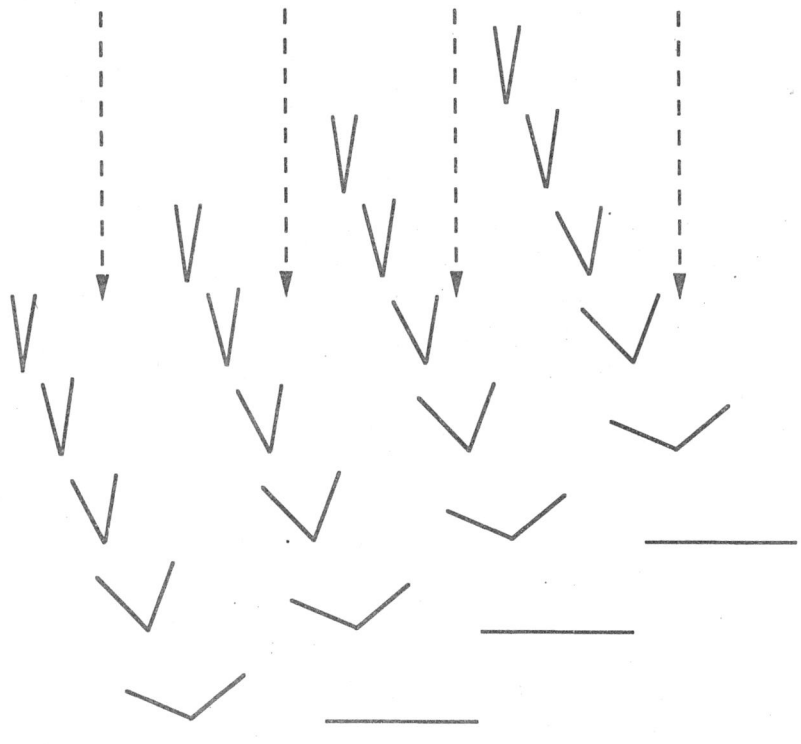

Figure 5.5. Second approximation to a qualitative description of the array of plant leaves that maximizes total canopy photosynthetic rate when illuminated by solar radiation incident from an arbitrary direction (viewed from a sun-based coordinate system).

nation angles of top canopy leaves may be affected by the asymmetry of the leaf array with respect to the solar beam. One-half of a leaf pair may be at the top of a larger number of leaf surfaces than the other half. But previous studies have shown that even moderate changes in the number of layers in a canopy affect top leaf inclination angles only slightly. Thus, it might be expected that top canopy layers would show nearly identical, large angles of leaf inclination, with leaves oriented symmetrically with respect to the direct solar beam and lower layers of leaves would show larger variance in orientation and a lead or lag of mean orientation with respect to the solar beam.

5.4   DISCUSSION

The model described in this paper predicts leaf orientations for a plant that shows solar-tracking leaf movement and for which the photosynthetic response to solar flux and orientation is instantaneous and without memory. Although incorporation of temporal dependence of photosynthetic rate into a complex model of light interception is a task for the future, it is instructive to compare results of the simple model presented here to data obtained from plants that show leaf heliotropism. Tropical forest trees include a large number of heliotropic species. Richards (1966) has noted that "pulvini are an extremely common feature of the leaves of tropical trees and lianes." Since the pulvinus is the region at the base of the leaf, leaflet, or pinnule which bends or twists and causes heliotropic movement, it might be expected that tropical forests are particularly rich in species that are solar-tracking. However, there have been almost no detailed experimental observations of leaf orientation and position in either solar tracking or non-tracking tropical forest trees. Many observations have been made on solar-tracking from small temperate and tropical plants, and it is upon these which we must rely for an immediate comparison with predictions of our model.

5.4.1   Leaves are most steeply inclined near the top of a plant

One prediction of the model presented in this paper is that the cosine of the angle of incidence, the cosine of an angle between a perpendicular to the leaf and the direction of the solar beam, will be small for top canopy leaves exposed to low incident solar flux and large for top canopy leaves illuminated by high incident solar flux. Experimental results from a wide range of solar tracking plant species show an increase in the cosine of angle of incidence in the morning, followed by a decrease in the cosine near noon (Herbert, 1984, 1992a,b; Herbert & Larsen, 1985). After noon, the cosine of angle of incidence increases to a second maximum in the early afternoon before decreasing steadily in the late afternoon. Low values of the cosine of angle of incidence in the early morning and afternoon represent a transition from night-time orientation of leaves or leaflets. However, the values of the cosine of angle of incidence for the period from mid-morning to mid-afternoon are characteristic of solar tracking and are qualitatively similar to the results predicted by our model. Low incident solar flux should result in a large cupping angle (the angle between the midribs of two leaves with azimuths which differ by 180°), and high incident flux should result in a small cupping angle (the angle between the midribs of two opposite leaves). Wainwright (1977) has shown that leaflets of *Lupinus arizonicus* have low values for the cupping angle near noon and much higher values in the morning and afternoon.

5.4.2   Leaf orientation leads and lags the sun

The asymmetry of leaf orientation with respect to sun position has been observed as a lead or lag in the azimuth of the steepest line on the leaf surface of several species of desert plants (Ehleringer & Forseth, 1980). Wainwright (1977) described a lead or lag of approximately 20° one hour after dawn and one hour before dusk for the solar-tracking angle, (the inclination of a perpendicular to the leaf plane with respect to the horizontal). Shell et al. (1974) have measured the angle between a perpendicular to the leaflet and the solar beam for *Phaseolus vulgaris*, obtaining a lead of 38° in the morning and a lag of 44° in the afternoon. The lead and lag of leaf orientation has been graphically demonstrated by Kawashima (1969a,b), who plotted the orientation of vectors perpendicular to leaflet surfaces of *Phaseolus vulgaris* on a polar projection of a hemisphere. The points representing the orientation of leaf perpendiculars clearly cluster in a position that leads the sun's orientation in the morning and lags in the afternoon. At solar noon, the leaf perpendiculars are oriented in a circle centered on the sun's position, with a radius representing about 60° of arc on the hemisphere. This radius corresponds to an angle between leaf midribs of about 30° (if there is no axial rotation or folding of the leaflet). Significant leaf or lag effects have not generally been shown in results reported by Herbert (1984; 1992a,b) and Herbert and Larsen (1985), but careful analysis shows some lead with respect to sun position for vectors oriented along the leaflet midribs for heliotropic orientation of leaflets of *Pithecellobium ungis-cati* (Figure 5.6). In this analysis of leaf orientation data, probability contours were plotted for the distribution on a sphere of leaf midribs using methods described by Diggle and Fisher (1985). The first two contours shown in Figure 5.6 represent probabilities of 0.2 and 0.8 that a midrib vector from the observed distribution will be found outside that contour. Since the sun position is just inside the 0.8 contour, it is clear that the lead angle is just barely significant at any reasonable confidence level.

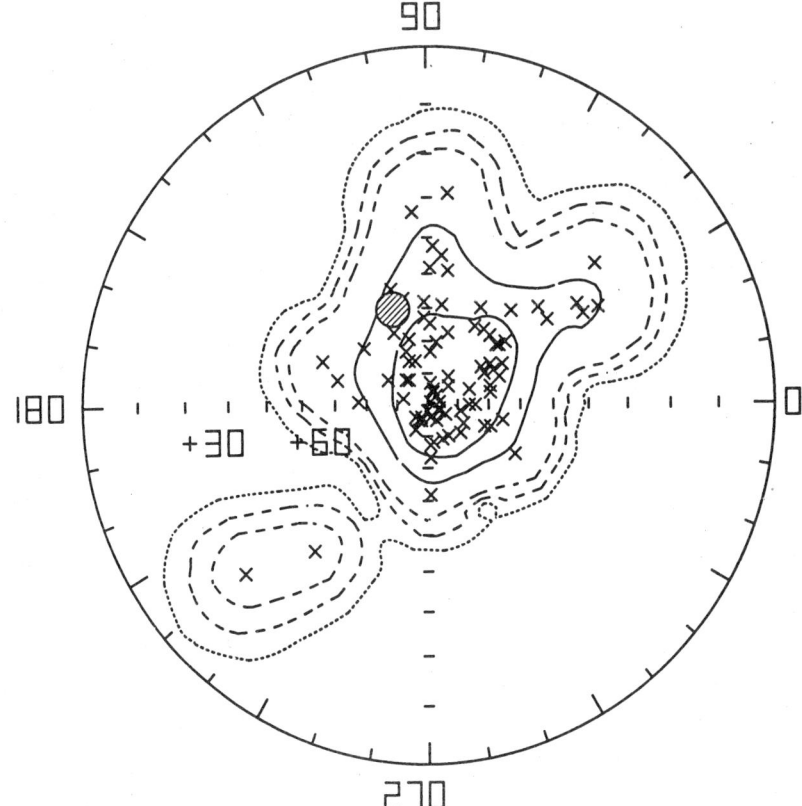

Figure 5.6. Orientation of the leaf midribs of Pithecellobium ungis-cati, *a solar-tracking plant, towards the sun for the period 9:29 – 10:30 h, solar time, on 3/3/90 in Miami, Florida. Midrib directions from leaf petiole to tip presented on an equal-angle projection of a hemisphere onto a circle (x) and solar position at 12:00 h (O). Angles represent compass directions, and solid lines represent contours of equal probability as calculated by the methods of Diggle and Fisher (1985).*

While plants show lead and lag phenomena, their relationship to the model described is not clear. Darwin (1881) described what seem to be independent effects of temperature and light flux upon leaf movement. Leaves do move into a vertical orientation as a result of high temperatures, independent of the direction of light. This movement to a vertical position as temperature increases could produce the appearance of the leaf movement leading or lagging sun position. A consequence of the temperature-induced lead or lag might be that the lead angle might equal zero at some time after solar noon because of the lag of temperature increase with respect to the elevation of the sun. Furthermore, many studies, including those of Herbert (1984; 1992a,b) and Herbert and Larsen (1985) are careful to measure leaf orientation to the direct solar beam only for top canopy leaves to avoid the influence of shading. Since the model described in this paper predicts the lead and lag phenomena primarily for leaves just below the top canopy layer, a full understanding and experimental verification of the relationship of lead and lag to maximization of total plant photosynthetic rate will require careful studies of leaf orientation as a function of time and position of the leaf in the canopy.

A major prediction of the model is that leaves should be steeply inclined in the upper layers of a plant canopy and nearly horizontal in the lower layers. Although this is often observed by workers in the field for both flowering plants and conifers, there has been very little work on the dependence of leaf angle on canopy depth for tropical forest species (Chazdon, 1985). Since a horizontally extended canopy effectively directs the penetration of incident light into a vertical direction, extended canopies should show a change in leaf angle with canopy depth, whether or not the leaves are solar tracking. Hopefully, future studies of extended monospecific plant stands and single trees with a large horizontal extent will be undertaken.

5.4.3   Inclination angles and leaf photosynthesis show variance within a plant

Experimental data on the variation of interception of light by solar tracking plants supports the model that we have described. Data obtained by Herbert (1991) show that although the mean inclination angle increases and the mean cosine of angle of incident decreases near solar noon, there is high variance of both inclination angle and cosine of angle of incidence. In fact, the distribution of these values is nearly uniform within a wide range and it appears that the change in mean value of the cosine of incidence is the result of a change in the maximum value of cosine of incidence, not in the distribution of values. The high variance in leaf angle and cosine of incidence is consistent with the model proposed in this paper. Although our model predicts an optimum set of leaf angles that maximizes total plant photosynthesis or minimizes inter-leaf variance in photosynthesis, even the simple two-dimensional model demonstrates that there are many different sets of leaf inclination angles that lead to a nearly identical result.

5.4.4   Spatial packing and inclination of leaves are interrelated

Honda and Fisher (1978) showed that for a model of *Terminalia catappa*, branching angles that maximize total unshaded leaf area are very close to observed values. A sequential model (Fisher & Honda, 1979) that added new branch units to a plant structure that had already been optimized to maximize total unshaded leaf area showed that new branch units do not grow out to produce optimal filling of the space. Thus, *Terminalia catappa* appears to have a deterministic branching pattern which results in maximization of light interception (Fisher, 1986).

The model of Honda and Fisher (1978) is closely related to models that fix the position of the leaf but vary the inclination angle. Higher values of leaf inclination, relative to the direction of the solar beam, result in a decrease in the solar flux intercepted by a leaf surface. Since a change in the amount of intercepted solar flux causes a change in the dimensions of the leaf as projected onto a plane perpendicular to the solar beam, changes in leaf inclination affect the packing of leaves with respect to this perpendicular plane. Thus, the determination of the set of leaf orientation angles or the packing of leaves in space required to maximize light interception or total plant photosynthesis is an equivalent solution to the problem of finding an optimal structure.

5.4.5   Maximization of carbon gain

This paper has concentrated on the relationship between the position and orientation of leaves in space and the instantaneous photosynthetic rate resulting from illumination of the leaf surface. However, the adaptive value of a particular geometry of the photosynthetic elements of plant structure is a function of photosynthetic activity integrated over time and the resulting carbon gain. Integration of photosynthetic rate over time requires a detailed time-dependent model of the biochemistry of photosynthesis such as those being developed by Farquhar (1989) and Laisk (Laisk & Walker, 1989; Laisk

& Eichelmann, 1989). Observations by Pearcy and Calkin (1983) estimate that sunflecks are responsible for as much as 60% of the total daily net carbon gain in two Hawaiian understory species. Although this is a reasonable first approximation, the kinetics of dynamic phenomena such as sunflecking are complicated and cannot be calculated simply by integrating static models over time (Chazdon et al., Chapter 1).

Chazdon (1985), working on rainforest understory palms, has begun the task of understanding the relationship between characteristics of the light environment, the geometry and spatial positioning of photosynthetic surfaces, and the resulting carbon gain. The rainforest understory presents a simplified geometrical environment since incident flux is directed downward through the forest canopy and does not change direction to the extent found in an environment illuminated for most of the day by the direct solar beam. However, the understory is subject to much dynamic variation in the light environment caused by sunflecks, making estimation of temporally integrated photosynthetic activity and carbon gain difficult.

The mechanistic, reductionist model presented in this paper predicts many interesting and important details of leaf angle and position in individual, isolated plants and in forest canopies. However, the model presented here is static and incorporates the dynamics of photosynthesis; thus, it cannot adequately predict integrated photosynthetic activity and carbon gain. New advances in computer hardware and programming techniques and in the modeling of photosynthesis will permit the detailed, dynamic modeling of parts of the complex geometrical and physiological structure of tropical forest canopies and can be expected to lead to accurate calculation of carbon gain and detailed understanding of many new phenomena important for tropical forest ecology.

REFERENCES

CHAZDON, R. L. (1985) Leaf display, canopy structure, and light interception of two understory palm species. *American Journal of Botany*, **72**, 1493–1502.

DARWIN, C. R. (1881) *The Power of Movement in Plants*. D. Appleton, New York.

DIGGLE P. J. FISHER, N. I. (1985) SPHERE: A contouring program for spherical data. *Computers and Geosciences*, **11**, 725 – 766.

EHLERINGER, J. R. & FORSETH, I. N. (1980) Solar tracking by plants. *Science* 210, 1094 – 1098.

FARQUHAR, G. D. (1989) Models of integrated photosynthesis of cells and leaves. *Philosophical Transactions of the Royal Society B (London)*, **323**, 357-367.

FISHER, J. B. (1986) Branching patterns and angles in trees. *On the Economy of Plant Form and Function.* (ed. T. J. GIVNISH,), Cambridge University Press, Cambridge, pp 493 – 523.

FISHER, J. B & HONDA, H. (1979) Ratio of tree branch lengths: the equitable distribution of leaf clusters on branches. *Proceedings of the National Academy of Sciences USA,* **76**, 3875 – 3879.

GELLER, G. N. & NOBEL, P. S. (1984) Cactus ribs: Influence on PAR interception and update. *Photosynthetica,* **18**, 482 – 494.

GELLER, G. N. & NOBEL, P. S. (1986) Branching patterns of columnar cacti: Influences on PAR interception and update. *American Journal of Botany,* **73**, 1192 – 1200.

GUTSCHICK, V. P. & WIEGEL, F. W. (1988) Optimizing the canopy photosynthetic rate by patterns of investment in specific leaf mass. *American Naturalist,* **132**, 67 – 86.

HERBERT, T. J. (1984) Axial rotation of *Erythrina herbacea* leaflets. *America Journal of Botany,* **71**, 76 – 79.

HERBERT, T. J. (1989) A model of daily leaf movement in relation to the radiation regime, in *Advances in Legume Biology,* (eds STIRTON, C.H., ZARUCCHI, J. L.) *Monographs in Systematic Botany from the Missouri Botanical Garden,* **29** pp. 629 – 643.

HERBERT, T.J. (1991) Statistical variation in interception of the direct solar beam by top canopy layers. *Ecology* **72**, 17 – 22.

HERBERT, T.J. (1992a) Geometry of heliotropic and nyctinastic leaf movements. *American Journal of Botany,* **79**, 547 – 550.

HERBERT, T. J. (1992b) Random wind-induced leaf orientation – Effect upon maximization of whole plant photosynthesis. *Photosynthetica,* **26**, 601 – 607.

HERBERT, T. J. & LARSEN, P. B. (1985) Leaf movement in *Calathea lutea* (Marantaceae). *Oecologia,* **67**, 238 – 243.

HERBERT, T. J. & NILSON, T. (1991) A model of variance of photosynthesis between leaves and maximization of whole plant photosynthesis. *Photosynthetica,* **25**, 597 – 606.

HONDA, H. E., FISHER, J. B. (1978) Tree branch angle: Maximizing effective leaf area. *Science* **199**, 888 – 890.

HORN, H. S. (1971) *The Adaptive Geometry of Trees.* Princeton University Press, Princeton.

KAWASHIMA, R. (1969a) Studies on the leaf orientation-adjusting movement in soybean plants. I. The leaf orientation-adjusting movement and light intensity on leaf surface. *Proceedings of the Crop Science Society of Japan,* **38**, 718 – 729.

KAWASHIMA, R. (1969b) Studies on the leaf orientation-adjusting movement in soybean plants. II. Fundamental pattern of the leaf orientation-adjusting movement and its significance for dry matter production. *Proceedings of the Crop Science Society of Japan,* **38,** 730 – 742.

KUROIWA, S. (1970) Total photosynthesis of a foliage in relation to inclination of leaves. *Prediction and Measurement of Photosynthetic Productivity.* PVDOC, Wageningen, pp. 79 – 89.

LAISK, A. EICHELMANN, H. (1989) Towards understanding oscillations: a Mathematical model of the biochemistry of photosynthesis. *Philosophical Transactions of the Royal Society B (London),* **323,** 369 – 384.

LAISK, A. & WALKER, D. A. (1989) A mathematical model of electron transport. Thermodynamic necessity for photosystem II regulation: 'Light Stomata'. *Philosophical Transactions of the Royal Society B (London),* **237,** 417 – 444.

LIETH, J. R. & REYNOLDS, J. F. (1987) The nonrectangular hyperbola as a photosynthetic light response model: geometrical interpretation and estimation of the parameter Θ. *Photosynthetica,* **21,** 363 – 366.

MYNENI, R. B. & IMPENS, I. (1985a) A procedural approach for studying the radiation regime of infinite and truncated foliage spaces. Part I. Theoretical considerations. *Agricultural and Forest Meteorology,* **33,** 327 – 337.

MYNENI, R. B. & IMPENS, I. (1985b) A procedural approach for studying the radiation regime of infinite and truncated foliage spaces. Part II. Experimental results and discussion. *Agricultural and Forest Meteorology,* **34,** 3 – 16.

MYNENI, R. B., ASRAR, G., KANEMASU, E. T., LAWLOR, D. J. & IMPENS, I. (1986a) Canopy architecture, irradiance distribution on leaf surfaces and consequent photosynthetic efficiencies in heterogeneous plant canopies. Part I. Theoretical considerations. *Agricultural and Forest Meteorology,* **37,** 189 – 204.

MYNENI, R. B., ASRAR, G., WALL, G. W., KANEMASU E. T. & IMPENS, I. (1986b) Canopy architecture, irradiance distribution on leaf surfaces and consequent photosynthetic efficiencies in heterogeneous plant canopies. Part II. Results and discussion. *Agricultural and Forest Meteorology,* 37, 205 – 218.

NIKLAS, K. J & KERCHNER, V. (1984) Mechanical and photosynthetic constraints on the evolution of plant shape. *Paleobiology* **19,** 79 – 101.

NILSON, T. (1968) On the optimum geometrical arrangement of foliage in the plant cover. *Investigations of Atmospheric Physics (Tartu),* **11,** 112 – 146.

OIKAWA, T. (1977a) Light regime in relation to plant population geometry II. Light penetration in a square-planted population. *Botanical Magazine (Tokyo),* **90,** 11 – 22.

OIKAWA, T. (1977b) Light regime in relation to plant population geometry III. Ecological implications of a square-planted population from the viewpoint of utilization efficiency of solar energy. *Botanical Magazine (Tokyo),* **90,** 301 – 311.

OIKAWA, T. & SAEKI, T. (1977) Light regime in relation to plant population geometry I. A Monte Carlo simulation of light microclimates within a random distribution foliage. *Botanical Magazine (Tokyo)*, **90**, 1 – 10.

PEARCY, R.W. & CALKIN, H. W. (1983) Carbon dioxide exchange of C_3 and C_4 tree species in the understory of a Hawaiian forest. *Oecologia*, **58**, 26 – 32.

RICHARDS, P. W. (1966) The *Tropical Rainforest*. Cambridge Univ. Press, Cambridge. p. 89.

SHELL, G. S. G., LANG, A. R. G. & SALE, P. J. M. (1974) Quantitative measures of leaf orientation and heliotropic response in sunflower, bean, pepper and cucumber. *Agricultural Meteorology*, **13**, 25 – 37.

WAINWRIGHT, G. M. (1977) Sun-tracking and related leaf movements in a desert lupine (*Lupinus arizonicus*). *American Journal of Botany*, **64**, 1032 – 1041.

6

Photosynthetic Characteristics of Tropical Trees Along Successional Gradients

Silvia Strauss-Debenedetti and Fakhri A. Bazzaz

Patterns of tropical succession and species replacement are shaped and determined by the dynamic nature of these forests. Gap-phase succession in tropical forests is driven by changes in resource availability arising from canopy disturbances and associated soil disturbances. Large gaps result primarily from major storms that cause the blowdown of numerous trees, while smaller gaps are created as individual trees lose branches, topple, or die standing (reviewed in Denslow, 1987). When such breaches in the canopy occur, light, temperature, humidity, and nutrients are altered (Chazdon & Fetcher, 1984a; Chiariello, 1984; Fetcher, Oberbauer & Strain, 1985; Ashton, 1992; Vázquez-Yanes & Orozco-Segovia, Chapter 18). Consequently, it is common to find all stages of forest succession occurring within the same forest at any given time along the understory-gap center continuum.

The ensuing differences in light quantity— and to a lesser extent, quality—between the undisturbed forest and the newly created gap environment play a crucial role in influencing species establishment, growth, and reproduction (Fetcher, Oberbauer & Chazdon, 1994; Denslow, 1987; Chazdon et al., Chapter 1; Fredeen & Field, Chapter 20). Tropical forest species differ in the extent to which gaps are required for successful regeneration. Based on their contrasting life-histories, tropical forest species have been categorized into two successional groups: light demanding or early successional, on the one hand, and shade tolerant or late successional, on the other. Many researchers have attempted to refine these criteria and make them more sensitive to the array of regeneration and life-history patterns found among tropical species (e.g., Martínez-Ramos, 1985; Whitmore, 1989; Swaine & Whitmore, 1989; Clark & Clark, 1992). In the absence of reliable data on species distributions along a successional sequence, however, no clear consensus has emerged on how to gauge the successional status of tropical species in relation to gap dynamics.

In this chapter, we evaluate the photosynthetic traits of tropical tree species to determine whether predictable patterns emerge between species broadly identified as early and late successional (pioneers vs. shade-tolerant species), within the framework of gap-phase succession. Our starting point in this analysis is the hypothesis that early successional, gap-requiring species exhibit a high degree of physiological flexibility (Bazzaz 1979 1984; Bazzaz & Pickett, 1980). We first describe the environment of early and late successional habitats within tropical forests, noting that these habitats require distinct physiological responses for survival and growth. We then focus on a comparison of the photosynthetic characteristics of seedlings of purported early and late successional species under an array of light conditions. Finally, we speculate on differences in responses of seedlings and mature trees and expand the original paradigm describing the photosynthetic performance of tropical trees to incorporate ontogenetic changes in performance. In concluding, we present a series of guidelines and suggestions for further research in the field.

6.1 PHOTOSYNTHESIS IN TROPICAL TREES: HYPOTHESES

The recognition that early and late successional species experience contrasting levels and scales of environmental heterogeneity has led to the formulation of hypotheses predicting their performance. The leading paradigm guiding much of the research on the ecophysiological patterns in tropical forest succession focuses on the assumption that resource variability is especially high in early successional environments. Indeed, levels of light, nutrients, and water differ within gaps, with distinct resource pools evident in the crown, bole and root zones of treefalls (Orians, 1982; Vitousek & Denslow, 1986). Gap shape, size, orientation, and the nature of the surrounding vegetation further influence the gap environment (Bazzaz, 1984; Barton, Fetcher & Redhead, 1989; Brandani, Hartshorn & Orians, 1988). Species specializing in these sites are therefore expected to be characterized by flexible response patterns (Bazzaz, 1979, 1984; Bazzaz & Pickett, 1980):

> ... permanent understory species may exhibit a "struggler" metabolism that is adapted to continuously low levels of resource flux, whereas species that depend on gaps have higher and more flexible metabolic rates capable of responding to resource pulses... (Bazzaz & Pickett, 1980, p. 303).

Researchers have extrapolated from this multiple-resource hypothesis to a one-dimensional focus on light, predicting that early successional species should exhibit wide acclimation of photosynthesis to changes in light levels, while a more constrained acclimation response pattern is expected among late successional species.

This approach is based on the assumption that constantly low light levels are characteristic of late successional environments, while highly heterogeneous conditions are prevalent in early successional ones. As we discuss later, both early and late successional environments experience light heterogeneity, albeit at different spatial and temporal scales. In fact, the often repeated (but not verified) adage that at least 75% of late successional (i.e., shade tolerant) species require exposure to openings to reach the canopy (Hartshorn, 1978) indicates that understory plants cope with changes in their light environment. Pearcy (1987) proposed that this requirement for frequent gap events implies that late successional species that eventually reach the canopy are especially sensitive to variations in the light environment and are thus characterized by high levels of photosynthetic acclimation.

We are thus left with two seemingly contradictory hypotheses. First, the multiple-resource model of Bazzaz and Pickett (1980) that predicts high levels of physiological flexibility among early successionals, and second, a single-resource model that predicts high levels of photosynthetic response among late successionals that experience frequent gap events (Pearcy, 1987). Here, we propose that these views are not contradictory, but rather reflect differences in the scale of response to environmental variation among early and late successionals and that such responses are, in part, mediated by ontogeny.

6.2 THE FOREST ENVIRONMENT

The gap dynamics model that has dominated much of the thinking in tropical biology brings to mind the so-called Swiss-cheese syndrome (*sensu* Lieberman & Lieberman, 1989) of a mosaic of open (gaps) versus closed (understory) forest habitats. The forest environment is spatially heterogeneous, with distinct microenvironments in temperature, vapor pressure deficit, and irradiance arising horizontally along the understory-gap continuum and vertically from the forest floor to the top of the canopy (Chazdon & Fetcher, 1984a, Chiariello, 1984; Fetcher, Oberbauer & Strain, 1985; Nuñez-Farfán & Dirzo, 1988; Chazdon et al., Chapter 1). Environmental heterogeneity also operates on a temporal scale, with high intensity photosynthetic photon flux density (PFD) sunflecks of short duration punctuating the understory (Chazdon & Pearcy, 1986; Chazdon, 1988; Chazdon et al., Chapter 1), and seasonal and diurnal changes in irradiance occurring in large gaps (reviewed in Bazzaz & Wayne, 1994; Chazdon et al., Chapter 1).

6.2.1 Forest gaps: Early successional environments

In all but very large clearings, total daily PFD in gaps is always less than full sunlight but significantly higher than in the understory (Bazzaz, 1984, 1991; Bazzaz & Wayne, 1994; Chazdon & Fetcher, 1984a; Moad, 1992; Chazdon et al., Chapter 1). Pioneers typically establish in gaps in the 150–500 m^2 size range (Brokaw, 1985), but gaps of different size experience strikingly different light levels. For example, Chazdon and Fetcher (1984a) found that total daily PFD in a 400 m^2 gap at La Selva was 20–25 times that of the understory, while in a 200 m^2 gap, PFD was only 9 times greater than in the understory.

Gaps are also spatially variable with resource heterogeneity arising along the root-crown zone of fallen vegetation (Orians,1982; Vitousek & Denslow, 1986), and with gap centers and edges experiencing distinct microenvironments (Bazzaz & Wayne, 1994). Plant growth further modifies the nature of the gap itself (Brandani, Hartshorn & Orians, 1988) and within-crown light heterogeneity can be high, being affected by leaf position and angle (Chazdon, Williams & Field, 1988). Seasonal changes in the light regime are also evident in gaps, and are largely a function of shifts in the sun's position and patterns of cloud cover and precipitation (Rich et al., 1993; Bazzaz & Wayne, 1994; Raich, 1989; Lee, 1989; Chazdon et al., Chapter 1).

6.2.2 The forest understory: Late successional environments

Light reaching the understory is altered both quantitatively and qualitatively (Chazdon et al., Chapter 1). This change in light quality may influence plant morphology and prevent germination of light-demanding species, but does not appear to have a direct effect on the carbon-gaining capacity of plants (Bazzaz, 1984, 1991; Vázquez-Yanes & Orozco-Segovia, 1984; Kwesiga & Grace, 1986; Ramos & Grace, 1990; Turnbull, 1991; Turnbull, Doley & Yates, 1993; Kitajima, 1994; Chazdon et al., Chapter 1; Vázquez-Yanes & Orozco-Segovia, Chapter 18).

Photosynthesis is instead limited by the quantity of light reaching the understory. Typically, this represents less than 3% of the light incident on the canopy or large clearings (Denslow, 1987; Chazdon et al., Chapter 1). For example, measurements at La Selva have indicated that the majority of 10-minute averages in the understory are under 10 $\mu mol\ m^{-2}\ s^{-1}$ (Chazdon & Fetcher, 1984b), and similar values have been reported for forests elsewhere (Denslow, 1987; Ashton, 1992). Consequently, short-duration sunflecks are an important source of PFD in the understory, where they can contribute over half of the daily PFD (Chazdon & Fetcher, 1984b; Chazdon et al., Chapter 1).

For plants growing in the understory, the creation of canopy gaps results in a sudden change in the quality and quantity of light, a potentially stressful situation. Not surprisingly, damage from photo-inhibition and photobleaching is likely to occur soon after the canopy opens with species differing in their capacity to recover from such damage (see Section 6.4.2.1). Despite similar maximum PFD of brief sunflecks and exposures of direct irradiance in gaps, photosynthetic responses to these two types of light regimes require distinct biochemical and stomatal specializations (Chazdon et al., Chapter 1).

6.2.3   Experimental light regimes

Testing whether early and late successionals differ in their responses to exposure to different light levels has led to a comparison of performance in plants, mostly seedlings, grown under real or simulated understory, and small and large gaps.

Greenhouse and controlled environment studies are particularly useful in this regard because they minimize confounding effects such as herbivore damage, competition, and variation in temperature and rainfall. Such studies, however, do not accurately mimic the forest light environment: high light in the greenhouse often exceeds what is encountered in gaps and it is typically too low in growth chambers; sunflecks are nearly impossible to duplicate, and light quality is rarely modified. Because of time and space constraints, most of these studies last for only a few months and focus only on the seedling-sapling phases. Still, studies performed under controlled conditions allow for the evaluation of one variable at a time and are particularly useful in assessing the interactions between environmental variables, such as light and nutrients.

Field studies, on the other hand, incorporate the complexity of the rainforest and allow for much longer periods of study. Field work, however, is complicated by the difficulty with site replication, weather unpredictability, ease of access and sampling. Recent advances in canopy access technology (cranes, balloons, towers, etc.) and in portable gas-exchange equipment will facilitate the acquisition of data at or near the canopy, thus expanding the data base beyond the seedling-sapling phase.

6.3   LEAF COMPONENTS OF RESPONSE

Growth in the light-limited understory presents a series of photosynthetic constraints which are clearly distinct from those in high irradiance gaps, calling for distinct adaptations. For example, species persisting in the understory are thought to benefit from adaptations that maximize light capture and minimize respiratory costs. Plants adapted to growth in gap or high light environments, on the other hand, face water, nutrient, high illumination, and photoinhibition stresses that can adversely interact with the maintenance of high photosynthetic rates (Björkman, 1981; Pearcy & Sims, 1994).

As expected, leaf structure and biochemistry are also altered with irradiance and leaf longevity increases in light-limited environments

(Reich et al., 1991; but see Mulkey, Wright & Smith (1993) for evidence of the reverse). In general, high-light or "sun" leaves tend to be thicker as the palisade component elongates. From a photosynthetic perspective, the main effect of these anatomical changes is to bring about a reduction in mesophyll resistance to carbon dioxide (Nobel, 1977). Increased irradiance also results in the development of leaves with increased concentration of carboxylation enzymes (Rubisco) and electron carriers and generally higher stomatal conductances (Björkman, 1981; Pearcy & Sims, 1994).

Canopy leaves of late successionals tend to be thicker compared to leaves of high-light grown seedlings of the same species (Oberbauer & Strain, 1984; 1986), but this trend is reversed for early successionals (Fetcher, Oberbauer, & Chazdon, 1994), indicating the presence of different developmental constraints. Patterns of variation in leaf thickness vary among tropical tree species. For example, high-light grown leaves of seedlings of the early successional *Cecropia obtusifolia* were 2.5 times thinner than those of the late successional *Poulsenia armata*, although the former photosynthesized at 50% higher rates (Strauss-Debenedetti & Berlyn, 1995). A similar lack of correspondence between photosynthesis and leaf thickness was found in the late successional *Pentaclethra macroloba* (Oberbauer & Strain, 1986).

6.4 PHOTOSYNTHESIS

Studies on the photosynthetic characteristics of tropical trees have largely focused on a comparison of photosynthetic rates of seedling leaves measured under saturating irradiances (A_{max}). These measurements are obtained from leaf (or plant) exposure to high light or from light-response curves, whereby photosynthesis is measured at various irradiances. Such curves provide much useful information, including saturating light level, quantum yield, maximum photosynthesis, and estimates of respiration rates.

In early successional environments, A_{max} rates are expected to be high, offering a competitive advantage to fast-growing pioneers and reflecting high resource availability (Chazdon et al., Chapter 1; Ackerly et al., Chapter 21). In the understory, on the other hand, survivorship appears to depend more on preventing herbivore and pathogen damage than on maintaining a high photosynthetic rate (Kitajima, 1994; Chapter 19). Light saturation for all species occurs at relatively low light levels, coincident with measured PFD in gaps and understory environments. Early successionals saturate at approximately 25% of full sunlight, while late successionals saturate at even lower levels (Langenheim et al., 1984; Oberbauer & Strain, 1984; Strauss-Debenedetti & Bazzaz, 1991; Newell et al., 1993). Although these results are based largely on results obtained for seedlings grown in greenhouse or growth-chamber light environments, Zotz and Winter (Chapter 3) present similar findings for intact canopy leaves of mature trees.

6.4.1 Photosynthetic plasticity

If early and late successional plants differ in their photosynthetic characteristics, then these differences are expected to arise when grown under similar conditions, especially in high light. Studies of intraspecific photosynthetic plasticity (*sensu* Bradshaw, 1965) form the bulk of our knowledge on the photosynthetic characteristics of tropical trees. Typically, seeds are germinated (or seedlings are transplanted shortly thereafter) and grown under constant, different light levels designed to emulate the forest understory and different-sized gaps (i.e. low, medium, and high light regimes). Observed differences in photosynthetic characteristics are generally viewed as being adaptive in nature, although they may instead simply reflect the constraints imposed by resource limitation.

6.4.1.1 Plasticity in early successional species

Early successionals are generally described as those species that germinate soon after gaps are formed, are fast growers, and reproduce before the gap closes or else become emergents themselves (Fetcher, Strain & Oberbauer, 1983; Augspurger, 1984; Denslow, 1987; Popma & Bongers, 1988; De Steven, 1988; King, 1991; Bazzaz, 1991; Ackerly, Chapter 21). Typical representatives in this group include members of the genera *Cecropia, Macaranga,* and *Trema.*

Early reports on the assimilation rates of tropical trees, while lacking standardization in techniques and sampling methods, indicate that early successionals have higher maximum photosynthetic rates than late successionals (reviewed in Bazzaz & Pickett, 1980). The results obtained in the last decade agree with these earlier reports. As can be seen from Table 6.1, species variously classified as pioneer or early successional have photosynthetic rates of about 10 μmol m^{-2} s^{-1}. For example, seedlings of *Cecropia obtusifolia* and *Ficus insipida* grown in uncovered temperate glasshouse benches exhibited A_{max} rates of 9.9 µmol m^{-2} s^{-1} and 9.4 µmol m^{-2} s^{-1}, respectively (Strauss-Debenedetti & Bazzaz, 1991), similar to the results obtained by Fetcher et al. (1987) for open-grown seedlings of *Heliocarpus appendiculatus* (8 µmol m^{-2} s^{-1}) and *Ochroma lagopus* (9 µmol m^{-2} s^{-1}) in Costa Rica. However, A_{max} rates of *Ochroma lagopus* grown under controlled conditions were three times higher (Oberbauer & Strain, 1984). Field-grown seedlings appear to have lower A_{max} rates, as evidenced from measurements of *Shorea parvifolia* (6.1 µmol m^{-1} s^{-1}) and *Shorea argentifolia* (4.3 µmol m^{-2} s^{-1}) in Sepilok, Sabah (Moad, 1992) and of *Milicia excelsa* (5 µmol m^{-2} s^{-1}) in Nigeria (Riddoch et al., 1991). These lower rates may, in part, reflect lower growth light conditions, although the exact nature of these differences in A_{max} remains unclear.

Early successional species exhibit a marked reduction in A_{max} when grown in low light, i.e., photosynthetic plasticity is high. A comparison of the light response curves of high and low light grown plants reveals a shift in the shape of the curve, with concomitant reductions in photosynthetic rates, and in light compensation and saturation points (Langenheim et al., 1984; Strauss-Debenedetti & Bazzaz, 1991; Turnbull, 1991; Newell et al., 1993). These responses are in agreement with the observation that the biochemical costs associated with maintaining a high photosynthetic rate in high light preclude equally high rates in low light (Pearcy & Sims, 1994; Fredeen & Field, 1991; Chazdon et al., Chapter 1; Fredeen & Field, Chapter 20).

Photosynthetic rates of early successionals decrease by an average factor of two or more in low light compared to high light, but generalizations are difficult to make due to differences in light regimes among studies (Table 6.1; contrast the data obtained for *Cordia alliodora* as reported by Fetcher et al., 1987 and Ramos & Grace, 1990). Large differences in A_{max} between light extremes have been obtained for *Cecropia obtusifolia* (Strauss-Debenedetti & Bazzaz, 1991) and *Ochroma lagopus* (Fetcher et al., 1987), while seedlings of *Solanum aviculare* exhibited very small reductions in A_{max} in the shade, where they experienced low survivorship (Turnbull, 1991; Table 6.1).

It is not immediately clear that these reductions in A_{max} are necessarily adaptive, although they support the notion that early successionals exhibit a high level of physiological flexibility (*sensu* Bazzaz & Pickett, 1980). Early successionals are rarely found in forest understories beyond the early seedling phase, where they quickly succumb to pathogens, herbivores, and lack of resources. On the other hand, plants growing in gaps may experience shading of some modules as

Table 6.1. *Maximum photosynthetic rates (A_{max}) of tropical trees grown under a variety of light levels. Trees are classified into early and late successional groupings based on the reported information.*

Early Successionals	A_{max} μmol m^{-2} s^{-1}			Growth Light Levels	References
	High	Medium	Low		
Cecropia obtusifolia	9.9	5.9	1.7	100%, 37%, 8%	Strauss-Debenedetti & Bazzaz, 1991
Ficus insipida	9.4	7.3	4.6	100%, 37%, 8%	Strauss-Debenedetti & Bazzaz, 1991
Heliocarpus appendiculatus	8	7.7	4	100%, 25%, 2%	Fetcher et al., 1987
Cordia alliodora	6	7.7	4	100%, 25%, 2%	Fetcher et al., 1987
Hampea appendiculatus	7	5.5	3	100%, 25%, 2%	Fetcher et al., 1987
Hampea appendiculatus	11.8	n.a.	n.a.	100%	Oberbauer & Strain, 1984
Dipteryx panamensis	8.1	n.a.	n.a.	100%	Oberbauer & Strain, 1984
Bursera simaruba	13.9	n.a.	n.a.	100%	Oberbauer & Strain, 1984
Ochroma lagopus	27.7	n.a.	n.a.	500 μmol m^{-2} s^{-1}	Oberbauer & Strain, 1984
Ochroma lagopus	~9	5.5	3	100%, 25%, 2%	Fetcher et al., 1987
Toona australis	7	n.a.	~3	canopy & understory	Pearcy, 1987
Toona australis	14	n.a.	n.a.	glasshouse seedlings	Pearcy, 1987
Cedrella odorata	~7.5	–	~5.5	33 & 3.3 mol m^{-2} d^{-1}	Ramos & Grace, 1990
Cedrella alliodora	~8.8	n.a.	~5	33 & 3.3 mol m^{-2} d^{-1}	Ramos & Grace, 1990
Ceiba pentandra	7	n.a.	n.a.	n.a.	Riddoch et al., 1991
Milicia excelsa	5	–	–	n.a.	Riddoch et al., 1991
Piper auritum	~12	n.a.	~6	15 & 2 mol m^{-2} d^{-1}	Walters & Field 1987
Piper hispidum	~10	n.a.	~1	~17 & 2 mol m^{-2} d^{-1}	Walters & Field, 1987
Omalanthus populifolius	~8	~6	~5	60%, 20%, 1%	Turnbull 1991
Solanum aviculare	~5.5	~5	~4.5	60%, 20%, 1%	Turnbull, 1991
Shorea parvifolia	6.07	n.a.	n.a.	~500-700 μmol m^{-2} s^{-1}	Moad, 1992

Table 6.1. (cont.).

Late Successionals	A_{max} $\mu mol\ m^{-2}\ s^{-1}$			Growth Light Levels	References
	High	Medium	Low		
Shorea leprosula	5.66	n.a.	n.a.	$\sim$500-700 $\mu mol\ m^{-2}\ s^{-1}$	Moad, 1992
Poulsenia armata	6.5	3.8	3.4	100%, 37%, 8%	Strauss-Debenedetti & Bazzaz, 1991
Brosimum alicastrum	5.9	4.1	4.2	100%, 37%, 8%	Strauss-Debenedetti & Bazzaz, 1991
Pseudolmedia oxyphyllaria	4.7	4.7	3.2	100%, 37%, 8%	Strauss-Debenedetti & Bazzaz, 1991
Pentaclethra macroloba	6.7	7	5.4	100%, 25%, 1%	Oberbauer & Strain, 1985
Pentaclethra macroloba	6.5	n.a.	$\sim$5	2000 & 100 $\mu mol\ m^{-2}\ s^{-1}$	Oberbauer & Strain, 1986
Pentaclethra macroloba	5.4	6.7	6	500, 300, 10 $\mu mol\ m^{-2}\ s^{-1}$	Oberbauer & Strain, 1986
Pentaclethra macroloba	5.6	n.a.	n.a.	500 $\mu mol\ m^{-2}\ s^{-1}$	Oberbauer & Strain, 1984
Blighia sapida	4	n.a.	n.a.	n.a.	Riddoch et al, 1991
Strombosia	4	n.a.	n.a.	n.a.	Riddoch et al, 1991
Brosimum alicastrum	6	n.a.	4.5	33 & 3.3 $mol\ m^{-2}\ d^{-1}$	Ramos & Grace, 1990
Swietenia	5.5	–	4	33 & 3.3 $mol\ m^{-2}\ d^{-1}$	Ramos & Grace, 1990
Argyrodendron peralatum	9.4	8.10	5.7	canopy, gap, understory	Pearcy, 1987
Castanospermum australe	$\sim$5	–	$\sim$3	canopy, understory	Pearcy, 1987
Hymenaea courbaril	$\sim$5	n.a.	$\sim$3	100% & 6%	Langenheim et al, 1984
Copaifera venezuelana	$\sim$7	n.a.	$\sim$7	100% & 6%	Langenheim et al, 1984
Agathis microstachya	$\sim$4	n.a.	$\sim$3	100% & 6%	Langenheim et al, 1984
Agathis robusta	$\sim$7	n.a.	$\sim$3.5	100% & 6%	Langenheim et al, 1984
Virola koschnyi	6.3	n.a.	n.a.	$\sim$24 $mol\ m^{-2}\ d^{-1}$	Oberbauer & Strain, 1984
Dipteryx panamensis	6	5	3	100%, 25%, 2%	Fetcher et al., 1987
Simarouba amara	5	4.5	4.5	100%, 25%, 2%	Fetcher et al., 1987
Carapa guianensis	6	4	3	100%, 25%, 2%	Fetcher et al., 1987

Table 6.1.   (cont.).

Late Successionals	A_{max} μmol m^{-2} s^{-1}			Growth Light Levels	References
	High	Medium	Low		
Minquartia guianensis	4	4	3	100%, 25%, 2%	Fetcher et al., 1987
Dubisia myoporodes	~8	~9	~3	60%, 20%, 1%	Turnbull, 1991
Enodia micrococca	~6	~6	~2.5	60%, 20%, 1%	Turnbull, 1991
Acmena ingens	~2.2	~3	~1.7	60%, 20%, 1%	Turnbull, 1991
Argyrodendron actinophyllum	~2.2	~3	~1.7	60%, 20%, 1%	Turnbull, 1991
Miconia affinis	~2.6	n.a.	~1.4	understory, treefall gap	Newell et al., 1993
Miconia gracilis	~3.0	n.a.	~1.0	understory, treefall gap	Newell et al., 1993
Miconia nervosa	~3.5	n.a.	~1.5	understory, treefall gap	Newell et al., 1993
Vatica oblongifolia	1.38	n.a.	n.a.	~500-700 μmol m^{-2} s^{-1}	Moad, 1992
Shorea maxwelliana	2.16	n.a.	n.a.	~500-700 μmol m^{-2} s^{-1}	Moad, 1992

the canopy closes. The ability to persist and outcompete other fast growing plants despite such encroachment suggests that distinct physiologies probably develop within a single tree. As we explain in a later section (6.4.2.1), it is the dynamic responses (i.e., acclimation) to such changes in light availability that are especially meaningful biologically.

6.4.1.2   Plasticity in late successionals

The vast majority of tropical tree species fall into the category of late successionals. Many of these persist in the understory, forming long-lived seedling banks until released by canopy gaps, while a smaller fraction complete their life-cycle in the understory (Hartshorn, 1978; Bazzaz, 1984; Martínez-Ramos, 1985; Denslow, 1987). Survival in the light-limited forest understory is usually accompanied by slow growth rates, long leaf life spans, and high allocation to defense structures that minimize herbivore damage (Augspurger, 1984; Popma & Bongers, 1988; Reich et al., 1991; Kitajima, 1994).

Compared to pioneers, seedlings of late successionals exhibit a comparatively narrow range of photosynthetic responses when grown under different light levels. A_{max} rates in high light are generally less than 7 µmol m^{-2} s^{-1} (Table 6.1), with rates as low as 1.4 µmol m^{-2} s^{-1} having been reported for field grown seedlings of *Vatica oblongifolia* in Sepilok, Sabah (Moad, 1992; Table 6.1). At the other end of the spectrum, canopy leaves of the Australian species *Argyrodendron peralatum* exhibited A_{max} rates as high as 9.4 µmol m^{-2} s^{-1} (Pearcy, 1987).

Many species fail to exhibit any differences in A_{max} between high and intermediate irradiances (e.g., *Pseudolmedia oxyphyllaria* [Strauss-Debenedetti & Bazzaz, 1991], *Pentaclethra macroloba* [Oberbauer & Strain, 1985], *Minquartia guianensis* [Fetcher et al., 1987], Table 6.1). Results such as these suggest that, as predicted by Bazzaz and Pickett (1980), late successionals have a comparatively restricted photosynthetic plasticity. A_{max} rates tend to decrease at low irradiances, converging in the 3–5 µmol m^{-2} s^{-1} range, similar to that of early successionals (Table 6.1). However, large interspecific differences and the lack of consistent experimental protocols caution against making any broad generalizations on plant performance.

For example, seedlings of *Poulsenia armata* (Strauss-Debenedetti & Bazzaz, 1991), *Agathis robusta* (Langenheim et al., 1984), and *Dipteryx panamensis* (Fetcher et al., 1987) nearly doubled in A_{max} between light extremes, while a much more restricted response pattern was reported for *Pseudolmedia oxyphyllaria* (Strauss-Debenedetti & Bazzaz, 1991), *Pentaclethra macroloba* (Oberbauer & Strain, 1986), and *Miconia affinis* (Newell et al., 1993) (Table 6.1).

These results confirm earlier reports that A_{max} and photosynthetic plasticity in late successionals are consistently lower than in early successionals. The low assimilation rates reported here are consistent with the low light level characteristic of forest understories but do not address either plant response to sunflecks or acclimation to changing light regimes (*sensu* Pearcy, 1987).

6.4.2 Photosynthetic acclimation

Comparisons of plant performance under different, constant light levels are especially useful in describing the range of phenotypes that a particular species can exhibit. These measurements of whole-plant plasticity deal with static responses, i.e. they address plant form and function under a given combination of resources. But because plants rarely encounter such constant conditions, it is questionable whether such measures of plasticity predict plant response when resource availability changes.

We are interested in assessing how tropical seedlings respond to the *change* in light levels that arise when gaps are created (or when they are shaded by faster growing plants). If, for example, plants with a shade-type phenotype are to respond to sudden increases in irradiance by increased growth and A_{max} rates, then changes in phenotype are expected. Switching experiments, whereby seedlings established under one light level are transferred to contrasting light regimes, specifically address these issues. We call such responses to changing environmental conditions acclimation, to be distinguished from plasticity in that they involve changes in an already expressed phenotype. As we discuss below, changes in phenotype can occur both in existing modules or in those developed under the new environmental conditions.

We acknowledge that there is a lack of consensus and much confusion with respect to the meaning and applicability of the terms plasticity and acclimation. For example, Grime and Campbell (1991) distinguish between morphological plasticity and cellular acclimation under stressful conditions, but these are distinct from what is being addressed here. Other researchers studying photosynthesis in tropical trees have used these terms interchangeably, or have implicitly or explicitly assumed *a priori* that plasticity responses predict switching responses (i.e., acclimation potential). However, as we have shown recently (Strauss-Debenedetti & Bazzaz, 1991), studies comparing performance under constant conditions (plasticity) may actually overestimate plant response to changes in irradiance (acclimation) for they do not accurately gauge the presence of carryover effects such as biochemical constraints or allocation patterns, which may interfere with the expression of full acclimation. We thus propose that acclimation refer to the responses arising in either existing or new modules after switching, whereas plasticity refer to the range of phenotypic expression under constant contrasting environmental conditions. This distinction applies most clearly when environmental conditions are experimentally controlled.

6.4.2.1 *Acclimation of existing modules*

There are a number of levels at which acclimation can be evaluated. In its simplest form, the responses of existing leaves or modules to sudden resource fluxes are monitored for signs of damage and recovery within a short time frame (hours to days) using chlorophyll fluorescence techniques and measuring photosynthetic rates. Adjustments to the new light conditions imply the coordinated efforts of various biochemical pathways (reviewed by Pearcy & Sims, 1994), but anatomical changes that enhance assimilation rates may also occur in fully developed leaves experiencing both increases or decreases in light levels (Kamaluddin & Grace, 1992a; 1992b; Turnbull, Doley & Yates, 1993).

Numerous studies have reported symptoms of photoinhibition, enzymatic breakdown or photobleaching in shade grown leaves following exposure to higher light levels (Strauss-Debenedetti & Bazzaz, 1991; Kamaluddin & Grace, 1992a; Mulkey & Pearcy, 1992; Castro, Fernández & Fetcher, 1991; Lovelock, Jebb & Osmond, 1994; Sims & Pearcy, 1992). While signs of photoinhibition may persist for over a year (Langenheim et al., 1984), there is no evidence to suggest that photoinhibition by itself leads to seedling death (Lovelock, Jebb & Osmond, 1994). In general, exposure to increased irradiance results in an immediate reduction in photosynthesis, with the time-course of recovery being species-specific and influenced by both leaf temperature (Mulkey & Pearcy, 1992) and water status of the plant (Lovelock, Jebb & Osmond, 1994). For example, in the late successional *Pentaclethra macroloba*, acclimation occurred at intermediate light levels but was less pronounced at high irradiances (Oberbauer & Strain, 1985).

Leaves developed in the shade or moderate light levels differ in their susceptibility to photoinhibition, coincident with what has been reported for temper. , plants (Anderson & Osmond, 1987). For example, shade grown eaves of the understory herb *Alocasia macrorrhiza* took several days to recover from brief (2 hour) exposures to high light, compared to overnight recovery among leaves developed at intermediate light levels (Mulkey & Pearcy, 1992).

Only two studies compare acclimation of existing leaves in tropical tree species of different shade tolerances or successional positions (Lovelock, Jebb and Osmond, 1994; Turnbull, Doley & Yates, 1993), so it is not possible to make large-scale predictions of performance at this time. In their Papua, New Guinea study, Lovelock, Jebb and Osmond, (1994) found that shade-tolerant plants exhibited the greatest degree of photoinhibition and the slowest levels of recovery to sudden increases in irradiance. At the other end of the spectrum, gap specialists acclimated rapidly in response to a close coupling between biochemical adjustments and changes in leaf orientation. Finally, they found a mixed array of response syndromes among shade-tolerant plants known to benefit from gap openings. The results of Turnbull, Doley and Yates (1993) with three Australian species, on the other hand, indicate that acclimation was restricted among preexisting leaves of early successional species transferred to high light.

6.4.2.2 Acclimation in new modules

A second way in which acclimation can be assessed involves a comparison of performance in leaves formed prior to and after exposure to new light conditions (usually simulating canopy openings). Such changes may not arise until several new leaves are formed (Sims & Pearcy, 1992), and thus the time-course of acclimation may depend on leaf production rates. Most studies have looked for evidence of a change in the phenotype from a morphological perspective, i.e., relative growth rate or biomass allocation (e.g. Oberbauer & Strain, 1985; Fetcher et al., 1987; Osunkoya & Ash, 1991; Popma & Bongers, 1991), and have implicitly assumed that the observed results were due to photosynthetic acclimation. This need not necessarily be the case, however, for increased irradiance *per se* may be sufficient to explain increased growth rates, even in the absence of changes in the biochemistry of photosynthesis (acclimation).

Few researchers have directly addressed the photosynthetic acclimation characteristics of tropical species, relying instead on measurements of plasticity. However, the assumption that photosynthetic acclimation can thus be inferred has recently been challenged (Strauss-Debenedetti & Bazzaz, 1991). There is, as yet, only a small body of data specifically addressing these issues, and there is no clear consensus whether species of different successional positions differ in their acclimation characteristics.

For example, Newell et al. (1993) found no differences in photosynthetic acclimation among seedlings of three *Miconia* species of different shade tolerances exposed to canopy openings, although this may have been due to the narrow range of light levels evaluated. A comparison of A_{max} rates prior to gap creation revealed no interspecific differences, and all species had very low assimilation rates. Within four months of exposure to the higher light levels, all species had nearly doubled their A_{max} rates to 3–4 μmol m^{-2} s^{-1} with *M. gracilis*, the most shade tolerant, exhibiting the lowest saturation PFD.

That photosynthetic acclimation may vary along a successional gradient among closely related species was shown in a study with greenhouse-grown Moraceae (Strauss-Debenedetti & Bazzaz, 1991). Seedlings of the early successional *Cecropia obtusifolia* exhibited more than an eight-fold increase in A_{max} rates when transferred from low to high irradiances, whereas a doubling in A_{max} was obtained for the pioneer *Ficus insipida*. These values compare favorably with those measured in plants grown continuously in high light (plasticity), suggesting that full acclimation had occurred. Among the late successionals studied, however, acclimation took place only in *Brosimum alicastrum*, with small adjustments arising in *Pseudolmedia oxyphyllaria*. On the other hand, and despite a near doubling in A_{max} between light extremes (plasticity), seedlings of *Poulsenia armata* did not acclimate.

Plasticity and acclimation studies do not lend supporting evidence to explain from a photosynthetic perspective alone the apparent requirement of a large fraction of tropical trees to exposure to canopy gaps (*sensu* Hartshorn, 1978). If increased irradiance *per se* rather than photosynthetic acclimation in late successionals results in increases in relative growth rates, then the original hypothesis of Bazzaz and Pickett (1980) is supported. Such a strategy is likely to be found among seedlings that will be quickly overtopped by faster growing plants (i.e., early successionals) and for whom acclimation would be prohibitively expensive. However, further research is needed to evaluate the acclimation characteristics of late successional species before such a conclusion is confirmed.

6.4.2.3 Acclimation across ontogenetic stages

Most of the above-mentioned studies focus almost exclusively on acclimation at the seedling phase, the assumption being that no differences exist between seedling and mature tree physiology. Not incorporated in this analysis are ontogeny-mediated changes which may arise independently of environmental conditions. If, for example, changes in leaf morphology and anatomy (and presumably photosynthetic traits) arise only once plants have attained a certain size, then it is possible that photosynthetic acclimation be delayed as well.

In this scenario, photosynthetic acclimation should be evident among those individuals that potentially experience higher light levels for longer periods of time and for whom acclimation is likely to pay larger growth dividends, i.e., fast-growing early successionals and post-sapling late successionals. If so, then Pearcy's (1987) contention that acclimation occurs in late successional canopy species would also be supported. As we discussed earlier, some late successionals exhibit large plastic responses but limited acclimation to changes in irradiance (Strauss-Debenedetti & Bazzaz, 1991). The possibility that seedling plasticity may be a good predictor of ontogeny-mediated changes needs to be explored.

There is evidence indicating distinct differences in leaf anatomy and morphology and photosynthetic traits both along the understory-canopy continuum and between seedlings and adults grown under similar light levels (Roth, 1984; Strauss-Debenedetti, unpub. data; Bongers & Popma, 1990; Oberbauer & Strain, 1986; Oberbauer & Strain, 1984; Fetcher, Oberbauer & Chazdon, 1994). For example, canopy leaves typically photosynthesize at higher rates than understory leaves, with intermediate rates being observed in gaps (Pearcy, 1987; Oberbauer & Strain, 1986). In part, such modifications can be expected to arise from changes in the microenvironment (e.g., light, water, humidity), similar to what has been observed in plants grown under different constant light treatments (e.g., plasticity). But one cannot exclude the possibility that ontogeny-mediated changes also explain some of these differences, irrespective of environmental conditions. Currently, few data have been gathered to support these ideas, but improvements in canopy access technology (e.g., the crane at the Smithsonian Tropical Research Institute in Barro Colorado Island, Panama) indicate that such data are forthcoming (Zotz & Winter, Chapter 3).

6.5 SUMMARY AND RECOMMENDATIONS

In this chapter we have variously assessed the leading paradigm in tropical forest ecophysiology that predicts a difference in the photosynthetic characteristics of trees of different successional positions

(Bazzaz & Pickett, 1980). Central to this hypothesis is the notion that the light environment of early successional habitats is highly heterogeneous, leading to higher levels of physiological flexibility. However, and as we have shown, it is now apparent that environmental heterogeneity also operates in the understory, albeit at a different, yet biologically meaningful, scale. Species persisting in the understory must contend with both short-term sunflecks and repeated canopy openings. Such recurrent exposure to gaps has been considered to be essential for the majority of late successional species, leading to the formulation of an alternative hypothesis conferring a high level of photosynthetic acclimation to this group of plants (Pearcy, 1987). Not explored in any detail, moreover, is the possibility that species may change their shade-tolerance characteristics with ontogeny, thus supporting both views.

The existing body of data does not unequivocally support either hypothesis, but rather points to a continuum of responses. Problems of interpretation arise primarily from experimental designs aimed at understanding static rather than dynamic species' responses. We have argued that the assumption that plasticity predicts or equals acclimation may in fact be erroneous or misleading (Strauss-Debenedetti & Bazzaz, 1991). Further, comparison of performance under different light levels is complicated by the lack of experimental standardization, resulting in a several-fold difference in light levels across studies. Hence, measurements of plasticity may be a function of the light levels used rather than a physiological trait.

There is an almost universal consensus that early successional species exhibit high photosynthetic rates in high light. The converse, however, that a species with a high photosynthetic rate is an early successional, is not necessarily true. Less easily understood is what accounts for their lack of persistence in the understory despite a marked reduction in assimilation. We have assumed that inherently high respiration rates are a necessary trade-off, limiting habitat breadth, but non-photosynthetic traits may be even more important (e.g., protection against herbivory, pathogens, etc.; Kitajima, 1994). Seedlings of late successionals, on the other hand, fail to exhibit large increases in photosynthesis when grown under high irradiances, but the extent to which this is due to experimentally-induced photoinhibition is unknown. In light-limited conditions, these species are capable of responding to rapidly fluctuating light flecks while maintaining respiratory losses at a minimum. Whether ontogenetic changes arise as plants grow through the canopy is unknown, but can be inferred from observed changes in leaf anatomy.

Unfortunately, there is a paucity of studies evaluating how plants respond to simulated or real canopy openings and almost none assessing ontogenetic changes, hence extrapolations to field conditions are impossible to make. The existing data are far from conclusive, indicating that acclimation occurs in pioneers and in some, but not all, late successionals. Again, differences in experimental technique and level at which acclimation has been studied (old vs. new modules) complicate the picture. The focus at the seedling level further limits the value of these studies.

Future work in this field needs to resolve the seemingly contradictory premises of the two hypotheses (Bazzaz & Pickett, 1980; Pearcy, 1987). First, it is necessary to expand the baseline data on species' acclimation characteristics. In so doing, strict guidelines on appropriate light levels must be set forth. Second, acclimation should be assessed for both existing and new modules, but needs to be integrated with whole plant growth, allocation characteristics, and knowledge of regeneration and survivorship in the field. Such studies must be accompanied by a thorough investigation of ontogenetic changes in the seedling-adult transition, especially among late successionals. Finally, models of plant performance need to incorporate biotic interactions (e.g., herbivory and susceptibility to diseases) to provide an integrative picture of species of different successional stages.

REFERENCES

ANDERSON, J. M. & OSMOND, C. B. (1987) Shade-sun responses: Compromises between acclimation and photoinhibition. *Topics in Photosynthesis, Vol. 9, Photoinhibition* (eds. D. J. KYLE, C. B. OSMOND & C. J. ARNTZEN) Elsevier, Amsterdam.

ASHTON, P. M. S. (1992) Some measurements of the microclimate within a Sri Lankan tropical rainforest. *Agricultural and Forest Meteorology,* **59,** 217–235.

AUGSPURGER, C. K. (1984) Light requirements of neotropical tree seedlings: A comparative study of growth and survival. *Journal of Ecology,* **72,** 777–795.

BARTON, A. M., FETCHER, N. & REDHEAD, S. (1989) The relationship between treefall gap size and light flux in a Neotropical rain forest in Costa Rica. *Journal of Tropical Ecology,* **5,** 437–439.

BAZZAZ, F. A. (1979) Physiological ecology of plant succession. *Annual Review of Ecology and Systematics,* **10,** 351–371.

BAZZAZ, F. A. (1984) Dynamics of wet tropical forest and their species strategies. *Physiological Ecology of Plants of the Wet Tropics.* (eds. E. MEDINA,

H. A. MOONEY, & C. VÁZQUEZ-YÁNES). Dr. W. Junk Publishers, The Hague, pp 233–243.

BAZZAZ, F. A. (1991) Regeneration of tropical forests: physiological responses of pioneer and secondary species. *Rain Forest Regeneration and Management* (eds. A. GÓMEZ-POMPA, T. C. WHITMORE, & M. HADLEY), The Parthenon Publishing Group, Parkridge, New Jersey, pp. 91–118.

BAZZAZ, F. A. & PICKETT, S. T. A. (1980) Physiological ecology of tropical succession: a comparative review. *Annual Review of Ecology and Systematics*, **11**, 287–310.

BAZZAZ, F. A. & WAYNE, P. M. (1994) Coping with environmental heterogeneity: The physiological ecology of tree seedling regeneration across the gap-understory continuum. *Exploitation of Environmental Heterogeneity in Plants* (eds. M. M. CALDWELL & R. W. PEARCY) Academic Press, San Diego, pp 349–390.

BJÖRKMAN, O. (1981) Responses to different quantum flux densities. *Physiological Plant Ecology I. Encyclopedia of Plant Physiology* (eds. O. L. LANGE, P. S. NOBEL, C. B. OSMOND & H. ZIEGLER) Springer-Verlag, N. Y. pp 57–107.

BONGERS, F. & POPMA, J. (1990) Leaf characteristics of the tropical rainforest flora of Los Tuxtlas, Mexico. *Botanical Gazette,* **151,** 354–365.

BRADSHAW, A. D. (1965) Evolutionary significance of phenotypic plasticity in plants. *Advances in Genetics,* **13,** 115–155.

BRANDANI, A., HARTSHORN, G. S. &. ORIANS, G. H. (1988) Internal heterogeneity of gaps and species richness in Costa Rican tropical wet forest. *Journal of Tropical Ecology,* **4,** 99–119.

BROKAW, N. V. L. (1985) Treefalls, regrowth, and community structure in tropical forests. *The Ecology of Natural Disturbance and Patch Dynamics* (eds. S. T. A. PICKETT, & P. S. WHITE) Academic Press, New York pp 53–69.

CASTRO, Y., FERNÁNDEZ, D. & FETCHER, N. (1991). Chronic photoinhibition in attached leaves of tropical tree species. *Plant Physiology* **96** (suppl.) 116.

CHAZDON, R. L. (1988) Sunflecks and their importance to forest understory plants. *Advances in Ecological Research,* **18,** 2–54.

CHAZDON, R. L. & FETCHER N. (1984a) Light environments of tropical forests. *Physiological Ecology of Plants of the Wet Tropics* (eds. E. MEDINA, H. A. MOONEY, & C. VÁZQUEZ-YÁNES) Dr. W. Junk Publishers, The Hague, pp 27–36.

CHAZDON, R. L. & FETCHER, N. (1984b) Photosynthetic light environments in a lowland tropical rainforest in Costa Rica. *Journal of Ecology,* **72,** 553–564.

CHAZDON, R. L. & PEARCY, R. W. (1986) Photosynthetic responses to light variation in rainforest species. I. Induction under constant and fluctuating light conditions. *Oecologia,* **69,** 517–523.

CHAZDON, R. L., WILLIAMS, K. & Field, C. B. (1988) Interactions between crown structure and light environment in five rain forest *Piper* species. *American Journal of Botany,* **75,** 1459–1471.

CHIARIELLO, N. (1984) Leaf energy balance in the wet lowland tropics. *Physiological Ecology of Plants of the Wet Tropics* (eds. E. MEDINA, H. A. MOONEY, & C. VÁZQUEZ-YÁNES) Dr. W. JUNK Publishers, The Hague,pp. 85–98.

CLARK, D. A. & CLARK, D. B. (1992) Life history diversity of canopy and emergent trees in a neotropical rain forest. *Ecological Monographs,* **62,** 315–344.

DENSLOW, J. S. (1987) Tropical rainforest gaps and tree species diversity. *Annual Review of Ecology and Systematics,* **18,** 431–451.

DE STEVEN, D. (1988) Light gaps and long-term seedling performance of a Neotropical canopy tree (*Dipteryx panamensis,* Leguminosae). *Journal of Tropical Ecology* **4,** 407–411.

FETCHER, N., OBERBAUER S. F. & CHAZDON R. L. (1994) Physiological ecology of plants. *La Selva: Ecology and Natural History of a Neotropical Rain Forest* (eds. L. A. MCDADE, K. S. BAWA, H. A. HESPENHEIDE, G. S. HARTSHORN), University of Chicago Press, Chicago and London, pp 128–141.

FETCHER, N., OBERBAUER, S. F. & STRAIN, B. R. (1985) Vegetation effects on microclimate in lowland tropical forest in Costa Rica. *International Journal of Biometeorology,* **29,** 145–155.

FETCHER, N., OBERBAUER, S. F., ROJAS, G. & STRAIN, B. R. (1987) Efectos del régimen de luz sobre la fotosíntesis y el crecimiento en plántulas de árboles de un bosque lluvioso tropical de Costa Rica. *Revista de Biologia Tropical,* **35 (suppl.**), 97–110.

FETCHER, N., STRAIN, B. R. &. OBERBAUER, S. F. (1983) Effects of light regime on the growth, leaf morphology and water relations of seedlings of two species of tropical trees. *Oecologia,* **58,** 314–319.

FREDEEN, A. L. & FIELD, C. B. (1991). Leaf respiration in *Piper* species native to a Mexican rainforest. *Physiologia Plantarum,* **82,** 85–92.

GRIME, J. P. & CAMPBELL, B. D. (1991). Growth rate, habitat productivity, and plant strategy as predictors of stress response. *Response of Plants to Multiple Stresses,* (eds. H. A. MOONEY, W. E. WINNER, & E. J. PELL) Academic Press, San Diego, pp 143–161.

HARTSHORN, G. S. (1978) Tree falls and forest dynamics. *Tropical Trees as Living Systems* (eds. P. B. Tomlinson & M. H. Zimmermann) Cambridge University Press, pp 617–638.

KAMALUDDIN, M. & GRACE, J. (1992a) Photoinhibition and light acclimation in seedlings of *Bischofia javanica,* a tropical forest tree from Asia. *Annals of Botany,* **69,** 47–52.

KAMALUDDIN, M. & Grace, J. (1992b) Acclimation in seedlings of a tropical tree, *Bischofia javanica,* following a stepwise reduction in light. *Annals of Botany,* **69,** 557–562.

KING, D. A. (1991) Correlations between biomass allocation, relative growth rate and light environment in tropical forest saplings. *Functional Ecology,* **5,** 485–492.

KITAJIMA, K. (1994). Relative importance of photosynthetic traits and allocation patterns as correlates of seedling shade tolerance of 13 tropical trees. *Oeeologia*, **98**, 419–428.

KWESIGA, F. K. & GRACE, J. (1986) The role of the red/far-red ratio in the response of tropical tree seedlings to shade. *Annals of Botany*, **57**, 283–290.

LANGENHEIM, J. H., OSMOND, C. B., BROOKS, A. & FERRAR, P. J. (1984) Photosynthetic responses to light in seedlings of selected Amazonian and Australian rainforest tree species. *Oecologia*, **63**, 215–224.

LEE, D. W. (1989) Canopy dynamics and light climates in a tropical moist deciduous forest in India. *Journal of Tropical Ecology*, **5**, 65–79.

LIEBERMAN, M. & LIEBERMAN, D. (1989) Forests are not just Swiss cheese: canopy stereogeometry of non-gaps in tropical forests. *Ecology*, **70**, 550–552.

LOVELOCK, C. E., JEBB, M. & OSMOND C. B. (1994). Photoinhibition and recovery in tropical plant species: Response to distuubance. *Oecologia*, **97**, 297–307.

MARTÍNEZ-RAMOS, M. (1985) Claros, ciclos vitales de los árboles tropicales y regeneración natural de las selvas altas perenifolias. *Investigaciones sobre la regeneración de selvals altas en Veracruz, Mexico II.* (eds. A. GOMEZ-POMPA & S. DEL AMO R.), Editorial Alhambra, Mexico, pp 191–240.

MOAD, A. S. (1992) *Dipterocarp Juvenile Growth and Understory Light Availability in Malaysian Tropical Forest.* Ph. D. Thesis, Harvard University.

MULKEY, S. S. & PEARCY, R. W. (1992) Interactions between acclimation and photoinhibition of photosynthesis of a tropical forest understory herb, *Alocassia macrorrhiza*, during simulated gap formation. *Functional Ecology*, **6**, 719–729.

MULKEY, S. S. , WRIGHT, S. J. & SMITH, A. P. (1993). Comparative physiology and demography of three neotropical forest shrubs: Alternative shade-adaptive character syndromes. *Oecologia*, **96**, 526–536.

NEWELL, E. A., McDONALD, E. P, STRAIN, B. R. & DENSLOW, J. S. (1993) Photosynthetic responses of *Miconia* species to canopy openings in a lowland tropical rainforest. *Oecologia*, **94**, 49–56.

NOBEL, P. S. (1977) Internal leaf are a and cellular CO_2 resistance: Photosynthetic implications of variations with growth conditions and plant species. *Physiologia Plantarum*, **40**, 137–144.

NÚÑEZ-FARFÁN J. & DIRZO, R. (1988) Within-gap spatial heterogeneity and seedling performance in a Mexican tropical forest. *Oikos*, **51**, 274–284.

OBERBAUER, S. F. & STRAIN, B. R. (1984) Photosynthesis and successional status of Costa Rican rain forest trees. *Photosynthesis Research*, **5**, 227–232.

OBERBAUER, S. F. & STRAIN, B. R. (1985) Effects of light regime on the growth and physiology of *Pentaclethra macroloba* (Mimosaceae) in Costa Rica. *Journal of Tropical Ecology*, **1**, 303–320.

OBERBAUER, S. F. & STRAIN, B. R. (1986) Effects of canopy position and irradiance on the leaf physiology and morphology of *Pentaclethra macroloba* (Mimosaceae). *American Journal of Botany,* **73,** 409–416.

ORIANS, G. H. (1982) The influence of treefalls in tropical forests in tree species richness. *Journal of Tropical Ecology,* **23,** 255–279.

OSUNKOYA, O. O. & ASH, J. E. (1991) Acclimation to a change in light regime in seedlings of six Australian rainforest tree species. *Australian Journal of Botany,* **39,** 591–605.

PEARCY, R. W. (1987) Photosynthetic gas exchange responses of Australian tropical forest trees in canopy, gap and understory micro-environments. *Functional Ecology,* **1,** 169–178.

PEARCY, R. W. & SIMS D. A., (1994) Photosynthetic acclimation to changing light environments: Scaling from the leaf to the whole plant. *Exploitation of Environmental Heterogeneity by Plants* (eds. M. M. Caldwell & R. W. Pearcy) Academic Press, San Diego, pp 145–174.

POPMA, J. & BONGERS, F. (1988) The effect of canopy gaps on growth and morphology of seedlings of rainforest species. *Oecologia,* **75,** 625–632.

POPMA, J. & BONGERS, F. (1991) Acclimation of seedlings of three Mexican tropical rainforest tree species to a change in light availability. *Journal of Tropical Ecology,* **7,** 85–97.

RAICH, J. W. (1989) Seasonal and spatial variation in the light environment in a tropical dipterocarp forest and gaps. *Biotropica,* **21,** 299–302.

RAMOS J. & Grace, J. (1990) The effects of shade on the gas exchange of seedlings of four tropical trees from Mexico. *Functional Ecology,* **4,** 667–677.

REICH, P. B. , UHL, C., WALTERS, M. B. & ELLSWORTH, D. S. (1991) Leaf lifespan as a determinant of leaf structure among 23 Amazonian tree species. *Oecologia,* **86,** 16–24.

RICH, P. M., CLARK, D. B. , Clark, D. A. & OBERBAUER, S. F. (1993) Long-term study of light environments in tropical wet forest using quantum sensors and hemispherical photography. *Agricultural and Forest Meteorology,* **65,** 107–127.

RIDDOCH, I., GRACE, J. , FASEHUN, F. E., RIDDOCH, B. & LADIPO, D. O. (1991) Photosynthesis and successional status of seedlings in a tropical semideciduous rainforest in Nigeria. *Journal of Ecology,* **79,** 491–503.

ROTH, I. (1984) *Stratification of Tropical Forests as Seen in Leaf Structure.* Dr. W. JUNK, The Hague.

SIMS, D. A. & PEARCY, R. W. (1992) Response of leaf anatomy and photosynthetic capacity in *Alocasia macrorhiza* (Araceae) to a transfer from low to high light. *American Journal of Botany,* **79,** 449–455.

STRAUSS-DEBENEDETTI, S. & BAZZAZ, F. A. (1991) Plasticity and acclimation in tropical Moraceae of different successional positions. *Oecologia,* **87,** 377–387.

STRAUSS-DEBENEDETTI, S. & BERLYN, G. P. (1995) Leaf anatomical responses to light in five tropical Moraceae of different successional status. *American Journal of Botany,* in press.

SWAINE, M. D., & WHITMORE, T. C. (1989) On the definition of ecological species groups in tropical rainforests. *Vegetatio,* **75,** 81–86.

TURNBULL, M. H. (1991) The effect of light quantity and quality during development on the photosynthetic characteristics of six Australian rainforest tree species. *Oecologia,* **87,** 110–117.

TURNBULL, M. H., DOLEY, D. & YATES, D. J. (1993) The dynamics of photosynthetic acclimation to changes in light quantity and quality in three Australian rainforest tree species. *Oecologia,* **94,** 218–228.

VÁZQUEZ-YANES, C. & OROZCO-SEGOVIA, A. (1984) Ecophysiology of seed germination in the tropical humid forests of the world: A review. *Physiological Ecology of Plants of the Wet Tropics* (eds. E. MEDINA, H. A. MOONEY, & C. VÁZQUEZ-YÁNES) Dr. W. Junk Publishers, The Hague, pp 37–50.

VITOUSEK, P. M. & DENSLOW, J. S. (1986). Nitrogen and phosphorous availability in treefall gaps of a lowland tropical rainforest. *Journal of Ecology,* **74,** 1167–1178.

WALTERS, M. B. & FIELD, C. B. (1987) Photosynthetic light acclimation in two rainforest *Piper* species with different ecological amplitudes. *Oecologia,* **72,** 449–456.

WHITMORE, T. C. (1989) Canopy gaps and the two major groups of forest trees. *Ecology,* **70,** 536–538.

7

Influence of Seasonal Drought on the Carbon Balance of Tropical Forest Plants

Stephen S. Mulkey and S. Joseph Wright

7.1 INTRODUCTION

Despite high annual rainfall, one of the principal determinants of resource availability in tropical forests is the seasonal occurrence of drought. Plant growth and reproduction in these forests are frequently coordinated with seasonal variation in rainfall (Richards, 1952). In this chapter, we review the role of seasonal drought in tropical plant demography and physiology, with particular emphasis on the relationships among water deficits, carbon balance, and plant survival. We begin with a review of physiological mechanisms permitting continued gas exchange during drought in tropical trees, understory shrubs, and herbs. Other drought responses of vines, hemiepiphytes, and epiphytes are considered in chapters by Holbrook and Putz (Chapter 13) and Medina (Chapter 2), and drought effects on tropical plant phenology are considered in detail by Wright (Chapter 15). Treefall gaps are a principal determinant of resource availability in the understory, and in the latter part of this review, we focus on the interaction between seasonal drought and gaps and their effect on plant demography and physiology. Central to this discussion is the role of declining plant water potentials in limiting carbon gain. We draw largely from our work with herbs and shrubs on Barro Colorado Island (BCI) in Panama because this experimental study included environmental variation in light and water. We emphasize the role of drought as a determinant of whole plant performance through its influence on the acquisition and allocation of carbon to competing functions.

7.2   PLANT WATER RELATIONS IN THE HUMID TROPICS

Water stress is an ecological factor closely linked to leaf, flower, and fruit phenology for a wide array of tropical species in seasonally dry forests (Alvim & Alvim, 1978; Augspurger, 1979; Bullock & Solis-Magallanes, 1990; Opler, Frankie & Baker, 1976; Reich & Borchert, 1984). Indeed, most tropical forests experience seasonal drought (reviewed by Wright, Chapter 15) and plants in these forests tolerate drought (*sensu* Jones, 1993) through mechanisms of water deficit avoidance and water deficit tolerance (Robichaux et al., 1984; Rundel & Becker, 1987). In all but the wettest of tropical forests, many woody plants avoid water deficits by becoming deciduous during the dry season. Leaf fall is programmed to occur after the cessation of rains, or it may be facultative in response to progressive drought (Borchert, 1983; Frankie, Baker & Opler, 1974). Plants that tolerate water deficits retain their leaves and maintain some level of gas exchange during the dry season. These plants face the possibility of reduced growth and death as a consequence of restricted carbon gain and negative carbon balance. Death of tissue from severe drought and lethal leaf temperatures produced by reduced transpirational cooling contribute to a restriction in the whole-plant carbon budget. Accordingly, this review focuses on leaf and canopy characteristics of plants that exhibit continued photosynthetic gas exchange in the face of drought. Although mortality can result from catastrophic xylem embolism during drought (Tyree & Sperry, 1989), this should be rare among species that are adapted to tolerate water deficits. High mortality from seasonal drought should occur only during extreme years, such as the catastrophic 1983 El Niño southern oscillation, which caused an extended dry season throughout Central America (Leigh et al., 1990). Moreover, underground processes play a crucial role in drought resistance, and the ecology of factors such as root-mediated stomatal closure (Schulze, 1993) and water transport by mycorrhizae (Davies, Potter & Linderman, 1993) is just beginning to be explored. Underground processes are considered here to the extent that they relate to whole-plant function during drought in tropical species.

7.2.1 Drought tolerance characteristics of leaves

The maintenance of leaf turgor is essential for uncompromised metabolic activity in the mesophyll during drought (Hanson & Hitz, 1982). The front line defense against turgor loss is stomatal closure in response to dry air, declining soil water potentials, or both (Tenhunen, Pearcy & Lange, 1987; Schulze, 1993). Compared to desert and mediterranean habitats where drought resistant leaves are common, water remains relatively abundant in air and soil in most tropical forests, even during the dry season (Wright, 1991). Nevertheless, tropical forest plants show diurnal and seasonal patterns of stomatal closure as pronounced as those of temperate zone plants (Robichaux et al., 1984) in response to water deficits in the atmosphere and soil (e.g., Mulkey, Wright & Smith, 1991; Oberbauer, Strain & Riechers, 1987). This is true for plants in the sunlit overstory canopy (Fetcher, 1979; Myers et al., 1987; Oberbauer, Strain & Riechers, 1987), as well as the more mesic shaded understory (Mulkey, Smith & Wright, 1991; Mulkey, Wright & Smith; 1991, Wright et al., 1991). Typically, as air humidity decreases and leaf water potentials become more negative diurnally, some degree of stomatal closure ensues. Midday stomatal closure becomes more pronounced as the dry season progresses (Figure 7.1). There is limited evidence that stomatal response to drought is more pronounced (Mooney et al., 1983) and that turgor loss occurs at higher leaf water potentials (Fetcher, Oberbauer & Chazdon, 1994) for plants from wet tropical forests than for those from drier forests. In contrast, leaves of young *Cordia alliodora* showed no stomatal closure under experimentally induced drought although the adults are drought deciduous (Fetcher, Oberbauer & Chazdon, 1994).

Leaf turgor is maintained by any mechanism that promotes water flow to the leaves. This can be accomplished through the generation of leaf water potentials more negative than those in the xylem stream by the active concentration of solutes in mesophyll cells during progressive drought (Tyree & Jarvis, 1982). Tolerance of water deficit through maintenance or adjustment of osmotic potential (Ψ) has been

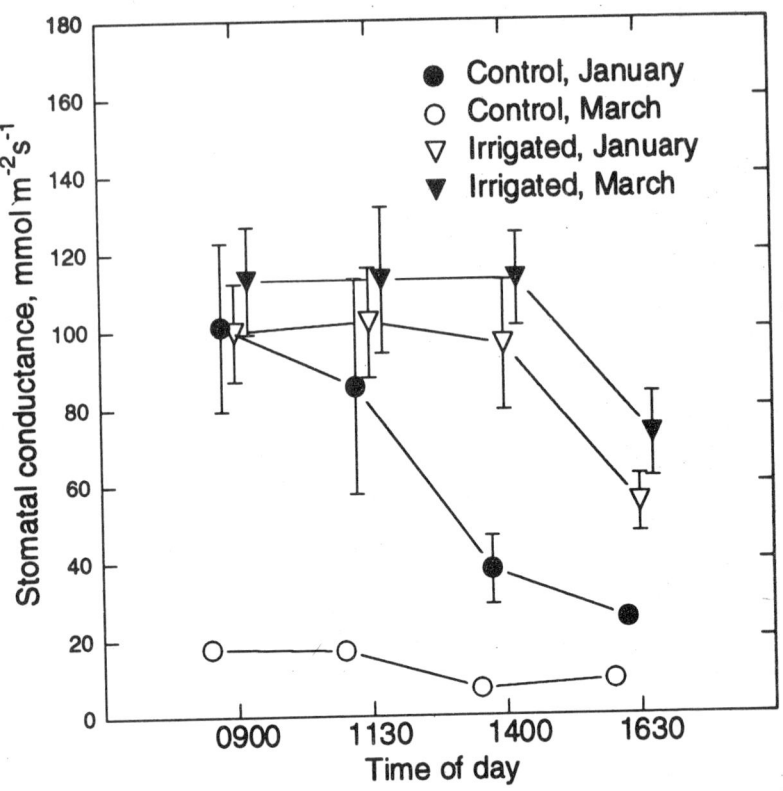

Figure 7.1.   Stomatal conductance in irrigated and control sites during early dry season (January) and late dry season (March) in Psychotria limonensis.

observed in a variety of tropical trees (Medina, 1983; Myers et al., 1987; Robichaux et al., 1984; Sobrado, 1986), shrubs (Ike & Thurtel, 1981; Mulkey, Wright & Smith, 1991; Wright et al., 1992), and herbs (Mulkey, 1986a,b). All else being equal, leaves with more negative leaf osmotic potentials lose turgor less readily than do leaves with less negative osmotic potentials. Three species of herbaceous bamboo that occur together in the shaded understory at BCI have varying sensitivity to drought stress largely as a function of their mesophyll osmotic potentials (Mulkey, 1986a; Figure 7.2). *Pharus latifolius* and *Streptochaeta sodiroana* maintain less negative osmotic potentials than *Streptochaeta spicata*, and thus lose turgor at higher water potentials than *S. spicata*. Relative to the other two, *S. spicata* is able to occupy a wider range of understory and gap habitats, and maintain gas

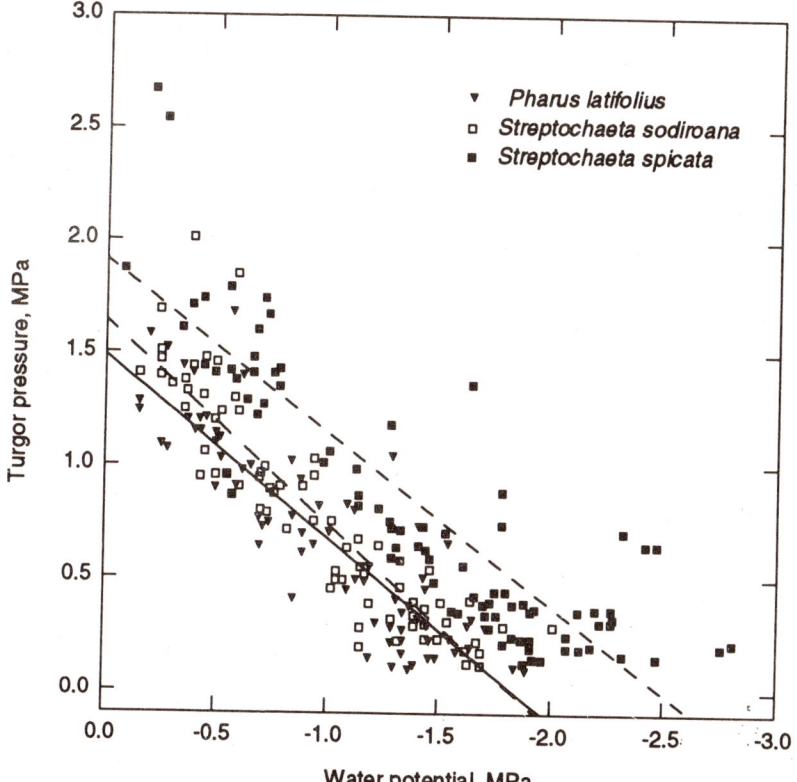

Figure 7.2.   Scatter plot of turgor pressure as a function of water potential for three species of herbaceous bamboo from BCI. Points for each species are derived from pressure-volume isoclines above zero turgor.

exchange at more constant levels throughout the day even during the dry season. Similar to results for tropical savannah tree species (Medina & Francisco, 1994), maximum photosynthesis is correlated with maximum stomatal conductance, but is inversely related to leaf water potentials among these bamboos. As determined from analysis of carbon stable isotopes, *S. spicata* has higher long-term integrated water-use efficiency (WUE, assimilation ÷ evapotranspiration) than the other two (Mulkey, 1986b). The canopy dominant *Pentaclethra macroloba* (Wild.) did not show osmotic adjustment during the short dry season in wet forest at La Selva (Oberbauer, Strain & Riechers, 1987).

The role of the volumetric modulus of elasticity (ε) in turgor maintenance is controversial (Cosgrove, 1988). With increasing drought

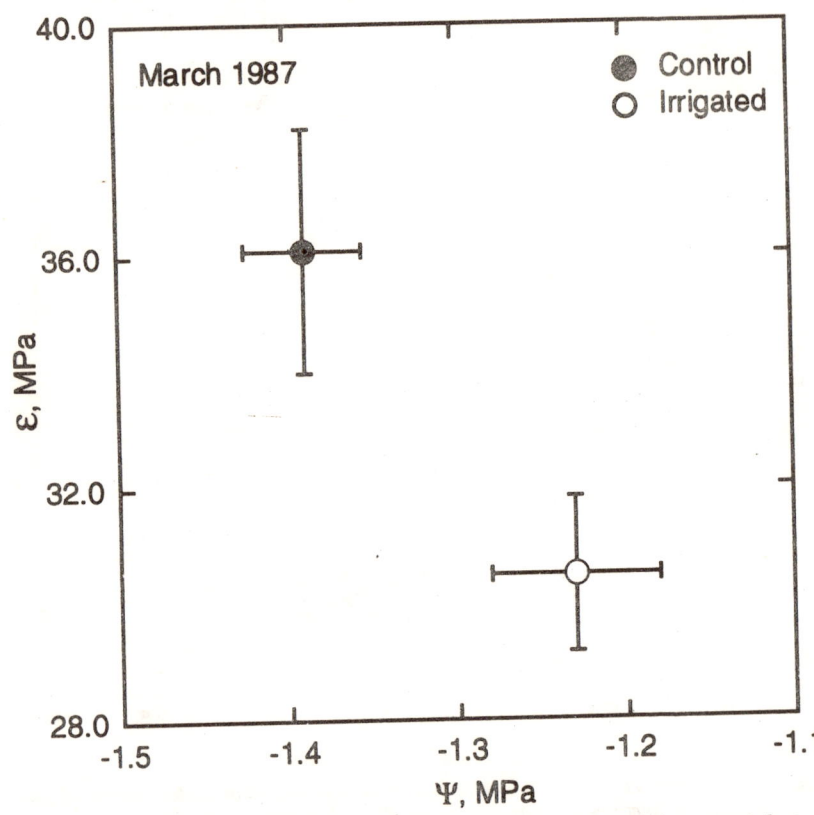

Figure 7.3. Differences in the elastic modulus (ε) and osmotic potential at full turgor (ψ) for Psychotria limonensis *growing in control and irrigated sites on BCI. Values were derived from pressure-volume isoclines.*

stress, ε has been shown to remain constant (e.g. Hsaio et al., 1976; Mulkey, 1986a), to decrease (Robichaux, 1984), and to increase (Bolaños & Longstreth, 1984; Mulkey, Wright & Smith, 1991). An inelastic cell wall (high ε) should act to increase the uptake of water from drying soil because more negative water potentials would result from a relatively small decrease in water content (e.g., Bowman & Roberts, 1985). Alternatively, a more flexible cell (low ε) should allow shrinkage during water loss, thus maintaining high turgor for a given change in water content (Robichaux et al. 1984). We suggest that as long as water can be extracted from the soil, plants utilizing osmotic adjustment would fare better if a small change in leaf water content resulted in a relatively large water potential gradient between the soil and the leaf (Mulkey, Wright & Smith, 1991). Under these circumstances, a stiffer cell wall should be an advantage. We have found that high ε is associated with osmotic adjustment in woody species (Figure 7.3; Mulkey, Wright & Smith, 1991). However, plants growing on thin soil (e.g., Hawaiian *Debautia*; Robichaux et al., 1984) or substrates without soil, such as epiphytes, should not be expected to adjust osmotically because there is little or no soil water to extract during drought. Such plants usually have low ε, which in theory helps to maintain turgor by permitting cell walls to shrink around a diminishing volume of symplasm (Holbrook & Putz, Chapter 13). The role of the volumetric modulus of elasticity in drought tolerance will likely become clear as tropical forest plants from diverse habitats are examined.

In addition to osmotic and elastic properties of the mesophyll, drought tolerant leaves of tropical forest plants also have characteristic morphologies that have been seen in drought tolerant plants from temperate regions. Leaves likely to experience high evaporative demand often have reduced stomatal density (Ball, Chapter 14; Nautiyal et al., 1994), high mass per area (Mooney, Field & Vásquez-Yanes, 1984; Nobel, 1980), greater thickness (Nautiyal et al., 1994), and a more waxy cuticle (Herbert & Larsen, 1985). Like their temperate cousins, drought-tolerant tropical plants often produce leaves that are compound rather than simple, and dissected rather than entire (Givnish, 1984; Givnish & Vermeij, 1976). These factors (and others) function to reduce either stomatal and cuticular transpiration per unit leaf area, or to minimize the vapor pressure gradient between the mesophyll and the air. Similarly, there has been some speculation that epiphytes and epiphylls growing on the surface of tropical leaves may reduce evaporation during the dry season (Coley & Kursar, Chapter 12).

7.2.2   Drought tolerance characteristics of canopies

Canopy characters are as important as leaf characters in determining drought tolerance. Collectively, the leaves of a canopy may minimize water loss during gas exchange by becoming less sensitive to evaporative demand during dry periods. In addition to seasonal changes in leaf-level stomatal regulation of water loss, one way to do this is through the production of a new leaf crop with drought resistant characters that are functional during the dry season. Production of leaves with differing tolerance to drought during periods of contrasting water availability occurs in desert and mediterranean habitats (e.g., Ehleringer, 1982). Heterophylly as a means of maintaining gas exchange during drought has recently been discovered in tropical forest plants (Mulkey et al., 1992). The understory shrub *Psychotria marginata* produces leaves with contrasting morphology and physiology in two seasonal flushes. Leaf production is bimodal, with a major peak at the beginning of the annual wet season and a secondary peak at the end of the wet season. This latter group has higher mass per area, and during the ensuing dry season has lower stomatal conductances than the group produced at the beginning of the rains. During drought, these leaves have twice the WUE of early wet-season leaves, even when under continuous irrigation. During extreme drought, these leaves maintain gas exchange while leaves of the other cohort close their stomata. Leaves produced during the early wet season have high mesophyll CO_2 concentrations, which may improve quantum yield and carbon gain during sunflecks (Pearcy, 1990) in the shaded understory during the wet season. Plants irrigated during two consecutive dry seasons continued to produce leaves with drought tolerant characters during the end of the wet season, suggesting that this ability has been strongly canalized by some environmental factor other than water availability (Mulkey et al., 1992). The bimodal leaf production results in canopy acclimation to changing conditions through co-occurring leaves that lack the physiological plasticity usually associated with acclimation. We have recently discovered a similar heterophylly in the canopy tree *Luehea seemannii* growing in dry forest in Panama (Mulkey, Kitajima & Wright, 1995).

Factors that reduce the surface area of a canopy exposed to strong vapor pressure gradients function to maintain gas exchange during drought. In addition to species that become completely deciduous during the dry season, a number of species drop only some of their leaves (Nautiyal et al., 1994; Wright, Chapter 15; Wright & van Schaik 1994). The extent to which gas exchange is maintained or enhanced in the remaining leaves over what it might have been without partial leaf drop is unknown. In contrast, plants may minimize direct irradiance and evaporative demand by reducing the leaf angle relative to the sun. Facultative wilting in the tropical pioneer tree, *Piper auritum*, in response to full midday sun results in increased water-use efficiency and the maintenance of stomatal conductance (Chiariello, Field & Mooney, 1987). Similarly, the understory herb *Calathea lutea*, which grows primarily in gaps, has solar tracking that limits exposure to the sun at midday, thus reducing leaf temperature (Herbert & Larsen 1985). *Pleiostachya pruinosa,* a gap colonizing herb, exhibits reversible upward curling around the leaf midvein during drought (A. P. Smith & S. S. Mulkey, pers. obs.). The effect of leaf orientation on leaf temperature and photosynthesis in tropical species has been reviewed by Herbert in Chapter 5 (Herbert, 1992a,b).

Every plant canopy creates a local climate that may enhance or impede gas exchange by virtue of the extent to which the canopy structure affects mixing with the atmosphere. The effectiveness of stomatal control of water loss is largely determined by the linkage between the canopy and the atmosphere as mediated by the canopy boundary layer ("the omega factor"). In Chapter 4, Meinzer and Goldstein demonstrate the role of the canopy boundary layer as one determinant of transpiration (Meinzer et al., 1993; Meinzer & Goldstein, Chapter 4). Theirs is one of the first attempts to quantify the boundary layer for a mature tropical tree, and it shows the large effect of differences in atmospheric linkage on bulk movement of water from the plant to the air during the dry season and wet season, respectively. This seasonal difference is a function of air humidity and wind velocity, as well as canopy structure. Similar within-canopy diurnal and spatial variation in linkage has been shown for temperate species (Jarvis, 1993). As our methods of canopy access improve, this approach holds considerable promise for understanding the landscape-level impact of forest gas exchange on the atmosphere (Parker, Smith & Hogan, 1992). Methods based on micrometeorology and energy flux, such as eddy correlation and the Bowen Ratio energy balance, permit the estimation of stand-level water balance (Dawson, 1993), but these techniques have yet to be widely applied to tropical canopies.

7.3. CARBON BALANCE DURING DROUGHT

For the last two decades, plant ecophysiologists have characterized plant performance by viewing the plant as an assemblage of functions competing for the same scarce resources, especially carbon (Mooney, 1972; Coley, Bryant & Chapin, 1985; Chapin, Schulze & Mooney, 1990). This view of the plant as a "balanced system" is based on the idea that carbon is the principal resource being partitioned among these functions, and that patterns of growth and reproduction reflect the outcome of selection on how carbon is allocated (Mooney & Chiarello, 1984). Drought can impair carbon gain through stomatal closure and reductions in photosynthesis through impaired metabolism. Once acquired, carbon can be allocated to a range of competing functions, and thus the impact of drought on carbon gain can be far-reaching, affecting multiple plant systems for possibly extended periods following the cessation of drought. For example, a single severe drought is suspected of causing several years of subsequent impaired performance and susceptibility to disease in a tropical forest tree (Gilbert, Hubbell & Foster, 1994).

7.3.1   Stomatal limitation of carbon gain

In order to assess the impact of stomatal closure on carbon gain, it is necessary to do a full accounting of carbon gains and losses during drought. Mooney, Field and Vázquez-Yanes (1984) point out the need to develop more complete studies of carbon budgets of tropical plants. Unfortunately, progress in this area has been slow because of the logistic difficulties of measuring allocation of carbon to competing functions. Despite the absence of a strict accounting, it is possible to assess the potential for the negative impact of stomatal closure on carbon budgets by looking for changes in gas exchange during drought. Although there was a strong negative effect of drought on survival (see section 7.3.4), drought had only minimal influence on photosynthesis in the field for three species of shrubs, *Psychotria* (Mulkey, Wright & Smith, 1993), and two species of herbs, *Calathea* and *Plieostachya* (Mulkey, Wright & Smith, 1991), grown in the experimental long-term irrigation plots on BCI. All five showed some degree of stomatal closure, but this resulted in only slightly lower photosynthetic rates at light saturation in control plots. By far the largest differences in photosynthetic rates at light saturation were due to habitat, rather than treatment (Figure 7.4).

For the three *Psychotria* species, it is possible to assess photosynthesis at light saturation in the field relative to maximal rates achieved by leaves taken to the laboratory and measured with an oxygen electrode at 5 percent CO_2 (Table 7.1). Because measurements were conducted under a standard, augmented CO_2 concentration, the rate of photosynthesis in the oxygen electrode is a benchmark reflecting minimal or no diffusion limitation against which assimilation in the field can be compared. Although rates in the laboratory should always be higher, the size of this increase could reflect water and sunlight availability during growth. The comparison in Table 7.1 is for opposite leaves of equivalent age at the same branch node. One leaf was used for field measures, while the other was taken during predawn hours directly to the laboratory for measures in the oxygen electrode. The difference was significant only for the effect of habitat ($F = 6.05$, $P < 0.02$; d.f. $= 1,44$) because gap-grown leaves showed the

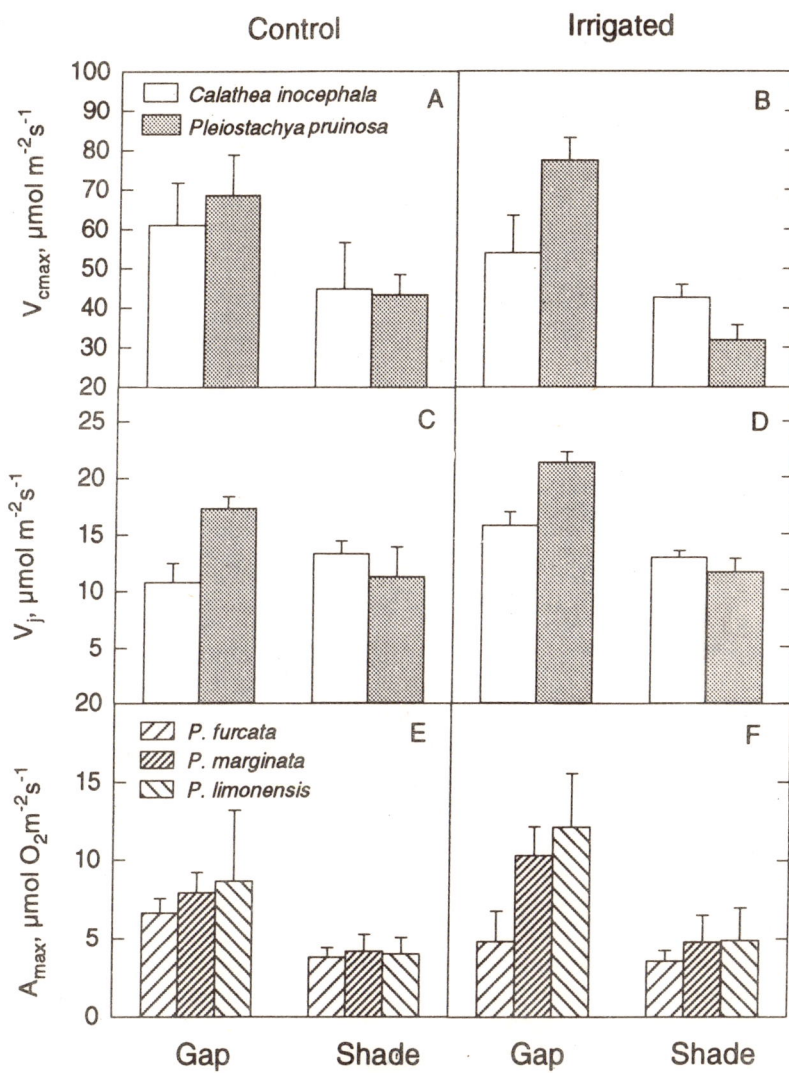

Figure 7.4. Variation in expression of photosynthetic capacity among species growing in gaps and shade in control and irrigated sites on BCI. V_j and V_{cmax} were derived from field constructed assimilation-CO_2 response curves and indicate electron transport limited substrate regeneration and carboxylation capacity, respectively. A_{max} is the maximum rate of oxygen evolution under 5 percent CO_2. All leaves were newly produced after gaps were created by a storm on 5 July 1989. Acclimation responses can be seen by comparing between bars of the same species in gaps and shade for a given treatment.

Table 7.1. *Difference between field and laboratory measures of A_{max} for the three species of Psychotria. Values are in $\mu mol\ m^{-2}\ s^{-1}$. Numbers in parenthesis are the proportional increase as a proportion of the maximum rate measured in the laboratory under conditions of augmented CO_2. The statistics are discussed in the text.*

Treatment	Habitat	Species		
		P. furcata	*P. marginata*	*P. limonensis*
Control				
	Gap	2.99 (0.49)	2.60 (0.30)	4.42 (0.51)
	Shade	1.41 (0.42)	2.43 (0.54)	0.48 (0.11)
Irrigated				
	Gap	0.94 (0.20)	3.45 (0.38)	3.12 (0.26)
	Shade	1.07 (0.30)	2.09 (0.44)	1.09 (0.20)

largest increase in A_{max} among all species, regardless of irrigation treatment. The effect of habitat was not found when the increase was calculated as a proportion of the maximum rate observed in the laboratory.

Epiphytes in the canopy often experience extreme drought during the dry season because of a lack of a soil water reservoir and high transpirational demand during frequent direct sunlight. Accordingly, one of the most compelling studies showing the link between CO_2 gas exchange and carbon balance is for epiphytes studied continuously over a year in the canopy of a large tropical emergent on BCI (Zotz & Winter, 1994a,b; Chapter 3). Gas exchange was strongly negatively affected by drought in two species with C_3 metabolism and one species with C_3-CAM intermediate metabolism. The evergreen fern *Polypodium crassifolium* showed a negative carbon balance and pronounced photoinhibition during the four months of the dry season. In all species, stomatal conductance was markedly higher during the wet season, but no mortality resulted from dry season restrictions on carbon gain. The orchid *Catasetum viridiflavum* produced new leaves during the second half of the dry season. *Clusia uvitana*, the C_3-CAM intermediate, showed high long-term WUE, but annual carbon gain was similar in all three species.

7.3.2 Influence of drought on photosynthetic capacity independent of stomatal function

A number of studies have shown that drought results in a loss of photosynthetic capacity (A_{max}) independent of stomatal limitation to gas exchange (Schulze, 1986). Although the cellular mechanisms of this remain an area of investigation (e.g., Gunasekera & Berkowitz, 1992), few studies examine the impact of seasonal drought on biochemical photosynthetic capacities of tropical plants. None of the five understory species studied by us on BCI showed a strong influence of drought on estimates of capacity derived from assimilation-CO_2 response curves (Figure 7.4; V_j and V_{cmax}; Mulkey, Smith & Wright, 1991) or from the oxygen-electrode light curves (A_{max}; Mulkey, Wright & Smith, 1993). All of these species were experiencing peak dry season drought stress at the time of measurement. In contrast, photoinhibition under strong light is known to be compounded by high temperature (Mulkey & Pearcy, 1992) and water stress (Cornic 1994; Havaux, 1992). Photoinhibition in such studies is typically associated with a reduction of dark-adapted fluorescence and biochemical limitations to both quantum yield and photosynthetic capacity (Chazdon et al., Chapter 1). Photoinhibition compounded by drought stress can be especially acute for canopy epiphytes during the dry season (Zotz & Winter, 1994b), and despite the present dearth of evidence, we expect this phenomenon to be common in the tropics whenever strong light and prolonged drought co-occur.

7.3.3   Effects of drought on leaf longevity and turnover

Evidence of the importance of drought for leaf longevity and production is suggested by the common association of leaf phenology with seasonal variation in water availability (Wright, Chapter 15). However, few studies have specifically associated leaf longevity with drought stress in tropical plants. In our irrigation study on BCI, drought negatively affected leaf longevity in all three species of *Psychotria*. The species with the shortest leaf life span, *P. furcata*, was by far the most strongly affected, showing a two-fold reduction in leaf life span in control plots relative to irrigated plots (Mulkey, Wright & Smith, 1993). In contrast, leaf longevity in the two herbs was unaffected by irrigation treatment (Mulkey, Wright & Smith, 1991). Instead, the species with the shortest leaf life span, *Plieostachya pruinosa*, was less able to replace dying leaves in plots without dry-season irrigation. Thus, the intrinsically high rate of leaf production in this species was impaired during drought. Kursar and Coley (unpublished) have shown that tropical understory species with long-lived leaves have greater capacity for osmotic adjustment than those with short-lived leaves.

7.3.4   Drought-related plant mortality

Although few controlled studies exist, drought has been shown to have a strong negative effect on plant growth and survival in tropical forests. This was found experimentally for very young seedlings of an understory shrub, *Hybanthus · prunifolius* (Augspurger, 1979), in which just a few days without water resulted in seedling death. Mortality of trees in moist forest in Panama increased by a factor of five following the 1983 El Niño-Southern Oscillation, the most severe drought in Central America in over 100 years (Leigh et al., 1990). This event accounted for much of the mortality of all woody species on BCI in a 50 ha census plot, with moisture-loving species growing on wet slopes and ravines showing the highest mortality (Hubbell & Foster, 1990). Although no mortality was reported, radial growth of trees in a Mexican dry forest was strongly correlated with the duration of the rainy season (Bullock, 1994). Drought has been an important source of mortality in the tropical heath forests of Brunei in Southeast Asia (Becker & Wong, 1994). Prolonged, severe drought substantially increased mortality of the smallest size classes of trees growing on these sandy, well-drained soils, while nearby dipterocarp forest, which grows on richer soils, did not exhibit this mortality.

Studies demonstrate that drought contributes to plant mortality not only during extreme periods such as El Niño, but also following more normal seasonal dry periods. Our work indicates that whole-plant carbon balance plays a role in this mortality. The dry season most affected survivorship of those species with the highest intrinsic rates of leaf turnover (Mulkey, Smith & Wright, 1991; Mulkey, Wright and Smith, 1993; Wright, Chapter 15). As noted above, *Pleiostachya pruinosa* showed reduced survivorship over five years of normal dry season conditions in control relative to irrigated treatments (Figure 7.5). Similarly, leaf turnover and mortality were higher for *Psychotria furcata* (Figure 7.5), than for *P. marginata* or *P. limonensis*. A study of seedlings of the canopy tree *Virola surinamensis* experimentally grown in these treatments showed that mortality from drought was inversely related to the amount of growth and root development after germination (Fisher et al., 1991).

Mortality during or following drought may be due to abiotic and biotic stresses associated with drought, or due to direct limitations to carbon gain and concomitant constriction of the carbon budget. An example of the first instance is mortality resulting from fires that frequently follow a prolonged dry season (Leighton & Wirawan, 1986). There is evidence that drought-related fires have affected tropical

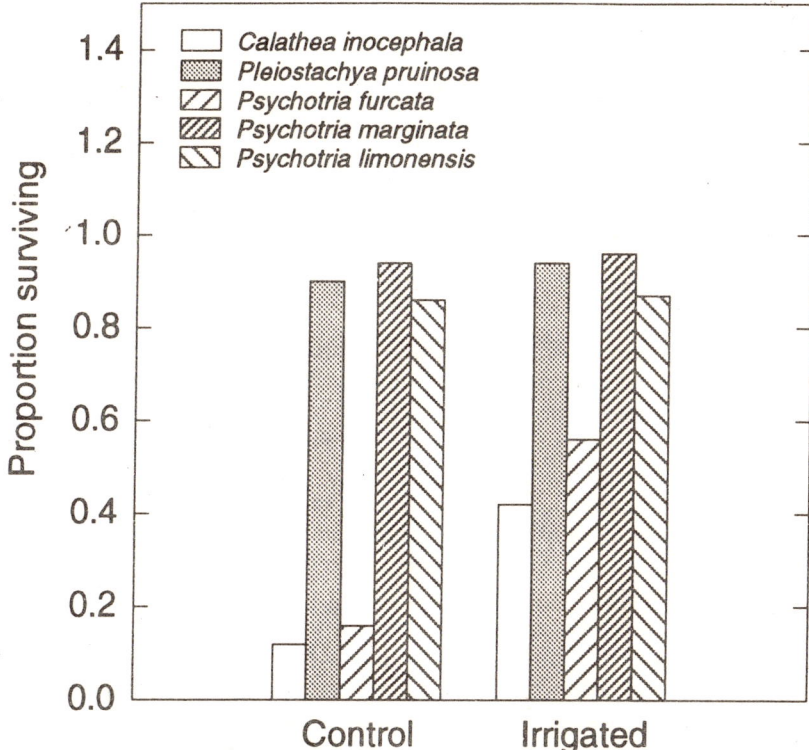

Figure 7.5.   Proportion of individuals of herbs and shrubs surviving for three years in control and irrigated sites on BCI.

forest composition over the last 4,000 years in tropical Africa (Hart et al., 1994). A wide range of plants have been shown to be more susceptible to attack by pathogens or insects during drought (e.g., Lewis, 1984; reviewed by Mattson & Haack, 1987), but there are few data for tropical species. A decade-long outbreak of canker disease in *Ocotea whitei* in Central Panama is likely to have been a consequence of the 1983 El Niño (Gilbert, Hubbell & Foster, 1994). Biochemical and physical features that promote resistance of plants to pathogens and insect attack are largely carbon dependent, especially for plants with long-lived leaves (Bazzaz et al., 1987), and we would expect allocation of carbon to tissue repair and pest resistance to be especially constrained during drought (Mattson & Haack, 1987). The data from the five-year irrigation experiment on BCI suggest that the effect of seasonal drought on the carbon budget is cumulative, requiring several years for mortality to ensue. In no case was death associated with a specific external factor other than the yearly occurrence of drought (Mulkey, Wright & Smith, 1993). However, as detailed above, leaf-level analysis of photosynthesis at light saturation during drought does not predict which species will suffer mortality.

7.4. INTEGRATED PLANT RESPONSES TO DROUGHT

Although the physiological literature focuses disproportionately on leaf-level phenomena, drought influences all aspects of plant function. It is perhaps understandable that the leaf should be the source of most of our knowledge given its enormous functional significance as the characteristic organ for vascular plants, and the logistic difficulties of studying processes in other plant organs, especially those below ground. Recent work shows that transport and storage functions can have great influence on plant responses to drought (e.g., Tyree & Ewers, Chapter 8). Another aspect of integrated plant function that has recently received attention is the way different plant characters vary ontogenetically in relation to each other (e.g., Reich, Walters & Ellsworth, 1992). Because there have been multiple evolutionary solutions for successful growth and reproduction in a given environment, it is more appropriate to examine a suite of functional characters than any single character alone in order to understand plant function. Similarly, it is also clear that environmental variation almost never occurs along a single axis. For example, the change in light availability that accompanies gap formation is only part of the host of environmental changes that accompanies a treefall, including changes in humidity, temperature, soil moisture, and nutrient availability. As our ability to make sophisticated measures in the field has improved, it has become possible to characterize integrated plant responses to multiple, simultaneous environmental challenges (e.g., Amthor and McCree, 1990; Chapin et al., 1987). Thus, we are approaching a state in which we can begin to define the finite group of functional character syndromes that exist for vascular plants. Here we consider selected aspects of integrated plant response as they relate to how drought affects tropical species.

7.4.1. The importance of underground structures and water-use efficiency

Functional aspects of water uptake and storage below ground are perhaps the most poorly studied area of plant response to drought. Although there is a considerable literature concerning crop plants, there is very little known about rooting depth, water storage, and root hydraulics in tropical forest species. At present, most models of water movement through the soil-plant-air continuum assume the region below ground to be a black box. The conventional wisdom that most tropical forests are made up of shallowly-rooted species (Richards, 1952) is certainly false for many tropical forests (e.g., Jordan & Herrera, 1981). Remarkably, we don't know the rooting depth for the most common tropical forest species (Sanford & Cuevas, Chapter 10). Variation in root length has obvious effects on the capacity to maintain metabolic function during drought, as well as determining diffusive resistance to water movement and the outcome of nearest-neighbor competition for soil water. Overstory trees in the large-scale experiment on BCI did not show great changes in phenology when irrigated (Wright, Chapter 15), possibly because their roots had access to the water table, which during the dry season may be only a few meters below the surface. Liana roots on BCI have been excavated to a depth of several meters (Tyree & Ewers, Chapter 8). Plants with high root-to-shoot ratios often survive better under resource limiting conditions (Chapin et al., 1987; Kitajima, Chapter 19). Although circumstantial evidence points to the paramount importance of root function during drought (e.g., Becker & Wong, 1994), there are no studies linking survivorship of tropical trees to specific mechanisms of below ground function.

Root function is especially germane to assessing the importance of water-use efficiency (WUE) for plant performance during drought. WUE has long been a controversial metric in the plant ecophysiology and agricultural literature. Expressed as the ratio of carbon assimilated per water lost (or vice versa in some literature), WUE is easily measured at the leaf level, and comparisons among species are common (e.g., Yoshie, 1986). Breeding programs have sought to improve WUE in crop plants that show genotypic variation in this parameter (e.g. Farquhar & Richards, 1984). Unfortunately, it is difficult to relate plant performance to WUE when plant performance is measured in a number of different ways, including instantaneous carbon gain, growth, and survival. All things being equal, high WUE should result in higher productivity and growth. Not surprisingly, high WUE is usually found in plants with low productivity because efficiency in water use is acquired through stomatal closure which partially limits carbon gain (Jones, 1993). From an evolutionary perspective, water conserving characteristics that confer high WUE would matter little if all competing neighbors have access to the same diminishing reserve of water. This is especially true for tropical understory plants if they compete with large overstory trees for the same water reserve, which would be the case if tropical forest species generally have shallow rooting depths. Accordingly, Jones (1993) concludes that "water-use efficiency is of only limited applicability to discussion of the fitness of plants for water-limited environments." Nevertheless, many studies link appropriate plant variation in WUE to ecological variation in atmospheric and soil drought (e.g., Tenhunen, Pearcy & Lange, 1987; Mulkey, 1986b; Mulkey, Smith & Wright, 1991; Mulkey et al., 1992). The recent use of carbon-and oxygen-stable isotopes has provided an integrated long-term estimate of WUE (e.g., Sternberg, Mulkey & Wright, 1989a,b; Goldstein et al., Chapter 9), and there is considerable circumstantial evidence that the distribution and abundance of some species are linked to this factor (e.g., Medina & Francisco, 1994).

Plants with the ability to store water after it is acquired would survive, and possibly grow better during drought, if assimilation was more efficient with regard to water use. Water may be stored in special organs such as tubers, or it may be in the various compartments of the xylem and connected mesophyll of the symplast. Plants with considerable ability to store water are said to have high capacitance, defined as the change in tissue water volume for a given change in tissue water potential (Koide et al., 1989). Desert plants with high WUE often have Crassulacean Acid Metabolism, which depends on a large internally maintained volume of water (Gibson & Nobel, 1986). In our study of *Pleiostachya* and *Calathea*, high WUE efficiency was associated with low mortality during drought in *Calathea*, which has a large underground tuber (Mulkey, Smith & Wright, 1991). The tuber no doubt contributes stored carbon as well as water to the maintenance of this species during seasonal drought. Indeed, this species seems to be exceptionally long-lived: individuals growing in deep shade that were marked in the early 1960s by H. Kennedy continue to thrive. In a pilot study monitoring stem flow (Baker & Van Bavel, 1987) in *Psychotria* during the dry season, we found that for plants in shade, water does not begin to move in the xylem near ground level until 3 or 4 hours after stomata are open and gas exchange is underway. Within a few minutes after water flow in the stem begins, the stomata begin to limit transpiration by closing. These species have no root storage organ and their capacitance is likely small. Regardless, small variation in capacitance among species may be the determinant of differences in the ability to maintain gas exchange through midday when evaporative demand increases. The cumulative effect of a few added minutes of carbon gain each day may be quite large for energy-limited plants in the shaded understory of tropical forests. Indeed, because the ability to make up for low carbon gain is most limited in shade, it may be that drought affects mortality more in the shade than in the sun, which is the case for *Pleiostachya pruinosa* (Mulkey, Smith & Wright, 1991).

7.4.2. Photosynthetic acclimation and drought tolerance

Most understory tropical forest plants will experience large variation in the availability of light and water during their lifetime. Because plants undergoing acclimation to changes in light that accompany treefall gap formation may experience seasonal drought, we would expect the allocation of carbon during this process to reflect the impact of both of these important environmental variables. As detailed above, the negative impact of photoinhibition can be amplified by drought, but most plants in newly-formed gaps continue to acclimate and produce new leaves with physiological and morphological characteristics suited to the high-light environment (Mulkey & Pearcy, 1992). Although there is evidence that soil water potentials are slightly higher in newly-formed gaps than in adjacent forest (Becker et al., 1988), we expect drought stress to be higher for gap plants because of lower humidity and higher evaporative demand under high light. Dry season water stress during the acclimation period may influence the entire carbon budget of the plant and could be a major determinant of how rapidly new leaves are produced. The process of acclimation is expensive for plants that produce sun leaves with substantially higher photosynthetic capacity than shade leaves because of the carbon and nutrients necessary for the more substantial photosynthetic machinery. This is especially true for new leaves with higher mass per area (leaf specific mass, LSM) than shade leaves. Respiratory costs for leaf construction during acclimation are considerably higher than that for maintenance construction in the understory rainforest herb, *Alocasia macrorrhiza* (Sims & Pearcy, 1991). Thus we might expect plants with high acclimation potential to be more negatively affected by drought than plants that maintain their leaves or produce leaves with photosynthetic capacity similar to shade leaves. The opposite may be the case if plants can quickly recoup the carbon lost due to new leaf production if those new leaves have significantly higher photosynthetic capacity and are functional during drought.

There is no consistent relationship between photosynthetic acclimation, as indicated by the ability to produce new leaves with higher A_{max}, and the ability to survive drought on BCI. *Calathea* shows little or no increase of photosynthetic capacity in newly-produced leaves in gaps and also shows little long-term effect of drought. *Pleiostachya* exhibits a large gap acclimation response and considerable drought-related mortality, and generally survives longer in gaps than in shade (Mulkey, Smith & Wright, 1991). In contrast, acclimation potential among the three *Psychotria* is lowest in the species most likely to die from the effects of drought (*P. furcata*; Figure 7.4 E, F), but note that this species has the highest leaf turnover. This species is twice as likely to die in gaps than in shade when not irrigated during the dry season (Mulkey, Wright & Smith, 1993). These results suggests that although carbon may be limiting during drought, acclimation potential to light *per se* is not a consistent indicator of the outcome of multiple stresses in gaps.

7.4.3   Suites of functional characters

Allocation theory, with its emphasis on the pivotal role of carbon, has done much to improve our understanding of plant form and function. However, because of the logistic difficulties of measuring carbon budgets, this approach rarely results in a complete picture. Recently, several workers have focused on suites of functional characters as indicators of how allocation patterns vary in response to the opportunity for carbon gain (Reich, Walters & Ellsworth, 1992; Mulkey, Kitajima & Wright, 1995). By comparing the variation among these characters across species with leaf displays that contrast in either time or space, it is possible to describe the tradeoffs that have resulted from selection to coordinate functions to maximize whole-plant carbon gain. Field (1991) has argued that A_{max} is the "master integrator" that reflects the allocation of nutrients to leaf function and the distribution of leaves in time and space. Building on this concept, Reich, Walters and Ellsworth (1992) have shown that there is a strong relationship among A_{max}, maximum conductance, leaf nitrogen concentration, LSM, leaf area ratio (leaf area per plant mass, LAR), and leaf longevity for a large group of woody plants from diverse ecosystems. To the extent that covariation among these characters reflects tight developmental linkage, they should also vary predictably among individuals of species with differing resource availability (e.g., wet vs. dry sites). Similarly, within-canopy variation should be predictable as the opportunity for carbon gain varies across seasons and between sunny and shady regions of the canopy. Work is underway to explore which characters covary in ecological time in the canopy of a dry forest in Panama (Mulkey, Kitajama & Wright, 1995). This approach may be a useful tool for scaling-up by linking leaf-level characters to whole-plant function.

It is clear that no single character, such as A_{max} of dry-season leaves, adequately reflects the influence of drought on the whole-plant carbon budget. Despite the fact that drought increases mortality in some species on BCI, the causal link with a plant's carbon budget cannot be established by looking at gas exchange alone (Mulkey, Smith & Wright, 1991; Mulkey, Wright & Smith, 1993). One piece of strong evidence linking carbon balance to survival during drought is the observation that those species with intrinsically high rates of leaf and whole-plant biomass turnover are most likely to experience increased mortality during drought (Mulkey, Smith & Wright, 1991; Mulkey, Wright & Smith, 1993). When several functional characters are analyzed together, we can begin to develop a picture of how species differences in leaf phenology and habitat affinity result in differences in the potential for carbon gain during drought. For example, *Calathea* has physiological and morphological characteristics consistent with persistence in the understory after gap closure and during repeated seasonal drought. Conservative gas exchange and controlled water loss in conjunction with water storage in the roots and long-lived leaves help this species persist under conditions inimical to *Pleiostachya* over a range of light and water availability. It is evident from the entire array of functional characters that *Pleiostachya* lacks efficient gas exchange and the ability to conserve water during drought. A similar approach has allowed us to see how a suite of functional characters permits *P. furcata* to have some life history characteristics of a weed, but little dependency on gaps for growth and reproduction. In particular, canopy architecture and high LAR result in low self-shading and contribute to high whole-plant efficiency in carbon uptake in this species. The influence of drought on the carbon budgets of these species is best understood by the negative effect on leaf longevity, which results in high carbon costs for leaf replacement.

In sum, it is only apparent that physiological features distinctly complement each life history when we expand the analysis to include a variety of leaf-level and whole-plant characters, including leaf demography.

7.5   A RESEARCH AGENDA

As stated in the preface to this volume, two areas of investigation for the future are the forest canopy and the rhizosphere. Development of these areas is crucial for understanding how carbon budgets are affected by drought. There should be two overarching goals for this research. First, we should seek to understand how drought has operated as a selective factor on carbon budgets by describing the regulatory mechanisms that have resulted in particular patterns of allocation. Our most important tool is the tried-and-true comparative study of carefully selected contrasting species which show alternative solutions to the same evolutionary problem. Second, an understanding of the role of tropical forest in regional and global hydrologic and carbon cycles requires that we develop techniques to quantify the effect of leaf-level and whole-plant drought responses for assemblages of individual plants and entire ecosystems. Because tropical forests constitute between 9 and 17 percent of the surface area of the earth's terrestrial ecosystems (Melillo et al., 1992) and are the largest of forest ecosystems containing most of the earth's biodiversity, such work is of critical importance.

7.5.1   The canopy

Progress in understanding the effect of drought on the forest canopy can be made through studies that detail the function of leaves through time and space in a way that is meaningful for the carbon budget of an entire tree. One important result to emerge from the past two decades of emphasis on the energy-limited shaded understory is that whole-plant characteristics, such as allocation to roots (Kitajima, Chapter 19), canopy architecture (Mulkey, Wright & Smith, 1993), and phenology (Wright, Chapter 15) are determinants of survival and growth in the face of drought. As detailed above, one promising avenue is through defining the finite set of physiological syndromes, including whole-plant features, that represent solutions to the water-loss carbon-gain evolutionary conundrum. Although work by Reich, Walters and Ellsworth (1992) and others has resulted in a list of features that serves as a descriptor across communities, we have found that this approach does not adequately describe the opportunity for carbon gain across the complex canopy of an individual tree (Mulkey, Kitajima & Wright, 1995). This is because seasonal drought and spatial variation in transpirational demand and light within a tree's canopy result in a complex mosaic of carbon gaining potential.

A first step in describing this mosaic is to build three-dimensional empirical models of the canopy in which small errors of measurement at the leaf level are not compounded when extrapolated to the branch and canopy level. Along with improved methods of accessing the canopy, we need architectural models that permit generalizations about the opportunity for carbon gain in particular regions of the canopy over the seasons. Using a canopy crane for access to the canopy of a tropical dry forest in Panama, we are applying an empirically-based model of photosynthesis and light interception to entire branches located in contrasting micro-habitats within the canopy (with K. Kitajima and R. W. Pearcy). The influence of drought on the local carbon budget of these branches can be compared across seasons and among species with contrasting phenologies. Ultimately, such a "bottom-up" approach must be quantitatively reconciled with the conclusions from "top-down" studies such as those described by Meinzer and Goldstein (Chapter 4).

7.5.2 The rhizosphere

Describing the role of roots during drought is just as important and possibly more daunting than studying the canopy. There are three areas that need substantial development. First, there are a number of evolved resource acquisition strategies that have resulted in species with roots with contrasting radial spread, depth, and growth rates. These strategies should be described through measures of factors such as capacitance and hydraulic conductance (Tyree & Ewers, Chapter 8). Any significance of water-use efficiency as a measure of plant water relations depends on a more detailed understanding of root function. Second, we need an understanding of the role played by mycorrhizae in the overall carbon budget of the plant and the possible acquisition of water. The vast majority of tropical trees maintain vesicular-arbuscular mycorrhizal associations that have hyphal connections among taxonomically unrelated tree species throughout the soil matrix. Although there has been pioneering work in this field (Janos, 1980), we know very little about the quantitative effects of mycorrhizal associations on the carbon budgets of tropical trees (Sanford & Cuevas, Chapter 10). Finally, we must develop models of root function based on their role as carbon sources during shoot growth, and carbon sinks during root growth. There is strong evidence that allocation to roots has compound influences over whole-plant resistance to pests, disease, and drought (e.g., Kitajima, Chapter 19), and that this is an essential component of shade tolerance. A full accounting of plant carbon budgets and adequate descriptions of physiological syndromes must include the role of carbon flux through the roots.

ACKNOWLEDGEMENTS

Our work on BCI has been supported by the University of Missouri and the Smithsonian Scholarly Studies Program. The irrigation study was funded by the Smithsonian Environmental Studies Program. Ongoing canopy studies have been supported by the National Science Foundation (IBN-9220759; BIR-9419994) and the Smithsonian Scholarly Studies Program. We would like to acknowledge the pivotal role of Alan P. Smith in providing inspiration, intellectual input, and logistic support during the decade of this research.

REFERENCES

ALVIM, P. de T. & ALVIM, R. (1978) Relation of climate to growth periodicity in tropical trees. *Tropical Trees as Living Systems* (eds. P. B. TOMLINSON & M. H. ZIMMERMANN) Cambridge University Press, London, pp. 445–464.

AMTHOR, J. S. & McCREE, K. (1990) Carbon balance of stressed plants: A conceptual model for integrating research results. *Stress Responses in Plants: Adaptation and Acclimation Mechanisms* (eds., R. G. ALSCHER & J. R. CUMMING) Wiley-Liss, New York, pp 1–16.

AUGSPURGER, C. K. (1979) Irregular rain cues and the germination and seedling survival of a Panamanian shrub (*Hybanthus prunifolius*) *Oecologia*, **44,** 53–59.

BAKER, J. M. & VAN BAVEL, C. H. M. (1987) Measurement of mass flow of water in the stems of herbaceous plants. *Plant, Cell and Environment*, **10,** 777–782.

BAZZAZ, F. A., CHARIELLO, N. R., COLEY, P. D. & PITELKA, L. F. (1987) Allocating resources to reproduction and defense. *Bioscience* **37,** 58–67.

BECKER, P. & WONG, M. (1994) Drought-induced mortality in tropical heath forest. *Journal of Tropical Sciences*, **5,** 416–417.

BECKER, P., RABENOLD P. E., IDOL J. R. & SMITH A. P. (1988) Water potential gradients for gaps and slopes in a Panamanian tropical moist forest's dry season. *Journal of Tropical Ecology*, **4**, 173–184.

BOLAÑOS, J. A. & LONGSTRETH, D. J. (1984) Salinity effects on water potential components and bulk elastic modulus of *Alternanthera philoxeroides* (Mart.) Griseb. *Plant Physiology*, **75**, 281–284.

BORCHERT, R. (1983) Phenology and control of flowering in tropical trees. *Biotropica*, **15**, 81–89.

BOWMAN, W. D. & ROBERTS, S. W. (1985) Seasonal changes in tissue elasticity in chaparral shrubs. *Physiologia Plantarum*, **65**, 233–236.

BULLOCK, S. H. (1994) Insensitivity of growth of deciduous trees to reduced drought. International Meeting of the Society for Conservation Biology and the Association for Tropical Biology, Guadalajara, Mexico.

BULLOCK, S. H. & SOLIS-MAGALLANES, J. A. (1990) Phenology of canopy trees of a tropical deciduous forest in Mexico. *Biotropica*, **22**, 22–35.

CHAPIN, F. S., BLOOM A., FIELD, C. B. & WARING, R. H. (1987) Plant response to multiple environmental factors. *Bioscience*, **37**, 49–57.

CHAPIN, F. S., SSHULZE E.-D. & MOONEY, H. A. (1990) The ecology and economics of storage in plants. *Annual Review of Ecology and Systematics*, **21**, 432–447.

CHIARIELLO, N. R., FIELD, C.B. & MOONEY, H. A. (1987) Midday wilting in a tropical pioneer tree. *Functional Ecology*, **1**, 3–11.

COLEY, P. D., BRYANT, J. P. & CHAPIN, F. S. III (1985) Resource availability and plant anti-herbivore defense. *Bioscience*, **37**, 110–118.

CORNIC, G. (1994) *Drought Stress and High Light Effects on Leaf Photosynthesis*. BIOS Scientific Publishers Ltd, United Kingdon

COSGROVE, D. J. (1988) In defense of the cell volumetric elastic modulus. *Plant Cell and Environment*, **11**, 67–69.

DAVIES, F. T., POTTER, J. R. & LINDERMAN, R. G. (1993) Drought resistance of mycorrhizal pepper plants independent of leaf P-concentration – Response in gas exchange and ater relations. *Physiologia Plantarum*, **87**, 45–53.

DAWSON T. E. (1993) Woodland water balance. *Trends in Ecology and Evolution*, **8**, 120–121.

EHLERINGER, J. (1982) The influence of water stress and temperature on leaf pubescence development in *Encelia farinosa*. *American Journal of Botany*, **69**, 670–675.

FARQUHAR, G. D. & RICHARDS, R. A. (1984) Isotopic composition of plant carbon correlates with water-use efficiency of wheat genotypes. *Australian Journal of Plant Physiology*, **11**, 539–552.

FETCHER, N. (1979) Water relations of five tropical tree species on Barro Colorado Island. *Oecologia*, **40**, 229–233.

FETCHER N., OBERBAUER, S. F. & CHAZDON, R. L. (1994) Physiological ecology of plants. *La Selva, Ecology and Natural History of a Neotropical Rain*

Forest (eds. L. A. MacDade, K. S. Bawa, H. A. Hespenheide & G. S. Hartshorn) University of Chicago Press, Chicago, pp 128–141.

Field, C. B. (1991) Ecological scaling of carbon gain to stress and resource availability. *Integrated Responses of Plants to Stress* (eds. H. A. Mooney, W. E. Winner & E. J. Pell) Academic Press, New York, pp 35–65.

Fisher, B., Howe, H. F., & Wright, S. J. (1991) Survival and growth of *Virola surinamensis* yearlings: Water augmentation in gap and understory. *Oecologia,* **86,** 292–297.

Frankie, G. W., Baker H. G. & Opler, P. A. (1974) Comparative phenological studies of trees in tropical wet and dry forests in the lowlands of Costa Rica. *Journal of Ecology,* **62,** 881–919.

Gibson, A. C. & Nobel, P. S. (1986) *The Cactus Primer.* Harvard University Press, Cambridge, Massachusetts.

Gilbert, G. S., Hubbell, S. P. & Foster, R. B. (1994) Density and distance-to-adult effects of a canker disease of trees in a moist tropical forest. *Oecologia,* **98,** 100–108.

Givnish, T. J. (1984) Optimal stomatal conductance, allocation of energy between leaves and roots, and the marginal cost of transpiration. *On the Economy of Plant form and Function* (ed. T. J. Givnish) Cambridge University Press, Cambridge, pp 171–214.

Givnish, T. J. & Vermeji, G. J. (1976) Sizes and shapes of liane leaves. *American Naturalist,* **110,** 743–778.

Gunasekera, D. & Berkowitz, G. A. (1992) Evaluation of contrasting cellular-level acclimation responses to leaf water deficits in three wheat genotypes. *Plant Science,* **86,** 1–12.

Hanson, A. D. & Hitz, W. D. (1982) Metabolic responses of mesophytes to plant water deficits. *Annual Review of Plant Physiology,* **33,** 163–203.

Hart, T. B., Hart, J. A., Dechamps, R., Fournier, M. & Atanolo, M. (1994) Changes in forest composition over the last 4000 years in the Ituri Basin, Zaire. *14th AETFAT Congress,* Nairobi, Association for the Taxonomic Study of the Flora of Tropical Africa.

Havaux, M. (1992) Stress tolerance of photosystem-II In vivo – Antagonistic effects of water, heat, and photoinhibition stresses. *Plant Physiology,* **100,** 424–432.

Herbert, T. J. (1992a) Random wind-induced leaf orientation – Effect upon maximization of whole plant photosynthesis. *Photosynthetica,* **26,** 601–607.

Herbert, T. J. (1992b) Geometry of heliotropic and nyctinastic leaf movements. *American Journal of Botany,* **79,** 547–550.

Herbert, T. J. & Larsen, P. B. (1985) Leaf movement in *Calathea lutea* (Marantaceae) *Oecologia,* **67,** 238–243.

Hsaio, T. C., Acevedo, E., Fereres, E. & Henderson, W. D. (1976) Water stress, growth, and osmotic adjustment. *Philosophical Transactions of the Royal Society London B,* **273,** 479–500.

HUBBELL, S. P. & FOSTER, R. B. (1990) Structure, dynamics, and equilibrium status of old-growth forest on Barro Colorado Island. *Four Neotropical Forests* (ed. A. H. GENTRY) Yale University Press, New Haven, pp 522–541.

IKE, I. F. & THURTEL, G. W. (1981) Water relations of cassava: water content, water, osmotic, and turgor potential relationships. *Canadian Journal of Botany*, **59**, 956–964.

JANOS, D. (1980) Vesicular-arbuscular mycorrhizae affect lowland tropical rain forest plant growth. *Ecology*, **61**, 151–162.

JARVIS, P. G. (1993) Water losses of crowns, canopies and communities. *Water Deficits: Plant Responses from Cell to Community* (eds. J. A. C. SMITH & H. GRIFFITHS) BIOS Scientific Publishers Limited, Oxford, United Kingdom pp 285–305.

JONES, H. G. (1993) Drought tolerance and water-use efficiency. *Water Deficits: Plant Responses from Cell to Community* (eds. J. A. C. SMITH & H. GRIFFITHS) BIOS Scientific Publishers Limited, Oxford, United Kindom, pp 193–204.

JORDAN, C. F. & HERRERA, R. (1981) Tropical rainforests: Are nutrients really critical? *American Naturalist*, **117**, 167–180.

KOIDE, R. T., ROBICHAUX, R. H., MORSE, S. R., & SMITH, C. M. (1989) Plant water status, hydraulic resistance, and capacitance. *Plant Physiological Ecology*, eds. R. W. PEARCY, J. EHLERINGER, H. A. MOONEY, and P. W. RUNDEL, Chapman and Hall, New York, pp 161–184.

LEIGH, E. G., WINDSOR, D. M., RAND, A. S. & FOSTER, R. B. (1990) The impact of the "El Niño" drought of 1982–83 on a Panamanian semideciduous forest. *Global Ecological Consequences of the 1982–83 El Niño-Southern Oscillation* (ed. P. W. GLYNN) Elsevier Oceanography Series, London, pp 473–486.

LEIGHTON, M. & WIRAWAN, D. M. (1986) Catastrophic drought and fire in Borneo tropical rain forest associated with the 1982–1983 El Niño Southern Oscillation event. *Tropical Rain Forests and the World Atmosphere* (ed. G. T. PRANCE) Westview Press, Boulder, Colorado, pp 75–102.

LEWIS, A. (1984) Plant quality and grasshopper feeding: Effects of sunflower condition on preference and performance in *Melanoplus differentialis*. *Ecology*, **65**, 836–843.

MATTSON, W. J. & HAACK, J. (1987) The role of drought in outbreaks of plant-eating insects. *Bioscience*, **37**, 110–118.

MEDINA, E. (1983) Adaptations of tropical trees to moisture stress. *Tropical Rain Forest Ecosystems* (ed. F. B. Golley) Elsevier Scientific Publishers Amsterdam, pp 225–237.

MEDINA, E. & FRANCISCO, M. (1994) Photosynthesis and water relations of savanna tree species differing in leaf phenology. *Tree Physiology*, **14**, 1367–1381.

MEINZER, F. C., GOLDSTEIN, G., HOLBROOK, N. M., JACKSON, P. & CAVALIER, J. (1993) Stomatal and environmental control of transpiration in a lowland tropical forest tree. *Plant Cell and Environment,* **16,** 429–436.

MELILLO, J. M., CALLAGHAN T. V., WOODWARD, F. I., SALATI, E. & SINHA, S. K. (1992) Effects on ecosystems. *Climate Change: The IPCC Scientific Assessment.* (eds. J. T. HOUGHTON, G. J. JENKINS, & J. J. EPHRAUMS), Cambridge University Press, Cambridge, United Kingdom pp 283–310.

MOONEY, H. A. (1972) The carbon balance of plants. *Ann. Rev. Ecol. Syst.,* **3,** 315–346.

MOONEY, H. A. & CHIARIELLO, N. R. (1984) The study of plant function – The plant as a balanced system. *Perspectives on Plant Population Ecology* (eds. R. DIRZO & J. SARUKHAN) Sinauer Associates, Sunderland, Massachusetts, pp 305–323.

MOONEY, H. A., FIELD, C. & VÁZQEZ-YANES, C. (1984) Photosynthetic characteristics of wet tropical forest plants. *Physiological Ecology of Plants of the Wet Tropics.* (eds. E. MEDINA, H. A. MOONEY & C. VÁZQUEZ-YANES), Dr. Junk Publishers, Hague, pp 113–128.

MOONEY, H. A., FIELD, C., VÁSQUEZ-YANES, C. & Chu, C. (1983) Environmental controls on stomatal conductance in a shrub of the humid tropics. *Proceedings of the National Academy of Sciences,* USA, **80,** 1295–1297.

MULKEY, S. S. (1986a) *Physiological ecology and demography of three species of neotropical herbaceous bamboo.* Ph.D. dissertation, The University of Pennsylvania.

MULKEY, S. S. (1986b) Photosynthetic acclimation and water-use efficiency of three species of understory herbaceous bamboo (Gramineae) in Panama. *Oecologia,* **70,** 514–519.

MULKEY, S. S. & PEARCY, R. W. (1992) Interactions between acclimation and photoinhibition of photosynthesis of a tropical forest understory herb, *Alocasia macrorrhiza,* during simulated canopy gap formation. *Functional Ecology,* **6,** 719–729.

MULKEY, S. S., SMITH, A. P. & WRIGHT, S. J. (1991) Comparative life history & physiology of two understory neotropical herbs. *Oecologia,* **88,** 263–273.

MULKEY, S. S., SMITH, A. P., WRIGHT, S. J., MACHADO, J. L., & DUDLEY, R. (1992) Contrasting leaf phenotypes control seasonal variation in water loss in a tropical forest shrub. *Proceedings of the National Academy of Sciences,* USA, **89,** 9084–9088.

MULKEY, S. S., WRIGHT, S. J. & SMITH, A. P. (1991) Drought acclimation of an understory shrub (*Psychotria limonensis*; Rubiaceae) in a seasonally dry tropical forest in Panama. *American Journal of Botany,* **78,** 579–587.

MULKEY, S. S., WRIGHT, S. J. & SMITH, A. P. (1993) Comparative physiology and demography of three neotropical forest shrubs: Alternative shade-adaptive character syndromes. *Oecologia,* **90,** 526–536.

MULKEY, S. S. & KITAJIMA, K., & WRIGHT, S. J. (1995) Photosynthetic capacity and leaf longevity in the canopy of a dry tropical forest. *Selbyana*, in press.

MYERS, B. J., ROBICHAUX, R. H., UNWIN, G. L. & CRAIG, I. E. (1987) Leaf water relations and anatomy of a tropical rainforest tree species vary with crown position. *Oecologia*, **74**, 81–85.

NAUTIYAL S., BADOLA, H. K., Pal, M. & NEGI, D. S. (1994) Plant responses to water stress changes in growth, dry matter production, stomatal frequency, and leaf anatomy. *Biologia Plantarum*, **36**, 91–97.

NOBEL, P. S. (1980) Leaf anatomy and water use efficiency. *Adaptation of Plants to Water and High Temperature Stress* (eds. N. C. TURNER & P. J. KRAMER) JOHN WILEY and Sons, New York, pp 43–52.

OBERBAUER, S. F., STRAIN, B. R. & RIECHERS, G. H. (1987) Field water relations of a wet-tropical forest tree species, *Pentaclethra macroloba* (Mimosaceae). *Oecologia*, **71**, 379–385.

OPLER, P. A., FRANKIE, G. W. & BACKER H. G. (1976) Rainfall as a factor in the release, timing, and synchronization of anthesis by tropical trees and shrubs. *Journal of Biogeography*, **3**, 231–236.

PARKER, G. G., SMITH, A. P. & HOGAN, K. P. (1992) Access to the upper forest canopy with a large tower crane. *Bioscience*, **42**, 664–670.

PEARCY, R. W. (1990) Sunflecks and photosynthesis in plant canopies. *Annual Review of Plant Physiology and Plant Molecular Biology*, **41**, 421–453.

REICH, P. B. & BORCHERT, R. (1984) Water stress and tree phenology in a tropical dry forest in the lowlands of Costa Rica. *Journal of Ecology*, **72**, 61–74.

REICH, P. B., WALTERS, M. B. & ELSWORTH, D. S. (1992) Leaf life-span in relation to leaf, plant and stand characteristics among diverse ecosystems. *Ecological Monographs*, **62**, 365–392.

RICHARDS, P. W. (1952) *The Tropical Rain Forest*. Cambridge University Press, London,

ROBICHAUX, R. H. (1984) Variation in the tissue water relations of two sympatric Hawaiian *Dubautia* species that differ in habitat and diploid chromosome number. *Oecologia*, **66**, 77–80.

ROBICHAUX, R. H., RUNDEL, P. W., STEMMERMANN, L. & CANFIELD, J. E. (1984) *Tissue Water Deficits and Plant Growth in Wet Tropical Environments*. Dr. W. Junk Publishers, The Hague.

RUNDEL, P. W. & BECKER, P. F. (1987) Cambios estacionales en las relaciones hidricas y en fenologia vegetative de plantas del estrato bajo del bosque tropical de la Isla de Barro Colorado, Panama. *Revista de Biologia Tropical*, **35 (Suppl. 1)**, 71–84.

SCHULZE, E.-D. (1986) Carbon dioxide and water vapor exchange in response to drought in the soil. *Annual Review of Plant Physiology*, **37**, 247–274.

SCHULZE, E.-D. (1993) Soil water deficits and atmospheric humidity as environmental signals. *Water Deficits: Plant Responses from the Cell to the Community* (eds. J. A. C. SMITH H. GRIFFITHS) BIOS Scientific Publishers Limited, Oxford, United Kingdom pp 129–146.

SIMS, D. A., & PEARCY, R. W. (1991) Photosynthesis and respiration in *Alocasia* following transfers to high and low light. *Oecologia,* **86,** 447–453.

SOBRADO, M. A. (1986) Aspects of tissue water relations and seasonal changes of leaf water potential components of evergreen and deciduous species coexisting in tropical dry forests. *Oecologia,* **68,** 413–416.

STERNBERG, L. d.-S. L., MULKEY, S. S. & WRIGHT, S. J. (1989a) Ecological interpretation of leaf carbon isotope ratios: Influence of respired carbon dioxide. *Ecology,* **70,** 1317–1324.

STERNBERG, L. d.-S. L., Mulkey, S. S. & WRIGHT, S. J. (1989b) Oxygen isotope ratio stratification in a tropical moist forest. *Oecologia,* **81,** 51–56.

TENHUNEN, J. D., PEARCY, R. W. & LANGE, O. L. (1987) Diurnal variations in leaf conductance and gas exchange in natural environments. *Stomatal Function* (eds. E. ZEIGER, G. D. FARQUHAR & I. R. Cowan) STANFORD University Press, Stanford, California, pp 323–352.

TYREE, M. T. & JARVIS, P. G. (1982) Water in tissues and cells. *Encyclopedia of Plant Physiology, New Series., Vol. 12b, Physiological Plant Ecology* (eds. O.L. LANGE, P.S. NOBEL, C.B. OSMOND, and H. ZIEGLER), Springer-Verlag, Berlin, pp 35–78.

TYREE, M. T. & SPERRY, J. S. (1989) Vulnerability of xylem to cavitation and embolism. *Annual Review of Plant Physiology and Plant Molecular Biology,* **40,** 19–38.

WRIGHT, S. J. (1991) Seasonal drought and the phenology of understory shrubs in a tropical moist forest. *Ecology,* **72,** 1643–1657.

WRIGHT, S. J. & VAN SCHAIK, C. P. (1994) Light and the phenology of tropical trees. *American Naturalist,* **143,** 192–199.

WRIGHT, S. J., MACHADO, J. L., MULKEY, S. S., & SMITH, A. P. (1992) Drought acclimation among tropical forest shrubs (*Psychotria*, Rubiaceae). Oecologia, **89,** 457–463,

YOSHIE, F. (1986) Intercellular CO_2 and water-use efficiency of temperate plants with different life-forms and from different microhabitats. *Oecologia,* **68,** 370–374.

ZOTZ, G. & WINTER, K. (1994a) Photosynthesis of a tropical canopy tree, *Ceiba pentandra*, in a lowland forest in Panama. *Tree Physiology,* **14,** 1291–1301.

ZOTZ G. & WINTER, K. (1994b) A one-year study on carbon, water and nutrient relationships in a tropical C_3-CAM hemiepiphyte, *Clusia uvitana* Pittier. *New Phytologist,* **127,** 45–60.

8

Hydraulic Architecture of Woody Tropical Plants

Melvin T. Tyree and Frank W. Ewers

This chapter reviews how hydraulic architectures of tropical trees and lianas (woody vines) influence the flow of water from roots to leaves. The hydraulic design potentially can limit plant water relations, gas exchange, successional distribution, and even the maximum height (or length, in the case of lianas) that a species can attain. Important parameters include vulnerability to drought-induced cavitation (since embolism reduces hydraulic conductance), root pressures (since these potentially could result in re-filling of conduits following cavitation events, thereby increasing conductance), leaf specific conductivity (which, together with transpiration rates, can predict pressure gradients throughout the plant), and water storage capacity (since this might determine the ability to survive water shortage). Some of the issues that are dealt with here are the impact of vessel diameter on drought- and freezing-induced embolism, the role of root pressures in the occasional removal of embolisms, and the ways in which the hydraulic architecture differs in different growth forms such as trees, shrubs, lianas, and hemiepiphytes. The ecological and physiological trade-offs of different architectures are discussed, and comparisons are made with temperate plants.

8.1 WATER RELATIONS AND HYDRAULIC ARCHITECTURE

Water deficits are created in plants when water loss by the leaves is greater than water absorption by the roots. The structure of the water conductive system (the hydraulic architecture) can potentially limit the flow of water to leaves and thus it can limit leaf water potential, stomatal behavior, and gas exchange. Since we recently reviewed hydraulic architecture in woody plants (Tyree & Ewers, 1991), as well as xylem structure and water flow in vines (Ewers, Fisher & Fichtner, 1991), in this review we provide only a brief overview of the concepts of hydraulic architecture. We focus our attention on tropical woody plants and provide some new information on the structure and function of root systems in relation to hydraulic architecture.

8.1.1 Parameters used to describe hydraulic architecture

8.1.1.1 Hydraulic conductivity (k_h)

Hydraulic conductivity per unit pressure gradient (k_h) is equal to the ratio between water flux (F, kg s^{-1}) through an excised stem segment and the pressure gradient (dP/dx, MPa m^{-1}) causing the flow:

$$k_h = F/(dP/dx) \qquad (1)$$

To determine the relationship of k_h to the hydraulic architecture of woody plants, it can be divided by the cross-section of the functional xylem (sapwood) in the segment or by the leaf area distal and/or attached to the segment.

8.1.1.2 Specific conductivity (k_s)

Specific conductivity (k_s), which is a measure of the porosity of the wood, is equal to k_h divided by the sapwood cross-sectional area (A_s, m^2). By Poiseuille's law for ideal capillaries, specific conductivity will be proportional in a linear manner to vessel number but in and when raised to the fourth power is linearly related to vessel diameter (Tyree and Ewers, 1991).

8.1.1.3 Leaf specific conductivity (k_l)

Leaf specific conductivity (k_l), also known as LSC (Tyree & Ewers, 1991), is equal to k_h divided by the leaf area distal to the segment (A_l, m^2). This is a measure of the hydraulic sufficiency of the segment to supply water to leaves distal to that segment. If we know the mean evaporative flux density (E, $kg\ s^{-1}\ m^{-2}$) from the leaves supplied by the stem segment and if we ignore water storage capacitance, then the pressure gradient through the segment is (dP/dx) = E/k_l. Therefore, the higher the k_l, the lower the dP/dx required to allow for a particular transpiration rate.

8.1.1.4 Huber value (HV)

The Huber value (HV) is defined as the sapwood cross-sectional area (or sometimes the stem cross-section) divided by the leaf area distal to the segment. Since the HV is in units of m^2 stem area per m^2 leaf area, it is often written without dimension. It is a measurement of the investment of stem tissue per unit leaf area fed. It follows from the above definitions that k_l = HV $\times$ k_s.

8.1.1.5 Water-storage capacitance (Q)

The water storage capacity (C) of plant tissue is defined as the mass of water (ω) that can be extracted per MPa change in water potential (Ψ) of the tissue (C = $\Delta\omega/\Delta\Psi$, $kg\ MPa^{-1}$). Since the size of C is proportional to the size of the tissue in question, tissue capacitance is defined as C per unit tissue volume (V) or per unit tissue dry mass or, for leaves, per unit area (A):

$$Q_{stem} = (\Delta\omega/\Delta\Psi)/V \qquad (2a)$$

$$Q_{leaf} = (\Delta\omega/\Delta\Psi)/A \qquad (2b)$$

Mechanisms of water storage include (1) elastic storage, associated with changes in the volume of tissue, (2) capillary storage associated with non-living elements and (3) cavitation, release of water (Zimmermann, 1983; Tyree & Yang, 1990; Tyree et al., 1991). Cavitation is the breakage of the water column within a non-living xylem element such as a fiber, a tracheid, or a vessel element (Tyree and Sperry, 1989).

8.1.1.6   Vulnerability to embolism

Water is normally under negative pressure (tension) as it moves through the xylem towards the leaves. The water is thus in a meta-stable condition and is vulnerable to cavitation due to air entry into the water columns. Cavitation results in embolism (air blockage), thus disrupting the flow of water (Sperry & Tyree, 1988; Tyree & Sperry, 1989). Cavitation in plants can result from water stress, and each species has a characteristic "vulnerability curve" which is a plot of the percent loss k_h in stems versus the xylem pressure potential, Ψ_{xp}, required to induce the loss. Vulnerability curves are typically measured by dehydrating large excised branches to known Ψ_{xp}. Stem segments are then cut under water from the dehydrated branches so that the air bubbles remain inside the conduits. An initial conductivity measurement is made and compared to the maximum k_h after air bubbles have been dissolved (Sperry, Donnelly & Tyree, 1987). The vulnerability curves of plants, in concert with their hydraulic architecture, can give considerable insight to drought tolerance and water relations "strategies" (see section 8.1.3 for more details about vulnerability curves)

8.1.1.7   Embolism repair

Embolisms may be dissolved in plants if Ψ_{xp} in the xylem becomes positive or close to positive for adequate time periods (Sperry et al., 1987; Borghetti et al. 1991; Tyree & Yang, 1990, 1992; Yang & Tyree, 1992; Edwards et al., 1994; Lewis, Harnden, & Tyree 1995). Embolisms disappear by dissolution of air into the water surrounding the air bubble. The solubility of air in water is proportional to the pressure of air adjacent to the water (Henry's Law). Water in plants tends to be saturated with air at a concentration determined by the average atmospheric pressure of gas surrounding plants. Thus, for air to dissolve from a bubble into water, the air in the bubble has to be at a pressure in excess of atmospheric pressure. If the pressure of water (P_w) surrounding a bubble is equal to atmospheric pressure (P_a), bubbles will naturally dissolve because surface tension (τ) of water raises the pressure of air in the bubble (P_b) above P_a. In general, $P_b = 2$ $\tau/r + P_w$, where r is the radius of the bubble. According to the cohesion theory of sap ascent, P_w is drawn below P_a during transpiration. Since $2\,\tau/r$ of a dissolving bubble in a vessel is usually < 0.03 MPa and since P_w is in the range of -0.1 to -10 MPa during transpiration, P_b is usually $\leq P_a$ and hence bubbles, once formed in vessels, rarely dissolve. Repair (= dissolution) occurs only when P_w grows large via root pressure or other mechanism (see section 8.2.3).

8.1.2 Relationship between parameters and xylem structure

All of the above parameters are influenced by xylem structure. If leaf area is held constant, increasing the number of conduits (vessels or tracheids) in the xylem should increase k_h and k_l in a linear manner. If xylem area is held constant as the conduits increase in number, k_s will increase linearly. Similarly, if conduits per cross sectional area remain constant, as conduits increase in number, HV will increase in a linear manner. Due to Poiseuille's law, increasing conduit diameter should increase k_h, k_s and k_l in a fourth power manner, with no effect on HV. The complicating factor is the effect of xylem structure on embolism, since nonconductance of conduits will directly reduce conductivity, potentially over-riding the impact of conduit size and number.

The relationship among conduit structure, water-stress-induced embolism, and conductive efficiency is complex. The size of the conduit lumen appears to have only a weak influence on drought-induced embolism (Tyree & Sperry, 1989, Cochard, Cruiziat & Tyree, 1992; Tyree, Davis & Cochard, 1995). By the air-seeding hypothesis, the spread of water-stress-induced embolism is controlled by the size of the pores on pit membranes that connect adjacent conduits. The smaller the pores on the pit membranes, the lower the water potential required to induce embolism (Crombie, Hipkins & Milburn, 1985; Sperry & Tyree, 1988; Cochard, Cruiziat & Tyree, 1992; Jarbeau, Ewers & Davis, 1994). Narrower pores on pit membranes could have the effect of reducing conductivity to well below the theoretical maximum predicted by Poiseuille's law. Although measured conductivity is typically 30 to 90% less than the theoretical (Zimmermann, 1983; Ewers & Cruiziat, 1990; Chiu & Ewers, 1992, 1993), there is no direct experimental evidence that membrane pore diameter limits conductivity through vessels. However, there is experimental evidence that pit resistance helps limit water flow through tracheids (Calkin, Gibson & Nobel, 1986). It is possible that there is an evolutionary trade-off between safety (promoted by narrow pores in pit membranes) and conductive efficiency (promoted by wider pores).

There is considerable evidence that, within a plant, wider conduits are more prone to water-stress-induced embolism than narrow conduits (Tyree & Dixon, 1986; Salleo & Lo Gullo, 1986, 1989; Sperry & Tyree, 1990; Lo Gullo & Salleo, 1991; Hargrave et al., 1994). Perhaps the wider diameter conduits in a plant consistently develop pores that are wider than for narrow conduits in the same plant (Sperry & Tyree, 1988). Alternately, it has been suggested that the bigger conduits may have many more pit membranes, thus increasing the statistical probability of having a particularly wide pore that is air-exposed (Hargrave et. al., 1994).

There is now fairly good evidence that wide and long conduits are more prone to freezing-induced embolism than short and narrow conduits; the correlation between conduit volume and vulnerability to embolism appears to hold true both within a plant and between unrelated taxa (Cochard & Tyree, 1990; Sperry & Sullivan, 1992). However, freezing temperatures do not occur in the tropics other than in extreme alpine situations (Krog et al., 1979; Beck, Scheibe & Hansen, 1987).

Whatever the cause of embolisms, it is important to know if plants are capable of reversing embolism, and whether such reversal can occur on a diurnal or seasonal basis. Other than refilling embolized conduits, the only known mechanism that a plant has to increase conductivity through an axis is the production of new xylem tissue.

Xylem is a complex tissue composed of various cell types, including vessels, tracheids, fibers and parenchyma cells. The living parenchyma cells in the xylem presumably play a major role in water capacitance. Likewise, fibers, which have a major role in mechanical support, may be involved with capacitance either through capillarity or through reversible embolism. The conduits (vessels and tracheids) would have a major impact on capacitance only if reversible embolism occurred in response to the range of Ψ that the plant experienced.

8.1.3 Root hydraulic architecture

Hydraulic architecture studies have been confined largely to the shoot system of plants. However, the root system may offer at least as much resistance to water flow as the shoot system (Tyree et al. 1994). New methods are now available to study hydraulic resistance of roots and shoots that show promise of advancing our knowledge of root hydraulic architecture (Yang & Tyree, 1994; Tyree et al. 1995). The concepts of hydraulic architecture are easiest to apply to those root systems that feed into a single stem. The leaf area being serviced by the roots can then be measured, and the total k_h and total k_l of the woody root system can then be measured for various distances from the stem insertion point. In contrast, a situation with many stems arising from a single root system, as in most shrubs, or with stems that are rooted many places along their length, as in some lianas, offers complexities in interpreting the pathways. Similar complexities are offered by trees that clonally propagate from root buds. Even in the simplest systems, fine roots are difficult to excavate for k_h measurements. In addition, the resistance offered by the root-soil interface and by the radial pathway of water movement through the cortex of fine roots involves a more complex analysis than that for axial movement through woody segments.

8.2   HYDRAULIC ARCHITECTURE OF TROPICAL WOODY PLANTS

The hydraulic architecture of woody plants has been reviewed recently (Tyree & Ewers, 1991). Hydraulic parameters have been compared and contrasted for gymnosperms and angiosperms and between plants with different growth forms (trees versus shrubs versus lianas). But these comparisons were of limited generality since hydraulic parameters were known for only 10 taxa at the time of the review. Sample sizes (number of stems measured) and statistics were also very limited for the taxa reviewed. Data are now available for 28 taxa, and some of the more recent studies are far more complete from a statistical standpoint (Patiño, Tyree & Herre, 1995).

Figure 8.1 summarizes the known ranges of hydraulic architecture parameters in 28 taxa covering a range of growth forms and phylogenies. Comparisons of mean values between species are difficult because k_l, k_s and HV often change significantly with stem diameter, D. For example, k_l and k_s can be 10 to 100 times greater when measured in the bole of trees (D = 300 mm) than when measured in young branches (D = 3 mm). Sometimes differences between species in k_l or k_s measured at D = 6 mm may be reversed at D = 60 mm; furthermore, stem morphologies may be such that the smallest segments bearing leaves were 20 mm diameter in one species but just 3 mm in another. The mean values computed for Figure 1 were computed from regression values at D = 15 mm, except for 3 tropical species with large stems for which we used D = 45 mm.

The rationale for comparing hydraulic parameters of small branches is that the pressure drop in small branches (say, from 50

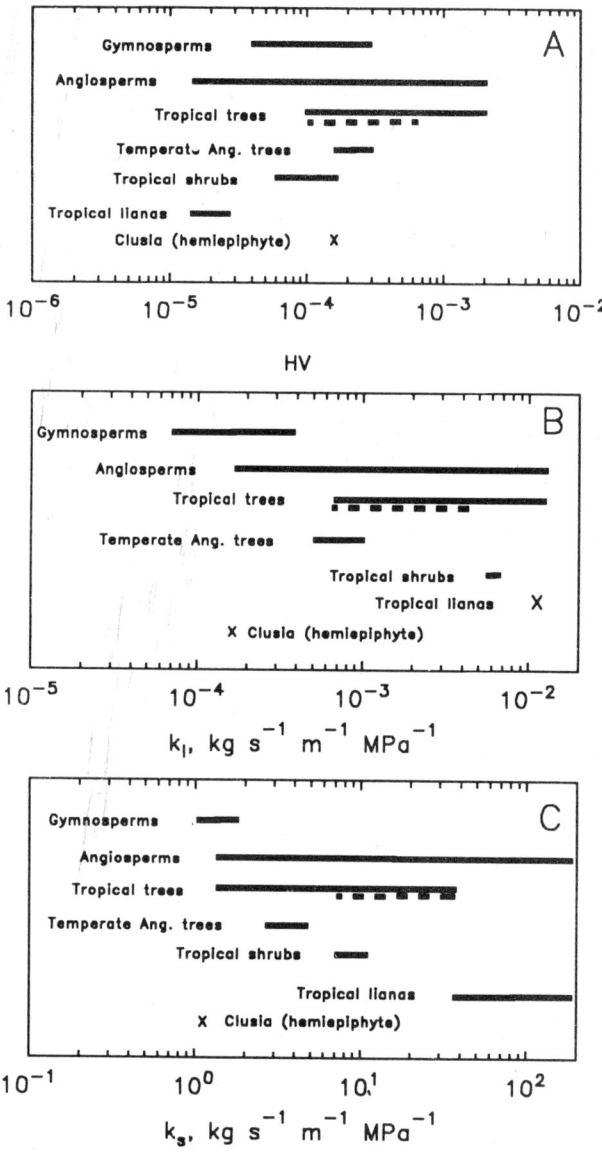

Figure 8.1. Ranges of hydraulic parameters (HV = Huber Value, k_l = leaf specific conductivity, k_s = specific conductivity) by phylogeny or growth form. Horizontal bars demark the ranges read from the bottom axis of each parameter. Among tropical trees, Ficus spp are indicated by a dashed line. Ranges too short to be represented by horizontal bars are indicated by 'X'.

mm stem diameter   the leaf base) accounts for more than half the overall pressure drop from the base of a large tree to the leaf, i.e., the main vascular resistance of shoots resides in small diameter branches.

8.2.1 Trees

The hydraulic architecture of rapidly-growing gap specialists has been well documented for *Schefflera morototoni* (Tyree et al., 1991) and for *Ochroma pyramidale* and *Pseudobambax septenatum* (Machado & Tyree, 1994). All species have unusually high k_l values due to high HV; the k_s values are comparable to those of temperate hardwoods, and greater than those in conifers. The hydraulic sufficiencies are 3 to 30 times higher than that of several temperate species. In the case of *Schefflera*, the dP/dx required to maintain midday transpiration rates is only 1.2 times that required to lift water against gravity in a 20 m tree. Rapid leader growth in many species is predicated on high hydraulic conductivity or low transpiration rates, which allow for maximum turgor potential in growing tips (Dixon et al., 1988), so high k_l values may be part of the necessary strategy for rapid growth in gap-specialists.

The hydraulic architecture of 7 species of *Ficus* has been studied (Patiño, Tyree & Herre, 1995); 3 species were free-standing and 4 were stranglers. The free-standing species presumably have a reliable water source once established as seedlings and must compete for light to grow into large trees. However, the hemiepiphytic species must be adapted to two distinct habitats (Holbrook & Putz, Chapter 13). When these species germinate high in the crown of tall trees, they trade higher light availability for much reduced water resources. Aerial roots gradually grow toward the ground over a number of years. When fully rooted, they have the advantage of both high light and water availability, with much less investment in biomass for support than required by the host tree to reach similar heights; essentially these species are structural parasites.

Patiño, Tyree & Herre (1995) found measurable differences in the hydraulic architecture of hemiepiphytic versus free-standing figs that might be related to challenges posed by the hemiepiphytic habit. Since hemiepiphytes invest less carbon in stems to establish and support themselves at great height, reduced values of k_l and HV were expected and, indeed, found. In stems 10–20 mm in diameter, the k_s were all higher (4 to 14 kg s^{-1} m^{-1} MPa^{-1}) than those of the gap-specialists (1.6 to 2.2), but the HV were all less. Consequently the k_l were in the same range as the gap specialists (5×10^{-4} to 30×10^{-4} kg s^{-1} m^{-1} MPa^{-1}), and thus much greater than in temperate trees.

8.2.2 Shrubs

The hydraulic architecture of shrubs has been studied less than the architecture of trees or lianas. Stems are somewhat less efficient per unit transport length than comparable diameter stems of trees. For instance, shrubs of *Bauhinia aculeata* and *B. galpinii* had k_l's about 50% less than in the trees *B. blakeana* and *B. variegata* (Ewers, Fisher & Fichtner, 1991). This was due to lower k_s values. Maximum vessel diameters were similar in the shrubs versus trees of *Bauhinia* (Ewers & Fisher, 1991), but the vessel frequency per mm^2 sapwood was about twice as great in the trees (Ewers, Fisher & Fichtner, 1991).

Transport distances from soil to leaves would normally be less in shrubs than in trees. Therefore, the requirements for transport efficiency per unit length of pathway should be proportionally reduced. An understory shrub would have reduced transpiration rates, further reducing requirements for transport efficiency (Mulkey & Wright, Chapter 7). The shrub *B. galpinii* showed an extremely variable relationship between stem xylem area and leaf area supplied by the stem (Ewers & Fisher, 1991). This variation in HV could be due to a plastic response to variation in light conditions. Perhaps plants grown under high light conditions produce more sapwood area per leaf, allowing for greater maximum transpiration rates.

8.2.3 Lianas

Lianas are a growth form in which external plants or objects provide mechanical support. As a result, the mechanical demands upon the xylem are much reduced compared to the situation in free-standing growth forms. However, the maximum transpiration and photosynthetic rates might be slightly greater for canopy lianas than for canopy trees (Teramura, Gold & Forseth, 1991; Ewers, Fisher & Fichtner, 1991; Holbrook & Putz, Chapter 13). This difference suggests that the hydraulic transport requirements are at least as great in canopy lianas as in trees. The situation in understory climbing species, which would be expected to have reduced transpiration demands, needs exploration.

8.2.3.1 Shoot hydraulic architecture of lianas

Compared to trees and shrubs, lianas tend to have lower HV's and higher k_s's. The higher k_s values in lianas are due to the fact that lianas have greater maximum vessel diameters in their stems (Ewers, 1985; Gartner et al., 1990; Gartner, 1991; Ewers and Fisher, 1991; Ewers, Fisher & Fichtner, 1991). The xylem flux density (water flow rate per xylem transverse area) in vines in Chamela, Mexico, was about 100 times greater than that reported for trunks of temperate trees (Fichtner & Schulze, 1990). However, the dP/dx values in liana stems appear to be similar to those reported for trees (Ewers, Fisher & Chiu, 1989; Ewers, Fisher & Fichtner, 1991).

The remarkable efficiency of the narrow liana stems can be attributed to a high ratio of sapwood to heartwood (Putz, 1983; Ewers, Fisher & Fichtner, 1991) and to vessels that are wider and often of a greater frequency (more vessels per transverse area) than in closely related trees or shrubs (Ewers, Fisher & Fichtner, 1991; Ewers & Fisher, 1991; Gasson & Dobbins, 1991). Recently we discovered vessels in stems of the "woody" monocotyledon vine *Smilax panamensis* to be up to 700 μm in diameter, which we believe is the largest vessel diameter ever reported in a plant (unpublished data). Gartner et al. (1990) found that in a tropical deciduous forest in Chamela, Mexico, lianas had significantly wider vessels and greater k_s values than the trees they were growing on. Within the genus *Bauhinia*, where liana species have higher k_s values than tree and shrub species, the wide vessels of lianas appear to hydraulically compensate for narrow stem diameters and low HV's (Ewers & Fisher, 1991; Ewers, Fisher & Fichtner, 1991).

Information is generally lacking on embolism in tropical lianas. In temperate species of the liana *Vitis*, however, there is evidence that the wide vessels of stems become 100% embolized in the winter, and are refilled via root pressures in the spring (Sperry et al., 1987). Examples of possible vessel refilling in tropical vines are discussed below.

8.3.2.2 Root hydraulic architecture and root pressures

An important issue is whether the root systems of lianas offer more or less resistance to water flow than does the shoot system. Equally relevant is the possible presence of "root pressures" in lianas, i.e., the osmotic water uptake caused by solute uptake into roots (Tyree et al.

1994). It is important to know whether embolized vessels of lianas are capable of being refilled by root pressures, or if the vessels, once embolized, remain permanently non-conductive. The water flow caused by root pressure is normally much less than that caused by transpiration-pull, and the osmotic force causing root pressure disappears during transpiration because the solutes are diluted by water drawn into the mesophyll. Root pressure effects are thus generally too small to measure, except when transpiration is minimum (at night or during rainstorms). The transition from push to pull has been demonstrated recently in a vinelike bamboo (Cochard, Ewers & Tyree, 1994). Other factors that would work against the root pressures being expressed in the stem are tissue water capacitance, possible lateral leakage of water, and gravity. The gravity gradient by itself dictates that there will be a loss of pressure of about 10 kPa per m of height in a standing water column. Four very different liana root systems are mentioned here to illustrate the variation in liana root morphology. This is followed by a discussion of root pressures in tropical climbing plants.

Enormous ligno-tubers of *Bauhinia fassoglensis* (Fabaceae, subfamily Caesalpinioideae) develop from the hypocotyl root axis. Tubers grow up to 2.5 m long and 0.5 m in width (Brenan, 1967), but no other large woody roots are produced. The widest root emerging from a tuber 1 m long and 0.25 m in diameter was 5 mm in xylem diameter; several stems up to 30 mm in xylem diameter and many narrow roots, most less than 1 mm in diameter, emerged from the tuber (Ewers & Fisher, unpublished data). The hydraulic capacitance of the tuber might play a major role in the water relations and resprouting capabilities of this species, as in other vines with large tubers (Mooney & Gartner, 1991). Modest root pressures develop in this species at predawn following periods of rain (Table 8.1).

A deep and extensive woody root system is illustrated by *Machaerium milleflorum* (Fabaceae, subfamily Papilionoideae). Excavations of several plants on Barro Colorado Island revealed that the total xylem transverse area of the woody roots was considerably greater than for the stem. One stem with a xylem diameter of 20 mm had four roots greater than 20 mm diameter emerging from the stem base, including a taproot of 50 mm diameter. The large roots were irregularly twisted and branched, reaching depths to 4.4 m. At 0.75 m from the root-stem junction, the total root transverse xylem area was almost 8X greater than for the stem (2438 mm^2 for 4 major roots plus 3 lateral roots versus 306 mm^2 for the stem). None of the individuals of this species exhibited positive stem xylem pressures during the wet or dry season (Ewers, Cochard & Tyree, submitted).

Table 8.1. Tropical plants capable of generating positive xylem pressure in the stem when shoots are excised.

Taxon	Max xylem $\Psi(kPa)$	Growth form	Source[1]
Schizaeaceae			
Lygodium venustrum	66	Vine	A
Poaceae			
Bambusa arundinaceae	200	Vine	B
Rhipidocladum racemiflorum	120	Vine	B
Saccharum spp.	300	Sugar cane	C
Fabaceae			
Bauhinia fassoglensis	5	Liana	D
Dilleniaceae			
Doliocarpus major	64	Liana	A
Tetracera hydrophilia	16	Liana	A
T. portabellensis	57	Liana	A

[1] A = Ewers, Cochard & Tyree (submitted)

B = Cochard, Ewers & Tyree (1994)

C = Meinzer et al. (1992)

D = Ewers, Fisher & Chiu (1989).

An extensive, widespreading root system is exhibited by *Tetracera hydrophilia* (Dilleniaceae). One stem 22 mm in xylem diameter had 10 major, mostly-horizontal, woody roots radiating from the stem base. These roots were 0.05 to 0.5 m deep and spread up to 7.5 m from the stem base. The horizontal roots had lateral woody sinker roots at about 1 m intervals. The deepest sinker root excavated penetrated 2.6 m. At 0.25 m from the root-stem junction, the ratio of root to stem xylem area was very close to 1:1 (483 mm^2 in the stem versus 475 mm^2 for all the roots). However, the root k_h at 0.25 m from the root-stem junction was almost 13 times greater than for the stem (58.9 x 10^{-3} versus 3.55 x 10^{-3} kg s^{-1} MPa^{-1} m). Modest root pressures developed in this species during both the wet and dry seasons (Table 8.1).

A shallow, fibrous root system is illustrated by *Rhipidocladum racemiflorum* (Poaceae), a viny bamboo that climbs up to 5 m in height along forest edges (Croat, 1978). The rhizomes branch frequently and several adventitious roots are associated with each aerial shoot. Positive root pressures of up to 120 kPa were recorded near the base of stems during the rainy season at night or during rainstorms, but not during the dry season (Cochard, Ewers & Tyree, 1994).

From the above examples and based upon a survey of root pressures at Barro Colorado Island, it appears that the build-up of positive root pressures is more likely with a shallow root system during the rainy season (e.g., *Rhipidocladum*) than with a deep rooting system during any time of the year (e.g., *Machaerium milleflorum*). Based upon the massive root system of *M. milleflorum* and the extensive, spreading root system of *Tetracera hydrophilia*, it appears as though the woody root system of some lianas may well offer less resistance to water flow than the woody shoot system. However, studies of the conductivity of intact root systems of vines are needed to confirm this.

It has been suggested that the wide vessels of tropical lianas remain conductive for many years due to positive root pressures that might refill embolized stem vessels on a diurnal or seasonal basis (Putz, 1983). Although there is anecdotal evidence that some tropical lianas exhibit positive root pressures, we know of only eight taxa of tropical plants where positive xylem pressures have been measured (Table 8.1). Of these, only the leaf-climbing fern, *Lygodium venustrum*, and three members of the Poaceae had root pressures of a magnitude such that they might be expected to refill embolized conduits throughout the shoots. The root pressures in the lianas in Table 8.1 (*Bauhinia fassoglensis* and three members of the Dilleniaceae) were not sufficient to refill vessels in the upper stems. For instance, the three members of the Dilleniaceae had stems to 18 m in height, but the positive pressures of up to 64 kPa were measured at 0.5 m above the soil level. Theoretically, the positive pressures would be dissipated by the gravity gradient in going up the stems and refilling of embolized vessels would not occur higher than 1.1 m above the ground.

In a survey of root pressures of vines of Barro Colorado Island, measurements were made on 65 individuals of 32 species of climbing plants, including 20 genera from 13 families (Ewers, Cochard & Tyree, submitted). None of the sampled 29 species of dicotyledonous vines exhibited root pressures that would be adequate to refill embolized vessels in canopy stems. The more likely scenario is that once the vessels of such vines become embolized, they remain permanently non-conductive.

As noted previously, wide vessels appear to be a crucial part of the water relations "strategy" of lianas. Since wide vessels in general are quite prone to freezing-induced embolism (Ewers, 1985; Cochard & Tyree, 1990; Sperry & Sullivan, 1992), and since in lianas the stem embolisms appear to be permanent, freezing could well help to limit the distribution of many tropical lianas. This would explain why there is such a strong inverse correlation between latitude and liana abundance (Gentry, 1991)

In contrast to dicotyledonous lianas, the climbing fern *Lygodium venustrum* and the viny bamboo *Rhipidocladum racemiflorum* exhibited root pressures that might be adequate to refill embolized conduits throughout the entire plant. For example, *Rhipidocladum racemiflorum* climbed to just 4.5 m, and Ψ_{xp} of up to 120 kPa were found near the stem base (Table 8.1). In addition, we found positive Ψ_{xp} even at the distal-most part of stems in this species (Cochard, Ewers & Tyree, 1994).

8.2.4 Hemiepiphytes

Some hemiepiphytes have shoots of the stature of large shrubs (up to 5 m tall) with aerial roots > 30 m and are dependent permanently on their host for support (e.g., *Clusia uvitana*), whereas others eventually become free-standing trees (e.g., several species of *Ficus*; Holbrook & Putz, Chapter 13). *Clusia* is intermediate in hydraulic architecture between a tree and a liana (Zotz, Tyree & Cochard, 1994). Aerial roots had very high k_s values (13 to 27 kg s^{-1} m^{-1} MPa^{-1}), but values for stems from the crown are 10 times less and similar to the values for many other tropical and temperate trees. HV and k_l values are low, but transpiration rates are also very low because *Clusia* has a C_3-CAM metabolism. The low transpiration and low k_l values combine to yield computed dP/dx values of *Clusia* that are comparable to that in the crowns of other tropical species. In the areal roots, dP/dx values are low and comparable to that of *Schefflera*. The hydraulic architecture of strangler figs (*Ficus* spp) is similar to that of other free-standing trees. To date, no studies have been done on strangler figs in the epiphytic stage.

8.3 CONCLUSIONS REGARDING HYDRAULIC ARCHITECTURE

Much more work has been done on hydraulic architecture of tropical plants since the review of Tyree and Ewers (1991), so it is appropriate to review the previous generalizations in light of these new data. One generalization that still applies to all woody plants is that k_l tends to decrease from base to apex, except in trees with strong apical dominance, where k_l may remain relatively constant or increase towards the dominant apex. A decline of k_l towards the apex permits a more or less equal competition for water in the crowns of trees without apical dominance. Other generalizations about k_s and k_l values are becoming less distinct: (1) Gymnosperms appeared previously to be unique in having lower k_s values than angiosperms because tracheids are smaller and less conductive than vessels. There are now examples of vessel bearing trees with k_s values comparable with gymnosperms. One example is *Clusia*, which has small diameter vessels and moderate vessel densities (vessels per unit cross section); other examples are rapidly-growing gap-specialist trees which have large vessels but extremely low densities, making k_s small. (2) Previously, angiosperms and gymnosperms fell into distinct classes in terms of ranges of k_l. Now, the ranges overlap with the inclusion of woody CAM species (*Clusia*).

Are there any interspecific patterns or generalizations expressed in terms of hydraulic parameters? It is difficult to be definitive with a data base of only 28 taxa, but we can make a tentative proposal. There is a rough correlation between k_l value and evaporative flux density (E) and, perhaps, plant size. Gymnosperms and *Clusia* have low E and k_l values, whereas many tropical species have high k_l and high E values. The results for *Ficus* were consistent with this trend. Hemiepiphyte species tend to be more conservative water spenders than free-standing species (especially when the hemiepiphytes are not rooted), and this lower E correlated with lower k_l in hemiepiphytes than in free-standing figs (Holbrook & Putz, Chapter 13.

The causal connection to explain this observation may be the role of k_l in determining the overall pressure potential (or water-potential) drop in the vascular system. E/k_l determines the pressure gradient in the vascular system so the overall pressure drop is given by

$$\Delta P = LE/k'_l, \tag{3}$$

where L is the path length from the base to the apex of a plant and k'_l is the average k_l over the path length. Plants with long path length and/or high E must have high k'_l to keep P within reasonable limits. Such plants must maintain relatively high water potentials either because they are very vulnerable to cavitation or because they must maintain high turgor pressure to grow rapidly. Another example might be very long plants, such as lianas, that can grow to be several-hundred meters long (Putz & Mooney, 1991). The high k_l values of lianas permits high E values with minimal pressure gradients in their long stems. This generalization would be strengthened if we had more examples of plants with low E. So far we have data on only three conifers and *Clusia*. The test of the generalization must await more studies of plants that grow in moisture-limited environments, e.g., epiphytes, CAM species, or chaparral shrubs.

There may also be a general relationship between k'_1 and plant size; the relationship certainly holds within a species. Yang and Tyree (1994) found that the absolute conductance per unit leaf area of whole maple shoots was nearly independent of plant size from saplings 1 m tall to young trees 12 m tall. Absolute conductance is defined here as $k_A = k'_1/L$ so k'_1 must increase in proportion to L for k_A to be constant. Short species may have lower k'_1 values than tall or long species. Preliminary unpublished data we have obtained on small woody plants and herbs support this hypothesis.

8.4   VULNERABILITY TO XYLEM DYSFUNCTION BY CAVITATION

It is presumed by many that there is a trade-off between large conduits with high hydraulic efficiency and the vulnerability of xylem to dysfunction (Zimmermann, 1983). This presumption is borne out of many anatomical surveys using both systematic and floristic approaches. These studies have shown the general trend in conduit diameter shown in Figure 8.2. Wet, warm environments tend to favor species with wide conduits, whereas cold or dry environments tend to favor species with narrow conduits. Space does not permit a full review of the vulnerability to cavitation in tropical plants; readers are referred to a recent review article (Tyree, Davis & Cochard, 1995) for details.

The vulnerability curves (VC) for a number of species are reproduced in Figure 8.3. These species represent the range of vulnerabilities observed so far, i.e., 50 percent loss k_h occurring at Ψ_{xp} values ranging from -0.7 to -11 MPa. We have access to VC data for about 60 species, representing plants from many different climates (temperate, Mediterranean, moist tropical, desert) and several growth forms (grasses, vines, trees, & shrubs). In order to assess the relationship between conduit size and vulnerability to cavitation, Tyree et al. (1994) plotted ψ_{50} = the drought-induced Ψ_{xp} needed to cause 50% loss k_h versus the average vessel diameter responsible for 95% of the hydraulic conductance (D_{95}). This relationship is reproduced in Figure

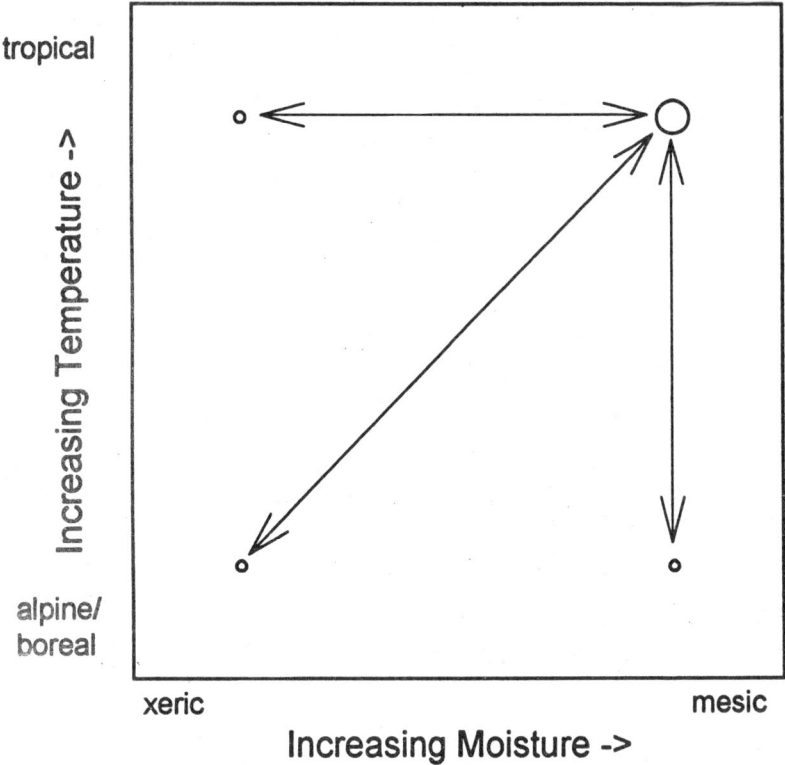

Figure 8.2. Diagram showing ecological gradients in temperature and moisture and the direction of evolution of vessel diameter in response to these gradients. The double-headed arrows indicate that plants evolve in both directions along the indicated gradients.

8.4. A log-log transform was used to look for both linear and nonlinear correlations. The regression line is shown in bold, and dotted lines are 95% confidence intervals. The weak, but statistically significant, regression accounts for only 21% of the variation. The weak correlation may be of use to evolutionary biologists, but is not of sufficient accuracy to be of predictive value to a comparative physiologist, i.e., a physiologist cannot predict the vulnerability of a species by measuring the mean conduit diameter.

We can classify tropical species into two categories; drought-evaders and drought-tolerators (Zotz, Tyree, & Cochard, 1994). The drought-evaders evade drought (low ψ_{xp} and high percent loss k_h) by having deep roots and a highly conductive hydraulic system. Taxa

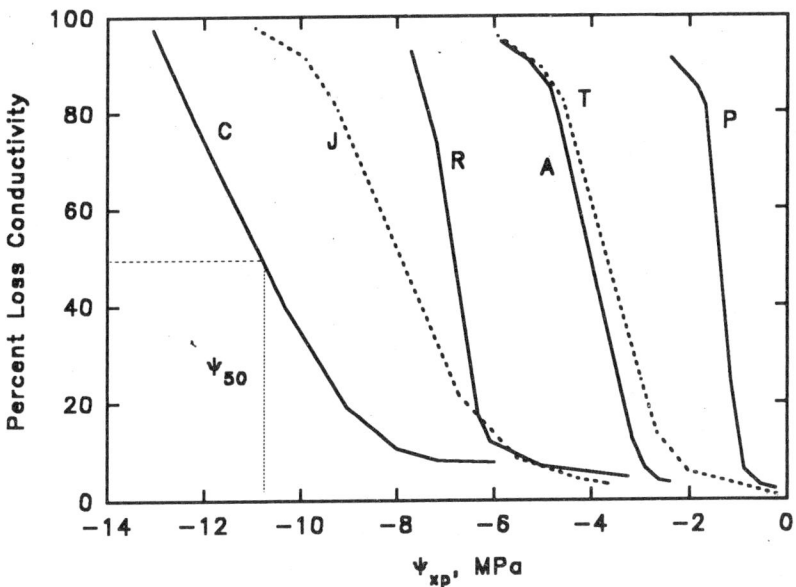

Figure 8.3. Vulnerability curves for various species. Y-axis is percent loss of hydraulic conductivity induced by the xylem pressure potential, ψ_{xp}, shown on the x-axis. C = Ceanothus megacarpus, J = Juniperus virginiana, R = Rhizophora mangle, A = Acer saccharum, T = Thuja occidentalis, P = Populus deltoides.

that fall into this category are *Ochroma* and *Schefflera* and perhaps *Ficus* and *Ouratea*. Alternatively they evade drought by being deciduous (*Pseudobombax*). The other species frequently reach very negative ψ_{xp} for part or all of the year and are shallow rooted or grow in saline environments; taxa in this category are *Rhipidocladum*, *Rhizophora* and *Psychotria*. Taxa that are of doubtful classification are *Cordia* and *Ficus* because we know nothing about rooting depths or typical dry season values of ψ_{xp}.

The large vessels of tropical species probably are not adapted to cold environments. In fact, freezing-induced xylem embolism could limit the geographic distribution of many tropical species. Conversely, why do conifers dominate in boreal and alpine habitats? There is a growing body of evidence that small conduits are less vulnerable than large conduits to freezing-induced embolism. Why should large conduits be more prone to freeze-induced dysfunction? This tendency probably has something to do with how long it takes air bubbles to dissolve rather than the tension when the ice first forms. This is because bubbles have

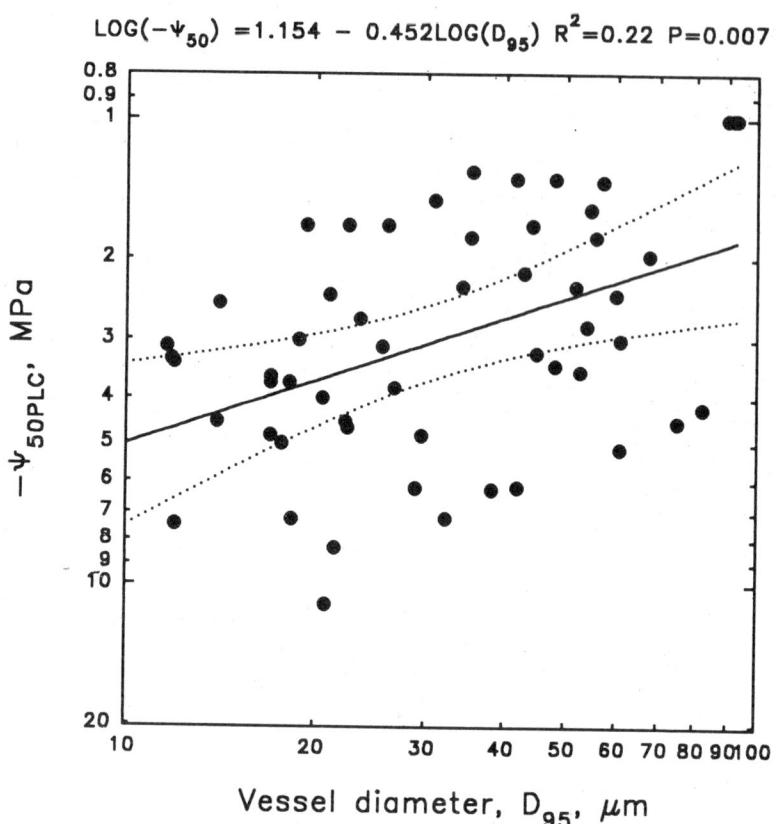

Figure 8.4. Log-log plot of xylem tension causing 50 percent loss of hydraulic conductivity (-ψ_{50PLC}) and mean diameter of the vessels that account for 95% of the hydraulic conductance (D_{95}).

to dissolve before the onset of a critical tension causing them to expand. The physics of air bubble dissolution is now well understood (Pickard, 1989; Yang & Tyree, 1992; Tyree & Yang, 1992). An analysis of the kinetics of bubble dissolution reveals that the time it would take for a bubble to dissolve increases approximately with the square of its initial diameter. If many small bubbles were formed when ice melted and if the bubbles were the same size regardless of the size of the conduit, then conduit size may not influence freezing-induced dysfunction. But Ewers (1985) studied bubble formation while freezing and thawing water in small glass capillary tubes and observed that bigger bubbles formed in large diameter tubes than in small tubes, and they took longer to redissolve in big versus small tubes. It seems likely that the same will happen in xylem conduits.

Hammel (1967) and Sucoff (1969) have shown that conifer stem segments can be frozen and thawed while the rest of the shoot is warm and transpiring without permanent loss of hydraulic conductivity, whereas the first frost event is known to induce > 90% loss k_h in wide-vessel oaks (Cochard & Tyree, 1990; Sperry & Sullivan, 1992) and some wide vessel chaparral shrubs (Langan & Davis, 1994).

Freezing should induce embolisms because air is not soluble in ice. So when water freezes, air comes out of solution. If water is saturated with air at 0 °C when it freezes, then approximately 28 ml of air will come out of solution for every liter of water frozen. What happens to this air when the ice melts? If the ice melts slowly and no tension develops in the tissue, then the air will dissolve. But if tensions develop beyond the critical tension given by the size of the bubble and the surface tension, then the bubbles will expand to make the conduit fully embolized and dysfunctional. Apparently this happens most in oak with vessels 100 μm in diameter, less in maple with vessels 30 μm in diameter, and least in conifers, with tracheids 10 μm diameter. Sperry and Sullivan (1992) have demonstrated a strong correlation with vulnerability to freezing-induced embolism and conduit volume. Figure 8.5 compares the ψ_{50} values (open circles) of 5 species versus conduit volume or specific conductivity; it can be seen that there is no trade-off between conduit size and drought-induced dysfunction (as stated previously), nor is there a trade-off between increased stem hydraulic efficiency and drought-induced dysfunction.

The solid circle (ψ_{50F}) values were obtained in the following way: shoots were dehydrated to various initial water potentials, ψ_i, and then wrapped in a bag, frozen to − 20 °C overnight, and then thawed to > 20 °C. This procedure induced a certain percent loss k_h. What is plotted is the ψ_i that induced 50% loss k_h after the freeze-thaw cycle. While there is a strong correlation, it is not clear why ψ_i measured before a freeze should influence percent loss k_h after a freeze-thaw cycle.

The ψ_i value *could not* influence the amount of air coming out of solution upon freezing—the same amount of air will come out regardless of the initial conditions because the solubility of air in ice is very low compared to water at 0 °C. Since the samples frozen were not frost tolerant, many living cells must have died; this would alter ψ_i an unknown amount, but certainly the value of ψ_{xp} will have risen from a value initially close to ψ_i to something much less negative. It is the magnitude of ψ_{xp} after the freeze and the capacity of the previously

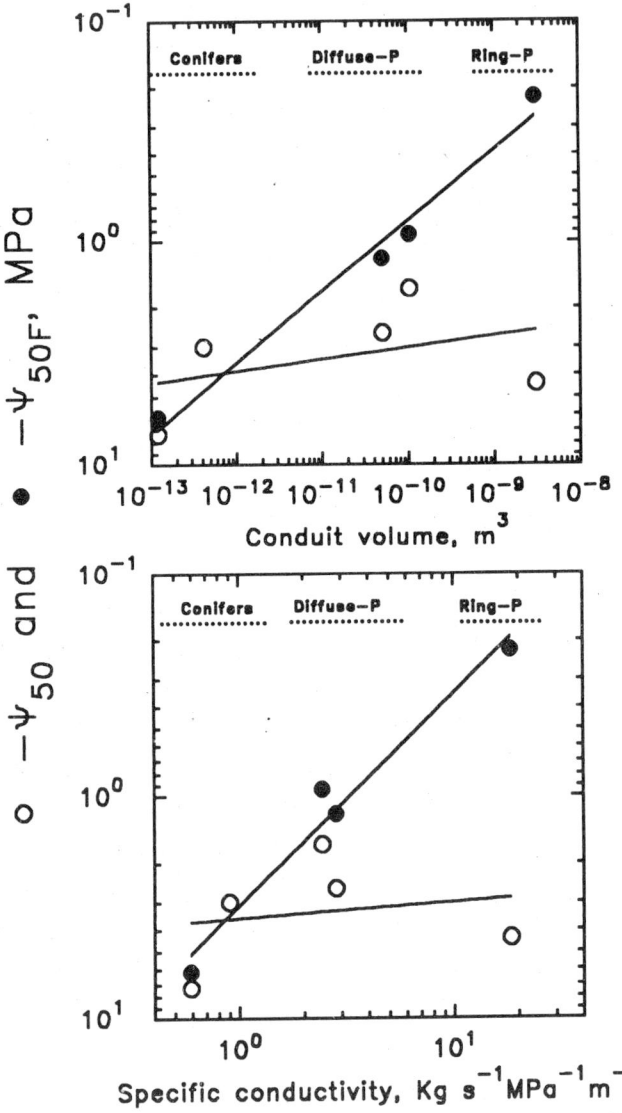

Figure 8.5.   Relationship between xylem tension and loss of conductivity. Open circles are tension (negative ψ_{xp}) needed to induce 50% loss $k_h = \psi_{50}$, closed 8circles are tension before a freeze required to induce 50% loss $k_h = \psi_{50F}$. The upper plot shows the relationship with conduit volume and the lower plot shows the relationship with specific conductivity. The lower curve suggests a trade-off between vulnerability to freezing-induced loss of conductance and specific conductivity of wood. (Replotted from Sperry & Sullivan, 1992).

frozen tissue to take up water from conduits after the freeze that will determine how much the embolisms expand and how much percent loss k_h results. If no air dissolved immediately after the thaw, only 2.8% of the tissue volume would be occupied by air (= the volume fraction of air in solution at the time of the freeze). Only if these bubbles are expanded to fill the entire conduits would we expect values of 50% loss of k_h or more.

8.5 FUTURE RESEARCH

From the above, the reader can see that there is a rapidly growing literature dealing with the hydraulic architecture of different growth forms in the tropics. A fair amount is known about the relationships between conduit diameter and air embolism in the xylem. Topics of particular interest for future research on tropical plants include the relative importance of fine root, woody root, and whole shoot conductance. Similarly, the impact of mycorrhizae on root hydraulic conductance is a crucial area for future investigation. An integrated view of hydraulic architecture will emerge from studies of root and shoot conductance during different parts of the life cycle, and comparative studies of early, middle, and late successional plants. We anticipate that xylem ecotypes will be discovered among widely distributed species.

REFERENCES

BECK, E., SCHEIBE, R. & HANSEN, J. (1987) Mechanisms of freezing avoidance and freezing tolerance in tropical alpine plants. *Plant Cold Hardiness* (ed. P. H. LI) ALAN R. LISS, New York.

BORGHETTI, M., EDWARDS, W. R. N., GRACE, J., JARVIS, P. G. & RASCHI A. (1991) The refilling of embolized xylem in *Pinus sylvestris L. Plant, Cell and Environment*, **14**, 357–369.

BRENAN, J. P. M. (1967) *Flora of Tropical East Africa. Part 2, Caesalpinioideae.* Crown Agents for Oversea Governments and Administrations, London, p 215.

CALKIN, H. W., GIBSON, A. C. & NOBEL, P. S. (1986) Biophysical model of xylem conductance in tracheids of the fern *Pteris vittata. Journal of Experimental Botany*, **37**, 1054–1064.

CHIU, S.-T. & EWERS, F. W. (1992) Xylem structure and water transport in a twiner, a scrambler, and a shrub of *Lonicera* (Caprifoliaceae). *Trees*, **6**, 216–224.

CHIU, S.-T. & EWERS, F. W. (1993) The effect of segment length on conductance measurements in Lonicera fragrantissima. Journal of Experimental Botany, **44**, 175–181.

COCHARD, H. & TYREE, M. T. (1990) Xylem dysfunction in Quercus: Vessel sizes, tyloses, cavitation, and seasonal changes in embolism. Tree Physiology, **6**, 393–407.

COCHARD, H., CRUIZIAT, P. & TYREE, M. T. (1992) Use of positive pressures to establish vulnerability curves: Further support for the air-seeding hypothesis and possible problems for pressure-volume analysis. Plant Physiology, **100**, 205–209.

COCHARD, H., EWERS, F. W. & TYREE, M. T. (1994) Water relations of a tropical vinelike bamboo (Rhipidocladum racemiflorum): Root pressures, vulnerability to cavitation, and seasonal changes in embolism. Journal of Experimental Botany, **45**, 1085–1089.

CROAT, T. B. (1978) Flora of Barro Colorado Island. Stanford University Press, Stanford, California.

CROMBIE D. S., HIPKINS, M. F. & MILBURN, J. A. (1985) Gas penetration of pit membranes in the xylem of Rhododendron as the cause of acoustically detectable sap cavitation. Australian Journal of Plant Physiology, **12**, 445–453.

DIXON, M. A., BUTT, J. A., MURR, D. P. & TSUJITA, M. J. (1988) Water relations of cut greenhouse roses: The relationship between stem water potential hydraulic conductance and cavitation. Scientific Horticulture, **63**, 109–118.

EDWARDS, W. R. N., JARVIS. P. G., GRACE, J. & MONCRIEFF, J. B. (1994) Reversing cavitation in tracheids of Pinus sylvestris L. under negative water potentials. Plant, Cell and Environment, **17**, 389–397

EWERS, F. W. (1985) Xylem structure and water conduction in conifer trees, dicot trees, and lianas. International Association of Wood Anatomists Bulletin, **6**, 309–317.

EWERS, F. W. & CRUIZIAT, P. (1990) Measuring water transport and storage. Techniques and Approaches in Forest Tree Ecophysiology (eds. J. P. LASSOIE & T. M. HINCKLEY) CRC Press, Boca Raton, Florida, 91–115.

EWERS, F. W. & FISHER, J. B. (1991) Why vines have narrow stems: Histological trends in Bauhinia (Fabaceae). Oecologia, **88**, 233-237.

EWERS, F. W., COCHARD, H. & TYREE, M. T. (submitted) A survey of root pressures in vines of a tropical lowland forest. Oecologia.

EWERS, F. W., FISHER, J. B. & CHIU, S. -T. (1989) Water transport in the liana Bauhinia fassoglensis (Fabaceae). Plant Physiology, **91**, 1625–1631.

EWERS, F. W., FISHER, J. B. & FICHTNER, K. (1991) Water flux and xylem structure in vines. The Biology of Vines (eds. F. E. PUTZ and H. A. MOONEY) Cambridge University Press, Cambridge, 127–160.

FICHTNER, K. & SCHULZE, E. D. (1990) Xylem flow in tropical vines as measured by a steady state heating method. Oecologia, **82**, 355–361.

GARTNER, B. L. (1991) : em hydraulic properties of vines vs. shrubs of western poison oak, *Toxicodendron diversilobum. Oecologia,* **87,** 180–189.

GARTNER, B. L., BULLOCK, S. H., MOONEY, H. A., BROWN, V. B. & WHITEBECK, J. L. (1990) Water transport properties of vine and tree stems in a tropical deciduous forest. *American Journal of Botany,* **77,** 742–749.

GASSON, P. & DOBBINS, D. R. (1991) Wood anatomy of the Bignoniaceae, with a comparison of trees and lianas. *International Association of Wood Anatomists Bulletin,* **12,** 389–417.

GENTRY, A. H. (1991) The distribution and evolution of climbing plants. *The Biology of Vines* (eds. F. E. Putz & H. A. Mooney) Cambridge University Press, Cambridge, pp 3–49.

HAMMEL, T. D. (1967) Freezing of xylem sap without cavitation. *Plant Physiology* **42,** 55–66.

HARGRAVE, K. R., KOLB, K. J., EWERS, F. W. & DAVIS, S. D. (1994) Conduit diameter and drought-induced embolism in *Salvia mellifera* (Labiatae). *New Phytologist,* **126,** 695–705.

JARBEAU, J., EWERS, F. W. & DAVIS, S. D. (1994) The mechanism of xylem dysfunction in two chaparral shrubs. *Plant, Cell and Environment,* in press.

KROG, J. O., ZACHARIASSEN, K. E., LARSEN, B. & SMIDSROD D. (1979) Thermal buffering in Afro-alpine plants due to nucleating agent-induced water freezing. *Nature,* **282,** 300–301.

LANGAN, S. J. & DAVIS, S. D. (1994) Xylem dysfunction caused by freezing and water stress in two species of co-occurring chaparral shrubs. *Bulletin of the Ecological Society of America,* in press.

LEWIS, A. M., HARNDEN, V. D. & TYREE, M. T. (1995). Collapse of water-stress emboli in the tracheids of *Thuja occidentalis* L. *Plant Physiology,* in press

LO GULLO, M. A. & SALLEO, S. (1991) Three different methods for measuring xylem cavitation and embolism: A comparison. *Annals of Botany,* **67,** 417–424.

MACHADO, J.-L. & TUREE, M. T. (1994) Patterns of hydraulic architecture and water relations of two tropical canopy trees with contrasting leaf phenologies: *Ochroma pyramidale* and *Pseudobombax septenatum. Tree Physiology,* **14,** 219–240.

MEINZER, F. C., GOLDSTEIN, G., NEUFELD, H. S., GRANTZ, D. A. & CRISOSTO, G. M. (1992) Hydraulic architecture of sugarcane in relation to patterns of water use during plant development. *Plant Cell and Environment.* **15,** 471-477.

MOONEY, H. A. & GARTNER B. L. (1991) Reserve economy of vines. *The Biology of Vines* (eds. F. E. PUTZ & H. A. MOONEY) Cambridge University Press, Cambridge, pp 166–179.

PATIÑO, S., TYREE, M. T., & HERRE, E. A. (1995) Comparison of hydraulic architecture of woody plants of differing phylogony and growth form with special reference to free-standing and hemiepiphytic *Ficus* species from Panama. *New Phytologist*, **129**, 125–134.

PICKARD, W. F. (1989) How might a tracheary element which is embolized by day be healed by night? *Journal of Theoretical Biology*, **141**, 259–279.

PUTZ, F. E. (1983) Liana biomass and leaf area of a "tierra firme" forest in the Rio Negro Basin, Venezuela. *Biotropica*, **15**, 185–189.

PUTZ, F. W. & MOONEY, H. A. (1991) *The Biology of Vines*. Cambridge University Press, Cambridge.

SALLEO, S. & LO GULLO, M. A. (1986) Xylem cavitation in nodes and internodes of whole *Chorisia insignis* H. B. and K. plants subjected to water stress: Relations between xylem conduit size and cavitation. *Annals of Botany*, **58**, 431–434.

SALLEO, S. & LO GULLO, M. A. (1989) Xylem cavitation in nodes and internodes of *Vitis vinifera* L. plants subjected to water stress. Limits of restoration of water conduction in cavitated xylem conduits. *Structural and Functional Responses to Environmental Stresses* (eds. K. H. KEEB, H. RICHTER & T. M. HINCKLEY) SPB Academic Publishing, The Hague, The Netherlands.

SPERRY, J. S. & SULLIVAN, J. E. M. (1992) Xylem embolism in response to freeze-thaw cycles and water stress in ring-porous, diffuse-porous, and conifer species. *Plant Physiology*, **100**, 603–613.

SPERRY, J. S. & TYREE, M. T. (1988) Mechanism of water stress-induced xylem embolism. *Plant Physiology*, **88**, 581–587.

SPERRY, J. S. & TYREE, M. T. (1990) Waterstress-induced xylem embolism in three species of conifers. *Plant, Cell and Environment*, **13**, 427–436.

SPERRY, J. S., DONNELLY, J. R. & TYREE, M. T. (1987) A method for measuring hydraulic conductivity and embolism in xylem. *Plant, Cell and Environment*, **11**, 35–40.

SPERRY, J. S., HOLBROOK, N. M., ZIMMERMANN, M. H. & TYREE, M. T. (1987) Spring filling of xylem vessels in wild grapevine. *Plant Physiology*, **83**, 414–417.

SUCOFF, E. (1969) Freezing of conifer xylem sap and the cohesion-tension theory. *Physiologia Plantarum*, **22**, 424–431.

TERAMURA, A. H., GOLD, W. G. & FORSETH, I. N. (1991) Physiological ecology of mesic, temperate woody vines. *The Biology of Vines* (eds. F. E. PUTZ & H. A. MOONEY) Cambridge University Press, Cambridge.

TYREE, M. T. & DIXON, M. A. (1986) Water stress-induced cavitation and embolism in some woody plants. *Physiologia Plantarum*, **66**, 397–405.

Tyree, M. T. and Ewers, F. W. (1991) The hydraulic architecture of trees and other woody plants. *New Phytologist*, **119**, 345–360.

TYREE, M. T., DAVIS, S. D., & COCHARD, H. (1995). Biophysical perspectives of xylem evolution: Is there a trade-off of hydraulic efficiency for vulnerability to dysfunction? *IAWA Journal*, **15**, in press.

TYREE, M. T. & SPERRY, J. S. (1989) The vulnerability of xylem to cavitation and embolism. *Annual Review of Plant Physiology and Molecular Biology,* **40,** 19–38.

TYREE, M. T. & YANG, S. (1990) Water-storage capacity of *Thuja, Tsuga,* and *Acer* stems measured by dehydration isotherms: Contributions of capillary water and cavitation. *Planta,* **182,** 420–426.

TYREE, M. T. & YANG, S. (1992) Hydraulic conductivity recovery versus water pressure in xylem of *Acer saccharum. Plant Physiology,* **100,** 669–676.

TYREE, M. T., PATIÑO, S., BENNINK, J., & ALEXANDER, J. (1995). Dynamic measurements of root hydraulic conductance using a high-pressure flowmeter for use in the laboratory and field. *Journal of Experimental Botany,* in press.

TYREE, M. T., SNYDERMAN, D. A., WILMOT, T. R., MACHADO J.-L. (1991) Water relations and hydraulic architecture of a tropical tree (*Schefflera morototoni*): Data, models, and a comparison with two temperate species (*Acer saccharum* and *Thuja occidentalis*). *Plant Physiology,* **96,** 1105–1113.

TYREE, M. T., YANG, S., CRUIZIAT, P., & SINCLAIR, B. (1994) Novel methods of measuring hydraulic conductivity of tree root systems and interpretation using AMAIZED: A maize-root dynamic model for water and solute transport. *Plant Physiology,* **104,** 189–199.

YANG, S. & TYREE, M. T. (1992) A theoretical model of hydraulic conductivity recovery from embolism with comparison to experimental data on *Acer saccharum. Plant, Cell and Environment,* **15,** 633–643.

YANG, S. & TYREE, M. T. (1994) Hydraulic architecture of *Acer saccharum* and *A. rubrum:* Comparison of branches to whole trees and the contribution of leaves to hydraulic resistance. *Journal of Experimental Botany,* **45,** 179–186.

ZIMMERMANN, M. T. (1983) *Xylem Structure and the Ascent of Sap.* Springer-Verlag, Berlin.

ZOTZ, G., TYREE, M. T. & COCHARD, H. (1994) Hydraulic architecture, water relations and vulnerability to cavitation of *Clusia uvitana:* A C_3-CAM tropical hemiepiphyte. *New Phytologist,* **127,** 287–295.

9

Evaluating Aspects of Water Economy and Photosynthetic Performance with Stable Isotopes from Water and Organic Matter

Guillermo Goldstein, Frederick C. Meinzer,
Leonel da Silveira Lobo Sternberg, Paula Jackson,
Jaime Cavelier, and N. Michele Holbrook

This chapter focuses on stable isotopes as a tool for investigating aspects of water economy and photosynthetic performance of tropical forest plants. We discuss stable isotope ratios in water as an index of spatial and temporal patterns of soil water utilization, and stable carbon isotope ratios as an integrated measure of internal conditions influencing carbon assimilation. These methods may help bridge the gap between instantaneous observations of plant responses and information obtained at extended temporal and spatial scales, and help us understand the contribution of different patterns of physiological regulation to processes evident at the whole organism, community, or ecosystem level.

Semideciduous tropical lowland forests, characterized by seasonal variations in rainfall and soil water and the presence of both evergreen and drought deciduous trees, are the most extensive of the world's lowland tropical forests. Most of the examples discussed here are taken from the forest on Barro Colorado Island (BCI, Smithsonian

Tropical Research Institute) in Panama where the dry season lasts three months. Water relations characteristics of trees in tropical savannas and deciduous forests, where the dry season may last five or six months, have been studied extensively (e.g., Goldstein & Sarmiento, 1987; Medina, 1982; Meinzer, Seymour & Goldstein, 1983; Sarmiento, Goldstein & Meinzer, 1985), and when appropriate they will be discussed for comparative purposes. The water economy of woody plants in lowland tropical forests where pronounced transient and seasonal water deficits occur frequently even though the dry season is shorter, is less well known (e.g., Oberbauer, Strain & Riechers, 1987; Mulkey, Wright & Smith, 1991; Mulkey, Wright & Smith, 1993).

We first consider the belowground aspects of tropical plants by focusing on water uptake by tropical plant root systems. We focus the second part of our discussion on aboveground processes, in particular, photosynthesis and transpiration. Although root systems can differ in many aspects of their growth dynamics, architecture, and function, the root biomass distribution with depth will ultimately determine the type of water source utilized. On the one hand, the growth of shallow roots involves a smaller investment of reduced carbon than does production of deep roots, but the former tap shallow water sources that fluctuate seasonally. On the other hand, deeper roots require a greater investment of reduced carbon but can access deeper water sources that are more stable throughout the year. Here we examine how root systems in a semideciduous forest partition soil water resources in time and space, and thus how they may minimize competition for water. The current dogma that roots of plants from tropical rainforests mainly exploit the uppermost few centimeters of the soil profile is discussed.

Leaves also vary considerably in their physical and physiological properties, particularly in their photosynthetic capacity (e.g., Chazdon et al., Chapter 1). Two components determine the intrinsic photosynthetic capacity of plants: the stomatal restriction of CO_2 movement to the site of carboxylation, and the biochemical capacity to transform CO_2 into organic products. Higher stomatal conductances will maximize the photosynthetic capacity of leaves, but may result in a high cost in terms of transpirational water loss, particularly if the water supply is limited because of low soil water availability or high resistance to water transport (Tyree et al., 1991; Mulkey & Wright, Chapter 7). Some of the mechanisms by which changes in photosynthetic CO_2 uptake (A) and stomatal conductance (g) regulate instantaneous and long-term water-use efficiency of canopy trees, high light demanding gap species, and understory plants are discussed.

9.1   BACKGROUND ON STABLE ISOTOPES

Most elements of biological and ecological interest have two or more stable (non-radioactive) isotopes (Ehleringer & Rundel, 1989). Of interest in this review are isotopes of carbon, hydrogen, and oxygen. Carbon-12, hydrogen, and oxygen-16 are lighter isotopes, while carbon-13, deuterium (^2H) and oxygen-18 are heavier. The lighter isotopes are present in the biosphere in far greater abundance (98.89% for carbon, 99.98% for hydrogen and 99.76% for oxygen) than the heavier ones. Because of the difficulties in directly measuring the abundance of stable isotopes, their abundance is usually measured relative to that in a standard (std) and expressed in delta (δ) units,

$$\delta X = [(R_{sample}/R_{std}) - 1]1000$$

where δX is the isotope ratio in delta units on a "per mil basis" (parts per thousand) relative to the standard, and R is the ratio of minor (in terms of absolute abundance) to major isotope measured for both the sample and the standard. $\delta^{13}C$ values are traditionally expressed relative to carbon dioxide derived from limestone of the Pee Dee Belemnite formation, whereas δD and $\delta^{18}O$ values are expressed relative to SMOW (Standard Mean Ocean Water) or V-SMOW (Vienna-SMOW) (White, 1989; Dawson, 1993a).

The stable carbon isotope composition of plant tissue ($\delta^{13}C$) is determined both by the isotopic composition of the CO_2 source, and by discrimination against the heavier isotope ^{13}C during photosynthetic CO_2 assimilation. $\delta^{13}C$ values of bulk atmospheric carbon dioxide are about $- 8.00‰$, which is largely in equilibrium with the carbonate pool of ocean water (Keeling, 1961). Because the biological effects on carbon isotope ratios of plant matter are usually of more interest than that of source CO_2 effects, carbon isotope ratios are frequently expressed in terms of discrimination (Δ), a standardized value that takes the variation in the $\delta^{13}C$ of the source CO_2 ($\delta^{13}C_{air}$) into account (Farquhar et al., 1989):

$$\Delta = (\delta^{13}C_{air} - \delta^{13}C_{plant})/(1 + \delta^{13}C_{plant}).$$

Both notations will be used throughout this chapter.

In C_3 plants, discrimination against ^{13}C during photosynthesis has two main components; one associated with diffusion of CO_2 through the stomata, and one associated with discrimination by the primary carboxylating enzyme. The overall discrimination during photosynthesis is determined largely by the ratio of intercellular to atmospheric CO_2 concentration (c_i/c_a) during biomass formation (Farquhar, O'Leary & Berry, 1982). The value c_i/c_a is also directly related to the ratio of the instantaneous rates of CO_2 assimilation and stomatal conductance (A/g), a measure of intrinsic water-use efficiency (Farquhar et al., 1989). Any environmental or plant factor that affects either stomatal conductance and/or the photosynthetic capacity may alter c_i/c_a and therefore the magnitude of the discrimination against ^{13}C.

Recently, measurements of the stable isotopes of hydrogen and oxygen (D and ^{18}O) in plant matter have been employed to characterize the utilization of soil and ground water by plants occurring in temperate, semi-arid, or coastal regions (e.g., White 1989; Ehleringer & Dawson, 1992; Ish-Shalom et al., 1992; Thorburn & Walker, 1993). This characterization is possible because there is usually a substantial difference in isotopic composition between ground water and surface water derived from recent precipitation. Unlike the fractionation of carbon isotopes that occurs during photosynthesis, no fractionation of isotopes occurs during water uptake by most plants. An exceptional case involves isotopic fractionation of hydrogen by roots during water uptake by mangroves (Lin & Sternberg, 1993). On the other hand, isotopic fractionation in water is maximal in those plant parts where transpiration is occurring. During the phase change from liquid to vapor in leaves and other transpiring organs, the water molecules containing the lighter isotopes of oxygen and hydrogen ($H_2{}^{16}O$) evaporate more readily than molecules containing the heavy isotopes ($D^{16}O$, $H_2{}^{18}O$) (Flanagan and Ehleringer, 1991; Flanagan, Comstock & Ehleringer, 1991). Thus, the remaining water becomes enriched in deuterium and ^{18}O. Therefore, δD and/or $\delta^{18}O$ values of leaf water are higher relative to soil and xylem water and will not reflect the values of the source water. The applicability of these methods in temperate environments has recently been reviewed by Ehleringer and Dawson (1992) and Dawson (1993a).

Of particular interest here is the measurement of the isotopic composition of xylem water in tropical forest plants to determine the depth in the soil profile from which the plants are extracting water. The paucity of studies involving application of stable isotope techniques to determine water sources of plants in tropical areas may result from the lack of isotopically distinct precipitation, such as that which occurs between the winter and summer months in temperate zones. However, because of the isotopic fractionation that occurs in soil water during evaporation, it should also be possible to use stable isotopes in tropical areas that experience a pronounced seasonality with respect to rainfall (Jackson et al., 1995). This is possible because evaporation of soil water causes water in the upper soil layers to become enriched in deuterium and oxygen-18 relative to deeper soil layers (Allison & Hughes, 1983). Application of this method is restricted to situations in which the isotopic composition of all of the potential water sources is known, and the differences in isotopic composition among sources are greater than the sampling error of the mass spectrometers currently used for stable isotope studies (Dawson, 1993a).

9.2 ISOTOPIC COMPOSITION OF XYLEM WATER IN RELATION TO EXPLOITATION OF WATER RESOURCES

Soil water availability in lowland tropical forests varies both temporally and spatially due to seasonality in rainfall and enhanced drying of upper soil layers during periods of low precipitation. On BCI, for example, both rainfall and water content in the upper portion of the soil profile decrease substantially between the months of January and April. Although this forest receives an average of over 2600 mm of rain a year, in more than half the years of record, less than 100 mm falls during the first 3 months of each year (Leigh, Rand & Windsor, 1982). Several important canopy species in the BCI forest, such as *Pseudobombax septenatum* and *Spondias radlkoferi* are drought-deciduous. These trees drop their leaves at the beginning of the dry season (December to January) and leaf out again in May. Evergreen tree species, on the other hand, remain active during the dry season, with species such as *Anacardium excelsum* producing new leaves 3 to 5 weeks after the beginning of the dry season. It has been hypothesized that, in order to maintain continuous physiological activity, these evergreen species may tap water from deeper parts of the soil profile where greater amounts of water are available.

Temporal and spatial patterns of belowground resource utilization remain largely unknown in lowland tropical forests. The utility of using excavations to study rooting depth is limited by the intricate nature of root systems and by the high species diversity and density of lowland tropical forests. Furthermore, these types of studies provide little insight into resource uptake rates or patterns of resource utilization because root presence is not a reliable indicator of active water uptake dynamics in either time or space (Ehleringer & Dawson, 1992). Stable isotopes provide an alternative way of evaluating integrated water uptake by forest plants. During the dry season, the iso-

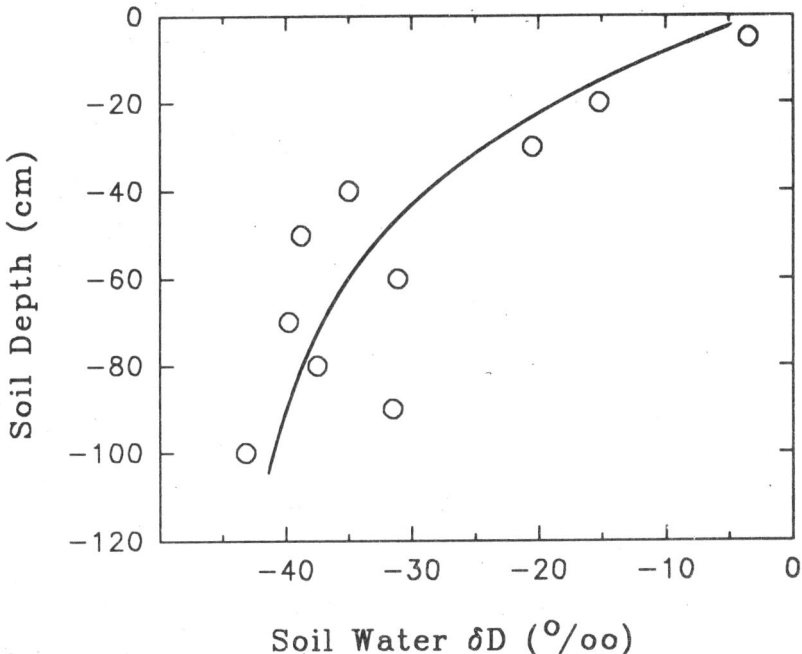

Figure 9.1. *The hydrogen isotope ratio (δD) of soil water as a function of soil depth. Samples were obtained from a tropical forest site near Barro Colorado Island, Panama, at the end of the dry season of 1992. From Jackson et al. (1995).*

topic composition of the soil water in the BCI forest varies as a function of depth (Figure 9.1). Water in the uppermost 30 cm of the soil profile becomes enriched in the heavier isotopes through evaporation. Average δD of the uppermost 30 cm is $-13.1‰$, whereas water in the deeper horizons is isotopically lighter with δD values ranging from -43.2 to $-31.1‰$ (Figure 9.1).

During the dry season, differences in the isotopic composition of xylem water may reflect patterns of soil water utilization between evergreen and deciduous canopy trees (Figure 9.2). Deciduous tree species have δD values for xylem water ranging from -5 to $-25‰$, suggesting that they extract water largely from the upper parts of the soil profile. Evergreen trees, on the other hand, have xylem water with lower δD values, and therefore appear to have access to soil water at greater depths. This is consistent with the relatively high levels of physiological activity exhibited by evergreen trees during the dry season (Meinzer et al., 1993). The root systems of the evergreen

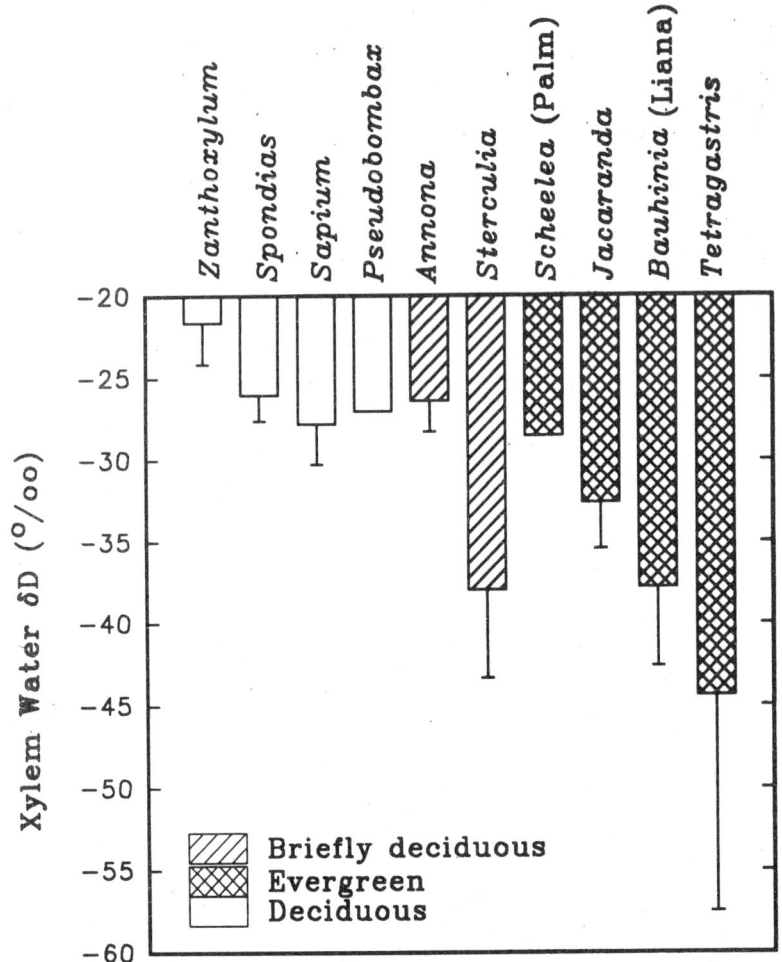

Figure 9.2.   The hydrogen isotope ratio of xylem sap (δD) for different species (categorized by life form) at a tropical forest site near Barro Colorado Island, Panama, at the end of the dry season of 1992. Bars are ± 1 SE of three to four samples for at least two individuals of each species. From Jackson et al. (1995).

species studied in the semideciduous forest of BCI may partition their exploitation of the soil water in such a way that no two species exhibit identical isotopic composition of xylem water (Figure 9.2). This is consistent with the high species diversity of tropical rainforests and with models accounting for the diversity of forest trees from niche differentiation through competitive displacement (Leigh, 1982).

Isotopic analysis of xylem and soil water can also be used to characterize soil water resources utilized by understory and gap species in semideciduous tropical forests. Differences in rates of water use between shade-tolerant shrubs and light-demanding species may be associated with vertical partitioning of soil water resources (Figure 9.3A). Gap species such as *Miconia argentea* and *Cecropia obtusifolia* that exhibit high transpiration rates apparently utilize deeper water sources (more negative δD values) than species such as *Piper cordulatum* and *Psychotria limonensis*, which are shade tolerant and have low rates of water consumption and less negative xylem water δD values. However, *Palicourea guianensis*, a gap species, had the lowest transpiration rates and also the highest δD values, indicating use of shallow water resources. As expected, leaf water potentials tended to be more negative if the plants' water source was shallower (more positive δD values) (Figure 9.3B). Nevertheless, the relationship between leaf water potential and the depth of water source was not as close as the one observed between depth of water source and transpiration (Figure 9.3A). Ehleringer and Dawson (1992) have suggested that a tight relationship between water source and plant water potential may not always occur if the individual species exhibit different water transport efficiencies. In such cases, plants exploiting similar water sources and subjected to the same evaporative demand conditions could operate at different leaf water potentials, depending on species-specific differences in water relations characteristics, such as stomatal regulation of water loss, leaf specific hydraulic conductivity, and elasticity of leaf tissues (Mulkey & Wright, Chapter 7).

9.3 SPATIAL AND TEMPORAL VARIATION IN CARBON ISOTOPE RATIOS OF THE SOURCE CO_2

Foliar carbon isotope ratios can be a powerful tool for understanding integrated responses of leaf gas exchange to water availability, light, and other environmental factors (e.g., Zimmerman & Ehleringer, 1990, Meinzer et al. 1992), provided variation in the isotopic composition of the source CO_2 is minor compared to variations in physiological processes in the leaf leading to discrimination against[13]C. In closed tropical forests, large spatial and temporal variations in the isotopic composition of gaseous CO_2 may confound the interpretation of foliar $\delta^{13}C$ values. For example, ambient carbon dioxide $\delta^{13}C$ values become progressively more negative from the upper canopy to the understory,

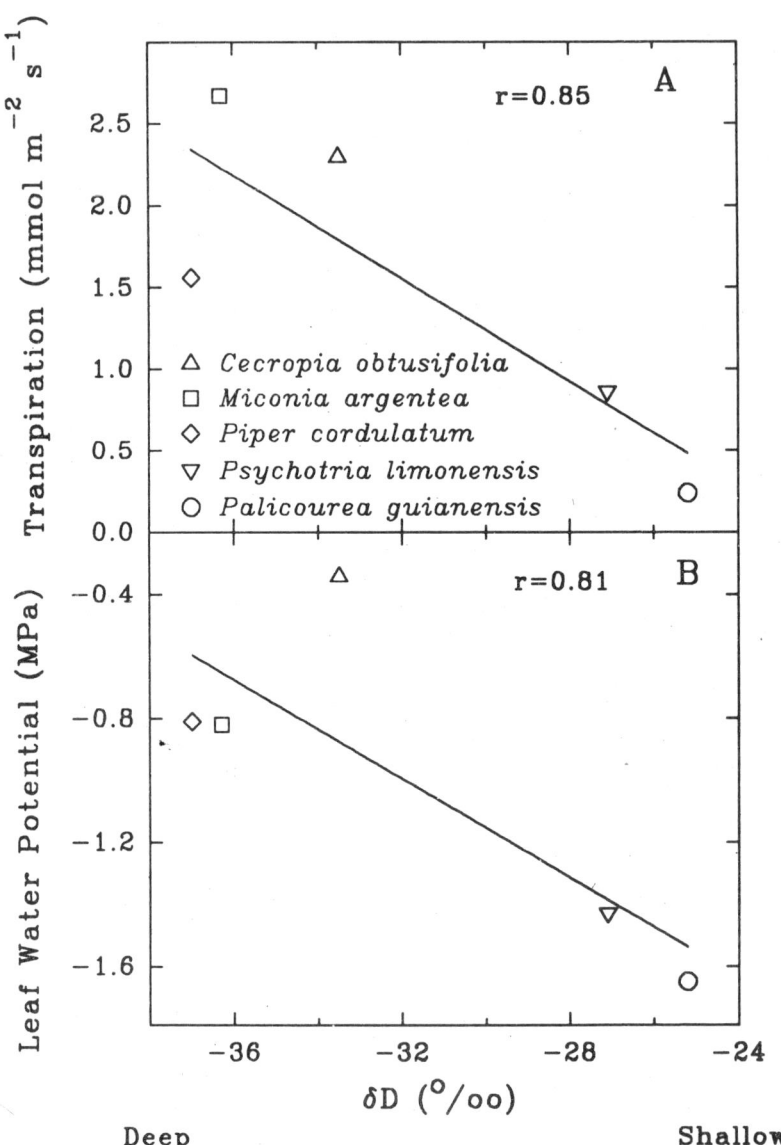

Figure 9.3. Relationship (A) between transpiration and hydrogen isotope ratios of xylem water (δD), and (B) between midday leaf water potential and δD for five species growing in gaps in a lowland tropical forest in Panama at the end of dry season. Transpiration was determined as mass flow of sap by the stem heat balance method. From Jackson et al. (1995).

particularly as the soil surface is approached. This vertical gradient in $\delta^{13}C$ values of the source CO_2 is partially the result of soil respiration and relatively poor ventilation at the forest floor (Vogel, 1978; Medina et al., 1986; Medina, Sternberg & Cuevas, 1991; Merwe & Medina, 1989). Since the CO_2 released from the forest floor is derived from organic material already depleted in ^{13}C, its isotopic composition should reflect the average isotopic composition of the decomposing material with isotope ratios ranging from -23 to $-36‰$

Horizontal variation in carbon isotope ratios of gaseous CO_2 between gap areas and adjacent understory sites in dense tropical forests showed differences of 0.5‰, which are consistent with the idea that gaps, particularly large ones, are more ventilated than the surrounding understory (Jackson et al., 1993). For small gaps (less than 10 m diameter), however, the $\delta^{13}C$ values of air are often indistinguishable from those in understory sites. Temporal variation of the isotopic composition of the air has been observed in the understory of a forest in Trinidad with values ranging from a 23% contribution from respired CO_2 in the morning to 8% in the evening (Broadmeadow et al., 1992).

Despite considerable spatial and temporal heterogeneity of $\delta^{13}C$ values in the air of tropical forests, several studies suggest that variation in $\delta^{13}C$ of ambient CO_2 can be predicted from measurements of ambient CO_2 concentration and $\delta^{13}C$ values of respired CO_2 (Quay et al., 1989; Keeling, 1961; Sternberg, 1989). This prediction assumes that the $\delta^{13}C$ value of ambient CO_2 is determined only by the relative proportions of bulk atmospheric carbon dioxide (with a $\delta^{13}C$ value in the range of -8 to $-7.8‰$), and respiratory CO_2 (with $\delta^{13}C$ values in the range of -28 to $-25‰$), and it is given by the following equation (Keeling, 1961)

$$\delta^{13}C_F = [(CO_2)_A/(CO_2)_F](\delta^{13}C_A - \delta^{13}C_R) + \delta^{13}C_R \qquad (1)$$

where $(CO_2)_F$ and $(CO_2)_A$ represent the CO_2 concentration in forest air and the bulk atmosphere, respectively; $\delta^{13}C_F$, $\delta^{13}C_R$, and $\delta^{13}C_A$ represent the $\delta^{13}C$ values of forest, respired, and atmospheric carbon dioxide, respectively. The relationship between the $\delta^{13}C$ value of ambient carbon dioxide and the inverse of its concentration is thus linear having a slope of $(CO_2)_A(\delta^{13}C_A - \delta^{13}C_R)$ and an intercept equal to the $\delta^{13}C$ value of respired carbon dioxide ($\delta^{13}C_R$).

In addition to dissipation of respired CO_2 by turbulent mixing with the atmosphere, some of the respired CO_2 may be refixed during photosynthesis. If this is the case, then the relationship described in eq. (1) no longer holds, since there is a large fractionation associated with photosynthetic CO_2 uptake. The following equation takes into account the influence on $\delta^{13}C_F$ of refixation of respired carbon dioxide by photosynthetic processes in the canopy (Sternberg, 1989):

$$\delta^{13}C_F = [(CO_2)_A/(CO_2)_F](\delta^{13}C_A - \delta^{13}C_R)(1\text{-}R) + \delta^{13}C_R + R^*E \quad (2)$$

where the abbreviations and symbols for concentration and isotopic composition are as previously defined, R represents the proportion of respired CO_2 refixed by photosynthesis relative to the total dissipation of carbon dioxide (photosynthesis and turbulent mixing), and E represents the fractionation factor associated with photosynthesis. Note that when a negligible proportion of respired CO_2 is refixed by photosynthesis, the equation reduces to its simpler form (Equation (1). Thus if the isotopic ratios and concentration of atmospheric, respiratory, and ambient CO_2 are known, together with the fractionation factor for photosynthesis, it should be possible to calculate the proportion of respired CO_2 refixed by photosynthesis. Using this equation, Sternberg (1989) calculated that in sites on BCI, Panama, about 7% of the carbon dioxide given off by respiration is recycled by photosynthesis. When the data of Quay et al. (1989), collected in the Amazon Basin, are fitted to Equation (2), the amount of respired CO_2 refixed by photosynthesis is estimated in the range of 5%, similar to the value reported for the forest on BCI. If further research indicates that the amount of respired CO_2 refixed by photosynthesis is generally small, then it should be possible to estimate the $\delta^{13}C$ values of ambient carbon dioxide based on its concentration alone.

Recent studies suggest that the $\delta^{13}C$ values of plants in the understory of tropical forests are lower (more negative) relative to those of plants with leaves in the upper canopy (Table 9.1). For example, the average difference in $\delta^{13}C$ between canopy and understory leaves was about 5.1‰ in a lowland wet forest in Puerto Rico (Medina, Sternberg & Cuevas, 1991), but only 2.8‰ in a semi deciduous forest in Trinidad (Broadmeadow et al. 1992). Foliar $\delta^{13}C$ values are also lower for understory plants of temperate forests relative to upper canopy trees (Garten & Taylor, 1992; Schleser, 1990). In these forests, the upper canopy leaves have $\delta^{13}C$ values similar to those of upper canopy leaves in tropical forests, but the understory plants of the temperate forests appear to exhibit more positive $\delta^{13}C$ values than those of understory plants in tropical forests (Table 9.1). Controversy exists as to the relative importance of factors that may lead to vertical differences in foliar $\delta^{13}C$. Some authors, such as Francey et al. (1985), suggest that physiological factors, such as stomatal and photosynthetic responses to light and their influence on the ratio of internal to ambient CO_2 concentration (c_i/c_a), can explain most of the vertical variation in foliar $\delta^{13}C$, while others, such as Merwe and Medina (1989), favor variations in isotopic composition of the source CO_2 as the primary cause. On BCI, Sternberg, Mulkey and Wright (1989) were able to calculate that 30 to 50% of the reduction in $\delta^{13}C$ values of understory plants was attributable to lower carbon isotope ratios of ambient CO_2 while the remaining percentage was due to a physiological response to lower light levels. Similar results were obtained by Jackson et al. (1993). Simultaneous measurements not only of carbon isotope ratios of the leaves and of the source CO_2 but also of diurnal variations in photosynthesis, stomatal conductance, and ambient CO_2 concentrations during leaf expansion would lead to a better understanding of factors responsible for the vertical gradient in foliar $\delta^{13}C$ values in forests. Nevertheless, it can be expected that the relative importance of physiological and environmental determinants of variation in $\delta^{13}C$ of leaves will depend on the type of forest where a given study is carried out.

9.4 INFERENCE OF LONG-TERM PHYSIOLOGICAL CHARACTERISTICS FROM LEAF CARBON ISOTOPE COMPOSITION

Values of $\delta^{13}C$ of leaves from evergreen and deciduous tropical forest species taken from the upper canopy are usually in the -26 to $-32\%_0$ range, while understory leaves from the same species tend to be in the -31 to $-36\%_0$ range (Medina, Sternberg & Cuevas, 1991; Broadmeadow et al., 1992; Jackson et al., 1993; and unpublished information). There do not appear to be substantial differences in foliar $\delta^{13}C$ between evergreen and deciduous trees. The $\delta^{13}C$ values for upper canopy leaves overlap with those of tropical savanna trees (-27 to $-30\%_0$; Goldstein et al., 1989). As in tropical forests, there is no substantial difference in foliar $\delta^{13}C$ between evergreen and deciduous savanna trees. The small differences in $\delta^{13}C$ among leaves of evergreen and deciduous species has been attributed to differences in nighttime respiration rates and differences in timing of leaf construction (dry season for the evergreen trees versus wet season for the deciduous trees, Goldstein et al., 1989).

One of the primary attractions of leaf carbon isotope analysis is that measurements made at a single point in time reflect an integration of the internal and external conditions influencing carbon uptake during

Table 9.1.   *Range of leaf carbon isotope ratios ($\delta^{13}C$) for upper canopy trees, lower canopy saplings, and understory shrubs in tropical and temperate forests. Data were obtained from different sources.*

Forest type	($\delta^{13}C$‰)		Source
	Upper Canopy	Saplings/Understory	
Tropical			
Semideciduous	-27.7 to -31.0	-34.8 to -35.0	Sternberg, Mulkey & Wright, (1989)
Semideciduous	-24.8 to -26.0	-28.9 to -29.1	Broadmeadow et al. (1992)
Deciduous	-28.9 to -29.1	-32.0 to -32.5	Broadmeadow et al. (1992)
Evergreen	-26.7 to -30.1	-33.2 to -34.8	Medina, Sternberg & Cuevas, (1991)
Caatinga	-28.8 to -32.6	-32.9 to -36.0	Medina et al. (1986)
Semideciduous	-25.5 to -29.5	-34.6 to -35.9	Jackson et al. (1993) and unpublished
Temperate			
Deciduous	-27.5 to -28.3	-29.1 to -30.4	Garten & Taylor (1992)
Deciduous	-26.1 to $-26.5.$	-30.1 to -30.4	Schleser (1990)

the formation of the leaf. Once variation in carbon isotope ratio of the source CO_2 has been factored out, the isotopic composition of the leaf tissues (measured as discrimination, Δ) represents an assimilation-weighted average of the ratio of internal to ambient CO_2 concentrations (Farquhar & Richards, 1984)

$$\Delta = a + (b - a)c_i/c_a \qquad (3)$$

where Δ is discrimination, and a and b are the isotopic fractionation factors associated with diffusion through the stomata (about 4.4‰) and carboxylation by Rubisco (29‰; Roeske & O'Leary, 1984), respectively. A limitation when using this equation is that the isotopic composition of the carbon source must be known. Because water loss from the leaf occurs along the same pathway as CO_2 uptake, instantaneous water-use efficiency (A/E) is equal to the ratio of the driving forces for the two gases multiplied by the ratio of their diffusivities in air

$$A/E = (c_a - c_i)/(e_i - e_a)\,1.56 \qquad (4)$$

As c_i is a factor in both Equations (3) and (4), the two can be combined to produce the following relationship

$$\Delta = 29 - ((A/E)(e_i - e_a)\,38.4/c_a) \qquad (5)$$

Note that the relationship between Δ and water-use efficiency is linear only if $(e_i - e_a)/c_a$ is constant. If one assumes that average c_a is

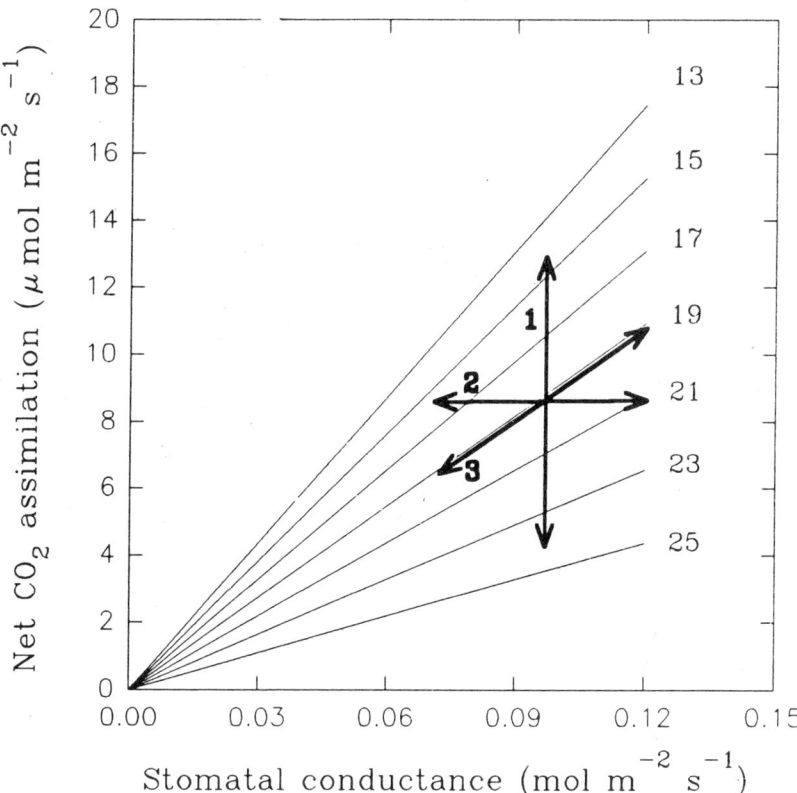

Figure 9.4. Hypothetical relationships between stomatal conductance (g), net CO_2 assimilation (A), and foliar carbon isotope discrimination (Δ) values for C_3 plants. All the lines representing constant values Δ intersect the origin. Arrows indicate three contrasting modes of regulation of photosynthetic gas exchange: 1) variations in A while maintaining g constant, 2) maintaining A constant while g changes, and 3) by similar, proportional variations in A and g such that A/g (and therefore c_i / c_a) remain constant. The first two responses lead to different values, whereas the last one leads to maintenance of constant Δ values.

350 ppm during the formation of the leaf tissue, then one can substitute $g (e_i - e_a)$ for E (where g is the total conductance to water vapor) to obtain the following linear relationship

$$\Delta = 29 - (A/g)\,0.11 \qquad (6)$$

where A/g is intrinsic water-use efficiency. When A is plotted against g, a family of lines connecting points of similar Δ (similar c_i/c_a) can be obtained (Figure 9.4).

Given the direct relationship between Δ and A/g, what do measurements of Δ tell us about the physiological mechanisms by which individual species adjust to changes in environmental conditions, or how different species experiencing the same environment balance carbon uptake and water loss? Similar intrinsic water-use efficiencies, and therefore Δ, may be attained by contrasting mechanisms, each having different ecological consequences. These are: 1) variations in photosynthetic capacity (as reflected in A) while maintaining g constant, 2) maintaining A constant while g changes, and 3) by similar, proportional variations in A and g such that A/g remains constant. Intermediate mechanisms are, of course, also possible. The first two mechanisms lead to different values, whereas the last one leads to maintenance of constant Δ values (Figure 9.4).

Some of these three modes of regulation have already been described. For example, in Hawaiian *Metrosideros polymorpha* populations growing at different levels of soil moisture availability, decreases in Δ values with declining water availability were associated with maintenance of nearly constant photosynthetic rates while g declined (Meinzer et al., 1992). Likewise, Geber and Dawson (1990) found that in an annual weed, variation in g led to strong genotypic differences in A/g and Δ. In contrast, genotypic variations in A/g (and therefore in Δ) in peanut resulted from variation in photosynthetic capacity at similar levels of g (Wright, Hubick & Farquhar, 1988). Similarly, the decrease in Δ in two shade-tolerant tropical species, *Piper cordulatum* and *Psychotria limonensis*, growing under higher irradiance in treefall gaps, resulted from an increase in A associated with a smaller increase in g (Jackson et al., 1993).

Analysis of the modes of regulation of photosynthetic gas exchange in relation to carbon isotope ratios, as described above, has typically been applied only to a single species in a given study. It is also of interest to know if seasonal variations in carbon isotope composition in plants of several different species growing in the same physical environment (e.g., canopy tree species) result from adjustments in g while A remains constant, variations in A while g remain constant, or variations in both A and g. Two of these modes of regulation of gas exchange have been observed in canopy evergreen species growing in a semideciduous forest near Panama City where leaf production occurs during both the rainy and dry seasons. Here, species-specific variation in Δ values should reflect intrinsic physiological characteristics of each species rather than the effect of external environmental factors because upper canopy leaves are subjected to similar irradiance, isotopic composition of the CO_2 source, and ambient water vapor pressure. Among these species, the magnitude of variation in carbon isotope composition differs seasonally (Figure 9.5). The range of Δ values is much smaller during the wet season (19‰ to 21‰) than during the dry season (16‰ to almost 21‰). Apparently, during the wet season, A and g co-vary in nearly a one to one relationship, which would limit the variation in Δ despite the nearly threefold variation among species in maximum photosynthetic rates (Figure 9.5). Thus, during the season in which water is not limiting, evergreen trees appear to modulate their stomatal conductance to match the maximum photosynthetic capacity. This results in leaves of different species having similar carbon isotope composition, and suggests that any differences in photosynthetic rates are determined primarily by differences in the intrinsic biochemical capacity for photosynthesis.

The preceding relationship in which A and g co-vary in nearly a one to one fashion does not hold for leaves mostly expanded in the dry season. During this season, the trees tend to produce leaves with lower Δ and with similar carbon-gaining capacity per unit leaf area (Figure 9.5). The values of A and g tend to be lower for leaves expanded during the dry season than for those expanded during the wet season, but intrinsic water-use efficiency for most canopy species is enhanced because the decrease in g is larger than the decrease in A. Thus, when water is limiting, it appears that the primary variable determining the differences in photosynthetic rate among species is the stomatal restriction of CO_2 transport to the sites of carboxylation rather than decreased carboxylating capacity per se. These patterns have important ecological implications. Evergreen species appear to produce leaves with similar carbon-gaining capacity during the dry season, but with substantial differences in water-use efficiency. Canopy species might be expected to exhibit large differences in water-use efficiency during the dry season, since they may differ in water transport capacity, internal water-storage capacity, and other water relations characteristics (Tyree & Ewers, Chapter 8). In the wet season, the water-use efficiency is more or less equal among species, but their carbon-gaining capacity is very different. When water is sufficient, the risk of maintaining a luxurious use of water is low, since transient water deficits are infrequent. When water is limiting, however, the adaptive value of increasing the efficiency of photosynthetic water use should be high, since the long-term maintenance of physiological activity may be more important than maximizing short-term carbon gain.

The interpretation of the differences in carbon isotope composition in canopy species is reasonably straightforward because the aerial

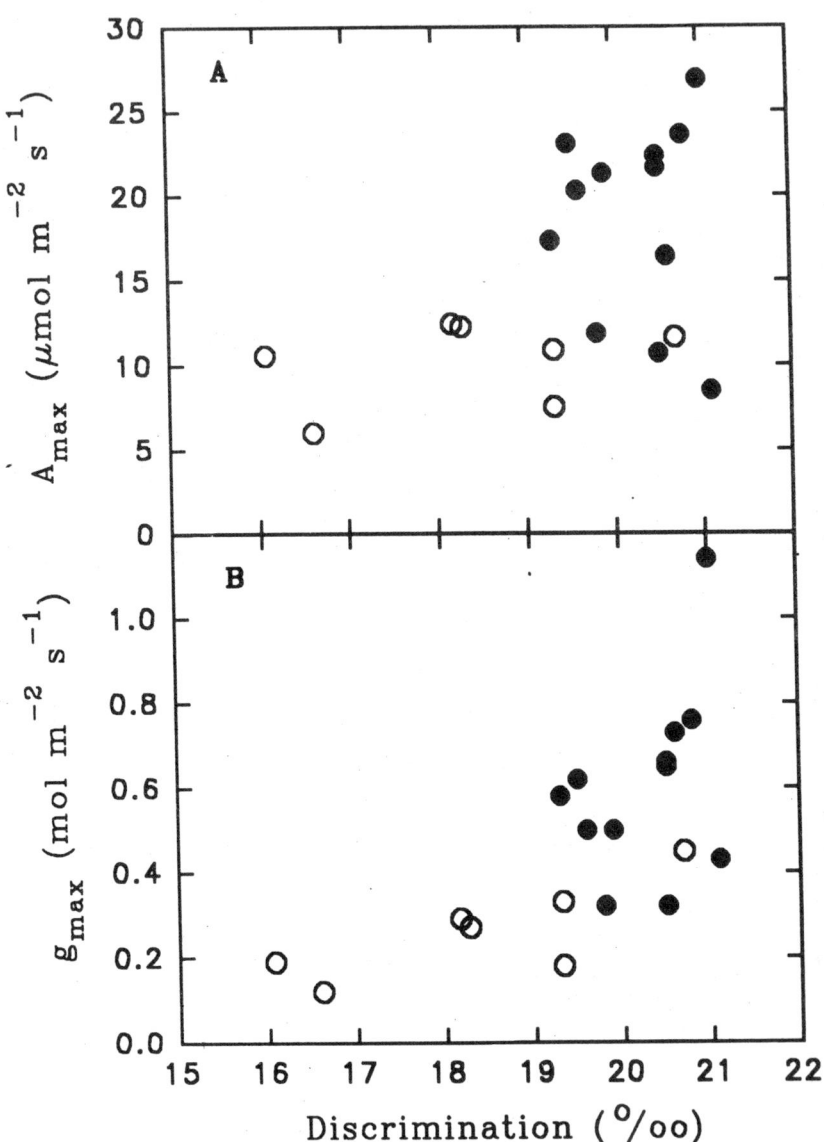

Figure 9.5. *Relationship between maximum net CO_2 assimilation (A_{max}) and discrimination (A), and between maximum stomatal conductance (g_{max}) and discrimination (B) for seven evergreen canopy species during the dry season (open symbols), and twelve evergreen or deciduous species during the wet season (closed symbols). Based on Cavelier et al., unpublished results.*

microenvironment is nearly identical for all species. In contrast, the interpretation of carbon isotope composition of leaf tissue in tropical forest floor plants is confounded by spatial variations in light and in the isotopic composition of the source CO_2. In a study carried out on BCI, species growing in gaps had $\delta^{13}C$ values 2–3‰ higher than plants of the same species growing in adjacent understory sites. It appeared that 65–80% of the difference in foliar $\delta^{13}C$ between gap and understory plants was the result of internal adjustments in leaf gas exchange characteristics, and the remaining 20–35% was attributable to differences in isotopic composition of source CO_2 (Jackson et al., 1993).

Shade-tolerant tropical plants have very negative $\delta^{13}C$ values (usually between -33 and -36‰). These low carbon isotope ratios can be explained in part by the substantial amount of respired CO_2 in the air near the forest floor, and in part by the extremely high c_i/c_a in leaves of these species. The maintenance of relatively high stomatal conductances even when light levels and photosynthetic rates are low is apparently of adaptive value for shade-tolerant plants that rely largely on diffuse light and short periods of direct sun (light flecks) for photosynthetic carbon gain in environments where evaporative demand is low (Chazdon et al., Chapter 1).

Overall, it appears that different lowland tropical forest life forms have been subjected to different evolutionary constraints in terms of regulation of water use and photosynthesis. For example, evergreen canopy trees appear to maintain a relatively high water-use efficiency during the dry season, in contrast to the shade-tolerant understory plants that exhibit high stomatal conductances despite their low photosynthetic capacity. Even within each of the different life forms, the individual species appear to differ in their photosynthetic capacity and water-use efficiency. It is expected that evergreen canopy species with deeper roots, and therefore access to more reliable soil water sources, will have lower intrinsic water-use efficiency than canopy evergreen shallow-rooted species. Similar degrees of species-specific variation in the regulation of water-use in relation to photosynthetic CO_2 assimilation apparently exist for other life forms, such as deciduous canopy trees, high light demanding gap species, and understory shrubs. The regulation of water use in other important groups of tropical forest species that were not included in this review, such as vines and hemiepyphytic plants, is also expected to be fine-tuned to the amount of resources to which they have access (Holbrook & Putz, Chapter 13). A picture of very narrow niche and resource partitioning may emerge as additional data on physiological and ecological characteristics of different lowland tropical forest species are obtained.

9.5 CONCLUSIONS

Some mechanisms of ecological importance related to resource acquisition by roots and leaves of lowland tropical forest plants that merit further research have been discussed in this chapter. Roots of different species appear to partition soil water in time and space, which can be taken as indirect evidence that species tend to minimize competition for water resources. The root systems of some evergreen tree species apparently obtain water from deeper soil layers, which is contrary to the current dogma that roots of plants from tropical forests explore mainly the uppermost few centimeters of the soil profile. Partitioning of soil water by roots also suggests that water is a resource available in limited amounts, at least during the dry season. The hypothesis that substantial niche partitioning driven by competition exists among trees and shrubs of tropical rainforests needs to be tested experimentally. Other stable isotope techniques, including analysis of stable N isotopes, which has the potential to provide information on N_2 fixation and to quantify inputs of fixed N to the system (Virginia et al., 1989), could be used to help falsify this long-standing hypothesis.

Foliar $\delta^{13}C$ values of tropical forest plants are the result of both physiological processes leading to discrimination against ^{13}C and of the magnitude of refixation of respiratory CO_2 emitted from the forest floor. Thus, the vertical stratification in the isotopic composition of the source CO_2 may confound the physiological and ecological interpretation of foliar $\delta^{13}C$ values. Despite considerable spatial and temporal heterogeneity of $\delta^{13}C$ values in air of closed tropical forests, recent studies suggest that variation in the $\delta^{13}C$ of air can be predicted from measurements of the ambient CO_2 concentration alone, providing reason for optimism concerning the utility of foliar $\delta^{13}C$ analysis in forest species (Sternberg, Mulkey & Wright, 1989; Broadmeadow et al., 1992). Nevertheless, the relationship between ambient CO_2 concentration and the isotopic composition of the source CO_2 needs to be better characterized for different types of tropical forests and during both the wet and dry seasons.

Results presented in this chapter suggest that during the wet season, stomatal conductance adjusts to the different intrinsic photosynthetic capacities of the canopy trees. However, during the dry season, stomatal conductance decreases and water-use efficiency increases by different amounts depending on the species, resulting in nearly constant carbon gain per unit leaf area across species. This behavior has important implications for predicting the effects of global climate change on the water and carbon balance of lowland tropical forest plants. It is expected, for example, that species that produce leaves during the dry season with high capacity for restricting water losses without a large decline in photosynthetic rates will be better able to tolerate the consequences of global temperature increase or more intense drought periods. The seasonal shift in photosynthetic strategies can be easily detected by variations in carbon isotope ratios of leaf samples obtained from leaves expanded during the dry and wet seasons. Larger root systems that can access more reliable water sources (inferred from δD or $\delta^{18}O$ values of water from stem and soil samples) should help in buffering changes in soil water availability. It thus seems clear that stable isotope techniques should help in predicting individual species responses to changing environmental conditions. For an appropriate interpretation of stable isotope composition of leaf and water samples, however, additional information on water relations and gas exchange characteristics of plants has to be obtained concurrently. For example, the consequences of shift in foliar Δ values cannot be evaluated fully unless the magnitudes of adjustments in A and g are known. Similarly, knowledge of the transpiration rates associated with given δD or $\delta^{18}O$ values for xylem water would confirm the differential availability of soil water pools having different isotopic composition.

In developing a research agenda for future work in the application of stable isotope techniques to tropical systems, we envision several other applications in addition to those identified in this chapter. Stable isotopes of C and H can potentially aid in scaling up processes from the leaf to the canopy level (see Meinzer & Goldstein, Chapter 4). Other ecological processes in tropical forests such as ecophysiological responses to forest fragmentation resulting in edge effects, can be examined with stable isotope techniques (Kapos et al., 1993). Dynamics of water movement in the soil and hydraulic lift are processes that can be examined rigorously with deuterated water. Hydraulic lift, soil water absorbed by deeply rooted plants and released into the upper soil layers, has been observed in deserts (Richards & Caldwell, 1987) and temperate deciduous forests (Dawson, 1993b). This phenomenon has not yet been documented in tropical plants. Still other possible research areas to which stable isotopes could be applied include the evaluation of species regeneration responses in relation to physiological status and stress, and the effects of internal water storage on the water economy of trees.

ACKNOWLEDGEMENTS

This chapter grew out of research supported by the Mellon Foundation and the Smithsonian Tropical Research Institute. The authors are grateful to the late A.P. Smith for his inspiration, support, and encouragement.

REFERENCES

ALLISON, G. B. & HUGHES, M. W. (1983) The use of natural tracers as indicators of soil-water movement in a temperate semi-arid region. *Journal of Hydrology,* **60,** 157–173.

BROADMEADOW, M. S. J., GRIFFITHS, H., MAXWELL, C. & BORLAND, A. M. (1992) The carbon isotope ratio of plant organic material reflects temporal and spatial variations in CO_2 within tropical forest formations in Trinidad. *Oecologia,* **89,** 435–441.

DAWSON T. E. (1993a) Water sources of plants as determined from xylem-water isotopic composition: Perspectives on plant competition, distribution, and water relations. *Stable Isotopes and Plant Carbon-Water Relations,* (eds. J. A. EHLERINGER, A. E. HALL & G. D. FARQUHAR) Academic Press, San Diego, pp 465–496.

DAWSON, T. E. (1993b) Hydraulic lift and water use by plants: Implications for performance, water balance, and plant-plant interactions. *Oecologia,* **95,** 565–574.

EHLERINGER, J. R. & DAWSON, T. E. (1992) Water uptake by plants: Perspectives from stable isotope composition. *Plant, Cell and Environment,* **15,** 1073–1082.

EHLERINGER, J. R. & RUNDEL P. W. (1989) Stable isotopes: History, units, and instrumentation. *Stable Isotopes in Ecological Research* (eds. P. W. RUNDEL, J. R. EHLERINGER & K. A. NAGY) Springer-Verlag, Heidelberg, ppp 1–15.

FARQUHAR, G. D. & RICHARDS, R. A. (1984) Isotopic composition of plant carbon correlates with water use efficiency of wheat genotypes. *Australian Journal of Plant Physiology,* **11,** 539–552.

FARQUHAR, G. D., HUBICK, K. T., CONDON, A. G., & RICHARDS, R. A. (1989) Carbon isotope fractionation and plant water-use efficiency. *Stable Isotopes in Ecological Research* (eds. P. W. RUNDEL, J. R. EHLERINGER, & K. A. NEGY) Springer Verlag, New York. pp 21–146.

FARQUHAR, G. D., O'LEARY M. H., & BERRY, J. A. (1982) On the relationship between carbon isotope discrimination and the intercellular carbon dioxide concentration in leaves. *Australian Journal of Plant Physiology,* **9,** 121–137.

FLANAGAN, L. B. & EHLERINGER, J. R. (1991) Stable isotopic composition of stem and leaf water: Applications to the study of plant water use. *Functional Ecology*, **5**, 270–277.

FLANAGAN, L. B., COMSTOCK, J. P. & EHLERINGER J. R. (1991) Comparison of modeled and observed environmental influences in stable oxygen and hydrogen isotope composition of leaf water in *Phaseolus vulgares*. *Plant Physiology*, **96**, 588–596.

FRANCEY, R. J., GIFFORD, R. M., SHARKEY, T. D., & WIER, B. (1985) Physiological influences on carbon isotope discrimination in huon pine (*Lagorostrobus franklinii*). *Oecologia*, **66**, 211–218.

GARTEN, C. T., & TAYLOR, G. E. (1992) Foliar $\delta^{13}C$ within a temperate deciduous forest: Spatial, temporal, and species sources of variation. *Oecologia*, **90**, 1–7.

GEBER, M. A., & DAWSON, T. E. (1990) Evolutionary responses of plants to global climate change. *Biotic Interactions and Global Change* (eds. P. M. KARIEVA, J. G. KINGSOLVER and R. B. HUEY) Sinauer Associates, Sunderland, Massachusetts, pp 179–197.

GOLDSTEIN, G. & SARMIENTO, G. (1987) Water relations of trees and grasses and their consequences for the structure of savanna vegetation. *Determinants of Tropical Savannas* (ed. B. H. WALKER) IUBS Monogr. Series No. 3, ICSU Press, Miami, pp 13–38.

GOLDSTEIN G., RADA, F., RUNDEL P., AZOCAR A., & OROZCO A. (1989) Gas exchange and water relations of evergreen and deciduous tropical savanna trees. *Annales des Sciences Forestieres.*, **46 (Suppl.)**, pp 448s–453s.

IISH-SHALOM, N., STERNBERG, L. da S. L., ROSS, M., O'BRIEN, J. & FLYN, L. (1992) Water utilization of tropical hardwood hammocks of the lower Florida Keys. *Oecologia*, **92**, 108–112.

JACKSON, P., CAVALIER, J., GOLDSTEIN, G., & MEINZER, F. C. (1995) Partitioning of water resources among plants of a lowland tropical forest. *Oecologia*, **101**, 197–203.

JACKSON P., MEINZER, F. C., GOLDSTEIN, G., HOLBROOK, N. M., CAVALIER, J., & RADA, F. (1993) Environmental and physiological influences on carbon isotope composition of gap and understory plants in a lowland tropical forest. *Stable Isotope and Plant Carbon-Water Relations* (eds. J. R. EHLERINGER, G. FARQUHAR & A. HALL), Academic Press, San Diego, pp 131–140.

KAPOS, V., GANADE, G., MATSUI, E. & VICTORIA, R. L. (1993) ^{13}C as an indicator of edge effects in tropical rainforest reserves. *Journal of Ecology*, **81**, 425–432

KEELING, C. D. (1961) The concentration and isotopic abundances of carbon dioxide in rural and marine air. Geochimica et Cosmochimica *Acta*, **24**, 227–298.

LEIGH, E. G. (1982) Why are there so many kinds of tropical trees? *The Ecology of a Tropical Forest; Seasonal Rhythms and Long-Term Changes* (eds. E. G. LEIGH, S. A. RAND, & D. A. WINDSOR) Smithsonian Institution Press, Washington, D. C., pp 63–66.

LEIGH, E. G., RAND, S. A., & WINDSOR, D. A., eds. (1982) *The Ecology of a Tropical Forest; Seasonal Rhythms and Long-Term Changes* Smithsonian Institution Press, Washington, D. C.

LIN, G. & STERNBERG, L. da S. L. (1993) Hydrogen isotopic fractionation by plant roots during water uptake in coastal wetland plants. *Stable Isotopes and Plant Carbon-Water Relations* (eds. J. R. EHLERINGER, G. FARQUHAR, & A. HALL) Academic Press. San Diego, pp 497–510.

MEDINA, E. (1982) Physiological ecology of neotropical savanna plants. *Ecology of Tropical Savannas* (eds. B. J. HUNTLEY, & B. H. WALKER) Springer-Verlag, Berlin. pp 308–335.

MEDINA, E., MONTES G., CUEVAS, E., & ROKZANDIC, A. (1986) Profiles of CO_2 concentration and $\delta^{13}C$ values in tropical rain forests of the upper Rio Negro basin, Venezuela. *Journal of Tropical Ecology*, **2**, 207–217.

MEDINA, E., STERNBERG, L., & CUEVAS, E. (1991) Vertical stratification of $\delta^{13}C$ values in closed natural and plantation forests in the Luquillo mountains, Puerto Rico. *Oecologia*, **87**, 369-372.

MEINZER, F. C., GOLDSTEIN, G., HOLBROOK, L. M., JACKSON, P., & CAVELIER, J. (1993) Stomatal and environmental control of transpiration in a lowland tropical forest tree. *Plant, Cell and Environment*, **16**, 429-436.

MEINZER, F. C., RUNDEL, P. W., GOLDSTEIN, G. & SHARIFI, M. R. (1992) Carbon isotope composition in relation to leaf gas exchange and environmental conditions in Hawaiian *Metrosideros polymorpha* populations. *Oecologia*, **91**, 305-311

MEINZER, F. C., SEYMOUR, V., & GOLDSTEIN, G. (1983) Water balance in developing leaves of four tropical savanna woody species. *Oecologia*, **60**, 237-243.

MERWE, N. J. & MEDINA, E. (1989) Photosynthesis and $^{13}C/^{12}C$ ratios in Amazonian rain forests. *Geochimica et Cosmochimica Acta*, **54**, 1091-1094.

MULKEY, S. S., WRIGHT, S. J. & SMITH, A. P. (1991) Drought acclimation of an understory shrub in a tropical forest. *American Journal of Botany*, **78**, 579-587.

MULKEY, S. S., WRIGHT, S. J. & SMITH, A. P. (1993) Comparative physiology and demography of three neotropical forest shrubs: Alternative shade adapted character syndromes. *Oecologia*, **46**, 526-537.

OBERBAUER, S. F., STRAIN, B. R., & RIECHERS, G. H. (1987) Field water relations of a wet-tropical forest tree species, *Pentaclethra macroloba* (Mimosaceae). *Oecologia*, **71**, 369-374.

QUAY, P., KING, S., WILBUR, D., WOFSY, S., & RICHEY, J. (1989) $^{13}C/^{12}C$ of atmospheric CO_2 in the Amazon basin: Forest and river sources. *Journal of Geophysical Research*, **94 (D15)**, 18327-18336.

RICHARDS, J. H. & CALDWELL, M. M. (1987) Hydraulic lift: Substantial nocturnal water transport between soil layers by *Artemisia tridentata* roots. *Oecologia*, **73**, 486-489.

ROESKE, C. A. & O'LEARY, M. (1984) Carbon isotope effects on the enzyme-catalyzed carboxylation of ribulose biphosphate. *Biochemistry,* **23,** 6275-6284.

SARMIENTO, G., GOLDSTEIN G., & MEINZER, F. (1985) Adaptive strategies of woody species in neotropical savannas. *Biological Review,* **60,** 315-355.

SCHLESER, G. H. (1990) Investigations of the $\delta^{13}C$ pattern in leaves of *Fagus sylvatica L. Journal of Experimental Botany,* **41,** 565-572.

STERNBERG, L. da S. L. (1989) A model to estimate carbon dioxide recycling in forests using $^{13}C/^{12}C$ ratios and concentrations of ambient carbon dioxide. *Agricultural and Forest Meteorology,* **48,** 163-173.

STERNBERG, L. da S. L., MULKEY, S. S., & WRIGHT, S. J. (1989) Ecological interpretation of leaf carbon isotope ratios: Influence of respired carbon dioxide. *Ecology,* **70,** 1317-1324.

THORBURN, P. J. & WALKER, G. R. (1993) The source of water transpired by *Eucaliptus camaldulensis*: Soil, grandwater, or stream. *Stable Isotopes and Plant Carbon-Water Relations* (eds. J. R. EHLERINGER, A. E. HALL & G. D. FARQUHAR) Academic Press, San Diego, pp 511-527.

TYREE, M. T., SYNDERMAN, D. A., WILMOT, T. A. & MACHADO J. L. (1991) Water relations and hydraulic architecture of a tropical tree (*Schefflera morototoni*): Data, models, and a comparison with two temperate species (*Acer saccharum* and *Thuja occidentalis*). *Plant Physioliology.* **96,** 1105-1113.

VIRGINIA, R. A., JARRELL, W. M., RUNDEL, P. W., SHEARER, G. & KOHL, D. H. (1989) The use of variation in the natural abundance of ^{15}N to assess symbiotic nitrogen fixation by woody plants. *Stable Isotopes in Ecological Research* (eds. P. W. RUNDEL, J. R. EHLERINGER & K. A. NAGY), Springer-Verlag, New York, pp 375-394.

VOGEL, J. C. (1978) Recycling of carbon in a forest environment. *Oecologia Plantarum,* **13,** 89-94.

WHITE, J. W. C. (1989) Stable hydrogen isotope ratios in plants: A review of current theory and some potential applications. *Stable Isotopes in Ecological Research* (eds. P. W. RUNDEL, J. R. EHLERINGER & K. A. NAGY), Springer-Verlag, New York.

WRIGHT, G. C., HUBICK, K. T., & FARQUHAR, G. D. (1988). Discrimination in carbon isotopes of leaves correlates with water-use efficiency of field-grown peanut cultivars. *Australian Journal of Plant Physiology* **15,** 815-825.

ZIMMERMANN, J. K., & EHLERINGER, J. R. (1990) Carbon isotope ratios are correlated with irradiance levels in the Panamenian orchid *Catasetum viridiflavum. Oecologia,* **83,** 247-249.

10

Root Growth and Rhizosphere Interactions in Tropical Forests

Robert L. Sanford, Jr. and Elvira Cuevas

10.1 INTRODUCTION

The challenge of working belowground, on the least understood portion of tropical forests, has been met by increasing numbers of studies over the last 25 years. For many basic questions concerning root physiology, however, only preliminary observations have been made, often in only one forest type or within a single species. In contrast to aboveground portions of tropical trees, roots are only beginning to be examined. Fine roots have been measured in numerous ways in recent years, yet large roots (sometimes called coarse roots) have been largely ignored. This situation is analogous to extensive studies of leaf activity without corollary studies of stem growth and respiration. A bottom-up physiological approach, synthesizing results of detailed studies of tropical tree roots, is not yet possible. Nevertheless, sufficient information is available to indicate trends in biomass and fine root allocation patterns for some tropical forests types in relation to either water and/or nutrient availability. Separating fine and coarse roots into discreet categories is a convenient methodological tool, especially in the measurement of biomass (Bohn, 1979), however, such dichotomies tend to disperse the focus that should be oriented toward whole-plant strategies. Likewise, the separation of tropical roots into descriptive morphological categories (Jenik, 1971) has allowed temperate zone-trained biologists to contend with the diversity of root forms in tropical forests, but has not provided a useful format for asking physiological questions. This approach may yet prove useful as new perspectives are incorporated, such as the architectural plasticity concept suggested by Fitter (1994).

Ultimately, tropical root physiology should be considered in the context of whole-plant carbon allocation patterns. In keeping with the purpose of this book, the focus and perspective of this chapter are on patterns belowground of biomass accumulation and allocation. Controls on allocation of carbon to root systems in temperate zone plants are controversial; tropical plants have received much less attention than temperate plants. Root biomass of tropical trees and forests is the most widely reported aspect of tropical tree roots, and a synthesis of that information is an appropriate starting point. Root response to disturbances of various intensities and scales and root adaptations to tropical environments are also the purview of this chapter. By understanding the ways that roots function in different tropical forests, it may be possible to recognize patterns that contribute to our overall understanding of carbon allocation in plants.

10.2   ROOT:SHOOT RATIOS AND SPECULATION ON THEIR CONTROLS

The best known and most widely investigated aspect of roots and root systems is the quantification of belowground plant mass. Root biomass has been measured for individual plants, for plant communities, and most often for representative ecosystems throughout the tropics. Early studies in tropical ecology, especially those intent on carbon and nutrient cycles, resulted in numerous measurements of the sizes of the various compartments (Golley, McGinnis & Clements, 1971; Kira, 1978). Root biomass is an obvious, if opaque, compartment. The compartment approach resulted from a static view of plant parts in which it was optimistically assumed that carbon allocation patterns to roots could be estimated based on aboveground biomass allometries (Bray,

1963). Yet, no general relationship between total above- and below-ground biomass or production has ever been established (Raich & Nadelhoffer, 1989). In recent years, attempts to estimate fine root allocation in relation to total carbon budgets have proven promising. In this context, a summary of total root biomass measurements is interesting for two reasons: 1) it allows for a perspective on below-ground plant mass in various tropical forest types, and 2) comparison with temperate zone forest systems is possible. In reviewing plant root biomass, it is important to remember that root biomass, like stem biomass, is the accumulated mass of a forest system, and it includes dead as well as live tissues.

Root plasticity and the regulation of the shoot:root ratio as a result of growing conditions are well known for temperate species (Russell, 1977), and considerable progress has been made over the last several decades in understanding the functional relationships between roots and shoots. Most of this progress has been from experiments on annual plants using solution culture, growth chambers, root boxes, non-destructive root observation techniques, and radioactive tracers (Klepper, 1991). How the results of that work can be transferred to tropical trees is not yet clear. In fact, there is general disagreement on the extent to which genetic vs. environmental influences determine root:shoot ratios. Moreover, it is poorly understood how temperature, water, fertility, and soil texture interact to exert controls on root:shoot ratios. Grime (1993) recognizes two alternative hypotheses for explaining root:shoot ratios. 1) the resource ratio hypothesis (Newman 1973, 1983; Iwasa & Roughgarden, 1984; Huston & Smith, 1987; Tilman, 1988, 1989), and 2) the root-shoot interdependence hypothesis (Donald, 1958; Grime, 1973, 1979; Chapin, 1980; Field & Mooney, 1986; Aerts, Boot & Van Der Aart, 1991; Campbell, Grime & MacKey, 1991). The resource ratio hypothesis asserts that trade-offs in allocation of captured resources will determine root:shoot ratios. When light is limiting, shoots will be relatively large, whereas on dry or infertile soils, root allocation will be largest. The interdependence hypothesis alleges that trade-offs will be severely constrained by the interdependence of root and shoot foraging activities. However, given the wide array of genera, rainfall, climate, and nutrient regimes in the tropics, one would expect a wide range of root biomass values. For example, if nutrient availability is a co-dominant control on root mass accumulation, then the range of nutrient use efficiencies in tropical forests would predicate a wide range of root:shoot ratios (Vitousek & Sanford, 1986).

Over the past 40 years, there have been more than 30 published studies of root and shoot biomass in tropical forests (Table 10.1). Most of these studies have been situated in lowland moist forests, with fewer in tropical deciduous forests and tropical montane forests. Few studies have examined mangrove, riparian, and Spodosol forests, which is unfortunate because these environments have produced the lowest and highest root biomass of all tropical forests.

The root:shoot ratio is related to the percentage contribution by roots to total community biomass (phytomass *sensu stricto*) by the following equation:

$$\% \text{ contribution} = ((R/S) \times 100)/((R/S) + 1).$$

Another way to consider % contribution is belowground biomass as a percent of total biomass (we include root mats, on the soil surface, as part of belowground biomass). Using these criteria, there are fairly distinct delineations of tropical forest ecosystems based on percent contribution (Figure 10.1). Tropical deciduous forests (also called tropical dry forests or seasonally dry forests), mangroves, and the forest on Spodosol all have the highest percentage contributions. These forests all occur in water stressed environments: the Spodosol forest is affected by alternating drought and flood events, the mangrove by salinity and tides, and the tropical deciduous forest by seasonal drought conditions. Tropical montane forest has the smallest range of % contribution, but the greatest amount of root biomass. Lowland moist forest, and the riparian forest, in particular, have low percent contribution. When compared to temperate coniferous and deciduous forests, tropical forests appear much more variable with regard to root biomass (Table 10.1). Temperate forests are most similar to tropical montane forest for percent contribution, although temperate forest above- and below ground biomass are much lower, on average, than tropical montane systems.

Perhaps the most striking distinction brought forward by this comparison is the large difference between tropical moist and tropical deciduous forest. Taken as a whole, it appears that tropical deciduous forests accumulate a greater fraction of their biomass in roots (percent contribution) than tropical moist forests, strengthening the hypothesis that with increasing drought conditions, relatively more carbon is allocated to the root system. Finally, the lowland riparian and Spodosol forests indicate that soil texture may be an important, though indirect, control on root biomass because mechanical impedance (subsoil hardpan) modifies water and aeration status, and nutrient content and availability (Glinski & Lipiec, 1990). The mangrove and Spodosol forest are the only examples of cases where belowground phytomass actually exceeds the aboveground.

Table 10.1.   Belowground and aboveground biomass, and percent contribution of belowground biomass for tropical forests globally.

Forest type	Country	Belowground Mg/ha	Aboveground Mg/ha	Total Mg/ha	% Contribution	Source
Spodosol	Venezuela	42	5.8	47.8	87.9	Bongers, Engelen & Kluge, 1985
Spodosol	Venezuela	69.8	29.8	99.6	70.1	Bongers, Engelen & Kluge, 1985
Spodosol	Venezuela	54.6	37	91.6	59.6	Sanford, 1989
Spodosol	Venezuela	127	182	309	41.2	Bongers, Engelen & Kluge, 1985
Mangrove	Panama	190	163	353	53.8	Golley et al., 1975
Mangrove	Puerto Rico	50	63	113	44.2	Golley, Odum & Wilson, 1962
TDF	Guanica	45	53	98	45.9	Murphy & Lugo, 1986
TDF	Senegal	58	82	140	41.4	Jung, 1969
TDF	India	17	28	45	37.8	Vyas, Gar & Vyas, 1977
TDF	Mexico	31	74	105	29.5	Castellanos, Maass & Kummerow, 1991
TDF	India	21	78	99	21.2	Golley, 1983
TDF	Congo	31	121	152	20.4	Bartholomew, Moyer & Laudelaut, 1953
TDF	India	11.7	52	63.7	18.2	Singh & Singh, 1981
Lowland	Venezuela	61	185	246	24.7	Sanford, 1989
Lowland	Cambodia	4	13.1	17.1	23.4	Hozumi et al., 1969
Lowland	Panama	40	150	190	21.1	Golley et. al., 1975
Lowland	Cambodia	70	345	415	16.9	Hozumi et al., 1969
Lowland	Ivory Coast	48	243	291	16.5	Muller & Nielsen, 1965
Lowland	Venezuela	61	335	396	15.4	Sanford, 1989
Lowland	Cambodia	51	297	348	14.7	Hozumi et al., 1969
Lowland	Sri Lanka	17.2	101	118.2	14.6	Koopmans & Andriesse, 1982
Lowland	Sarawak	6.3	49	55.3	11.4	Koopmans & Andriesse, 1982
Lowland	Cambodia	19	153	172	11.0	Hozumi et al. 1969
Lowland	Ghana	25	233	258	9.7	Greenland & Kowal, 1960
Lowland	Thailand	31	295	326	9.5	Ogawa, Yoda & Kora, 1961
Lowland	Ivory Coast	49	513	562	8.7	Muller & Nielsen, 1965

Table 10.1. (cont).

Forest type	Country	Belowground Mg/ha	Aboveground Mg/ha	Total Mg/ha	% Contribution	Source
Lowland	Thailand	31	331	362	8.6	Kira et al., 1967
Lowland	Thailand	33	371	404	8.2	Ogawa et al. 1961
Lowland	Brazil	32	406	438	7.3	Klinge, 1986
Lowland	Costa Rica	14.4	221	235.4	6.1	Raich, 1980
Lowland	Malaysia	20.5	475	495.5	4.1	Kato, Tadaki & Ogawa, 1978
Lowland	Panama	11	316	327	3.4	Golley et al., 1975
Lowland	Panama	10	366	376	2.7	Golley et al., 1975
Montane	Puerto Rico	75	226	301	24.9	Scatena & Larsen, 1991
Montane	Jamaica	54	209	263	20.4	Tanner, 1985
Montane	Brazil	328	1397	1725	19.0	Basilevich & Rodin, 1968
Montane	Brazil	201	860	1061	18.9	Basilevich & Rodin, 1968
Montane	Venezuela	56.4	348	404.4	13.9	Grimm & Fassbender, 1981
Montane	Nuova Guinea	40	310	350	11.4	Edwards & Grubb, 1982
Riparian	Panama	12	1175	1187	1.0	Golley et al., 1975
Average		49.8	296.0	345.8	18.0	
Mangrove		120	113	233	49.0	
Tropical Deciduous Forest		34	73	107	32.7	
Spodosol		55	37	925	9.6	
Temperate Coniferous		33	126	159	20.9	
Temperate Deciduous		42	190	232	18.2	
Mantane		125	559	684	18.0	
Lowland		32	270	302	11.9	
Riparian		12	1175	1187	1.0	

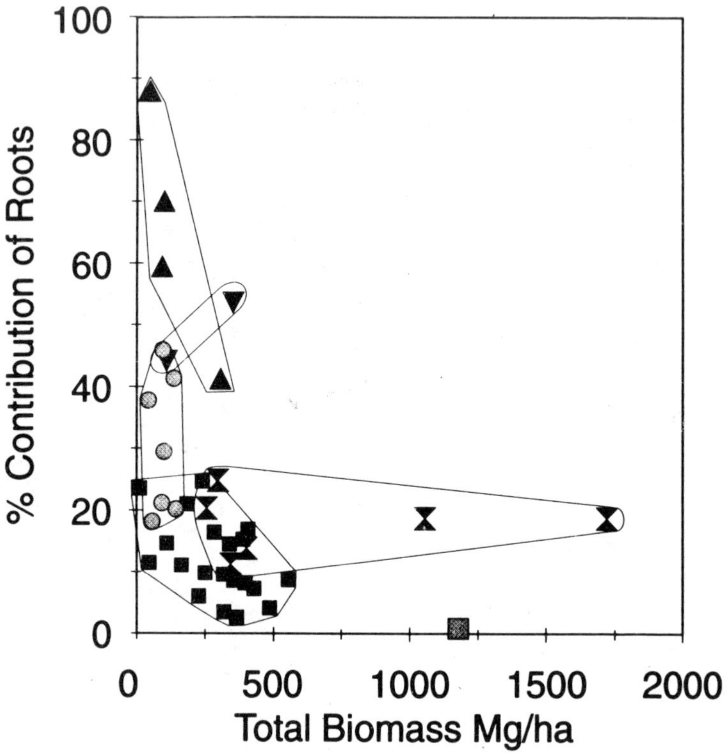

Figure 10.1. Total biomass and percent contribution of belowground biomass for tropical forests of the world. Six tropical forest vegetation types are shown; ▇ lowland, ▼ mangrove, ✕ montane, ▨ riparian ▲ Spodosols, ● tropical dry forest.

10.3 GROWTH PATTERNS

10.3.1 Specialized roots

Roots of tropical trees are by far the most diverse in morphology and habitat than for any other forest type. The diversity of tropical root growth forms has inspired an architectural approach to root system classification (Jenik, 1978) that uses criteria similar to the architectural classification models of shoot systems in tropical trees (Halle &

Oldeman, 1970). Rather than explore each of the 25 organization models of tropical root systems, only the most extreme and most common root systems are addressed here. Aerial stilt roots, adventitious roots, canopy roots, strangling roots, and apogeotropic roots are examples of specialized roots that are seldom or never found in other parts of the world.

Adventitious roots and apogeotropic roots add an above-soil, vertical dimension to root distribution in tropical forests. Adventitious roots in the canopy of tropical trees proliferate beneath the moist bryoflora and accumulated organic matter on branches and stems (Nadkarni, 1981). Although there are no estimates of fine root mass, respiration, or turnover in tropical forest canopies, these roots have been shown to be highly efficient in cycling several essential ions (Nadkarni, 1986; Herwitz, 1991). Similarly, apogeotropic roots, those that grow or orient against gravity, can extend root distribution (in some tropical forests) from the mineral soil upward, as high as 13 m into the sub-canopy (Sanford, 1987). Apogeotropic roots have been measured growing upward as fast as 1.9 cm per day (Sanford, 1987). These roots grow in response to nutrient gradients due to stem-flow (portion of rainfall that flows down stems and is usually enriched in nutrient elements relative to rain fall) or in response to sources of decomposing litter that have been trapped in the canopy (Raich, 1983; de Foresta & Kahn, 1984; Holbrook, Putz & Chai, 1985), or organic matter accumulation of different nutrient quality on the forest floor (St. John, 1983; Cuevas & Medina, 1988).

Little is known regarding measures of mass, production, mortality, or respiration of apogeotropic roots. Apogeotropic roots constitute an important fraction of ecosystem structure in the root mat found in tierra firme forests in San Carlos de Rio Negro, Venezuelan Amazon. Between 36 and 48% of the total mass of roots are found in this layer growing above the mineral soil surface (Stark & Spratt, 1977; Sanford, 1989). Fine root production (0-2 mm diameter) in the root mat has been estimated to be 0.85 g m-2 day-1 (Cuevas & Medina, 1988) and root mat respiration accounted for 67-82% of total soil respiration (Medina et al., 1980). Additionally, a substantial amount of litter mineralization, nutrient retention, and nutrient absorption takes place in the root mat (Stark & Jordan, 1978; Herrera et al., 1978; Cuevas & Medina, 1988; Tiessen, Chacon & Cuevas, 1994; Tiessen, Cuevas & Chacon, 1994).

Strangling roots are formed by epiphytes and hemiepiphytes and eventually fuse to form a cylinder around the host tree (Holbrook & Putz, Chapter 13). Eventually, if the host tree dies, the fused roots serve as a trunk and the strangler remains standing on its own. The genus *Ficus* (Moraceae) provides the best known examples, some of which have had rooting habits described in detail (Putz & Holbrook, 1989), but species of the following genera also have the same habit: *Metrosideros* (Myrtaceae), *Griselinia* (Grisilinaceae), *Nothopanax* = *Polyscias* (Araliaceae), and *Clusia* (Guttiferae) (Walter, 1971). In a study of two hemiepiphytic, strangler *Ficus* species in the Venezuelan llanos, Putz and Holbrook (1989) found upward and downward growing roots, although only the downward growing roots fused. The xylem structure (vessel diameter and density) of upward growing roots did not differ from downward growing roots, leading them to conclude that moisture availability was similar for the two types of roots. Another interesting aspect of strangler roots is that their structure and function shifts as the strangler matures. Initially, strangler roots provide a holdfast and acquire resources from the host crown. Eventually, these roots extend downward to the soil, thicken, and anastomose until support by the host is no longer necessary. Growth rates, respiration, and mortality have yet to be measured for strangling roots.

Stilt (prop) roots are heavily sclerified, stout aerial roots that originate from an aerial stem with mechanical support as a primary function (Benzing, 1991). These roots are commonly found on mangrove species (where they comprise at least 24% of the aboveground biomass; Golley, Odum & Wilson, 1962), on swamp forest species (notably *Anthocleista nobilis* [Potaliaceae] and *Voacanga thouarsii* [Apocynaceae]; Jenik, 1971, 1978), on numerous palm species (Dransfield, 1978), and on *Pandanus* spp. (Pandanaceae). In mangrove species, stilt roots tend to emerge perpendicular to the stem. Aboveground, they have numerous peculiarities, such as some anatomical features normally associated with stems, elongation by cell formation rather than by cell elongation, and no root apical geotropic or phototropic response (Gill & Tomlinson, 1975). Once mangrove roots penetrate the soil, they undergo marked physiological and anatomical changes including the loss of chlorophyll, initiation of lateral roots without stimulation from injury, and extensive development of aerenchyma just above the soil surface. Taken together, these changes produce a root system where aerial roots have only 5% gas space before penetration into the soil, compared to 50% gas space in the large subterranean roots and branches after soil penetration (Gill & Tomlinson, 1975), allowing for respiration to continue even when the roots are in an anaerobic environment, such as during high tide. Growth rates of stilt roots have been measured in the case of *Pan-*

danus baptistii, where growth follows a seasonal pattern with rates of elongation at 1 cm per month or less during November–December, then increasing to a maximum elongation rate of 1 cm per day sometime between January and May (Gill & Tomlinson, 1975). Respiration, photosynthesis (when it occurs), and mortality rates have not been measured for stilt roots.

10.3.2   Vertical and horizontal distribution of roots

Non-specialized, lateral and tap roots occur in (and sometimes on top of) the mineral soil. These are the most abundant roots in terms of mass or surface area in tropical forests. Roots in some humid tropical forests form dense mats on top of the soil surface (Stark & Spratt, 1977). In Amazon tierra firme forest, the average depth of the surface root mat is 20 cm (+ 0.9 cm SD) above the surface of the mineral soil (Sanford, 1987). Elsewhere, especially on Spodosols, root mats have also been observed in tropical Asia (Brunig & Sander, 1983) and in the Amazon (Klinge, Medina & Herrera, 1977; Bongers, Engelen & Kling, 1985; Sanford, 1989). A high proportion of fine roots in the superficial soil layers is characteristic of tropical forests, although degree of rooting depth varies according to soil type, type of forest, and species. Recent excavations have uncovered fine roots at up to 18 m depth in evergreen forests in northeastern state of Para, Brazil (Nepstad et al., 1995). Using a soil water balance approach, Nepstad et al. (1995) assessed the hydrologic role of deeply-penetrating roots and contend that these roots help explain why Amazonian evergreen forests extend well into the seasonally dry, eastern and southern regions of Amazonia (1.8×10^6 km^2) where long, dry seasons would otherwise preclude an evergreen habit. Taproots and sinker roots of many species reach a depth of 3-5 m in some well drained, deep soils (Mensah & Jenik, 1968; Forster, 1970; Lawson, Armstrong-Mensah & Hall, 1970; Huttel, 1975). Mulkey and Wright (Chapter 7), Goldstein et al. (Chapter 9), and Wright (Chapter 15) emphasize the importance of rooting depth in understanding plant water relations and phenological patterns.

Although there has been considerable curiosity and some research on the diversity of root forms, little information is available on the horizontal extent of tree roots. Horizontal root distribution is species and age dependent. Singh and Srivastava (1984) found that young (< 6 yr) teak (*Tectona grandis*) had most fine root biomass within 50 cm of the tree stem, and that lateral distribution increased with increasing age. In older trees, higher fine root density was found again at shorter distances from the trees, probably as a result of decreased competition with other species' roots, larger nutrient availability via stemflow, and better moisture conditions. Kummerow et al. (1990) also found similar tendencies in horizontal distribution of fine roots for two species in the dry deciduous forest in Chamela, Mexico. Pickett (1976) reported the horizontal extent of roots in mineral soil for the neotropical lowland tree *Castilla elastica*. Roots of the same tree were found far outside the canopy drip-line, as far as 5 times the radial extent of the tree crown (crown radius = 7 m; root distance = 35 m). *C. elastica* is considered to be a fast-growing, disturbance-response species, so these results may not be easily extrapolated to slower growing species. (For further information on growth of pioneer species, see Ackerly, Chapter 21).

10.3.3 Fine root growth rates

High fine-root biomass is a common feature in many tropical forests, generally occurring as root mats on the soil surface or concentrated in the uppermost 10–20 cm of mineral soil in tropical moist forest (Stark & Spratt, 1977), in tropical dry forest (Kummerow et al.,1990; Castellanos, Maass & Kumerow, 1991), in tropical montane forest (Grimm & Fassbender, 1981), and in forest on Spodosols (Klinge & Herrera, 1978; Bongers, Engelen & Klinge, 1985). The large amount of fine root biomass near the soil surface has been the subject of many recent studies on growth rates, seasonality patterns, and carbon costs (maintenance and structural). The observations that fine root production may exceed leaf production in some tropical forests are intriguing and may well lead tropical plant physiologists to consider examining belowground processes much more thoroughly.

Several recent studies in intact forests (moist lowland and tropical dry forest) focus on growth of fine roots. Working with sequential cores and mini-rhizotrons, Sanford (1995) measured fine root (< 2 mm) growth in the Amazon for intact tierra firme forest. In this forest system, approximately 50% (1267 g m^{-2}) of total fine root mass is located in the uppermost 10 cm of the mineral soil. Repeated monthly observations of fine roots indicated that there were two peaks in production, one in September and another in December, with root mortality greatest in October. Although there are no data for aboveground leaf production, leaf litterfall peaks between November and

January (Cuevas & Medina, 1986). Hence, in this forest, it appears that root mortality and leaf senescence are sequential, occurring just before the onset of the dry season (January – March). This trend should be tested at other sites and for longer than 1 year to provide the basis for a robust, general pattern.

Using similar techniques in lowland moist forest in Costa Rica (across a 2 km gradient of soil fertility), Sanford and Vitousek (1995) measured fine root (< 2 mm) and leaf litter at three intact forest sites ranging from recent alluvium to highly weathered soils (0-10 cm depth). On the most nutrient poor sites, total fine root litter was 1483 g m^{-2} yr^{-1}, which far exceeded leaf litter (763 g m^{-2} yr^{-1}). In contrast, total fine root litter on the alluvial sites (412 g m^{-2} yr^{-1}) was far less than leaf litter (718 g m^{-2} yr^{-1}). Average monthly fine root turnover in these two forests was 40% for the forest on weathered soils, compared to 27% for the alluvial forests. Fine root mortality was highest during the dry season (January – April), followed by a peak in fine root production in July during the wet season. Leaf litter reached a peak in March – April.

Fine root measurements in tropical dry forests also indicate seasonality of fine root production. Kummerow et al. (1990) found that fine root biomass decreased 160 g m^{-2} between wet and dry seasons in western Mexico. Singh and Singh (1981) have estimated the annual turnover of fine roots to be 37 percent, with most mortality concentrated in the dry season. Srivastava et al. (1986), working in teak forest plantations in the same tropical dry region of India, found that fine root biomass increased 71 percent from the dry season to the wet season maximum. Kavanaugh and Kellman (1992) used an experimental approach with root ingrowth bags to measure fine root proliferation on tropical dry forest on sand dunes on the east coast of Mexico. They measured enormous seasonal fluxes of fine root growth that coincided with the onset of the wet season. The reduction of root growth that occurs late in the wet season is independent of soil moisture and soil nutrients. Kavanaugh and Kellman (1992) speculate that carbohydrate reserves that had accumulated in roots over the course of the dry season were depleted in the early (wet season) period of rapid growth. Singh and Srivastava (1985), working with changes in fine root biomass and total non-structural carbohydrates (TNC) as related to size, depth, and seasonality in teak plantations, had already found an inverse relation between live fine root biomass and carbohydrate reserves. Although these seasonal changes were observed at all soil depths, the higher concentrations and bigger changes were measured at 0-20 cm depth, where the highest proportion of fine roots are found. Singh and Singh (1981) also found lower caloric values in fine roots of deciduous trees during the rainy season when compared to dry season values. This pattern of fine root carbohydrate depletion has been observed previously in temperate forest species (Ford & Deans, 1977; Ericsson & Persson, 1980; Marshall & Waring, 1985).

Root turnover can also be an important source of nutrients in tropical ecosystems. Total fine root litter was 1536 g m^{-2} yr^{-1} and exceeded leaf litter (757 g m^{-2} yr^{-1}) (Cuevas & Medina, 1986) by a factor of 2 in a Venezuelan tierra firme forest, where fine root turnover was 26 percent per month. Nitrogen and phosphorus quantities added to the soil by fine root turnover were very large – 190 kg N ha^{-1} yr^{-1} and 9 kg P ha^{-1} yr^{-1}, respectively (assuming no retranslocation from the roots before death). Fluxes of N and P from litterfall amounted to 121 kg ha^{-1} yr^{-1} and 2 kg P ha^{-1} yr^{-1} (Cuevas & Medina, 1986), indicating that fine root production and turnover in forests on very infertile soils may exceed leaf litterfall as a source of nutrient input.

10.3.4   Disturbance responses

Fine root growth and mortality in lowland moist forests have been measured following various disturbances, including hurricane, forest clearings, and light gap (canopy) openings. Working in the subtropical wet forest life zone in Puerto Rico, Parrotta and Lodge (1991) followed fine root production and mortality changes that occurred in response to Hurricane Hugo, a 1989 storm with moderate intensity and a recurrence interval of 50–60 yr (Scatena & Larsen, 1991). Parrotta and Lodge (1991) found an extremely low, live fine root (< 3 mm) biomass (0–2 g m^{-2}) in the upper 10 cm of mineral soil 4 weeks after passage of the storm; two months later, live root mass was reduced to 0 g m^{-2}. Over the following 10 months, live fine root biomass fluctuated, with peaks in February (29 g m^{-2}) and in June (49 g m^{-2}). Sharp decreases (in April and August) followed each of the peaks. Parrotta and Lodge (1991) suggest that more than one year is required for fine root recovery following hurricane disturbance, and they attribute this slow rate of recovery to three factors: physical disturbance, moisture stress, and low non-structural carbohydrate reserves that contributed to low levels of fine roots following hurricane Hugo.

Forest clearing is another form of disturbance in lowland tropical forests that has been studied with respect to fine roots. Silver and

Vogt (1993) worked in the same forest type in Puerto Rico as Parrotta and Lodge (1991), where they created small forest clearings, removing all of the aboveground biomass from half of their experimental plots. Within 6 weeks of felling, fine root biomass increased significantly from 95 g m^{-2} to 160 g m^{-2}, followed by a 40 percent decline in the plots with aboveground material removed, but no change in the control plots. The disturbance created by hurricane Hugo resulted in a sharp decrease (by nine months following the hurricane) to 20–25 g m^{-2} in all experimental plots, representing a 75% loss of the fine roots that were in the plots prior to creating the clearings. Silver and Vogt (1993) speculate that the short, rapid (within 5 days) increase in fine root mass immediately following forest clearing was a stress response to aboveground damage. Other studies of forest clearing effects on fine roots in Costa Rica found no significant differences in fine root biomass 11 weeks following forest clearing (Ewel et al., 1981), or even after 1 year (Raich, 1980). However, in Amazon forests, Sanford (1995) reported large differences in fine root biomass and mortality in forest clearings that were used for slash and burn agriculture and subsequently abandoned (Table 10.2). Fine root biomass (< 2 mm) in traditional agricultural plots (0–10 cm depth) was only 7% percent of intact forest fine root biomass 2 years after forest clearing. Following 3 years of succession, fine root biomass was 32%, and after 7 years, fine root biomass was nearly 70% of intact forest. This pattern parallels leaf biomass accumulation during succession; after 3 years, standing leaf biomass is 35% of intact forest leaf biomass and after 7 years, 71% (Jordan & Uhl, 1978; Uhl & Jordan, 1984; Cuevas & Medina, 1986). These results indicate that following slash and burn succession, carbon is accumulated in roughly equal proportions by fine roots and leaves.

Light-gap openings are the most ephemeral and least extensive of all disturbances addressed here. Two studies have examined fine roots in gaps and their results are contradictory. In lowland moist forest in Costa Rica, Sanford (1989) found fine root (< 2 mm diameter) biomass significantly less than in adjacent disturbed forest. This reduction in fine root biomass was observed in the canopy zone and in the root up-turn zone of the treefall but not along the bole. In contrast, Sanford (1990) found no difference in fine root biomass accumulation among gap zones, but did measure significant differences in root biomass in gaps of different sizes. These later results suggest that the species of tree forming the gap opening and the size of the gap may control fine root growth more than any inherent growth patterns within zones of a gap.

*Table 10.2.   Fine root biomass, mortality, and turnover calculated from se-
quential cores and mini-rhizotrons in successional and old-
growth tierra firme forest, San Carlos de Rio Negro, Venezuela.
All values are for fine roots (> 2 mm) in the surface (0-10 cm)
horizon. Leaf litter was collected from litter traps at the same
sites over the same (1 yr) time period.*

	Fine root mass (g m^{-2})	Average monthly mortality (%)	Fine rotto turnover g m^{-2} yr^{-1}	Leaf litter g m^{-2} yr^{-1}
Agriculture	36.4	20.1	87.8	–
1 yr succession	198.6	17.0	405.1	241.3
3 yr succession	157.9	29.4	557.1	266.8
7 yr succession	341.3	19.4	806.8	538.7
old-growth	494.1	25.9	1535.7	757.0

Studies on fine root growth following perturbance of lowland tropical moist forest are also interesting with regard to root growth during
secondary succession. Spatial distribution of root mass varies with the
successional stage of the vegetation and root distribution follows nutrient distribution in both space and time (Myers et al., 1994). In early
successional stages when litter layer formation is minimal, fine roots
extend deepest into the soil. After a litter layer develops, nutrient
distribution is more concentrated near the soil surface and more fine
roots are found in the superficial layers (Berish, 1982; Berish & Ewel,
1988; Bowen, 1984).

Difficulties involved with fine root production estimates have been
reviewed previously (Singh et al., 1984; Kurz & Kimmons, 1987;
Nadelhoffer & Raich, 1992). In general, estimates derived from sequential core methods tend to yield large overestimates when fine root
biomass is large and/or spatially variable, and underestimates if
growth and mortality occur simultaneously. The ingrowth core
method has also been used in tropical forests and it appears to be
promising, although there are too few studies to examine the relationship between fine root production and aboveground production.
Cuevas, Brown & Lugo, (1991) used both methods simultaneously to
measure fine root production and found no significant differences
between the two methods.

Cuevas, Brown & Lugo, (1991) found similar total production
(above-plus belowground) in a *Pinus caribea* plantation and a secondary forest (dominated by *Tabebuia heterphylla*) of the same age growing on similar soils and climate. Interestingly, belowground allocation
differed according to species dominance: the *Pinus caribaea* plantation

allocated 6% of total production belowground, whereas the secondary forest allocated 44% belowground. Cuevas, Brown & Lugo, (1991) concluded that a community level response can be determined by water and nutrient constraints, although the degree of response is species-specific.

10.3.5 Large roots

In contrast to fine roots, large roots in tropical forests have been largely ignored. Studies of large root biomass have not been complemented by growth, mortality, and respiration measurements in lowland moist forest. Singh and Singh (1981) provide the only measure of coarse root production in a tropical dry forest in India, where net production of coarse tree roots ($40 \text{ g m}^{-2} \text{ yr}^{-1}$) and coarse shrub roots ($20 \text{ g m}^{-2} \text{ yr}^{-1}$) account for only 7 percent of total net production ($860 \text{ g m}^{-2} \text{ yr}^{-1}$).

The only other aspect of large roots that has received systematic attention is root grafting. As long ago as 1952, LaRue reported intraspecific root grafting in 34 genera of tropical trees and suggested that natural root grafting is more common in tropical trees than in temperate ones. Recently, Basnet et al. (1993) observed extensive root grafts in lower montane forest in Puerto Rico, where over 60% of all stems of the canopy-dominant tree tabonuco (*Dacryodes excelsa*) grafted to form unions. Most grafts occurred at the soil surface, and none were found further than 20 cm depth. Grafting proclivity depended on life stage in this species; no grafts were encountered in seedlings and saplings. In addition, there were no interspecific root grafts observed.

Two hypotheses have been generated to explain the abundance of root grafts in lowland tropical moist forests: 1) inter-tree translocation of water and nutrients (Bormann, 1966; Graham & Bormann, 1966), and 2) root grafting increases the anchorage, support, and stability of trees growing in shallow soils (Kumar, Kulkarmi & Srimathi, 1985; Keeley, 1988). The recent work by Basnet et al. (1993) supports the later hypothesis because they found no evidence for inter-tree competition, and because tabonuco trees with root grafts were significantly less likely to be damaged by hurricanes. Basnet et al. (1993) further suggested that root grafting may also promote the dominance of tabonuco over other species because dominant trees in a union are known to supply photosynthates to suppressed trees (Kuntz & Riker, 1956; Bormann, 1966), and numerous "dead" tabonuco trees were observed to recover after hurricane Hugo battered tabonuco unions.

The advent of further studies on root grafting in tropical trees may prove important in understanding the persistence of single-species stands in the moist lowland tropics. Root grafts are especially intriguing in the sense that individual tree growth patterns and carbon allocation patterns are not likely to be independent.

10.4   CARBON ALLOCATION TO ROOTS

Direct measurements of root respiration are usually difficult due to the fact that in intact forest soils, CO_2 produced by soil organisms is mixed with CO_2 produced by living roots. Hence, studies of soil respiration include heterotrophic and autotrophic sources of carbon and, until recently, have been used exclusively for soil carbon budgets (with an occasional guess made for root respiration proportions). Raich and Nadelhoffer (1989) have provided the basis (and the assumptions) for estimating root respiration based on the following regression equation:

$$R_s - P_a = P_b + R_r$$

where R_s is soil respiration, P_a is aboveground detritus collection, P_b is belowground detritus production, and R_r is root respiration. Using this equation, total annual carbon allocation to roots ($P_b + R_r$) can be made from measurements of annual rates of soil respiration (R_s) and aboveground detritus production (P_a), both of which have been measured in tropical (and other) forests around the world (Figure 10.2). Following the same protocols as Raich and Nadelhoffer (1989), we do not include soil respiration rates that were obtained with unreliable techniques. Hence, for tropical forests, we estimate that soil respiration (R_s) is directly related to aboveground litterfall (P_a) by the least squares regression: $R_s = (2.74\,P_a) + 233$ ($r^2 = 0.85$, $P < 0.001$, $n = 6$).

The most important of the assumptions required to use this equation, are: 1) use of litterfall (exclusive of coarse woody debris) to estimate aboveground detritus production and 2) that total carbon allocation to roots is equal to the sum of carbon allocated to root detritus production plus respiration ($P_b + R_r$); this is equivalent to what most investigators refer to as fine root production.

Tropical forest carbon allocation to roots is strikingly similar to the global forest pattern for forests described by Raich and Nadelhoffer (1989). Subsequent analyses of global patterns of fine root production (Nadelhoffer & Raich, 1992) suggest that the carbon allocation to fine root production can be reliably derived from the nitrogen budget

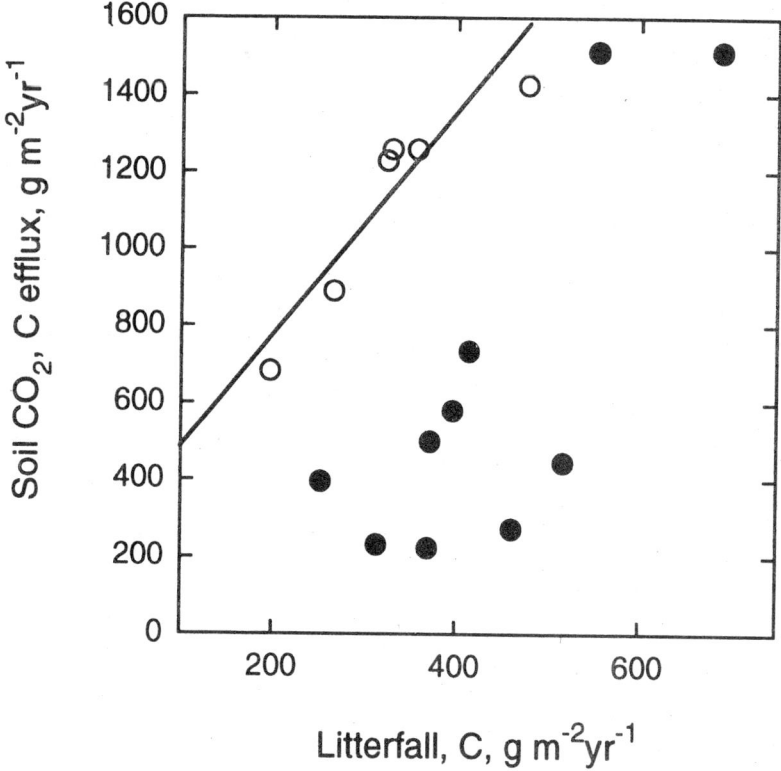

Figure 10.2. The relationship between soil respiration (measured as CO_2 carbon) and litterfall carbon in tropical forests of the world. Soil respiration values that were obtained with reliable techniques are shown as open circles (O); all other values are shown as solid circles (●). The solid line shows the linear regression on open circles (reliable techniques).

method, where carbon allocated to fine root production = (0.76 P_a) + 5.52. Following this scenario, approximately one-third of total carbon allocation to roots should be utilized for fine root production, with the rest allocated toward respiration. For the only tropical forests where it has been tried, the N budget method provides a large over estimation of fine root production (Sanford & Vitousek, 1995). For tropical forests, where N is not limiting to tree growth, the N budget method may not be applicable.

10.5 RHIZOSPHERE INTERACTIONS

The soil immediately around a root is generally referred to as the rhizosphere (Hiltner, 1904). More recently, Russell (1977) defined the rhizosphere as an area equal to the diameter of the soil through which fine roots and root hairs explore. The flora of the rhizosphere includes bacteria, fungi, and actinomycetes. It has been known for many years that the abundance of bacteria is much greater in the rhizosphere than in soil further from the root (Newman, 1985). Fungi are also usually more abundant in the rhizosphere, but less so than bacteria, and occasionally they show no clear difference in abundance between the rhizosphere and the bulk soil (Papavizas & Davey 1961). The causes for rhizosphere microbial population gradients are the organic exudates produced by roots and generally referred to as rhizodeposition (Newman, 1985). Rhizodeposition is strongly influenced by moisture stress (Martin, 1977); thus microbial activity in the rhizosphere depends on soil water availability to the plant. Rhizodeposition includes sugars, organic acids, amino acids, sloughed root cells, mucigel, and unidentified insoluble, carbon rich compounds (Rovira, 1969; Barber & Martin, 1976; Whipps & Lynch, 1983; Newman, 1985). Most of the data on rhizodeposition come from a few soils and plant species, mainly from annual, temperate zone crop plants. Exceptions are the exploratory studies that have been conducted on a dozen temperate woody species, and two tropical woody species: *Persia gratissima* (Zentmeyer, 1961) and *Zamia* sp. (Mishra & Kanaujia, 1973/74). Although rhizodeposition has been widely demonstrated, there are no estimates of total carbon costs for any tropical forest species.

In contrast to rhizosphere inquiry, there have been numerous studies on mycorrhizae in tropical forests. Both vesicular-arbuscular mycorrhizae (VAM) and ectomycorrhizae (EM) are present in tropical forests, with the former predominant in tropical host taxa worldwide, and the later most abundant in areas of the lowland paleotropics such as miombo woodland in Africa or Asian dipterocarp forests (Janos, 1983). Species with ectomycorrhizal associations are also known to occur in groves in lowland moist forest in Cameroon (Newbery et al., 1988), as fairly common taxa in extremely nutrient poor Amazon forests (St. John & Uhl, 1983; Moyersoen, 1993) and as the dominant association in tropical moist forest in Liberia (Germain & Evrard, 1956; Gerard, 1960) and Zaire (Voorhoeve, 1964). The ability of EM to mineralize soil organic matter, to absorb inorganic phosphorus, and to effectively "mine" the soil for nitrogen has long been controversial

(Alexander, 1983), but recent work has confirmed that EM in some temperate forest species can obtain nitrogen directly from organic matter (Read, Leake & Langdale, 1989; Hogberg, 1990).

VAM have been shown to improve phosphorus absorption in tropical moist forests as well as tree seedling/sapling survival (Janos, 1985). More recently, Lodge (1993) has proposed that VAM are important in controlling nutrient availability to plants in lowland moist forest and in tropical dry forest. The degree of importance of VAM varies with soil nutrient availability. Salcedo, Elliot and Sampaio (1991), working in an Atlantic coastal forest in Recife, Brazil, showed that phosphorus from the litter/fermentation layer is cycled back to the vegetation via VAM mediated mechanisms. However, 61% of the added ^{32}P moved down into the surface mineral soil, where P in the soil solution is controlled by microbial biomass activity. In contrast, Stark and Jordan (1978), working on a P-deficient, upland Amazon forest (Cuevas & Medina, 1988), found that nearly 100% of the added ^{32}P was retained in the root mat associated with the litter layer, with less than 1% moving down to the surface mineral soil. Fungal and other microbial biomass fluctuate rapidly in response to wetting and drying cycles, even in non-seasonal tropical forests, leading to net nutrient immobilization followed by pulses of mineralization (Lodge, 1993). This pulsed nutrient hypothesis (Lodge, McDowell & McSwiney, 1994) may be an important process in nutrient conservation in many tropical forests, and may help explain the results of Behera, Pati and Basu (1991) and Yang and Insam (1991), who found highest immobilization of nutrients by fungi in the rain season, thereby conserving nutrients against leaching.

The carbohydrate physiology of VAM has not been studied in great detail, however there is clear evidence, based on $^{14}CO_2$, that the carbon requirements of VAM are supplied by the host plant (Ho & Trappe, 1973). The magnitude of the drain on photosynthesis from VAM infection is hard to estimate (Gianinazzi-Pearson & Gianinazzi, 1983). Based on studies of temperate crop-plants, enhanced carbon demand by VAM-infected roots is based on the fungus itself as well as increased metabolic activity following infection (Pang & Paul, 1980; Tinker, 1978; Cox & Tinker, 1976). However, VAM-infected plants can compensate for increased carbon demand by increased photosynthesis (Allen et al., 1981a,b). Studies of tropical woody plants have yet to examine carbon costs associated with mycorrhizal infection.

Actinorhizal plants are rare in the neotropics. Exceptions include *Myrica* and *Alnus*, species of Central and South American highlands (Paschke & Dawson, 1992). The exotic actinorhizal tree *Casuarina equisetifolia* is planted for windbreaks in the highlands. The presence of infective and effective *Frankia* in five tropical forest soils of Costa Rica, including lowland wet and dry forest and premontane rainforest, suggests that there may be unknown actinorhizal host species in these forests (Paschke & Dawson, 1992).

10.6   ROOT LONGEVITY

Classical studies of fine root growth and turnover have been conducted under laboratory conditions, usually on herbaceous crop plants (Gregory, 1987), or in the field with temperate deciduous and conifer species (Vogt & Bloomfield, 1991). Temperate zone deciduous tree roots tend to be shorter lived (< 1 yr) than their coniferous counterparts (1–12 yrs), however there is very little agreement as to the causes of root longevity and average longevity (Vogt & Bloomfield, 1991). The only study of fine root growth for a tropical tree species was for a teak (*Tectona grandis*) plantation in a seasonally dry environment (Srivastava, Singh & Upadhyay, 1986) where, by estimating fine root biomass at monthly intervals, it was concluded that 50% of all very fine roots (< 1 mm) die during the dry season.

As with temperate trees, there is considerable speculation on genetic vs. environmental controls on root longevity for tropical trees. Cuevas, Brown and Lugo (1991), working with a 12-yr-old *Pinus caribaea* plantation and a secondary forest of the same age growing under the same edaphic and climatic conditions, estimated that mean longevity of fine roots was 0.27 and 0.81 years, respectively, thus pointing toward a genetic control at the species level. Seasonal variation in precipitation controls root turnover both quantitatively and qualitatively in lowland moist and tropical dry forests, and presumably affects both structural and maintenance costs to root systems. Increased growth of roots probably follows peak leaf abscission in moist forests, whereas in lowland tropical deciduous forests, root production commences with the onset of the wet season, simultaneous with leaf emergence (Singh & Singh, 1981; Kummerow et al., 1990; Kavanaugh & Kellman, 1992).

The extent of soil moisture controls on tropical tree root turnover as a whole is unknown. Limited data for temperate trees indicate that fine roots live longer with increased moisture availability (Vogt & Bloomfield, 1991).

The effect of soil nutrients on root turnover has received some attention in tropical forests, although clear patterns have not yet emerged. Fitter (18.) summarized data from temperate plants, finding that fine roots we. e most abundant in young root systems, in roots grown in nutrient-poor soils, and in the parts of split root systems supplied with the highest nutrient levels. St. John (1983) found that root proliferation may be related to zones of organic carbon accumulation and higher nutrient availability. Vitousek and Sanford (1986) report that most lowland tropical forests appear to be phosphorus limited. Phosphorus solubility, availability, and absorption kinetics differ from nitrogen. Hence, it is possible that tropical tree roots, in general, respond more to P than to N availability, especially on non-flooded soils. This hypothesis is supported by the work of Cuevas and Medina (1988), where a significant response in fine root proliferation to a specific nutrient addition was used as an index of nutrient limitation to dystrophic systems. Fine roots in non-flooded, upland Amazon forests grew in response to P and Ca additions, whereas a positive response to N addition was the only response in tall Amazon caatinga on flooded soils. The nutrient limitations observed in the soil/litter system of two dystrophic soils (Tiessen, Cuevas & Chacon, 1994; Tiessen, Chacon & Cuevas, 1994) confirms the reutilization patterns of N and P observed by Cuevas and Medina (1986, 1988) and Medina and Cuevas (1989).

Vogt and Bloomfield (1991) found that higher nutrient soil status (especially N) promotes temperate tree fine root longevity. Although minimal data support the role of limiting nutrients in tropical forests on shoot/root distribution and longevity, Gower (1987) found that fine root biomass was inversely related to phosphorus and calcium availability in lowland moist forest. He also found that fine root biomass had no relationship with nitrogen availability. Working in the same forest, Sanford and Vitousek (1995) reported that fine root turnover (average monthly mortality) was inversely related to phosphorus availability; 27% on high P soils and 40% on low P soils. These two studies report low fine root biomass and low turnover rates on the sites richer in soil P, indicating that phosphorus availability may influence root longevity for tropical trees, while N availability influences fine root longevity for temperate trees.

10.7   DIRECTIONS FOR FUTURE RESEARCH

Root biomass, production, mortality and respiration patterns are just beginning to be revealed in tropical forests. As a general rule, lowland forests (both wet and dry) have received more attention than montane tropical forests, and lowland moist forests are more explored than dry forests or forests on stress-inducing soils such as Spodosols, riparian soils, or saline environments. The extremely low root:shoot ratio of the riparian forest and the extremely high root:shoot ratios of the forest on Spodosols are interesting and deserve more attention. Although both of these environments are very limited in a global sense, they mark the extremes for root:shoot ratios and perhaps for carbon allocation belowground.

Specialized roots may provide some of the best initial sources for direct measurement of physiological processes. Apogeotropic and adventitious roots are readily observed in numerous lowland moist and montane tropical forests, but they have seldom, if at all, been sampled for respiration or mortality rates. These roots are reported to be morphologically and anatomically indistinguishable from fine roots in mineral soils, so they represent a likely starting point for direct measurement in the field. The lack of information on large roots (aside from biomass) is striking, especially in light of recent studies on root grafting. In some tropical forests, carbon allocation patterns may be substantially modified by root grafts.

One promising approach to resolving root carbon allocation involves soil carbon budgets. In contrast to temperate zone forests, where dozens of reliable soil respiration measurements have been made, there are only six reliable soil respiration values (and accompanying litterfall measurements) for tropical forests globally. Not only is this inadequate to determine the reliability of the technique, it is also impossible to determine if differences exist between tropical forest types. This potentially robust approach needs to be further investigated and tested.

Investigation of controls on tropical tree root turnover are just beginning. Promising areas of research include construction of carbon allocation budgets at the individual (seedling and mature tree stage) and at the community level, and tests of moisture and nutrient modification on carbon allocation patterns. Initial results from field studies indicate that both biomass accumulation and fine root turnover are controlled, in part, by these environmental variables.

Finally, further study is warranted on seasonality patterns of fine root growth and mortality. Additional studies of root:shoot allocation patterns should extend for at least 2 years to establish patterns of seasonal trade-offs between roots and shoots. For future research, whole-tree aspects of carbon allocation should result in further integration and study of tropical tree root physiology.

REFERENCES

AERTS, R., BOOT, R. G. A. & VAN DER AART, P. J. M. (1991) The relation between above- and belowground biomass allocation patterns and competitive ability. *Oecologia*, **87**, 551– 559.

ALEXANDER, I. J. (1983) The significance of ectomycorrhizas in the nirogen cycle. *Nitrogen as an Ecological Factor* (eds. J. A. LEE, S. MCNEILL & I. H. RORISON) Blackwell Scientific, Oxford, UK.

ALLEN, M. J., SEXTON, J. C., MOORE, T. S. & CHRISTENSEN, M. (1981a) The influence of phosphate source on vesicular-arbuscular mycorrhizae of *Bouteloua gracilis.* *New Phytologist.*, **87**, 687–694.

ALLEN, M. J., SMITH, W. K., MOORE, T. S. & CHRISTENSEN, M. (1981b) Comparative water relations and photosynthesis of mycorrhizal and non-mycorrhiozal *Bouteloua gracilis*. *New Phytologist.*, **88**, 683–93.

BARBER, D. A. & MARTIN, J. K. (1976) The release of organic substances by cereal roots into the soil. *New Phytologist.*, **76**, 69–80.

BARTHOLOMEW, W. V., MEYER, J. & LAUDELOUT, H. (1953) Mineral nutrient immobilization under forest and grass fallow in the Yangambi (Belgian Congo) region. *Publication de L'institut National Pour L'etude Agronomique du Congo Belge. Serie Scientifique* 57.

BASILEVICH, N. I. & RODIN, L. E. (1968) Reserves of organic matter in the underground sphere of terrestrial phytocoenoses. *Methods of Productivity Studies in Root Systems and Rhizosphere Organisms* (ed. M. S. GHILAROV) USSR Academy of Sciences, Leningrad.

BASNET, K., SCATENA, F. N., LIKENS, G. E. & LUGO, A. E. (1993) Ecological consequences of root grafting in tabonuco (*Dacroydes excelsa*) trees in the Luquillo experimental forest, Puerto Rico. *Biotropica*, **25**, 28–35.

BEHERA, N., PATI, D. P. & BASU, S. (1991) Ecological studies of soil microfungi in a tropical forest soil of Orissa, India. *Journal of Tropical Ecology*, **32**, 136–143.

BENZING, D. H. (1991) Aerial roots and their environments. *Plant Roots: the Hidden Half.* (eds. Y. WAISEL, A. ESHEL, & U. KAFKAFI) Marcel Dekker, New York.

BERISH, C. (1982) Root biomass and surface area in three successional tropical forests. *Canadian Journal of Forest Research*, **12**, 699–704.

BERISH, C. & EWEL, J.J. (1988) Root development in simple and complex tropical successional ecosystems. *Plant and Soil*, **106**, 73–84.

BOHN, W. (1979) *Methods of Studying Root Systems.* Springer-Verlag, Berlin.

BONGERS, F., Engelen, D. & Klinge, H. (1985) Phytomass structure of natural plant communities on Spodosols in southern Venezuela: The Bana woodland. *Vegetatio*, **63**, 13–34.

BORMANN, F. H. (1966) The structure, function, and ecological significance of root grafts in *Pinus strobus* L. *Ecological Monographs*, **36**, 1–26.

BOWEN, G. D. (1984) Tree roots and the use of soil nutrients. *Nutrition of Plantation Forests* (eds. G. D. BOWEN & E. K. S. NAMBIAR) Academic Press, New York.

BRAY, J. R. (1963) Root production and the estimation of net productivity. *Can. J. Bot.,* **41**, 65–72.

BRUNIG, E. F. & SANDER, N. (1983) Ecosystem structure and functioning: Some interactions of relevance to agroforestry. *Plant Research and Agroforestry* (ed. P. A. HUXLEY) ICRAF, Nairobi, Kenya.

CAMPBELL, B. D., GRIME, J. P. & MACKEY, J. M. L. (1991) A trade-off between scale and precision in resource foraging. *Oecologia,* **87**, 532–538.

CASTELLANOS, J., MAASS, M. & KUMMEROW, J. (1991) Root biomass of a dry deciduous tropical forest in Mexico. *Plant and Soil,* 131, 225–228.

Chapin, F. S. III (1980) The mineral nutrition of wild plants. *Annual Review of Ecology and Systematics,* **11**, 233–260.

COX, G. & TINKER, P. B. (1976) Translocation and transfer of nutrients in vessicular- arbuscular mycorrhizas I. The arbuscule and phosphorus transfer: A quantitative ultrastructural study. *New Phytologist.,* **77**, 371–378.

CUEVAS, E. & MEDINA, E. (1986) Nutrient dynamics within Amazonian forests. I. Nutrient flux in fine litterfall and efficiency of nutrient utilization. Oecologia, **68**, 466–472.

CUEVAS, E. & MEDINA, E. (1988) Nutrient dynamics within Amazonian forests. II. Fine root growth, nutrient availability, and leaf litter decomposition. *Oecologia,* **76**, 222–235.

CUEVAS, E., BROWN, S. & LUGO, A.E. (1991) Above- and belowground organic matter storage and production in a tropical pine plantation and a paired broadleaf secondary forest. *Plant and Soil,* **135**, 257–68.

DE FORESTA, H. & KAHN, F. (1984) Un systeme racinaire adventif dans un tronc creux d'*Eperua falcata. Revue d' Ecologie: La Terre et Vie,* **39**, 347–50.

DONALD, C. M. (1958) The interaction of competition for light and for nutrients. *Australian Journal of Agricultural Research,* **9**, 421–32.

DRANSFIELD, J. (1978) Growth forms of rainforest palms. *Tropical Trees as Living Systems* (eds. P. B. TOMLINSON & M. H. ZIMMERMANN) Cambridge University Press, UK.

EDWARDS, P. J. & GRUBB, P. J. (1982) Studies of mineral cycling in a montane rainforest in New Guinea. IV. Soil characteristics and the division of mineral elements between the vegetation and soil. *Journal of Ecology,* **70**, 649–666.

ERICSSON, A. & PERSSON, H. (1980) Seasonal changes in starch reserves and growth of fine roots of 20-year-old Scots pine. *Structure and Function of a Northern Coniferous Forest – An Ecosystem Study* (ed. T. PERSSON) *Ecological Bulletins* N. **32**, Swedish National Science Research Council, Stockholm.

EWEL, J., BERISH, C., BROWN, B., PRICE, N. & RAICH, J. (1981) Slash and burn impacts on a Costa Rican wet forest site. *Ecology,* **62,** 816–829.

FIELD, C. & MOONEY, H. A. (1986) The photosynthesis-nitrogen relationship in world plants. *On the Economy of Plant Form and Function* (ed. T.V. GIVNISH) Cambridge University Press, Cambridge.

FITTER, A. H. (1991) Characteristics and functions of root systems. *Plant Roots: The Hidden Half* (eds. Y. WAISEL, A. ESHEL, & U. KAFKAFI) Marcel Dekker, New York.

FITTER, A. H. (1994) Architecture and biomass allocation of root systems. *Exploitation of Environmental Heterogeneity by Plants* (eds. M. M. CALDWELL & R. W. PEARCY) Academic Press, San Diego, California.

FORD, E. D. & DEANS, J. D. (1977) Growth of a sitka spruce plantation: Spatial distribution and seasonal fluctuations of lengths, weights, and carbohydrate concentrations of fine roots. *Plant Soil,* **47,** 463–85.

FORSTER, M. (1970) Einige Beobachtungen zur Ausbildung des Wurzelsystems tropischer Waldbaume. *Allgemeine Forst-und Jagzeitung,* **141,** 185–188.

GERARD, P. (1960) Etude ecologique de la foret dense *Gilbertiodendron dewevrei* dans la region de l'Vele. *Publication de L'institut National Pour L'etude Agronomique du Congo Belge, Serie Scientifique,* **87,** 1–159.

GERMAIN, R. & EVRARD, C. (1956) Etude ecologique et phytosociologique de la forest a Brachystegia laurentii. *Publication de L'Institut National Pour L'Etude Agronomique du Congo Belge,* **67,** 1–105.

GIANINAZZI-PEARSON, V. & GIANINAZZI, S. (1983) The physiology of vesicular-arbuscular roots. *Plant and Soil,* **71,** 197–209.

GILL, A. M. & TOMLINSON, P. B. (1975) Aerial roots: An array of forms and functions. *The Development and Function of Roots* (eds. J. G. TORREY & D. T. CLARKSON) Academic Press, New York.

GLINSKI, J. & LIPIEC, J. (1990) *Soil Physical Conditions and Plant Growth.* CRC Press, Boca Raton, Florida.

GOLLEY, F. B. (1983) Tropical rainforest ecosystems: Structure and function. *Ecosystems of the World,* **14A,** Elsevier Scientific, Amsterdam.

GOLLEY, F. B., MCGINNIS, J. T., CLEMENTS, R. G., CHILD, G. I. & DUEVER, M. J. (1975) *Mineral Cycling in a Tropical Moist Forest,* University Georgia Press, Athens, GA.

GOLLEY, F. B., MCGINNIS, J. T. & CLEMENTS, R. C. (1971) La biomasa y estructura de algunos bosques de Darien, Panama. *Turrialba,* **21,** 189–196.

GOLLEY, F. B., ODUM, H. T. & Wilson, R. F. (1962) The structure and metabolism of a Puerto Rican red mangrove forest in May. *Ecology,* **43,** 9–19.

GOWER, S. T. (1987) Relations between mineral nutrient availability and fine root biomass in two Costa Rican tropical wet forests: A hypothesis. *Biotropica,* **19,** 171–175.

GRAHAM, B. F. Jr. & BORMANN, F. H. (1966) Natural root grafts. *Botanical Review,* **32,** 255–292.

GREENLAND, D. J. & KOWAL, J. M. L. (1960) Nutrient content of the moist tropical forest of Ghana. *Plant Soil*, **12**, 154–173.

GREGORY, P. J. (1987) Development of root systems in plant communities. *Root Development and Function*, (eds. P. J. GREGORI, J. V. LAKE & D. A. ROSE) Cambridge University Press, Cambridge.

GRIME, J. P. (1973) Competitive exclusion in herbaceous vegetation. *Nature*, **242**, 344–347.

GRIME, J. P. (1979) *Plant Strategies and Vegetation Processes*. Wiley & Sons, Chichester, UK.

GRIME, J. P. (1993) The role of plasticity in exploiting environmental heterogeneity. *Exploitation of Environmental Heterogeneity by Plants* (eds. M. M. CALDWELL & R. W. PEARCY) Academic Press, San Diego, California.

GRIMM, V. & FASSBENDER, H. W. (1981) Ciclos bioquimicos en un ecosistems forestalde los Andes Occidentales de Venezuela. III. Inventario de las reservas organicos y minerales (N, P, K, Ca, Mg, Mn, Al, Na). *Turrialba*, **31**, 27–37.

HALLE, F. & OLDEMAN, R. A. A. (1970) *Essai sur L'architecture et la Dynamique de Croissance des Arbres Tropicaux*. Masson, Paris.

HERWITZ, S. R. (1991) Aboveground adventitious roots and stemflow chemistry of *Ceratopetalum virchowii* in an Australian montane tropical rainforest. *Biotropica*, **23**, 210–218.

HERRERA, R., MERIDA, T., STARK, N. & JORDAN, C. F. (1978) Direct phosphorus transfer from leaf litter to roots. *Naturwissenschaften*, **65**, 208.

HILTNER, L. (1904) Uber neuere Erfarungen und Probleme auf dem Gebeit der Bodenbakteriologie und unter besonderer Beruck sichtigung der Grundungung und Brache. *Arbeiten Duetsche Landwirtschafts-Gesellschaft*, **98**, 59–78.

HO, I. & TRAPPE, J. M. (1973) Translocation of ^{14}C from Festuca plants to their endomycorrhizal fungi. *Nature*, **244**, 30–31.

HOGBERG, P. (1990) ^{15}N natural abundance as possible marker of the ectomycorrhizal habit of trees on mixed African woodlands. *New Phytologist*, **115**, 483–86.

HOLBROOK, N. M., PUTZ, F. E. & CHAI, P. (1985) Aboveground branching of the stilt-rooted palm *Eugeissona minor*. *Principes*, **29**, 142–46.

HOZUMI, K., YODA, K., KOKAWA, & KIRA, T., (1969) Production ecology of tropical rainforests in Southeast Cambodia. I. Plant biomass. *Nature and Life in South-east Asia* (Kyoto) **6**, 1–51.

HUSTON, M.A. & SMITH, T. A. (1987) Plant succession: Life history and competition. *American Naturalist*, **130**, 168–98.

HUTTEL, C. (1975) Root distribution and biomass in three Ivory Coast rainforest plots. *Tropical Ecological Systems*. (eds. F. B. GOLLEY & E. MEDINA) Springer-Verlag, Berlin.

IWASA, Y. & ROUGHGARDEN, J. (1984) Shoot/root balance of plants: optimal growth of a system with many vegetative organs. *Theoretical Population Biology,* **25,** 78–104.

JANOS, D. P. (1983) Tropical mycorrhizas, nutrient cycles and plant growth. *Tropical Rain Forest: Ecology and Management* (eds. S. L. SUTTON, T. C. WHITMORE), & A. C. Chadwick Blackwell Scientific Publications, Oxford.

JANOS, D. P. (1985) Mycorrhizal fungi: Agents or symptoms of tropical community composition? *Proceedings of the 6th North American Conference on Mycorrhizae* (ed. R. MOLINA) Forest Research Laboratory, Corvallis, Oregon.

JENIK, J. (1971) Root structure and underground biomass in equatorial forests. *Productivity of Forest Ecosystems* (Proceedings, Brussels, Symposium, 1969) UNESCO, Paris.

JENIK, J. (1978). Roots and root systems in tropical tres: Morphologic and ecologic aspects. *Tropical Trees as Living Systems,* (eds. P. B. TOMLINSON & M. H. ZIMMERMANN) Cambridge University Press, UK.

JORDAN, C. F. & UHL, C. (1978) Biomass of a tierra firme forest of the Amazon basin. *Oecologia Plantarum,* **13,** 255–268.

JUNG, G. (1969) Cycles biogeochimiques dans un ecosysteme de region tropicale seche *Acacia albida* (Del.). *Oecologia Plantarum,* **4,** 195–210.

KATO, R., TADAKI, Y, & OGAWA, H. (1978) Plant biomass and growth increment studies in Pasoh forest. *Malaysian Nature Journal,* **30,** 14–18.

KAVANAUGH, T. & KELLMAN, M. (1992) Seasonal pattern of fine root proliferation in a tropical dry forest. *Biotropica,* **24,** 157–65.

KEELEY, J. E. (1988) Population variation in root grafting and a hypothesis. *Oikos,* **52,** 364– 66.

KIRA, T., OGAWA, H., YODA, K., & OGINO, K. (1967) Comparative ecological studies on three main types of forest vegetation in Thailand. IV. Dry matter production with special reference to the Khao Chong rainforest. *Nature and Life in Southeast Asia, Kyoto,* **6,** 149–174.

KIRA, T. (1978) Community structure and organic matter dynamics in tropical lowland forests of southeast Asia with special reference to pasoh forest, West Malaysia. *Tropical Trees as Living Systems* (eds. P. B. TOMLINSON & M. H. ZIMMERMANN) Cambridge University Press. UK.

KLEPPER, B. (1991). Root-shoot relationships. *Plant Roots: The Hidden Half.* (eds. Y. WAISEL, A. ESHEL, & U. KAFKAFI) Marcel Dekker, New York.

KLINGE, H. & HERRERA, R. (1978) Biomass studies in Amazon caatinga forest in southern Venezuela. 1. Standing crop of composite root mass in selected stands. *Tropical Ecology,* **19,** 93–110.

KLINGE, H., MEDINA, E. & HERRERA, R. (1977). Studies on the ecology of Amazon caatinga forest in southern Venezuela. *Acta Scientifica Venezolana,* **28,** 270–76.

KOOPMANS, T. Th. & ANDRIESSE, J. P. (1982) Baseline study monitoring project of nutrient cycling in shifting cultivation. *Department of Agricultural Research Internal Report BO 82–6, Koninklijk Instituut voor de Tropen, Amsterdam.*

KUMAR, H., KULKARMI, D. & SRIMATHI, R. A. (1985) Natural grafts in sandal. *Indian Forester,* **8, 153–154.**

KUMMEROW, J., CASTELLANOS, J., MASS, M. & LARIGAUDERIE A. (1990) Production of fine roots and the seasonality of their growth in a Mexican deciduous dry forest. *Vegetatio* **90,** 73–80.

KUNTZ, J. E. & RIKER, A. J. (1956) The use of radioactive isotopes to ascertain the role of root grafting in the translocation of water nutrients and diseases among forest trees. *Proceedings of the International Conference on the Peaceful Uses of Atomic Energy (Geneva, Switzerland)* **12,** 144–48.

KURZ, W. A. & KIMMONS, J. P. (1987) Analysis of some sources of error in methods used to determine fine root production in forest ecosystems: A simulation approach. *Canadian Journal of Forest Research,* **17,** 909–912.

LaRUE, E.D. (1952) Root grafting in tropical trees. *Science* **115,** 296.

LAWSON, G. W., ARMSTRONG-MENSAH, K. O. & HALL, J. B. (1970) A catena in tropical moist semideciduous forest near Kade, Ghana. *Journal of Ecol.,* **58,** 371–398

LODGE, D.J. (1993) Nutrient cycling by fungi in wet tropical forests. *Aspects of Tropical Mycology* (eds. S. ISAAC, J.C. FRANKLAND, R. WATLING & A. J. WHALLEY) Cambridge University Press, Cambridge, UK.

LODGE, D. J., McDOWELL, W. H. & McSWINEY, C. P. (1994) The importance of nutrient pulses in tropical forests. *Trends in Ecology and Evolution,* **9,** 384–387.

MARSHALL, J. D. & WARING, R. H. (1985) Predicting fine root production and turnover by monitoring root starch and soil temperature. *Canadian Journal of Forest Research,* **15,** 791–800.

MARTIN, J. K. (1977) Factors influencing the loss of organic carbon from wheat roots. *Soil Biology and Biochemistry,* **9,** 1–17.

MEDINA, E. & CUEVAS, E. (1989) Patterns of nutrient accumulation and release in Amazonian forests of the upper Rio Negro basin. *Mineral Nutrients in Tropical Forest and Savanna Ecosystems* (ed. J. PROCTOR) Blackwell Scientic, Oxford, UK.

MEDINA, E., KLINGE, H., JORDAN, C. & HERRERA, R. (1980) Soil respiration in Amazonian rain forests in the Rio Negro basin. *Flora,* **170,** 240–250.

MENSAH, K. O. & JENIK, J., (1968) Root systems of tropical trees. 2. Features of the root system of Iroko (*Chlorophora excelsa* Benth. et. Hook.). *Preslia Prague,* **40,** 21–27.

MISHRA, R. R. & KANAUJIA, R. S. (1973/74) Investigations into the rhizosphere microflora XII. Seasonal variation in the microflora of certain gymnosperms. *Sydowia,* **27(1/6),** 302–311.

MOYERSOEN, B. (1993). Ectomicorrizas y Micorrizas Vesculo-arbusculares en Caatinga Amazonica del Sur de Venezuela. *Scientia Guaianae* Caracas, Venezuela.

MULLER, D & NIELSEN, J. (1965) Production brute, pertes par respiration et production nettee dans la foret ombriphile tropicale. *Det Forstlige Forsogsvaesen i Danmark*, **29**, 69–160.

MURPHY, P. G. & LUGO, A. E. (1986) Structure and biomass of a subtropical dry forest in Puerto Rico. *Biotropica* **18**, 89–96.

MYERS, R.J.K., PALM, C.A., CUEVAS, E., GUNATILLEKE, I.U.N. & BROSSARD, M. (1994) Synchronization of nutrient mineralization and plant nutrient demand. *Biological Management of Tropical Soil Fertility* (eds. P. WOOMER, & M. J. Swift) John Wiley & Sons, West Sussex, U. K.

NADELHOFFER, K. J. & RAICH, J. W. (1992) Fine root production estimates and belowground carbon allocation in forest ecosystems. *Ecology,* **73**, 1139–1147.

NADKARNI, N. (1981). Canopy roots: Convergent evolution in rainforest nutrient cycles. *Science,* **214**, 1023–1024.

NADKARNI, N. M. (1986) The nutritional effects of epiphytes on host trees with special reference to alteration of precipitation chemistry. *Selbyana,* **9**, 44–51.

NEPSTAD, D. C., DE CARVALHO, C. R., DAVIDSON, E. A., JIPP, P. H., LEFEBVRE, P. A., NEGREIROS, G. H., da SILVA, E. D., STONE, T. A., TRUMBORE, S. E. & VIEIRA, S. (1995) The deep-soil link between water and carbon cycles of Amazonian forests and pastures. *Nature* **372**, 666–667.

NEWBERY, D. M., ALEXANDER, I. J., THOMAS D. W. & GARTLAN, J. S. (1988) Ectomycorrhizal rainforest legumes and soil phosphorus in Korup National Park, Cameroon. *New Phytologist.,* **109**, 433–450.

NEWMAN, E. I. (1973) Competition and diversity in herbaceous vegetation. *Nature,* **224**, 310.

NEWMAN, E. I. (1983) Interactions between plants. Physiological plant ecology. III. Responses to the chemical and biological environment. *Encyclopedia of Plant Physiology, New Series* **12C.** Springer-Verlag, Berlin.

NEWMAN, E. I. (1985) The rhizosphere: Carbon sources and microbial populations. *Ecological Interactions in the Soil* (ed. A. H. FITTER) Blackwell Scientific, Oxford.

OGAWA, H., YODA, K. & KORA, T. (1961) A preliminary survey of the vegetation of Thailand. *Nature and Life in Southeast Asia (Kyoto),* **5**, 49–80

PANG, P. C. & PAUL, E. A. (1980) Effects of vesicular-arbuscular mycorrhizae on ^{14}C and 15N distribution in nodulated faba beans. *Canadian Journal of Soil Science,* **60**, 241–250.

PAPAVIZAS, G.C. & DAVEY. C. B. (1961). Extent and nature of the rhizospere of *Lupinus. Plant and Soil,* **14**, 215–236.

PARROTTA, J. A. & LODGE, D. J. (1991) Fine root dynamics in subtropical wet forest following hurricane disturbance in Puerto Rico. *Biotropica,* **23 (Suppl.)** 343–347.

PASCHKE, M. W. & DAWSON, J. O. (1992) The occurrence of *Frankia* in tropical forest soils of Costa Rica. *Plant and Soil,* **142,** 63–67.

PICKETT, S. T. A. (1976) Distribution and interactions of surface roots of *Castilla elastica* (Moraceae) in lowland Costa Rica. *Turrialba,* **26,** 156–159.

PUTZ, F. E. & HOLBROOK, M. (1989) Strangler fig rooting habits and nutrient relations in the llanos of Venezuela. *American Journal of Botany,* **76,** 781–788.

RAICH, J. (1980) Fine roots grow rapidly after forest felling. *Biotropica,* **12,** 231–232.

RAICH, J. (1983) Understory palms as nutrient traps: A hypothesis. *Brenesia,* **21,** 119–129.

RAICH, J. W. & NADELHOFFER, K.J. (1989) Belowground carbon allocation in forest ecosystems: Global trends. *Ecology,* **70,** 1346–1354.

READ, D. J., LEAKE, J. R. & LANGDALE, A. R. (1989) The nitrogen nutrition of mycorrhizal fungi and their host plants. *Nitrogen, Phosphorus and Sulphur Utilization by Fungi* (eds. L. BODDY, R. MARCHANT, & D. J. READ) Cambridge University Press, Cambridge, UK.

ROVIRA, A. D. (1969) Plant root exudates. *Botanical Review,* **35,** 35–57.

RUSSELL, E. W. (1977) *Soil Conditions and Plant Growth, 10th edition.* Longman Group Ltd. London.

SALCEDO, I. H., ELLIOTT, E .T. & SAMPAIO, E. V. S. B. (1991) Mechanisms controlling phosphorus retention in the litter mat of Atlantic coastal forests. *Phosphorus Cycles in Terrestrial and Aquatic Ecosystems. Regional Workshop 3: South and Central America* (eds. H. TIESSEN, D. LOPEZ-HERNANDEZ & I. H. SALCEDO) SCOPE, UNEP & Saskatchewan Institute of Pedology, Saskatoon, Canada.

SANFORD, R. L. Jr., (1987) Apogeotropic roots in an Amazon rainforest. *Science,* **235,** 1062–1064.

SANFORD, R. L. Jr., (1989) Root systems of three adjacent, old growth Amazon forests and associated transition zones. *Journal of Tropical Forest Research,* **1,** 268–279.

SANFORD, Jr., R. L. (1990) Fine root biomass under light gap openings in an Amazonian rainforest. *Oecologia,* **83,** 541–545.

Sanford, R. L., Jr., (1995) Fine root production, mortality and nutrient turnover in successional and old-growth Amazon rain forests. *Journal Tropical Ecology.* (submitted)

SANFORD, R. L. Jr., & VITOUSEK, P. M. (1995) Root and leaf litter production, and nutrient turnover in lowland tropical moist forest, Costa Rica. *Ecology* (submitted)

SCATENA, F. N. & LARSEN M. C. (1991) Physical aspects of hurricane Hugo in Puerto Rico. *Biotropica*, **23**, 317–323.

SILVER, W. L. & VOGT, K. A. (1993) Fine root dynamics following single and multiple disturbances in subtropical wet forest ecosystems. *Journal of Ecology*, **81**, 729–738.

SINGH, K. P. & SINGH, R. P. (1981) Seasonal variation in biomass and energy of small roots in tropical dry deciduous forest, Varanasi, India. *Oikos*, **37**, 88–92.

SINGH, K. P. & SRIVASTAVA, S. K. (1984) Spatial distribution of fine roots in young trees (*Tectona grandis*) of varying girth sizes. *Pedobiologia*, **27**, 161–170.

SINGH, K. P. & SRIVASTAVA, S. K. (1985) Seasonal variations in the spatial distribution of root tips of teak (*Tectona grandis* Linn F.) plantations in the Varanasi Forest Division, India. *Plant and Soil*, **84**, 93–104.

SINGH, J. S., LAURENROTH, W. K., HUNT, H. W. & SWIFT, D. W. (1984) Bias and random errors in estimators of net root production: a Simulation approach. *Ecology*, **65**, 1760–1764.

SRIVASTAVA, S. K., SINGH, K. P. & UPADHYAY, R. S. (1986) Fine root growth dynamics in teak (*Tectona grandis* Linn. F.). *Canadian Journal of Forest Research*, **16**, 1360–1364.

STARK, N. & JORDAN, C. F. (1978) Nutrient retention by the root mat of an Amazonian rain forest. *Ecology*, **59**, 434–37.

STARK, N. & SPRATT, M. (1977) Root biomass and nutrient storage in rainforest Oxisols near San Carlos de Rio Negro. Journal of *Tropical Ecology*, **18**, 1–9.

ST. JOHN, T. V. (1983) Response of tree roots to decomposing organic matter in two lowland Amazonian rain forests. *Canadian Journal of Forest Research*, **13**, 346–349.

ST. JOHN, T. V. & UHL, C. (1983) Mycorrhizae in the rainforest of San Carlos de Rio Negro, Venezuela. *Acta Scientifica Venezolana*, **34**, 233–237.

TANNER, E. V. J. (1985) Jamaican montane forests: Nutrient capital and cost of growth. Journal of *Ecology* **73**, 553–568.

TIESSEN, H., CHACON, P. & CUEVAS, E. (1994b) Phosphorus and nitrogen status in soils and vegetation along a toposequence of dystrophic rainforests on the upper Rio Negro. *Oecologia*, **99**, 145–150.

TIESSEN, H. CUEVAS, E. & CHACON, P. (1994a) The role of organic matter in sustaining soil fertility. *Nature*, **371**, 783–785.

TILMAN, D. (1988) *Plant Strategies and the Structure and Dynamics of Plant Communties. Princeton University Press*, Princeton, New Jersey.

TILMAN, D. (1989) Competition, nutrient reduction and the competitive neighborhood of a bunchgrass. *Functional Ecology*, **3**, 215–219.

TINKER, P. B. (1978) Effects of vesicular-arbuscular mycorrhizas on plant nutrition and plant growth. *Physiologie Vegetale*, **16**, 743–751.

UHL, C. & JORDAN, C. F. (1984) Succession and nutrient dynamics following forest cutting and burning in Amazonia. *Ecology,* **65,** 1476–1490.

VITOUSEK, P. M. & SANFORD, R. L. (1986) Nutrient cycling in moist tropical forest. *Annual Review of Ecology and Systematics,* **17,** 137–167.

VOGT, K. A. & BLOOMFIELD, J. (1991) Tree root turnover and senescence. *Plant Roots; The Hidden Half* (eds. Y. WAISEL, A. ESHEL & U. KAFKAFI) Marcel Dekker, New York.

VOQRHOEVE, I. A. G. (1964) Some notes on the tropical rainforest of the Yoma-Gola national forest near Bomi Hills, Liberia. *Commonwealth Forestry Review,* **43,** 17–24.

VYAS, L. N., GAR, R. K. & VYAS, N. L. (1977) Stand structure and aboveground biomass in dry deciduous forests of Aravalli Hills at Udipar (Rajasthan) India. *Biologia* (Bratislava) **32** 265–270.

WALTER, H. (1971) *Ecology of Tropical and Subtropical Vegetation.* Oliver and Boyd, Edinburgh.

WHIPPS, J. M. & LYNCH, J. M. (1983) Substrate flow and utilization in the rhizosphere of cereals. *New Phytologist,* **95,** 605–623.

YANG, J. C. & INSAM, H. (1991) Microbial biomass and relative contributions of bacteria and fungi in soil beneath tropical rainforest, Hainan Island, China. *Journal Tropical Ecology,* **7,** 385–395.

ZENTMEYER, G. A. (1961) Chemotaxis of zoospores for roots exudates. *Science,* **133,** 1595–1596.

II

Ecophysiological Aspects of Species Interactions

Introduction

The complex species interactions within tropical forests are as fascinating as the species themselves. In these forests, plants are confronted with a multitude of challenges imposed by their neighbors. Threats to survival come from many agents, such as physical damage from falling canopy debris, loss of leaf tissue to herbivores, or attack by pathogens. These selection pressures have shaped patterns of plant growth and development. The following three chapters describe new areas of ecophysiological research that are linked by their focus on species interactions.

Tropical leaves suffer higher rates of herbivory compared to temperate leaves. The wide array of plant defenses against herbivores has already attracted considerable attention from ecophysiologists. Young, expanding leaves are more vulnerable to herbivory than older leaves.

In Chapter 11, Coley and Kursar discuss anti-herbivore defenses of young leaves, taking a developmental, organismal approach. They describe biochemical and structural changes during leaf development in relation to ecological patterns in herbivory among species from tropical forests throughout the world. Their synthesis suggests that delayed greening is a common anti-herbivore defense in species with slowly expanding leaves in shaded environments. Species whose leaves expand quickly, in contrast, are favored in high-light conditions. Fast-expanding leaves have higher concentrations of nitrogen and lower concentrations of defensive compounds than slow-expanding leaves—and suffer greater rates of leaf area loss due to herbivory.

In wet tropical forests, conditions are appropriate for the development of entire communities of nonvascular plants (primarily liverworts) that live on leaf surfaces. Coley and Kursar (Chapter 12) explore the world of epiphylls from an ecophysiological perspective. They examine the influence of abiotic factors on epiphyll physiology, colonization, and community structure. From the host leaf's point of view, evidence for both positive and negative effect of epiphylls is presented. Cyanobacteria associated with the epiphyll species can fix substantial quantities of nitrogen, which may be taken up by the host plant. Other effects of epiphylls include delayed leaf senescence, increased susceptibility to pathogens, protection from herbivores, and shading. Leaf surface characteristics, as well as longevity, appear to influence patterns of epiphyll colonization. Relatively long-lived leaves have greatly reduced rates of epiphyll accumulation. But since leaves can be colonized over a longer time period, epiphyll cover does not vary consistently with leaf lifespan at the time of leaf senescence. The ecophysiology of epiphylls in relation to host plants is an exciting new research direction.

The focus on plants that grow on plants continues in Chapter 13 as Holbrook and Putz describe particular ecophysiological features of tropical climbers (vines and lianas) and hemiepiphytes. These plants are structural parasites of trees; they are otherwise physiologically independent from their hosts. The structure and function of vine stems is described in detail, with an emphasis on their unique xylem and phloem anatomy. Leaf-level physiology shows no consistent differences between trees and vines, although leaf morphology may be a more variable feature within vine species. Lianas hold the record for the most deeply rooted species in tropical forests.

Hemiepiphytes, particularly those that are stranglers, face unique functional challenges as they move from their establishment sites on branches in the upper canopy down to the soil. Two diverse tropical forest genera, *Ficus* and *Clusia*, provide excellent examples for comparative ecophysiological studies between epiphytic and terrestrially-rooted individuals. Epiphytic individuals exhibit a variety of mechanisms for tolerating a greater degree of water stress. *Clusia* species show remarkable plasticity in photosynthetic metabolism and are currently being actively investigated on many fronts.

As these chapters so aptly demonstrate, the ecophysiological underpinnings of species interactions are well deserving of continued detailed study.

11

Anti-Herbivore Defenses of Young Tropical Leaves: Physiological Constraints and Ecological Trade-offs

Phyllis D. Coley and Thomas A. Kursar

Damage to leaves by herbivores can have a significant and extensive impact on growth and reproduction of plants (Marquis, 1984, 1992a, b; Marquis & Braker, 1993), which in turn can influence competitive outcomes and community composition (Janzen, 1970; Dirzo, 1984; Clark & Clark, 1985; Dirzo & Miranda, 1991). In tropical forests, approximately 11% of the annual leaf area produced is consumed by herbivores and pathogens (Coley & Aide, 1991), a resource loss equivalent to investments in reproduction (Bazzaz et al., 1987). This loss would be substantially higher except for the fact that plants allocate considerable resources to physical, chemical, and phenological defenses. Most of our understanding of the costs and benefits of defenses and of the interplay between defenses and herbivores is based on research on mature leaves. However, young expanding leaves are the most vulnerable stage during the life of a leaf. In tropical shade-tolerant species of plants, expanding leaves suffer 5-100 times the rates of damage from pathogens and herbivores as mature leaves (Coley &

Aide, 1991). Seventy percent of the lifetime damage can occur during this small window of vulnerability. Many leaf developmental traits may therefore be the result of selection by herbivores and pathogens. These defensive traits include rapid expansion, ant defense, secondary compounds, and synchronous leaf production to satiate herbivores. In addition, delayed chloroplast development postpones input of resources until the leaf is mature and better defended. We discuss the physiological processes and costs behind these defenses. We also present data showing that although all these defensive traits may be effective to varying degrees, they are not all present in a single species. In a survey of over 250 species from lowland rainforests worldwide, we found subsets of traits that consistently co-occurred. We argue that these patterns of co-occurrence suggest that either the adaptive value of each trait is dependent on the ecological setting, or there are physiological constraints such that certain defense options are incompatible. In order to understand particular leaf developmental traits, it is therefore important to view all of them together and to consider them within an ecological context.

11.1   HERBIVORY

In most tropical species, young leaves suffer more damage from herbivores and pathogens than mature leaves. This is particularly marked in shade-tolerant rainforest species, where damage rates to young leaves average 20 times higher than damage to mature leaves (Table 11.1). Rainforest species that specialize on gaps also have high rates of damage to young leaves. These data are underestimates of actual damage, as sucking insects are abundant on young leaves, but are rarely measured. Young leaves that are completely eaten may also escape measurement (Coley, 1982; Lowman, 1984, 1985a,b).

Table 11.1.   Rates of damage to young and mature leaves of gap specialists and shade-tolerant species of humid rainforests. % / day = % of leaf area removed per day. Data are from Coley (1983), Wint (1983), Cooke et al. (1984), Newberry & de Foresta (1985), Aide (1993), Marquis & Braker (1993).

	Herbivory on Young Leaves		Herbivory on Mature Leaves	
	# species	%/day	# species	%/day
Shade-tolerant	130	0.616	98	0.028
Gap specialist	30	0.633	30	0.186

Table 11.2.   *The percentage of the total lifetime herbivore damage that occurs while leaves are young and expanding. Temperate data are a pooled estimate for an undetermined number of species in a Liriodendron forest (Reichle et al. 1973). Dry forest data are for 24 species in Costa Rica and Brazil (Stanton, 1975, Nascimento & Hay, 1993, Ribeiro et al., unpubl. ished data). Wet forest data are for 25 gap species in Panama and Costa Rica (Coley, 1983; Waltz, 1984), and 29 shade-tolerant species in Panama, Papua New Guinea, and Sarawak (Coley, 1983, Wint, 1983, Cooke et al., 1984).*

	#species	% of damage
Temperate forest	?	27.0
Tropical dry forest	24	59.8
Tropical wet forest:		
Gap specialist	25	40.4
Shade-tolerant	29	70.1

Rates of damage to young expanding leaves are so high that the majority of lifetime damage occurs during this short period of the lifespan (Table 11.2). In dry forest species and rainforest gap specialists, young leaf herbivory comprises 60% and 40%, respectively, of the lifetime damage. In shade-tolerant species of rainforests, 70% of the lifetime damage occurs during expansion. This is particularly surprising, since shade-tolerant leaves live 2.5 years on average. The 2-5 week period of leaf expansion therefore represents less than 4% of the lifespan, but the vast majority of the damage.

Young leaf herbivory in broad-leaved species appears to be of much greater significance in tropical than in temperate forests. Young tropical leaves have higher absolute rates of herbivory as well as a greater percentage of the lifetime damage compared to young temperate leaves (Table 11.2; Coley & Aide, 1991). This suggests that if we are to understand plant/herbivore interactions in the tropics, we need to have a much greater understanding of young leaf defenses.

11.2   NUTRITIONAL PROPERTIES

Two main proximate reasons have been identified that may explain why young leaves tend to have higher herbivory than mature leaves. One reason may be that young leaves are less tough and fibrous and therefore easier to chew and digest (Coley, 1983; Lowman & Box,

1983; Juniper & Southwood, 1986). For most species, herbivory drops dramatically at the end of leaf expansion during a period of 3-5 days when the leaf begins to toughen. In a detailed study of 5 species on Barro Colorado Island (BCI) Panama, herbivory dropped by a factor of 4 over this period (Kursar & Coley, 1992a). Toughness may be the most effective defense (Coley, 1983), yet it is not an option for expanding leaves because the formation of a lignified cell wall is not compatible with cell division and cell expansion.

The other reason young leaves are preferred may be their high nutritional value. Young leaves typically have higher water contents and 2-4 times the nitrogen content per mass of mature leaves (Feeny, 1970; Mattson & Scriber, 1987; Coley & Aide, 1991; Kursar & Coley, 1991). Diets with high nitrogen and water increase herbivore fitness, and are frequently positively correlated with food choice of herbivores (Scriber & Slansky, 1981; Slansky & Rodriguez, 1987). Any changes in the developmental patterns of young leaves that reduce the concentration of nitrogen might therefore reduce rates of herbivory (Moran & Hamilton, 1980; Kursar & Coley, 1991).

11.3 DEFENSES

Perhaps because of the greater nutritional value and vulnerability of young leaves, a remarkable variety of defensive traits have been identified in expanding leaves of tropical species. In this section, we summarize the major defenses of young leaves from a physiological perspective. In the following section (11.4), we discuss the trade-offs and physiological constraints limiting these defense options.

11.3.1 Rapid leaf expansion

Individual leaves that expand rapidly and toughen early will shorten the period when leaves are most vulnerable to herbivores (Orians & Janzen, 1974; McKey, 1979; Hay et al., 1988; Kursar & Coley, 1991). There is some evidence that rapid expansion reduces the total damage while leaves are young. Rates of herbivory and total damage are more than twice as high on slow- versus fast-expanding *Pentagonia* (Rubiaceae) leaves (Ernest, 1989). Aide and Londono (1989) showed that because of the rapid expansion of *Gustavia superba* (Lecythidaceae), its major herbivore has only a three day window in which oviposition can be successful. Larvae that hatch a few days later suffer high mortality due to the increase in toughness and decline in leaf quality.

Species differ dramatically in the rates of leaf expansion. We measured rates of leaf expansion in 208 of the most common shade-tolerant woody species at four lowland rainforest sites in Borneo, Africa, and Central America (Kursar & Coley, unpublished data, see Appendix 11.1 for site descriptions). On average, expanding leaves double in size every 5 days, with rates ranging over an order of magnitude. Four percent of the species surveyed had leaves that doubled in size each day. Rapid expansion of this magnitude would require rapid translocation and use of stored reserves rather than current photosynthate.

11.3.2 Secondary metabolites

Young tropical leaves are better defended chemically than mature leaves, in marked contrast to the temperate zone, where mature leaves have higher concentrations of many secondary metabolites (Coley & Aide, 1991). However, we still have an incomplete picture of chemical defense, as most work has focused on phenolic and terpenoid compounds, two widespread and abundant defensive compounds.

Initially, workers hypothesized that defense of young leaves by phenolic compounds would not be feasible because of problems of sequestering phenolic compounds away from cell machinery. However, this is apparently not the case. For 81 tropical species, levels of total phenols (measured by the Folin-Denis assay) and proanthocyanidins (condensed tannins, measured by the BUOH/HCl assay) were almost twice as high as in mature leaves (Coley & Aide, 1991). Mono-, sesqui-, and diterpenes are also generally higher in young as compared to mature leaves (Crankshaw & Langenheim, 1981; Langenheim et al., 1986)

Both terpenoid and phenolic compounds are carbon-based secondary metabolites containing no nitrogen. In environments where a plant's carbon pool is high compared to its nutrient pool, the "excess" carbon is generally shunted into carbon-based defenses (Bryant, Chapin & Klein, 1983). This is a plastic phenotypic response to imbalances in the source/sink relationships. It is therefore a mechanistic explanation for defense changes within a species or even within a single genotype. Defense increases in the sun have been shown for carbon-based defenses in a variety of tropical species (Langenheim et al., 1981; Coley, 1986;

Feibert & Langenheim, 1988; Mole, Ross & Waterman, 1988; Denslow et al., 1990; Nichols-Orians, 1991; Sagers, 1992).

Young tropical leaves undoubtedly contain a battery of other chemical defenses including saponins, alkaloids, cyanogenic compounds, monoterpenes, defensive proteins, and toxic amino acids. Although there have been few studies of these secondary metabolites, we suggest that this diverse group of low molecular weight compounds may be particularly well represented in young leaves. First, unlike condensed tannins, these compounds can probably be reclaimed and the resources used for other purposes. Reclamation may be advantageous when the leaf becomes protected by toughness and no longer needs the high levels of chemical protection required while expanding (McKey, 1979). Second, these low molecular weight compounds may have high turnover rates (Coley, Bryant & Chapin, 1985; but see Mihaliak, Gershenzon & Croteau, 1991; Baldwin, Karb & Ohnmeiss, 1994). Compounds with high turnover require continual synthesis to maintain a constant pool size. This cost of synthesis may be reasonable during the short period of leaf expansion, but could accumulate to prohibitively high levels if continued throughout the 2-3 year leaf lifespan. Thus, we predict that low molecular weight defenses will be most-effective in expanding leaves but will be much less common in long-lived mature leaves. These leaves will instead rely on toughness and tannins, defenses that do not have continued maintenance costs.

The actual turnover rates for different compounds are extremely poorly understood. Evidence from alkaloids are conflicting, with some authors measuring half lives of a few days (Robinson, 1974), and others detecting only minor (33%) turnover in 36 days (Baldwin, Karb & Ohnmeiss, 1994). Monoterpenes were also thought to turn over on the order of hours or days, however, these high rates have recently been questioned (Mihaliak, Gershenzon & Croteau, 1991; Gershenzon, Murtagh & Croteau, 1993). Clearly, a great deal more data on turnover rates of secondary compounds are necessary before we can accurately assess costs.

11.3.3   Anthocyanins

Many tropical species flush entire canopies of red, white or light green young leaves (Richards, 1952; Burgess, 1969; Opler, Frankie & Baker, 1980). The visually striking red coloration is due to anthocyanins (Harborne, 1979). The role of anthocyanins has been hotly debated since the turn of the century (Smith, 1909), with two major hypotheses emerging. Anthocyanins may screen harmful UV radiation (Caldwell, 1981). This could be important in the canopy, but UV screening is not a likely function in the shaded understory (Lee & Lowry, 1980; Lee, Brammeier & Smith, 1987). More recently, bioassays with leaf-cutter ants suggested that anthocyanins have antifungal properties which can protect young leaves against attack by fungal pathogens (Coley & Aide, 1989). Although leaf-cutters are also detered by anthocyanins, we suggest that they evolved in response to selection by pathogens. Leaf-cutters are less common in mature forest, the habitat where young red leaves are the most common. Furthermore, leaf-cutters do not occur in southeast Asia and Africa, yet young red leaves are as abundant as in the neotropics. In contrast, fungal pathogens are frequent in all forest understories of the humid tropics. Young expanding leaves have poorly developed cuticles and are particularly vulnerable to attack by pathogens (Garcia-Guzman & Dirzo, 1991; Coley & Kursar, unpublished data). Anthocyanins disappear from the leaf when it stops expanding and the cuticle develops.

Although the two hypotheses for the function of anthocyanins, defense and UV protection, are not mutually exclusive, a comparison of canopy and understory leaves would help separate them. UV levels are clearly higher in the canopy. Data on relative rates of pathogen attack are unfortunately lacking. However, we speculate that the possibility of pathogen infection is lower in the canopy due to shorter leaf life spans and a drier environment. If anthocyanin levels are reduced in the canopy, this would support an antifungal role and suggest that screening UV is not important. Elevated levels in the canopy imply that anthocyanins also function as a UV screen.

11.3.4   Ant defense

Extra-floral nectary secretion and food body rewards are found almost exclusively on young leaves (Bentley & Elias, 1983; Beattie, 1985; Huxley & Cutler, 1991). Food bodies typically contain sugar or lipid and occasionally protein. Nectar or food rewards are presented to ants that in turn provide partial protection from herbivores (Janzen, 1966, 1967; Schupp, 1986; Koptur, 1984; Bentley, 1977; Oliveira, Silva & Martins, 1987). Ant defense is particularly common in the tropics (Coley & Aide, 1991). Thirty-two percent of the 243 species surveyed on BCI produced ant rewards on the young leaves (Schupp & Feener, 1991). There was a strong phylogenetic component, with the vast majority of genera having only ant-defended species or only species without ant defense. Ant rewards were twice as common in gap specialists as compared to shade-tolerant species. This pattern also persisted after controlling for phylogeny. Vines had ant rewards more often than trees or shrubs, but this was because vines tended to be gap species and to occur in families rich in ant rewards. Evidently the vine habit alone does not favor the evolution of ant defense.

What explains the predominance of ant rewards in gap species? One hypothesis was that ants are more common in forest gaps (Bentley, 1977). However, in an extensive study on BCI, Schupp and Feener found no difference in either abundance or composition between understory and gap ant communities, suggesting that the availability of ants does not drive the presence of ant rewards (Schupp & Feener, 1991; Feener & Schupp, 1995).

Alternatively, we suggest that the predominance of ant rewards in gap species is related to differences in leafing phenology. The more continuous production of young leaves in gap as opposed to understory species (Coley, 1983) should favor ant defense. Continuous and predictable presence of ant rewards would lead to a higher probability of continual ant guarding, which increases the effectiveness of defense against herbivores (Schupp & Feener, 1991; McKey, 1984, 1989).

In addition to phenology, gap and understory species differ in their overall investment in defenses. Gap species invest less in physical and chemical defenses and suffer higher rates of damage (Coley & Aide, 1991; Coley, 1983; Marquis & Braker, 1994). Although solid evidence is lacking, it is generally assumed that ant defenses are relatively inexpensive (O'Dowd, 1979, 1980; Beattie, 1985; Keeler, 1981). Furthermore, the potentially greater rates of carbon gain in light gaps make the carbon-based sugar and lipid rewards relatively cheap. Based on these resource and phenological arguments, it is not surprising that gap species are commonly defended by ants.

11.3.5   Synchronous leaf production

Synchronous production of young leaves has been shown to satiate available herbivores and thus reduce leaf damage (Leigh & Smythe, 1978; McKey, 1979; Aide, 1988). In an African dry forest, young leaf damage was significantly higher on species with prolonged leaf flushing and on individuals that flushed near the end of a peak (Lieberman & Lieberman, 1984). Similarly, for 12 species on Barro Colorado Island, leaves that were produced in synchrony with the population averaged half the rates of herbivory as leaves produced at other times (Aide, 1991, 1993). Although this represents a substantial savings in damage, the cost of synchrony may also be considerable. Leaf flushes occur once or twice a year, so resources must be stored in the interim. There are direct costs of storage and defense of the stored resources, as well as the opportunity cost of lost photosynthesis associated with postponed leaf production. However, the fact that approximately one-third of the common species on BCI exhibit periodic flushing as saplings suggests that synchrony can be cost-effective (Aide, 1993).

It should be noted that, although it has been strongly argued that herbivore pressure selects for synchrony even in aseasonal environments (Aide, 1993; Coley & Aide, 1991), abiotic factors may also be important. Major periods of leaf flush are frequently tied to the onset of rains in dry forests, and even in humid forests (Frankie, Baker & Opler, 1974; Reich & Borchert, 1984; Lieberman & Lieberman, 1984; Wright, Chapter 15). The extent to which this pattern represents a physiological constraint or simply a convenient cue is debated.

11.3.6 Delayed greening

The above-mentioned defenses reduce damage by making leaves less attractive or less available to herbivores. Another approach is to minimize the impact of inevitable herbivory by delaying the input of valuable resources until the leaf is defended by toughness. One strategy that is extremely common in rainforest species worldwide is to delay the greening process until after full leaf expansion (Kursar & Coley, 1992a).

Young leaves of many tropical species have little chlorophyll and instead appear white, red, or pink. Although the visually dramatic red coloration has captured the attention of scientists for decades (Smith, 1909; Burgess, 1969; Opler, Frankie & Baker, 1980), little work focused on the more remarkable fact that these species are not green due to a delay in chloroplast development (Baker & Hardwick, 1973; Baker, Hardwick & Jones, 1975). In studies on BCI, Kursar and Coley (1992a,b,c) found that the input of chlorophyll, light harvesting proteins, and photosynthetic enzymes is delayed until the leaf is fully expanded and effectively protected by the increase in toughness (Figure 11.1). Chlorophyll and rubisco contents are only 10-20% of the levels seen in normally greening young leaves. The delay in the development of the lipid-rich chloroplast membranes means energy contents are also less than in green young leaves. Delayed greening thus results in substantially lower leaf nitrogen and energy invested in the chloro-

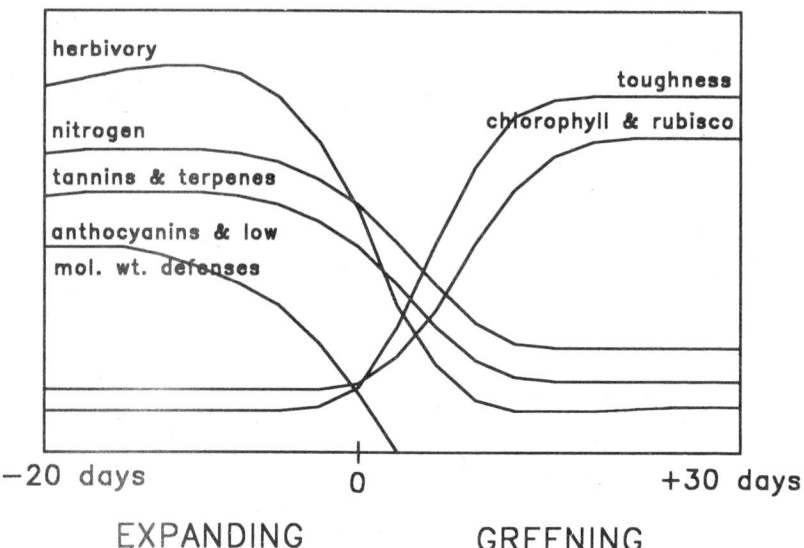

Figure 11.1.   Schematic representation of changes occurring during leaf development for species with delayed greening. Age 0 indicates full leaf expansion. Negative ages are for expanding leaves, and positive ages are for days since full expansion. Based on data in Kursar and Coley (1991, 1992a,b,c Coley & Kursar, unpublished data). We suspect that low molecular weight secondary compounds decrease during expansion, but no data exist to test this.

plasts than in typical green young leaves. For a given amount of herbivory, leaves with delayed greening lose fewer resources.

Clearly, the cost of delayed greening is forfeited photosynthesis during leaf development. Respiration exceeds photosynthesis throughout expansion, and maximal rates of net photosynthesis are not achieved until approximately 30 days after the leaf reaches full size (Kursar & Coley, 1992b,c). During expansion, light use efficiency for incident PPFD is about half that of green young leaves (Kursar & Coley, 1992c). Furthermore, during the 60 day period of leaf expansion and maturation, leaves with delayed greening only absorb 65% as much radiation as green leaves (Kursar & Coley, 1992c). The reduced photosynthetic capacity coupled with reduced light absorption translate to a much lower carbon gain during leaf expansion for species with delayed greening.

We can compare the costs and benefits of normal and delayed greening under different environments to determine if delayed greening is ever advantageous (Kursar & Coley, 1992a, 1995a). Photosynthesis, respiration, import, and tissue loss to herbivores must be considered in cost accounting and can all be measured in terms of glucose equivalents. For a given light level, we can then calculate the cost of delayed greening. For white leaves to be favored, the benefits of reduced loss to herbivores must exceed the cost of forfeited photosynthesis. For example, under saturating light, rates of damage would have to exceed 80 percent in order to favor white leaves. Therefore, in the high-light conditions of treefall gaps, the cost of forfeited photosynthesis is high and is not compensated for by the reduced loss of resources to herbivores. However, in the low light conditions of the understory, the benefits of delayed greening outweigh the costs. On BCI, the average damage rate to young leaves is about 39 percent. At these damage rates, the net cost of producing young leaves with normal vs delayed greening balances at 5 μ mol m^{-2} s^{-1}, a typical light level for the dark understory microsites (Chazdon & Fetcher, 1984; Denslow et al., 1990; Chazdon et al., Chapter 1). At higher herbivory, or lower light, leaves with delayed greening are most cost-effective.

Our cost analysis suggests that species with delayed greening are at a selective advantage under conditions of low light and high herbivory (Kursar & Coley, 1995a). In fact, it is only at PPFD levels less than approximately 20 μmol m^{-1} s^{-1} (1% of full sunlight) that we would expect to find species with delayed greening. These conditions are likely to be met only in tropical lowland rainforests. Casual observations suggest that delayed greening is rare in forests with higher light, such as temperate and tropical dry forests. Within the lowland rainforest of BCI, data from 175 of the most common tree species confirm that delayed greening occurs overwhelmingly in shade-tolerant species (Kursar & Coley, 1992a). Furthermore, the syndrome is common and geographically and phylogenetically widespread. In a survey of 250 species from lowland forests of southeast Asia, Africa, and Panama, we found one third of the species had delayed greening (< 0.5 mg dm^{-2} chlorophyll). Sixty-one percent of the 44 surveyed dicot families had species with delayed greening, suggesting that delayed greening has evolved independently many times. Delayed greening is particularly common in the Caesalpiniaceae, Sapindaceae, and Connaraceaè, and we have seen it in lycopods, ferns, and monocots as well.

There is no correlation between the extent of delayed greening and the leaf life span of 31 shade-tolerant species on BCI (ANOVA p = 0.074), although the trend is towards shorter life spans in species with normal greening. We argue that this trend is because higher light levels independently favor both shorter leaf life spans (Grime, 1977; Chabot & Hicks, 1982) and normal greening. This is clearly true comparing gap and understory species (Coley, 1983; Kursar & Coley, 1992a), and may also apply to more subtle variations in light in the understory. An alternative explanation is that longer life spans allow a greater period to pay back leaf construction costs. However, green and white leaves have similar mature leaf construction costs (Kursar & Coley, 1995), and differ primarily in the timing of investment. Species with delayed greening simply postpone part of the investment until after full leaf expansion. Furthermore, in the shade, forfeited photosynthesis is balanced by reduced losses to herbivores so that production of white leaves may, in fact, cost less than green leaves (Kursar & Coley, 1992a). We therefore find it unlikely that any correlation between life span and greening pattern is due to payback times.

Delayed greening provides an advantage only under conditions of high herbivory and low PPFD. Because the canopy and understory light environments differ substantially, we suggest that delayed greening may not be appropriate in the canopy. Although saplings do not green more quickly in high light (Kursar & Coley, 1992), the more profound ontogenetic changes associated with reaching the canopy might allow adult trees to green rapidly. Alternatively, greening of adult and juvenile leaves could be similar. The lack of plasticity, despite obvious advantages of early greening in the canopy, would suggest that these developmental changes are difficult to switch on and off. Intriguingly, this implies that selection for characters useful in the understory dominates. To date, these hypotheses remain untested.

It is interesting to note that not only is delayed greening common in tropical rainforests worldwide, but it also occurs in tropical marine algae. Species of the green alga *Halimeda* expand new, non-green segments at night when herbivory by reef fish is considerably lower (Hay et al., 1988). Greening occurs at dawn. Several other genera of siphonous green algae also appear to delay greening in young tissue until dawn (Kursar, Coley & Hay, unpublished data). Convergence of marine and terrestrial plants is strong evidence for the adaptive role of delayed greening under conditions of low light and high herbivory. On reefs, light varies temporally (day versus night), while in the rainforest it varies spatially (gap versus understory). Despite these differences, the combination of high herbivory and low light appears to have selected for delayed greening in two entirely different environments.

11.3.7 Tropical and temperate comparisons

The preceding survey of defense and herbivory points out that young tropical leaves suffer higher relative and absolute amounts of damage as compared to their temperate counterparts. They also have, on average, higher levels of defense and a greater variety of defense options. Given these patterns, can we speculate as to the implications for plant/herbivore interactions in temperate and tropical forest communities? First, depending on such an ephemeral food source may put ecological and evolutionary constraints on host-finding by tropical herbivores, forcing their life cycles to be very tightly coupled with the phenology of leaf flushing. Secondly, the greater diversity of defense approaches in tropical young leaves may mean greater diversity in dynamical relationships with herbivores.

11.4 TRADE-OFFS AND PHYSIOLOGICAL CONSTRAINTS

Given the extremely high rates of herbivory to young leaves, one might expect selection to favor the use of the entire battery of defenses described above. However, we rarely see the full range of defenses in a single species. In fact, subsets of these defenses tend to consistently co-occur. In this section, we present data from a survey of defenses for the most common species at 4 lowland rainforest sites in Borneo, Africa, and Panama (see Appendix 11.1 for site descriptions). We describe the patterns of co-occurrence of defenses and suggest that there are trade-offs and physiological constraints that limit the defensive possibilities for any given species.

11.4.1 Expansion and nitrogen

Both rapid expansion and low nitrogen reduce herbivory damage to young leaves. However, there is a negative relationship between the two such that fast expanders have higher concentrations of nitrogen (Figure 11.2). Presumably this is because rapid expansion requires high levels of enzymes in order to accomplish leaf construction. It appears to be physiologically impossible to expand rapidly with low levels of protein.

11.4.2 Expansion and toughness

Toughness is one of the best defenses (Tanton, 1962; Raupp, 1985; Coley, 1983), and although young leaves are much more tender than

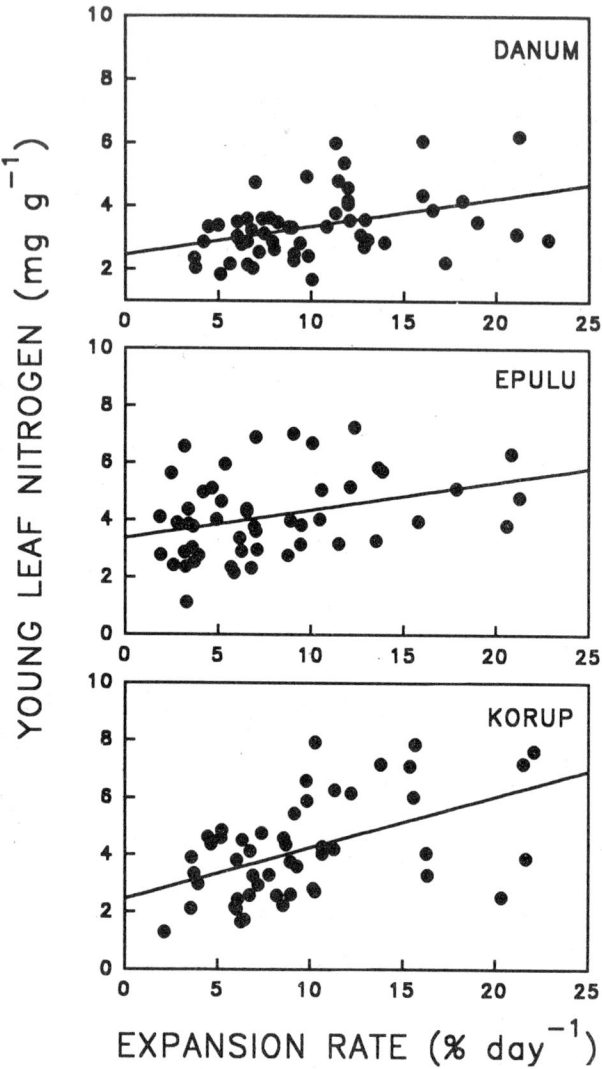

Figure 11.2.   Regression of expansion rate of young leaves versus leaf nitrogen content for the most common species in three lowland tropical rainforests (see Appendix 11.1 for site descriptions). Leaves were collected at 10% of full expansion, dried, and analyzed using a CHN analyzer. Expansion rate was measured as the % change in size d^{-1} for leaves between 20% and 80% of full size. Each point is a different species and is the average of several samples. Danum: $r^2 = 0.16$ $p < 0.01$; Epulu: $r^2 = 0.12$ $p < .02$; Korup: $r^2 = 0.26$ $p < 0.001$.

mature leaves, there is still considerable variation among young leaves of different species (Figure 11.3). In species with higher toughness, the greater stiffening of the cell wall is apparently sufficient to slow expansion. This difference in toughness between slow and fast expanders could have an impact on very young instars of herbivores, although it is unlikely to affect larger insects.

11.4.3 Expansion and herbivory

Surprisingly, fast-expanding young leaves suffered both higher rates of herbivory as well as larger amounts of damage than slow-expanding species (Figure 11.4). This pattern is even stronger if one controls phylogeny by looking within families (Figure 11.5). Rapid expansion should reduce herbivory by shortening the period of greatest vulnerability. However, fast-expanding leaves are also less tough and have higher nitrogen, both characteristics that would make them more attractive to herbivores. So to directly examine the relationship between expansion and herbivory, we statistically removed the effects of toughness and nitrogen using partial regressions. The positive correlation between expansion and herbivory remained essentially unchanged (Table 11.3). This pattern suggests that slow-expanding species must be better defended chemically.

Table 11.3. Herbivory as a function of expansion rate by rainforest site and by plant family. Values of r^2 are compared for simple regression and for a partial regression where the effects of young leaf nitrogen and toughness are removed. Values for simple regression come from Figures 11.4 and 11.5. There are no data on nitrogen and toughness for the BCI species (nd).

	Simple regression		Partial regression	
	r^2	P value	r^2	P value
By site:				
Danum	0.28	0.01	0.31	0.01
Korup	0.10	0.03	0.13	0.09
Epulu	0.07	0.08	0.09	0.24
B.C.I.	0.28	0.01	nd	
By family/site:				
Annonaceae/Danum	0.31	0.08	0.36	0.07
Dipterocarpaceae/Danum	0.45	0.07	0.38	0.10
Euphorbiaceae/Epulu	0.70	0.04	0.76	0.02
Sapotaceae/Epulu	0.73	0.06	0.28	0.29
Rubiaceae/Korup	0.00	0.98	0.00	0.32
Leguminosae/Korup	0.85	0.03	0.87	0.02

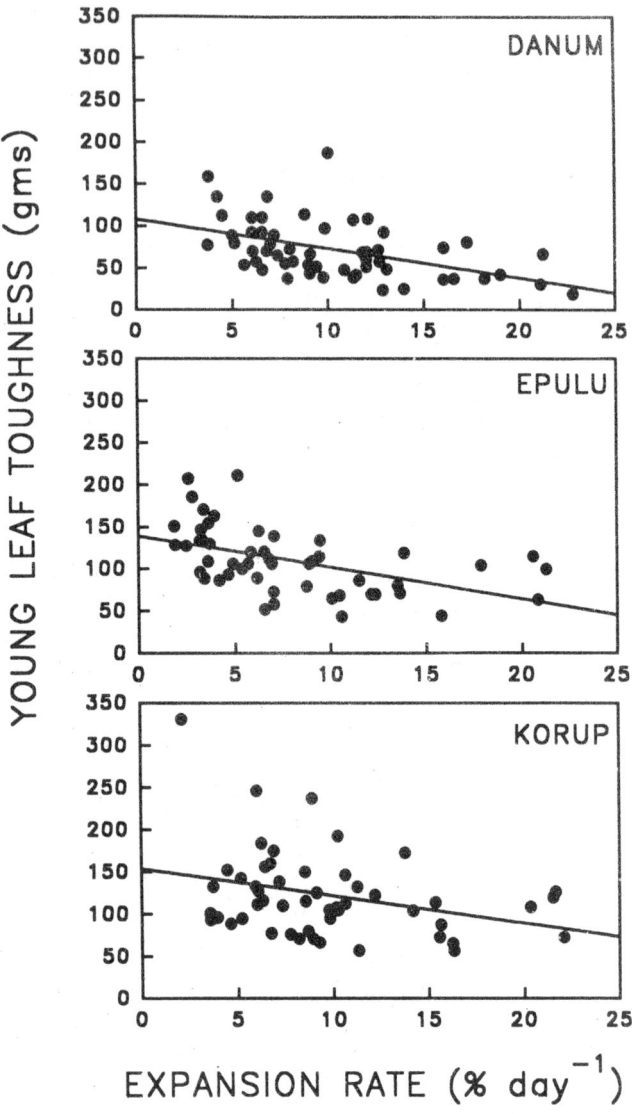

Figure 11.3.. Regression of expansion rates of young leaves versus toughness for the most common species in three tropical rainforests (see Appendix 11.1 for site descriptions). Toughness was determined on leaves at 75% of full expansion with a Chatillon penetrometer and reported in grams. This is the weight necessary to punch a rod 3mm in diameter through the leaf. Each point is a different species and is the average of several samples. Danum: $r^2 = 0.23$ $p < 0.001$; Epulu: $r^2 = 0.24$ $p < .001$; Korup: $r^2 = 0.10$ $p < 0.03$.

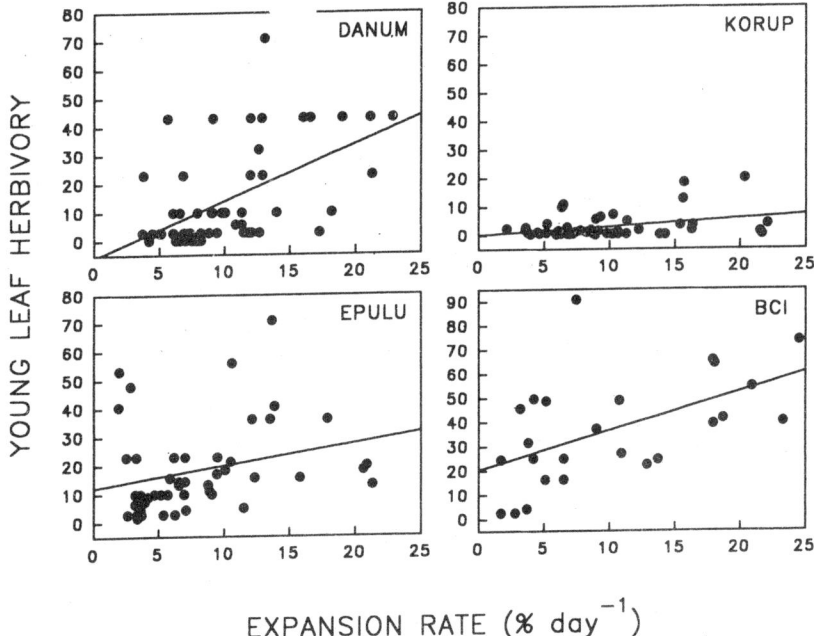

Figure 11.4. Regression of expansion rates of young leaves versus the amount of leaf damage at the end of expansion (herbivory) for the most common species in three tropical rainforests (see Appendix 11.1 for site descriptions). Each point is a different species and is the average of several samples. Danum: $r^2 = 0.28$ p < 0.001; Epulu: $r^2 = 0.07$ p < .08; Korup: $r^2 = 0.10$ p < 0.03; BCI: $r^2 = 0.28$ p < 0.01.

High levels of defense may actually be incompatible with rapid expansion (Richards 1952; McKey 1979). First, production of defense compounds in young leaves may decrease the rate of leaf development by competing for the materials and/or energy available for growth (Mooney & Chu, 1974; Orians & Janzen, 1974). Simply building the structural components of a leaf that doubles in size every day may require most of the metabolic machinery. Second, it has been suggested that sequestration of secondary metabolites away from cell machinery may be more difficult or costly in fast-expanding leaves (Rhoades & Cates, 1976; McKey, 1979). This would lead to a higher probability of autotoxicity. Although this is certainly plausible, fairly high levels of condensed tannins have been measured in fast-expanding young leaves (Coley, 1983).

Rapid expansion and chemical defense are probably the two most effective defenses for young leaves. Yet, because of the possibilities of

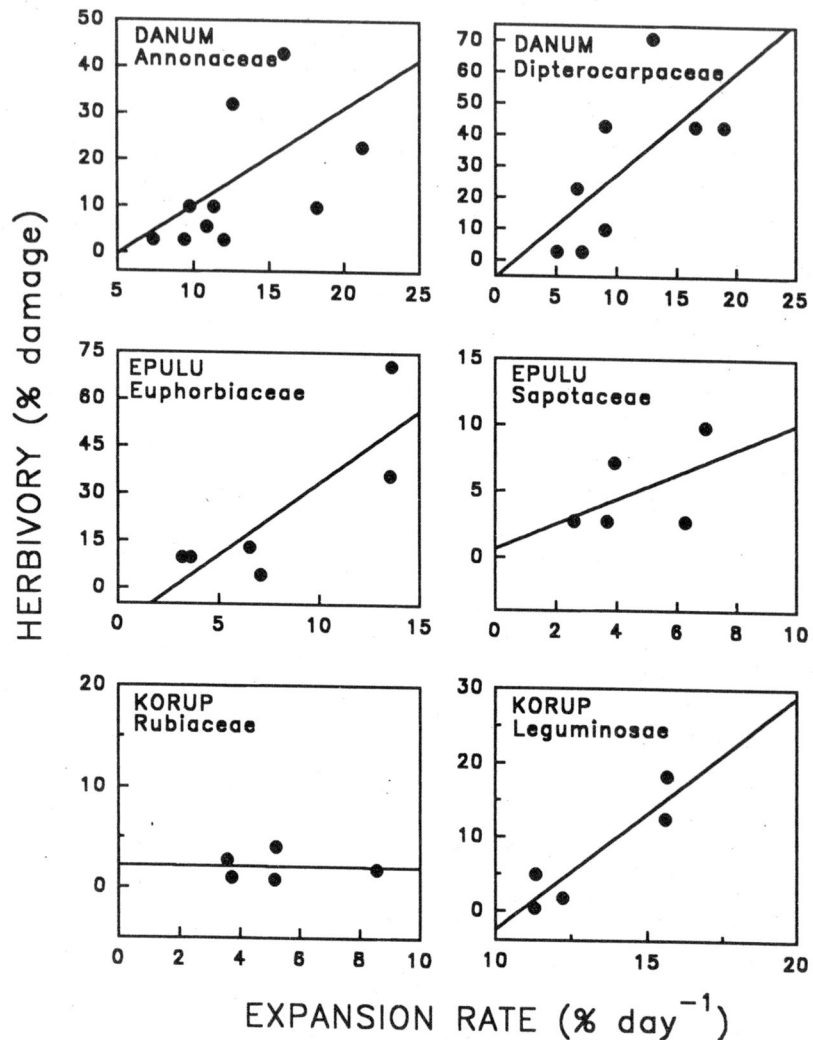

Figure 11.5.   Regression of expansion rates of young leaves versus the amount of leaf damage at the end of expansion (herbivory) for the six families containing 5 or more species. (see Appendix 11.1 for site descriptions). Each point is a different species and is the average of several samples. Danum, Annonaceae: $r^2 = 0.31$ $p < 0.08$; Danum, Dipterocarpaceae: $r^2 = 0.45$ $p < 0.07$; Epulu, Euphorbiaceae: $r^2 = 0.70$ $p = 0.04$; Epulu, Sapotaceae: $r^2 = 0.73$ $p < .06$; Korup, Rubiaceae: $r^2 = 0.00$ n.s.; Korup, Leguminosae: $r^2 = 0.85$ $p < 0.03$.

autotoxicity and competition for resources, the limited data available suggest that it is physiologically impossible to have both rapid expansion and high levels of secondary metabolites. Chemically unprotected shoot tips of ant acacias grow much faster than those of chemically well-protected non-ant acacias (Janzen, 1966; Rehr, Feeny & Janzen, 1973). Better defended shoots of species in families such as Asclepiadaceae, Apocynaceae, Passifloraceae, and Vitaceae tend to elongate more slowly than less well defended shoots of species in families such as Convolvulaceae, Leguminosae and Bignoniaceae (Orians & Janzen, 1974).

We speculate that the trade-off between expansion rate and chemical defense may have implications for herbivore population dynamics and life history traits. Species with rapidly expanding leaves may more easily escape detection by specialist herbivores, and instead, because of their higher palatability, be fed on primarily by generalists. In contrast, slow expanders, with their variety of chemical defenses, would more likely be targets of specialists that are capable of handling the secondary metabolites. Third, higher levels of chemical defense in slow expanders may mean that larval development of herbivores is slowed. This prolonged period of larval vulnerability opens the possibility for predators and parasitoids to have greater control of herbivore populations. Increased pressure from the third trophic level may, in turn, dampen herbivore outbreaks.

11.4.4 Expansion and delayed greening

Rapidly expanding leaves delay greening, whereas more slowly expanding leaves tend to green normally (Figure 11.6). There are at least two possibilities to explain this pattern. First, the negative correlation may mean that fast-expanding leaves cannot simultaneously develop the photosynthetic machinery because of within-leaf competition for limited resources. We suspect that limitation might arise from constraints on import into the expanding leaf. Rates of phloem transport are affected by the size and number of sieve tubes (Canney, 1975). We hypothesize that the most rapidly expanding species must have anatomical traits, such as a large cross-sectional area of phloem in the petiole or flushing twig, in order to accomodate such high rates of resource delivery. It seems less likely that resource limitation would result from insufficient resources at the whole plant level. Although delayed greening spreads construction costs over a slightly longer period, the additional time (approximately 10-20 days) appears insufficient to supply the resources from current photosynthate, especially

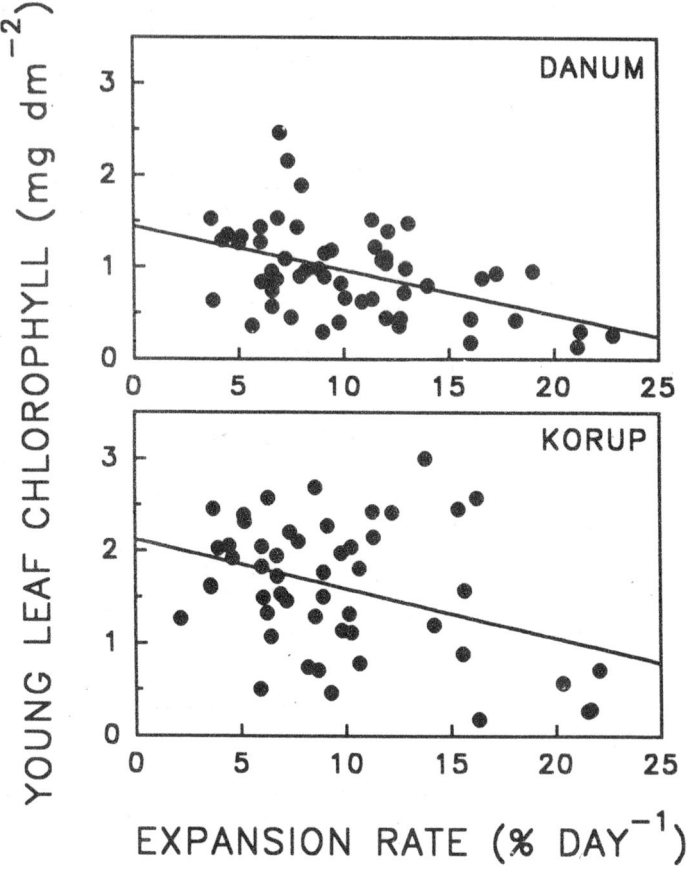

Figure 11.6.   Regression of expansion rates versus chlorophyll content for two lowland tropical forests (see Appendix for site descriptions). Chlorophyll was determined spectrophotometrically on fresh leaves collected at 75% of full expansion. Each point is a different species and is the average of several samples. For reference, a value of 2.2 mg dm^{-2} is typical of a young green leaf.

in the understory. We therefore favor the hypothesis that physical constraints on resource delivery into the leaf may put an upper limit on resource availability. Data for evaluating these physiological constraints are lacking.

A second possibility for the negative relationship between greening and expansion is that the high herbivory on rapid expanders selects for delayed greening as a way of minimizing the impact of this damage. Although there are no physiological constraints prohibiting de-

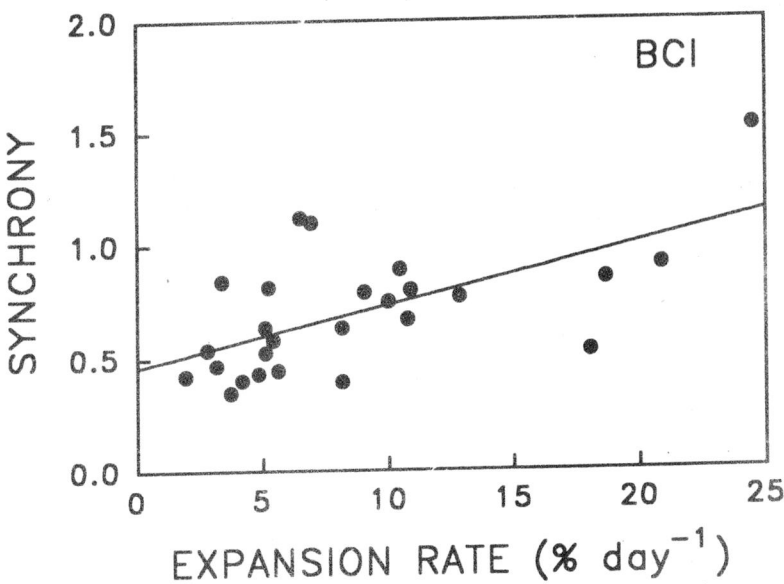

Figure 11.7.   Regression of expansion rates versus synchrony of leaf production for 25 species on BCI. (See Appendix 11.1 for site description). Each point is a different species. The synchrony index is the coefficient of variation for leaf production measured on a monthly basis. A higher value indicates a more synchronous production of leaves. Data on synchrony are from Coley (1983) and Aide 1993. $r^2 = 0.36$, $p < 0.001$.

layed greening in slow expanders, it would not be advantageous. Since slow-expanding leaves suffer little damage, there would be no point to paying the costs of delayed greening without the benefits. Thus, the correlation we see between expansion and delayed greening may be the combined result of both physiological constraints and ecological tradeoffs.

11.4.5   Expansion and synchrony

The degree of intraspecific synchrony in production of young leaves is significantly positively correlated with the speed of leaf expansion (Figure 11.7). Since fast expanders suffer high levels of herbivory and are apparently less well defended chemically, it is not surprising that selection has favored the added protection of synchrony. Unlike greening and the production of secondary metabolites, synchrony should not interfere with expansion rates. Synchronous production of young leaves involves mobilization of large amounts of stored resources, but this is at the whole plant level and would not compete with resource allocation within a leaf.

11.4.6   Expansion, ant defense, and synchrony

Ant rewards, such as extra-floral nectaries and food bodies, are produced on expanding leaves. Does investment in these rewards compete with resources for rapid expansion? Combining data from several studies on BCI, we found no difference in expansion rates for ant-defended versus non-ant-defended species (T-test, $p < 0.084$; $n = 46$; Coley, 1983; Schupp & Feener, 1991; Aide, 1993; Kursar & Coley, unpublished data). This result is consistent with the argument that ant rewards are relatively cheap (O'Dowd, 1979, 1980; Beattie, 1985; Keeler, 1981) and would therefore not interfere significantly with rapid expansion.

Since effective ant defense relies on continuous guarding by ants, it has been suggested that continuous leaf production would favor the evolution of ant defense (Schupp & Feener, 1991). For 60 species on BCI, those that had ant defense were significantly less synchronous (T-test, $p < 0.027$, $n = 60$; Coley, 1983; Schupp & Feener, 1991; Aide, 1993; Kursar & Coley, unpublished data). Continuous leaf production thus permits ant defense.

11.5   RESOURCE AVAILABILITY AND DEFENSE STRATEGIES

Although lowland tropical rainforests are similar with respect to many abiotic and biotic factors, subtle differences in resource availability appear to shift the balance between defense strategies. BCI, Danum, and Epulu all have forests typical of relatively rich soils, whereas both forest structure and soil analyses suggest that Korup is a more nutrient-poor site (see Appendix). Interestingly, young leaves at Korup are significantly tougher (ANOVA $p < 0.0001$) and greener (ANOVA $p < 0.0001$) and suffer lower herbivory (ANOVA $p < 0.0001$). Although there is no significant difference among sites in expansion rates, Korup has the highest frequency of slow-expanding leaves if one corrects for the higher elevation and cooler temperatures (see Appendix for correction factor). These patterns suggest a shift in the nutrient-poor site away from leaves that rely on minimizing the impact of herbivory through rapid expansion and delayed greening towards leaves that are better protected chemically from herbivores. It is well established that mature leaves of nutrient-poor sites are effectively defended by chemicals and toughness, as lost resources are difficult to replace (Janzen, 1974; McKey et al., 1978; Coley, 1987). Apparently a similar pattern is seen with the young leaves. It is tempting to speculate that species that might otherwise be able to persist on nutrient-poor soils are excluded because of inappropriate patterns of young leaf defense. And indeed, families that consistently have rapid expansion, such as Connaraceae, Sapindaceae, and to a lesser extent Caesalpiniaceae, are under-represented at Korup.

11.6 CONCLUSIONS

The correlations between different suites of defensive characteristics are consistent in all the lowland forests we have studied. This suggests that there are fundamental underlying causes for the patterns of co-occurring traits. We argue that it is productive to identify two extreme syndromes of defense: fast and slow expanders (Table 11.4). In order to expand rapidly, species must have high nitrogen, presumably in metabolically important enzymes. Perhaps because of competition for resources, greening and the synthesis of secondary metabolites would be delayed. We predicted a similar pattern for ant defense, but data suggest it is equally common in fast and slow expanders. There could also be problems of autotoxicity from secondary metabolites in rapidly dividing cells. Synchrony would add further protection without interfering with expansion. At the other extreme, slow-expanding leaves would not require the high concentrations of enzymes, and because herbivores prefer leaves with high nitrogen,

Table 11.4. Defensive characteristics of fast- and slow-expanding young leaves.

	Fast-expanders	Slow-expanders
Nitrogen content	high	low
Toughness	low	medium
Chlorophyll	low	high
Photosynthetic capacity	low	moderate
Chemical defenses	low	high
Extrafioral nectaries	common	common
Synchrony	common	rare

selection should favor reduced levels. Production of secondary metabolites would not only be physiologically possible in leaves with slow expansion, but would be strongly selected for by herbivores. Once leaves are effectively protected, delayed greening and synchrony become disadvantageous.

These data suggest that the defenses of young tropical leaves are diverse, and that the only way to understand individual defenses is by looking at them in the ecological context of other co-occurring traits. It seems that ecological trade-offs and physiological constraints may play a major role in shaping defenses of young leaves.

ACKNOWLEDGEMENTS

We are especially grateful to the people who shared their knowledge of different field sites and allowed us to work there: Terese and John Hart in Zaire, Clive Marsh and Elaine Gasis in Sabah, and John Hazam, Andrew Allo Allo, and Ferdinand Namata in Cameroon. The Smithsonian Tropical Research Institute provided research facilities in Panama. We thank Julie Glick, Ann Dickinson, and Denise Dearing for technical assistance. The manuscript was improved by the editors and an anonymous reviewer. This research was supported by a J.S. Guggenheim Fellowship (to P.D.C.) and NSF grants BSR-8407712, BSR-8806080, and BSR-9119619 (to T.A.K. and P.D.C.).

REFERENCES

AIDE, T. M. (1988) Herbivory as a selective agent on the timing of leaf production in a tropical understory community. *Nature*, **336**, 574–575.

AIDE, T. M. (1991) Synchronous leaf production and herbivory in juveniles of *Gustavia superba*. *Oecologia*, **88**, 511–514.

AIDE, T. M. (1993) Patterns of leaf development and herbivory in a tropical understory community. *Ecology*, **74**, 455–466.

AIDE, T. M. & LONDOÑO, E. C. (1989) The effects of rapid leaf expansion on the growth and survivorship of a lepidopteran herbivore. *Oikos*, **55**, 66–70.

ANONYMOUS (1987) *The Korup Project Soil Survey and Land Evaluation* Prepared by WWF in collaboration with The Land Resources Development Centre of the British Government for the Government of Cameroon.

BAKER, N. R. & HARDWICK, K. (1973) Biochemical and physiological aspects of leaf development in cacao (*Theobroma cacao*). I. Development of chlorophyll and photosynthetic activity. *New Phytologist*, **72**, 1315–1324.

BAKER, N. R., HARDWICK, K. & JONES, P. (1975) Biochemical and physiological aspects of leaf development in cacao (*Theobroma cacao*). II. Development of chloroplast ultrastructure and carotenoids. *New Phytologist,* **75,** 513–518.

BALDWIN, I. T., KARB, M. J. & OHNMEISS, T. E. (1994) Allocation of ^{15}N from nitrate to nicotine: Production and turnover of a damage-induced mobile defense. *Ecology* **75,** 1703–1713.

BAZZAZ, R. A., CHIARIELLO, N. R., COLEY, P. D. & PITELKA, L. F. (1987) Allocating resources to reproduction and defense. *Bioscience,* **37,** 58–67.

BEATTIE, A. J. (1985) *The Evolutionary Ecology of Ant-Plant Mutualisms.* Cambridge University Press, Cambridge.

BENTLEY, B. (1977) Extrafloral nectaries and protection by pugnacious bodyguards. *Annual Review of Ecology and Systematics,* **8,** 407–427.

BENTLEY, B. and ELIAS, T. (1983) *The Biology of Nectaries.* Columbia University Press, New York.

BRYANT, J. P., CHAPIN, F. S. III., & KLEIN, D. R. (1983) Carbon/nutrient balance of boreal plants in relation to vertebrate herbivory. *Oecologia,* **40,** 357–368.

BURGESS, P. F. (1969) Color changes in the forest. *Malayan Nature Journal,* **22,** 171–173.

CALDWELL, M. M. (1981) Plant responses to ultraviolet radiation. *Physiological Ecology,* **I,** 179–197. Springer-Verlag, Heidelberg.

CANNEY, M. J. (1975) Mass transfer. *Transport in Plants I: Encyclopedia of Plant Physiology,* (eds. M. H. ZIMMERMAN, & J. A. MILBURN) pp 291–300.

CHABOT, B. F. & HICKS, D. J. (1982) The ecology of leaf life spans. *Annual Review of Ecology and Systematics,* **13,** 229–259.

CHAZDON, R. L. & FETCHER, N. (1984). Photosynthetic light environments in a lowland tropical rain forest in Costa Rica. *Journal of Ecology,* **72,** 553–564.

CLARK, D. B. & CLARK D. A. (1985) Seedling dynamics of a tropical tree: Impacts of herbivory and meristem damage. *Ecology,* **66,** 1884–1892.

COLEY, P. D. (1982) Rates of herbivory on different tropical trees. *The Ecology of a Tropical Forest* (eds. E. G. LEIGH; Jr., A. S. RAND, & D. M. WINDSOR, D. M.) Smithsonian Institution Press, Washington, D. C., pp 123–132.

COLEY, P. D. (1983) Herbivory and defensive characteristics of tree species in a lowland tropical forest. *Ecological Monographs,* **53,** 209–233.

COLEY, P. D. (1986) Costs and benefits of defense by tannins in a neotropical tree. *Oecologia,* **70,** 238–241.

COLEY, P. D. (1987) Patterns of plant defense: Why herbivores prefer certain species. *Revista de Biologia Tropical,* **35 (Suppl.** 1), 251–263.

COLEY P. D. & AIDE, T. M. (1989) Red coloration of tropical young leaves: A possible antifungal defense? *Journal of Tropical Ecology,* **5,** 293–300.

COLEY P. D. & AIDE, T. M. (1991) Comparison of herbivory and plant defenses in temperate and tropical broad-leaved forests. *Plant-Animal Interactions: Evolution Ecology in Tropical and Temperate Regions*. (eds. P. W. PRICE, T. M. LEWINSOHN, T. M., WILSON FERNANDES, G. W. & W. W. BENSON, W. W.) John Wiley and Sons, N. Y., pp. 25–49.

COLEY, P. D., BRYANT, J. P., & CHAPIN, F. S. III. (1985) Resource availability and plant anti-herbivore defense. *Science*, **230**, 895–899.

COOKE, F. P., BROWN, J. P. & MOLE, S. (1984) Herbivory, foliar enzyme inhibitors, nitrogen and leaf structure of young and mature leaves in a tropical forest. *Biotropica*, **16**, 257–263.

CRANKSHAW, D. R. & LANGENHEIM, J. H. (1981) Variation in terpenes and phenolics through leaf development in *Hymenaea* and its possible significance to herbivory. *Biochemical Systematics and Ecology*, **9**, 115–124.

CROAT, T. B. (1978) *Flora of Barro Colorado Island*. Stanford University Press, Stanford, California.

DENSLOW, J. S, SCHULTZ, J. C., VITOUSEK, P. M.& STRAIN, B. R. (1990) Growth responses of tropical shrubs to treefall gap environments. *Ecology*, **71**, 165–179.

DIRZO, R. (1984) Herbivory, a phytocentric overview. Perspectives in *Plant Population Biology* (eds. R. DIRZO & J. SARUKHAN, J.) Sinauer, Sunderland, MA. pp 141–165.

DIRZO, R. & MIRANDA, A. (1991) Altered patterns of herbivory and diversity in the forest understory: A case study of possible consequences of contemporary defaunation. *Plant-Animal Interactions: Evolution Ecology in Tropical and Temperate Regions* (eds. PRICE, P. W., T. M. LEWINSOHN, G. W. WILSON FERNANDES, & W. W. BENSON) JOHN WILEY and SONS, N. Y., pp. 273–287.

ERNEST, K. A. (1989) Insect herbivory on a tropical understory tree: Effects of leaf age and habitat. *Biotropica*, **21**, 194–199.

FEENER, D. H., Jr. & SCHUPP, E. W. (1995) Effect of treefall gaps on the patchiness and species richness of neotropical ant assemblages. (in press). *Ecology*.

FEENY, P. P. (1970) Seasonal changes in oak leaf tannins and nutrients as a cause of spring feeding by winter moth caterpillars. *Ecology*, **51**, 565–581.

FEIBERT, E. B. & LANGENHEIM, J. H. (1988) Leaf resin variation in *Copaifera langsdorfii*: relation to irradiance and herbivory. *Phytochemistry*, **27**, 2527–2532.

FOSTER, R. B. & BROKAW, N. V. L. (1982) Structure and history of the vegetation of Barro Colorado Island. *The Ecology of a Tropical Forest* (eds. E. G. LEIGH, Jr., A. S. RAND & D. M. WINDSOR) Smithsonian Institution Press, Washington, D. C., pp 67–81.

FRANKIE, G. W., BAKER, H. G. & OPLER, P. (1974) Comparative phenological studies of trees in tropical wet and dry forests in the lowlands of Costa Rica. *Journal of Ecology*, **62**, 881–919.

GARCIA-GUZMAN, M. & DIRZO, R. (1991) *Plant-pathogen-animal interactions in a tropical rain forest.* AIBS abstract.

GARTLAN, J. S., NEWBERY, D. M., THOMAS, D. W. & WATERMAN, P. G. (1986) The influence of topography and soil phosphorus on the vegetation of Korup Forest Reserve, Cameroun. *Vegetatio*, **65**, 131–148.

GERSHENZON, J., MURTAGH, G. J. & CROTEAU, R. (1993) Absence of rapid terpene turnover in several diverse species of terpene-accumulating plants. *Oecologia*, **96**, 583–592.

GRIME, J. P. (1977) Evidence for the existence of three primary strategies in plants and its relevance to ecological and evolutionary theory. *American Naturalist*, **111**, 1169–1194.

HARBORNE, J. B. (1979) Function of flavonoids in plants. *Chemistry and Biochemistry of Plant Pigments* (ed. T. W. GOODWIN) Academic Press, New York, pp 736–788.

HART, T. B., HART, J. A. & MURPHY, P. G. (1989) Monodominant and species-rich forests of the humid tropics: Causes for their co-occurrence. *American Naturalist*, **133**, 613–633.

HAWKINS, P. & M. BRUNT. (1965) *The Soils and Ecology of West Cameroon.* FAO Expanded Program of Technical Assistance No 2083, Food and Agriculture Organization, Rome.

HAY, M. E., PAUL, V. J., LEWIS, S. M., GUSTAFSON, K., TUCKER, J. & TRINDELL, R. N. (1988) Can tropical seaweeds reduce herbivory, by growing at night? Diel patterns of growth, nitrogen content, herbivory and chemical versus morphological defenses. *Oecologia*, **75**, 233–245.

HOLDRIDGE, L. R., GRENKE, W. C., HATHEWAY, W. H., LIANG, T. & TOSI, J. A., Jr. (1971). *Forest Environments in Tropical Life Zones: A Pilot Study.* Pergamon, Oxford, England.

HUXLEY C. R. & CUTLER, D. F. (1991) *Ant-Plant Interactions*, Oxford University Press, Oxford.

JANZEN, D. H. (1966) Coevolution of mutualism between ants and acacias in Central America. *Evolution*, **20**, 249–275.

JANZEN, D. H. (1967) Interaction of the bull's-horn acacia (*Acacia cornigera* L.) with an ant inhabitant (*Pseudomyrmex ferruginea* F. Smith) in Eastern Mexico. *University of Kansas Science Bulletin*, **47**, 315–558.

JANZEN, D. H. (1970) Herbivores and the number of tree species in tropical forests. *American Naturalist*, **104**, 501–528.

JANZEN, D. H. (1974) Tropical blackwater rivers, animals, and mast fruiting by the Dipterocarpaceae. *Biotropica*, **6**, 69–103.

JUNIPER, B. & SOUTHWOOD, T. R. E. (1986) *Insects and the Plant Surface,* Arnold, London.

KEELER, K. H. (1981) A model of selection for facultative nonsymbiotic mutualism. *American Naturalist*, **118**, 488-498.

KOPTUR, S. (1984) Experimental evidence for defense of *Inga* (Mimosoideae) saplings by ants. *Ecology*, **65**, 1787–1793.

KURSAR, T. A. & COLEY, P. D. (1991) Nitrogen content and expansion rate of young leaves of rainforest species: Implications for herbivory. *Biotropica* **23**, 141–150.

KURSAR, T. A. & COLEY, P. D. (1992a) Delayed greening in tropical leaves: An anti-herbivore defense? *Biotropica*, **24**, 256–262.

KURSAR, T. A. & COLEY, P. D. (1992b) Delayed development of the photosynthetic apparatus in tropical rainforest species. *Functional Ecology*, **6**, 411–422.

KURSAR, T. A. & COLEY, P. D. (1992c) The consequences of delayed greening during leaf development for light absorption and light use efficiency. *Plant, Cell and Environment*, **15**, 901–909.

KURSAR, T. A. & COLEY, P. D. (1995) The cost of leaf construction in shade-tolerant rainforest plants. (in press) *American Naturalist*.

LANGENHEIM, J. H., ARRHENIUS, S. P. & NASCIMENTO, J. C. (1981) Relationship of light intensity to leaf resin composition and yield in the tropical leguminous genera *Hymenaea* and *Copaifera*. *Biochemical Systematics and Ecology*, **9**, 27–37.

LANGENHEIM, J. H., MACEDO, C. A., ROSS, M. K. & STUBBLEBINE, W. H. (1986) Leaf development in the tropical leguminous tree *Copaifera* in relation to microlepidopteran herbivory. *Biochemical Systematics and Ecology*, **14**, 51–59.

LEE, D. W. & LOWRY, J. B. (1980) Young leaf anthocyanin and solar ultraviolet. *Biotropica*, **12**, 75–76.

LEE, D. W., BRAMMEIER, S., & SMITH, A. P. (1987) The selective advantages of anthocyanins in developing leaves of mango and cacao. *Biotropica*, **19**, 40–49.

LEIGH, E. G., Jr., & SMYTHE, N. (1978) Leaf production, leaf consumption, and the regulation of folivory on Barro Colorado Island. *The Ecology of Arboreal Folivores* (ed. G. G. MONTGOMERY) Smithsonian Institution Press, Washington, D. C. pp 35–50.

LEIGH, E. G., Jr., & WRIGHT, S. J. (1990) Barro Colorado Island and tropical biology. *Four Neotropical Rainforests* (ed. A. H. GENTRY) Yale University Press, pp 28–47.

LEONG, K. M. (1974) The geology and mineral resources of the upper Segama Valley and Darvel Bay, Sabah, Malaysia. *Geological Survey of Malaysia Memoir 4*, Kuala Lumpur, Malaysia.

LIEBERMAN, D. & LIEBERMAN, M. (1984) The causes and consequences of synchronous flushing in a dry tropical forest. *Biotropica*, **16**, 193–201.

LOWMAN, M. D. (1984) An assessment of techniques for measuring herbivory: Is rainforest defoliation more intense than we thought? *Biotropica*, **16**, 264–268.

LOWMAN, M. D. (1985a) Spatial and temporal variation in insect grazing of the canopies of five Australian rainforest trees. *Australian Journal of Ecology,* **10,** 7–24.

LOWMAN, M. D. (1985b) Insect herbivory in Australian rainforests – is it higher than in the neotropics? *Proceedings of the Ecological Society of Australia,* **14,** 109–119.

LOWMAN, M. D. & BOX, J. R. (1983) Variation in leaf toughness and phenolic content among five species of Australian rainforest trees. *Australian Journal of Ecology,* **8,** 17–25.

MARQUIS, R. J. (1984) Leaf herbivores decrease fitness of a tropical plant. *Science,* **226,** 537–539.

MARQUIS, R. J. (1992a) Selective impact of herbivores. *Plant Resistance to Herbivores and Pathogens* (eds. R. S. FRITZ, & E. L. SIMMS). University of Chicago Press, Chicago, Illinois. pp 301–325.

MARQUIS, R. J. (1992b) A bite is a bite is a bite? Constraints on response to folivory in *Piper arieianum* (Piperaceae). *Ecology,* **73,** 143–152.

MARQUIS, R. J. & BRAKER, H. E. (1994) Plant-herbivore interactions: Diversity, specificity, and impact. La Selva: *Ecology and Natural History of a Neotropical Rainforest* (eds. L. McDADE, G. H. HARTSHORN, H. HESPENHEIDE & K. BAWA) University of Chicago Press, Chicago, pp 263–281.

MATTSON, W. J. & SCRIBER, J. M. (1987) Nutritional ecology of insect folivores of woody plants: Nitrogen, water, fiber, and mineral considerations. *Nutritional Ecology of Insects, Mites, Spiders and Related Invertebrates* (eds. F. SLANSKY, & J. G. RODRIGUEZ) John Wiley and Sons, New York pp 105–146.

McKEY, D. D. (1979) The distribution of secondary compounds within plants. *Herbivores: Their Interactions with Secondary Plant Metabolites.* (eds. G. A. ROSENTHAL & D. H. JANZEN). Academic Press, New Jersey, pp 55–133.

McKEY, D. D. (1984) Interaction of the ant-plant *Lenardoxa africana* (Caesalpiniaceae) with its obligate inhabitants in a rainforest in Cameroon. *Biotropica,* **16,** 81–99.

McKEY, D. D. (1989) Interactions between ants and leguminous plants. *Advances in Legume Biology* (eds. J. ZARUCCHI, & C. STIRTON). Missouri Botanical Gardens, St Louis.

McKEY, D. D., WATERMAN, P. G., MBI, C. N., GARTLAN, S. J. & STRUHSAKER, T. T. (1978) Phenolic content of vegetation in two African rainforests: Ecological implications. *Science,* **202,** 61–64.

MIHALIAK, C. A., GERSHENZON, J. & CROTEAU, R. (1991) Lack of rapid monoterpene turnover in rooted plants: Implications for theories of plant chemical defense. *Oecologia,* **87,** 373–376.

MOLE, S., ROSS, J. A. M. & WATERMAN, P. G. (1988) Light-induced variation in phenolic levels in foliage of rainforest plants. I. Chemical changes. *Journal of Chemical Ecology,* **14,** 1–21.

MOONEY, H. A. & CHU, C. (1974) Seasonal carbon allocation in *Heteromeles arbutifolia*, a California shrub. *Oecologia*, **14**, 295–306.

MORAN, N. & HAMILTON, D. W. (1980) Low nutritive quality as a defense against herbivores. *Journal of Theoretical Biology*, **86**, 247–254.

NASCIMENTO, M. T. & HAY, J. D. (1993) Intraspecific variation in herbivory on *Metrodorea pubescens* (Rutaceae) in two forest types in central Brazil. *Revista Brasiliera de Biologia*, **53**, 143–153.

NEWBERY, D. M. & DE FORESTA, H. (1985) Herbivory and defense in pioneer gap and understory trees in tropical rainforests in French Guiana. *Biotropica*, **17**, 238–244.

NEWBERY, D. M., ALEXANDER, I. J., THOMAS, D. W. & GARTLAN, J. S. (1988) Ectomycorrhizal rain-forest legumes and soil phosphorus in Korup National Park, Cameroon. *New Phytologist*, **109**, 433–450.

NICHOLS-ORIANS, C. M. (1991) The effects of light on foliar chemistry, growth and susceptibility of seedlings of a canopy tree to an attine ant. *Oecologia*, **86**, 552–560.

O'DOWD, D. J. (1979) Foliar nectar production and ant activity on a neotropical tree, *Ochroma pyramidale*. *Oecologia*, **43**, 233–248.

O'DOWD, D. J. (1980) Pearl bodies of a neotropical tree, *Ochroma pyramidale*: Ecological implications. *American Journal Botany*, **67**, 543–549.

OLIVEIRA, P. S., SILVA, A. F. & MARTINS, A. B. (1987) Ant foraging on extrafloral nectaries of *Qualea grandiflora* (Vochysiaceae) in cerrado vegetation: Ants as potential anti-herbivore agents. *Oecologia*, **74**, 228–230.

OPLER, P. A., FRANKIE, G. W. & BAKER, H. G. (1980) Comparative phenological studies of treelet and shrub species in tropical wet and dry forest in the lowlands of Costa Rica. *Journal of Ecology*, **68**, 167–188.

ORIANS, G. H. & JANZEN, D. H. (1974) Why are embryos so tasty? *American Naturalist*, **108**, 581–592.

RAUPP, M. J. (1985) Effects of leaf toughness on mandibular wear of the leaf beetle *Plogiaodera versicolora*. *Ecological Entomology*, **10**, 73–79.

REHR, S. S., FEENY, P. P. & JANZEN, D. H. (1973) Chemical defense in Central American non-ant acacias. *Biochemical Systematics*, **1**, 63–67.

REICH, P. & BORCHERT, R. (1984) Water stress and tree phenology in a tropical dry forest in the lowlands of Costa Rica. *Journal of Ecology*, **72**, 61–74.

REICHLE, D. E. & CROSSLEY, D. A. (1967) Investigation of heterotrophic productivity in forest insect communities. *Secondary Productivity of Terrestrial Ecosystems*. (ed. ʼK. PETRUSEWICS) Panstwowe Wydawnietwo Naukowe, Warsaw, pp. 563–587.

REICHLE, D. E., GOLDSTEIN, R. A., VAN HOEK, R. I., Jr., & DODSON, G. D. (1973) Analysis of insect consumption in a forest canopy. *Ecology*, **54**, 1076–1084.

RHOADES D. F. & CATES, R. G. (1976) Toward a general theory of plant antiherbivore chemistry. Biochemical interactions between plants and insects. (eds. J. WALLACE, & R. L. MANSELL) *Recent Advances in Phytochemistry,* **10,** 168–213, Plenum Press, New Yersey.

RICHARDS, P. W. (1952) *The Tropical Rainforest.* Cambridge University Press, Cambridge.

ROBINSON, T. (1974) Metabolism and function of alkaloids in plants. *Science,* **184,** 430–435.

SAGERS, C. L. (1992) Manipulation of host plant quality: Herbivores keep leaves in the dark. *Functional Ecology,* **6,** 741–743.

SCHUPP, E. W. (1986) *Azteca* protection of *Cecropia*: ant occupation benefits juvenile trees. *Oecologia,* **70,** 379–385.

SCHUPP, E. W. & FEENER, D. H., Jr. (1991) Phylogeny, lifeform, and habitat dependence of ant-defended plants in a Panamanian forest. *Ant-Plant Interactions* (eds. C. R. HUXLEY & D. F. CUTLER) Oxford University Press, Oxford. pp 175–197.

SCRIBER, J. M. & SLANSKY, F., Jr. (1981) The nutritional ecology of immature insects. *Annual Review of Entomology,* **26,** 183–211.

SLANSKY, F., Jr., & RODRIQUEZ, J. G. (1987) *Nutritional Ecology of Insects, Mites, Spiders, and Related Invertebrates.* John Wiley and Sons, New York.

SMITH, A. M. (1909) On internal temperature of leaves in tropical insolation, with special reference to the effect of their color on temperature. *Annals of the Royal Botanical Garden Peradeniya,* **4,** 229–297.

STANTON, N. (1975) Herbivore pressure on two types of tropical forests. *Biotropica* **7,** 8–12.

TANTON, M. T. (1962) The effect of leaf "toughness" on feeding of the larvae of the mustard beetle, *Phaedon cochleariae* Fabricius. *Entomologia Experimentalis et Applicata,* **5,** 74–78.

WALTZ, S. A. (1984) *Comparative study of predictability, value and defenses of leaves of tropical wet forest trees (Costa Rica).* Ph. D. thesis, University of Washington, Seattle.

WINDSOR, D. M. (1988) *Climate and moisture availability of Barro Colorado Island, Panama: long-term environmental records from a lowland tropical forest.* Smithsonian Institution Press, Washington D. C., USA.

WINT, G. R. W. (1983) Leaf damage in tropical rainforest canopies. *Tropical Rainforest: Ecology and Management* (eds. S. L. SUTTON, WHITMORE, & A. C. CHADWICH) Blackwell Publishers, Oxford, pp. 229–239.

APPENDIX 11.1   SITE DESCRIPTIONS

Danum Valley Field Station, Sabah, Malaysia: Danum Valley is a lowland forest in northern Borneo. (Latitude: 4° 54' N; elevation: 200 m; rainfall: 260 cm yr^{-1} with two mild dry seasons). It is the most diverse of the four study sites, and is dominated by trees in the Dipterocarpaceae and Euphorbiaceae. Soils are Tertiary sediments and are moderately rich in nutrients (Leong, 1974). Plant identifications were made with the help of Dr. Elaine Gasis, the Danum Valley Herbarium, and Dr. K.M. Kochummen of the Forest Research Institute of Malaysia. Data were collected in February and March, 1990.

Epulu, Zaire: The site was a lowland forest near the town of Epulu in the Ituri forest of north eastern Zaire. (Latitude: 1° 25′ N; elevation: 750 m; rainfall: 180 cm yr^{-1} with a mild dry season in January). The study area has been extensively studied by Drs. John and Terese Hart (Hart, Hart & Murphy, 1989). Dominant families are the Leguminosae, Sapotaceae, and Meliaceae. Soils are oxisols and are moderately poor in nutrients (Hart, Hart & Murphy, 1989). The Harts identified the most common plant species for our study and provided plant identifications. Mr. Semeke helped in the field. Data were collected in July and August, 1990. Because Epulu was 600 m higher than the other sites, we temperature-corrected expansion rates for interspecific comparisons (section 11.5). Assuming an adiabatic lapse rate of 6° C/km^{-1} and a $.Q_{10}$ of 2.0, we multiplied expansion rates by 1.28.

Korup National Park, Mundemba, Cameroon: This lowland forest site is on the border with Nigeria. (Latitude: 4° 80′ N; elevation: 150m; rainfall: 550 cm yr^{-1}). Soils are weathered from precambrian gneiss and are relatively poor in nutrients, particularly in extractable phosphorus (Hawkins & Brunt, 1965, Gartlan et al., 1986; Anonymous, 1987; Newbery et al., 1988). As a consequence, the forest structure was primarily thin polls of low stature. The study species were chosen from a list of the most common species compiled by Dr. Duncan Thomas, previously of the Missouri Botanical Garden. Plant identifications were done by Mr. Ferdinand Namata, a local botanist who had worked with Dr. Thomas. Data were collected in May and June, 1990.

Barro Colorado Island (BCI), Panama: BCI is a protected island in Lake Gatun. (Latitude: 9° 09′ N; elevation: 150 m; rainfall: 260 cm yr^{-1} with a 3–4 month dry season; Windsor 1988). The vegetation on BCI is classified as tropical moist forest (Holdridge et al., 1971) and has been extensively studied (Croat, 1978; Foster & Brokaw, 1982). Soils are varied, ranging from nutrient-poor oxisols to fertile alfisols (Leigh & Wright, 1990). Vegetation is typical of tropical forest on fertile soil (Foster & Brokaw, 1982). Plant identifications were done by P. Coley and T. Kursar with the help of the BCI Herbarium. Herbivory data were collected from January through March, 1993. Other data were collected over the years by D. Feener, T.M. Aide, E. Schupp, and ourselves.

12

Causes and Consequences of Epiphyll Colonization

Phyllis D. Coley and Thomas A. Kursar

Leaves, particularly those in humid tropical regions, are frequently colonized by epiphylls (Richards, 1954; Pocs, 1982). Epiphyllous communities are generally dominated by lichens and liverworts, although mosses, algae, fungi, and bacteria can also occur (Winkler, 1971; Smith, 1982; Andrews & Hirano, 1991). The interactions between epiphylls and host leaves have not been well studied, but here we suggest that epiphylls may have significant ecological and evolutionary effects on their hosts. We briefly summarize physiological performances of lichens and liverworts. We examine the influence of abiotic factors on the extent of epiphyll colonization and on competitive outcomes among different epiphyllous species. We review the evidence for both positive and negative effects of epiphyll occupation on the host leaves. And finally, we present evidence that host leaf characteristics can influence the rates of epiphyll colonization.

12.1 COMPOSITION OF EPIPHYLLOUS COMMUNITIES

Liverworts are the major structural component of the epiphyllous community, and although they occur in temperate regions, they are exceptionally abundant and speciose in the humid tropics (Richards, 1954; Pocs, 1982). Liverworts are in the class Hepaticae (division Bryophyta), with epiphyllous liverworts being dominated by members of the family Lejeuneaceae. Asian epiphyllous liverworts are the most diverse, but there are at least 175 neotropical species (Pocs, 1982), with over 70 species having been collected at a single site (Winkler, 1967). A partial list of 23 species has been reported from La Selva Biological Station in Costa Rica (Bien, 1982), and 15 species have been found on a single shrub species on Barro Colorado Island (BCI) in Panama (Marino & Salazar Allen, 1993). Perhaps because of the transient nature of their substrate (leaves vs. rocks or trunks), epiphyllous species have relatively rapid life cycles (Schuster, 1980). They can reproduce sexually within a few months, although reproduction and dispersal are dominated by asexual propagules. The conspicuous part of the life cycle, the leafy thallus, is the gametophyte.

Lichens are also common epiphylls belonging to many different families. Almost 500 epiphyllous species have been described, the majority being from the tropics (Farkas & Sipman, 1993). As many as 50 species have been found on a single leaf (Gradstein, 1992). Most epiphyllous lichens have a crustose growth form, with foliose forms being more common on trunks or at higher elevations (Gradstein, 1992). Fungi, cyanobacteria, mosses, and occasionally algae are also found growing on leaves (Richards, 1954; Smith, 1982; Andrews & Hirano, 1991; Galloway, 1991).

12.2 ABIOTIC EFFECTS ON EPIPHYLL COLONIZATION

Light intensity, precipitation, temperature, and humidity have all been suggested as factors that control the distribution of epiphylls (Richards, 1954, 1984; Pocs, 1982; Frahm, 1987; van Reenen, 1987; Marino & Salazar Allen, 1993). Unfortunately, these factors are rarely examined experimentally and are frequently confounded in nature.

12.2.1 Rainfall

12.2.1.1 Site comparisons

The amount and seasonality of rainfall appear to be the most important correlates of liverwort cover. In Nigerian forests, growth of liverworts was more rapid at wetter sites and during wetter seasons (Olarinmoye, 1974). In El Salvador, growth of liverworts dropped by one-third during the dry season (Winkler, 1967). Numerous non-quantitative distributional studies show that liverworts are associated with more humid sites (Fulford, Crandall & Stotler, 1970; Smith, 1982; Pocs, 1982; Richards, 1984; Frahm, 1987; van Reenen, 1987; Thiers, 1988).

In a more quantitative study (Coley, Kursar & Machado, 1993), rates of epiphyll colonization were studied at four sites in Panama that differed in rainfall and elevation. One site is in lowland moist forest on BCI (9°N, 100m elevation). Annual rainfall averages 260 cm/yr, most of it occurring during an 8 month rainy season (Windsor, 1990). The other three sites are in montane rainforest on the continental divide at the Fortuna Watershed in Chiriqui Province, Panama (9°N, 1200 m elevation) and are therefore cooler than BCI. Annual rainfall is approximately 500 cm/yr with no dry season and a substantial additional contribution from fog drip (Cavelier, 1989; Victoria, unpublished data). The three Fortuna sites are separated by only 10 km, but differ by a factor of two in rainfall (368 cm/yr at Filo Hornito vs. 664 cm/yr at Arena, Cavelier, 1989). To measure rates of epiphyll colonization, young leaves (3-6 wks) were marked (586 leaves, 2 leaves/plant). At this age, no leaves had visible signs of epiphylls. After one and two years, marked leaves were remeasured for percent cover by liverworts and lichens.

Liverwort cover after two years was 6 times higher on the wetter La Fortuna sites compared to the drier and more seasonal forest of BCI (Figure 12.1). Liverwort cover ranged from 2% at the driest site to 20% at the wettest. In contrast, lichens covered approximately 25% of the leaf and showed no relationship with rainfall. Because BCI and La Fortuna differ in many factors, such as temperature and host species, a comparison among the three sites at La Fortuna more directly addresses the role of rainfall. Sites were only a few kilometers apart, and appeared similar in terms of temperature and host species composition. Even at this scale, leaves in the wetter sites had significantly greater liverwort cover (Figure 12.1).

On an even finer scale, Marino and Salazar Allen (1993) examined liverwort cover on two shrub species growing on ridges and ravines

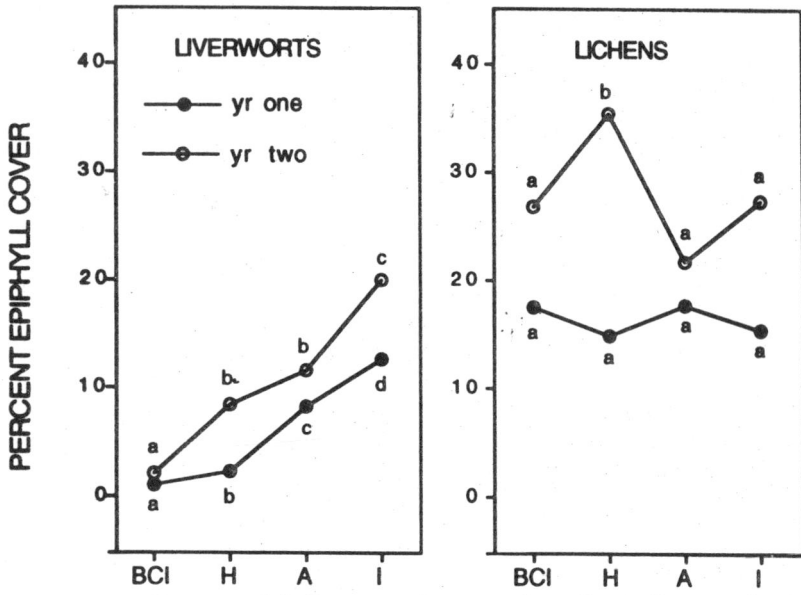

Figure 12.1. *Percent epiphyll cover on leaves at four sites in Panama for one-year-old (n = 586) and two-year-old leaves (n = 324 due to mortality). Values are means + / − one standard error. Sites are compared within years for lichen or liverwort cover (ANOVA on arcsin-square root transformed data with Duncan's multiple-range test). Values followed by different letters differ significantly at P < 0.05. The four sites are arranged by increasing rainfall / water availability and are abbreviated as follows: BCI = Barro Colorado Island, 9°N, 100 m elevation, 260 cm rain / yr; H = Filo Hornito, La Fortuna, Chirique, 9°N, 1200 m elevation, 368 cm rain / yr; A = Quebrada Arena, La Fortuna, Chirique, 9°N, 1200 m elevation, 664 cm rain / yr and I = IRHE, La Fortuna, Chirique, 9°N, 1200 m elevation. Rainfall data are not available for the IRHE site. However, IRHE probably has similar rainfall to Quebrada Arena and is even better protected from wind, suggesting that it has the highest overall water availability. Epiphyll cover was determined by running 5 transects (100 mm × 1 mm) on each leaf. Percent cover was calculated as the percent of the 1 mm² grid squares that were at least 10% covered by epiphylls. Data from Coley, Kursar & Machado (1993).*

on BCI. Both microsites had similar light regimes, but ravines were presumably more humid. They found no difference in the total percent cover, although different species dominated in different microsites.

12.2.1.2   Irrigation experiments

To assess directly the effects of rainfall and to control for host species and site effects, rates of epiphyll colonization were quantified on *Hybanthus prunifolius* (Violaceae), a common understory shrub growing in irrigated and control plots on BCI (Coley, Kursar & Machado, 1993). Irrigated plots were sprayed daily with water from 1.8m tall sprinklers, which wetted leaf surfaces and maintained high soil moisture throughout the year (Wright & Cornejo, 1990). Liverwort cover in control plots was less than 2%, a value similar to that found in the community survey (Figure 12.1). This was significantly less than the 20.5% cover observed in the irrigated plots (Figure 12.2; ANOVA, $P < 0.001$). The two treatments differed only in the dry season relative humidity, leaf wetting and soil moisture, yet the magnitude of the liverwort response was similar to the differences observed between BCI and the wettest site at La Fortuna (2% vs 20% cover, Figure 12.1). This result suggests that differences in the amount and seasonality of rainfall could be sufficient to account for the differences among sites in liverwort cover.

In contrast to liverworts, lichen cover on *Hybanthus* leaves was significantly higher in the control plots that experienced the normal 4 month dry season as compared to irrigated plots (ANOVA, $P < 0.05$, Figure 12.2). These data suggest that epiphylls respond differently to abiotic factors, and that, compared to liverworts, lichens do relatively better in drier habitats. However, from this experiment, it is not possible to determine if the poor growth response of lichens in watered plots was a direct response to humidity or was due to increased competition from liverworts.

12.2.2 Light

The effects of light on epiphyll colonization have received much less attention than rainfall, presumably because light effects are more subtle. However, liverworts may grow best in intermediate light levels. In a study on BCI, Marino and Salazar Allen (1993) measured liverwort distribution on two common species of understory shrub, *Psychotria horizontalis* (Rubiaceae) and *Hybanthus prunifolius* (Violaceae). They found that cover was approximately 20% in small gaps as compared to 2% in the understory. In Monteverde, Costa Rica, liverwort cover on *Piper* (Piperaceae) leaves was twice as high in gaps as in the understory (Monge-Najera, 1989). At the other extreme, liverworts are encountered only infrequently on the same hosts growing in large rainforest gaps (Frahm, 1990; Marino & Salazar Allen, 1993) or on canopy leaves (P. D. C., personal observation). Laboratory studies on liverworts suggest that the combination of high temperature and low light are detrimental

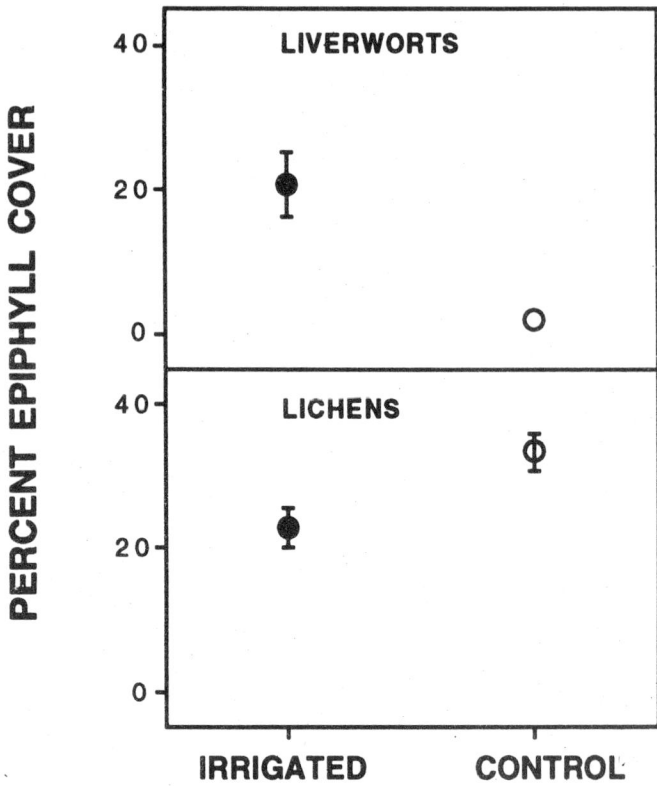

Figure 12.2.   Percent epiphyll cover on one-year-old Hybanthus prunifolius *leaves growing in control and irrigated plots on BCI. Values are means + / − one standard error, n = 20 leaves per treatment. Lichen cover is significantly greater in control vs watered plots (ANOVA on transformed data, P < 0.02). Liverwort cover is significantly greater in watered vs control plots (ANOVA on transformed data, P < 0.0001). Data from Coley, Kursar & Machado (1993).*

(Frahm, 1987). In contrast, lichens seem to prefer drier, brighter habitats (Gradstein, 1992).

12.2.3   Temperature

Richards (1984) and Frahm (1990) suggest that high temperatures may be responsible for the relative rarity of bryophytes in lowland forests. The high temperatures cause elevated respiration rates that may not be balanced by photosynthesis in the shaded understory. They argue that the cooler temperatures and higher light of elfin forests favor bryophyte growth, although more detailed physiological studies are necessary to draw firm conclusions.

12.3 PHYSIOLOGICAL CHARACTERISTICS OF LICHENS AND LIVERWORTS

It is not yet possible to relate the distributional patterns outlined above to the physiological characteristics of tropical epiphylls. Unfortunately, little work has been done on the physiology of tropical lichens or liverworts, and essentially nothing on epiphylls. The following conclusions are primarily extrapolated from work on temperate or arctic organisms growing in drier and colder habitats.

12.3.1 Photosynthesis and water relations

12.3.1.1 *Liverworts*

Bryophytes and lichens differ from vascular plants primarily in their water relations. Leafy liverworts lack a water-repellent cuticle, so water is absorbed and lost over the entire leaf surface. Water movement within the leaf is diffuse, occurring along cell walls or from cell to cell (Proctor, 1982; Chopra & Kumra, 1988). As a consequence of conducting gas exchange across the wet leaf surface, CO_2 diffusion is 10^4 times slower than in air, and photosynthesis may frequently be diffusion limited (Proctor, 1982). Measured rates of photosynthesis in bryophytes range from 1–40 nmol $g^{-1}s^{-1}$ (Proctor, 1982). Most saturate at 200–400 µmol photons m^{-2} s^{-1}, although we suspect that shade-tolerant species, such as rainforest epiphylls, may saturate at much lower light levels.

Water contents of liverworts vary widely, with wet weights frequently being 10 times the dry weights (Chopra & Kumra, 1988). In wet-habitat species, most water is stored in the vacuoles of large cells, with water loss leading to loss of turgor. Although some bryophytes can photosynthesize at extremely low water potentials, -8 MPa (equivalent to a thallus at equilibrium with air of 94% relative humidity), photosynthesis in more drought-sensitive species typically drops at water potentials lower than -1.0 MPa (equivalent to 99.3% relative humidity) (Proctor, 1982; Sveinbjornsson & Oechel, 1992). Desiccation alters membrane structures, which must be reconstituted upon rehydration before resumption of normal cell function (Stewart

& Bewley, 1982). After rewetting, there is frequently a burst of respiration (Proctor, 1982). Cell membranes are also initially leakier, such that nutrients and solutes can leach out. The apparent restriction of liverworts to wet sites is consistent with these physiological responses, particularly the energy and nutrient costs associated with rehydration and the drop in photosynthesis at lower water contents.

12.3.1.2 Lichens

In lichens, most water is absorbed by cell walls and intercellular spaces (Hale, 1983), rather than by large vacuoles. Wet weights of crustose lichens are 1–3 times the dry weights, a water content considerably lower than for liverworts. Many species of lichens can withstand long periods of desiccation, with desert species surviving for up to 4 years (Rogers, 1977). The ability of lichens to take up water vapor from the atmosphere depends on the phycobiont (Lange, Kilian & Ziegler, 1986; Lange, 1988). Lichens with green algal symbionts are capable of high rates of photosynthesis after rehydration with only water vapor. Lichens with blue-green bacterial symbionts require liquid water in order to rehydrate and carry out net positive photosynthesis. Lichen morphology can also affect the ability to take up different forms of water, which in turn influences habitat requirements (Larson, 1984).

Photosynthetic rates for lichens are expected to be lower than for liverworts, as both algal and fungal components contribute to dark respiration (Hale, 1983). Maximal rates of photosynthesis for foliose lichens range from 6–30 nmol $g^{-1} s^{-1}$ or 0.25-1.4 μmol $m^{-2} s^{-1}$ (Lange & Matthes, 1981; Hale, 1983; Kershaw, 1985), with 15 nmol $g^{-1}s^{-1}$ being reported for a tropical species, *Leptogium azureum* (Collemataceae) (G. Zotz, personal communication). Crustose species, the more common form of epiphyllous lichens, generally have lower rates of photosynthesis, approximately 3 nmol $g^{-1} s^{-1}$ (Kershaw, 1985). After desiccation and rewetting, the general pattern for lichens is an initial increase in photosynthesis with increasing thallus water content, followed by a decrease at higher water contents (Lange & Matthes, 1981; Lange & Ziegler, 1986). In a terrestrial lichen, *Ramalina maciformis*, photosynthesis in water-saturated tissue (after soaking for several hours) was only 10 to 25% of maximal rates (Lange & Tenhunen, 1981). In general, photosynthesis is highest at water contents of 35-70% of saturation (Hale, 1983). There is even evidence that prolonged periods of saturation can cause death (Farrar, 1976).

12.3.1.3   Differences between lichens and liverworts

The above data suggest that the extent and periodicity of rainfall will have different effects on the photosynthetic performance of lichens and liverworts. Following rainfall, water-saturated epiphylls will begin drying until they equilibrate with ambient humidity. In rainforest understories, ambient relative humidities are frequently between 90% and 95% (Windsor, 1990). Humidity at the leaf surface may be even higher due to boundary layer effects, however, even with these high humidity levels, liverworts may not exhibit peak photosynthetic activity. Liverworts suffer reduced photosynthesis at tissue water potentials of less than -1.0 MPa, which is equivalent to 99.3% relative humidity. Furthermore, during rehydration after desiccation, liverworts experience a burst of respiration and lose solutes through leaky membranes. Because liverworts perform best at relatively high and constant tissue water contents, they may thrive in cloud forest sites with frequent rewetting by rain and slow drying due to high ambient humidities. In contrast, lichens can carry out photosynthesis at much lower water potentials, but experience substantial depression of photosynthesis at higher tissue water contents. As a consequence, they should be favored in drier sites.

12.3.2   Epiphylls as indicators of atmospheric pollution

Lichens, and to a lesser extent liverworts, are increasingly being used as bioindicators of atmospheric pollutants since they are generally more sensitive than vascular plants (Gilbert, 1973; Nash & Wirth, 1988). Lichens have been particularly useful in monitoring elevated sulfur dioxide emissions around urban areas, as the effects are cumulative and species differ in their sensitivities (Kershaw, 1985). Lichens and liverworts also concentrate heavy metals to a greater extent than vascular plants (Nash & Wirth, 1988). Heavy metal concentrations in epiphyllous liverworts were analyzed in an Amazonian lowland rainforest to determine if there was long range transport of pollutants (Montagnini, Neufeld & Uhl, 1984). Although acid rain had previously been detected, levels of six heavy metals in epiphyllous liverworts, including cadmium, lead, and nickel, were lower than in temperate areas. Similarly, low levels of lead were measured in a lichen on BCI (Lawrey & Hale, 1979). Montagnini, Neufeld and Uhl (1984) suggest that the levels of heavy metals found in rainforest epiphylls represent background levels with little input from pollution.

Long-term monitoring of heavy metal concentrations in epiphylls may allow us to sample pollutant deposition inexpensively in remote areas.

12.4 CO-OCCURRENCE OF LICHENS AND LIVERWORTS

We might expect to see competitive interactions among epiphyllous species given the structurally simple nature of the host-leaf surface and the fact that space becomes limiting on older leaves. And indeed, several investigators have found evidence suggestive of competition. In lowland forests in Costa Rica (Bien, 1982) and Nigeria (Olarinmoye, 1975), liverwort species were always found overgrowing crustose lichens and algae. On *Hybanthus* leaves growing in irrigated plots on BCI, lichens and liverworts were equally abundant (approximately 22%). However, there was a significant negative correlation between lichen and liverwort cover ($r^2 = 0.58$ p < 0.001; Figure 12.3),

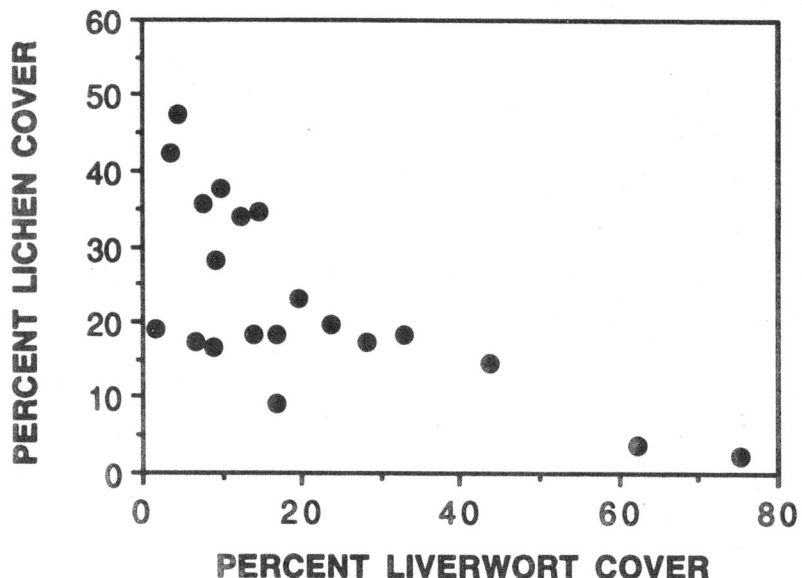

Figure 12.3. Co-occurrence of lichen and liverworts on one year-old Hybanthus prunifolius *leaves growing in irrigated plots on BCI. Plotted values are the percent cover (with a potential range from 0%–100%). The mean percent cover was 20.5% for liverworts and 22.8% for lichens. For regression analysis, values were transformed using the arcsin-square root of proportional cover ($R^2 = 0.58$, P < 0.0001). Data from Coley, Kursar & Machado (1993).*

suggesting that competition may be important. Liverworts appeared to be dominant as they could overgrow lichens. We never saw lichens overgrowing liverworts. Competitive dominance by liverworts suggests that lichens may be maintained in the community due to better colonizing ability. We were unable to find any direct evidence on rates of colonization, although a comparison across years suggests that lichens are more rapid colonists (Figure 12.1)

There is also evidence for competitive dominance among liverworts. On BCI, Marino and Salazar Allen (1993) suggest that the liverwort *Aphanolejeunea sicaefolia* is less common at wet sites because of competitive exclusion by another liverwort, *Leptolejeunea elliptica*. They did not indicate if competitive dominance was due to physiological or morphological characteristics. In Nigeria, liverwort species with appressed shoots were at a competitive disadvantage compared to liverworts with leafy shoots (Olarinmoye, 1975).

An alternative explanation for the negative association between liverworts and lichens is that epiphylls are sorting out along microsite differences. Microsite variation in both light and humidity could contribute to differential success of liverworts and lichens. The light environment in the understory varies spatially, especially during the dry season when some canopy tree individuals become deciduous. On BCI, liverworts grew more quickly in high light than in shade (Marino & Salazar Allen, 1993), and in wet as opposed to dry sites (Figure 12.2). Epiphyllous lichens are thought to prefer drier sites with higher light (Gradstein, 1992; Galloway, 1991). Lichens were more common in drier control plots (Figure 12.2). Opposite responses to small differences in light and humidity by liverworts and lichens would contribute to the observed negative association (Figure 12.3).

12.5   EPIPHYLL EFFECTS ON HOST LEAVES

A few workers have considered the consequences of epiphyll occupation for the host plant. Here we review the evidence for both positive and negative effects of epiphylls on hosts, and point out that this interesting area deserves much more scientific attention.

12.5.1   Nitrogen fixation

Many studies have shown that cyanobacteria associated with the epiphyllous community can fix substantial quantities of nitrogen (Ruinen, 1975; Sengupta et al., 1981; Nandi & Sen, 1981; Goosem &

Lamb, 1986). In greenhouse experiments, growth of crop seedlings was increased when leaves were sprayed with suspensions of nitrogen-fixing micro-organisms (Sengupta et al., 1981; Nandi & Sen, 1981). There was no growth effect when soil alone was sprayed. Bentley and co-workers argue that the nitrogen fixed by epiphyllous cyanobacteria on rainforest leaves can be taken up by the host plant leaves and can contribute significantly to their nitrogen budget (Bentley & Carpenter, 1984; Bentley, 1987). However, in a detailed field study of coffee, uptake of epiphyllous nitrogen was negligible (Roskoski, 1980, 1981).

This discrepancy in results from crops vs rainforest species may be due to the different growth requirements of the host leaves. Fast-growing, light-demanding crop species may be nitrogen limited and may have thinner cuticles, facilitating nitrate uptake from epiphyllous sources on the leaf surface. In contrast, most shaded rainforest species are light limited, have thick cuticles, and are unlikely to show growth responses to elevated nitrogen. Furthermore, nitrogen fixation in the understory may be considerably lower than in the high light of the experimental conditions. We therefore suggest that fast-growing species in high-light microsites such as treefall gaps are more likely to benefit from epiphyllous nitrogen than understory individuals.

12.5.2   Delayed senescence and hormone production

There is the intriguing possibility that hormones produced by the epiphyllous community may influence their host plants. High levels of indole-acetic acid (IAA) are produced by epiphyllous yeast and bacteria (Wichner & Libbert, 1968a,b; Diem, 1971). IAA on leaf surfaces can be taken up by the plant (Sethunathan, 1970), and at least one report suggests that the majority of IAA in pea and corn plants was produced by epiphyllous bacteria (Libbert et al., 1966). Maintenance of a high IAA concentration is known to delay leaf abscission (Addicott, 1982), suggesting a possible impact of the epiphyllous community on host leaf longevity. In analogous systems, fungi, galls, and leaf-mining caterpillars cause elevated levels of cytokinins in the host leaves (Engelbrecht, 1971). This delays senescence and creates "green islands" around sites of attack. It is tempting to speculate that epiphylls may also prolong the suitability of their host leaf through auxin or cytokinin production, although no studies have addressed this possibility. Manipulating senescence would be advantageous to the epiphylls, yet would be detrimental to the host, as it interferes with normal patterns of senescence and reuse of reabsorbed resources from the leaf.

12.5.3 Increased susceptibility to pathogens

Epiphylls may enhance the probability of host colonization by pathogenic bacteria and fungi, although evidence is primarily circumstantial. First, epiphylls keep the host leaf surface wet for long periods, which may increase the probability of infection (Gregory, 1971). For temperate crops, the duration of leaf wetness positively influences colonization, infection, and sporulation (Huber & Gillespie, 1992). Second, leaching from epiphylls may increase surface levels of nutrients, which are known to be important microbial substrates (Andrews, 1992). Third, for infection to be successful, spores must stick successfully, so physical trapping by uneven leaf surfaces enhances colonization (Andrews, 1992). We suggest that epiphylls would greatly increase the potential for trapping of spores. All these effects of epiphylls should, theoretically, increase rates of infection and hasten senescence of the host leaf. Accelerated leaf senescence has been shown for crop species with epiphyllous fungi (Jachmann & Fehrmann, 1989), but no studies have directly addressed any of these issues for rainforest plants.

Although pathogens are an important source of damage to rainforest leaves, successful infection is frequently dependent on previous leaf damage by herbivores (Garcia-Guzman & Dirzo, 1991). With the possible exception of *Radula* (Berrie & Eze, 1975; Eze & Berrie, 1977), most epiphyllous liverworts do not penetrate the host cuticle (Winkler, 1967). Lichens may more frequently damage the cuticle (Winkler, 1967). Therefore, unless epiphylls damage leaf surfaces, increased occurrence of pathogens on epiphyll-covered leaves may not necessarily lead to increased infection of the host.

12.5.4 Protection from herbivores

Liverworts are thought to be well defended against herbivores, and contain high concentrations of terpenoids as well as phenolic compounds (Chopra & Kumra, 1988; Zinsmeister & Mues, 1990; Yoshimoto, Katoh & Takeda, 1990). Only one species of butterfly is known to feed on liverworts (DeVries, 1988). Lichens are also known to contain a variety of secondary metabolites with antibiotic and antiherbivore properties (Emmerich et al., 1993; Lawrey, 1980; Hale, 1983; Vartia, 1973). In an experiment with grapefruit (*Citrus*) and an understory cyclanth (*Cyclanthus*), removal of epiphylls (primarily liverworts) increased palatability of leaves to leaf cutter ants by 2–3 times (Mueller & Wolf-Mueller, 1991). It is possible that host leaves with epiphylls may be less attractive to other herbivores as well, although this has never been documented.

12.5.5 Light interception

Perhaps the most obvious effect of epiphylls on hosts is shading (Richards, 1966). We measured the amount of photosynthetically active radiation (PAR) transmitted through several types of epiphyll cover typically seen in the field (Coley, Kursar & Machado, 1993). Percent transmittance through a single layer of non-overlapping liverwort leaves was 45% of ambient, decreasing to 15% under dense growth. Measurements in a lowland forest in Costa Rica showed a 20–30% reduction in light by liverworts on the palm *Welfia georgii* (Bien, 1982). Light reduction of this magnitude is particularly significant in environments such as the rainforest understory, where the ambient light intensity is well below levels needed to saturate photosynthesis. In rainforest understories, ambient light levels are between 10–30 µmol photons m^{-1} s^{-1} (0.5–1.5% of full sun) and may limit plant growth more than nutrients (Chazdon & Fetcher, 1984; Chazdon, 1988; Oberbauer et al., 1988; Chazdon et al., Chapter 1). Our data show a 55-85% reduction in light incident on the portions of the leaf covered with epiphylls. Furthermore, after two years, epiphyll cover averaged 45%, with some leaves nearly completely covered. We estimate that this could reduce lifetime photosynthesis by 20%. Studies on coffee (Roskoski, 1981) and eelgrass (Sand-Jensen, 1977) also suggest that shading by epiphylls can reduce host leaf photosynthesis by 20–30%.

12.5.6 Advantages and disadvantages of epiphyll cover

In the last sections, we presented evidence for both positive and negative effects of epiphylls on host leaves. What is the net effect?

Does epiphyll occupation change leaf productivity or longevity? Unfortunately, we have no data on the lifetime productivity of colonized vs. epiphyll-free leaves, so we shall speculate. First, we suggest that epiphylls reduce instantaneous rates of photosynthesis because positive effects of increased nitrogen fixation are negligible and are unlikely to outweigh the negative effects of light interception. Since understory leaves will be operating below saturating light levels, we estimate that shading by epiphylls will reduce photosynthesis by as much as 20%. Second, we suggest that decreased photosynthesis will not be offset by any increased protection from herbivores. Rates of herbivory on mature leaves are quite low, accumulating to an average of 20% leaf area damaged during the 2–3 year life span (Coley, 1983; Coley & Aide, 1991).

Associational protection by epiphylls would therefore have to be substantial to balance the reduction in photosynthesis. Third, we suggest that epiphylls increase leaf mortality either by reducing light and photosynthesis or by increasing pathogen infection. To test this, we used the data set from BCI and La Fortuna (Figure 12.1; Coley, Kursar & Machado, 1993), and compared mortality rates within a single species for leaves with differing amounts of epiphyll cover. Sample sizes were small (approximately 10 leaves per species), however, in 5 of the 6 most common species, two-year-old leaves that died had higher (P > 0.05) epiphyll cover after one year than those that lived. Finally, while we favor the idea that epiphyllous communities have detrimental effects on host leaves overall, we emphasize that a great deal more work is needed before any conclusions can be drawn fairly.

12.6   HOST CHARACTERISTICS

In this section, we explore the possibility that, in addition to abiotic factors, host leaf properties may also influence colonization and success by epiphylls. Both physical and chemical characteristics of the leaf surface are known to vary among host species. However, the potential for interspecific variation in suitability of host leaves for epiphylls is generally ignored, and it is assumed that colonization is simply a function of time. As a result, the idea that epiphylls grow only on fairly long-lived leaves is frequently seen in the literature (Richards, 1954; Pocs, 1982; Bien, 1982; Bentley, 1987). Here we suggest that host properties can deter epiphyll colonization, and that long-lived leaves are in fact better defended.

12.6.1　Physical defenses

12.6.1.1　Leaf surface texture

It is generally thought that smooth, waxy leaf surfaces present difficulties for successful attachment by epiphylls, particularly facultative species of epiphylls (Pocs, 1978; Bischler, 1968). Obligate epiphylls have elaborate structures for attachment, suggesting that physical barriers of host leaves exist (Bischler, 1968). Despite the appeal of this idea, no systematic data link surface properties with colonization. In an attempt to test this, we altered the texture of the upper epidermis by gently scraping a subset of the leaves. Leaves were scraped with a kitchen scrubber wrapped in several layers of cloth. This caused a fine-scale scratching of the cuticle which was visible with a dissecting microscope. We scraped one leaf per plant for the individuals studied in the BCI/La Fortuna comparison (explained in Figure 12.1; Coley, Kursar & Machado, 1993). Our technique for scraping leaf surfaces did visually alter the surface texture, but colonization rates were not significantly affected for either lichens or liverworts at any site (ANOVA, $P > 0.50$ for lichens, $P > 0.79$ for liverworts).

12.6.1.2　Drip tips

In addition to surface properties, leaf shape has been suggested as influencing epiphyll colonization. Species with drip tips, the acuminate tips of leaves, are common in rainforests. Drip tips facilitate water run-off from leaf surfaces (Dean & Smith, 1978) and hasten leaf drying. More rapid drying was thought to reduce colonization by epiphylls (Jungner, 1891; Richards, 1966). Despite widespread appeal of this idea, recent studies suggest that it is not true. Bien (1982), working at La Selva, Costa Rica, found no correlation between length of drip tips and epiphyll cover. Martin Burd (unpublished data.) removed drip tips from two common understory species on BCI, *Faramea occidentalis* and *Psychotria marginata* (Rubiaceae). He paired altered and control leaves and followed rates of epiphyll colonization for one year. Epiphyll cover was highly variable, but was unrelated to the presence of the drip tip.

12.6.2　Leaf lifetime affects colonization

Typically it is assumed that all leaves are equally suitable for colonization by epiphylls, so that leaf age is the key element determining the extent of cover (Richards, 1954; Pocs, 1982; Bentley, 1987). Although epiphylls accumulate through time on any given host leaf, our data suggest that not all species of hosts support similar rates of epiphyll colonization and growth. We examined epiphyll cover on nine host species for which we had large populations of marked leaves on BCI (Coley, Kursar & Machado, 1993; Coley & Kursar, unpublished data). We knew the exact age of each marked leaf, as well as the average lifespan for that species. Leaf life span was calculated as the time until 50% survivorship of marked leaves. All species were shade-tolerant and exhibited a range of leaf life spans, from less than one year to more than five years.

Species with longer lived leaves had greatly reduced rates of epiphyll accumulation (Figure 12.4). After one year, short-lived *Acalypha* leaves had 40% cover, whereas long-lived *Ouratea* leaves had only 2% cover. On average, the 4 species with short-lived leaves had 5 times the percent cover as the 5 species with long-lived leaves (28.3% vs 5.5%, ANOVA P < 0.001). As was found in the community survey, most of the epiphyll cover for these nine species was composed of lichens. Rates of colonization by liverworts were also affected by leaf life span. After one year, liverworts had colonized 45% of leaves with life spans of one year and only 5% of the longer-lived leaves. Since we did not identify individual epiphyll species, we cannot say if species composition as well as abundance varied with host leaf lifespan.

Although the rates of epiphyll colonization and accumulation were considerably lower for species with long-lived leaves, these leaves are available for colonization for many more years. The final percent cover near senescence, however, was indistinguishable for long- and short-lived leaves (29.9% vs 28.3% respectively, ANOVA P > 0.05).

This interspecific variation in colonization rates could arise due to differences in both texture and chemical defenses. Although we have no data directly addressing chemical defenses against epiphylls, species with long-lived leaves are better defended against herbivores and pathogens (Coley & Aide, 1991; Coley, 1988).

12.7 ACTIVE DEFENSE AGAINST EPIPHYLLS

Although previous studies claimed that long-lived leaves had higher epiphyll cover (Richards, 1954; Pocs, 1982; Bentley, 1987), they did not control for the age of the measured leaves. In contrast, our data show that longer-lived leaves have both lower rates of colonization and

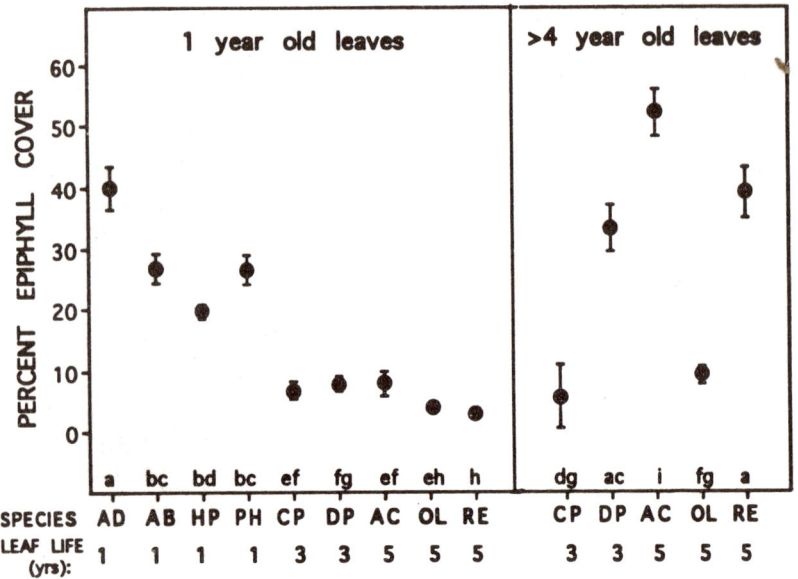

Figure 12.4.   Percent epiphyll cover for host species with different average leaf life spans. Measured leaves were either one year old, or more than four years old. Means + / − one standard error are plotted. Differences among species were determined by ANOVA on arcsin-squareroot transformed values using Duncan's multiple-range test. Values that do not share letters are significantly different at P < 0.05. Species and the number of leaves measured are as follows: AD = Acalypha diversifolia (Euphorbiaceae) 42 (1 yr); AB = Alseis blackiana (Rubiaceae) 33 (1 yr); HP = Hybanthus prunifolius (Violaceae) 124 (1 yr); PH = Psychotria horizontalis (Rubiaceae) 101 (1 yr), DP = Desmopsis panamensis (Annonaceae) 28 (1 yr) 47 (> 4 yr), CP = Connarus panamensis (Connaraceae) 46 (1 yr) 12 (> 4 yr), AC = Aspidosperma cruenta (Apocynaceae) 30 (1 yr) 40 (> 4 yr), OL = Ouratea lucens (Ochnaceae) 78 (1 yr) 74 (> 4 yr), RE = Rheedia edulis (Clusiaceae) 91 (1 yr) 42 (> 4 yr). The species can be divided into 3 groups, A: 1-year-old leaves on species with short-lived leaves, B: 1-year-old leaves on species with long-lived leaves, and C: > 4-year- old leaves on species with long-lived leaves. Groups A and C are not signficantly different, but both differ from group B (P < 0.001, ANOVA on arcsin-square root using Duncan's multiple range test).

accumulate similar cover throughout the entire leaf life span (Figure 12.4). The lower susceptibility of long-lived leaves could simply be a passive consequence of physical or chemical characters which protect the leaf from herbivory and the environment. However, we have argued that most evidence points towards detrimental effects of epiphyll cover, suggesting that selection may have favored characters in long-lived leaves that specifically deter epiphylls.

Our data suggest that there may be two alternative strategies to reduce the negative impact of epiphylls on host leaves. Leaves could be long-lived and well protected physically or chemically against epiphylls, or they could be poorly protected and avoid excess colonization by being short-lived. Although, as we have shown, defense against epiphyll colonization is possible, it may not always be cost-effective. For example, in particularly humid environments such as La Fortuna, epiphyll colonization and growth can be extremely high, nearly covering leaves in 2–3 years. Therefore, in very humid environments, it may be too costly to defend leaves against epiphylls, and species with shorter leaf life spans would be at a selective advantage (Grubb, 1977).

12.8 OVERVIEW AND CONCLUSIONS

The ecophysiology of epiphyllous lichens and liverworts is an area much in need of more research. Although lichens and liverworts show different distributional patterns, the extent to which this is due to physiological constraints is not known. In particular, we need to know more about rates of photosynthesis under different light and humidity conditions. Liverworts are most successful in humid habitats, possibly due to their limited ability to retain a high tissue water content. The greater abundance of lichens in drier habitats most likely arises from a greater physiological tolerance of desiccation, however, lichens could be excluded from wetter habitats by competition with liverworts rather than by their own physiological limits.

A greater understanding of water relations and gas exchange in epiphylls would not only help explain broadscale patterns of distribution, but might also shed light on mechanisms for coexistence of epiphyll species on a single leaf. Are there subtle differences in physiology that allow many species to partition a given habitat and coexist? We have already mentioned the possibility that microscale variation in light, temperature, and humidity can influence the competitive ability of different species. In addition, other factors that have received no attention include interspecific differences among epiphylls in rates of colonization of new hosts, or in the ability to circumvent host defenses.

The substantial interspecific differences in host suitability suggest that leaves are not a passive substrate for colonization. Are there physical or chemical properties that affect epiphyll success, and are these specific for epiphylls, or simply a consequence of defense against other enemies, such as pathogens and herbivores? If defenses against epiphylls exist, how are they correlated with particular life history characteristics of host species or with particular habitats?

A major controversy continues to surround the nature of the interaction between epiphylls and host leaves. Is it beneficial or detrimental to a host leaf to be colonized by epiphylls? A quantitative approach to the costs and benefits of colonization should examine light interception and the reduction of photosynthesis, incidence of pathogen attack, protection from herbivores, and rates of nitrogen fixation and uptake by the host.

The study of community structure, ecophysiology, and species coexistence in epiphyllous lichens and liverworts is particularly tractable because of their small size and abundance. Each host leaf is a replicated assemblage of epiphyll species on which one can easily manipulate environmental factors as well as species composition. Furthermore, host leaves live an average of 2.5 yrs (Coley & Aide, 1991), so colonization, growth, and species interactions are taking place over a relatively short time span. All of these attributes make the epiphyllous community an excellent system for experimental and field studies.

ACKNOWLEDGMENTS

We thank Jose Luis Machado for invaluable help and ideas and for invention of the Acme Medidor, Diana Valencia, Julie Glick and Luis Guttierez for field and laboratory assistance, Noris Salazar Allen for advice and epiphyll identification, S. Joseph Wright for use of irrigated plots, Jaime Cavelier, Jose Felix Victoria, and Martin Burd for access to unpublished data, and S.R. Gradstein and an anonymous reviewer for comments on the manuscript. We thank Franklin Gonzalez and the Departamento de Manejo de Cuencas for generous help and permission to work at the Fortuna Watershed. We are grateful to the Smithsonian Tropical Research Institute for logistical support and Exxon Fellowships. The research was supported by NSF grants BSR-8407712, BSR-8806080, and BSR-9119619.

REFERENCES

ADDICOTT, F. T. (1982) *Abscission*. University of California Press, Berkeley.

ANDREWS, J. H. (1992) Biological control in the phyllosphere. *Annual Review of Phytopathology,* **30,** 603–635.

ANDREWS, J. H. & HIRANO, S. S. (1991) *Microbial Ecology of Leaves.* Springer.

BENTLEY, B. L. (1987) Nitrogen fixation by epiphylls in a tropical rainforest. *Annals of the Missouri Botanical Garden,* **74,** 234–241.

BENTLEY, B. L. & CARPENTER, E. J. (1984) Direct transfer of newly-fixed nitrogen from free-living epiphyllous microorganisms to their host. *Oecologia,* **63,** 52–56.

BERRIE, G. K. & EZE, J. M. O. (1975) The relationship between an epiphyllous liverwort and host leaves. *Annals of Botany,* **39,** 955–963.

BIEN, A. R. (1982) Substrate specificity of leafy liverworts (Hepaticae: Lejeuneaceae) in a Costa rican rainforest. M. Sc. Thesis, SUNY at Stony Brook.

BISCHLER, H. (1968) Notes sur l'anatomie des amphigastres et sur le develop-pement du paramphigastre et des rhizoides chez *Drepano-, Rhaphido-* et *Leptolejeunea. Revue Bryologique et Lichenologique,* **36,** 45–55.

CAVELIER, J. (1989) Root biomass, production and the effect of fertilizers in two tropical rain forests. Dissertation, University of Cambridge, England.

CHAZDON, R. L. (1988) Sunflecks and their importance to forest understorey plants. *Advances in Ecological Research,* **18,** 1-63.

CHAZDON, R. L. & FETCHER, N. (1984) Photosynthetic light environments in a lowland tropical rain forest in Costa Rica. *Journal of Ecology,* **72,** 553–564.

CHOPRA, R. N. & KUMRA, P. K. (1988) *Biology of Bryophytes.* John Wiley & Sons, New York.

COLEY, P. D. (1983) Herbivory and defensive characteristics of tree species in a lowland tropical forest. *Ecological Monographs,* **53,** 209–233.

COLEY, P. D. (1988) Effects of plant growth rate and leaf lifetime on the amount and type of anti-herbivore defense. *Oecologia,* **74,** 531–536.

COLEY, P. D. & AIDE, T. M. (1991) A comparison of herbivory and plant defenses in temperate and tropical broad-leaved forests. *Ecology of Plant-Animal Interactions: Tropical and Temperate Perspectives* (eds P. W. PRICE, T. M. LEWINSOHN, G. W. FERNANDES & W. W. BENSON), John Wiley and Sons, New York, pp 25–49.

COLEY, P. D., KURSAR, T. A. & MACHADO, J.-L. (1993) Colonization of tropical rainforest leaves by epiphylls: Effects of site and host plant leaf lifetime. *Ecology,* **74,** 619–623.

DEAN, J. M. & SMITH, A. P. (1978) Behavioral and morphological adaptations of a tropical plant to high rainfall. *Biotropica,* **19,** 152–154.

DEVRIES, P. (1988) The use of epiphylls as larval hostplants by the neotropical riodinid butterfly, *Sarota gyas. Journal of Natural History,* **22,** 1447–1450.

DIEM, M. H. (1971) Production de l'acide indolyl-3-acetique par certaines levures epiphylles. *Comptes Rendus Hebdomadaires des Seances de L'aca-demie de Sciences. Serie D Sciences Naturelles,* **272,** 941–943.

EMMERICH, R., GIEX, I., LANGE, O. L. & PROKSCH, P. (1993) Toxicity and antifeedant activity of lichen compounds against the polyphagous herbivororous insect *Spodoptera littoralis. Phytochemistry,* **33,** 1389–1394.

ENGELBRECHT, L. (1971) Cytokinin activity in larval-infected leaves. *Biochemie und Physiologie der Pflanzen,* **162,** 9–27.

EZE, J. M. O. & BERRIE, G. K. (1977) Further investigations into the physiological relationship between an epiphyllous liverwort and its host leaves. *Annals of Botany,* **41,** 351–358.

FARKAS, E. E. & SIPMAN, H. J. M. (1993) Bibliography and checklist of foliicolous lichenized fungi up to 1992. *Tropical Bryology,* **7,** 93–148.

FARRAR, J. F. (1976) Ecological physiology of the lichen *Hypogymia physodes I.* Some effects of constant water saturation. *New Phytologist,* **77,** 99–103.

FRAHM, J.-P. (1987) Which factors control the growth of epiphytic bryophytes in tropical rainforests? *Symposia Biologica Hungarica,* **35,** 639–648.

FRAHM, J.-P. (1990) The ecology of epiphytic bryophytes on Mt. Kinabalu, Sabah (Malaysia). *Nova Hedwigia,* **51,** 121–132.

FULFORD, M., CRANDALL, B. & STOTLER, R. (1970) The ecology of an elfin forest in Puerto Rico. II. The leafy hepaticae of Pico del Oeste. *Journal of the Arnold Arboretum,* **51,** 56–69.

GALLOWAY, D. J. (1991) *Tropical Lichens: Their Systematics, Conservation, and Ecology.* Clarendon Press, Oxford.

GARCIA-GUZMAN, G. & DIRZO, R. (1991) *Plant-pathogen-animal interactions in a tropical rainforest.* AIBS abstract.

GILBERT, O. L. (1973) Lichens and air pollution. *The Lichens,* (eds. V. AHMADJIAN & M. E. HALE) Academic Press, New York pp 443–472.

GOOSEM, S. & LAMB, D. (1986) Measurements of phyllosphere nitrogen fixation in a tropical and two sub–tropical rainforests. *Journal of Tropical Ecology,* **2,** 373–376.

GRADSTEIN, S. R. (1992) The vanishing tropical rainforest as an environment for bryophytes and lichens. *Bryophytes and Lichens in a Changing Environment* (eds. J. W. BATES & A. M FARMER), Clarendon Press, Oxford, pp. 234–258.

GREGORY, P. H. (1971) The leaf as a spore trap. *Ecology of Leaf Surface Microorganisms,* (eds. R. F. PREECE & C. H. DICKINSON) Academic Press, New York, pp 239–244.

GRUBB, P. J. (1977) Control of forest growth and distribution on wet tropical mountains; with special reference to mineral nutrition. *Annual Review of Ecology and Systematics,* **8,** 83–107.

HALE, M. E. (1983) *The Biology of Lichens.* Edward Arnold Press, Baltimore, Manyland.

HUBER, L., & GILLESPIE, T. J. (1992) Modeling leaf wetness in relation to plant disease epidemiology. *Annual Review of Phytopathology,* **30,** 553–577.

JACHMANN, H. T. & FEHRMANN, H. (1989) Effects of phyllosphere microorganisms on the senescence of wheat leaves. *Zeitschrift fuer Pflanzenkrankheiten und Pflanzenschutz,* **96,** 124–133.

JUNGNER, J. R. (1891) Anpassungen der Pflanzen an das Klima in den Gegenden der regenreichen Kamerungebirge. *Botanisches Centralblatt,* **47,** 353–360.

KERSHAW, K. A. (1985) *Physiological Ecology of Lichens.* Cambridge University Press, Cambridge.

LANGE, O. L. (1988) Ecophysiology of photosynthesis: Performance of poikilohydric lichens and homoiohydric mediterranean sclerophylls. *Journal of Ecology,* **76,** 915–937.

LANGE, O. L. & MATTHES, U. (1981) Moisture-dependent CO_2 exchange of lichens. *Photosynthetica,* **15,** 555–574.

LANGE, O. L. & TENHUNEN, J. D. (1981) Moisture content and CO_2 exchange of lichens. II. Depression of net photosynthesis in *Ramalina maciformis* at high water content is caused by increased thallus carbon dioxide diffusion resistance. *Oecologia,* **51,** 426–429.

LANGE. O. L. & ZIEGLER, H. (1986) Different limiting processes of photosynthesis in lichens, *Biological Control of Photosynthesis* (eds. R. MARCELLE, H. CLIJSTERS & M. VAN POUKE) Martinus Nijhoff Publishers, Dordrecht, pp. 147–161.

LANGE, O. L., KILIAN, E. & ZIEGLER, H. (1986) Water vapor uptake and photosynthesis of lichens: Performance in species with green and blue-green algae as phycobionts. *Oecologia,* **71,** 104–110.

LARSON, D. W. (1984) Habitat overlap/niche separation in two *Umbilicaria* lichens: A possible mechanism. *Oecologia,* **62,** 118–125.

LAWREY, J. D. (1980) Correlations between lichen secondary chemistry and grazing activity by *Pallifera varia. The Bryologist,* **83,** 328–334.

LAWREY, J. D. & HALE, M. E. (1979) Lichen growth responses to stress induced by automobile exhaust pollution. *Science,* **204,** 423–424.

LIBBERT, E., WICHNER, S., SCHIEWER, U., RISCH, H. & KAISER, W. (1966) The influence of epiphytic bacteria on auxin metabolism. *Planta,* **68,** 327–334.

MARINO, P. C. & SALAZAR ALLEN, N. (1993) Tropical epiphyllous hepatic communities growing on two species of shrub in Barro Colorado Island: the influence of light and microsite. *Lindbergia,* **17,** 91–94.

MONGE-NAJERA, J. (1989) The relationship of epiphyllous liverworts with leaf characteristics and light in Monte Verde. *Cryptogamie, Bryologie-Lichenologie,* **10,** 345–352.

MONTAGNINI, F., NEUFELD, H. S. & UHL, C. (1984) Heavy metal concentrations in some nonvascular plants in an Amazonian rainforest. *Water, Air and Soil Pollution,* **21,** 317–321.

MUELLER, U. G. & WOLF-MUELLER, B. (1991) Epiphyll deterrence to the leaf-cutter ant *Atta cephalotes. Oecologia,* **86,** 36–39.

NANDI, A. S. & SEN, S. P. (1981) Utility of some nitrogen-fixing microorganisms in the phyllosphere of crop plants. *Plant and Soil*, **63**, 465–476.

NASH, T. H., III. & WIRTH, V. (1988) *Lichens, Bryophytes, and Air Quality*. J. Cramer, Berlin.

OBERBAUER, S. F., CLARK, D. B., CLARK, D. A. & QUESADA, M. (1988) Crown light environments of saplings of two species of rainforest trees. *Oecologia*, **75**, 207–212.

OLARINMOYE, S. O. (1974) Ecology of epiphyllous liverworts: Growth in three natural habitats in Western Nigeria. *Journal of Bryology*, **8**, 275–289.

OLARINMOYE, S. O. (1975) Ecological studies of epiphyllous liverworts in Western Nigeria. II. Notes on competition and successional change. *Revue Bryologique et Lichenologique*, **41**, 457–463.

POCS, T. (1978) Epiphyllous communities and their distribution in East Africa. *Congres International de Bryologie*, **13**, 681–713.

POCS, T. (1982) Tropical forest bryophytes. *Bryophyte Ecology* (ed. A. J. E. SMITH) Chapman and Hall, London, pp 59–104.

PROCTOR, M. C. F. (1982) Physiological ecology: Water relations, light, and temperature responses, carbon balance. *Bryophyte Ecology*, (eds. A. J. E. SMITH), Chapman and Hall, New York, pp 333–381.

RICHARDS, P. W. (1954) Notes on the bryophyte communities of lowland tropical rainforest with species reference to Moraballi Creek, British Guiana. *Vegetatio*, **5–6**, 319–328.

RICHARDS, P. W. (1966) *The Tropical Rain Forest*. Cambridge University Press, Cambridge.

RICHARDS, P. W. (1984) The ecology of tropical forest bryophytes, *New Manual of Bryology, Volume 2* (ed. R. M. SCHUSTER) Nichinan, Miyazaki, pp 1233–1270.

ROGERS, R. W. (1977) Lichens of hot and semi-arid lands. *Lichen Ecology* (ed. M. R. D. SEAWARD), Academic Press, pp 211–252.

ROSKOSKI, J. P. (1980) N_2 fixation (C_2H_2 reduction) by epiphylls on coffee, *Coffea arabica*. *Microbial Ecology*, **6**, 349–355.

ROSKOSKI, J. P. (1981) Epiphyll dynamics of a tropical understory. *Oikos*, **37**, 252–256.

RUINEN, J. (1975) Nitrogen fixation in the phyllosphere. *Fixation by Free-Living Microorganisms* (ed. W. D. P. STEWART) Cambridge University Press, Cambridge, pp 85–100.

SAND-JENSEN, K. (1977) Effect of epiphytes on eelgrass photosynthesis. *Aquatic Botany*, **3**, 55–63.

SAS (1987) *SAS/STAT Guide for Personal Computers. Version 6*. SAS Institute, Cary, North Carolina.

SCHUSTER, R. M. (1980) *The Hepaticae and Anthocerotae of North America*. Vol. IV. Columbia University Press, New York.

SENGUPTA, B., NANDI, A. S., SAMANTA, R. K., PAL, D. SENGUPTA, D. N. & SEN, S. P. (1981) Nitrogen fixation in the phyllosphere of tropical plants: Occurrence of phyllosphere nitrogen-fixing microorganisms in Eastern India and their utility for the growth and nitrogen nutrition of host plants. *Annals of Botany*, **48**, 705–716.

SETHUNATHAN, N. (1970) Foliar sprays of growth regulators and rhizosphere effect in *Cajanus cajan* Millsp. I. Quantitative changes. Plant and Soil, **33**, 62–70.

SMITH, A. J. E. (1982) *Bryophyte Ecology*. Chapman and Hall, London.

STEWART, R. R. C. & BEWLEY, J. D. (1982) Stability and synthesis of phospholipids during desiccation and rehydration of a desiccant-tolerant and a desiccation-intolerant moss. *Plant Physiology*, **69**, 724–727.

SVEINBJORNSSON, B. & OECHEL, W. C. (1992) Controls on growth and productivity of bryophytes: Environmental limitations under current and anticipated conditions. *Bryophytes and Lichens in a Changing Environment*, (eds. J. W. BATES and A. M. FARMER) Oxford Science Publishers, Oxford, pp 77–102.

THIERS, B. M. (1988) Morphological adaptations of the Jungermanniales (Hepaticae) to the tropical rainforest habitat. *Journal of the Hattori Botanical Laboratory*, **64**, 5–14.

VAN REENEN, G. B. A. (1987) Attitudinal bryophyte zonation in the Andes of Colombia: A preliminary report. *Symposia Biologica Hungarica*, **35**, 631–636.

VARTIA, K. O. (1973) Antibiotics in lichens. *The Lichens* (eds. V. AHMADJIAN & M. E. HALE) Academic Press, New York, pp 547–561.

WICHNER, S. & LIBBERT, E. (1968a). Interactions between plants and epiphytic bacteria regarding their auxin metabolism. I. Detection of IAA-producing epiphytic bacteria and their role in long duration experiments on tryptophan metabolism in plant homogenates. *Physiclogia Plantarum*, **21**, 227–241.

WICHNER, S. & LIBBERT, E. (1968b). Interactions between plants and epiphytic bacteria regarding their auxin metabolism. II Influence of IAA-producing epiphytic bacteria on short-term IAA production from tryptophan in plant homogenates. *Physiologia Plantarum*, **21**, 500–509.

WINDSOR, D. M. (1990) Climate and moisture availability in a tropical forest, long term records for Barro Colorado Island, Panama. *Smithsonian Contributions to the Earth Sciences*, **29**, 1–145.

WINKLER, S. (1967) Die epiphyllen Moose der Nevelwalder von El Salvador, S. C. *Revue Bryologique et Lichenologique*, **35**, 303–369.

WINKLER, S. (1971) Okologische Beziehungen zwischen den epiphyllen Moosen der Regenwalder des Choco (Colombia, S. A.). *Revue Bryologique et Lichenologique*, **37**, 949–959.

WRIGHT, S. J. & CORNEJO, F. H. (1990) Seasonal drought and leaf fall in a tropical forest. *Ecology,* **71,** 1165–1175.

YOSHIMOTO, O., KATOH, K. & TAKEDA, R. (1990) Growth and secondary metabolite production in cultured cells of liverworts. *Bryophyte Development: Physiology and Biochemistry*, (eds. R. N. CHOPRA & S. C. BHATLA) CRC Press, Boca Raton, Florida, pp 209–223.

ZINSMEISTER, H. D. & MUES, R. (1990) *Bryophytes: Their Chemistry and Chemical Taxonomy*. Clarendon Press, Oxford.

13

Physiology of Tropical Vines and Hemiepiphytes: Plants that Climb Up and Plants that Climb Down

N. Michele Holbrook and Francis E. Putz

Plants become climbers, in order ... to reach the light, and to expose a large surface of their leaves to its action and to that of the free air. This is effected ... with wonderfully little expenditure of organized matter ...

> — DARWIN, 1867, *The Movements and Habits of Climbing Plants*

And the whole life history of these outlandish trees seems beautifully contrived to accomplish this objective: — to seize a place in the sun in the midst of a dense tropical forest.
> — DOBZHANSKY & MURCA-PIRES, 1954, *Strangler Trees*

13.1 INTRODUCTION

Vines climb up. From a starting point in the forest understory, vines use the mechanical competence of neighboring plants in their ascent towards the bright sun of the forest canopy. Hemiepiphytes climb down. They begin life in the treetops and extend earthward to form permanent and often substantial connection with the soil (Figure 13.1). Vines and hemiepiphytes hold in common a period of mechanical dependence, yet the physiological consequences and constraints associated with this reliance upon external support differs substantially between the two groups. In vines, adaptations essential for effective climbing constrain their outward form and thus influence the uptake, transport, and storage of energy and materials. For hemiepiphytes, in contrast, the major constraints may not be the particular rigors associated with life as either an epiphyte or a tree, but rather in the plasticity required to succeed as both. Our discussion of tropical vine physiology is morphologically organized, considering in turn the function of stems, leaves, and roots. With hemiepiphytes, we focus primarily on the nature of the transition between the epiphytic and tree growth forms. We draw upon the roles of vines and hemiepiphytes in tropical forest communities only as such discussion provides insights into their physiology. For clarity we will refer to all climbing plants as vines, reserving the term lianas for woody vines (including climbing palms). The category of herbaceous vines includes all climbers that lack secondary growth, even if their stems are perennial. We use the term hemiepiphyte to refer exclusively to species that establish as epiphytes and later form a connection with the soil (primary hemiepiphytes; Putz & Holbrook, 1986). Plants that germinate on the ground, ascend trunks of large trees using adhesive roots, and then later lose connection with the soil (secondary hemiepiphytes, such as many Araceae) are, in many ways, more vine-like in their physiology and morphology than the stranglers on which we focus. We emphasize that all of these plants are only structural parasites and acquire neither nutrients nor water directly from their host.

13.2 TROPICAL VINES

13.2.1 Stem function

Vine stems are narrow relative to the mass of foliage for which they serve as conduits for water, mineral nutrients, hormones, and photosynthate. Although this observation was clearly stated more than 100 years ago by Schenck (1893), questions still remain about xylem and phloem function, mechanical properties, and behavior of the vascular cambium in vines (see Tyree & Ewers, Chapter 8).

Figure 13.1.   Characteristic growth habits of epiphytic and terrestrially-rooted hemiepiphytes. (A) Ficus tuerkheimii *growing on the side of a large tree in Costa Rica; (B)* F. tuerkheimii *seedling; (C)* F. trigonata *seedling growing on a* Copernicia tectorum *palm tree in Venezuela; (D)* F. trigonata *surrounding a* C. tectorum *palm; (E)* F. nymphaeifolia *growing on an* Attalea *palm in Venezuela; (F) large* Ficus *sp. in lowland rainforest in Malaysia.*

A

B

C

D

E

F

Vine stems are long and thin because they grow slowly in diameter but rapidly in length. Herbaceous vines (i.e., those incapable of secondary thickening) continue to grow in length after primary diameter growth has ceased. Among tropical lianas, there are occasional individuals that are very thick (the record may be a 50.8 cm dbh [diameter at 1.4 m] *Entada monostachya* in central Panama; Putz, 1984). Some tropical lianas display moderately fast diameter growth rates (e.g., mean growth rate of 17 *E. monostachya* was 5.82 mm/y over an 8 y period; Putz, 1990), but most canopy lianas grow much more slowly in diameter than canopy trees in the same forest (mean growth rate of 15 species was 1.37 mm/y). As expected, and as observed in trees, lianas with low density wood tend to grow more quickly in diameter than species with denser wood. Rapid stem diameter growth, however, may be disadvantageous to plants that rely on trees for mechanical support; even with stems that weigh little for their length, lianas commonly fall from the canopy (Putz, 1990).

Although stem morphology is the structural feature that distinguishes vines from other plants, little is known of the developmental patterns that give rise to this growth form. Some normally scandent species can grow to be free-standing adults (e.g., Gartner, 1991), suggesting that there may be a developmental switch associated with the hormonal control of stem thickening. Perhaps this developmental pattern is simply an expression of neoteny insofar as most lianas are free-standing until they are 0.3–1.5 m tall (Putz, 1984). Alternatively, stem thickening may be triggered by attachment to a stationary support, as reported for leaf expansion on leader shoots (Raciborski, 1900). For vines, the transition from being self-supporting to climbing is due to internode lengthening. It is also often accompanied by marked changes in stem anatomy, including increased vessel diameter and an increased ratio of soft-to-hard tissues in the xylem (Dobbins, 1981). Neither the proximate nor the ultimate causes of these allometric and anatomical changes are understood.

Vines are flexible because they are long and slender. In addition, flexibility may result from peculiar stem anatomies. Anomalous secondary growth — concentric bands of phloem alternating with xylem bands, phloem that intrudes far into the xylem, isolated bands of phloem embedded in xylem, massive proliferation of xylem parenchyma, and other cambial variants (Carlquist, 1988, 1991) — is common among, but not unique to, lianas. The juxtaposition of hard and soft tissues and the isolation of strands of xylem in cable-like arrays confers flexibility to liana stems and may lessen the likelihood of mechanically-induced xylem cavitation (Putz & Holbrook, 1991). The presence of thin-walled and potentially meristematic tissues in the xylem promotes healing of mechanically damaged stems (Dobbins & Fisher, 1986; Fisher & Ewers, 1991). Fenestration of stems and compartmentation of the xylem path (sectorial ascent) of some species of lianas (e.g., *Serjania* spp. and other Sapindaceae) may also help mechanically damaged liana stems avoid loss of conductive capacity.

Information on the relationship between xylem anatomy and water flux in lianas is limited (for a review, see Ewers, Fisher & Fichtner, 1991; Tyree & Ewers, Chapter 8). Lianas and herbaceous vines tend to have very large diameter vessels. Because flow rates in vessel-like capillaries increase with aperture radius to the fourth power, they have substantial conductive capacities. In a comparison of water flow rates through excised sections of vines (10 species) and trees (10 species) from a dry forest in Mexico, Gartner et al. (1990) estimated that, under similar water potential gradients, water flow velocities would be 24 times greater in the vines. While the large diameter vessels found in many lianas have attracted much attention from researchers, Carlquist (1991) points out that most lianas also have very narrow vessels. Although high xylem conductivities are due to the large vessels, small vessels provide a limited conductive capacity if the large vessels become dysfunctional.

Maintenance of hydraulic conductivity after cavitation of many large xylem vessels may be particularly important if cavitation and recovery of vessel function are commonplace in lianas. To date, there is some limited evidence in support of this idea (Ewers, Fisher & Fichtner, 1991; Tyree & Ewers, Chapter 8). Given the slow rates of stem diameter increment and long life spans of lianas, their xylem vessels must remain functional for many decades. The frequent absence of obvious heartwood in even large diameter lianas (Putz, 1983) also argues for prolonged functional life spans of liana vessels. Positive root pressures have been reported for a number of tropical lianas (Tyree & Ewers, Chapter 8). That temperate species of *Vitis* refill xylem vessels in the spring after being drained during the winter suggests that refilling of cavitated vessels may occur in other lianas (Sperry et al., 1987). The critical experiments for testing for vessel refilling, however, have apparently not been conducted on tropical lianas. A good place to start might be in determining whether or not tropical species of *Vitis* refill cavitated vessels. Although the anomalous secondary growth and huge xylem vessels of many lianas have justifiably drawn the attention of researchers, lianas with dense wood and uniformly narrow vessels are not particularly rare.

The narrowness of vines' stems relative to their leaf areas creates a potential problem related to storage of water, photosynthate, and other materials (Mooney & Gartner, 1991). At least some arid zone vines have belowground storage organs (Rundel & Franklin, 1991) as do some temperate zone species (e.g., *Clematis, Campsis, Pueraria, Apios*). Some tropical vines (e.g., *Bauhinia* sp. and *Serjania* sp.) also have belowground storage organs, but tropical vine root biology remains largely unexplored. The narrowness of vine stems may also necessitate increased efficiency of phloem transport. Ewers (personal communication) has shown that sieve elements are wider in some vines than in congeneric trees. At a more macroscopic level, the intruded phloem, concentric phloem rings, and other cambial variants involving the distribution of phloem in vines serve to increase the proportional representation of phloem in stem cross sections without simply adding it to the outer circumference, where it may be more prone to mechanical disruption.

13.2.2 Leaf physiology

What characteristics of vines might lead us to expect them to have leaf physiologies that differ from trees? We discuss in this section how mechanical dependence might influence leaf size, leaf structure, phenology, water and nutrient relations, and photosynthetic properties of climbing plants. Given rather limited data on tropical vines, we will base much of the discussion on readily observable characteristics of vines, ecological patterns of vine abundance, and speculation based on vine natural history. Physiological characteristics of temperate zone vines were recently reviewed by Teramura, Gold and Forseth (1991); Castellanos (1991), in the same volume, characterizes the photosynthetic properties of the tropical vines for which data are available. Based on these studies, we conclude that there are no consistent differences in leaf-level physiology between trees and vines (Teramura, Gold & Forseth, 1991; Castellanos, 1991).

One possible contrast between leaves of vines and trees is that vines may experience more heterogeneous light environments than trees. Castellanos (1991) presents data on photosynthetic rates at different light levels that suggests that vines perform better than trees in heterogeneous environments. This assumption about vine light environments, however, deserves further inspection. We examine in this section characteristics of the light environments experienced by canopy vines and canopy trees growing in a wet tropical lowland forest in the context of the degree of heterogeneity experienced by the two growth forms. Presumably, most of the points made will also be relevant to understory vines and climbing plants in dry and montane forests.

Vines and trees both grow to the canopy from the forest floor where, except in treefall gaps, light strongly limits photosynthesis. For a tree seedling, the pathway to the canopy is vertical. Vines, in contrast, cannot grow straight up except during the upright juvenile stage (rarely beyond 1 m height) or when they encounter upright supports suitable for attachment (e.g., root climbers on tree trunks or tendril climbers that ascend the narrow stems of vines already in the canopy). Most vines follow a much more circuitous path upwards and may cover as much distance horizontally as vertically.

On their way to the canopy, trees encounter a variety of light environments. While we as pedestrians may envision forest light environments as two-dimensional patchworks, the heterogeneity is three-dimensional. Within the small crowns of tree saplings of two species, the coefficient of variation of daily total PPFD (photosynthetically-active photom flux density) averaged more than 120% (Oberbauer, et al., 1989). Although similar data are not available for large trees, the variation is likely to be substantial, at least until they emerge from the canopy. Individual trees adjust to spatial heterogeneity in PPFD by leaf- and branch-specific adjustments of physiological responses as well as by varying the size, number, and longevity of leaves. Trees may grow up through patches of dense shade by translocating photosynthate from better-lit areas of the crown. There are species-specific limits of shade tolerance, however, so the range of light environments to which a tree is likely to be exposed is somewhat constrained.

Canopy vines, like trees, experience a range of light environments during their climb to the treetops. This range is limited by the need to maintain a generally positive carbon balance. To a greater extent than in trees, however, vines seem to be able to increase internode lengths or suppress leaf expansion in response to local shade (Peñalosa, 1983). Leaf production and turnover rates in vines also seem to be higher than in most trees (Peñalosa, 1984; Putz & Windsor, 1987; Hegarty, 1990). Rapid changes in leaf orientation also may characterize climbing plants, at least in the north temperate zone (e.g., *Pueraria lobata*; Teramura, Gold & Forseth, 1991). These characteristics associated with the climbing habit could result in vine leaves being exposed to less variation in light environments than tree leaves in the same forest. It should be noted, however, that due to most vines' ability to spread vegetatively, the range of environmental conditions experienced by a vine over its lifetime may exceed that of most trees.

Leaf shapes that appear unusual to temperate zone botanists are common among all types of climbers. This, however, may be attributed to phylogeny rather than any particular adaptive significance. For example, the unusual leaf shapes of many Passifloraceae, Menispermaceae, Dioscoreaceae, and Marcgraviaceae are unlikely to be related to their climbing habits. Vines that climb trunks of trees with the aid of adventitious roots or adhesive tendrils often display heteroblasty—marked ontogenetic variation of leaf size, shape, and other characteristics (for a recent review, see Lee & Richards, 1991 or Ray, 1990). While mild heteroblasty is expressed in most woody tendril climbers and twiners as upright juveniles develop into climbing adults, pronounced variation in leaf size and shape of the sort discussed by Givnish and Vermeij (1976) is more characteristic of herbaceous root climbers. Lee and Richards (1991) suggest that the high frequency of heteroblasty in vines is due to the temporal variability in their light environments; again, data are inadequate to test this notion.

Whether or not there are general phenological differences between vines and trees is unresolved, nor is it clear why such differences would be expected. The data available on vines and tree-leaf phenology do not suggest consistent trends. In comparison to canopy trees on Barro Colorado Island, Panama, for example, canopy vines have longer leaf production periods each year, and vines are less likely to go deciduous (Putz & Windsor, 1987). In the drier and more seasonal forests of Guanacaste, Costa Rica, in contrast, many vines are deciduous and seasonality in leaf production is pronounced (Opler, Baker & Frankie, 1991).

13.2.3 Roots and nutrient relations

Given their different growth habits, we expect substantial morphological differences between the root systems of tropical vines and those of free-standing plants in the same forests. We hypothesize that in the tropics, vines have smaller maximum diameter roots compared to trees of the same stem diameter or leaf area. A lack of large diameter roots should translate into root systems that are more spatially extensive for a given investment in belowground biomass. This does not lead us to any predictions about fine root characteristics, root proliferation rates in response to localized resource patches, or plasticity of uptake-kinetic mechanisms. The expected expansiveness of liana root systems would be augmented by their general capacity to propagate vegetatively by rooting of fallen stems.

In a greenhouse experiment with four species of temperate zone lianas and four species of sympatric trees, Putz (in preparation) found that liana roots grew up nutrient gradients much more rapidly than tree roots. The lianas, however, did not show higher rates of biomass increment or higher root:shoot ratios. Similarly, the lianas also failed to show the expected greater efficiency in exploring soil volumes, measured either as horizontal expansion per unit root biomass or the ratio between root growth rates in the fertilized and unfertilized portions of the soil. The four liana species maximized horizontal expansion through root length growth at the expense of lateral branching. Interpreting these results is complicated insofar as lateral root growth in some vines serves both to increase the soil volume exploited for nutrients and the area over which vegetative propagation by root sprouts is possible.

Lianas may be among the most deep-rooted species in tropical forests. Leaf persistence and expansion during dry periods, when many trees are deciduous (Putz & Windsor, 1987), suggest access to water sources not tapped by trees. Likewise, Longino (1986) reported that leaf production by a liana species in Costa Rica was negatively correlated with rainfall. Stable hydrogen isotope composition of water extracted from the xylem of a liana (*Bauhinia* sp.) indicates that its roots had access to deeper portions of the soil profile (Jackson et al., 1995). Direct evidence of deep rooting by vines in the Brazilian Amazon is provided by research investigating vine root functions at 4–6 m depth (T. G. Reston, personal communication). Furthermore, well diggers in the same region report that roots 10–12 m belowground are predominantly from lianas (D. C. Nepstad, personal communication).

Vine abundance seems to decline with decreasing soil fertility. Although we are aware of no fundamental reason why vines should be more nutrient-demanding than trees, this trend has been reported over vast geographical areas (e.g., Gentry, 1991), as well as in sites with both nutrient-rich and nutrient-poor soils (e.g., Putz, 1983; Putz & Chai, 1987). That alkaloids and other highly mobile defenses seem to be particularly well represented in vine-rich families (e.g., Hegarty, Hegarty & Gentry, 1991) is also consistent with vines' increased occurrence in nutrient-rich soils; cause and effect, however, are not at all clear. Furthermore, in light of the rapid leaf turnover rates in many vines, mobile defenses would also be expected.

13.3 HEMIEPIPHYTES

There are far fewer species of hemiepiphytes (< 2000) than of either holoepiphytes (approximately 25,000) or trees (Putz & Holbrook, 1986; Gentry & Dodson, 1987; Benzing, 1990). Hemiepiphytes are reported in 26 families. The majority are members of either the Moraceae (notably *Ficus*, with approximately 500 hemiepiphytic species) or Clusiaceae (about 200 species; Willis, 1973; Madison, 1977). In most hemiepiphytes, mature individuals develop only modest connections with the ground. The best known hemiepiphytes, however, are the stranglers, whose descending roots anastomose and fuse to form a cylinder that surrounds and ultimately may kill the host tree (Dobzhansky & Murca-Pires, 1954). In this way, stranglers can go on to become large, free-standing trees. Strangling species occur in the genera *Ficus* and *Clusia*, as well as in *Schefflera* (Araliaceae), *Coussapoa* (Cecropiaceae), *Posoqueria* (Rubiaceae), and *Metrosideros* (Myrtaceae). Although most hemiepiphytes do not strangle their hosts, stranglers (particularly *Ficus* and *Clusia* species) are of interest because they grow in a wide range of tropical habitats from aseasonal wet forests to seasonally deciduous forests and savannas. In addition, they can grow to be large, canopy trees (Putz & Holbrook, 1986; Richards, 1952), and they are often important food sources for birds and mammals (e.g., Lambert & Marshall, 1991). Stranglers are intriguing physiologically because of their transition in growth form from epiphyte to tree. We focus here on stranglers and begin with a discussion of establishment and germination requirements. We then turn to the question of the transition from epiphyte to tree and contrast morphological and physiological characteristics of epiphytic and tree-phase individuals. Finally, we consider the morphology of large hemiepiphytes with respect to the strangling growth form.

13.3.1 Seed germination and seedling establishment

Little is known about the establishment site requirements of hemiepiphytes (see Williams-Linera & Lawton, 1995). *Ficus* and *Clusia* seedlings seem to germinate wherever substantial humus accumulates in the canopy. Successful establishment in the canopy, however, appears limited to sites in which the roots have access to a substantial volume of soil or the decaying interior of the supporting tree (Laman, 1994; Putz & Holbrook, 1986). Large stranglers are generally restricted to the base of large branches (Todzia, 1986).

Hemiepiphytes are enigmatic in terms of their relationship with the soil. The epiphytic phase appears to be obligatory for most hemiepiphytic species of *Ficus*. Why should strangler figs be so plastic in terms of growth form, but not opportunistic in terms of establishment sites? In contrast, many *Clusia* species appear to be more flexible in terms of germination sites, with seedlings thriving both when growing in forest soil as well as in the canopy (N. M. H., personal observation, Ball et al., 1991; Popp et al., 1987).

Hemiepiphytes, particularly stranglers, occupy a more limited range of establishment sites than do holoepiphytes. Stranglers often grow such that at least some of their roots are protected from the atmosphere, i.e., inside hollow trunks or behind persistent leaf bases of palm trees (Putz & Holbrook, 1986, 1989; Kelly, 1985; Vanderdyst, 1922; Bessey, 1908). Architectural characteristics of candidate host species, including deep branch crotches or openings into hollow stems, may be more important than surface characteristics such as texture (Todzia, 1986) or chemicals leached from the bark (Titus, Holbrook & Putz, 1990). Most hemiepiphytes are animal dispersed (Putz & Holbrook, 1986) and food preferences of the dispersal agents may also influence the range of host species. In a survey of Barro Colorado Island, Panama, Todzia (1986) found that hemiepiphytes occur in all areas of trees' canopies, from the main trunk to the small branches near the edge of the crown. Species that grow to be quite large (*Ficus* spp.) were more common on the host's main trunk, while species that remain mechanically dependent on the host tree throughout their life (e.g., *Cosmibuena skinneri* [Rubiaceae], *Havetiopsis flexilis* [Clusiaceae], and *Clusia uvitana* [= *Clusia odorata*]) were more evenly distributed throughout tree crowns.

13.3.1.1   *Ficus*

Germination of *Ficus* species seems to be most limited by moisture. In *F. aurea*, a strangling species, germination was substantially reduced by low substrate water potential (Swagel & Ellmore, 1992). After two weeks at −0.25 MPa, germination was 30% lower than in the 0:0 MPa controls and radical growth rate was similarly reduced. Germination studies of *F. religiosa* in Israel (Galil & Meiri, 1981) also indicate the need for constant (and high) humidity. Germination in petri plates that were intermittently covered and exposed at four day intervals was reduced 25% compared to the continuously covered controls, and the seedlings appeared weak and wrinkled. Greater exposure to drying (two day intervals) prevented germination altogether. Continuous irrigation beneath an adult *F. religiosa* resulted in germination of numerous seedlings on the soil (Galil & Meiri, 1981). In contrast, in montane cloud forest at Monteverde, Costa Rica, daily watering of *F. pertusa* and *F. tuerckheimmii* seeds placed in experimental soil plots, did not result in germination (Titus, Holbrook & Putz, 1990). Seeds collected from the same individuals, however, did germinate when kept moist in small pots and petri plates indoors. In the few observations of Ficus seeds germinating beneath an adult, all were heavily infected with fungi and died soon after (Titus, Holbrook & Putz, 1990). All the *Ficus* species examined to date require light for germination (Titus, Holbrook & Putz, 1990; Galil & Meiri, 1981; Bessey, 1908).

13.3.1.2   *Clusia*

We know of no published reports of germination requirements in *Clusia*. *Clusia* seeds are much larger than those of *Ficus*, appear to have little or no dormancy, and may even occasionally be viviparous. Cloud forest *Clusia* species may reproduce vegetatively through the rooting of fallen branches (Lawton & Putz, 1988).

13.3.2   Comparisons between epiphytic and terrestrially-rooted individuals

Hemiepiphytes sidestep many of the establishment constraints faced by normal trees (e.g., light limitations, mechanical damage from falling debris, root competition, trampling and consumption by terrestrial herbivores, fires, flooding), only to confront other limitations associated with the epiphytic habit. Foremost among these are the potential shortfalls in water or nutrient supply caused by growing on a limited substrate volume, and the danger of dislodgment. Because hemiepiphytes tend to grow in microsites with substantial accumulation of organic humus, nutrient deficiency may be less important. Indeed, such substrates are often nutrient rich (Putz & Holbrook, 1989; Nadkarni, 1984), and epiphytic individuals may have equal or higher concentrations of foliar nutrients (per leaf area) than conspecific trees (Putz & Holbrook, 1989, Holbrook and Putz, unpublished data). Low water content of the substrate, however, remains a frequent, severe threat in the epiphytic phase. Many differences between epiphytic and terrestrially-rooted hemiepiphytes suggest moisture availability as a major limiting factor during the epiphytic phase (Holbrook & Putz, 1995b; Putz and Holbrook, 1986; Ting, et al., 1985). An important inter-generic difference is that while *Ficus* exclusively utilizes the C_3 photosynthetic pathway (Holbrook and Putz, 1995b; Ting et al., 1985, 1987), the water-conserving crassulacean acid metabolism (CAM) occurs in many epiphytic and hemiepiphytic *Clusia* species (Franco, Ball and Lüttge, 1990; Popp et al., 1987; Ting et al., 1987; Tinoco Ojanguren and Vazquez-Yanes, 1983; Medina, Chapter 2).

13.3.2.1 Ficus

(i) Gas exchange and water use

Strangler figs use water more conservatively when epiphytic than when firmly rooted in the ground. This pattern is most evident during extended rainless periods, when the stomata of epiphytic plants open only for a few hours each morning and then close for the rest of the day. During such dry periods, stomata of conspecific, co-occurring trees remain open throughout the day (Holbrook & Putz, 1995b; Ting et al., 1987). Epiphytes that had been watered for several days during the dry season had diurnal patterns of stomatal conductance similar to those of the trees. This indicates that the low stomatal conductance of epiphytic plants was related to substrate drying rather than to the greater leaf-to-air vapor pressure difference that also occurs during the dry season (Holbrook & Putz, 1995b). During the rainy season, stomatal conductance in epiphytic individuals markedly increased (to values similar to those of irrigated epiphytes, although never exceeding those of the trees; Holbrook & Putz, 1995b).

Epiphytes' lower stomatal conductance during drought periods suggests that they experience greater water stress. During the dry season, however, midday leaf water potentials of epiphytic *F. trigonata* and *F. pertusa* were significantly higher than in conspecific trees (Holbrook & Putz, 1995b). Short-term irrigation of epiphytic *Ficus pertusa* plants increased pre-dawn values, but midday leaf water potentials remained at pre-irrigation levels (Holbrook & Putz, 1995b). During the rainy season, epiphytes and trees of both *F. trigonata* and *F. pertusa* had similar pre-dawn and mid-day leaf water potentials.

Water potential isopleths (pressure-volume curves) indicate substantial differences in the foliar dehydration parameters (cell wall elasticity, cell sap osmotic potential) of epiphytic and terrestrially-rooted individuals (Holbrook & Putz, 1995a). Epiphytes in all of the five *Ficus* species examined had higher (less negative) osmotic potentials and lower moduli of elasticity (more elastic cell walls) than the conspecific trees. Compared with differences between the two growth forms, seasonal differences in either of these parameters were either non-existent or relatively minor. Thus, the turgor loss point is estimated to occur at a significantly higher leaf water potential in epiphytes (5 species), but at the same (3 species) or higher (2 species) relative water content than in conspecific trees. This means that although midday leaf water potentials were higher in epiphytic than in tree-phase individuals during the dry season, estimated leaf turgor pressures were similar (Holbrook & Putz, 1995b).

During the dry season, epiphytic *Ficus* plants limit stomatal opening to the early morning. Carbon uptake, thus, occurs during the period most advantageous (for a C_3 plant) in terms of minimizing water loss (Holbrook & Putz, 1995b). Epiphytic strangler figs have fleshy thickenings at the base of their stem that may store water; strong stomatal control may allow epiphytes to protect such water reserves from rapid depletion (Holbrook & Putz, 1995b, Putz & Holbrook, 1986). This storage may be of critical importance if the water supply is sporadic (e.g., heavy dew-fall or occasional light showers) and substrate drying rapid (Holbrook & Putz, 1995b). The less negative leaf water potentials (and associated cellular water relations) of the epiphytes than conspecific trees also supports the idea of reliance on stored water (Holbrook, 1995).

Many questions remain concerning aspects of stomatal regulation and water relations of strangler figs. For example, what is the physiological basis for the differences in stomatal behavior in the two forms? Epiphytic and terrestrially-rooted individuals of *F. nymphaeifolia* have substantially different stomatal conductances during the dry season, but they show similar responses to changes in the leaf-to-air vapor pressure gradient (Holbrook & Putz, unpublished data). Preliminary reports of xylem hydraulic conductivity of Ficus species suggest that the specific conductivity ($kg\ s^{-1}\ m^{-1}\ MPa^{-1}$) may be extremely high compared to co-occurring woody species (S. Patino, unpublished data, cited in Zotz, Tyree & Cochard, 1994). Further studies are needed of xylem structure and hydraulic architecture of epiphytic and tree-phase hemiepiphytes in order to understand the relationships among water uptake, transport, and stomatal conductance in the two growth forms.

(ii) Leaf structure

Marked morphological and anatomical differences exist between leaves of epiphytic and tree *Ficus* individuals (Table 13.1). In particular, epiphytic stomatal density is several times lower than in conspecific trees (Holbrook & Putz, 1995a). Leaves excised from epiphytic *Ficus* dehydrate much more slowly than do leaves of conspecific trees (Holbrook & Putz, 1995a). *Anthurium bredemyeri*, a herbaceous plant which grows both as an epiphyte and on the forest floor, showed a similar pattern of lower stomatal density in leaves of epiphytic individuals (Rada & Jaimez, 1992). Other morphological differences between the two growth forms include pubescence in epiphytic, but not tree, leaves (2 of 5 *Ficus* species examined), thinner leaves in epiphytic plants (4 of 5 *Ficus* species), and 2 to 4 times higher specific leaf area (cm^2 g^{-1}) in epiphytic individuals (5 *Ficus* species; Holbrook & Putz, 1995a, Table 13.1). Differences in leaf thickness were primarily due to the palisade parenchyma layer, which was thinner in epiphytic leaves of all five species (Holbrook & Putz, 1995a).

(iii) Leaf turnover

Members of the genus *Ficus* typically show population-level asynchrony of leaf turnover, paralleling the asynchronous flower production that permits local persistence of their pollinator (e.g., Kjellberg & Maurice, 1989). Although a number of phenological studies have included mature hemiepiphytes, we know of only one that included epiphytes and focused on differences in leaf turnover between the two growth forms (Putz, Romano & Holbrook, 1995). This study considered two species (*Ficus pertusa* and *F. trigonata*) growing in a seasonally dry region of Venezuela, with observations recorded at two-week intervals over a two year period. Trees of both species are functionally evergreen. Leaf fall occurs over a short period with individuals remaining fully leafless only briefly (1–2 weeks). Despite highly seasonal rainfall, leaf production in the trees was independent of precipitation, and individuals in every phenological stage could be found at any time of the year. Epiphytes followed the same overall pattern, with the production of new leaves in some individuals occurring at the height of the six-month dry season. Nevertheless, production and loss of leaves by epiphytic individuals were more tied to recent rainfall events, and epiphytes had shorter leaf life spans and higher frequency of deciduousness than conspecific trees. Mortality was also higher in the epiphytes; 17% died during the two year study period, compared to no mortality of trees.

These differences are consistent with physiological data indicating frequent and severe water stress in epiphytic individuals. Nevertheless, epiphytic plants retain some leaves during the dry season and may even produce new ones (Putz, Romano & Holbrook, 1995). Dur-

Table 13.1.   Relative differences between epiphytic and terrestrially-rooted (tree) individuals of Ficus and Clusia hemiepiphytes.

	Ficus			Clusia		
	epiphyte	tree	source	epiphyte	tree	source
photosynthetic pathway	C_3	C_3	9	C_3-CAM primarily at night	C_3-CAM night, early AM and late afternoon	8,9,10
dry season stomatal opening	low, limited to early morning	high, open throughout day	4, 5			8, 9
dry season midday leaf water potential	more negative	less negative	4, 5	comparative values not available, both appear to be high compared with Ficus		2, 5
leaf osmotic potential	less negative	more negative	3	more negative	less negative trees similar to Ficus epiphytes; epiphytes similar to Ficus epiphytes in dry season and to Ficus trees in wet season	3
cell wall properties (leaves)	elastic	stiff	3	seasonal elastic values lower (more elastic) than Ficus	stiff	3
specific leaf area (area/mass)	high	low	3, 7	epiphytes higher, but actual values lower than in Ficus		3
stomatal density	low	high	3	no difference		
epidermal water loss rate	low	high	3			
foliar nitrogen (N/mass)	high	low	5, 7	approximately equal (5) or lower in epiphytes (1)		1, 5
leaf thickness	thin	thick	3	no difference (3) or thinner in epiphytes (1), overall thicker than Ficus trees		1, 3
leaf life span	short	long	6			

(1) Ball et al., 1991
(2) Borland et al., 1992
(3) Holbrook & Putz, 1995a
(4) Holbrook & Putz, 1995b
(5) Holbrook & Putz, unpublished data
(6) Putz, Romano & Holbrook, 1995
(7) Putz & Holbrook, 1989
(8) Sternberg et al., 1987
(9) Ting et al, 1987
(10) Ting et al, 1985

ing the dry season, the organic material collected from behind persistent leaf bases of host palms showed moisture contents that were variable, but substantially higher than soil collected from the top 10 cm of soil directly beneath the host tree (Holbrook & Putz, 1995b). The source of this moisture is unknown. Dewfall and intermittent rain showers represent the most likely sources of water. In addition, termites known to humidify their nests with water carried up from deep in the soil (*Nausutitermes*; Lee & Wood, 1971) are common on the most common host tree, *Copernicia tectorum* (Putz & Holbrook, 1989).

13.3.2.2 Clusia

(i) Gas exchange

Patterns of gas exchange and carbon acquisition in *Clusia* are dominated by their ability to modulate between C_3 and CAM. Although some *Clusia* species are primarily C_3 (e.g., C. *venosa*, [Franco, Ball & Lüttge, 1990]; C. *valerii* [Ting et al., 1987]), most of those studied, including all of the hemiepiphytic species so far examined, are highly plastic in their photosynthetic metabolism (Zotz & Winter, 1993; Borland et al., 1992; Winter et al., 1992; Ball et al., 1991; Franco, Ball & Lüttge, 1990; Sternberg et al., 1987; Ting et al., 1987; Ting et al., 1985). Early studies of CAM in *Clusia* (first reported by Tinoco-Ojanguren & Vazquez-Yanes,1983) focused on the relationship between limitations in substrate water during the epiphytic phase and the high water-use efficiency typically associated with nocturnal CO_2 uptake (Popp et al., 1987; Sternberg et al., 1987; Ting et al., 1987; Ting et al., 1985). Subsequent investigations have explored the effect of environmental factors such as light, vapor pressure, and nitrogen availability on the balance between day-time and night-time CO_2 uptake (Zotz & Winter, 1993; Franco, Ball & Lüttge, 1991, 1992; Haag-Kerwer, Franco & Lüttge, 1992; Winter et al., 1992; Lee, Schmitt & Lüttge, 1989; Schmitt, Lee & Lüttge, 1988).

Field measurements demonstrate that the capacity for and expression of CAM are not limited to epiphytic individuals (Zotz & Winter, 1993; 1994; Sternberg et al., 1987; Ting et al., 1987). Nevertheless, gas exchange measurements reflect lower levels of water in the substrate available to epiphytes. During the dry season, epiphytic individuals acquire CO_2 from the atmosphere only during the night (i.e. no phase II or IV uptake, *sensu* Osmond, 1978) and have smaller diel fluctuations in titratable acidity than conspecific trees (Sternberg et al., 1987; Ting et al., 1987). "Tree" individuals show high rates of carbon uptake during the early light period (phase II), late afternoon (phase IV), and at night (phase I). With the transition to the rainy season, a greater fraction of daily carbon uptake occurs during the daytime in both epiphytes and trees. In *Clusia* epiphytes most daily carbon uptake may occur during a brief stomatal opening in the early light period (Sternberg et al., 1987; Ting et al., 1987). *Clusia* trees, on the other hand, may shift to CO_2 uptake throughout the day with little or none at night (Ting et al., 1987) or continue all four CAM phases, but with a greater emphasis on daytime uptake (Sternberg et al., 1987).

Measurements of $\delta^{13}C$, an index of the proportion of CO_2 fixed through the C_3 vs. CAM pathway (Osmond, 1978), range between −14 and −27 in *Clusia* species, with somewhat higher values in the epiphytes (Borland et al., 1992; Popp et al., 1987; Sternberg et al., 1987; Ting et al., 1987). Although an integrated measure, these values will be weighted towards conditions prevailing during periods of leaf development. Zotz & Winter (1994) report that most leaf area expansion in *C. uvitana* takes place during the rainy season; furthermore, leaves less than 12 weeks old acquire CO_2 only during the light period, regardless of the environmental conditions. On the other hand, high intercellular CO_2 concentrations during midday malic acid decarboxylation (phase III) may lead to a net efflux of CO_2 and thus an overall enrichment in C^{13} (Sternberg et al., 1987 Goldstein et al., Chapter 9).

Several aspects of photosynthetic metabolism in *Clusia* distinguish it from other CAM or C_3-CAM species. Foremost is the rate at which *Clusia* individuals are able to shift their photosynthetic behavior between typical C_3 patterns and CAM. Reversible shifts within 24 hours have been reported (Zotz & Winter, 1993). Hemiepiphytic *Clusia* species exhibit such variability in their photosynthetic behavior that simple characterization as a C_3-CAM intermediate may be misleading (Schmitt, Lee & Lüttge, 1988). Patterns of organic acid accumulation in *Clusia* contrast with those of other CAM plants due to high diel fluctuation (exceeding 1000 mol m^{-3}), substantial accumulation of citric acid as well as malic acid, and maintenance of high $[H^+]$ even when all CO_2 uptake occurs during the light (Borland et al., 1992; Franco, Ball & Lüttge, 1990, 1991, 1992; Haag-Kerwer, Franco & Lüttge, 1992; Popp et al., 1987; Ting et al., 1985; see also Medina, Chapter 2; for a more detailed discussion of organic acid accumulation patterns in *Clusia*). The latter has been suggested to play an important role in the flexibility and sensitivity with which dark CO_2 uptake in Clusia responds metabolically to environmental conditions (Zotz & Winter, 1993).

Table 13.2.   Effect of environmental conditions on the partitioning of CO_2 uptake between light and dark periods in hemiepiphytic species of Clusia.

	Change in % daytime CO_2 uptake[1]	Reference
↓ substrate water content	decrease	Zotz & Winter, 1993 Franco, Ball & Lüttge, 1992 Winter et al., 1992 Franco, Ball & Lüttge, 1991 Schmidt, Lee & Lüttge, 1988
↑ VPD	decrease	Winter et al., 1992 Lee, Schmidt & Lüttge, 1989 Schmidt, Lee & Lüttge, 1988
↓ PPFD	increase[2]	Zotz & Winter, 1993 Franco, Ball & Lüttge, 1992 Haag-Kerwer, Franco & Lüttge 1992 Winter et al., 1992 Franco et al., 1991
↓ Nitrogen	increase	Franco, Ball & Lüttge, 1991
↓ ΔT (day – night)	increase	Haag-Kerwer et al., 1992.
↑ ambient CO_2	increase	Winter et al., 1992
↓ ambient CO_2	decrease	Winter et al., 1992

[1] Relative to a well-watered and fertilized plant grown at moderate VPD, high PPFD, ambient CO_2, and $T_{day} = 30\,°C$, $T_{night} = 20\,°C$.
[2] But see Schmidt, Lee & Lüttge, 1988.

Changes in environmental conditions influence not only total (24 h) carbon gain in *Clusia*, but also the proportion of CO_2 uptake in daylight and at night. Water availability greatly affects diurnal patterns of stomatal opening, as do a number of other factors (Table 13.2). Decreases in substrate water availability may increase the fraction of nocturnal CO_2 uptake before any measurable decline in bulk leaf water potential or impaired chloroplast function. Reductions in leaf water potential and afternoon fluorescence can occur, however, with prolonged drought (Winter et al., 1992). Reduced overall CO_2 uptake, however, does not necessarily coincide with a greater proportion of nocturnal uptake (Table 2). Thus, a shift towards a more CAM-like metabolism cannot be interpreted as simply a stress response. Zotz & Winter (1993) suggest that partitioning between daytime and nighttime CO_2 uptake is controlled by leaf [H⁺] at the end of the light period. Nocturnal CO_2 fixation will, thus, be enhanced by conditions that increase organic acid decarboxylation during the light period (Zotz & Winter, 1993).

The rapidity and flexibility with which hemiepiphytic species of *Clusia* modulate their photosynthetic metabolism in relation to environmental factors far exceeds that of any other group studied to date. Whether or not this allows them to simultaneously optimize water, carbon, and photon utilization is not known (Zotz & Winter, 1995; Borland et al., 1992). Estimates of annual productivity in *C. uvitana* are surprisingly high, approaching values estimated for the host tree, *Ceiba pentandra* (Bombacaceae), in lowland Panama and exceeding those reported for temperate regions (Zotz & Winter, 1994; Zotz & Winter, Chapter 2). Changes between C_3 and CAM photosynthesis within a single species have been interpreted primarily in light of the increased water-use efficiency associated with nocturnal stomatal opening (e.g., Osmond, 1978). The fact that photosynthetic plasticity is retained in the tree form further suggests that it represents an adaptation to more than reductions in substrate water availability.

(ii) Water relations

Differences in leaf water relations between epiphytic and tree forms of *C. minor* were small compared to those reported for *Ficus* species (Holbrook & Putz, 1995a). Osmotic potentials were less negative in *C. minor* than in *Ficus* trees (with values for *C. minor* trees being similar to *Ficus* epiphytes). The bulk moduli of cell wall elasticity for *C. minor* epiphytes and trees were also substantially lower (more elastic) than in either *Ficus* growth form. At full saturation *C. minor* had significantly different osmotic potentials between growth forms (epiphyte vs. tree), season, and their interaction (season x growth form). The epiphytes had lower (more negative) osmotic potentials, particularly in the wet season—the opposite of what was found in *Ficus*. Higher values of apparent cell wall elasticities occurred in trees and during the dry season. These differences, however, were small. Furthermore, *C. minor* epiphytes and trees did not differ in either relative water content or leaf water potential at the turgor loss point (Holbrook & Putz, 1995a).

Clusia species maintain plasticity in photosynthetic metabolism even after they become well-rooted in the soil and have relatively high leaf water potentials which exhibit little diurnal or seasonal variation (Zotz, Tyree & Cochard, 1994; Zotz & Winter, 1994; Holbrook, unpublished data). This raises the question whether *Clusia* trees, with access to soil water reserves often exhibit primarily nocturnal CO_2 uptake, whereas co-occurring *Ficus* trees exhibit high rates of daytime

CO_2 uptake (Ting et al., 1987). Studies of the hydraulic architecture and vulnerability to cavitation in C. uvitana indicate that the branch hydraulic conductivity per supported leaf area of this C_3-CAM hemiepiphyte is substantially lower than reported for other tropical woody species (0.93 to 2.39 vs. 5 to 30 kg s^{-1} m^{-1} MPa^{-1}, respectively; Zotz, Tyree & Cochard, 1994). This difference can be primarily attributed to a high ratio of leaf area to branch cross-sectional area; the specific hydraulic conductivity (normalized by cross-sectional area) of leaf-bearing branches was similar to values reported for co-occurring trees (Zotz, Tyree & Cochard, 1994). Stems of C. uvitana are extremely vulnerable to cavitation, with 50% loss of hydraulic conductivity occurring at a water potential of −1.3 MPa (Zotz, Tyree & Cochard, 1994). This susceptibility to cavitation may restrict the high rates of water loss associated with daytime stomatal opening to periods of high soil water availability (Zotz & Winter, 1994; Zotz, Tyree & Cochard, 1994). In addition, the aerial roots of C. uvitana have extremely high specific conductivity which may play an important role in preventing low stem xylem water potentials for individuals located high in the canopy (Zotz, Tyree & Cochard, 1994)

(iii) Leaf structure

Differences in leaf structure between epiphytic and terrestrially rooted *Clusia* were also minor compared with those reported for *Ficus* (Holbrook & Putz, 1995a). For example, while epiphytic *Ficus* had specific leaf areas two to three times greater than the conspecific tree form, in *Clusia* there was only a 30% difference in the same direction. Furthermore, *Clusia* showed no significant differences in stomatal density or leaf thickness (Holbrook & Putz, 1995a, Ball et al., 1991). We are not aware of any detailed study of leaf production or duration, but apparently all species of *Clusia* are evergreen.

13.3.3   To strangle or not to strangle?

The most commonly asked question regarding stranglers is: Do they really strangle their host? While death of the host tree due to the constriction of its stem may occur, separating this source of mortality from the deleterious effects of competition between strangler and host for light and soil resources remains problematic. Furthermore, strangulation could be either active or passive. The descending roots of stranglers have a circumferential cambium and thus are capable of inward growth; when the host tree has died and rotted away stranglers gradually reduce the size of the central cavity (Smith, 1956; personal observation). Passive strangulation occurs when the host tree's diameter growth is constrained by the unyielding cylinder of the strangler. Dobzhansky and Murca-Pires (1954) report that strangler figs growing on palms frequently kill their hosts and state that, as palms lack a vascular cambium for growth in diameter, active strangulation must occur. Competitive effects, however, cannot be ruled out. On the other hand, we have observed many palms which remained healthy and thriving even when nearly completely enveloped by a strangler and have only rarely observed instances in which the death of the palm might be attributed to the presence of a hemiepiphyte.

How stranglers kill their host retains an illicit charm, but perhaps the more biologically interesting question is why they do so. Given that hemiepiphytes go to a fair bit of trouble to capitalize on their position as structural parasites, why should they then outgrow it? Again there may be no single answer; strangling and non-strangling hemiepiphytes exist in both *Ficus* and *Clusia*. One potential issue is that of allometry. The conducting tissue needed to supply a large leaf area may require a substantial cross-sectional area of stem; a secondary consequence of this may be restriction of the host tree's growth. That vine stems support vast leaf areas in relation to their cross-sectional area suggests that this is not a necessary constraint, although hemiepiphytes may lack plasticity to alter their xylem structure. Competition for resources may also influence host-hemiepiphyte interactions. Hemiepiphyte and host compete directly for sunlight, water, and soil nutrients. Elimination of this competition might outweigh the strangler's biomass costs of producing enough wood to be self-supporting.

A final consideration, and the one that we feel may be most important, is that of longevity. Hemiepiphytes occur most often on older trees. Host age increases both tree size and, more importantly, the probability of an access point into a deep recess in the trunk. If canopy trees are replaced at an average rate of about 100 years (e.g., as occurs on BCI; Putz & Milton, 1982), and hemiepiphytes are constrained to succumb with their support, then the strangler's objective of "seizing a place in the sun" may be very short indeed. Furthermore, the host's demise may be hastened by competition for resources and the added mechanical load of the hemiepiphyte. Todzia (1986) notes that hemiepiphytic species that do not become mechanically self-sufficient appear to have shoot morphologies that reduce shading of the host, whereas those that become free-standing trees do not. Vines overcome their mechanical dependence by being tough enough to survive the fall of their supporting tree (Putz & Holbrook, 1991; Putz, 1984). Fallen branches of some *Clusia* species are able to take root and survive (Lawton & Putz, 1988), and we have observed fallen strangler figs in Florida and *Metrosideros* in Hawaii that have survived and grown back to the forest canopy. It is unlikely, however, that large *Ficus* individuals would be able to adapt readily to a complete severing of rooting connections with the soil. Biomass expended in becoming mechanically self-sufficient may be a small price to pay for a substantially increased lifespan.

13.3.4 Non-strangling hemiepiphytes

Little is known about the physiology and ecology of non-strangling hemiepiphytes. In general, non-strangling primary hemiepiphytes (ones that begin as epiphytes and later establish a root connection to the ground) appear more restricted to wet habitats than true stranglers. They also span the range from basically epiphytic plants with only a minor root connection to the ground to very large individuals having one or more very large connections with the forest soil. Many *Clusia* and *Ficus* species fall into the latter class, with these large roots (up to 20 cm diameter) descending parallel with the bole of the host tree and often attached to it by horizontal roots that clasp the host's trunk. It would be interesting to know the degree to which non-strangling hemiepiphytes rely on water and nutrients from roots in the canopy and in stem flow compared with those established in the forest soil. With only a single connection with the ground, non-strangling hemiepiphytes are susceptible to damage (e.g., fire; Putz & Susilo, 1995) and we suspect that they remain more epiphytic in their physiology than stranglers so as to survive temporary loss of connection with the soil.

Secondary hemiepiphytes germinate on the ground and ascend into the canopy using adhesive roots to attach to the sides of trees (Ray, 1990, 1992). As noted above, these plants are in many ways more vine-like in their growth form than primary hemiepiphytes, and their physiology may be better understood in relation to that of lianas. Nevertheless, because of the intermittent nature of their connection with the soil, we expect these plants to be as physiologically flexible as they are plastic in their shoot morphology (Ray, 1988, 1990).

Finally, we consider those plants that are true hemiepiphytes in that they lack roots in the ground for a portion of their life span, but are difficult to categorize in terms of growth form. For example, the tree fern *Cibotium glaucum* (Dicksoniaceae) is common in the understory of forests in Hawaii where it may begin life as an epiphyte and extend roots down to the ground (i.e., a primary hemiepiphyte) or grow up directly from the soil (C.P. Lund, personal communication). As roots are initiated just below the fronds and grow down along the outside of the central stem, however, tall individuals of *C. glaucum* may reach a point at which no living roots are actually in contact with the soil. Is this an example of a plant being a secondary hemiepiphyte upon itself? Clearly the physiological consequences of the tree fern growth form warrant greater investigation. A second hemiepiphyte that exhibits a diversity of growth forms is *Ficus crassiuscula*. Seedlings of this primary hemiepiphyte germinate on stumps or in the canopy, where they can form vine-thickets up to 5 m across consisting of many sprawling shoots (Daniels & Lawton, 1991). Eventually one shoot undergoes a transformation into an erect self-supporting form, which eventually may grow to form, a large canopy tree. What triggers this developmental change is not known. *F. crassiuscula* is a member of the subgenus *Pharmacosycea* and has thus evolved the strangling habit independently of the better known species of the subgenus *Urostigma*.

13.4 FUTURE DIRECTIONS

Vines and hemiepiphytes encompass a wide variety of growth forms and exhibit a high degree of morphological and physiological plasticity. Many of the consequences of this flexibility, however, have yet to be investigated. As techniques for investigation of phloem structure and function improve, we predict that studies of carbohydrate movement in vines will provide insight into characteristics that increase rates of phloem transport. This prediction is based on the simple observation that for plants with large leaf areas but necessarily small stem cross-sections, the phloem is subjected to the same constraints as the xylem. We suspect that major insights on vine nutrient and water relations will derive from studies of the architecture of vine root systems. Observations of extremely deep rooting by some vines and apogeotropic roots in others suggests that vines may display belowground versatility on par with their aboveground habits. The hemiepiphytic life history forms a natural experiment in which developmental and physiological responses to environmental parameters may be examined. Studies of photosynthetic metabolism in *Clusia* species have yielded important insights into the mechanisms underlying their remarkable ability to adjust patterns of CO_2 fixation in relation to environmental conditions (e.g., Zotz & Winter, 1993), while examination of *Ficus* species has demonstrated a high degree of developmental plasticity (e.g., Holbrook & Putz, 1995a). Further studies of a wide variety of hemiepiphytes, including non-strangling species, are needed to better understand the range of physiological behaviors associated with the transition between epiphytic and ground-rooted forms. As with vines, studies of hemiepiphytes should be extended into the rooting zone (both above- and belowground). Finally, there is an overwhelming need for long-term studies involving both intensive physiological observations and experimental manipulations (e.g., Zotz & Winter, 1994; Laman, 1994; Putz, Romano and Holbrook, 1995).

ACKNOWLEDGEMENTS

We would like to thank C. Malmström, R. O. Lawton, C. P. Lund, and the reviewers for helpful comments. We thank G. Zotz for generously providing access to work in press.

REFERENCES

BALL, E., HANN, J., KLÜGE, M., LEE, H. S. J., LÜTTGE, U., ORTHEN, B., POPP, M., SCHMITT, A. K. & TING, I. P. (1991) Ecophysiological comportment of the tropical CAM-tree *Clusia* in the field. I. Growth of *Clusia rosea* Jacq. on St. John, US Virgin Islands, Lesser Antilles. *New Phytologist,* **117,** 473–481.

BENZING, D. (1990) *Vascular Epiphytes.* Cambridge University Press, Cambridge.

BESSEY, E. A. (1908) The Florida strangling figs. *Annual Report of the Missouri Botanical Garden,* **19,** 25–34.

BORLAND, A. M., GRIFFITHS, H., MAXWELL, C., BROADMEADOW, M. S. J., GRIFFITHS, N. M., & BARNES, J. D. (1992) On the ecophysiology of the Clusiaceae in Trinidad: Expression of CAM in *Clusia minor* L. during the transition from wet to dry season and characterization of three endemic species. *New Phytologist,* **122,** 349–357.

CARLQUIST, S. (1988) *Comparative Wood Anatomy: Systematic, Ecological and Evolutionary Aspects of Dicotyledon Wood.* Springer-Verlag, Berlin.

CARLQUIST, S. (1991) Anatomy of vine and liana stem: A review and synthesis. *The Biology of Vines* (eds. F. E. PUTZ & H. A. MOONEY) Cambridge University Press, Cambridge.

CASTELLANOS, A. E. (1991) Photosynthesis and gas exchange in vines. *The Biology of Vines* (eds. F. E. PUTZ & H. A. MOONEY) Cambridge University Press, Cambridge.

DANIELS, J. D. & LAWTON, R. O. (1991) Habitat and host preferences of *Ficus crassiuscula*: A neotropical strangling fig of the lower-montane rainforest. *Journal of Ecology,* **79,** 129–142.

DARWIN, C. (1867) The movements and habits of climbing plants. *Journal of the Linnean Society,* **9,** 1–118.

DOBBINS, D. R. (1981) Anomalous secondary growth in lianas of the Bignoniaceae is correlated with the vascular pattern. *American Journal of Botany,* **68,** 142–144.

DOBBINS, D. R. & FISHER, J. B. (1986) Wound response in girdled stems of lianas. *Botanical Gazette* **147,** 278–289.

DOBZHANSKY, T. & MURCA-PIRES, J. (1954) Strangler trees. *Scientific American,* **190,** 78–80.

EWERS, F. W., FISHER, J. B., & FICHTNER, K. (1991) Water flux and xylem structure in vines. *The Biology of Vines* (eds. F. E. PUTZ & H. A. MOONEY) Cambridge University Press, Cambridge.

FISHER, J. B. & EWERS, J. B. (1991) Structural responses to stem injury in vines. *The Biology of Vines* (eds. F. E. PUTZ & H. A. MOONEY) Cambridge University Press, Cambridge.

FRANCO, A. C., BALL, E. & LÜTTGE, U. (1990) Patterns of gas exchange and organic acid oscillations in tropical trees of the genus *Clusia. Oecologia,* **85,** 108–114.

FRANCO, A. C., BALL, E. & LÜTTGE, U. (1991) Influence of nitrogen, light, and water stress on CO_2 exchange and organic acid accumulation in the tropical C_3-CAM tree, *Clusia minor. Journal of Experimental Botany,* **42,** 597–603.

FRANCO, A. C., BALL, E. & LÜTTGE, U. (1992) Differential effects of drought and light levels on accumulation of citric and malic acids during CAM in *Clusia. Plant, Cell and Environment,* **15,** 821–829.

GALIL, J. & MEIRI, L. (1981) Druplet germination in *Ficus religiosa* L. *Israel Journal of Botany,* **30,** 41–47.

GARTNER, B. L. (1991) Relative growth rates of vines and shrubs of western poison oak, *Toxicodendron diversilobum* (Anacardiaceae). *American Journal of Botany,* **78,** 1345–1353.

GARTNER, B. L., BULLOCK, S. H., MOONEY, H. A., BROWN, V. B., & WHITBECK, J. L. (1990) Water transport properties of vine and tree stems in a tropical deciduous forest. *American Journal of Botany,* **77,** 742–749.

GENTRY, A. H. (1991) The distribution and evolution of climbing plants. *The Biology of Vines* (eds. F. E. PUTZ & H. A. MOONEY) Cambridge University Press, Cambridge.

GENTRY, A. H. & DODSON, C. H. (1987) Contribution of non-trees to species richness of a tropical rain forest. *Biotropica*, **19**, 149–156.

GIVNISH, T. J. & VERMEIJ, G. J. (1976) Sizes and shapes of liana leaves. *American Naturalist*, **110**, 131–160.

HAAG-KERWER, A., FRANCO, A. C. & LÜTTGE, U. (1992) The effect of temperature and light on gas exchange and acid accumulation in the C_3-CAM plant *Clusia minor* L. *Journal of Experimental Botany*, **43**, 345–352.

HEGARTY, E. E (1990) Leaf life-span and leafing phenology of lianas and associated trees during a rainforest succession. *Journal of Ecology*, **78**, 300–312.

HEGARTY, M. P., HEGARTY, E. E., & GENTRY, A. H. (1991) Secondary compounds in vines with an emphasis on those with defensive function. *The Biology of Vines* (eds. F. E. PUTZ & H. A. MOONEY) Cambridge University Press, Cambridge.

HOLBROOK, N. M. (1995) Stem water storage. *Plant Stems: Physiology and Functional Morphology* (ed. B. L. GARTNER) Academic Press, San Diego, pp. 151–174.

HOLBROOK, N. M.. & PUTZ, F. E. (1995a) From epiphyte to tree: Differences in leaf structure and water relations associated with the transition in growth form. *Plant, Cell and Environment,* in press.

HOLBROOK, N. M. & PUTZ, F. E. (1995b) Water relations of hemiepiphytes: Field measurements of two species of strangler figs growing in a Venezuelan palm savanna. *Oecologia* (submitted).

JACKSON, P. C., CAVELIER, J., GOLDSTEIN, G., MEINZER, F. C., & HOLBROOK, N. M. (1995) Partitioning of water resources among plants of a lowland tropical forest. *Oecologia* **101**, 197–203.

KELLY, D. L. (1985) Epiphytes and climbers of a Jamaican rainforest: Vertical distribution, life forms and life histories. *Journal of Biogeography*, **12**, 223–241.

KJELLBERG, F. & MAURICE, S. (1989) Seasonality in the reproductive phenology of *Ficus*: Its evolution and consequences. *Experientia*, **45**, 653–660.

LAMAN, T. G. (1993) Seedling establishment of the hemiepiphyte *Ficus stupenda* in the Bornean rainforest canopy. *Bulletin of the Ecological Society of America*, **74** (**Suppl.**), 321.

LAMAN, T. G. (1994) The ecology of strangler figs (hemiepiphytic *Ficus* spp.) in the rainforest canopy of Borneo. Ph.D. dissertation. Harvard University.

LAMBERT, F. R. & MARSHALL, A. G. (1991) Keystone characteristics of bird-dispersed *Ficus* in an Malaysian lowland rainforest. *Journal of Ecology*, **79**, 793–809.

LAWTON, R. O. & PUTZ, F. E. (1988) Natural disturbance and gap-phase regeneration in a wind-exposed tropical cloud forest. *Ecology,* **69,** 764–777.

LEE, D. W. & RICHARDS, J. (1991) Heteroblastic development in vines. *The Biology of Vines* (eds. F. E. PUTZ & H. A. MOONEY) Cambridge University Press, Cambridge, pp 205–243.

LEE, H. S. J., SCHMITT, A. K. & LÜTTGE, U. (1989) The response of the C_3-CAM tree, *Clusia rosea,* to light and water stress. II. Internal CO_2 concentration and water use efficiency. *Journal of Experimental Botany,* **211,** 171–179.

LEE, K. E. & WOOD, T. G. (1971) *Termites and Soils.* Academic Press, New York.

LONGINO, J. T. (1986). A negative correlation between growth and rainfall in a tropical liana. *Biotropica,* **18,** 195–200.

MADISON, M. (1977) Vascular epiphytes: Their systematic occurrence and salient features. *Selbyana,* **2,** 1–13.

MOONEY, H. A. & GARTNER, B. L. (1991) Resource economy of vines. *The Biology of Vines* (eds. F. E. PUTZ & H. A. MOONEY) Cambridge University Press, Cambridge.

NADKARNI, N. M. (1984) Epiphyte biomass and nutrient capital of a neotropical elfin forest. *Biotropica,* **16,** 249–256.

OBERBAUER, S. F., CLARK, D. A., CLARK, D. B., & QUESADA, M. (1989) Comparative analysis of photosynthetic light environments within the crown of juvenile rainforest trees. *Tree Physiology,* **5,** 13–23.

OPLER, P. A., BAKER, H. G. & FRANKIE, G. W. (1991) Seasonality of climbers: A review and example from Costa Rican dry forest. *The Biology of Vines* (eds. F. E. PUTZ & H. A. MOONEY) Cambridge University Press, Cambridge.

OSMOND, C. B. (1978) Crassulacean Acid Metabolism: A curiosity in context. *Annual Review of Plant Physiology,* **29,** 379–414.

PEÑALOSA, J. (1983) Shoot dynamics and adaptive morphology of *Ipomoea phillomega* (Vell.) House (Convolvulaceae) a tropical rainforest liana. *Annals of Botany,* **52,** 737–754.

PEÑALOSA, J. (1984) Basal branching and vegetative spread in two tropical rainforest lianas. *Biotropica,* **16,** 1–9.

POPP, M., KRAMER, D., LEE, H., DIAZ, M., ZIEGLER, H. & LÜTTGE, U. (1987) Crassulacean acid metabolism in tropical dicotyledonous trees of the genus *Clusia. Trees,* **1,** 238–247.

PUTZ, F. E. (1983) Liana biomass and leaf area of a "tierra firme" forest in the Rio Negro Basin, Venezuela. *Biotropica,* **15,** 185–189.

PUTZ, F. E. (1984) The natural history of lianas on Barro Colorado Island, Panama. *Ecology,* **65,** 1713–1724.

PUTZ, F. E. (1990) Liana stem diameter growth and mortality rates on Barro Colorado Island, Panama. *Biotropica,* **22,** 103–104.

PUTZ, F. E. & CHAI, P. (1987) Ecological studies of lianas in Lambir National Park, Sarawak, Malaysia. *Journal of Ecology,* **75,** 523–531.

PUTZ, F. E. & HOLBROOK, N. M. (1986) Notes on the natural history of hemiepiphytes. *Selbyana,* **9,** 61–69.

PUTZ, F. E. & HOLBROOK, N. M. (1989) Strangler fig rooting habits and nutrient relations in the llanos of Venezuela. *American Journal of Botany,* **76,** 781–788.

PUTZ, F. E. & HOLBROOK, N. M. (1991) Biomechanical studies of vines. *The Biology of Vines* (eds. F. E. PUTZ & H. A. MOONEY) Cambridge University Press, Cambridge.

PUTZ, F. E. & MILTON, K. (1982) Tree mortality rates on Barro Colorado Island. *The Ecology of a Tropical Forest: Seasonal Rhythms and Long Term Changes* (eds. E. G. LEIGH, Jr., A. S. RAND, & D. M. WINDSOR) Smithsonian Institution Press, Washington, D. C.

PUTZ, F. E. & SUSILO, A. (1994) Figures and fire. *Biotropica,* **26,** 468–469.

PUTZ, F. E., ROMANO, G. B. & HOLBROOK, N. M. (1995) Phenology of epiphytic and tree-phase strangler figs in a Venezuelan palm savanna. *Biotropica,* **27,** 183-189.

PUTZ, F. E. & WINDSOR, D. M. (1987) Liana phenology on Barro Colorado Island, Panama. *Biotropica,* **19,** 334–341.

RACIBORSKI, M. (1900) Über die Vorläuferspitze. *Flora,* **87,** 1–25.

RADA, F., & JAIMEZ, R. (1992). Comparative ecophysiology and anatomy of terrestrial and epiphytic *Anthurium bredemyeri* Schott in a tropical Andean cloud forest. *Journal of Experimental Botany,* **43,** 723–727.

RAY, T. S. (1988) Survey of shoot organization in the Araceae. *American Journal of Botany,* **75,** 56–84.

RAY, T. S. (1990) Metamorphosis in Araceae. *American Journal of Botany,* **77,** 1599–1609.

RAY, T. S. (1992) Foraging behavior in tropical herbaceous climbers (Araceae). *Journal of Ecology,* **80,** 189–203.

RICHARDS, P. W. (1952) *The Tropical Rain Forest.* Cambridge University Press, Cambridge.

RUNDEL, P. W. & FRANKLIN, T. (1991) Vines in arid and semi-arid ecosystems. *The Biology of Vines* (eds. F. E. PUTZ & H. A. MOONEY) Cambridge University Press, Cambridge.

SCHENCK, H. (1893) Beiträge zur Biologie und Anatomie der Lianen, im besonderen der in Brasilien einheimischen Arten, 2. Beiträge zur Anatomie der Lianen. *Botanische Mitteilungen aus den Tropen 5* (ed. A. F. W. SCHIMPER) G. Fischer, Jena.

SCHMITT, A. K., LEE, H. S. L. & LÜTTGE, U. (1988) The response of the C_3–CAM tree, *Clusia rosea,* to light and water stress. I. Gas exchange characteristics. *Journal of Experimental Botany,* **39,** 1581–1590.

SMITH, D. (1956) Untitled. *Principes,* **1,** 3.

SPERRY, J. S., HOLBROOK, N. M, ZIMMERMANN, M. H., & TYREE, M. T. (1987) Spring filling of xylem vessels in wild grapevines. *Plant Physiology*, **83**, 414–417.

STERNBERG, L. da S. L., TING, I. P., PRICE, D. & HANN, J. (1987) Photosynthesis in epiphytic and rooted *Clusia rosea* Jacq. *Oecologia*, **72**, 459–460.

SWAGEL, E. N. & ELLMORE, G. S. (1992) Germination constraints imposed by substrate water potential upon achenes of the strangling figure *Ficus aurea*. *American Journal of Botany*, **79**, 185.

TERAMURA, A. H., GOLD, W. G. & FORSETH, I. N. (1991) Physiological ecology of mesic, temperate woody vines. *The Biology of Vines* (ed. F. E. PUTZ & H. A. MOONEY) Cambridge University Press, Cambridge.

TING, I. P., LORD, E. M., STERNBERG, L. da S. L., & DENIRO, M. J. (1985) Crassulacean acid metabolism in the strangler *Clusia rosea* Jacq. *Science*, **229**, 969–971.

TING, I. P., HANN, J., HOLBROOK, N. M., PUTZ, F. E., STERNBERG, L. da S. L., PRICE, D., & GOLDSTEIN, G. (1987) Photosynthesis in hemiepiphytic species of *Clusia* and *Ficus*. *Oecologia*, **74**, 339–346

TINOCO-OJANGUREN, C. & VAZQUEZ-YANES, C. (1983) Especies CAM en la selva humeda tropical de Los Tuxtlas, Veracruz. *Boletin de la Sociedad Botanico de Mexico*, **45**, 150–153.

TITUS, J. H., HOLBROOK, N. M. & PUTZ, F. E. (1990) Seed germination and seedling distribution of *Ficus pertusa* and *F. tuerckheimii*: Are strangler figs autotoxic? *Biotropica*, **22**, 425–428.

TODZIA, C. (1986) Growth habits, host tree species, and frequency of hemiepiphytes on Barro Colorado Island, Panama. *Biotropica*, **18**, 22–27.-

VANDERDYST, R. P. H. (1922) Nouvelle contribution a l'etude de *Ficus* epiphitiques su l'*Elaeis*. *Revue de Zoologie et de Botanique Africaines*, **10**, 65–74.

WILLIAMS-LINERA, G. & LAWTON, R. D. (1995) The ecology of hemiepiphytes in forest canopies. *Forest Canopies* (eds. M. D. LOWMAN & N. M. NADKARNI) Acdemic Press, Orlando.

WILLIS, J. C. (1973) *A Dictionary of the Flowering Plants and Ferns*. Cambridge University Press, Cambridge.

WINTER, K., ZOTZ, G., BAUR, B. & DIETZ, K-J. (1992) Light and dark CO_2 fixation in *Clusia uvitana* and the effects of plant water status and CO_2 availability. *Oecologia*, **91**, 47–51.

ZOTZ, G., TYREE, M. T., & COCHARD, H. (1994). Hydraulic architecture, water relations, and vulnerability to cavitation of *Clusia* uvitana Pittier: A C_3-CAM tropical hemiepiphyte. *New Phytologist*, **127**, 287–295.

ZOTZ, G. & WINTER, K. (1994) A one-year study on carbon, water, and nutrient relationships in a tropical C_3-CAM hemiepiphyte, *Clusia uvitana*. New Phytologist, **127**, 45–60.

ZOTZ, G. & WINTER, K. (1993) Short-term regulation of Crassulacean acid metabolism activity in a tropical hemiepiphyte, *Clusia uvitana*. *Plant Physiology*, **102**, 835–841.

III

Ecophysiological Patterns across Tropical Forest Communities

INTRODUCTION

Ecophysiological tolerances of species interact with environmental factors, such as climate and soil properties to influence which species occur at a given location. Assemblages of species occupying discrete climatic and edaphic regimes are recognized as communities. The tremendous diversity of species within tropical forests has made it challenging to define the relative importance of biotic vs. abiotic factors in determining the composition of communities. Moreover, seasonal variation at a particular location can determine local differences among communities, such as the difference between tropical dry forest and wet forest, and genotypic differences among the individuals of the same species living in these communities. Chapters in this section explore how tropical forest communities and members of these communities respond to large-scale environmental variation. Specific topics include climatic altitudinal gradients, seasonal variation in rainfall, the effects of salinity in mangrove forests, and ecotypic differentiation of species across communities.

Tropical montane forests are instantly recognizable as different from lowland forest in many respects. Four types of rainforest are widely recognized based on altitudinal variation in rainfall, cloud cover and available light, mist, soil leaching, CO_2 partial pressure, and seasonal and diurnal temperature regimes. Cavelier in Chapter 14 summarizes the major differences among these communities, and how specific environmental factors dominate the physiognomy of each. This chapter explores how leaf morphology and function vary as a function of available light, temperature, and soil nutrient content. In montane forests, high rates of soil leaching and slow litter decomposition are often associated with slow growth and short stature. This chapter details the importance of soil nutrient relations for the characteristics of these communities over altitudinal gradients.

Seasonality of rainfall and light are tightly correlated with leaf, flowering, and fruiting phenology in tropical forests. Many species produce new leaves at the beginning of the rainy season, when light may be limiting due to cloud cover, but water is abundant. Some species are leafless during the dry season, while many flower during the early dry season. Wright (Chapter 15) reviews the seasonality of tropical forests and identifies environmental factors that affect the timing of leaf production and flowering. Proximate cues and ultimate selective factors that may have caused patterns of phenology are discussed. Correlations between radial growth of trees and irradiance suggest that trees may be light limited during the rainy season. Wright examines the predictions that (1) plants without dry-season water supplies must produce leaves and flowers in the wet season, and (2) when water is available, selection will favor leaf and flower production in the dry season when sunlight is high and herbivory low.

Mangrove forests are unique among tropical forests because these communities experience tidal inundation with sea water. Vines, palms, ferns and epiphytes that are so characteristic of other lowland tropical forests are poorly represented in mangrove forests. Ball (Chapter 16) describes how the physiological characteristics of plants in mangrove forests differ from rainforest species because of edaphic conditions. Tidal inundation strongly influences the availability of soil nutrients, while subjecting trees to varying degrees of waterlogging. Maintaining favorable water relations in a saline soil requires that mangroves maintain strongly negative xylem water potentials. Ball

explores the notion that conservative water use has evolved as an adaptive response to salinity. It is clear that there are large differences among mangrove forests, depending on the frequency of inundation.

Intraspecific ecotypic differentiation can account for the presence of the same species in communities with contrasting environments. As discussed by Hogan (Chapter 17), the population structure of tropical plant species suggests that genetic differentiation of ecotypes should be rare. However, he shows that differentiation among populations has been found whenever it has been looked for in tropical species. Overall, there is very little known about the relative importance of ecotypic differentiation as opposed to phenotypic plasticity as determinants of the distribution of plant species across tropical forest communities. Chapter 17 emphasizes the importance of understanding this phenomenon to provide a solid basis for conservation, restoration, and plant responses to climate change.

14

Environmental Factors and Ecophysiological Processes along Altitudinal Gradients in Wet Tropical Mountains

Jaime Cavelier

14.1 INTRODUCTION

Forests of tropical mountains occupy a wide range of environmental conditions as determined by altitudinal variation in temperature, water, light, and soils. Janzen (1967) noted that tropical lowland climates have relatively little overlap in temperature variation with tropical montane climates. As such, these steep altitudinal gradients offer a unique evolutionary laboratory for the study of how plant characters vary in response to environmental variables. In this chapter, I describe how environmental factors and leaf-level ecophysiological processes vary along altitudinal gradients in wet tropical mountains, and discuss the environmental factors that control the distribution of forests types. I begin with a description of the effect of altitude on air and soil temperature, light quality and quantity, water supplies, decomposition rates, and nutrient availability. Information is provided for small and large mountains around the world, with emphasis on the Andes, the Caribbean, and Southeast Asia. Some information is provided for the African mountains. Next, I review elevational trends in leaf morphology, anatomy, and ecophysiology, with special emphasis on canopy trees. This includes analyses of the reduction in leaf size with increasing altitude, the adaptive significance of leaf thickness in pachyphylls, and the relationships between leaf structure and nutrient availability. Water availability is generally high in these wet forests and plant water relations reflect this. Stomatal responses to environmental factors and daily variation of leaf water potentials are discussed. Over altitudinal gradients in wet tropical forests there is considerable variation in the opportunity for carbon gain. Accordingly, I review what is known of photosynthesis and the relationship between leaf nitrogen content and photosynthetic performance in these forests. Finally, I address the "Massenerhebung effect" and potential edaphic, climatic and biological factors that may control the distribution of forest types on both small and large tropical mountains.

14.2 ALTITUDINAL VARIATION IN CLIMATIC AND EDAPHIC FACTORS

14.2.1 Soil and air temperatures

With increasing altitude in tropical mountains, there is a decrease in the mean soil temperature and air temperature (Figure 14.1). The soil temperature gradient varies with the size of the mountain range, with steeper gradients for small mountains (Grubb, 1971; Grubb & Whitmore, 1966). Mean annual air temperature decreases with altitude due to a drop in atmospheric pressure and air density with increasing elevation (Barry, 1981). At the Sierra Nevada de Santa Marta (SNSM) along the Caribbean coast of Colombia, the mean air temperature gradient is 0.60°C/100 m (van der Hammen, 1984a). In the middle part of the gradient (1300-3300 m), stabilized soil temperatures (at depths of 50 to 100 cm) and mean air annual temperatures are the same (Figure 14.1). This altitudinal range corresponds to areas of high rainfall and cloud cover. In contrast, soil temperature between 500 m and 1300 m in altitude is 0.4–1.0°C lower than air temperatures, while between 3300 m and 4100 m, soil temperature is 1.6-2.6°C higher than air temperature. At the SNSM, these altitudes roughly correspond to the limits between lowland rainforest and lower montane rainforest (1000–1200 m) and the limits between upper

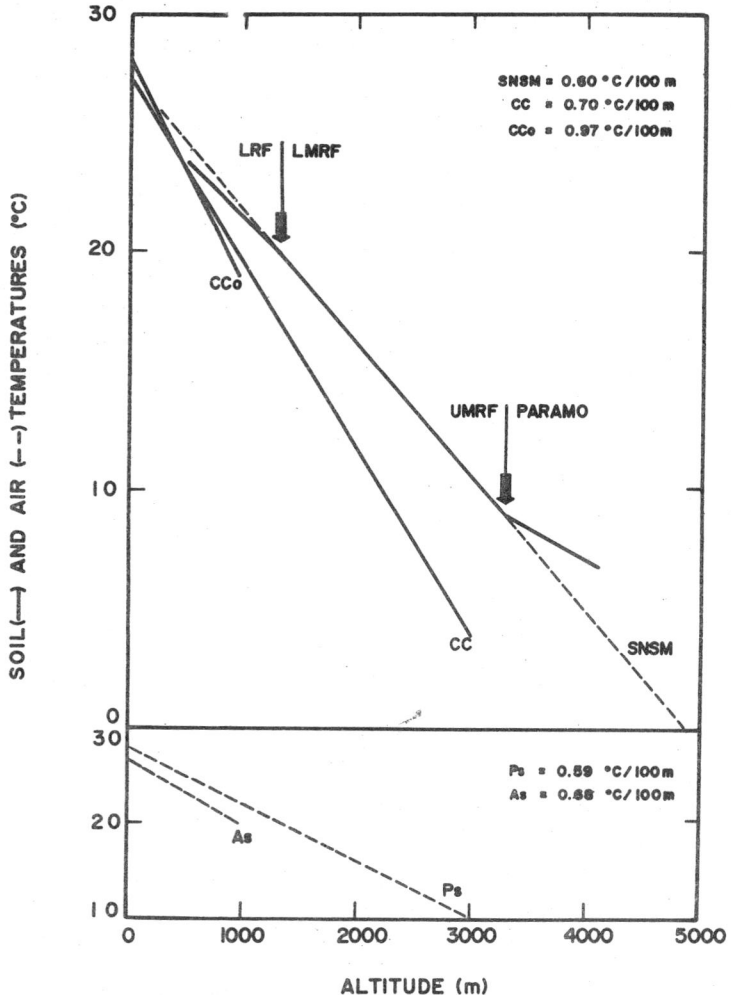

*Figure 14.1. Relationship between stabilized soil temperature and altitude
(—), and between mean air annual temperature and altitude (---), for small
and large tropical mountains; Cerro Copey (CCo), Margarita Island,
Venezuela (Sugden, 1986), Cordillera de la Costa (CC), Dominican Republic
(Schubert & Medina, 1982), Sierra Nevada de Santa Marta (SNSM),
Colombia (van der Hammen, 1984a) and Pacific (Ps) and Atlantic side (As),
Costa Rica (Holdridge et al. 1971). The air temperature gradient at SNSM
was plotted knowing the mean annual temperature of 27°C at sea level and
0°C at 4850 m. The limits between Lowland Rainforest (LRF) and Lower
Montane Rainforest (LMRF), and between UMRF and paramo vegetation are
also marked.*

montane rain forest and paramo vegetation (3300 m), suggesting a temperature-water control for the distribution of forest types in large wet tropical mountains.

The decrease in mean annual air temperature with increasing altitude also varies with the rainfall regime of the slope, with lower temperatures for wetter slopes (Figure 14.1). For instance, the Atlantic watershed (wetter) in Costa Rica runs about 2°C cooler than the Pacific side; that is, equivalent temperatures occur approximately 330 m lower in elevation on the Atlantic side (Holdridge et al., 1971). These differences in air temperature, as well as in rainfall and wind exposure between slopes, may help explain the distribution of montane forests on the windward and leeward side of tropical mountain ranges (Beard, 1949).

14.2.2 Altitudinal changes of light quality and quantity

Because of cloud cover, the amount of sunlight reaching montane forests can be quite small. Solar radiation increases exponentially with increasing altitude, both in an ideal (pure, dry) atmosphere, and as observed at mountain stations in the Alps (Barry, 1981; Körner, Allison & Hilscher, 1983). Nevertheless, a decline in the radiation levels with increasing altitude, reported for the mountains of New Guinea, shows the effect of cloud cover in montane environments (Körner, Allison & Hilscher, 1983). Indeed, spot measurements in tropical mountains show that cloud cover reduces radiation flux density (RFD) in comparison to that received in the lowlands. The attenuation ranges from 16% in the Blue Mountains of Jamaica (Aylett, 1985), to 40% in the elfin cloud forest of Pico del Oeste in Puerto Rico (Baynton, 1968), to 53% in the elfin cloud forest of Serranía de Macuira (Cavelier & Mejia, 1990). Research is needed to measure the annual reduction in RFD caused by cloud cover, particularly under the bases of cumulus clouds where attenuation of radiation seems to be greatest and where the elevational limit of montane forests occurs.

Along altitudinal gradients, radiation at the ground level under forest canopy increases from 0.05–0.2% (500–1600m) to 0.4–1.0% (3000–3400 m) of the total radiation arriving above the canopy regardless of cloud cover (van der Hammen & van Reenen, 1984). These increments in the radiation at the ground level are correlated with a reduction of the total coverage of trees (i.e., lower leaf area index) and with changes in canopy structure along the altitudinal gradient. More open canopies in the mountains, resulting partly from steep slopes, would result in understory plants receiving longer sunflecks during cloud-free periods (usually during the morning) and diffuse light during foggy periods (usually during the afternoon). More open canopies would mean that the red/far red shift in spectral composition due to filtering by leaves should be less in montane forests than in lowland forests. More work is needed on the characterization of the light regime in the understory of montane forests (e.g., Grubb & Whitmore, 1967) and the potential effects on seed germination and seedling growth (Samper, 1992).

14.2.3 Water supplies

In temperate mountains, precipitation generally increases with altitude, while in tropical mountains, rainfall increases from the base of the mountain to a certain altitude and then decreases (Hastenrath, 1967, 1991; Lauer, 1975). Less often, rainfall decreases with altitude, with a maximum at the base (Lauer, 1975). Rarely, there is a second peak of rainfall in the upper part of the mountains. The altitude of maximum rainfall raises with decreasing annual total, from 1500 m on the wet eastern slopes of the Colombian Andes, to 2500 m along the dry Northern Rift in Ethiopia (Lauer, 1975). This sets the conditions for the occurrence of montane rainforests at different elevations on wet than on dry mountains. Rainfall at different altitudes is correlated with the presence of "cloud belts" located at 1000–1400 m (the lower limit of montane forests) and at 3600–4000 (the upper limit of montane forests) for the northern Andes (Cleef, 1981; Lauer, 1976).

Although rainfall and fog interception are the principal sources of water for montane forests, there is little information on the relative dependence of the forests on these two water sources (Baynton, 1969; Cavelier & Goldstein, 1989a, Ekern, 1964; Juvik & Ekern, 1978; Vogelmann, 1973). In the elfin cloud forest of Serrania de Macuira, Colombia, fog interception represents 48% of the annual water sources (Cavelier & Goldstein, 1989a). Similar results have been obtained in Hawaii, where fog-drip amounts to 750 mm (50% of annual rainfall) between 1500 and 2500 m on the windward slope of Mauna Loa (Juvik & Ekern, 1978). Fog is not an important direct water source in all cloud forests. For instance, in the elfin cloud forest of El Zumbador at 3100 m in the Andes of Merida, Venezuela, fog interception accounts for only 3.5% of the water inputs (Cavelier & Goldstein, 1989a). The importance of fog as a water source in tropical montane rainforest diminishes as rainfall increases. In very rainy montane forests, clouds maintain high humidity of the air and high soil water content of the soil, reducing the effect of seasonal variations in rainfall (Medina, 1986; Sugden, 1986). Water can also enter the forest in the form of hail or dew, but these two forms of water have been poorly documented (Veneklaas, 1991).

14.2.4   Decomposition of organic matter and nutrient availability

The reduction of air and soil temperature and the concomitant increase in humidity and soil water content with altitude are partly responsible for low rates of decomposition of organic matter in high tropical mountains (Grubb, 1977). Slower rates of organic matter decomposition are reflected in the higher carbon content and the C/N ratios of the soil (Swift, Heal & Anderson, 1979), as shown for the mountains of Colombia (Alexander & Pichott, 1979; Jenny, Binghsm & Padilla-Saravia, 1948) and Thailand (Yoda & Kira, 1969). Altitudinal gradients in Hawaiian mountains exhibit a reversed trend, with total carbon decreasing with increasing elevation on a young lava flow (Vitousek, Matson & Turner, 1988). The reduction in the rates of disappearance of litter with increasing altitude is also the result of low "substrate quality" of the decomposing litter. Litter with low N and P concentrations (Enriquez, Duarte & Sand-Jensen, 1993) and high concentrations of carbon-based compounds such as lignin, cellulose, hemicellulose (Anderson & Swift, 1983), and other cell wall materials (Niemann et al., 1992) is harder for microorganisms to utilize. Besides a general decline in the concentration of N per unit of leaf mass along altitudinal gradients (Grubb, 1977; Heaney & Proctor, 1989; Tanner, 1977a; Vitousek, 1982, 1984), and higher concentrations of lignin, cellulose, and fibre acid detergent in montane forest litter (Cavelier, 1989), little information has been gathered on the quality of leaf litter (see also Bruijnzeel et al., 1993).

   Breakdown of organic matter in montane rainforest soils, particularly on mor (C/N > 25) and mull soils (C/N < 15), may be inhibited by phenolic compounds (Coulson, Davis & Lewis, 1960; Davis, 1971). Phenolics are usually synthesized under low nitrogen availability (Horner, Gosz & Cates, 1988), and their concentration is negatively correlated with the total nitrogen concentration in leaf tissues (Bruijnzeel et al., 1993). Experimental evidence with grand fir (*Abies grandis*), a temperate tree species, showed that a significant decrease in production of total phenolics can be induced with nitrogen fertilization (Muzika & Pregitzer, 1992). Thus, when nitrogen is in short supply in the soils of montane forests (Heaney & Proctor, 1989; Marrs et al., 1988), foliar polyphenols are synthesized, causing a positive feedback system during the decomposition of litter, reducing N availability (Bruijnzeel et al., 1993). In addition, decomposition of leaf litter can be reduced in montane forest soils by anoxia in waterlogged soils such as peat (Whitmore, 1984) and low pH in "mor" soils (Grubb & Tanner, 1976; Tanner 1977b).

14.3 CHANGES IN LEAF MORPHOLOGY AND FUNCTION WITH CHANGES IN ALTITUDE

14.3.1 Leaf Size

In spite of the well documented decline of leaf size with increasing altitude in tropical mountains (Cleef et al., 1984; Cuatrecasas, 1934; Grubb et al., 1963; Körner, Allison & Hilscher, 1983; Leigh, 1975; Sugden, 1985; Tanner & Kapos, 1982; Vareschi, 1966), there is little agreement on the mechanisms underlying this variation. Several authors have suggested that air temperature is the cause of decreasing leaf size with increasing altitude (Cuatrecasas, 1958; Körner, Allison & Hilscher, 1983; Leigh, 1975; Parkhurst & Loucks, 1972; Pittier, 1935; Richards, 1954). Experimental evidence suggests that low temperatures reduce cell growth (Nobel, 1980) and that temperatures at night (i.e. minimum temperatures) play a major role in determining leaf growth rates in grasses (Busso & Richards, 1992). Further experimental work is needed to understand the role of air temperature in determining leaf size along altitudinal gradients (see also Woodward, Körner & Crabtree, 1986).

Besides temperature, four other environmental factors may be responsible for the reduction in leaf size with increasing altitude. (1) Although soil water seems to be readily available thought the year (Kapos & Tanner, 1985), the hydraulic conductance of stems may not be high enough to compensate for the high water losses that occur during sunny periods in montane forests. A water supply limited by the hydraulic conductance of the stems would impose a reduction in leaf size and leaf area index. Slow growth of montane trees (Tanner, 1977a) as a result of a limited mineral supply is correlated with lignified, short, and narrow vessels. These characteristics could result in a reduction of hydraulic conductance (Carlquist, 1975; Tyree & Ewers, Chapter 8). If the hydraulic conductance of species with high wood density (i.e. slow growing species) is lower than that of fast growing species, the reduction in leaf size in montane rainforest trees could be interpreted as a byproduct of mineral shortage. It is interesting to observe that there are fast growing secondary trees in montane environments with large leaves (i.e., several Solanaceae) rooted in the same soil as slow growing trees of the primary forest with characteristic small leaves and low leaf area index. (2) The frequency of frosts increases with height (Troll, 1968), and small leaves may suffer less frost damage than broad leaves (Parkhurst & Loucks, 1972). (3) Large canopy leaves at high altitude would require larger amounts of nitrogen and other nutrients that may not be available. Indeed, limitations in N availability may reduce leaf expansion, resulting in small leaves (Motta & Medina, 1978). Larger leaves would mean an increase in leaf area index (LAI) above a value that cannot be economically sustained with the given radiation and nutrient availability at higher altitudes (Marrs et al., 1988). (4) Windy environments of elfin cloud forests (Lawton, 1982) may select for small leaves which are not easily removed by wind (Parkhurst & Loucks, 1972). In general, however, montane rainforests seem not to be particularly windy (Grubb & Whitmore, 1966), and this cannot be a general explanation of the reduction of leaf size with increasing altitude.

14.3.2   Leaf size and anatomy in the strata of the forest

Leaf size changes along the vertical gradient (i.e., vegetation layers) in montane rainforests as well as in lowland rainforests (Richards, 1954). Givnish and Vermeij (1976) explain this variation in leaf size along the vertical gradient in the cloud forest of Rancho Grande, Venezuela (Huber, 1986) in terms of the balance between the photosynthetic gain of the whole plant and the energetic requirements to provide water for transpiration (Medina, 1986). Variation in leaf size as a function of foliage height was also analyzed by Parkhurst and Loucks (1972) using an ecophysiological model that assumes that leaf size tends to optimize the efficiency of water use (WUE). The model predicts smaller leaves for the higher strata as a result of high temperature and high radiation, and larger leaves in the lower strata due to high temperature and low radiation. The model does not predict relatively large leaves in the understory of montane forests where radiation and air temperatures are relatively low. In any case, it seems unlikely that water-use efficiency plays any role in the selection of leaf characteristics in an environment where water seems readily available throughout the year.

Among vertical strata in tropical forests, light and humidity seem to have a strong influence not only on leaf size, but on the anatomical structure of the leaves (Roth, 1990). Beadle (1954) and Loveless (1961, 1962) have proposed that sclerophyllous leaves are advantageous under conditions of nutrient deficiency. Typically, canopy leaves are more sclerophyllous than understory leaves, but it seems unlikely that this is related to nutrient availability because both canopy and understory plants are growing in the same soil. It seems more likely that the decrease in leaf area per mass (specific leaf area, SLA) along the vertical gradient from the forest floor to the canopy is caused by changes in humidity and/or light (Mott, Gibson & O'Leary, 1982; Roth, 1990).

14.3.3 Leaf thickness: The role of dry matter, air and water

The term "pachyphyll" was introduced by Grubb (1974) to describe the anatomical characteristics of leaves of tropical montane rainforests, and to distinguish them from "pycnophylls" (Grubb, 1986), the dense scleromorphic leaves found in areas with mediterranean climates. One of the most conservative characteristics of pachyphylls is the presence of a hypodermis. In montane forests, this tissue is more common in canopy than in understory leaves (Roth, 1990) and has been suggested to function as a water supply for the mesophyll (Cavelier & Goldstein, 1989b; Grubb, 1977). Because UV-B is known to increase with increasing altitude in tropical mountains (Robberecht, Caldwell & Billings, 1980), hypodermis could also protect the chlorophyll from UV radiation as shown for Saguaros (Darling, 1989).

The increase in leaf thickness in pachyphylls (largely a function of increases in air and water volumes rather than increases in dry matter, Figure 14.2), tends to increase the ratio of internal to external leaf surface (i.e., the mesophyll cell surface area per unit of leaf area, A^{mes}/A). The increase in A^{mes}/A will occur if the size of the mesophyll cells remains the same or decreases with increasing lamina thickness

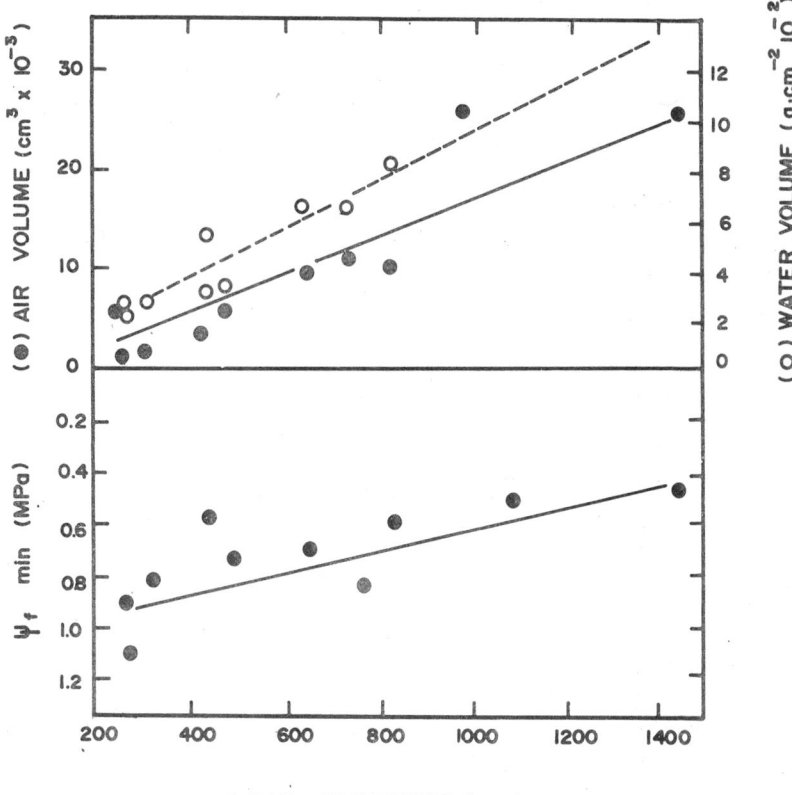

Figure 14.2. The upper panel shows the relationship between leaf thickness and air volume (r = 0.97, p < 0.01, Y = 0.019X − 1.668) and water volume (r = 0.77, p < 0.01, Y = 4.89·10⁻⁵X + 0.0189) in the mesophyll of 10 tree species of the elfin cloud forest of Cerro Jefe, Panama. The lower panel shows the relationship between leaf thickness and minimum tissue water potential measured in the field (r = 0.59, p < 0.05, Y = 4.129·10⁻⁴ X − 1.046) for the same group of species. Modified from Cavelier & Goldstein (1989b).

(Nobel, 1991), a feature that has not been measured but is likely to occur in leaves of montane environments characterized by low temperatures. High A^{mes}/A should result in reduced total resistance for CO_2 movement per unit of leaf area, and thus in higher photosynthetic rates (Nobel, Zaragoza & Smith, 1975). Nevertheless, limited measurements of gas exchange rates in montane rainforest trees showed that photosynthetic rates are not particularly high (Aylett, 1985). Furthermore, because nitrogen content per unit of mass decreases with increasing altitude (Grubb, 1977), and N content (g g^{-1}) positively correlates with photosynthetic capacity (Medina, 1984), we expect CO_2 uptake rates of montane rainforest trees to be low. Perhaps a high A^{mes}/A ratio in pachyphylls partially compensates for the reduction in the partial pressure of CO_2 at high altitudes (Körner & Diemer, 1987). Indeed, a reduction in the p_i/p_a ratio and in the discrimination at the leaf level against the heavy isotope of carbon (^{13}C) along altitudinal gradients suggest a CO_2 limitation for photosynthesis with increasing altitude (Körner, Farquhar & Wong, 1991). Further research is needed to understand the causes and adaptive significance of the internal leaf structure of pachyphylls.

14.3.4   Leaf characters and nutrient availability

14.3.4.1   Pachyphylly and sclerophylly

The term "sclerophylly" was introduced at the beginning of the century by Schimper (1903) to describe the small, thick, and leathery leaves that he thought were an adaptation to physiological drought. It has been suggested that sclerophylly is an adaptation not only to drought, as originally proposed by Schimper, but also as protection against insect damage (Coley, 1983; Grubb, 1986) or an adaptation to mineral shortage (Beadle, 1954; Loveless, 1961, 1962). Sclerophylly (*sclero* = the degree of hardness of a leaf; Loveless 1961) has been traditionally quantified as SLA (cm^2 g^{-1}; Evans, 1972; Medina, 1984; Stocker, 1931;), or as the ratio of crude fiber dry weight / crude protein dry weight (Loveless, 1961, 1962). Based on a plot of the fiber/protein ratio as a function of phosphorus content (% dry weight), Loveless (1961, 1962) suggested that sclerophylly decreases with increasing phosphorus content up to a certain level (about 0.3%), above which increased phosphorus content does not result in further proportional decrease in the fiber/protein content. However, in a group of 26 tree species of 4 elfin cloud forests in Colombia and Venezuela, phosphorus concentration was well above 0.3% and nevertheless, there were both sclerophytic and mesophytic (*sensu* Loveless, 1961) leaves (Cavelier, 1986). Thus, if phosphorus plays an important role in determining sclerophylly in montane forest trees, it does so at much higher concentrations than in those studied by Loveless.

In a limited number of studies of montane rainforest trees, SLA is positively correlated with leaf nitrogen (Cavelier, 1986; Cuenca, 1976) and leaf phosphorus content per unit leaf weight (Cuenca, 1976).

Medina (1984) suggests that this relationship would be expected if leaf expansion is inhibited by deficiencies of N and P. The apparent limitation of N and P in leaves of montane rainforest trees and its relationship to sclerophylly (measured as SLA) can be explored at the ecosystem level by looking at the circulation of total N and P and within-stand nutrient use efficiencies (*sensu* Vitousek, 1982, 1984). As detailed above, most montane rainforests are characterized by low levels of nitrogen circulation and low concentration of nutrients in litterfall (Medina, Sobrado & Herrera, 1978; Tanner, 1977b) resulting in a large amount of organic matter produced per unit of nutrient taken up, which is the definition of an efficient nutrient economy.

14.3.4.2 Nutrient retranslocation

If N and P availabilities in montane forest soils are lower than in the lowlands, and retranslocation of nutrients before leaf shedding is an important mechanism for nutrient conservation, percentage retranslocation should be higher in montane than in lowland forests (Table 14.1). In general, P retranslocation is higher with increasing altitude (Table 14.1), with retranslocation being inversely correlated to foliar concentration (Veneklaas, 1991). If N availability in soils decreases with elevation in all mountain ranges as described for Volcan Barva in Costa Rica (Marrs et al., 1988), one should expect an increase in N reabsorption with increasing altitude. The available information shows otherwise. Percentage of N reabsorption is very similar for all montane forests, with the exception of low values for very tall and productive lower montane rainforest in New Guinea (16%) and the relatively high values for the stunted and low productivity elfin cloud forest of Puerto Rico (22%).

14.3.5   Leaf water relations

14.3.5.1 Stomatal conductance and transpiration rates

Odum (1970) suggested that small trees of montane forests result when the saturation deficit is low without interruption, and dry air is not available for transpiration, reducing transport of minerals in the transpiration stream (see also Leigh, 1975). However, lower and upper montane rainforests are not always cloudy and usually experience clear skies in the mornings when plants can transpire (Cavelier & Goldstein, 1989a; Cavelier & Mejia, 1990; Grubb & Whitmore, 1966;

Table 14.1. Concentration (% dry weight) of nitrogen and phosphorus in canopy leaves and in fresh litterfall in montane rainforests (upper montane rainforest (UMRF) and lower montane rainforest (LMRF)) and lowland rainforests (LRF). Retranslocation was calculated as (concentration of nutrients in canopy leaves – concentration in fresh litter) / (concentration in fresh litter).

Montane rainforests	Concentrations (% dry weight)		Retranslocation (%)	
	N	P	N	P
Colombia (1)				
UMRF				
Canopy leaves	1.47	0.09		
Fresh litter	0.90	0.03	39	67
LMRF				
Canopy leaves	1.78	0.127		
Fresh litter	1.08	0.070	39	45
Jamaica (2)				
Canopy leaves	1.30	0.068		
Fresh litter	0.76	0.030	42	56
New Guinea (3)				
Canopy leaves	1.32	0.086		
Fresh litter	1.11	0.059	16	31
Puerto Rico (4)				
Canopy leaves	0.99	0.063		
Fresh litter	0.77	0.024	22	62
Sabah (5)				
at 610m				
Canopy leaves	1.45	0.047		
Fresh litter	1.05	0.017	28	64
at. 790m				
Canopy leaves	1.72	0.054		
Fresh leaves	1.05	0.017	39	69
at 870m				
Canopy leaves	1.34	0.046		
Fresh litter	0.83	0.020	38	57
Venezuela (6)				
Canopy leaves	1.74	0.08		
Fresh litter	1.20	0.06	31	25
Venezuela (7)				
Canopy leaves	1.64	0.111		
Fresh litter	1.15	0.062	30	44
AVERAGE			32 (16-42)	52 (69-25)

Table 14.1.   (cont.).

Lowland rainforests	Concentrations (% dry weight)		Retranslocation (%)	
	N	P	N	P
Columbia (8)				
Canopy leaves	1.93	0.070		
Fresh litter	1.30	0.035	33	50
Ghana (9)				
Canopy leaves	2.52	0.140		
Fresh litter	2.10	0.09	16	36
Venezuela (10)				
Canopy leaves	1.78	0.06		
Fresh litter	1.56	0.03	11	50
Caatinga				
Canopy leaves	1.08	0.07		
Fresh litter	0.70	0.05	35	29
Bana				
Canopy leaves	0.89	0.04		
Fresh litter	0.58	0.02	35	50

(1) Veneklaas (1991), (2) Tanner (1977a,b), (3) Grubb & Edwards (1982), Edwards (1982), (4) Medina, Cuevas & Weaver (1981), Weaver et al. (1986), (5) Proctor et al. (1989), (6) Fassbender & Grimm (1981), (7) Steinhardt (1979), (8) Fölster & De las Salas (1976), (9) Nye (1961), (10) Cuevas & Medina (1986).

Odum, Drewry & Kline, 1970; van der Hammen, 1984a). Significant transpiration rates were calculated for cloud forest trees in the Luquillo Mountains of Puerto Rico by Weaver, Byer & Bruck (1973), Medina, Cuevas and Weaver (1981), and Gates (1969). Daily transpiration rates for trees during cloudy and sunny days at the elfin cloud forest of Macuira were similar in spite of the great differences in leaf-to-air water concentration gradients (Cavelier, 1990). Arguments against Odum's hypothesis have been proposed by Grubb (1977) and Medina, Cuevas and Weaver (1981).

Although water availability is high and constant in montane rainforests (Kapos & Tanner, 1985), local plants seem to be very sensitive to water shortage. Measurements of stomatal conductance (g_s) under clear skies have shown that increases in VPD (the vapor pressure difference between the leaf and air) result in marked decreases in g_s through stomatal closure (Aylett, 1985; Cavelier, 1990; Körner, Allison & Hilscher, 1983; Medina, Cuevas & Weaver, 1981; Meinzer, Goldstein & Jaimes, 1984). Decreases in g_s with increases in VPD are probably a response to avoid further water deficits (Medina, Sobrado & Herrera, 1978) and suggest a stomatal response to changes in relative humidity (Lange et al., 1971; Schulze et al., 1972; Mulkey & Wright, Chapter 7). It is possible that there is a limitation to water transport in the stems of these trees. This would be clarified by studies of the patterns of hydraulic architecture and water relations similar to those carried out for lowland trees with contrasting leaf phenologies (Machado & Tyree, 1994; Tyree & Ewers, Chapter 8).

Stomatal conductance and stomatal density decrease across species with elevation in the mountains of New Guinea (Körner, Allison & Hilscher, 1983) and Puerto Rico (see Cintrón, 1970 for stomatal density), while an opposite trend has been observed in the Alps (Bonnier, 1895 cited in Friend & Woodward, 1990; Körner & Mayr, 1980; Wagner, 1892 cited in Friend & Woodward, 1990) and in Australia (Körner & Cochrane, 1985). The decrease in these two characteristics in tropical mountains is interpreted by Körner, Allison and Hilscher (1983) as a result of decreasing irradiance with increasing altitude. Apparently, stomatal conductance and stomatal density are determined by climatic conditions not necessarily coupled with altitudinal gradients, like photosynthetic active radiation received by the leaf, which decreases in tropical but not in temperate mountains (Körner, Allison & Hilscher, 1983). Stomatal size does not change with elevation in New Guinea, but increases in the Luquillo mountains in Puerto Rico (Cintrón, 1970) in spite of the reduction of PAR with increasing altitude. This difference may be due to the growth forms used in the study in New Guinea (sclerophyll shrubs, dwarf shrubs, tussock grasses and herbs) and in Puerto Rico (trees). Leaf anatomy in canopy trees remains to be studied along a single altitudinal gradient in a wet tropical mountain range, a task that is increasingly difficult due to deforestation and fragmentation of montane forests worldwide.

14.3.5.2   Leaf water potential and its components

Values of minimum tissue water potentials measured in the field are relatively high in montane rainforests (Cavelier, 1990; Kapos & Tanner, 1985; Medina, Cuevas & Weaver, 1981) when compared with values of lowland rainforest trees (Myers et al., 1987; Oberbauer, Strain & Riechers, 1987). Less information is available for the components of tissue water potentials (Table 14.2). Values for the osmotic potentials at full hydration in montane rainforest trees (Cavelier, 1990) are similar to those reported in the wet lowlands (Oberbauer, Strain & Riechers, 1987; Myers et al., 1987), but higher than those in dry deciduous forests (Fanjul & Barradas, 1987; Sobrado, 1986). These results suggest that montane rainforest trees experience lower

Table 14.2.   *Components of tissue water potential and other values related to the water status in tropical cloud forest tree species. Osmotic potential at full hydration ($\Psi\pi^{100}$) and at turgor loss point ($\Psi\pi^{tlp}$), relative water content at turgor loss point (r_{tlp}), and weight-average bulk tissue elastic modulus (E). Data from Serrania de Macuira from Cavelier (1990) and from Cerro Santa Ana, La Montaña (Cable Car Station at the Sierra Nevada de Mérida, Venezuela), and El Zumbador (Estado Táchira, Venezuela) from Cavelier (unpublished data).*

Species	$\Psi\pi^{100}$ (Mpa)	$\Psi\pi^{tlp}$ (Mpa)	R_{tlp} (%)	E (Mpa)
Serrania de Macuira (865m)				
Clusia rosea	-1.23	-1.60	0.90	13.56
Guapira fragrans	-1.61	-1.73	0.93	21.30
Myrcia splendens	-1.63	-2.00	0.87	13.07
Myrciantes fragrans	-1.96	-2.36	0.84	10.12
Rapanea guianensis	-1.70	-2.06	0.86	10.20
Rugea marginata	-2.13	-2.60	0.89	20.21
Cerro Santa Ana (815m)				
Actinostemon concolor	-2.50	-3.40	—	8.00
Ardicia cuneata	-1.60	-2.20	—	9.90
Coccolaba coronata	-2.30	-3.50	—	4.10
Guapira opposita	-1.80	-2.50	—	11.50
Myrcia splendens	-2.30	-3.30	—	5.10
Weinmannia pinnata	-1.80	-2.80	—	6.00
La Montaña (2700m)				
Vaccinium floribundum	-2.45	-2.73	0.93	29.80
Weinmannia aff. glabra	-1.56	-2.07	0.91	9.22
Podocarpus oleifolius	-1.57	-2.22	0.88	10.53
Clusia minor	-1.69	-2.02	0.87	7.17
Freziera subintegrifolia	-1.74	-1.96	0.94	49.34
Hdyosmus glabratuma	-1.52	-1.83	0.87	9.47
El Zumbador (3100m)				
Brunellia aff. goudotii	-1.70	-1.90	—	3.30
Clusia cf. articulata	-1.70	-2.00	—	10.80
Ocotea calophylla	-1.50	1.60	—	10.40
Persea ferruginea	-1.40	-1.60	—	5.20
Podocarpus oleifolius	-1.10	-1.70	—	6.30
Symbolanthus sp.	-1.10	-1.30	—	6.70
Symplocos suaveolens	-0.90	-1.40	—	3.10
Weinmannia pinnata	-1.50	-1.70	—	2.30

or similar water deficits than those in the lowlands (see also Fetcher, 1979; Mulkey & Wright, Chapter 7).

Pachyphylls (*sensu* Grubb, 1977) are not characterized by their ability to withstand water shortage (Buckley, Corlett & Grubb, 1980;

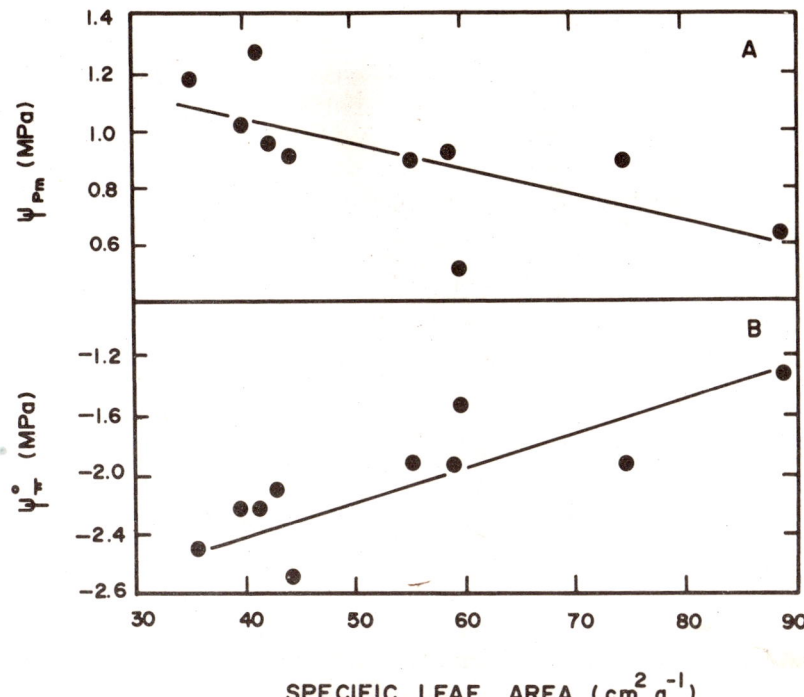

Figure 14.3. The upper panel shows the relationship between specific leaf area and the minimum turgor pressure in the field (estimated form PV curves and daily courses of tissue water potential) in a group of 10 tree canopy species of the elfin cloud forest of Cerro Jefe, Panama (r = − 0.68, p < 0.05, Y = 8.86· 10^3X + 1.4). The lower panel shows the relationship between specific leaf area and the osmotic potential at turgor loss point (r = 0.81, p < 0.01, Y = 0.019X − 3.10) for the same group of species. Modified from Cavelier & Goldstein (1989b).

Kapos & Tanner, 1985). Nevertheless, when the components of tissue water potential are compared among species within the same cloud forest but with different degrees of sclerophylly (measured as SLA), an interesting picture emerges (Figure 14.3). In a cloud forest in Panama, scleromorphic leaves (low SLA) have lower osmotic potentials at turgor loss point and experience higher turgor pressure in the field than do mesophyll leaves (high SLA). Furthermore, thick leaves experience more positive leaf water potentials than thin leaves, suggesting that thick and low SLA leaves have more physiological characteristics of drought tolerance than thin and high SLA leaves (Cavelier & Goldstein, 1989b).

The modulus of elasticity is positively correlated with the concentration of calcium in leaf tissues (Cavelier, 1986), supporting the idea that calcium in pectins and proteins in the cell wall decreases elasticity (Bennet-Clark, 1956; Burstrom, 1968). This correlation has important implications for the water relations of cloud forest trees that grow in highly leached soils. If the concentration of calcium in leaf tissues is determined by the availability in the soil (Grubb, 1977), the elasticity of cell walls, an important mechanism for turgor maintenance, may be partly determined by a soil property. Thus, in a low calcium environment, elasticity of cell walls would tend to be high (low E), and water deficits would be transduced into instantaneous decreases in cell volume rather than reduced hydrostatic pressure.

14.3.6 Photosynthesis

14.3.6.1 *Leaf nitrogen content and photosynthetic rates*

Leaf nitrogen content per unit leaf weight is highly correlated with photosynthetic rate at saturating light (Field & Mooney, 1986; Medina, 1984; Mooney & Gulmon, 1979). Since N content is such a strong determinant of photosynthesis, and leaf N content is low in montane rainforest trees, net CO_2 uptake is likely to be low. Limited measurements of maximum photosynthetic rates show that net CO_2 uptake in montane rainforest trees ranges from 5.0 µmol CO_2 m^{-2} s^{-1} for *Hedyosmum arborescens* (in the subcanopy) to 6.6 µmol CO_2 m^{-2} s^{-1} for *Clethra occidentalis* (in the upper canopy) (Aylett, 1985). These values are in the lower range of values reported for fast growing secondary species (13.3-16.4 µmol CO_2 m^{-2} s^{-1}) and for primary forest tree species in tropical lowland rainforests (3.9-15.4 µmol CO_2 m^{-2} s^{-1}, Aylett, 1985; 15-20 µmol CO_2 m^{-2} s^{-1} for *Ceiba pentandra*, Zotz & Winter, 1993; Zotz & Winter, Chapter 3).

There is a general decline in foliar concentration of N when passing from lowland rainforests (1.66% dry weight for 8 forests) to upper montane rainforests (1.13 % dry weight for 9 forests) (after Grubb, 1977). In Hawaii, nitrogen concentrations per unit leaf weight in *Metrosideros polymorpha* decreased with increasing elevation on both old and young lava flows (Vitousek, Matson & Turner, 1988). When N concentration is expressed per unit leaf area, a different picture emerges: montane rainforest trees show higher N concentration than lowland rainforest trees (Medina, Cuevas & Weaver, 1981), or there is no difference with altitude, as in the case of *Metrosideros polymorpha* in Hawaii (Vitousek, Matson & Turner, 1988). N content per unit leaf area increases with altitude in trees in mountains in the temperate zone (Korner, 1989; Körner, Bannister & Mark, 1986) where there is an inverse relationship between nitrogen content per unit leaf area and specific leaf area (Körner & Cochrane, 1985). This relationship was also found for lowland rainforest species (Reich et al., 1994; Raich & Walters, 1994). Greater nitrogen content per unit leaf area partially explained the higher efficiency of carbon dioxide uptake (i.e., the linear slope of the relationship between assimilation and internal partial pressure of CO_2) of high-altitude herbaceous plants in the Alps (Körner & Diemer, 1987). Similar photosynthesis measurements are needed along a tropical altitudinal gradient to see if greater N content per unit leaf area correlates with high assimilation rates. As proposed by Chapin (1980), plants growing in infertile soils, in both tropical and temperate mountains, may respond to the stress by simply maintaining higher concentrations through luxury consumption and/or reduced growth (Körner, Bannister & Mark, 1986) . In the mean time, correlations of leaf N content per unit leaf weight and assimilation, and the variation of the carbon isotopic composition of plant material, can provide insight into the relationship between leaf nitrogen and rates of carbon assimilation along altitudinal gradients in tropical mountains.

Field and Mooney (1986) suggested that the ratio of photosynthetic capacity (A_{max}) to N content for evergreen sclerophylls is low because proportionally less of the nitrogen is allocated to photosynthetic reactions. Low maximum photosynthetic rate in *H. arborescens*, with relatively high N concentration (1.85%), could be related to the preferential allocation of N to compounds required for longevity (e.g., herbivory deterrents like alkaloids) rather than to photosynthetic compounds (e.g., Rubisco). It is possible that in montane habitats, where nutrient availability is low (Marrs et al., 1988; Grubb, 1971, 1977) and growth conditions are generally unfavorable (i.e. low temperature and acid soils), plants devote most of their resources to survival rather than to productivity (see also Grubb, 1977). To test this hypothesis, information is needed on gas exchange rates and on the nitrogen allocation of these and other species of montane rainforest trees to test this hypothesis.

Trunk-growth increases in response to fertilization with N (Vitousek et al., 1987; Tanner et al., 1990; Tanner, Kapos & Franco, 1992; Gerrish & Bridges, 1984), which suggests that productivity of montane forests is limited by nitrogen. At the ecosystem level, there is a decline in productivity with increasing altitude. For instance, annual leaf litterfall declines from 10.5 ton ha^{-1} in a lowland rainforest in Ghana to 5.5 ton ha^{-1} in an upper montane rain forest on a wet slope in Jamaica (Grubb, 1977). Aylett (1985) concluded that this decline in productivity may be partly due to a reduction in solar radiation in montane environments. Indeed, although montane forests have lower LAI than most lowland rainforests, 95% of the incoming short-wave radiation was intercepted before reaching the ground level in Jamaican forests. Thus, an increase in LAI would likely not result in significantly increasing interception of radiation and forest productivity (Aylett, 1985). The value of the LAI is probably the maximum a forest can sustain given the available radiation and nutrient levels.

14.3.6.2   Altitudinal gradients and leaf carbon isotope ratios

A reduction in the discrimination against the heavy isotope of carbon (^{13}C) has been found with increasing altitude (Körner, Farquhar & Wong, 1991; Korner, Farquhar & Roksandic, 1988; Tieszen et al., 1979; Vitousek, Matson & Turner, 1988). Reduced discrimination against ^{13}C, is probably the result of differences in discrimination by plants, rather than atmospheric variation in ^{13}C (Körner, Farquhar & Roksandic, 1988). Indeed, a decrease in the p_i/p_a (intercellular/ambient partial pressure of CO_2) with increasing elevation has been recently documented by Körner, Farquhar and Roksandic (1988), Körner and Diemer (1987), and F. Rada (personal communication) in the high Andes of Venezuela, suggesting a CO_2 limitation for photosynthesis with increasing altitude. This CO_2 limitation with altitude could partially explain the reduction in productivity, biomass, and stature in montane rain forests in comparison with most lowland rainforests.

If the carbon isotopic composition of leaf tissues is a measure of WUE (Farquhar & Richards, 1984; Hubick, Farquhar & Shorter, 1986a, 1986b; Goldstein et al., Chapter 9), reduced discrimination against ^{13}C with increasing altitude suggests that there is an increase in the WUE along elevational gradients. This variation correlates with changes in leaf anatomy and morphology of pachyphylls, which have thick leaves with high mesophyll cell surface area per unit leaf area. An increase in the A^{mes}/A could partially compensate for the decrease in partial pressure of CO_2 and the reduction in leaf size in montane environments. Regardless, the idea of an increase in WUE with increasing altitude must be interpreted very carefully since water vapor pressure changes with altitude and WUE *per se* may not have been selected for.

14.4   FACTORS CONTROLLING THE DISTRIBUTION OF MONTANE FORESTS

14.4.1   Montane rain forest formations

Along altitudinal gradients in wet tropical mountains there are several vegetation types that were recognized in the neotropics as early as the beginning of the 19th century by Alexander von Humboldt (Troll, 1968). The most general classification system of forest types has been proposed by Grubb (1974; 1977) with the names of Lowland Rainforest (LRF), Lower Montane Rainforest (LMRF), Upper Montane Rainforest (UMRF), and Subalpine Rainforest (SARF). The physiognomical characteristics of these formation types are summarized in Grubb (1977), and information on epiphytic flora is found in Cleef et al. (1984). The altitudinal limits of the formation types vary with the size of the mountain. Forest formations extend to higher elevations on taller, massive mountains than on smaller mountains, outline ridges, and mountains close to the sea. This effect is known as the Massenerhebung effect (Grubb, 1977; Richards, 1954; van Steenis, 1961, 1972). This concept was introduced at the beginning of the 20th century to describe the fact that treeline and snowline occur at higher elevations in the central Alps than on mountains in their outer margins (Barry, 1981). In temperate mountains, the increase in air temperature above land masses results in a decrease in the incidence of frost, which permits a longer growing season and thus an elevational extension of treeline (Whitmore, 1984). This cannot explain the Massenerhebung effect for tropical mountains because the incidence of frosts is not related to seasonality, but to altitude, with mountains less than 3000 m elevation receiving only occasional frosts.

14.4.2.   The Massenerhebung effect in tropical mountains

14.4.2.1   Temperature lapse rate

The most obvious environmental factor that could control the distribution of forests along altitudinal gradients is temperature (Cuatrecasas, 1958; Hedberg, 1951; Richards, 1954). For the limit of a given forest type on a small mountain to be at the same temperature as on a large mountain, the lapse rate would need to be significantly steeper on small mountains. For instance, for the lower limit of the LMRF in Trinidad (700–900 m) to have the same temperature as in the main ranges of the Colombian Cordilleras (1200–1500 m), the adiabatic lapse rate in Trinidad would need to be between 0.97 to 1.00°C/100 m. In fact, this gradient is 0.97°C/100 m for Cerro Copey, a small mountain (930 m) on Margarita Island, Venezuela, not far from Trinidad (Sugden, 1986). Temperature could be an important environmental factor to control the lower limit of the LMRF either by itself or in combination with other factors, such as those related to cloud cover at the condensation level (i.e., lower radiation and higher humidity).

14.4.2.2 Condensation level: Land masses and distance to the coast

Mossy forests and other montane rainforest types occur at lower altitudes on small isolated peaks on islands or near the sea rather than on large mountain masses inland in the Caribbean (Beard, 1944, 1949; Cavelier & Mejia, 1990; Sugden, 1982, 1986; Tanner, 1977b) and Southeast Asia (van Steenis, 1972, 1984). According to van Steenis (1972), the occurrence of montane forests at low elevation (as low as 400 m on Gunung Payung, Java; Bruijnzeel et al., 1993) is the result of low clouds over the sea. The altitude at which mossy forests occur on mountains in Indonesia is significantly correlated with the distance to the sea (Flenley, 1979). This may be due to increased atmospheric humidity above the sea and/or a steeper adiabatic lapse rate associated with small mountains as mentioned before.

In summary, the Massenerhebung effect in tropical mountains is probably related to the occurrence of cloud condensation at lower altitudes on islands or in small mountains close to the sea, and at high altitudes on large mountains inland. Because the Massenerhebung effect has been reported in mountains with very different bedrock and geological histories, the mechanism that explains the effect must lie in a direct or indirect effect of cloud cover (i.e. rain, cloud water, higher humidity, lower radiation, lower temperature).

14.4.2.3 Nutrient availability, soil water, and temperature

Grubb and Whitmore (1966) first suggested that the most important factor determining the distribution of montane forests in tropical mountains is the frequency of fog. Grubb (1971, 1977) noted that the zonation of montane forests could be explained in terms of the availability of certain nutrients that become less available as their rates of mineralization from humus decrease with either lower mean temperature (high altitudes) or higher soil water content. High soil water content, and sometimes waterlogged soil conditions, would result from frequent cloud cover (Grubb, 1977, Whitmore, 1984). Because clouds tend to form at a particular altitude, increased water availability would start rather abruptly at this elevation, as has been reported in Borneo (Kitayama, 1992) and Colombia (Cavelier & Goldstein, 1989b). Experimental evidence gathered at Volcan Barva (2906 m) in Costa Rica, shows that when temperature was increased and soil water content decreased simultaneously in laboratory incubations (with a concomitant increase in aeration), nitrogen-mineralization rates increased. A positive relationship was found with altitude of origin. Similar measurements at Serrania de Macuira (865 m) produced the same type of results (Cavelier & Tanner, unpublished data), suggesting that high soil water content may be the proximal cause of the Massenerhebung effect in tropical mountains.

Although waterlogging is commonly associated with montane forests of Southeast Asia (Whitmore, 1984) and very stunted montane thicket in Puerto Rico (Lyford, 1969), waterlogging is not a requirement for montane forests to occur. The formation of peat is controlled primarily by waterlogging (Grubb & Tanner, 1976; Whitmore, 1984; Cuevas, 1987), but mor soils (C/N > 25) are formed on strongly leached upland sites with poor parent materials and very low pH (usually less than 4.0; Grubb & Tanner, 1976). Soil water content in mull soils, even during the rainy season, is well below saturation (Figure 14.4). In spite of this difference, soil water content is constantly high in mor and mull soils in montane forests. Under this condition, decomposition of organic matter is slower than under wetting and drying cycles (Birch 1958, 1964). Constant wet conditions favor microbial nutrient immobilization, making it difficult for plants to compete for limiting nutrients (Vitousek, 1984). Under the cloud cap, forests experience nearly constant soil water conditions due to high relative humidity and continuously available water in the form of rain and/or mist. These changes are probably the cause of the sudden increase in the carbon content of the soils at ca. 1200, the altitude of the first cloud belt in large mountains (Figure 14.5).

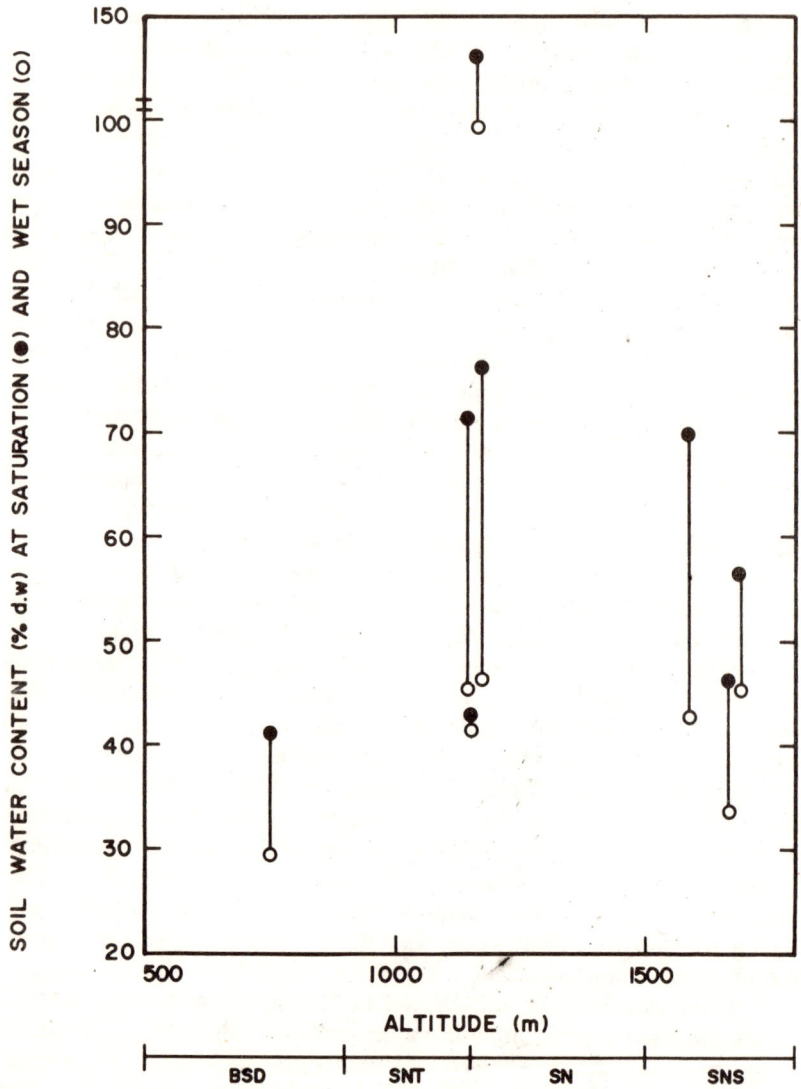

Figure 14.4. Soil water content (% dry weight), at saturation and during the wet season, along the altitudinal gradient of Cordillera de la Costa, Parque Nacional "Henry Pittier," Venezuela (after Zinck, 1986). Forest types are: BSD = Bosque Seco Deciduo (dry deciduous forest), SNT = Selva Nublada de Transición (transitional cloud forest), SN = Selva Nublada (cloud forest) and SNS = Selva Nublada Superior (approx. montane thicket). For the description of the forest types see Huber (1986).

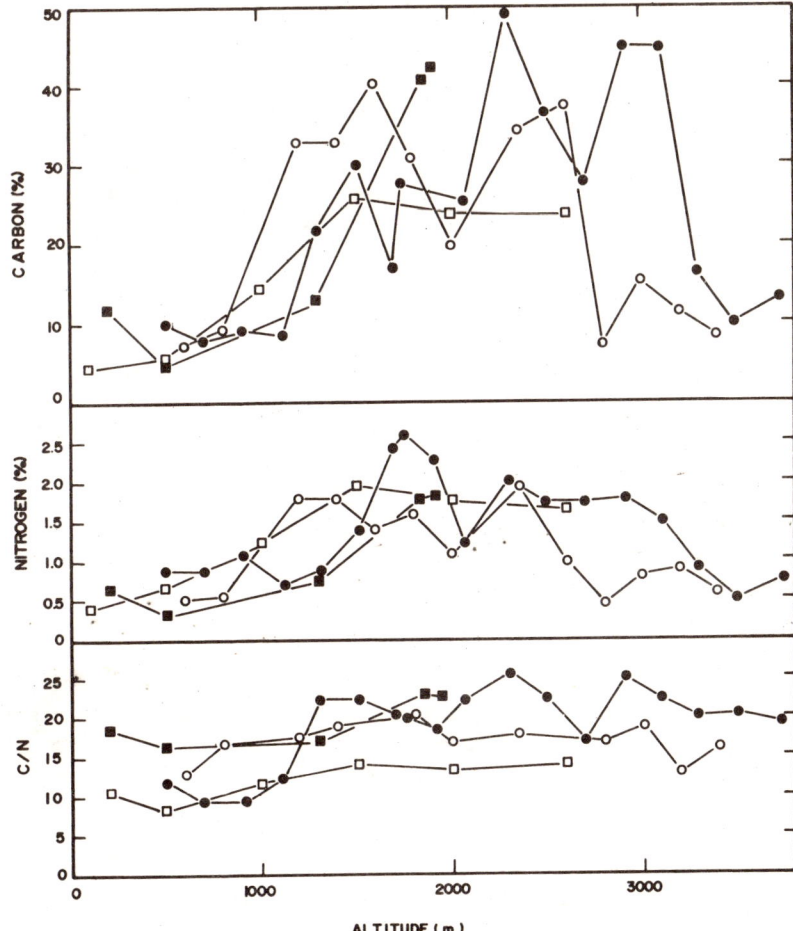

Figure 14.5.   Soil carbon, total nitrogen, and the C/N ratio for the surface soils along the altitudinal gradients of the Sierra Nevada de Santa Marta (●), Colombia (van der Hammen, 1984b), Mount Kinabalu (○), Borneo (Kitayama, 1992), Volcán Barva (□), Costa Rica (Marrs et al., 1988; Heaney & Proctor, 1989), and Gunung Mulu (■), Sarawak (Proctor, Anderson & Vallack, 1983).

14.4.3 The limits between forests types: Controlling factors

14.4.3.1 The limit between lowland and lower montane rainforests

It has been suggested that the upper limit of each formation is a function of temperature, and that the lower limit is probably determined by competition rather than by high temperatures (Grubb, 1977). Although in the Holdridge Life Zone System the premontane and lower montane altitudinal belts are separated by the "frost line or critical temperature line" (18°C mean annual biotemperature and approximately 1500 m elevation; Holdridge et al., 1971), there is little information on the actual temperature threshold that separates the occurrence of tropical lowland from tropical montane forest flora. On Mount Kinabalu (Borneo), the upper boundary of the LRF coincides with the threshold of Koppen's tropical climate (18 °C mean daily minimum temperature; Kitayama, 1992). This correlation is of little use in understanding the cause-and-effect relationships between climate and vegetation, since Koppen's climatic boundaries where taken from empirical studies of vegetation (Holdridge et al., 1971; Thornthwaite, 1943).

The limit between LRF and LMRF correlates with the base of the "first cloud belt" and a rainfall peak. The increase in the soil water content at around 1200 m created by cloud water, rainfall, and increased humidity due to the presence of clouds can by itself set the limit between the LRF and the LMRF, or induce changes in soil chemistry which in turn influence which species grow there. As noted above, at this altitude there is a sharp increase in organic carbon and total nitrogen in the Andes (van der Hammen, 1984b), Borneo (Kitayama, 1992), Costa Rica (Marrs et al., 1988), and Sarawak (Proctor, Anderson & Vallack, 1983). Under the influence of higher soil water content and/or changes in one or more of the chemical properties of the soils, competitive exclusion is a possible mechanisim contribuiting to the sudden changes in species abundance along altitudinal gradients (Grubb, 1977). One example of such competitive exclusion is *Shorea platyclados*, the common montane dipterocarp that becomes successful in competition with lowland dipterocarps as conditions become moister and cooler (Whitmore, 1984).

14.4.3.2 The limit between lower and upper montane rainforests

The limit between lower and upper montane rainforests in the mountains of Southeast Asia is very sharp (Whitmore, 1984). The UMRF has an abrupt lower boundary that correlates well with the altitude of cloud formation. Frequent cloud cover induces waterlogged soils that result in the development of peat (Proctor et al., 1988, 1989). Under these conditions, certain species of the UMRF with slow-decomposing litter would be favored and the process of peat development would be reinforced, sharpening the limit with the LMRF (Whitmore, 1984). Not all the limits between LMRF and UMRF are correlated with sharp differences in soil chemical properties. For instance, the limits between these two forest formations (ca. 2400 m), at SNSM, and on Mt. Kinabalu in Borneo seem to be associated with excess water (Kitayama, 1992), as indicated by the epiphytic flora (Wolf, 1989) and 30% soil surface cover by mosses, including *Sphagnum.* Some of the soil characteristics that change at this limit at the SNSM include thicker surface root layer, lower pH and earthworm activity in the UMRF (van der Hammen, 1984b).

14.4.3.3 The upper limit of upper montane rainforests

The upper limit of the UMRF coincides with the appearance of a few days of frost (Troll, 1968) that can occur any time of the year (Sarmiento, 1986). Although giant rosette plants and a tree species growing above the continuous tree line in the Northern Andes of Venezuela are known to avoid freezing temperatures by means of supercooling (Rada et al., 1985a, 1985b), and Afroalpine giant rosettes are known to tolerate extracellular freezing (Beck et al., 1982), no studies have been carried out to investigate freezing avoidance or tolerance of trees of the upper montane rainforests. These trees are likely to have physiological mechanisims to deal with occasional sub-zero temperatures and frosts. Frost rarely occurs as low as 1500 m in tropical mountains (van Steenis, 1961).

Flenley (1979, 1992, 1993) hypothesized that vegetation zonation on the upper parts of tropical mountains is related to high levels of ultraviolet light, particularly UV-B (280-320 nm). This idea is supported by three pieces of evidence. First, there is a positive relationship between UV-B and altitude, with the highest values in high tropical mountains (Caldwell, 1971; Caldwell, Robberecht & Billings, 1980). Second, experimental effects of UV-B on plant growth show that plants under UV-B become stunted and develop small and thick leaves with a hypodermis (Murali & Teremura, 1986; Teremura, 1983). Indeed, leaves are smaller and the occurrence of a hypodermis (% species) is greater in subalpine and UMRF than in LRF (Tanner & Kapos, 1982). Plants under UV-B also develop extra flavonoid pigments (Caldwell, Robberecht & Flint, 1983; Teremura, 1983; Murali & Teremura, 1986), the most abundant secondary compounds in *Polylepis quadrijuga* (Rosaceae), a tree species above the continuous tree line in the Paramos on the Eastern Cordillera of the Andes, Colombia (Velez, 1994). Third, during the Late Pleistocene (characterized by lowering of the forest formations to c. 2000 m), the UMRF apparently disappeared completly due to the absence of a habitat with a suitable combination of temperature and UV-B insolation.

14.5 AN AGENDA FOR FURTHER RESEARCH

Not all environmental factors are equally well studied over altitudinal gradients in tropical mountains. We have relatively good information on air and soil temperature and rainfall regimes. However, there are only scattered data and observations on cloud cover, radiation regimes, light quality and quantity, and the direct and indirect effects of cloud water on plants with increasing altitude. Because the environment of montane rainforest understory is obviously very different from those in the lowlands, we need data to characterize more fully these differences.

There is good understanding of the general environmental and biological factors that control the decomposition rates of organic matter, but with the exception of N content, there is still little information on the quality of montane rainforest litter. Further experimental work is needed to understand the importance of anoxia in waterlogged soils (peat), high and constant soil water content in mull and mor soils, and the presence of carbon-based compounds (e.g., polyphenols) in reducing decomposition rates of montane rainforest litter.

The possibility that other environmental factors besides temperature may select for leaf functional characteristics in canopy trees of montane rainforests should be explored. Further research is needed to understand the causes and adaptive significance of leaf thickness in pachyphylls, a feature that is apparently more related to air and water volumes than to mass per unit of leaf area. *In situ* gas exchange measurements of montane rainforest trees are few. Information is needed on nitrogen allocation in pachyphylls to test the hypothesis that the ratio of photosynthetic capacity (A^{max}) to N content is low because proportionally less of the nitrogen is allocated to photosynthetic reactions as it is in other evergreen sclerophylls. Detailed information on the rates of nitrogen mineralization and nitrification should be collected along single altitudinal gradients for a better understanding of the N retranslocation in montane rainforest trees. Observations and experiments are needed on the variation of UV-B insolation along the altitudinal gradients in tropical mountains and the effect of exclusion of UV-B radiation on leaf morphology (e.g., leaf size), anatomy (e.g., occurrence of hypodermis), and physiology (e.g., epidermal transmittance and the synthesis of flavonoids) of canopy leaves along single altitudinal gradients.

Some experimental evidence (e.g., soil/plants transplants along altitudinal gradients and plant competition experiments in temperature-controlled environments) is needed to determine if the temperature limit between LRF and LMRF, and the environmental factors associated with occurrence of the cloud base, have a measurable effect on the growth and survival of lowland and lower montane rainforest tree species. Experiments in which soil water content is reduced by trenching would help elucidate the effect of high and constant soil water content and/or waterlogged conditions on the growth of montane rainforest trees. These experiments and measurements should provide useful information on the proximal causes and mechanisms of the Massenerhebung effect in tropical mountains.

ACKNOWLEDGMENTS

I would like to thank the people that have encouraged me to study montane rainforests in Central and South America: Guillermo Goldstein (University of Hawaii), Guillermo Sarmiento and Maximina Monasterio (Universidad de los Andes, Merida, Venezuela), Alan Smith and Egbert Leigh, Jr. (Smithsonian Tropical Research Institute), Edmund Tanner, Valerie Kapos and Peter Grubb (University of Cambridge), and Carlos A. Mejia (Universidad de los Andes, Bogota, Colombia). Jose Luis Machado and Santiago Madriñan provided useful references for this review. I appreciate the editors' comments on this chapter.

REFERENCES

ALEXANDER, E. B. & PICHOTT, J. (1979) Soil organic matter in relation to altitude in equatorial Colombia. *Turrialba*, **29**, 183–188.

ANDERSON, J. M. & SWIFT, M. J. (1983) Decomposition in tropical forests. *Tropical Rain Forest: Ecology and Management* (eds. S. L. SUTTON, T. C. WHITMORE, & A. C. CHADWICK). Special Publication Number 2 of the British Ecological Society, Blackwell Scientific Publications, Oxford, pp 287–309.

AYLETT, G. P. (1985) Irradiance interception, leaf conductance and photosynthesis in Jamaican upper montane rainforest trees. *Photosynthetica*, **19**, 323–327.

BARRY, R. G. (1981) *Mountain Weather and Climate*. Methuen, London and New York.

BAYNTON, H. W. (1968) The ecology of an elfin cloud forest in Puerto Rico, 2. The microclimate of Pico del Oeste. *Journal of the Arnold Arboretum*, **49**, 419–430.

BAYNTON, H. W. (1969) The ecology of an elfin forest in Puerto Rico, 3. Hilltop and forest influence on the microclimate of Pico del Oeste. *Journal of the Arnold Arboretum*, **50**, 80–92.

BEADLE, N. C. W. (1954) Soil phosphate and the delimitation of plant communities in eastern Australia. *Ecology*, **35**, 370-375.

BEARD, J. S. (1944) Climax vegetation in tropical America. *Ecology*, **25**, 127–158.

BEARD, J. S. (1949) The natural vegetation of the windward and leeward Islands. *Oxford Forestry Memoirs*, **21**, 192.

BECK, E., SENSER, M., SCHEIBER, R., STEIGER, H. & PONGRATZ, P. (1982) Frost avoidance and freezing tolerance in Afroalpine "giant rosette" plants. *Plant, Cell and Environment*, **5**, 215–222.

BENNET-CLARK, T. A. (1956) Salt accumulation and mode of action of auxin. A preliminary hypothesis. *The Chemistry and Mode of Action of Plant Growth Substances* (eds. R. A. WAIN & F. WIGHTMAN) Butterworth Scientific Publications, London.

BIRCH, H. F. (1958) The effect of soil drying on humus decomposition and nutrient availability. *Plant and Soil*, **10**, 9–31.

BIRCH, H. F. (1964) Mineralization of plant nitrogen following alternate wet and dry conditions. *Plant and Soil*, **20**, 43–49.

BONNIER, G. (1895) Recherches expérimentales sur l'adaptation des plantes au climat alpin. *Ann. des Science Naturelles, Botanique*, Ser. VII, **20**, 217–360.

BRUIJNZEEL, L. A., WATERLOO, M. J., PROCTOR, J., KUITERS, A. T. & KOTTERINK, B. (1993) Hydrological observations in montane rainforests on Gunung

Silam, Sabah, Malaysia, with special reference to the "Massenerhebung" effect. *Journal of Ecology*, **81**, 145–167.

BUCKLEY, R. C., CORLETT, R. T. & GRUBB, P. J. (1980) Are the xeromorphic trees of tropical upper montane rainforests drought-resistant? *Biotropica*, **12**, 124–136.

BURSTROM, H. G. (1968) Calcium and plant growth. *Biological Review*, **43**, 287–316.

BUSSO, C. A. & RICHARDS, J. H. (1992) Diurnal variation in the temperature response of leaf extension of two bunchgrasses species in the field. *Plant, Cell and Environment*, **15**, 855–859.

CALDWELL, M. M. (1971) Solar U. V. irradiation and the growth and development of higher plants. *Photophysiology*, Vol. VI, (ed. A. C. GIESE) Academic Press, New York, pp 131–268.

CALDWELL, M. M., ROBBERECHT, R. & BILLINGS, W. D. (1980) A steep latitudinal gradient of solar ultraviolet-B radiation in the arctic-alpine zone. *Ecology*, **61**, 600–611.

CALDWELL, M. M., ROBBERECHT, R. & FLINT, S. D. (1983) Internal filters: Prospects for UV-acclimation in higher plants. *Physiologia Plantarum*, **58**, 445–450.

CARLQUIST, S. (1975) *Ecological Strategies of Xylem Evolution*. University of California Press, Berkely.

CAVELIER, J. (1986) *Relaciones hídricas y de nutrientes en bosques enanos nublados*. Tesis de Maestría, Universidad de los Andes, Mérida, Venezuela.

CAVELIER, J. (1989) *Root biomass, production, and the effect of fertilization in two tropical rainforests*. Ph. D. thesis, University of Cambridge, UK.

CAVELIER, J. (1990) Tissue water relations in elfin cloud forest tree species of Serranía de Macuira, Guajira, Colombia. *Trees*, **4**, 155–163.

CAVELIER, J. & GOLDSTEIN, G. (1989a) Mist and fog interception in elfin cloud forests in Colombia and Venezuela. *Journal of Tropical Ecology*, **5**, 309–322.

CAVELIER, J. & GOLDSTEIN, G. (1989b) Leaf anatomy and water relations in tropical elfin cloud forest tree species. *Structural and Functional Responses to Environmental Stresses* (eds. K. H. KREBB, H. RICHTER & T. M. HINCKLEY) SPB Academic Publishing, The Hague, pp 243–253.

CAVELIER, J. and MEJIA, C. A. (1990) Climatic factors and tree stature in the elfin cloud forest of Serranía de Macuira, Colombia. *Agricultural and Forest Meteorology*, **53**, 105–123.

CHAPIN, F. S. III, (1980) The mineral nutrition of wild plants. *Annual Review of Ecology and Systematics*, **11**, 233–260.

CINTRÒN, G. (1970) Variation in size and frequency of stomata with altitude in the Luquillo Mountains. *A Tropical Rain Forest* (eds. H. T. ODUM & R. F. PIGEON) Atomic Energy Commission, Springfield, pp H 133–135.

CLEEF, A. M. (1981) The vegetation of the paramos of the Colombian Cordillera Oriental. *Dissertationes Botanicae Band*, **61**, Cramer, Vaduz.

CLEEF, A. M., RANGEL, O., VAN DER HAMMEN, T. & JARAMILLO, R. (1984) La vegetación de las selvas del transecto Buritaca. *Studies on Tropical Andean Ecosystems, Volume 2, La Sierra Nevada de Santa Marta (Colombia) Transecto Buritaca-La Cumbre* (eds. T. VAN DER HAMMEN & P. M. RUIZ) J. CRAMER, Berlin, pp 267–395.

COLEY, P. D. (1983) Herbivory and defensive characteristics of tree species in a lowland tropical forest. *Ecological Monographs*, **53**, 209–233.

COULSON, C. B., DAVIES, R. I. & LEWIS, D. A. (1960) Polyphenols in plants, humus, and soil. I. Polyphenols of leaves, litter, and superficial humus from mull and mor sites. *The Journal of the Soil Science*, **11**, 20–29.

CUATRECASAS, J. (1934) Observaciones geobotánicas en Colombia. *Trabajos del Museo Nacional de Ciencias Naturales, Madrid*, Serie Botánica, **27**, p 144.

CUATRECASAS, J. (1958) Aspectos de la vegetación natural de Colombia. *Revista de la Academia Colombiana de Ciencias Físicas y Naturales*, **10**, 221–264.

CUENCA, G. (1976) *Balance nutricional de algunas leñosas de dos ecosistemas contrastantes: Bosque nublado y bosque deciduo.* Tesis de Licenciatura, Escuela de Biología, Universidad Central de Venezuela, Caracas.

CUEVAS, E. (1987) Perfil nutricional de la vegetación de turberas en el maciso del Chimanta, Edo. Bolivar, Venezuela. Resultados preliminares. *Acta Científica Venezolana*, **38**, 366–375.

CUEVAS, E. & MEDINA, E. (1986) Nutrient dynamics in Amazonian forest ecosystems. I. Nutrient flux in fine litter fall and efficiency of nutrient utilization. *Oecologia*, **68**, 466–472.

DARLING, M. S. (1989) Epidermis and hypodermis of the Saguaro cactus (*Cereus giganteus*): Anatomical and spectral properties. *American Journal of Botany*, **76**, 1698–1706.

DAVIS, R. I. (1971) Relation of polyphenols to decomposition of organic matter and to pedogenetic processes. *Soil Science*, **111**, 80–85.

EDWARDS, P. J. (1982) Studies of mineral cycling in a montane rainforest of New Guinea. V. Rates of cycling in throughfall and litter fall. *Journal of Ecology*, **70**, 807–827.

EKERN, P. C. (1964) Direct interception of cloud water on Lanaihale, Hawaii. *Proceedings of the Soil Science Society of America*, **28**, 419–421.

ENRIQUEZ, S., DUARTE, C. M., & SAND-JENSEN, K. (1993) Patterns in decomposition rates among photosynthetic organisms: The importance of detritus C:N:P content. *Oecologia*, **94**, 457–471.

EVANS, G. C. (1972) *The Quantitative Analysis of Plant Growth.* University of California Press, Berkeley.

FANJUL, L. & BARRADAS, V. L. (1987) Diurnal and seasonal variations in the water relations of some deciduous and evergreen trees of a deciduous dry forest of the western coast of Mexico. *Journal of Applied Ecology*, **24**, 289–303.

FARQUHAR, G. D. & RICHARDS, R. A. (1984) Isotopic composition of plant carbon correlates with water use efficiency of wheat genotypes. *Australian Journal of Plant Physiology*, **11**, 539–552.

FASSBENDER, H. W. & GRIMM, V. (1981) Ciclos biogeoquímicos en un ecosistema forestal de los Andes Occidentales de Venzuela. II. Producción y descomposición de los residuos vegetales. *Turrialba*, **31**, 39–47.

FETCHER, N. (1979) Water relations of five tropical tree species on Barro Colorado Island, Panama. *Oecologia*, **40**, 229–233.

FIELD, C., & MOONEY, H. A. (1986) The photosynthesis-nitrogen relationship in wild plants. *On the Economy of Plant Form and Function* (ed. T. J. GIVNISH). Cambridge University Press, pp. 25–49.

FLENLEY, J. R. (1979) *The Equatorial Rain Forest: A geological history.* Butterworths, London.

FLENLEY, J. R. (1992) Ultraviolet-B insolation and the altitudinal forest limit. *Nature and Dynamics of Forest-Savanna Boundaries* (eds. P. A. FURLEY, J. PROCTOR & J. A. RATTER) Chapman and Hall, London, pp 273–282.

FLENLEY, J. R. (1993) Cloud forest, the Massenherbebung effect, and ultraviolet insolation. *Tropical Montane Cloud Forests* (eds. L. S. HAMILTON, J. O. JUVIK & F. N. SCATENA) East-West Center Program on Environment, Hawaii, pp 94–96.

FÖLSTER, H. & DE LAS SALAS, G. (1976) Litterfall and mineralization in three tropical evergreen stands, Venezuela. *Acta Científica Venezolana*, **27**, 196–202.

FRIEND, A. D. & WOODWARD, F. I. (1990) Evolutionary and ecophysiological responses of mountain plants to the growing season environment. *Advances in Ecological Research*, **20**, 59–124.

GATES, D. E. (1969) The ecology of an elfin forest in Puerto Rico, 4. Transpiration rates and temperatures of leaves in cool humid environments. *Journal of the Arnold Arboretum*, **50**, 93–98.

GERRISH, G. & BRIDGES, K. W. (1984) A thinning and fertilizer experiment in *Metrosideros* dieback stands in Hawaii. *Hawaii Botanical Science Paper*, **43**, 1–107.

GIVNISH, T. J. & VERMELJ, G. J. (1976) Sizes and shapes of liane leaves. *American Naturalist*, **110**, 743–778.

GRUBB, P. J. (1971) Interpretation of the "Massenerhebung" effect on tropical mountains. *Nature*, **229**, 44–45.

GRUBB, P. J. (1974) Factors controlling the distribution of forests on tropical mountains: New facts and a new perspective. *Altitudinal Zonation in Malesia* (eds. J. R. FLENLEY) *University of Hull, Geography Department Miscellaneous Series*, **16**, 1–25.

GRUBB, P. J. (1977) Control of forest growth and distribution on wet tropical mountains, with special reference to mineral nutrition. *Annual Review of Ecology and Systematics*, **8**, 83–107.

GRUBB, P. J. (1986) Sclerophylls, pachyphylls and pycnophylls: The nature and significance of hard leaf surfaces. *Insects and the Plant Surface* (eds. B. JUNIPER & R. SOUTHWOOD) Edward Arnold, London.

GRUBB, P. J. & EDWARDS, P. J. (1982) Studies on mineral cycling in a montane rainforest in New Guinea. III. The distribution of mineral elements in the above ground material. *Journal of Ecology*, **70**, 623–648.

GRUBB, P. J. & TANNER, E. V. J. (1976) The montane forests and soils of Jamaica: A reassessment. *Journal of the Arnold Arboretum*, **57**, 313–368.

GRUBB, P. J. & WHITMORE, T. C. (1966) Comparison of montane and lowland rain forests in Ecuador. II. The climate and its effect on the distribution and physiognomy of forests. *Journal of Ecology*, **54**, 303–333.

GRUBB, P. J. & WHITMORE, T. C. (1967) A comparison of montane and lowland forest in Ecuador. II. The light reaching the ground vegetation. *Journal of Ecology*, **55**, 33–57.

GRUBB, P. J., LLOYD, J. R., PENNINGTON, T. D. & WHITMORE, T. C. (1963) A comparison of montane and lowland rainforest in Ecuador. I. The forest structure, physiognomy, and floristics. *Journal of Ecology*, **51**, 567–601.

HASTENRATH, S. (1967). Rainfall distribution and regime in Central America. *Archives for Meteorology Geophysics and Bioclimatology Series B.*, **15**, 201–241.

HASTENRATH, S. (1991). *Climate Dynamics of the Tropics*. Kluwer Academic Publishers, Dordrecht.

HEANEY, A. & PROCTOR, J. (1989) Chemical elements in litter in forests on Volcán Barva, Costa Rica. *Mineral Nutrients in Tropical Forest and Savanna Ecosystems* (ed. J. PROCTOR), Special Publication Number 9 of the British Ecological Society, Blackwell Scientific Publications, Oxford, pp 255–271.

HEDBERG, O. (1951) Vegetation belts of the East African Mountains. *Svensk Botanisk Tidskrift*, **45**, 140–202.

HOLDRIDGE, L. R., GRENKE, W. C., HATHEWAY, W. H., LIANG, T. & TOSI, J. A. (1971) *Forest Environments in Tropical Life Zones. A Pilot Study*. Pergamon Press, Oxford.

HORNER, J. D., GOSZ, J. R. & Cates, R. G. (1988) The role of carbon-based plant secondary metabolites in decomposition in terrestrial ecosystems. *The American Naturalist*, **136**, 869–883.

HUBER, O. (ed.) (1986) *La Selva Nublada de Rancho Grande Parque Nacional "Henry Pittier"*. Fondo Editorial Acta Científica Venezolana, Caracas.

HUBICK, K. T., FARQUHAR, G. D. & SHORTER, R. (1986a) Correlation between water-use efficiency and carbon isotope discrimination in diverse peanut

(*Arachis*) germplasm. *Australian Journal of Plant Physiology*, **13**, 803–816.

JANZEN, D. H. (1967) Why mountain passes are higher in the tropics. *American Naturalist*, **101**, 233–249.

JENNY, H., BINGHSM, F. & PADILLA-SARAVIA, B. (1948) Nitrogen and organic matter contents of equatorial soils of Colombia, South America. *Soil Science*, **66**, 173–186.

JUVIK, J. O. & EKERN, P. C. (1978) *A Climatology of Mountain Fog on Mauna Loa, Hawaii Island*. University of Hawaii, Water Resources Research Center, Technical Report No. 18.

KAPOS, V. & TANNER, E. V. J. (1985) Water relations of Jamaican upper montane rainforest trees. *Ecology*, **66**, 241–250.

KITAYAMA, K. (1992) An altitudinal transect study of the vegetation on Mount Kinabalu, Borneo. *Vegetatio*, **102**, 149–171.

KÖRNER, CH. (1989) The nutritional status of plants from high altitudes. *Oecologia*, **81**, 379–391.

KÖRNER, CH. & COCHRANE, P. M. (1985) Stomatal responses and water relations of *Eucalyptus pauciflora* in summer along an elevational gradient. *Oecologia*, **66**, 443–455.

KÖRNER, CH. & DIEMER, M. (1987) *In situ* photosynthetic response to light, temperature, and carbon dioxide in herbaceous plants from low and high altitude. *Functional Ecology*, **1**, 179–194.

KÖRNER, CH. & MAYR, R. (1980) Stomatal behavior in alpine plant communities between 600 and 2600 meters above sea level. *Plants and Their Atmospheric Environment* (eds. J. GRACE, E. D. FORD & P. G. JARVIS) Blackwell, Oxford, pp 205–218.

KÖRNER, CH., ALLISON, A. & HILSCHER, H. (1983) Altitudinal variation of leaf diffuse conductance and leaf anatomy in heliophytes of montane New Guinea and their interrelation with microclimate. *Flora*, **174**, 91–135.

KÖRNER, CH., BANNISTER, P. & MARK, A. F. (1986) Altitudinal variation in stomatal conductance, nitrogen content and leaf anatomy in different plant life forms in New Zealand. *Oecologia*, **69**, 577–588.

KÖRNER, CH., FARQUHAR, G. D. & ROKSANDIC, Z. (1988) A global survey of carbon isotope discrimination in plants from high altitude. *Oecologia*, **74**, 623–632.

KÖRNER, CH., FARQUHAR, G. D. & WONG, S. C. (1991) Carbon isotope discrimination by plants follows latitudinal and altitudinal trends. *Oecologia*, **88**, 30–40.

KÖRNER, CH., SCHEEL, J. A. & BAUER, H. (1978) Maximum leaf diffusive conductance in vascular plants. *Photosynthetica*, **13**, 45–82.

LANGE, O. L., LOSCH, P., SCHULZE, E. -D. & KAPPEN, L. (1971) Responses of stomata to changes in humidity. *Planta*, **100**, 76–86.

LAUER, W. (1975) *Klimatische Grundzüge der Höhenstufung tropischer Gebirge.* Tagungsbericht und wissenschaftliche Abhandlungen, 40 Deutscher Geographentag, Innsbruck, F. Steiner.

LAUER, W. (1976) Zur hygrischen Hohenstufung tropischer Gebirge. *Neotropische Okosysteme* (ed. J. SCHMITHUSEN) Biogeografica 7, The Hague.

LAWTON, R. O. (1982) Wind stress and elfin stature in a montane rainforest tree: An adaptive explanation. *American Journal of Botany*, **69**, 1224–1230.

LYFORD, W. H. (1969) The ecology of an elfin forest in Puerto Rico. 7. Soil, root, and earthworm relationships. *Journal of the Arnold Arboretum*, **50**, 210–224.

LEIGH, E. G., Jr. (1975) Structure and climate in tropical rainforest. *Annual Review of Ecology and Systematics*, **6**, 67–86.

LOVELESS, A. R. (1961) A nutritional interpretation of sclerophylly based on differences in the chemical composition of sclerophyllous and mesophytic leaves. *Annals of Botany*, **25**, 168–183.

LOVELESS, A. R. (1962) Further evidence to support a nutritional interpretation of sclerophylly. *Annals of Botany*, **26**, 551–561.

MACHADO, J. L. & TYREE, M. T. (1994) Patterns of hydraulic architecture and water relations of two tropical canopy trees with contrasting leaf phenologies: *Ochroma pyramidale* and *Pseudobombax septenatum*. *Tree Physiology*, **14**, 219–240.

MARRS, R. H., PROCTOR, J., HEANEY, A. & MOUTHFORD, M. D. (1988) Changes in soil nitrogen-mineralization and nitrification along an altitudinal transect in tropical rainforest in Costa Rica. *Journal of Ecology*, **76** , 466–482.

MEDINA, E. (1984) Nutrient balance and physiological processes at the leaf level. *Physiology of Plants of the Wet Tropics* (eds. E. MEDINA, H. A. MOONEY & C. VASQUEZ-YANES) W. JUNK, The Hague, pp 139–154.

MEDINA, E. (1986) Aspectos ecofisiológicos de plantas de bosques nublados tropicales: el bosque nublado de Rancho Grande. *La Selva Nublada de Rancho Grande, Parque Nacional "Henri Pittier"* (ed. O. HUBER) Fondo Editorial Acta Científica Venezolana, Caracas, pp 189–196.

MEDINA, E., CUEVAS, E. & WEAVER, P. L. (1981) Composición foliar y transpiración de especies leñosas del Pico del este, Sierra de Luquillo, Puerto Rico. *Acta Científica Venezolana*, **32**, 159–165.

MEDINA, E., SOBRADO, M. & HERRERA, R. (1978) Significance of leaf orientation for leaf temperature in an Amazonian sclerophyll vegetation. *Radiation and Environmental Biophysics*, **15**, 131–140.

MEINZER, F., GOLDSTEIN, G. & JAIMES, M. (1984) The effect of atmospheric humidity on stomatal control of gas exchange in two tropical coniferous species. *Canadian Journal of Botany*, **62**, 591–595.

MOONEY, H. A. & GULMON, S. L. (1979) Environmental and evolutionary constraints on the photosynthetic characteristics of higher plants. *Topics in Plant Population Biology* (eds. O. T. SOLBRIG, S. JAIN, G. B. JOHNSON & P. H. RAVEN) Columbia University Press, New York, pp 316–337.

MOTT, K. A., GIBSON, A. C. & ÒLEARY, J. W. (1982) The adaptive significance of amphistomatic leaves. *Plant, Cell and Environment*, **5**, 455–460.

MOTTA, N. & MEDINA, E. (1978) Early growth and photosynthesis of tomato (*Lycopersicon esculentum*) under nutritional deficiencies. *Turrialba*, **28**, 131–141.

MURALI, N. S. & TERAMURA, A. H. (1986) Intraspecific differences in *Cucumis sativus* sensitivity to ultraviolet-B radiation. *Physiologia Plantarum*, **58**, 673–677.

MUZIKA, R. M. & PREGITZER, K. S. (1992) Effect of nitrogen fertilization on leaf phenolic production of grand fir seedlings. *Trees*, **6**, 241–244.

MYERS, B. J., ROBICHAUX, R. H., UNWIN, G. L. & CRAIG, I. E (1987) Leaf water relations and anatomy of a tropical rainforest tree species vary with crown position. *Oecologia*, **74**, 81–85.

NIEMANN, G. J., PUREVEEN, J. B. M., EIJKEL, G. B., POORTER, H. & BOON, J. J. (1992) Differences in relative growth rate in 11 grasses correlate with differences in chemical composition as determined by pyrolysis mass spectrometry. *Oecologia*, **89**, 567–573.

NOBEL, P. S. (1980) Leaf anatomy and water use efficiency. *Adaptation of Plants to Water and High Temperature Stress* (eds. N. C. TURNER & P. J. KRAMER) John Wiley and Sons, New York, pp 43–55.

NOBEL, P. S. (1991) *Physicochemical and Environmental Plant Physiology*. Academic Press, San Diego.

NOBEL, P. S., ZARAGOZA, L. J. & SMITH, W. K. (1975) Relationship between mesophyll surface area, photosynthetic rate, and illumination level during development for leaves of *Plectranthus parviflorus* Henckel. *Plant Physiology*, **55**, 1067–1070.

NYE, P. H. (1961) Organic matter and nutrient cycles under moist tropical forest. *Plant and Soil*, **13**, 333–346.

OBERBAUER, S. F., STRAIN, B. R. & RIECHERS, G. H. (1987) Field water relations of a wet-tropical forest tree species, *Pentaclethra macroloba* (Mimosaseae). *Oecologia*, **71**, 369–374.

ODUM, H. T. (1970) Rainforest structure and mineral cycling homeostasis. *A Tropical Rain Forest* (eds. H. T. ODUM & R. F. PIGEON) Atomic Energy Commission, Springfield, pp H 3–52

ODUM, H. T., DREWRY, G. & KLINE, J. R. (1970) Climate at El Verde, 1963–1966. *A Tropical Rain Forest* (eds. H. T. ODUM & R. F. PIGEON) Atomic Energy Commission, Springfield, pp 347–417.

PARKHURST, D. & LOUCKS, O. (1972) Optimal leaf size in relation to environment. *Journal of Ecology*, **60**, 505–537.

PITTIER, H. (1935) Apuntaciones sobre geobotánica de Venezuela. *Boletín de la Sociedad Venezolana de Ciencias Naturales*, **3**, 93–114.

PROCTOR, J., ANDERSON, J. M., & VALLACK, H. W. (1983) Comparative studies on forests, soils and litterfall at four altitudes on Gunung Mulu, Sarawak. *The Malaysian Forester*, **46**, 60–75.

PROCTOR, J., LEE, Y. F., LANGLEY, A. M., MUNRO, W. R. C. & NELSON, T. (1988) Ecological studies on Gunung Silam, a small ultrabasic mountain in Sabah, Malaysia. I. Environment, forest structure and floristics. *Journal of Ecology*, **76**, 320–340.

PROCTOR, J., PHILLIPS, C., DUFF, G. K., HEANEY, A. & ROBERTSON, F. M. (1989) Ecological studies on Gunung Silam, a small ultrabasic mountain in Sabah, Malaysia. II. Some forest processes. *Journal of Ecology*, **77**, 317–331.

RADA, F., GOLDSTEIN, G., AZOCAR, A. & MEINZER, F. (1985a) Freezing avoidance in Andean giant rosette plants. *Plant, Cell and Environment*, **8**, 501–507.

RADA, F., GOLDSTEIN, G., AZOCAR, A., & MEINZER, F. (1985b) Daily and seasonal osmotic changes in a tropical treeline species. *Journal of Experimental Botany*, **36**, 987–1000.

REICH, P. B. & WALTERS, M. B. (1994) Photosynthesis-nitrogen relations in Amazonian tree species. II. Variation in nitrogen vis-a-vis specific leaf area influences. *Oecologia*, **97**, 73–81.

REICH, P. B., WALTERS, M. B., ELLSWORTH, D. S. & UHL, C. (1994) Photosynthesis-nitrogen relations in Amazonian tree species. I. Patterns among species and communities. *Oecologia*, **97**, 62–72.

RICHARDS, P. W. (1954) *The Tropical Rain Forest*. Cambridge University Press, Cambridge.

ROBBERECHT, R., CALDWELL, M. M. & BILLINGS, W. D. (1980) Leaf ultraviolet optical properties along a latitudinal gradient in the arctic-alpine life zone. *Ecology*, **61**, 612–619.

ROTH, I. (1990) Leaf structure of a Venezuelan cloud forest in relation to microclimate. *Encyclopedia of Plant Anatomy*, **XIV**, Gebrüder Borntraeger, Berlin, 244 p.

SAMPER, C. (1992) *Natural disturbance and plant establishment in an Andean cloud forest*. Ph. D thesis, Harvard University, Cambridge.

SARMIENTO, G. (1986) Ecological features of climate in high tropical mountains, in *High Altitude Tropical Biogeography* (eds. F. VUILLEUMIER & M. MONASTERIO) Oxford University Press, New York, pp 11–45.

SCHIMPER, A. F. W. (1903) *Plant Geography upon a Physiological Basis*. Clarendon Press, Oxford.

SCHUBERT, G. & MEDINA, E. (1982) Evidence of quaternary glaciation in the Dominican Republic: Some implications for Carribean paleontology. *Paleogeography, Paleoclimatology and Paleoecology*, **39**, 281–294.

SCHULZE, E. -D., LANGE, O. L., BUSCHBOM, U., KAPPEN, L. & EVENARI, M. (1972) Stomatal responses to changes in humidity in plants growing in the desert. *Planta*, **108**, 259–270.

SOBRADO, M. A. (1986) Aspects of tissue water relations and seasonal changes of leaf water potential components of evergreen and deciduous species coexisting in tropical dry forest. *Oecologia*, **68**, 413–416.

STEINHARDT, U. (1979) Untersuchungen uber den Wasser- und Nahrstoffhaushalt eines andien Wolkenwaldes in Venezuela. *Gottinger Bodenkundliche Berichte*, **56**, 185.

STOCKER, O. (1931) Transpiration und Wasserhaushalt in verschiedenen Klimazonen. I. Untersuchungen and der artischen Baumgrenze in Schwedisch Lappland. *Jahrb. wiss. Bot.* **75**, 494–549.

SUGDEN, A. M. (1982) The vegetation of the Serrania de Macuira, Guajira, Colombia: A contrast of arid lowlands and an isolated cloud forest. *Journal of the Arnold Arboretum*, **63**, 1–30.

SUGDEN, A. M. (1985) Leaf anatomy in a Venezuelan montane forest. *Botanical Journal of the Linnean Society*, **90**, 231–241.

SUGDEN, A. M. (1986) The montane vegetation and flora of Margarita Island, Venezuela. *Journal of the Arnold Arboretum*, **67**, 187–232.

SWIFT, M. J., HEAL, O. W., & ANDERSON, J. M. (1979) *Decomposition in Terrestrial Ecosystems* (*Studies in Ecology, Vol. 5*) University of California Press, Berkeley.

TANNER, E. V. J. (1977a) *Mineral cycling in montane rain forests in Jamaica.* Ph. D. thesis, University of Cambridge, UK.

TANNER, E. V. J. (1977b) Four montane rainforests of Jamaica: A quantitative characterization of the floristics, the soils, and the foliar mineral levels, and a discussion of the interrelations. *Journal of Ecology*, **65**, 883–918.

TANNER, E. V. J. & KAPOS, V. (1982) Leaf structure of Jamaican upper montane rainforest trees. *Biotropica*, **14**, 16–24.

TANNER, E. V. J., KAPOS, V. & FRANCO, W. (1992) Nitrogen and phosphorus fertilization effects on Venezuelan montane forest trunk growth and litterfall. *Ecology*, **73**, 78–86.

TANNER, E. V. J., KAPOS, V., FRESCOS, S., HEALEY, J. R. & THEOBALD, A. M. (1990) Nitrogen and phosphorus fertilization of Jamaican montane forest trees. *Journal of Tropical Ecology*, **6**, 231–238.

TERAMURA, A. H. (1983) Effects of ultraviolet-B radiation on the growth and yield of crop plants. *Physiologia Plantarum*, **58**, 415–427.

THORNTHWAITE, C. W. (1943) Problems in the classification of climates. *Geographical Review*, **33**, 233–255.

TIESZEN, L. L., SENYIMBA, M. M., IMBAMBA, S. K. & TROUGHTON, J. H. (1979) The distribution of C_3 and C_4 grasses and carbon isotope discrimination along an altitudinal and moisture gradient in Kenya. *Oecologia*, **37**, 337–350.

TROLL, C. (1968) The cordilleras of the Tropical Americas. Aspects of climate, phytogeographical, and agrarian ecology. Geo-ecología de las Regiones Montañosas de las Americas Tropicales. *Colloquium Geographicum*, **9**, 163–186.

VAN DER HAMMEN, T. (1984a) Temperaturas de suelo en el transecto Buritaca-La Cumbre. *Studies on Tropical Andean Ecosystems, Volume 2, La Sierra Nevada de Santa Marta (Colombia) Transecto Buritaca-La Cumbre* (eds. T. VAN DER HAMMEN & P. M. RUIZ) J. Cramer, Berlin, pp 67–74.

VAN DER HAMMEN, T. (1984b) Tipos de suelos en relación con ecosistemas en el transecto Buritaca-La Cumbre. *Studies on Tropical Andean Ecosystems, Volume 2, La Sierra Nevada de Santa Marta (Colombia) Transecto Buritaca-La Cumbre* (eds. T. VAN DER HAMMEN & P. M. RUIZ) J. Cramer, Berlin, pp 139–154.

VAN DER HAMMEN, T. & VAN REENEN, G. B. A. (1984) Intensidad de la luz a nivel del suelo en el transecto Buritaca. *Studies on Tropical Andean Ecosystems, Volume 2, La Sierra Nevada de Santa Marta (Colombia) Transecto Buritaca-La Cumbre* (eds. T. VAN DER HAMMEN & P. M. RUIZ) J. Cramer, Berlin, pp 93–98.

VAN STEENIS, C. G. G. J. (1961) An attempt towards an explanation of the effect of mountain mass elevation. *Proceedings Koninklike Nederlandse Akademie van Wetenschappen*, Series C, **64**, 435–442.

VAN STEENIS, C. G. G. J. (1972) The effect of mountain mass elevation. *The Mountains of Java* (ed. C. G. G. J. VAN STEENIS) E. J. Brill, Laiden, pp 19–20.

VAN STEENIS, C. G. G. J. (1984) Floristic altitudinal zones in Malesia. *Botanical Journal of the Linnean Society*, **89**, 289–292.

VARESCHI, V. (1966) Sobre las formas biológicas de la vegetación tropical. *Boletín de la Sociedad Venezolana de Ciencias Naturales*, **110**, 508–518.

VELEZ, V. (1994) Fenología, herbivoría y fitoquímica de *Polylepis quadrijuga* (Rosaceae) en el Páramo de Chingaza. Tesis de Biología, Universidad de los Andes, Bogotá.

VENEKLAAS, E. J. (1991) Litterfall and nutrient fluxes in two montane tropical rainforests, Colombia. *Journal of Tropical Ecology*, **7**, 319–336.

VITOUSEK, P. (1982) Nutrient cycling and nutrient use efficiency. *The American Naturalist*, **119**, 553–572.

VITOUSEK, P. (1984) Litterfall, nutrient, cycling, and nutrient limitation in tropical forests. *Ecology*, **65**, 285–298.

VITOUSEK, P. M., MATSON, P. A & TURNER, R. (1988) Elevational and age gradients in Hawaiian montane rainforest: Foliar and soil nutrients. *Oecologia*, **77**, 565–570.

VITOUSEK, P. M., WALKER, L. R., WHITEAKER, L. D., MULLER-DOMBOIS, D. & MATSON, P. A. (1987) Biological invasion by *Myrcia faya* alters ecosystem development in Hawaii. *Science*, **238**, 802–804.

VOGELMANN, H. W. (1973) Fog interception in the cloud forest of eastern Mexico. *Bioscience*, **23**, 96–100.

WAGNER, A. (1892) Zur Kenntniss des Blattbaues der Alpenflanzen und dessen biologischer Bedeutung. *Sitzungsberichte der Kaiserlichen Akademie der Wissenschaften, in Wein, Mathematisch-naturwissenschaftliche Klasse,* **100**, 487–547.

WEAVER, P. L., BYER, M. D. & BRUCK, D. L. (1973) Transpiration rates in the Luquillo mountains of Puerto Rico. *Biotropica,* **5**, 123–133.

WEAVER, P. L., MEDINA, E., POOL, D., DUGGER, K., GONZALES-LIBOY, J. & CUEVAS, E. (1986) Ecological observations in the dwarf cloud forest of the Luquillo Mountains in Puerto Rico. *Biotropica,* **18**, 79–85.

WHITMORE, T. C. (1984) *Tropical Rain Forests of the Far East. Clarendon Press, Oxford.*

WOLF, J. H. (1989) Comunidades epífitas en un transecto altitudinal en la Cordillera Central, Colombia: datos iniciales sobre la cantidad de especies de briófitos y líquenes. *Studies on Tropical Andean Ecosystems, Volume 3, La Cordillera Central Colombiana, Transecto Parque los Nevadao (Segunda Parte)* (eds. T. VAN DER HAMMEN, S. DIAZ-PIEDRAHITA & V. J. ALVAREZ) J. Cramer, Berlin, pp 455–459.

WOODWARD, F. I., KÖRNER, CH. & CRABTREE, R. C. (1986) The dynamics of leaf extension in plants with diverse altitudinal ranges. *Oecologia,* **70**, 222–226.

YODA, K. & KIRA, T. (1969) Comparative ecological studies on three main types of forest vegetation in Thailand. V. Accumulation and turnover of soil organic matter, with notes on the altitudinal soil sequence on Khao (Mt.) Luang peninsular Thailand. *Nature and Life in Southeast Asia,* **6**, 83–112.

ZINCK, A. (1986) Propiedades y estabilidad mecánicas de los suelos en ambiente de selva nublada. *La Selva Nublada de Rancho Grande, Parque Nacional "Henri Pittier"* (ed. O. HUBER) Fondo Editorial Acta Científica Venezolana, Caracas, pp 91–105.

ZOTZ, G. & WINTER, K. (1993) Short-term photosynthesis measurements predict leaf carbon balance in tropical rainforest canopy species. *Oecologia,* **191**, 409–412.

15

Phenological Responses to Seasonality in Tropical Forest Plants

S. Joseph Wright

Tropical rainforests include more plant species than any other habitat (Gentry, 1988). An unmatched variety of phenological patterns contributes to this diversity. This variety is illustrated by the one tropical forest for which the reproductive phenologies of most plants are known. On Barro Colorado Island, Panama, most species flower and fruit during short, predictable periods; still more than 600 species have been recorded in flower and in fruit during each calendar month (Croat, 1975, 1978).

Two largely independent research disciplines have sought explanations for the wide variety of phenological patterns observed among tropical forest plants. Ecologists have studied biotic interactions that select for timing (Janzen, 1967; Bawa, 1983). Examples include seasonal changes in levels of herbivory, pollination, seed predation, and seed dispersal. Physiologists, on the other hand, have studied abiotic factors that limit plants or cue plant responses (Schimpfer, 1903). Examples include wetting after drought, changes in photoperiod, seasonal changes in temperature, and short-term changes in temperature associated with heavy rainfall or invasions by cold air masses (Njoku, 1963; Wycherley, 1973; Opler, Frankie & Baker, 1976; Buttrose & Alexander, 1978; Ashton et al., 1988). A number of environmental factors that influence the phenologies of tropical forest plants have been identified. The approach has been to study single factors, however, and an integrative understanding of tropical forest plant phenologies has yet to emerge.

The control of plant phenology in tropical forests is of practical interest because deforestation is rapidly changing tropical landscapes and modifying tropical climates, which may in turn disrupt the phenologies of tropical forest plants. Forest evapotranspiration contributes to high tropical rainfall (Salati et al., 1979), and global climate models predict that deforestation will diminish rainfall and intensify tropical seasonality (Shukla, Nobre & Sellers, 1990). On a smaller scale, deforestation changes environmental conditions in surviving forest patches (Kapos, 1989). Changing tropical climates may soon disrupt the mechanisms that control the timing of plant production and alter interactions among species. The potential consequences are illustrated by the frequent failure of introduced agricultural crops when the mechanisms that control timing fail in a new environment, and critical life history events occur at the wrong time (Evans, 1980).

The purpose of this article is to review the seasonality of tropical forests, to identify environmental factors that vary seasonally and have consistent effects on plants, and to integrate those factors to predict the timing of leaf production and flowering. A clear distinction will be drawn between the proximate cues of plant phenologies and the ultimate selective factors that have shaped phenologies over evolutionary time. The timing of fruit maturation and seed dispersal will not be considered due to the additional complications imposed by germination and seedling requirements. As an example, more than 200 plant species delay fruit maturation by at least one full season after flowering on Barro Colorado Island (Croat, 1975), presumably so that dispersal coincides with conditions that are optimal for seedling establishment (Garwood, 1983).

15.1 TROPICAL FOREST SEASONS

The movements of the intertropical convergence zone (ITCZ) influence seasonality over large parts of the tropics (Hastenrath, 1985). The ITCZ develops when air warmed by the zenithal sun rises and cools adiabatically to form clouds and rain. Outside the ITCZ, the surface tradewinds rush to replace air rising within the ITCZ, and cloud cover and rainfall are reduced. The ITCZ moves latitudinally some two months after the zenithal sun, and its movements bring wet and dry seasons to large parts of the tropics. Rainfall, cloud cover, irradiance, atmospheric saturation deficits, windspeed, and potential evapotranspiration all covary seasonally. In addition to direct effects on plant growth, these seasonal patterns may affect populations of animals and microbes that interact with plants. As a consequence, most tropical forest plants experience simultaneous seasonal change in several environmental factors. These environmental factors will now be grouped into four broad classes, and evidence of their effects on tropical forest plant phenologies will be evaluated.

15.1.1 Moisture availability

In many tropical forests, low dry-season rainfall reduces moisture availability just as increases in atmospheric saturation deficits, irradiance, and leaf temperatures increase transpirational demand. The resultant moisture shortfalls are a primary determinant of tropical vegetation types.

Evergreen forests predominate where rainfall exceeds evapotranspiration for 11 or 12 months (Köppen, 1936; Medina, 1983; Whitmore, 1984). Plant water deficits have been recorded in evergreen tropical forests both during unusual drought events and also as an adaptive response to high irradiance (Chiariello, Field & Mooney, 1987; Oberbauer, Strain & Riechers, 1987). Predictable seasonal water deficits are absent from most tropical evergreen forests, however, and moisture seasonality is unlikely to be an important factor influencing plant phenologies (Frankie, Baker & Opler, 1974; Newstrom et al., 1994).

Semideciduous tropical forests occur where rainfall exceeds evapotranspiration for eight to 10 months (Köppen, 1936; Medina, 1983; Whitmore, 1984). Dry-season drought is predictable in semideciduous forests, and in some years, plant moisture stress can be severe (Leigh et al., 1990; Wright et al., 1992; Condit, Hubbell & Foster, 1995). Nonetheless, many plants maintain year-round growth. For example, 90% of the plant species are evergreen and 20% of the plant species flower, fill, and disperse fruit during the four-month dry season on Barro Colorado Island, Panama (Croat, 1975, 1978). Moisture seasonality is only likely to be an important factor influencing the phenologies of drought-sensitive species in semideciduous forests (Reich & Borchert, 1984; Wright & Cornejo, 1990a,b; Wright et al., 1992).

In contrast, deciduous tropical forests predominate where evapotranspiration exceeds rainfall for five or more months (Köppen, 1936; Medina, 1983; Whitmore, 1984). Dry-season water deficits are potentially severe, and the dominant phenological pattern is to avoid dry-season water stress by becoming deciduous during the dry season (Frankie, Baker & Opler, 1974; Lieberman & Lieberman, 1982; Bullock & Solis-Magallanes, 1990). Deciduous species may renew growth during the dry season, but constraints imposed by water deficits are again evident. For example, many deciduous species flower after a heavy dry-season rain (Opler, Frankie & Baker, 1976). More perplexing are deciduous species that renew growth under continuous drought. Reich and Borchert (1982, 1984, 1988) hypothesized that these species experience water stress early in the dry season because senescent leaves have lost the ability to control transpiration. As a consequence, leaf abscission occurs before soil water is completely depleted, and the leafless tree is able to take up water and renew growth under continuous drought.

At this time, predictions of the relationship between phenology and water stress for deciduous and semideciduous forests depend on species-by-species information on adaptations for drought resistance (Reich & Borchert, 1984; Borchert, 1994). The possible adaptations are few in number (Jones, Turner & Osmond, 1981). Transpirational water loss can be reduced through reductions in leaf area, cuticular conductance, and leaf temperatures via reductions in the amount of radiation absorbed. Water uptake can be maintained through deep, extensive root systems and low resistances to xylem water flow. Tissue tolerance to moisture stress can be enhanced through increases in tissue osmotic concentrations and the rigidity of cell walls. These mechanisms of drought resistance have all been observed among tropical forest plants (Medina, 1983). The critical question here is whether growth is possible during the drier season, as growth and cell expansion are among the first plant processes to be adversely affected by small water deficits and reductions in tissue turgor pressures (Bradford & Hsiao, 1982).

Most tropical deciduous forests include a few evergreen plants (Frankie, Baker & Opler, 1974; Lieberman & Lieberman, 1982; Bullock & Solis-Magallanes, 1990). In contrast to their deciduous neighbors, evergreen species maintain positive leaf turgor potentials in the dry season (Sobrado, 1986; Fanjul & Barradas, 1987). Similar dry-season reductions in osmotic potentials occur in both deciduous and evergreen species, however, evergreen species have deep root systems and inelastic cell walls relative to their deciduous neighbors (Sobrado & Cuenca, 1979; Sobrado, 1986; Fanjul & Barradas, 1987). Deep root systems also characterize evergreen trees in tropical savannahs (Sarmiento, Goldstein & Meinzer, 1985). A few deciduous forest species may even reverse the normal pattern of leaf phenology. For example, the shrub *Jacquinia pungens* is deciduous in the wet season and foliated in the dry season in the deciduous forests of Mesoamerica (Janzen, 1972). The depth of the root system may again be critical. The deep tap root of *J. pungens* is exceptional for a deciduous forest shrub and allows the maintenance of high dry-season water potentials (Fanjul & Barradas, 1987; Janzen, 1972; Oberbauer, 1985). Recent data for trees from semideciduous and deciduous tropical forests suggest that analyses of hydraulic architecture and consequent dry-season water relations can predict phenologies (Borchert, 1994; Machado & Tyree, 1994; Tyree & Ewers, Chapter 8).

15.1.2 Light

Most tropical forests experience substantial seasonal variation in irradiance. In the wet tropics, global radiation averages 50% greater in the highest month than in the lowest month [compiled for the 24 tropical sites with annual rainfall > 1000 mm in Müller (1982)]. Seasonal changes in cloud cover, daylength and solar elevation contribute to seasonal variation in irradiance (Chazdon & Fetcher, 1984; Chazdon et al., Chapter 1). The importance of day length and solar elevation increase with latitude, and the month of maximum global radiation falls close to the summer solstice at higher latitudes within the tropics (Figure 15.1). The quantitative importance of cloud cover can be illustrated by a comparison of the equinoxes for a 17-year record from central Panama (Windsor, 1990). Global radiation averages 31% greater on the March equinox when dry-season conditions with negligible cloud cover prevail. Cloud cover increases toward the equator (Hastenrath, 1985), and within 10° of the equator, the month of maximum global radiation is less tightly constrained by solar elevation (Figure 15.1) and reflects the local seasonality of cloud cover.

Predictable seasonal changes in light levels will be a potent selective agent on the phenologies of light limited plants. Light limits many tropical forest plants. Photosynthetically active radiation (PAR) in the shaded understory of tropical forests is as low as in any other habitat

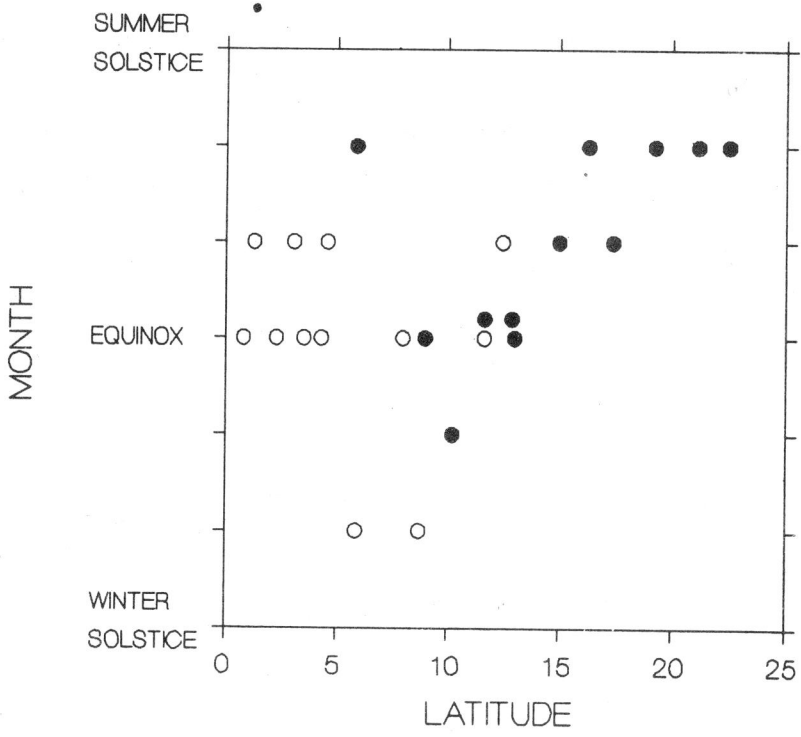

Figure 15.1. *The relation between latitude and the month of maximum global radiation within the wet tropics. Closed and open symbols represent sites where the ratio of maximum to minimum mean monthly global radiation is above and below the median for the wet tropics, respectively. Annual rainfall was greater than 1000 mm at all sites. Data compiled from Müller (1982) and Windsor (1990).*

occupied by autotrophic plants (Björkman & Ludlow, 1972). The growth of understory plants increases dramatically in response to short sunflecks that pass through small openings in the canopy (Pearcy, 1983) and after treefalls open the canopy overhead (Brokaw, 1985; Fisher, Howe & Wright, 1991).

Light may also limit taller plants. Radiant energy is extinguished exponentially with distance beneath forest canopies (Kira, Shinozaki & Hozumi, 1969). Above-canopy PAR was reduced by 94% just 5 m into the canopy of a wet forest in Puerto Rico (Johnson & Atwood, 1970), and global radiation was reduced by 53% just 6 m into the canopy of a Malaysian rainforest (Aoki, Yabuki & Koyama, 1975; Yoda, 1974). PAR is extinguished more rapidly than global radiation due to differential absorption by leaves. Limiting levels of PAR predominate directly below the uppermost leaves of tropical forests (Chazdon et al., Chapter 1).

Correlations between plant performance and irradiance suggest that tropical forest trees are light limited. The radial growth of trees from evergreen forests increases with irradiance in Surinam and possibly Costa Rica (Schulz, 1960; Clark & Clark, 1994). Seed set increases with irradiance for several trees in Borneo (Wycherley, 1973), and the likelihood of a mast flowering increases with irradiance in peninsular Malaysia (Van Schaik, 1986). These correlations are consistent with limitation by irradiance, but shifts in allocation and effects of covarying environmental factors cannot be discounted.

Physiological measurements provide stronger evidence that light limits tropical trees. Light often limits photosynthesis by canopy leaves. Canopy leaves measured *in situ* become saturated with light at photosynthetic photon flux densities (PFD) of 450 to 600 μmol m^{-2} s^{-1} for trees from a wide variety of tropical forests (Oberbauer & Strain, 1986; Pearcy, 1987; Zotz and Winter, Chapter 3). Lower light levels will limit photosynthesis. PFD varied by three orders of magnitude and limited photosynthesis in most determinations made between 0800 and 1000 h for leaves from above 29 m in an Australian rainforest tree. PFD was less than 400 μmol m-2 s-1 in 71% of diurnal course measurements for canopy leaves of *Pentaclethra macroloba* in an evergreen forest in Costa Rica (Oberbauer & Strain, 1986). Low light levels associated with heavy cloud cover often limit photosynthesis by the emergent tree *Ceiba pentandra* during the wet season in central Panama (Zotz & Winter, Chapter 3). Self shading, lateral shading by neighbors, and shading by competing lianas affect all but the uppermost leaves in tropical forests, and light interception by clouds affects all leaves. Thus, predictable, seasonal changes in irradiance may be an important selective agent on the phenologies of tropical forest plants.

15.1.3   Biotic interactions

Seasonal changes in the activities of interacting animals and microbes also influence tropical forest plant phenologies. Possibilities include selection to coincide with the seasonality of mutualists, such as pollinators, and selection to avoid the seasonality of pests, such as seed predators and herbivores. Many tropical forest animals specialize on a particular host plant or on a small group of related host plants

(Janzen, 1980; Frankie et al., 1983). It is well established that host specialists can cause strong selection on plant phenologies (Augspurger, 1981). However, specialists must track the seasonality of their hosts, and stabilizing selection or directional selection for earlier or later production may result depending on the autecologies of the particular species involved (Wright, 1990; Van Schaik, Terborgh & Wright, 1993). Host specialists are unlikely to have consistent, predictable effects on plant phenology.

Consistent effects are more likely when a variety of host generalists share a common seasonality. For example, in seasonal tropical forests, many herbivorous insects are inactive or restricted to moist microsites during the drier season (Janzen, 1973; Wolda, 1978). This shared seasonality creates the potential for phenological selection.

This potential will be illustrated by a comparison of seasonal changes in leaf herbivory across three tropical forests with very different seasonalities. Tropical forest is near its distributional limit in the Accra Plains of Ghana where rainfall seasonality is extreme. Insect herbivory was largely restricted to the wet season at this site, and strong selection for dry-season leaf production resulted (Lieberman & Lieberman, 1982). The seasonal fluctuations of herbivorous insects decrease in tropical forests with milder dry seasons (Wolda, 1988). Relative to Accra, seasonal changes in herbivory were modest in the semideciduous forests of Barro Colorado Island, Panama, averaging 20–22% in the three driest months and 25–42% in the nine wetter months (Aide, 1988). Finally, seasonal changes in herbivory were absent for an understory shrub in the evergreen forests of La Selva, Costa Rica (Marquis, 1987). The potential for selection by seasonal insect populations increases with rainfall seasonality. Less is known about seasonal changes in tropical forest microbes, but a similar pattern is to be expected (G. Gilbert, personal communication).

15.1.4 Mineral nutrients

Temporal variation in nutrient availability has been documented for several tropical forests. Pulses of high nitrogen and phosphate availability are often associated with rapid changes in moisture availability that cause lysis of soil and litter microbes (reviewed by Lodge, McDowell & McSwiney, 1994). Microbial death simultaneously frees nutrients and reduces microbial competition for those nutrients. The type of change in moisture availability that induces microbe death varies among forests. Microbe death is caused by droughts in normally aseasonal, everwet forests, and by the rapid increase in moisture availability at the beginning of the wet season in monsoonal and other seasonally dry forests.

There may also be long-term seasonal changes in nutrient availability that are unrelated to rapid changes in moisture availability. Magnesium and calcium availability increased steadily throughout the wet season, reached their annual peak in the final month of the wet season, and then crashed rapidly to low dry-season levels for three consecutive years on Barro Colorado Island (Yavitt & Wright, unpublished data). This seasonal pattern of availability may characterize nutrients whose concentrations remain high in recalcitrant litter fractions until final release from decayed litter (Cornejo, Verala & Wright, 1994). Other nutrients that are released more rapidly during decomposition had different seasonal patterns on Barro Colorado Island. Phosphate and ammonium increased in the dry season then decreased in the wet season, while potassium and sulfate did not vary seasonally. Likewise, phosphate and ammonium availabilities did not vary among months in the relatively aseasonal, evergreen forests of La Selva, Costa Rica (Vitousek & Denslow, 1986).

Predictable changes in nutrient availability, like those outlined above, may select for plant phenologies. Many wild plants are able to store and recycle nutrients effectively (Chapin, 1980); however, this would tend to minimize the impact of temporal variation in nutrient availabilities. Perhaps for this reason the potential effect of temporal variation in nutrient availability on plant phenologies has generally been overlooked.

15.2 SEASONALITY AND PHENOLOGY

Seasonal changes in moisture availability, irradiance, and insect activity can be expected to have consistent effects on large numbers of plant species in seasonal tropical forests. These three factors will often covary, with the drier season having higher irradiance and reduced insect activity. We have yet to integrate the potential effects of these three factors in order to predict tropical forest plant phenologies.

Rainfall seasonality will take precedence over other seasonal changes whenever water deficits are sufficient to arrest plant cell expansion and growth (Reich & Borchert, 1984). Given adequate water supplies, selection will favor leaf production during seasons of low herbivore activity and high irradiance. This will be true because most herbivory occurs in the first month of leaf life before leaves

become fully lignifie. .nd also because photosynthetic capacities decline with age once lea res complete expansion (Coley, 1980; Wright &
Van Schaik, 1994). Flowering will be selected to coincide with maximum photosynthesis because it is energetically most efficient to transfer assimilates directly to growing organs rather than to store them
in different tissues first and to mobilize and translocate them later
(Chapin, Schulze & Mooney, 1990; Wright & Van Schaik, 1994). Thus,
there are two general predictions. First, plants without dry-season
water supplies must produce leaves and flowers in the wet season
(Reich & Borchert, 1984). Second, when water is available, selection
will favor leaf and flower production in seasons when irradiance is
high and herbivory is low (Wright & Van Schaik, 1994). Specific
predictions will vary with site seasonality and plant drought sensitivity.

In evergreen tropical forests, where moisture deficits are absent or
occur irregularly and fluctuations in herbivore pressure are modest,
the timing of leaf and flower production is predicted to coincide with
peak irradiance. This hypothesis was tested for four evergreen forests
where minimum mean monthly rainfall was greater than 60 mm
(Wright & Van Schaik, 1994). On average, twice as many tree species
as expected by chance centered leaf and flower production on the three
sunniest months in evergreen forests from Central America, the
Guyana shield, central Amazonia and Atlantic coastal Brazil (Figure
15.2). This result is consistent with the hypothesis that tree phenologies have been selected to coincide with peak irradiance (Wright &
Van Schaik, 1994).

Specific predictions for more seasonal forests must incorporate an
independent assessment of dry-season access to water. Such an analysis was possible for monsoon forests from East Java, where roots were
excavated, and for deciduous and semideciduous forests from
Venezuela and Panama, where dry-season tissue water potentials
were measured (Coster, 1932; Sobrado, 1986; Wright & Van Schaik,
1994). Irradiance peaked in the dry season in all three forests, and 12
of the 17 species with access to dry-season water supplies concentrated leaf and flower production in the dry season. In contrast, the
15 species with limited access to dry-season water all concentrated
leaf and flower production in the wetter season (Wright & Van Schaik,
1994). Additional evidence that water stress limits leaf and flower
production in deciduous and semideciduous tropical forests was reviewed in section 15.1.1.

Dry-season peaks in irradiance and dry-season lows in pest activity
will both select for dry-season leaf and flower production in most

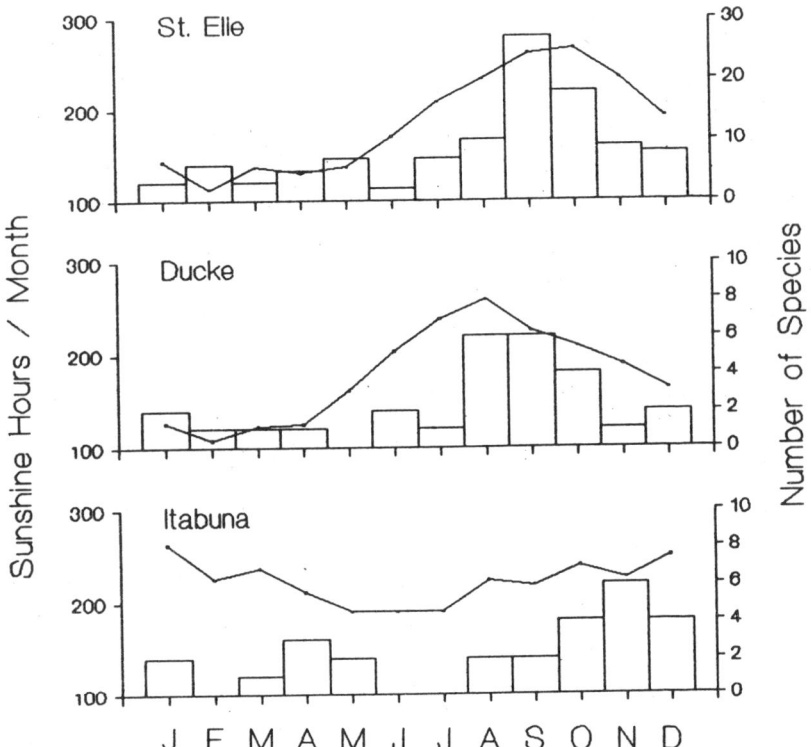

Fig. 15.2. The relation between flowering and sunshine for three evergreen tropical forests. The solid line represents mean numbers of sunshine hours for each month. The histogram represents numbers of tree species in flower. Flowering data for St. Elie, French Guyana; Ducke, Brazil; and Itabuna, Brazil from Sabatier (1985), Alencar et al. (1979), and Alvim and Alvim (1978), respectively. Sunshine hours taken from these sources or from standard meteorological references (see Wright & Van Schaik, 1994).

seasonal tropical forests. The seasonal forests of Makakou, Gabon, offer an opportunity to evaluate the relative importance of seasonal changes in insect abundance and irradiance. Insect biomass follows the expected seasonal pattern at Makokou and averages 3.9 times greater in the wet season. The three dry months are exceptionally cloudy, however, and the number of sunshine hours actually averages 2.1 times greater during the wetter months (Charles-Dominique, 1977). More than 95% of the tree and liana species examined at Makakou concentrated leaf and flower production during the wetter, sunnier months, and the dry-season peak of leaf and flower production characteristic of other tropical forests with a three-month dry season was absent (Hladik, 1973, 1978; Charles-Dominique, 1977). The Makokou forest phenologies are also consistent with the hypothesis that seasonal patterns of irradiance have been an important selective agent on tree seasonality in tropical forests.

Specific predictions are also possible for plant life forms for two reasons. First, there are consistent differences in rooting depths among life forms. Mean maximum rooting depths are less than 20 cm for mature herbs, between 30 and 90 cm for mature shrubs, between 70 and 120 cm for 2 m to 3 m tall tree saplings, and potentially much deeper for mature canopy trees and lianas on Barro Colorado Island, Panama (Becker & Castillo, 1990; Mulkey, Smith & Wright 1991; Wright et al., 1992; personal observation). All else equal, dry-season water deficits should be maximal among understory herbs, intermediate among shrubs, and least for mature trees and canopy lianas. A second difference among plant life forms that might contribute to differences in phenologies concerns irradiance seasonality at sites with significant seasonal changes in cloud cover. Cloud cover scatters and reduces direct beam radiation and increases diffuse radiation. Cloud cover will cause large reductions in PAR in the canopy, where direct beam radiation is important, and smaller reductions in PAR in the understory, where diffuse radiation is more important (Björkman & Ludlow, 1972). All else equal, understory herbs and shrubs should experience relatively modest dry-season increases in irradiance, while canopy trees and lianas should experience large dry-season increases in irradiance. Both rooting depths and irradiance seasonality lead to the prediction that selection for dry-season production should be weakest among understory herbs, intermediate among understory shrubs, and strongest for canopy trees, and lianas. On Barro Colorado Island, Panama, the proportion of forest species that concentrate both flowering and fruit maturation in the dry season is least for understory herbs and greatest for trees, as predicted (Table 15.1).

15.3 PROXIMATE CUES

The proximate cues that initiate plant production may differ from the ultimate selective factors. For example, photoperiod is an excellent predictor of freezing winter temperature at higher latitudes, and photoperiod is the proximate cue for the phenologies of many temperate zone plants even though potential damage from freezing is the

Table 15.1. *The number of forest herb, shrub, and tree species that flower and disperse fruit in the same season on Barro Colorado Island, Panama. Data compiled from Croat (1987).*

	Season	
	Dry	Wet
Herbs	3	33
Shrubs	5	21
Trees	22	39

ultimate selective factor. Likewise, in the tropics, the proximate factors that initiate phenological changes may differ from the ultimate selective factors.

Physical cues that anticipate seasonality will vary with latitude within the tropics. At higher tropical latitudes, there is significant annual variation in photoperiod and temperature, and changes in both factors anticipate movements of the intertropical convergence zone and seasonal changes in rainfall, irradiance, and animal and microbial activity. Two examples will be used to illustrate the potential importance of photoperiod and temperature as proximate cues for tropical forest plant phenologies. The native range of the avocado (*Persea americana*) falls between 15° and 20° N, and reproductive buds differentiate in response to seasonal changes in temperature (Buttrose & Ambrose, 1978). The forest tree *Hildegardia barteri* occurs at 7° N, where photoperiod changes by 53 min, and a 1 hr decrease in photoperiod suppresses seedling leaf production (Njoku, 1963). Photoperiod and temperature cannot be uniformly important cues throughout the tropics; however, because photoperiod is constant at the equator, and seasonal temperature variation is negligible at the meteorological equator at about 5°N (Hastenrath, 1985), plants at these latitudes must respond to other proximate cues.

Rainfall and the restoration of plant water status after drought is the proximate cue for renewed growth in many deciduous and semideciduous forest plants (Opler, Frankie & Baker, 1976; Augspurger, 1982; Reich & Borchert, 1984). However, a large-scale forest irrigation experiment demonstrated that seasonal changes in soil moisture and drought were generally unimportant as phenological cues for canopy trees and lianas from semideciduous forest on Barro Colorado Island. Irrigation maintained soil moisture near field capacity for five consecutive dry seasons. Nonetheless, the timing of phenological events associated with the transition to the dry season (flowering and leaf fall) and also the transition to the wet season

(flowering) were unaffected in more than 95% of the species considered (Wright, 1991; Wright & Cornejo, 1990a,b). The absence of an
effect of forest irrigation on tree and liana phenologies suggests that
changes in atmospheric conditions that were not altered by irrigation
may be important phenological cues.

A comparison of the timing of leaf fall from three sites in western
Africa again implicates atmospheric conditions as an important phenological cue. Forests in the Ivory Coast, Nigeria, and Gabon experience two dry seasons (Hopkins, 1966; Bernhard-Reversat, Huttel &
Lemee, 1972; Hladik, 1978). In the Ivory Coast and Nigeria, rainfall
reaches its annual low from December through February. In Gabon,
mean monthly rainfall exceeds 80 mm at this time, but falls to just 22
mm between June and August. If rainfall and soil drought were the
proximate cue for leaf fall, then leaf fall would track rainfall, and the
timing of peak leaf fall would differ between these three forests. In
contrast, irradiance and potential evapotranspiration peak between
December and February in all three forests. If atmospheric conditions
were the proximate cue for leaf fall, then peak leaf fall would occur
between December and February in all three forests, and this was, in
fact, the case (Hopkins, 1966; Bernhard-Reversat, Huttel & Lemee,
1972; Hladik, 1978). In the low latitude tropics where variation in
most physical factors is limited, changes in atmospheric conditions
may become important cues for plant phenologies.

At least three possible mechanisms might allow atmospheric conditions to cue tropical forest plant phenologies. First, small temperature
drops associated with heavy rainfall and with tropical incursions by
cold air masses are both well established proximate cues for tropical
forest plant phenologies (Wycherley, 1973; Ashton, Givnish & Appanah
1988). A second possibility is that seasonal increases in irradiance
might cue production. A similar response is shown by many understory
plants when production is initiated in response to increased irradiance
shortly after the creation of a treefall gap. In this case, increased photosynthesis and increased tissue carbohydrate concentrations might trigger growth. The third possibility is that increases in potential evapotranspiration may create temporary water deficits that hasten the abscission of senescent leaves with limited control of transpiration (Reich
& Borchert, 1984). This could trigger bud break, particularly in species
where bud break is suppressed by the presence of old leaves (Borchert,
1983). Further studies of the mechanisms that control the timing of
plant production in the low latitude tropics are clearly needed to explore
these possibilities and identify other, as yet unanticipated, cues.

15.4 CONCLUSIONS

The timing of production by tropical forest plants can be predicted from seasonal patterns of rainfall and irradiance and mechanisms of drought resistance. Many more species produce leaves and reproduce in the season of greatest irradiance than expected by chance in evergreen forests and in seasonal forests, where peak irradiance occurs in the wet season. In seasonal forests where peak irradiance occurs in the dry season, disproportionately large numbers of species with adaptations that maintain dry-season water uptake produce leaves and reproduce in the drier, sunnier seasons, whereas species that lack these adaptations are limited to the wetter, cloudier season.

The strength of these community-level patterns and the universal effects of low light and water on plants suggest that these may be the primary evolutionary determinants of plant seasonality in tropical forests. Interacting organisms also create strong selective pressures on plant seasonality. For many tropical forest plants, these effects may fine-tune the timing of production within constraints imposed by the seasonal pattern of rainfall and irradiance.

The proximate mechanisms that control the phenologies of low-latitude tropical forest plants are largely unexplored. Evidence reviewed here suggests that atmospheric conditions, including cloud cover and atmospheric saturation deficits may be critical. Tropical deforestation increases atmospheric saturation deficits in surviving forest patches (Kapos, 1989) and on a larger scale, reduces evapotranspiration and convectional cloud cover (Salati et al., 1979). This raises the possibility that environmental conditions may change in tropical forest reserves as surrounding forest is removed (Shukla, Nobre & Sellers, 1990). These changes may disrupt the timing of plant production and alter interactions among species because very few reserves will be large enough to maintain the natural hydrological cycle. Tropical forests are being destroyed at a rapid pace. Global climate models indicate that deforestation will intensify seasonality in the tropics (Shukla, Nobre & Sellers, 1990). The possible consequences for forest plants in remnant reserves lends urgency to the study of tropical plant seasonality.

REFERENCES

AIDE, T. M. (1988) Herbivory as a selective agent on the timing of leaf production in a tropical understory community. *Nature*, **336**, 574–575.

ALENCAR, J. DA C., DE ALMEIDA, R. A. & FERNANDES, N. P. (1979) Fenologia de especies florestais em floresta tropical umida de terra firme na Amazonia Central. *Acta Amazonica,* **9**, 163–198.

ALVIM, P. DE T. & ALVIM, R. (1978) Relation of climate to growth periodicity in tropical trees. *Tropical Trees as Living Systems* (eds. P. B. TOMLINSON & M. H. ZIMMERMANN) Cambridge University Press, London, pp 445–464.

AOKI, M., YABUKI, K. & KOYAMA, H. (1975) Micrometeorology and assessment of primary production of a tropical rainforest in West Malaysia. *Journal of Agricultural Meteorology,* **31**, 115–124.

ASHTON, P. S., GIVNISH, T. J. & APPANAH, S. (1988) Staggered flowering in the Dipterocarpaceae, new insights into floral induction, and the evolution of mast fruiting in the aseasonal tropics. *American Naturalist,* **132**, 44–66.

AUGSPURGER, C. K. (1981) Reproductive synchrony of a tropical shrub, experimental studies on effects of pollinators and seed predators on *Hybanthus prunifolius* (Violaceae). *Ecology,* **62**, 775-788.

AUGSPURGER, C. K. (1982) A cue for synchronous flowering. *The Ecology of a Tropical Forest* (eds. E. G. LEIGH, JR., A. S. RAND & D. M. WINDSOR) Smithsonian Institution Press, Washington, D. C., pp 133–150.

BAWA, K. S. (1983) Patterns of flowering in tropical plants. *Handbook of Experimental Pollination Biology,* (eds. C. E. JONES & R. J. LITTLE) Van Nostrand Reinhold Co., New York, pp 394–410.

BECKER, P. & CASTILLO, A. (1990) Root architecture of shrubs and saplings in the understory of a tropical moist forest in lowland Panama. *Biotropica,* **22**, 242–249.

BERNHARD-REVERSAT, F., HUTTEL, C. & LEMEE, G. (1972) Quelques aspects de la periodicite ecologique et de l'activity vegetate saisoniere en foret ombrophile sempervirente de Cote-d'Ivoire. *Papers from a Symposium on Tropical Ecology with an Emphasis on Organic Production* (eds. F. B. GOLLEY & R. MISRA) University of Georgia Press, Athens, Georgia, pp 219–234.

BJÖRKMAN, O. & LUDLOW, M. M. (1972) Characterization of the light climate on the floor of a Queensland rainforest. *Carnegie Institution of Washington Yearbook,* **71**, 85–94.

BORCHERT, R. (1983) Phenology and control of flowering in tropical trees. *Biotropica,* **15**, 81–89.

BORCHERT, R. (1994) Soil and stem water storage determine phenology and distribution of tropical dry forest trees. *Ecology,* **75**, 1437–1449.

BRADFORD, K. J. & HSIAO, T. C. (1982) Physiological responses to moderate water stress. *Physiological Plant Ecology II. Water Relations and Carbon Assimilation* (eds. O. L. LANGE, P. S. NOBEL, C. B. OSMOND & H. ZIEGLER) *Encyclopedia of Plant Physiology, New Series, Volume 12B.* Springer-Verlag, Berlin, pp 263–324.

BROKAW, N. (1985) Gap-phase regeneration in a tropical forest. *Ecology*, **66**, 682–687.

BULLOCK, S. H. & SOLIS-MAGALLANES, J. A. (1990) Phenology of canopy trees of a tropical deciduous forest in Mexico. *Biotropica*, **22**, 22–35.

BUTTROSE, M. S. & ALEXANDER, D. McE. (1978) Promotion of floral initiation in 'fuerte' avocado by low temperature and short daylength. *Scientia Horticulturae*, **8**, 213–217.

CHAPIN, F. S., III. (1980) The mineral nutrition of wild plants. *Annual Review of Ecology and Systematics*, **11**, 233–260.

CHAPIN, F. S., III, Schulze, E. -D. & MOONEY, H. A. (1990) The ecology and economics of storage in plants. *Annual Review of Ecology and Systematics*, **21**, 423–447.

CHARLES-DOMINIQUE, P. (1977) *Ecology and Behaviour of Nocturnal Primates.* Columbia University Press, New York.

CHAZDON, R. L. & FETCHER, N. (1984) Light environments of tropical forests. *Physiological Ecology of Plants of the Wet Tropics* (eds. E. MEDINA, H. A. MOONEY & C. VAZQUEZ-YANES) DR. W. Junk Publishers, The Hague, pp 27–36.

CHIARIELLO, N. R., Field C. B. & MOONEY H. A. (1987) Midday wilting in a tropical pioneer tree. *Functional Ecology*, **1**, 3–11.

CLARK, D. A. & CLARK, D. B. (1994) Climate induced annual variation in canopy tree growth in a Costa Rican tropical rainforest. *Journal of Ecology*, **82**, 865–872.

COLEY, P. D. (1980) Effects of leaf age and plant life history patterns on herbivory. *Nature*, **284**, 545–546.

CONDIT, R., HUBBELL, S. P. & FOSTER, R. B. (1995) Changes in tree species abundance in a Neotropical forest over eight years, impact of climate change. *Journal of Tropical Ecology*, in press.

CORNEJO, F. H., VERALA, A. & WRIGHT, S. J. (1994) Tropical forest litter decomposition under seasonal drought: Nutrient release, fungi and bacteria. *Oikos*, **70**, 183–190.

COSTER, C. (1932) Wortelstudien in de tropen. I. De jeugdontwikkeling van een zeventigtal boomen en groenbemesters. *Tectona*, **25**, 828–872.

CROAT, T. B. (1975) Phenological behavior of habit and habitat classes on Barro Colorado Island (Panama Canal Zone). *Biotropica*, **7**, 270–277.

CROAT, T. B. (1978) *Flora of Barro Colorado Island.* Stanford University Press, Stanford, California.

DOLEY, D., YATES, D. J. & UNWIN G. L. (1987) Photosynthesis in an Australian rainforest tree, *Argyrodendron peralatum*, during the rapid development and relief of water deficits in the dry season. *Oecologia*, **74**, 441–449.

EVANS, L. T. (1980) The natural history of crop yield. *American Scientist*, **68**, 388–397.

FANJUL, L. & BARRADAS, V. L. (1987) Diurnal and seasonal variation in the water relations of some deciduous and evergreen trees of a deciduous dry forest of the western coast of Mexico. *Journal of Applied Ecology*, **24**, 289–303.

FISHER, B., HOWE, H. F. & S. J. WRIGHT. (1991) Survival and growth of *Virola surinamensis* yearlings: Water augmentation in gap and understory. *Oecologia*, **86**, 292–297.

FRANKIE, G. W., BAKER, H. G. & OPLER, P. A. (1974) Comparative phenological studies of trees in tropical wet and dry forests in the lowlands of Costa Rica. *Journal of Ecology*, **62**, 881–919.

FRANKIE, G. W., HABER, W. A., OPLER, P. A. & BAWA, K. S. (1983) Characteristics and organization of the large bee pollination system in the Costa Rican dry forest. *Handbook of Experimental Pollination Biology* (eds. C. E. Jones & R. J. LITTLE) Van Nostrand Reinhold, New York, pp 441–447.

GARWOOD, N. C. (1983) Seed germination in a seasonal tropical forest in Panama: A community study. *Ecological Monographs*, **53**, 159–181.

GENTRY, A. H. (1988) Changes in plant community diversity and floristic composition on environmental and geographical gradients. *Annals of the Missouri Botanical Garden*, **75**, 1–34.

HASTENRATH, S. (1985) *Climate and Circulation of the Tropics*. D. Reidel Publishing Co., Dordrecht, Holland.

HLADIK, A. (1978) Phenology of leaf production in rainforest of Gabon: Distribution and composition of food for folivores. *The Ecology of Arboreal Folivores* (ed. G. G. MONTGOMERY), Smithsonian Institution Press, Washington, D. C., pp 51–72.

HLADIK, C. M. (1973) Alimentation et activite d'un groupe de chimpanzes reintroduits en foret Gabonaise. *Terre et Vie*, **27**, 343–413.

HOPKINS, B. (1966) Vegetation of the Olokemeji forest reserve, Nigeria IV. The litter and soil, with special reference to their seasonal changes. *Journal of Ecology*, **54**, 687–703.

JANZEN, D. H. (1967) Synchronization of sexual reproduction of trees within the dry season in Central America. *Evolution*, **21**, 620–637.

JANZEN, D. H. (1972) *Jacquinia pungens*, a heliophile from the understory of tropical deciduous forest. *Biotropica*, **2**, 112–119.

JANZEN, D. H. (1973) Sweep samples of tropical foliage insects: Effects of seasons, vegetation types, elevation, time of day and insularity. *Ecology*, **54**, 687–708.

JANZEN, D. H. (1980) Specificity of seed-attacking beetles in a Costa Rican deciduous forest. *Journal of Ecology*, **68**, 929–952.

JOHNSON, P. L. & ATWOOD, D. M. (1970) Aerial sensing and photographic study of the El Verde rainforest. *A Tropical Rain Forest* (ed. H. T. ODUM), Division of Technical Information, U. S. Atomic Energy Commission, Washington, D. C., pp B–63–B–78.

JONES, M. M., TURNER, N. C. & OSMOND, C. B. (1981) Mechanisms of drought resistance. The Physiology and Biochemistry of Drought Resistance in Plants (eds. L. G. PALEG & D. ASPINALL), Academic Press, Sydney, Australia, pp 15–38.

KAPOS, V. (1989) Effects of isolation on the water status of forest patches in the Brazilian Amazon. Journal of Tropical Ecology, 5, 173–185.

KIRA, T., SHINOZAKI, K. & HOZUMI, K. (1969) Structure of forest canopies as related to their primary productivity. Plant and Cell Physiology, 10, 129–142.

KÖPPEN, W. (1936) Das geographische system der Klimate. Handbuch der Klimatologie, Vol. I (eds. W. KOPPEN & W. GEIGER) Teil C. Gebruder Borntraeger, Berlin.

LEIGH, E. G. JR., WINDSOR, D. M., RAND, S. A. & FOSTER, R. B. (1990) The impact of the El Nino drought of 1982–1983 on a Panamanian semideciduous forest. Global Ecological Consequences of the 1982–1983 El Nino-Southern Oscillation (ed. P. W. GLYNN) Elsevier Press, New York, pp 473–486.

LIEBERMAN, D. & LIEBERMAN, M. (1982) The causes and consequences of synchronous flushing in a dry tropical forest. Biotropica, 16, 193–201.

LODGE, D. J., McDOWELL, W. H. & McSWINEY, C. P. (1994) The importance of nutrient pulses in tropical forests. Trends in Ecology and Evolution, 9, 384–387.

MACHADO, J. -L. & TYREE, M. T. (1994) Patterns of hydraulic architecture and water relations of two tropical canopy trees with contrasting leaf phenologies: Ochroma pyramidale and Pseudobombax septenatum. Tree Physiology, 14, 219–240.

MARQUIS, R. J. (1987) Variacion en la herbivoria foliar y su importancia selectiva en Piper arieianum (Piperaceae). Revista de Biologia Tropical, 35 (supp. 1), 133–150.

MEDINA, E. (1983) Adaptations of tropical trees to moisture stress. Tropical Rain Forest Ecosystems (ed. F. B. GOLLEY), Elsevier, Amsterdam, pp 225–237.

MULKEY, S. S., SMITH, A. P. & WRIGHT, S. J. (1991) Comparative life history and physiology of two understory neotropical herbs. Oecologia, 88, 263–273.

MÜLLER, M. J. (1982) Selected climate data for a global set of standard stations for vegetation science. Tasks for Vegetation Science 5. W. Junk Publishers, The Hague.

NEWSTROM, L. E., FRANKIE, G. W., BAKER, H. G. & COLWELL, R. K. (1994) Diversity of long-term flowering patterns. La Selva (ed. L. A. McDADE, K. S. BAWA, H. A. HESPENHEIDE & G. S. HARTSHORN) University of Chicago Press, Chicago, Illinois, pp 142–160.

NJOKU, E. (1963) Seasonal periodicity in the growth and development of some forest trees in Nigeria. *Journal of Ecology*, **51**, 617–624.

OBERBAUER, S. F. (1985) Plant water relations of selected species in wet and dry tropical lowland forests in Costa Rica. *Revista de Biologia Tropical*, **33**, 137–142.

OBERBAUER, S. F. & STRAIN, B. R. (1986) Effects of canopy position and irradiance on the leaf physiology and morphology of *Pentaclethra macroloba* (Mimosaceae). *American Journal of Botany*, **73**, 409–416.

OBERBAUER, S. F., STRAIN, B. R., & RIECHERS, G. H. (1987) Field water relations of a wet-tropical forest tree species, *Pentaclethra macroloba* (Mimosaceae). *Oecologia*, **71**, 369–374.

OPLER, P. A., FRANKIE, G. W. & BAKER, H. G. (1976) Rainfall as a factor in the release, timing, and synchronization of anthesis by tropical trees and shrubs. *Journal of Biogeography*, **3**, 231–236.

PEARCY, R. W. (1983) The light environment and growth of C_3 and C_4 tree species in the understory of a Hawaiian forest. *Oecologia*, **58**, 19–25.

PEARCY, R. W. (1987) Photosynthetic gas exchange responses of Australian tropical forest trees in canopy, gap, and understory microenvironments. *Functional Ecology*, **1**, 169–178.

REICH, P. B. & BORCHERT, R. (1982) Phenology and ecophysiology of the tropical tree, *Tabebuia neochrysantha* (Bignoniaceae). *Ecology*, **63**, 294–299.

REICH, P. B. & BORCHERT, R. (1984) Water stress and tree phenology in a tropical dry forest in the lowlands of Costa Rica. *Journal of Ecology*, **72**, 61–74.

REICH, P. B. & BORCHERT, R. (1988) Changes with leaf age in stomatal function and water status of several tropical tree species. *Biotropica*, **20**, 60–69.

SABATIER, D. (1985) Saisonalite et determinisme du pic de fructification en foret guyanaise. *Terre et la Vie*, **40**, 289–320.

SALATI, E., DALL'OLIO, A., MATSUI, E. & BAT, J. A. (1979) Recycling of water in the Amazon basin: An isotopic study. *Water Resources*, **15**, 1250–1258.

SARMIENTO, G., GOLDSTEIN, G. & MEINZER, F. (1985) Adaptive strategies of woody species in neotropical savannas. *Biological Reviews*, **60**, 315–355.

SCHIMPFER, A. F. W. (1903) *Plant Geography Upon a Physiological Basis*. Clarendon Press, Oxford.

SCHULZ, J. P. (1960) *Ecological Studies in Northern Suriname*. Noord-Hollandse Uitgeversmaatschappij, Amsterdam.

SHUKLA, T., NOBRE, C. & SELLERS, P. (1990) Amazon deforestation and climate change. *Science*, **247**, 1322–1325.

SOBRADO, M. A. (1986) Aspects of tissue water relations and seasonal changes of leaf water potential components of evergreen and deciduous species coexisting in tropical dry forests. *Oecologia*, **68**, 413–416.

SOBRADO, M. A. & CUENCA, G. (1979) Aspectos del uso de agua de especies deciduas y siempreverdes en un bosque seco tropical de Venezuela. *Acta Cientifica Venezolana*, **30**, 302–308.

VAN SCHAIK, C. P. (1986) Phenological changes in a Sumatran rainforest. *Journal of Tropical Ecology*, **2**, 327–347.

VAN SCHAIK, C. P., TERBORGH, J. W. & WRIGHT, S. J. (1993) The phenology of tropical forests: Adaptive significance and consequences for primary consumers. *Annual Review of Ecology and Systematics*, **24**, 353–377.

VITOUSEK, P. M. & DENSLOW, J. S. (1986) Nitrogen and phosphorus availability in treefall gaps of a lowland tropical forest. *Journal of Ecology*, **74**, 1167–1178.

WHITMORE, T. C. (1984) *The Tropical Rain Forests of the Far East*. Oxford University Press, Oxford.

WINDSOR, D. M. (1990) Climate and moisture variability in a tropical forest, long-term records for Barro Colorado Island, Panama. *Smithsonian Contributions to Earth Sciences*, **29**, 1–145.

WOLDA, H. (1978) Seasonal fluctuations in rainfall, food, and abundance of tropical insects. *Journal of Animal Ecology*, **47**, 369–381.

WOLDA, H. (1988) Insect seasonality: Why? *Annual Review of Ecology and Systematics*, **19**, 1–18.

WRIGHT, S. J. (1990) Cumulative satiation of a seed predator over the fruiting season of its host. *Oikos*, **58**, 272–276.

WRIGHT, S. J. (1991) Seasonal drought and the phenology of shrubs in a tropical moist forest. *Ecology*, **72**, 1643–1657.

WRIGHT, S. J. & CORNEJO, F. H. (1990a) Seasonal drought and the timing of flowering and leaf fall in a neotropical forest. *Reproductive Ecology of Tropical Forest Plants* (eds. K. S. BAWA & M. HADLEY), Man and the Biosphere Series, UNESCO, Paris and Parthenon Publishing, Carnforth, England, pp 49–61.

WRIGHT, S. J. & CORNEJO, F. H. (1990b) Seasonal drought and leaf fall in a tropical forest. *Ecology*, **71**, 1165–1175.

WRIGHT, S. J. & VAN SCHAIK, C. P. (1994) Light and the phenology of tropical trees. *American Naturalist*, **143**, 192–199.

WRIGHT, S. J., MACHADO, J. L., MULKEY, S. S. & SMITH, A. P. (1992) Drought acclimation among tropical forest shrubs (*Psychotria, Rubiaceae*). *Oecologia*, **89**, 457–463.

WYCHERLEY, P. R. (1973) The phenology of plants in the humid tropics. *Micronesica*, **9**, 75–96.

YODA, K. (1974) Three-dimensional distribution of light intensity in a tropical rainforest of west Malaysia. *Japanese Journal of Ecology*, **24**, 247–254.

16

Comparative Ecophysiology of Mangrove Forest and Tropical Lowland Moist Rainforest

Marilyn C. Ball

"Mangrove" is an ecological term referring to a taxonomically diverse association of woody trees and shrubs that form the dominant vegetation in tidal, saline wetlands along tropical and subtropical coasts (Tomlinson, 1986). There, moist lowland rainforest gives way to mangrove vegetation where the forest experiences tidal inundation with saline water. There is an abrupt transition from rainforest, with its high diversity of tree species, to mangrove forest of relatively few species. The diverse assemblage of life forms so common in rainforest gives way to forest where vines, palms, ferns, and epiphytes are poorly represented and conifers are absent (Tomlinson, 1986). For example, Tomlinson (1986) conservatively recorded 114 species from 66 genera in his treatment of the floristics of mangrove forests worldwide, with species richness being greatest in the Indo-Pacific region. Thus, fewer mangrove species are found worldwide than one might encounter in a few hectares of moist tropical forest, particularly in areas supporting the greatest biodiversity (Whitmore, 1992).

Tropical lowland moist rainforest and mangrove forest also differ in structure. The complex, multilayered canopy of rainforest contrasts with the relatively simple canopy and sparse understory vegetation of mangrove forest. In addition, the mangrove species are often distributed differentially along tidally maintained environmental gradients such that a banded zonation pattern forms normal to shore (Hutchings & Saenger, 1987). These bands, which may be monospecific or dominated by a mixture of two to three tree species, are well correlated with the frequency and duration of tidal inundation (Watson, 1928; Chapman, 1944; Macnae, 1968). However, correlations between mangroves and tidal characteristics tend to be site specific because tidal characteristics mainly influence vegetation through intermediate factors. These factors, which either directly affect growth or are resources required for growth, define the "environmental space" in which species are distributed (Austin, 1985). In the case of a mangrove forest, such factors would include the form and availability of nutrients (Boto, 1982), the degree of soil saturation, and the salinity of surface and soil water (Thom, 1967; Clarke & Hannon, 1970).

This review considers how physiological attributes of mangroves might differ from those of rainforest species given inherent differences in their edaphic environments, and how physiological attributes might relate to mangrove forest structure along tidally maintained environmental gradients.

16.1 NUTRIENTS

Tropical moist lowland rainforest and mangrove forest appear to differ in nutrient relationships in many respects, including pathways of nutrient cycling and the relative importance of phosphate and nitrogen in limiting productivity. In rainforests, mycorrhizae play an important role in the rapid cycling of nutrients required to sustain lush growth (Vitousek, 1984; Whitmore, 1992). Much less is known of nutrient cycling in mangrove forests. In Australia, leaf processing by crabs may enhance rates of nutrient cycling as crabs break down litter at > 75 times the rate of microbial decay alone, thereby facilitating high rates of productivity by sedimentary bacteria (Robertson & Daniel, 1989). In addition to their role in decomposition, some sedimentary bacteria may contribute directly to the supply of available nitrogen. Rates of nitrogen fixation by bacteria are higher in the rhizosphere of mangrove roots than in adjacent plant-free sediments, implying that the bacteria may benefit from root exudates while the mangroves benefit from an enhanced supply of combined inorganic nitrogen (Zuberer & Silver, 1978). In general, however, little is known of the possible involvement of mycorrhizal or other symbiotic systems in satisfying demands for nutrients in nutrient-poor coastal environments.

The distribution and availability of nutrients in mangrove forests are strongly influenced by tidal regimes in several ways. First, tidal waters transport dissolved and particulate-bound nutrients and plant litter into and out of the forests. Second, patterns of tidal inundation influence the degree of soil saturation and hence, redox potential of the soil, which has major implications for the availability of both micro- and macro-nutrients (Boto, 1982; Valiela, 1984).

Soils in infrequently flooded areas near the landward limits of mangrove vegetation undergo periodic changes in redox potential (Boto, 1982). During periods of low tidal amplitude, these soils change from flooded, anaerobic conditions to a partially aerobic state. Under aerobic conditions, phosphorus readily adsorbs onto clay and precipitates with calcium, magnesium, and iron. These reactions could cause phosphorus to be largely unavailable in mineral, oxidized soils at infrequently flooded sites (Boto, 1982). In contrast, the availability of phosphorus can be much higher in anaerobic sediments of frequently flooded sites where bacteria and H_2S reduce ferric iron (Fe^{+3}) to ferrous iron (Fe^{+2}) (Valiela, 1984). This process increases availability of dissolved phosphate in interstitial water because ferrous iron is much less effective at adsorbing phosphate than ferric iron. In addition, the presence of abundant organic matter in anaerobic soils may contribute to accumulation of dissolved phosphate by inhibiting its adsorption onto clay surfaces (Valiela, 1984). The net result is that the ratio (by weight) of adsorbed phosphate to the equilibrium concentration of dissolved phosphate ranges between 1 and 5 in reduced sediments, and between 25 and 5000 in aerobic sediments (Krom & Berner, 1980). These differences between aerobic and anaerobic sediments in availability of phosphorus may explain why mangrove growth was enhanced by the addition of phosphate at an infrequently flooded site (Boto & Wellington, 1983, 1984), whereas both mangrove (Boto & Wellington, 1983, 1984) and saltmarsh vegetation (Valiela, 1984) failed to respond to addition of phosphate at frequently flooded sites.

Nitrogen is another nutrient whose availability is affected by tidal flooding regimes (Boto, 1982; Valiela, 1984). At high tidal elevations, concentrations of nitrate and ammonia are very low as they are readily leached from mineral, oxidized soils. At low tidal elevations, ammonia, which is the major form of combined inorganic nitrogen under reducing conditions, can reach relatively high concentrations because it adsorbs onto organic particles. However, anaerobic conditions inhibit the uptake of nitrogen by saltmarsh plants (Howes et al., 1981; Mendelssohn, McKee & Patrick, 1981). This may explain why, despite relatively high concentrations of ammonia, both mangrove (Boto & Wellington, 1983, 1984) and saltmarsh vegetation (Valiela, 1984) respond to the addition of nitrogen when growing in waterlogged and anaerobic soils at frequently flooded sites. These results are consistent with nitrogen being, in general, limiting to productivity in marine systems (Valiela, 1984). This contrasts with observations that phosphate limits growth in tropical moist lowland rainforest (Vitousek, 1984; Whitmore, 1992).

As discussed above, tidal flooding regimes can affect the distribution and availability of nutrients. Levels of nutrients increase with increasing tidal inundation (Boto, 1982). Mangrove growth may be limited by lack of phosphate in aerobic, infrequently flooded soils and by lack of nitrogen in anaerobic, frequently flooded soils, as demonstrated by Boto and Wellington (1983, 1984). These results are consistent with predictions based on chemical transformations of nutrients in aerobic and anaerobic marine sediments (Valiela, 1984). There may be interspecific differences in nutrient requirements for growth and in the physiological capacity to satisfy these requirements among species that characteristically occupy different positions along gradients of tidal inundation. However, little is known of species responses to differences in the form and availability of nutrients.

16.2 WATERLOGGING

Mangrove forests are subject to varying degrees of waterlogging, depending on the frequency of tidal inundation and the drainage characteristics of the soil. The plants do not respond to waterlogging itself, but rather to a range of factors associated with it. McKee (1993) noted that plants growing in waterlogged soils may be adversely affected by root oxygen deprivation, strongly reducing conditions, and accumulation of soil phytotoxins, including reduced forms of iron and manganese, organic acids, and gases such as methane. In addition, species growing in anaerobic marine sediments must also cope with toxic concentrations of H_2S, a potent inhibitor of cytochrome oxidase in respiratory metabolism. Thus, interspecific differences in both the capacity to maintain aerobic metabolism in roots and sensitivity to soil phytotoxins, partic rly H_2S, may be major determinants of relative differences in tolera e to waterlogged soils (McKee, 1993). Most research on mechanisms of survival in waterlogged mud has focused on aeration of root systems. Although the roots may cope with temporary periods of anaerobiosis (McKee & Mendelssohn, 1987), root functioning depends on maintenance of aerobic conditions in tissues growing in an otherwise anaerobic environment. The structure of mangrove roots increases the availability of oxygen to roots growing in oxygen-deficient sediments (Tomlinson, 1986). In general, the roots are shallow and have numerous lenticels and extensive aerenchyma. In *Avicennia marina*, for example, the aerenchyma accounts for as much as 70% of the total root volume (Curran, 1985). Many species also have aerial roots which occur in a wide variety of forms, including pneumatophores (e.g., *Avicennia marina*), knee roots (e.g. *Bruguiera exaristata*), stilt roots (e.g., *Rhizophora stylosa*), and plank roots (e.g., *Xylocarpus granatum*). These structural attributes provide an effective means of root ventilation (Curran, 1985; Nickerson & Thibodeau, 1985; Thibodeau & Nickerson, 1986; McKee & Mendelssohn, 1987; McKee, 1993), provided that the roots are exposed to the atmosphere at least during low tides. Such exposure is critical because aeration of the root systems depends on diffusion of oxygen down partial pressure gradients from the atmosphere to sites of respiration in the roots, and diffusion of oxygen occurs 10,000 times more rapidly through air than through water (Scholander, Dam & Scholander, 1955). The presence of aerobic conditions within roots is indicated by leakage of oxygen from roots, which causes surrounding soils to be more oxidized than would be expected in anaerobic sediments (Nickerson & Thibodeau, 1985; Thibodeau & Nickerson, 1986; McKee & Mendelssohn, 1987; McKee, 1993). Such oxidation of the rhizosphere may have major implications for nutrient uptake and plant/microbe interactions under waterlogged conditions.

Despite apparently effective systems for root ventilation, waterlogging can adversely affect growth and functioning of mangroves (Naidoo, 1983, 1985; Pezeshki, DeLaune & Patrick, 1990; McKee, 1993). Adverse effects of waterlogging undoubtedly result from a multitude of processes, some of which may have interactive effects on plant growth. For example, rates of water uptake are lower in waterlogged mangroves than in those grown under well drained conditions (Naidoo, 1985). Tolerance to increasing salinity is reduced by anaerobic conditions caused by waterlogging (van der Moezel et al., 1988), possibly because anoxia interferes with both salt exclusion and selectivity for K^+ over Na^+ (Kriedemann & Sands, 1984; Drew &

Dikumwin, 1985). Indeed, concentrations of K^+ are lower in shoot tissues in mangroves (Naidoo, 1985) and other salt-tolerant plants (van der Moezel et al., 1988) grown in waterlogged than in well-drained conditions. As relatively high tissue concentrations of K^+ are required for protein synthesis and other metabolic processes, limitations in the supply of K^+ can lead to reduction in growth, and under extreme deficiency, to metabolic dysfunction (Ball, Chow & Anderson, 1987). Finally, high levels of H_2S can inhibit nitrogen uptake (Bradley & Morris, 1990) and growth (McKee, 1993). This process might explain intriguing results of Smith et al. (1991), who showed that burrowing by crabs improved soil aeration, which in turn reduced levels of sulfide and enhanced productivity and reproductive output in a mangrove forest dominated by the Rhizophoraceae. Interspecific differences in the ability to cope with waterlogged conditions, including effects of anoxia on uptake of water and nutrients, may well contribute to characteristic segregation of species along tidal inundation gradients.

16.3 SALINITY

Salinity in the mangrove environment is due largely to NaCl and can vary in time and space from freshwater to hypersaline conditions. The salt concentration in seawater is approximately 35 g 1^{-1} which in solution includes 483 mM Na^+ and 558 mM Cl^- (Harvey, 1966). Seawater also contains relatively high concentrations of other ions, and very little is known of their individual or collective effects on plant growth and functioning in coastal wetlands.

Responses of mangroves to salinity are as varied as the environments in which they are found. Some species, such as *Sonneratia lanceolata* (Figure 16.1), grow well in freshwater and sustain maximal growth rates only over a limited range of low salinities (Ball & Pidsley, 1995). Most mangrove species can grow in freshwater, but growth is stimulated by saline conditions, with optimal salinities for growth ranging from 5% to 50% seawater depending on the species and the growth conditions (Ball, 1988a). Finally, some species, such as *Rhizophora mangle* (Werner & Stelzer, 1990) and *Sonneratia alba* (Ball & Pidsley, 1995) appear unable to grow to maturity in freshwater. Thus, interspecific differences in salt tolerance reflect a physiological continuum ranging from moderately salt-tolerant glycophytes to highly salt-tolerant and apparently obligate halophytes (Ball, 1988a).

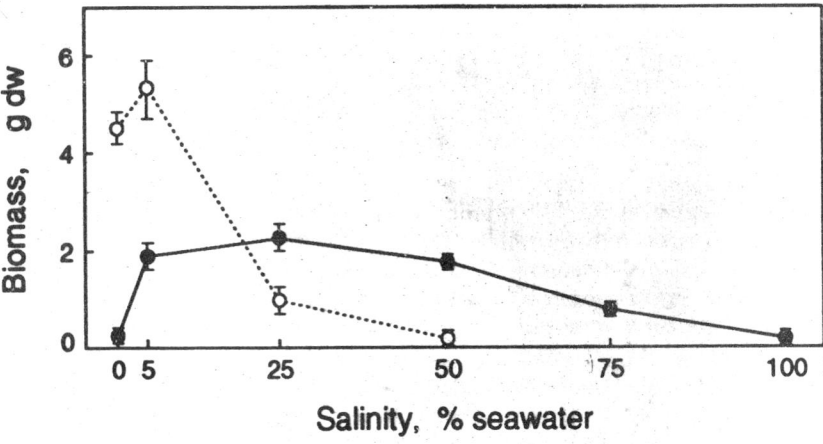

Figure 16. 1. Growth of Sonneratia alba *and* S. lanceolata *in response to salinity. Plants were grown from seed for three months in dilutions of seawater enriched with nutrients. Biomass refers to whole plants. Values are mean ± SE, n = 5, except n = 2 for* S. lanceolata *grown in 50% seawater. Solid and dashed lines link mean data from* S. alba *and* S. lanceolata, *respectively. Redrawn from Ball and Pidsley (1995).*

16.3.1 Physiological bases of salt tolerance

Coping with saline environments places three requirements on plant functioning: maintenance of favorable water relations, of selective ion uptake and compartmentation, and of a balance between the supply of ions to the shoot and the capacity to accomodate the salt influx.

16.3.1.1 Maintenance of favorable water relations

Habitats of moist tropical forest and mangrove forest place very different demands on plant water relations. A positive water balance can only be maintained if tissue water potentials are lower than the osmotic potentials of the substrate in which the plant is growing. A tree in a moist tropical rainforest may rarely experience water potentials more negative than -2 MPa, whereas a mangrove growing in seawater would never have a water potential less negative than the osmotic potential of seawater, approximately -2.5 MPa (Sperry, Tyree & Donnelly, 1988). Typical values of leaf water potential in field-grown mangroves range from -2.5 to -6 MPa (Scholander et al., 1964; Scholander, 1968; Rada et al., 1989; Naidoo, 1989; Smith et al., 1989; Sternberg et al., 1991). The water potential of a plant cell is determined by the hydrostatic pressure and osmotic potential of the cell sap, with osmotic potential being generated by solutes dissolved in the cell sap. Thus, intracellular concentrations of these solutes must increase with increasing salinity if cellular osmotic potentials are to be maintained at levels lower than those in the root environment.

Most plants accumulate inorganic ions for osmotic adjustment. In salt-sensitive species, like those of moist lowland rainforest, K^+ plays a major biophysical role in osmotic adjustment of the vacuole, whereas its osmotic role is largely replaced by Na^+ and Cl^- in salt-tolerant species (Leigh et al., 1986) such as mangroves (Popp, 1984a). Salt concentrations in leaves of the mangrove Avicennia marina, for example, increase from 300 to 600 mM NaCl (bulk leaf water basis) with increases in the salinity in which the plants were grown from 50 to 500 mM NaCl, respectively (Downton, 1982; Ball & Farquhar, 1984b, Clough, 1984). Most of these ions accumulate in the vacuole, with ion concentrations in other intracellular compartments probably differing little from those in salt sensitive species (Flowers, Hajibagheri & Clipson, 1986). In non-vacuolar regions of the cell, osmotic adjustment is apparently achieved by synthesis of compatible solutes, i.e. organic compounds of low molecular weight that do not interfere with metabolism such as proline, quaternary ammonium compounds, mannitol and cyclitols (Popp, 1984a, b; Popp, Larher & Weigel, 1984; Popp, Polania & Weiper, 1993). Intracellular compartmentation of ions is essential because metabolic processes in mangroves (Ball & Anderson, 1986; Sommer, Thonke & Popp, 1990) and other halophytes (Greenway & Osmond, 1972) are just as sensitive in vitro to high concentrations of NaCl as those isolated from salt-sensitive species. Accumulation of ions in the cytoplasmic compartments would lead to metabolic dysfunctioning and cell death even in highly salt tolerant species. Thus, the use of inorganic ions for osmotic adjustment is a common feature of plant water relations, but glycophytes and halophytes differ in both the ions used for adjustment and in the extent to which high concentrations can be sequestered in the vacuole.

16.3.1.2 Maintenance of favorable ion relations

Mangroves exclude at least 90% of the external salt in solution from entry into the transpiration stream (Scholandér et al., 1962; Scholander et al., 1966). Salt uptake is not directly coupled with water uptake. The salt concentrations in the xylem decrease hyperbolically with increase in the volume flux such that the flux of salt to the leaves does not increase with increase in transpiration rates (Ball, 1988b). Indeed, the salt flux to leaves can even decrease at very high rates of water loss (Munns, 1985; Ball, 1988b). Mechanisms of salt exclusion by the roots are unknown, but the major means of ion entry into the transpiration stream is apparently via symplastic pathways (Mallery & Teas, 1984; Moon et al., 1986; Ball, 1988b, Werner & Stelzer, 1990; Lin & Sternberg, 1993).

Salt tolerance depends on balancing the transport of ions to the shoot with the requirements or capacities of the shoot to accommodate the salt influx. Some species have salt secretion glands in the leaves. In these species, some of the salt carried to the leaves in the transpiration stream is absorbed for osmotic adjustment of growing tissues, with ion concentrations maintained within physiologically acceptable levels by salt secretion (Atkinson et al., 1967; Ball, 1988b). Most mangrove species lack salt secretion glands and must accommodate the salt influx to the shoot by growth, increasing succulence, and re-export from the leaves. The demand for ions is presumably greatest during leaf area expansion as ions play a major role in the water relations of developing tissues (Kriedemann, 1986). As leaf area expansion ceases, the concentrations of Na^+ and Cl^- ions per unit dry matter increase with leaf age (Atkinson et al., 1967; Werner & Stelzer, 1990; Popp, Polania & Weiper, 1993). However, the concentrations of these ions on a bulk leaf water basis are maintained relatively constant by increase in succulence, i.e. water content per unit area (Atkinson et al., 1967; Werner & Stelzer, 1990; Popp, Polania & Weiper, 1993). The ionic composition of the leaves also changes as they age. The increase in Na^+ is accompanied by decrease in K^+, which Werner and Stelzer (1990) interpreted as reflecting retranslocation of K^+ from older leaves to young expanding leaves by exchange of vacuolar K^+ for Na^+ combined with selective phloem transport of K^+ (Jeschke, 1984; Jeschke, Pate & Atkins, 1987). Thus, senescing leaves become depleted of K^+ and have high concentrations of Na^+ (Werner & Stelzer, 1990). Failure to maintain favourable ion concentrations would cause premature death of old leaves, and eventually affect growth of new leaves (Munns & Termaat, 1986). The great longevity of the leaves of many mangrove species (Saenger & Moverly, 1985) implies that the salt balance of the leaves is well controlled.

16.3.2   On the importance of water to plant form and function along salinity gradients

16.3.2.1   Water relations

Rada et al. (1989) conducted an intriguing study of seasonal changes in leaf water relations in three species (*Coccoloba uvifera, Conocarpus erectus* and *Rhizophora mangle*) with different distributions along a salinity gradient from inland to seashore environments, respectively, in Venezuela. There was a linear relationship between ground water potential and leaf osmotic potentials at full turgor. *Coccoloba uvifera*, which grows in soils with a relatively high proportion of fresh water, had the least negative osmotic potentials, whereas *R. mangle*, which grows in the most saline soils, had the most negative osmotic potentials. During the dry season, leaf osmotic potential decreased in all three species, but there were interspecific differences in the way osmotic adjustment was achieved. The decrease in osmotic potential in response to increase in salinity resulted from a net increase in osmotically active solutes inside the leaf cells in *C. uvifera*, from a decrease in symplasmic water content causing concentration of intracellular solutes in *R. mangle*, and from a combination of both mechanisms in *C. erectus*. Rada et al. (1989) commented that these interspecific differences in mechanisms of osmotic adjustment may be important if, as suggested by Meinzer et al. (1986), an active net increase in osmotically active solutes per cell is more costly than sequestering the same amount of solutes in a smaller cell volume. Under this scenario, *C. uvifera* would expend more energy than *R. mangle* to maintain its leaf water potential below that of the soil substrate during periods of high salinity (Rada et al., 1989).

Rada et al. (1989) also studied relationships between cell wall elasticity and turgor maintenance in the three species growing naturally along a salinity gradient. Elastic properties of cell walls are an important component of plant water relations because they affect the change in volume for a given change in pressure. In other words, differences in cell wall elasticity affect the relationship between relative water content and turgor pressure. With a decrease in relative water content, leaves with more elastic cell walls can maintain higher turgor pressures whereas those with more rigid cell walls can generate greater water potential differences between leaves and soil.

Cell wall elasticity decreased along the salinity gradient from the inland to the seashore sites such that *C. uvifera* had the most elastic cell walls and *R. mangle* the most rigid ones. Leaves of all species showed seasonal decreases in cell wall elasticity with lowering of soil water potentials during the dry season. Decrease in elasticity coupled with decrease in osmotic potential during periods of high soil salinity would contribute to maintenance of turgor because a large difference in water potential between soil and leaves (which would increase water uptake) can be produced with relatively little water loss in cells with rigid walls (Bolaños & Longstreth, 1984). Indeed, large decreases in water potential result from relatively small losses of intracellular water in mangrove leaves (Scholander et al., 1966; Scholander, 1968). Despite changes in osmotic potential and cell wall elasticity, Rada et al. (1989) found that turgor loss was not completely avoided during the dry season as leaves of two species had zero turgor pressure for a few hours during midday, but the reduction in turgor was much less than it would have been if there had been no acclimation to increasing salinity. Increasing water uptake through establishment of large water potential gradients between the soil and leaves may therefore be more important for the water economy of whole plants than midday turgor maintenance (Rada et al., 1989).

16.3.2.2 Hydraulic architecture

Maintenance of water flow into the plant during times of drought may contribute to avoidance of embolism of xylem vessels. In trees, embolism accounts for a 10–20% loss in hydraulic conductivity over the normal range of xylem pressures, but can become catastrophic when the xylem pressure becomes more negative (Sperry, 1986; Tyree & Dixon, 1986; Tyree & Ewers, Chapter 8). Generation of positive xylem pressures can refill embolized vessels (Tyree et al., 1986), but Sperry, Tyree and Donnelly (1988) pointed out that this option would not be available to a mangrove growing in seawater because the osmotic potential of the xylem sap is always near that of pure water (Scholander et al., 1962). Given this high osmotic potential, the xylem pressure must be negative in order to balance the osmotic potential of seawater (-2.5 MPa) bathing the roots. Thus, mangrove species are potentially more susceptible than inland species to drought-induced embolism, which could impair water movement through the xylem.

In an elegant study, Sperry, Tyree and Donnelly (1988) contrasted the hydraulic architecture of two tree species in the Rhizophoraceae: *Cassipourea elliptica*, which occurs in moist tropical rainforest, and the mangrove, *Rhizophora mangle*. Vulnerability of xylem conduits to cavitation (water column breakage) and subsequent embolism were consistent with the range of xylem pressures experienced by these species in their natural habitats. In *R. mangle*, the more tolerant species, stems lost little hydraulic conductivity at xylem pressures more positive than -6 MPa, but loss of conductivity was complete at a xylem pressure of -7 MPa. Embolisms were apparently caused by air-seeding at intervessel pit membranes, with the greater permeability of intervessel pit membranes in *C. elliptica* than in *R. mangle* accounting for the greater vulnerability of *C. elliptica* to embolism. Two anatomical features of the xylem could contribute to greater resistance to embolism in *R. mangle* (Sperry, Tyree & Donnelly, 1988). First, thick vessel walls could reduce the probability of air-seeding where these walls contact air-filled extracellular spaces. Second, shorter vessel length could promote compartmentalisation of potential embolisms during periods of drought or high salinity. In contrast, the highly permeable and relatively long vessels in *C. elliptica* may enable greater xylem conductance with minimal loss in safety given the characteristically high soil water potentials in its environment (Sperry, Tyree and Donnelly, 1988). These anatomical differences between *R. mangle* and *C. elliptica* may contribute to differences in their vulnerability to embolism, but the carbon costs of constructing xylem to minimize the occurrence and severity of embolism may be greater in the saline environment which is subject to lower soil water potentials.

16.3.2.3   Water-use characteristics

Both moist lowland rainforest and mangrove forest occur in environments with an abundant supply of water, but, unlike rainforest trees, mangroves have low stomatal conductances, transpire slowly and maintain unusually high water-use efficiencies for C_3 species (Ball, 1988a). For comparison, non-mangrove species growing in nonsaline sites can have stomatal conductances in excess of 400 mmol $m^{-2} s^{-1}$, whereas stomatal conductances in adjacent mangroves generally do not exceed 200 mmol $m^{-2} s^{-1}$ (Björkman, Demmig & Andrews, 1988). Under field conditions, Clough and Sim (1989) found stomatal conductances to range from 79 to 271 mmol $m^{-2} s^{-1}$, with corresponding assimilation rates ranging from 5.8 to 19.1 μmol CO_2 $m^{-2} s^{-1}$ in mangroves growing in a low salinity environment. These values contrast with an even lower range of stomatal conductances (i.e. 18 to 84 mmol $m^{-2} s^{-1}$) and assimilation rates (i.e. 2.5 to 10.3 μmol CO_2 $m^{-2} s^{-1}$) found in plants growing in highly saline environments (Clough & Sim, 1989). These observations accord well with other studies on mangroves showing that stomatal conductance, and hence also transpiration, decreases with increases in salinity (Ball & Farquhar, 1984a,b; Björkman, Demmig & Andrews, 1988; Lin & Sternberg, 1992c).

Low stomatal conductance restricts the efflux of water, but also limits the influx of CO_2, causing a leaf to operate with low intercellular CO_2 concentrations and correspondingly low assimilation rates, but with a high water-use efficiency. These gas exchange characteristics are reflected in the carbon isotope composition of leaves because isotope fractionation occurs during diffusion and carboxylation (Farquhar et al., 1982). Values of $\delta^{13}C_{PDB}$ in mangroves, ranging from -19.6 to -26.5, are consistent with fractionation predicted from their low intercellular CO_2 concentrations and high water-use efficiencies (Farquhar et al., 1982). These values are less negative than $\delta^{13}C_{PDB}$ values of -30.5 to -35.9 in less water-use efficient leaves of tropical moist lowland rainforest species (Jackson et al., 1993; Goldstein et al., Chapter 9).

Mangroves maintain or increase water-use efficiency with increased salinity, at least over the range of salinities at which growth is vigorous (Ball & Farquhar, 1984a,b; Ball, 1988b). Extremely stressed plants may show a loss of water-use efficiency when salinities are excessive. In general, water-use efficiency increases with increase in the salt tolerance of the species (Ball, 1988a). Interspecific differences in water-use characteristics are apparent from differences in the relationship between assimilation rate and stomatal conductance in response to variation in leaf temperature, irradiance, and the vapor pressure difference (vpd) between the leaf and air (Table 16.1). In the Rhizophoraceae, the slope, dA/dg, increases with increasing salinity tolerance of the species. For comparison, these values of dA/dg are greater than the range of values, 9.5 to 60.4 mol/mol, reported for less water-use efficient leaves of tropical moist lowland rainforest species (Jackson et al., 1993). The data in Table 1 show that, for a wide range of environmental factors affecting photosynthesis over the course of a day, stomatal conductance at a given assimilation rate is lower (and water use is more conservative) the greater the salinity tolerance of the species (Ball, Cowan & Farquhar, 1988).

In a recent review, Ball and Passioura (1993) concluded that conservative water use may be a consequence of many factors, but may particularly reflect the dangers of runaway embolism and inhibitory accumulation of salt around the roots. Mangrove roots take up water very slowly, primarily via symplastic pathways (Moon et al., 1986; Lin & Sternberg, 1993b). It appears that whatever mechanisms are involved in controlling the entry of ions into the transpiration stream

Table 16.1. Interspecific variation in gas exchange characteristics and in the display and properties of leaves fully exposed to the sun in relation to salinity tolerance in the Rhizophoraceae. Species are listed in order of increasing salinity tolerance, with Bruguiera gymnorrhiza being the least salt tolerant. Values for dA/dg were calculated by linear regression of the assimilation rate ($mol\ CO_2\ m^{-2}\ s^{-1}$) as a function of stomatal conductance to water vapor ($mmol\ m^{-2}\ s^{-1}$) with variation in irradiance, leaf temperature and leaf-to-air vapor pressure difference; dw = dry weight (after Ball, Cowan & Farquhar, 1988).

Species	$dA\ dg^{-1}$ ($\mu mol\ mol^{-1}$)	Rosette area (cm^2)		Individual leaf area (cm^2)		Specific leaf weight ($g\ dw\ m^{-2}$)	Succulence ($g\ water\ m^{-2}$)
		Total	Projected	Total	Projected		
Bruguiera gymnorrhiza	72	635	356	58	32	133.1	262.5
Rhizophora apiculata	96	553	196	69	25	148.8	348.4
Rhizophora stylosa	101	419	126	44	13	169.3	387.9
Ceriops tagal	113	102	39	8	3	189.2	463.2

may also restrict water flow through the roots. Indeed, hydraulic conductances of roots of *Avicennia marina*, even of plants grown in freshwater, are as much as two orders of magnitude lower than those of salt sensitive species and decrease with increasing salinity (Field, 1984). Given a large resistance to water flow, rapid transpiration rates would then induce such a low water potential in the leaves that an impossibly high concentration of solutes in the cells would be required to maintain turgor. Furthermore, rapid induction of very low water potentials would strain the capabilities of the xylem to limit the occurrence and severity of embolism. Mangroves may have evolved to operate with a sufficiently low stomatal conductance to maintain the water potential above a value that is on the threshold of inducing substantial embolism (Sperry, Tyree & Donnelly, 1988; Sperry & Tyree, 1988; Tyree & Sperry, 1988; Jones & Sutherland, 1991).

Low transpiration rates may also minimize the accumulation of salt around the roots of plants growing in waterlogged soils which are not well flushed by the ebb and flow of tidal water (Passioura, Ball & Knight, 1992). When roots extract water from saturated soil, the water must flow downwards from the soil surface towards the roots. Salt is carried, by convection, in this flow of water, but is excluded by the roots as they absorb the water. Thus, the salinity rises in the soil occupied by the roots. Salinity continues to rise until a sufficiently large concentration gradient develops to diffuse the salt back to the soil surface (where the concentration remains at approximately that of tidal water) as fast as it enters the soil by convection. Rapid rates of transpiration could cause salt to become so concentrated in the soil water that it would severely limit the rate of water uptake by the roots. For trees grown in sediments bathed by seawater, Passioura, Ball and Knight (1992) calculated the limiting rate of transpiration to be about 1 mm day^{-1}. This estimate accords well with measured values of 0.67 mm day^{-1} in a coastal stand of *Rhizophora mangle* in Florida (Miller, 1972) and 2 mm day^{-1} in an estuarine mangrove forest in northern Australia (Wolanski & Ridd, 1986). Indeed, most of the water lost from a mangrove forest comes from direct evaporation from the soil (Miller, 1972; Wolanski & Ridd, 1986). In contrast, a well-developed canopy of crop plants growing under similar climatic conditions would transpire at the potential rate set by the environment, which often exceeds 7 mm day^{-1}. Given the potential hazards of salinization of the soil (Passioura, Ball & Knight, 1992) and the difficulties of coping with waterlogged conditions (McKee, 1993), it is not surprising that mangrove roots are relatively shallow (Gill & Tomlinson, 1977) and primarily extract water near the soil surface (Lin & Sternberg, 1994).

16.3.2.4   Implications of conservative water use for photosynthesis

Salinity-dependent limitations in rates of photosynthesis may cause leaves of plants in highly saline environments to be more subject to photoinhibition than those with higher photosynthetic capacities in less saline habitats (Björkman, Demmig & Andrews, 1988). Photoinhibition is light-dependent loss in photosynthetic functioning of photosystem II, which is manifest in whole leaves as a decline in the quantum efficiency of photosynthesis (i.e., mol CO_2 fixed or mol O_2 evolved per mol photons absorbed) under limiting light intensities (Osmond, 1994). Photoinhibition can result from direct photodamage to photosystem II and from photoprotection, in which excessive excitation energy is deflected away from PS II and dissipated harmlessly, primarily as heat (Osmond, 1994). Protective dissipation occurs mainly by means of the transthylakoid pH gradient (Krause & Behrend, 1986) and xanthophyll cycle pigments (Demmig-Adams, 1990), enabling a so-called "down regulation" to balance the light energy received by PS II with its capacity to use it (Chow, 1994).

It follows from the low photosynthetic rates of mangroves that light requirements for maximal photosynthesis are considerably less than the amounts of light available on bright, sunny days. Rates of photosynthesis in field-grown mangrove leaves generally become light-saturated at incident quantum flux densities ranging from 25 to 50% full sunlight (Ball & Critchley, 1982; Björkman, Demmig & Andrews, 1988; Cheeseman et al., 1991), consistent with the light levels received by the leaves in their normal, almost vertical, orientations (Ball, Cowan & Farquhar, 1988). Low photosynthetic rates under saturating irradiance inevitably result in an excess of excitation energy when leaves are exposed to direct sunlight. This is a potentially hazardous situation, particularly for photosystem II, which becomes photoinhibited when more light is absorbed than can be used in photosynthetic photochemistry (Osmond, 1981).

Interactive effects of salinity and light on photosynthesis occur when high salinities reduce the capacity for photosynthetic carbon assimilation under high irradiance (Björkman, Demmig & Andrews, 1988). Under shaded conditions, the quantum yield of mangrove leaves was as high as in leaves of non-mangrove species, and analysis of fluorescence characteristics showed that energy conversion was unaffected by high salinity. However, fully-exposed sun leaves of mangroves had markedly depressed quantum yields and fluorescence was severely quenched, such that the efficiency of photosystem II photochemistry decreased with increasing radiation (Björkman, Demmig & Andrews, 1988). Other studies have also found that naturally displayed sun leaves can have lower quantum yields than shaded leaves (Cheeseman et al., 1991; Lovelock & Clough, 1992). This depression in quantum yield of mangrove leaves naturally receiving high irradiances was correlated with the concentration per unit leaf area of zeaxanthin pigment (Lovelock & Clough, 1992), implying that the loss of photosynthetic activity at limiting irradiances is due to protective dissipation of excessive excitation energy through the xanthophyll cycle (Demmig-Adams et al., 1989). In contrast, no such depression in photochemical efficiency was detected in sun leaves of non-mangrove species growing in adjacent non-saline sites (Björkman, Demmig & Andrews, 1988). Interspecific differences in the capacity of leaves to maintain relatively high photosynthetic rates and to provide protection from excessive irradiance with spatial and temporal variation in environmental factors affecting photosynthesis may influence the relative performance of species along salinity gradients.

Salinity varies both temporally and spatially in a mangrove swamp, but the soil salinity around the roots changes more slowly than does the microclimate surrounding the leaves, which has a direct and immediate effect on diurnal water use in relation to carbon gain. The transpiration rate depends on both the leaf conductance to water vapor and the vapor pressure gradient between the leaf and air (vpd). Diurnal variation in vpd is caused mainly by variation in leaf temperature because ambient vapor pressure changes little over the course of a day. The smaller the difference between leaf temperature and air temperature, the closer the evaporative demand of the leaf will reflect the saturation vapor deficit of the air. Thus, maintenance of leaf temperature close to air temperature is essential to minimizing water loss and maximizing water-use efficiency.

Optimal leaf temperatures for photosynthesis in mangroves are very close to the average air temperatures in the tropical and subtropical environments in which the plants are grown. Both assimilation rate and stomatal conductance are maximal at leaf temperatures ranging from 25 to 30°C, and decline precipitously with increase in leaf temperature above 35°C (Moore et al., 1972, 1973; Andrews, Clough & Muller, 1984; Andrews & Muller, 1985; Ball, Cowan & Farquhar, 1988). The transpiration rates, even at optimal leaf temperatures, are not sufficient to prevent heating of the leaves above ambient air temperatures during periods of intense insolation. Leaves operating with high transpiration rates can take advantage of the high irradiances required for maintenance of high photosynthetic rates with minimal increase in leaf temperature over air temperature. In contrast, leaves with more conservative water use must avoid high irradiances if leaf temperatures are to be kept within physiologically acceptable limits. In such leaves, avoidance of high light intensities in the middle of the day, when the heat load on the leaf is greatest, would allow the leaves to maintain fairly constant, but low, assimilation rates throughout the day, thus achieving a greater net gain of carbon than if the leaves were horizontal and subject to temperature-dependent inhibition of photosynthesis for extended periods (Cowan, 1982).

It follows that increase in leaf angle is a compromise between requirements for illumination and for maintenance of favorable leaf temperatures with minimal evaporative cooling. For example, when exposed canopy leaves of *Rhizophora apiculata* were held in a horizontal position, leaf temperatures increased from 4 to 11°C above ambient air temperatures of approximately 30°C, while incident irradiation increased from 1430 to 2085 μmol m^{-2} s^{-1}. In contrast, leaves left in their natural almost vertical orientation avoided the maximum heat load during midday when irradiance and air temperatures are greatest. During midday, these leaves received only 20% of available sunlight and were approximately 10°C cooler than they would have been if fully exposed to the sun. Earlier and later in the day, the leaves received about 500 μmol m^{-2} s^{-1} and leaf temperatures were 30°C, conditions nearly optimal for photosynthesis (Ball, Cowan & Farquhar, 1988). Such maintenance of leaf temperature close to air temperature by avoidance of high irradiance in leaves of *Rhizophora stylosa* was critical to maximizing the total integrated gain of carbon for a minimum expenditure of water during a day (Andrews & Muller, 1985).

16.3.2.5.   Implications of conservative water use for display and properties of leaves

Interspecific differences in the display and properties of foliage reflect the increasingly conservative water-use characteristics associated with increasing salinity tolerance. This is shown by variation in three major characteristics of leaves that contribute to maintenance of favorable leaf temperatures with minimal evaporative cooling. As discussed above, leaf angle affects the radiant heat loading on the leaf. The greater the leaf angle, the lower the proportion of leaf area projected on a horizontal surface. In the Rhizophoraceae, leaf angle in foliage fully exposed to the sun is greater, and hence the proportion of projected leaf area is smaller, the greater the salinity tolerance of the species (Table 16.1). Thus, the species that are more conservative in water use are those that tend most to avoid intense radiation (Ball, Cowan & Farquhar, 1988).

A second leaf property influencing leaf temperature is that of leaf size. Heat convection between a leaf and its environment depends on resistance to transfer imposed by a boundary layer, the characteristics of which are a function of leaf geometry and wind speed. Decrease in leaf size enhances boundary layer conductance and results in the temperature of the leaf being closer to ambient air temperature without putting the leaf at a disadvantage in terms of light interception. In other words, a smaller leaf can intercept more light with less increase in temperature than a larger leaf. Leaf size in the Rhizophoraceae is smallest in the most salt-tolerant (and most water conservative) species (Table 16.1), and decreases with increasing exposure to sunlight (Ball, Cowan & Farquhar, 1988). It follows that leaves of mangrove species that dominate low salinity wetlands are much larger than those of species that dominate hypersaline environments. Apparently, mangrove leaves are smallest under conditions in which, due to intense radiation or limitations to evaporative cooling, they experience the greatest heat load (Ball, Cowan & Farquhar, 1988).

Heat capacity per unit area, which increases with dry weight and water content per unit area, is a third leaf property influencing leaf temperature. Among the Rhizophoraceae, specific leaf weight and succulence, and thus also heat capacity, increase with salinity (Camilleri & Ribi, 1983), exposure to sunlight (Ball, Cowan & Farquhar, 1988; Medina et al., 1990; Lovelock, Clough & Woodrow, 1992), and with increase in the salinity tolerance of the species (Ball, 1988a). The heat capacities of the leaves in Table 16.1 range from 1.1 to 2.2 × 10³ J m^{-2} °C^{-1} in Bruguiera gymnorrhiza and Ceriops australis, respectively. During a lull in air movement, leaf temperatures would increase because of reduction in boundary layer conductance. However, the rate of temperature increase would be slower in the leaf with a greater heat capacity. Thus, there is a tendency for mangrove leaves to have a greater mass per unit area under conditions in which, due to intense irradiation or limitations to evaporative cooling, they are most vulnerable to rapid fluctuations in leaf temperature (Ball, Cowan & Farquhar, 1988).

The display and properties of foliage contribute to conservative water use, but not without costs to the plant. Increasing the angle of inclination reduces heat loading on a leaf, but is at the expense of light harvesting in that a larger leaf area index is required to intercept a given amount of light. Decrease in leaf size enhances heat transfer rates, but this requires greater investment in supportive and conductive tissue per unit of exposed leaf area than in large leaves. Increase in heat capacity of leaves buffers against rapid changes in temperatures, but at the expense of leaf carbon, which might otherwise be invested in expansion of leaf area. Thus, maintenance of favorable leaf temperatures with minimal evaporative cooling is at the expense of the assimilative capacity of the plant, with the expense increasing as water use becomes more conservative (Ball, Cowan & Farquhar, 1988).

16.3.3 Growth of mangroves in response to salinity

16.3.3.1 Growth characteristics

Like most halophytes (Flowers, Hajibagheri & Clipson, 1986), mangroves typically show maximum growth under relatively low salinity conditions, but differ in the range of salinities over which high growth rates are sustained (Ball, 1988a). For example, Figure 16.1 shows growth responses of two closely related species, *Sonneratia alba* and *S. lanceolata,* to increase in salinity from freshwater to seawater conditions. Maximal growth of *S. alba* and *S. lanceolata* occurred in salinities ranging from 5 to 50% seawater and from 0 to 5% seawater, respectively. Under optimal conditions for growth of both species (i.e., 5% seawater), the less salt tolerant species achieved twice the height, leaf area and total biomass of the more salt-tolerant species (Ball & Pidsley, 1995). These results are consistent with results of other studies on mangroves comparing growth of both closely related species, i.e., *Ceriops australis* and *Ceriops tagal* (Smith, 1988), and unrelated but sympatric species, i.e. *Aegiceras corniculatum* and *Avicennia marina* (Ball, 1988b). It appears that increasing salt-tolerance is at the expense of growth and competitive ability under the low salinity conditions in which most species grow best (Ball, 1988a).

16.3.3.2. Increasing salt tolerance at the expense of growth

There are many attributes associated with increasing salt tolerance, but two in particular have major implications for interspecific differences in architecture and growth (Ball, 1988a, b). First, the carbon cost of water uptake increases with increasing salinity and is greater in the more salt-tolerant species (Ball, 1988b). This is manifest in the field by increase in root biomass along gradients of increasing soil salinity (Soto, 1988). Indeed, the root biomass of mangrove forests is generally greater than that measured in other forest systems (Komiyama et al., 1987). Second, water use becomes increasingly conservative with increasing salinity and with increase in the salt tolerance of the species (Ball & Farquhar 1984a,b; Ball, 1988b; Ball, Cowan & Farquhar, 1988). Conservative water use may reflect the high carbon cost of water uptake (Ball, 1988b) and contribute to maintenance of favorable ion (Ball, 1988b) and water relations (Sperry, Tyree & Donnelly, 1988; Rada et al., 1989), but is manifest in the field by decreasing light interception along gradients of increasing salinity (Ball, Cowan & Farquhar, 1988; Davie, 1988). These two attributes, the high carbon cost of water gain by the roots and conservative water use by the leaves, may contribute to enhancement of salt tolerance but at the expense of the growth rate, such that species tolerant of broad ranges of salinity tend to grow more slowly than less tolerant species even under optimal salinities for growth (Ball, 1988a).

16.3.3.3. Causes of reduction in growth in response to salinity

What causes growth to decline with increase in salinity above optimal levels? The relative growth rate (mg g^{-1} day^{-1}) is the product of the net assimilation rate (g m^{-2} leaf area day^{-1}) and the leaf area ratio (m^2 leaf area kg^{-1} plant mass). This relationship can be used to quantify relative contributions of changes in net assimilation rate and carbon partitioning to changes in relative growth rates.

Reduction in net assimilation rate can result from increase in total plant respiration and/or decrease in rates of photosynthetic CO_2 assimilation per unit leaf area. Long and Baker (1986) noted that total plant respiration might be reduced by salinity because suppression of growth would decrease the growth component of respiration more than salinity would increase the maintenance component. Indeed, studies of whole plant respiration have found differences in carbon losses between plants grown under optimal and salt-stressed conditions to be either small or negligible (McCree, 1986). In contrast, photosynthetic rates of CO_2 fixation commonly decline with increasing salinity in many species (Long & Baker, 1986), including mangroves under both field (Clough & Sim, 1989) and laboratory conditions (Ball & Farquhar, 1984a,b; Björkman, Demmig & Andrews, 1988; Lin &

Sternberg, 1992c). Thus, decreased photosynthetic rates appear to account for most of the decline in net assimilation rate in response to adverse salinities.

Leaf growth generally declines with increase in salinity above optimal levels (Kriedemann, 1986). This decline has long been attributed to inadequate turgor for cell expansion, but Termaat et al. (1985) have shown that decrease in rate of leaf area expansion induced by increase in salinity is independent of turgor. Mechanisms controlling changes in leaf growth with increasing salinity remain elusive (Munns, 1993).

There have been few analyses of changes in relative growth rates of halophytes in response to salinity. Two studies on mangroves yielded similar results. In *Aegiceras corniculatum*, *Avicennia marina* (Ball, 1988b), *Sonneratia alba*, and *S. lanceolata* (Ball & Pidsley, 1995), leaf area ratio was relatively insensitive to salinity despite major decreases in both rates of leaf initiation and leaf area expansion. Such maintenance of a ratio between leaf area and plant biomass would tend to conserve, at the whole-plant level, the water-use efficiencies determined by stomatal behavior at the leaf level (Ball, 1988b). In *Aegiceras corniculatum* and *Avicennia marina*, leaf area ratios decreased only at the highest salinity, and most of the decline in relative growth rate with increase in salinity from 50 to 500 mM NaCl could be attributed to decrease in the net assimilation rate due to reduction in rates of photosynthesis (Ball, 1988b). In *Sonneratia alba* and *S. lanceolata*, the changes in growth shown in Figure 16.1 were largely due to changes in the net assimilation rate because leaf area ratio changed little with increase in salinity from 0 to 100% sea water (Ball & Pidsley, 1995). Thus, the results of these two studies emphasize the importance of net assimilation rate as the primary determinant of changes in growth in response to salinity, but more studies are required before these findings can be generalized.

16.3.3.4.   Implications of interspecific differences in salt tolerance for mangrove forest structure along salinity gradients

One of the most striking freatures of mangrove forest is the tendency for species to become distributed differentially in a banded zonation pattern oriented roughly parallel to shore (e.g., Saenger et al., 1977). This segregation of species into zones, which are either monospecific or strongly dominated by only one or two species, contrasts sharply with the more complex structure of floristically diverse moist lowland rainforest. Many biological processes contribute to the development of zonation patterns in mangrove forest, as recently reviewed by Smith (1992). These biological processes include dispersal (Rabinowitz, 1978), herbivory (Smith, 1987), and competition (Clarke & Hannon, 1970; Ball, 1980; Jimenez & Sauter, 1991), which can differentially affect mangrove dynamics along tidal gradients (Clarke & Myerscough, 1993). Much less attention has been given to understanding ecophysiological processes which also play a key role in development of forest structure. It is the physiological attributes of species that form the framework for their responses to environment and interactions with other species.

A model (Figure 16.2) has been proposed to explain how interspecific differences in salt tolerance might contribute to segregation of species along a salinity gradient (Ball, 1988a,b). Species intolerant of high salinity, such as *Bruguiera gymnorrhiza*, operate with relatively high transpiration rates and low water-use efficiencies. These species can maintain larger leaves with greater projected leaf areas than more salt tolerant species. In low-salinity environments, stands of *B. gymnorrhiza* have dense canopies that permit little penetration of direct sunlight to the forest floor. In contrast, highly salt-tolerant species, such as *Ceriops australis*, are very conservative in their use of water and maintain small leaves with a low proportion of projected leaf area. This slowly growing species typically forms stands with porous canopies, which under low salinity regimes could not exclude the more rapidly growing and densely canopied species characteristic of low-salinity environments. Thus, despite growing maximally under relatively low-salinity conditions, the very attributes that enable *C. australis* to tolerate highly saline conditions may reduce its competitive ability under the low-salinity conditions where it grows optimally. Individual *C. australis* occasionally occur in low-salinity environments, but the species is limited largely to highly saline habitats where it is a superior competitor or where competition with other species may be absent. Thus, species from an available pool could become distributed differentially along a salinity gradient because of differences in tolerance limits and because of the ways in which physiological attributes associated with differences in salt tolerance might affect competitive interactions for resources along the gradient.

Physiological studies of species responses to static salinity gradients contribute to our understanding of salt tolerance, but salinity regimes under natural field conditions vary in time and space, with the time scale and magnitude of fluctuations dependent on the climate and hydrological characteristics of the coastal environment. Seasonal

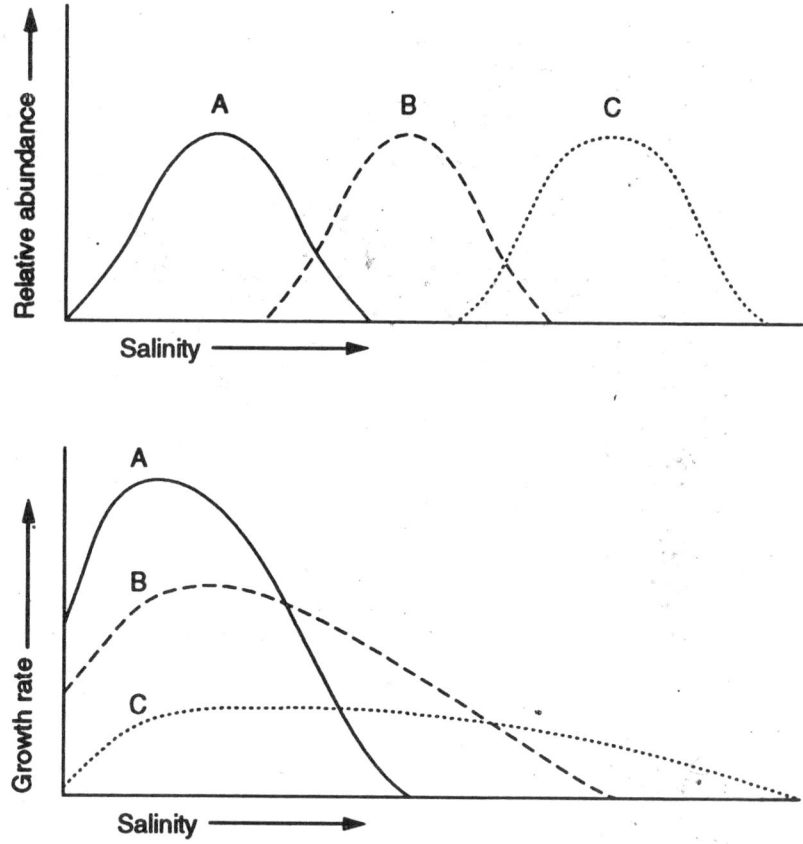

Fig. 16.2. Proposed model linking ecological and physiological responses of three hypothetical mangrove species to a salinity gradient. The top diagram shows the distribution of three species (A, B, C) along a natural salinity gradient. Each species is dominant over a different range of salinities. The bottom diagram shows the physiological responses of the same species to the salinity gradient. Species A is dominant in salinities optimal for its growth, whereas species B and C are dominant in salinities that are sub-optimal for their growth, but which limit or prevent the growth of other species. (after Ball, 1988a).

fluctuations in salinity may be an important determinant of the distribution and relative abundance of species along spatial salinity gradients, yet very little is known of relationships between environmental heterogeneity and the population biology of mangroves. Recently, Lin and Sternberg (1993a) contrasted growth of seedlings of *Rhizophora mangle* under constant and fluctuating salinity regimes in which the mean salinities were the same as those in constant salinity regimes. Growth in a constant salinity regime exceeded that in a fluctuating regime at all salinity levels. Lin and Sternberg (1993a) interpreted the results as being consistent with the occurrence of *R. mangle* in a dwarf growth form in highly saline environments subject to large fluctuations in salinity (Lin and Sternberg, 1992a,b,c). Much more attention needs to be given to understanding growth processes with temporal variation in salinity.

In another study, Ball and Pidsley (1995) considered how interspecific differences in salt tolerance between two closely related species, *Sonneratia alba* and *S. lanceolata* (Figure 16.1), might relate to their differential distribution along salinity gradients that fluctuate in time and space in northern Australia. Seedlings of *Sonneratia alba* grow in salinities ranging from freshwater to seawater, with growth being maximal in 5 to 50% seawater. These characteristics are consistent with widespread distribution of *S. alba* along coastal wetlands in the Northern Territory of Australia, where the average annual rainfall is at least 1000 mm. In these areas, river discharge seepage, and surface runoff presumably are sufficient to lower seawater salinities during periods critical to seedling establishment and growth (Semeniuk, 1983). This may be essential to successful establishment because growth of juvenile *S. alba* in seawater is so poor that it is unlikely to grow to maturity without exposure to lower salinities at least during the wet season. This may explain the absence of *S. alba* from more arid shores along northern Australia. Conversely, it is evident from the poor growth of *S. alba* in freshwater that this species is unlikely to develop to maturity in the absence of exposure to saline conditions, at least during the dry season. An apparent requirement for some minimal level of salinity is consistent with the occurrence of *S. alba* in seasonally brackish areas that experience prolonged exposure to freshwater, and the absence of *S. alba* from areas that are permanently inundated with freshwater (Ball & Pidsley, 1995).

In contrast, *S. lanceolata* behaves like a salt-tolerant glycophyte in that it shows a fair degree of salt tolerance, with maximum germination and growth occurring in salinities ranging from freshwater to 5% seawater (Figure 16.1). These characteristics might be expected to limit *S. lanceolata* to environments experiencing prolonged periods of freshwater inundation, and a relatively low-salinity regime. Indeed, vigorous trees of *S. lanceolata* grow along the margins of freshwater billabongs in northern Australia, and they occur primarily in low salinity environments in areas receiving high rainfall. However, *S. lanceolata* also occurs in dense stands along the upstream reaches of the South Alligator River system, where salinities are favorable for growth for a maximum of seven months per year. Fruits mature and seeds are released late in the wet season when salinities are minimal and within the optimal range for growth. As salinities approach maximal levels late in the dry season, progressive increase in senescence of older leaves, coupled with decrease in leaf initiation, cause the canopies to become depleted of foliage. *Sonneratia lanceolata* is not drought deciduous, and excessive loss of foliage with prolonged exposure to high salinities is associated with death of branches as well as trees. Survival in such marginal habitats apparently depends on the balance between rapid growth in the wet season and die-back during the dry season (Ball & Pidsley, 1995).

It is apparent that seasonal fluctuation in salinity may contribute to survival of both *Sonneratia alba* and *S. lanceolata* in many environments. These data support recent theory showing how temporal variation in environmental factors could contribute to persistence of rare species and maintenance of biodiversity (Chesson, 1990). Linking ecophysiological attributes such as salt tolerance of seedlings to population biology will require an understanding of plant responses to dynamic salinity gradients, particularly during regeneration.

16.4 CONCLUSION

Mangrove forest and tropical moist lowland rainforest occur in very different edaphic environments. In mangrove forest, tidal flooding with saline water produces gradients in nutrients, waterlogging, and salinity. Of these three factors, salinity is probably the most important in defining the boundary between mangrove and rainforest. Adaptations to saline environments are complex and involve all aspects of plant functioning. The present review has emphasized the most studied of those functions, namely plant water relations. In contrast to the behavior of plants in adjacent moist lowland rainforest, mangroves typically have low transpiration rates and high water-use efficiencies despite growing in an environment with an abundance of water. These water-use characteristics become increasingly conservative with increase in salinity and with increase in the salt tolerance of the species. Maintenance of high water-use efficiency may be a consequence of both high carbon costs of water uptake by the roots and restrictions in the rates of water usage by the leaves to maintain favorable water and salt relations. Conservative water use, which may have adaptive significance for survival in saline environments, has far-reaching consequences for plant functioning. Maximizing carbon gain relative to water use is a whole-plant phenomenon involving a complex balance between several levels of plant functioning: stomatal behavior in relation to photosynthesis, variation in leaf properties in relation to light interception and evaporative demand, and the partitioning of carbon between structures supplying and those consuming carbon-based assimilates (Cowan & Farquhar, 1977; Cowan, 1986). Differences in water-use characteristics thus find expression at all levels of plant form and function, and indeed are major determinants of mangrove forest structure along natural salinity gradients (Ball, 1988a).

Salinity gradients, however, do not exist in isolation from other edaphic and climatic factors. For example, combined effects of salinity and waterlogging are important determinants of species distributions in estuarine wetlands of southeastern Australia (Clarke & Hannon, 1970). Understanding how mangroves respond to salinity in combination with other factors, and how such responses influence forest dynamics, remain major challenges to ecophysiologists.

REFERENCES

ANDREWS, T. J. & MULLER, G. J. (1985) Photosynthetic gas exchange of the mangrove, *Rhizophora stylosa* Griff., in its natural environment. *Oecologia*, 65, 449–455.

ANDREWS, T. J., CLOUGH, B. F. & MULLER, G. J. (1984) Photosynthetic gas exchange and carbon isotope ratios of some mangroves in North Queensland. *Physiology and Management of Mangroves* (ed. H. J. TEAS) Tasks for Vegetation Science, Vol. 9, W. Junk Publishers, The Hague, pp 15–23.

ATKINSON, M. R., FINDLAY, G. P., HOPE, A. B., PITMAN, M. G., SADDLER, H. D. W. & WEST, K. R. (1967) Salt regulation in the mangrove *Rhizophora mucronata* Lam and *Aegialitis annulata* R. Br. *Australian Journal of Biological Sciences*, 20, 589–599.

AUSTIN, M. P. (1985) Continuation concept, ordination methods, and niche theory. *Annual Review of Ecology and Systematics*, 16, 39–61.

BALL, M. C. (1980) Patterns of secondary succession in a mangrove forest of southern Florida. *Oecologia*, 44, 226–234.

BALL, M. C. (1988a) Ecophysiology of mangroves. *Trees*, 2, 129–142.

BALL, M. C. (1988b) Salinity tolerance in the mangroves *Aegiceras corniculatum* and *Avicennia marina*. I. Water use in relation to growth, carbon partitioning, and salt balance. *Australian Journal of Plant Physiology*, 15, 447–464.

BALL, M. C. & ANDERSON, J. M. (1986) Sensitivity of photosystem II to NaCl in relation to salinity tolerance. Comparative studies with thylakoids of the salt-tolerant mangrove, *Avicennia marina*, and the salt-sensitive pea, *Pisum sativum.* Australian Journal of Plant Physiology, **13**, 689–698.

BALL, M. C. & CRITCHLEY, C. (1982) Photosynthetic responses to irradiance by the grey mangrove, *Avicennia marina*, grown under different light regimes. *Plant Physiology*, **74**, 7–11.

BALL, M. C. & FARQUHAR, G. D. (1984a) Photosynthetic and stomatal responses of two mangrove species, *Aegiceras corniculatum* and *Avicennia marina*, to long-term salinity and humidity conditions. *Plant Physiology*, **74**, 1–6.

BALL, M. C. & FARQUHAR, G. D. (1984b) Photosynthetic and stomatal responses of the grey mangrove, *Avicennia marina*, to transient salinity conditions. *Plant Physiology*, **74**, 7–11.

BALL, M. C. & PASSIOURA, J. B. (1993) Carbon gain in relation to water use: Photosynthesis in mangroves. *Ecophysiology of Photosynthesis* (eds. E. -D. SCHULZE & M. M. CALDWELL) Springer-Verlag, Berlin, pp 247–259.

BALL, M. C. & PIDSLEY, S. M. (1995) Growth responses to salinity in relation to distribution of two mangrove species, *Sonneratia alba* and *S. lanceolata*, in northern Australia. *Functional Ecology*, **9**, 77–85.

BALL, M. C., CHOW, W. S. & ANDERSON, J. M. (1987) Salinity-induced potassium deficiency causes loss of functional photosystem II in leaves of the grey mangrove, *Avicennia marina*, through depletion of the atrazine-binding polypeptide. *Australian Journal of Plant Physiology*, **15**, 351–361.

BALL, M. C., COWAN, I. R. & FARQUHAR, G. D. (1988) Maintenance of leaf temperature and the optimization of carbon gain in relation to water loss in a tropical mangrove forest. *Australian Journal of Plant Physiology*, **15**, 263–276.

BJÖRKMAN, O., DEMMIG, B. & ANDREWS, T. J. (1988) Mangrove photosynthesis: Response to high irradiance stress. *Australian Journal of Plant Physiology*, **15**, 43–61.

BOLAÑOS, J. & LONGSTRETH, D. (1984) Salinity effects on water potential components and bulk elastic modulus of *Alternanthera philoxeroides* (Mart.) Griseb. *Plant Physiology*, **75**, 281–284.

BOTO, K. G. (1982) Nutrient and organic fluxes in mangroves. *Mangrove Ecosystems in Australia* (ed. B. F. CLOUGH) Australian National University Press, Canberra, pp 239–257.

BOTO, K. G. & WELLINGTON, J. T. (1983) Phosphorus and nitrogen nutritional status of a northern Australian mangrove forest. *Marine Ecology Progress Series*, **11**, 63–69.

BOTO, K. G. & WELLINGTON, J. T. (1984) Soil characteristics and nutrient status in a northern Australian mangrove forest. *Estuaries*, **7**, 61–69.

BRADLEY, P. M. & MORRIS, J. T. (1990) Influence of oxygen and sulfide concentration on nitrogen uptake kinetics in *Spartina alterniflora. Ecology,* **71**, 282–287.

CAMILLERI, J. C. & RIBI, G, (1983) Leaf thickness of mangroves (*Rhizophora mangle*) growing in different salinities. *Biotropica,* **15**, 139–141.

CHAPMAN, V. J. (1944) 1939 Cambridge University expedition to Jamaica. I. A study of the botanical processes concerned in the development of the Jamaican shoreline. *Journal of the Linnean Society of London B,* **52**, 407–447.

CHEESEMAN, J. M., CLOUGH, B. F., CARTER, D. R., LOVELOCK, C. E., EONG, O. J. & SIM, R. G. (1991) The analysis of photosynthetic performance in leaves under field conditions: A case study using *Bruguiera* mangroves. *Photosynthesis Research,* **29**, 11–22.

CHESSON, P. L. (1990) Geometry, heterogeneity, and competition in variable environments. *Philosophical Transactions of the Royal Society of London, Series B,* **330**, 165–173.

CHOW, W. S. (1994) Photoprotection and photoinhibition. *Molecular Processes of Photosynthesis* (ed. J. BARBER) Vol. 10, Advances in Molecular and Cell Biology, JAI Press, Inc., Greenwich, Connecticut, pp 151–196.

CLARKE, L. D. & HANNON, N. J. (1970) The mangrove swamp and salt marsh communities of the Sydney district. III. Plant growth in relation to salinity and waterlogging. *Journal of Ecology,* **58**, 351–369.

CLARKE, P. J. & MYERSCOUGH, P. J. (1993) The intertidal distribution of the grey mangrove (*Avicennia marina*) in southeastern Australia: The effects of physical conditions, interspecific competition, and predation on propagule establishment and survival. *Australian Journal of Ecology,* **18**, 307–315.

CLOUGH, B. F. (1984) Growth and salt balance of the mangroves *Avicennia marina* (Forsk.) Vierh. and *Rhizophora stylosa* Griff. in relation to salinity. *Australian Journal of Plant Physiology,* **11**, 419–430.

CLOUGH, B. F. & SIM, R. G. (1989) Changes in gas exchange characteristics and water use efficiency of mangroves in response to salinity and vapor pressure deficit. *Oecologia,* **79**, 38–44.

COWAN, I. R. (1982) Regulation of water use in relation to carbon gain in higher plants. *Physiological Plant Ecology II. Water Relations and Carbon Assimilation* (eds. O. L. LANGE, P. S. NOBEL, C. B. OSMOND, & H. ZIEGLER) Springer-Verlag, Berlin, pp 589–614.

COWAN, I. R. (1986) Economics of carbon fixation in higher plants. *On the Economy of Plant Form and Function* (ed. T. J. GIVNISH) Cambridge University Press, Cambridge, pp 133–170.

COWAN, I. R. & FARQUHAR, G. D. (1977) Stomatal function in relation to leaf metabolism and environment. *Integration of Activity in the Higher Plant* (ed. D. H. JENNINGS) Cambridge University Press, Cambridge, pp 471–505.

CURRAN, M. (1985) Gas movements in the roots of *Avicennia marina* (Forsk.) Vierh. *Australian Journal of Plant Physiology*, 12, 97–108.

DAVIE, J. D. S. (1988) Differences in shoot architecture in mature communities of the tropical mangrove *Avicennia marina* (Forsk.) Vierh. *Proceedings of the Ecological Society of Australia*, 15, 213–219.

DEMMIG-ADAMS, B. (1990) Carotenoids and photoprotection in plants: A role for the xanthophyll zeaxanthin. *Biochimica et Biophysica Acta*, 1020, 1–24.

DEMMIG-ADAMS, B., WINTER, K., KRÜGER, A. & CZYGAN, F. -C. (1989) Zeaxanthin and the induction and relaxation kinetics of the dissipation of excess excitation energy in leaves in 2% O_2, 0% CO_2. *Plant Physiology*, 90, 887–893.

DOWNTON, W. J. S. (1982) Growth and osmotic relation of the mangrove *Avicennia marina* as influenced by salinity. *Australian Journal of Plant Physiology*, 9, 519–528.

DREW, M. C. & DIKUMWIN, E. (1985) Sodium exclusion from the shoots by roots of *Zea mays* (cv. LG 11) and its breakdown with oxygen deficiency. *Journal of Experimental Botany*, 36, 55–62.

FARQUHAR, G. D., BALL, M. C., VON CAEMMERER, S. & ROKSANDIC, Z. (1982) Effect of salinity and humidity on ^{13}C values of halophytes – evidence for diffusional isotope fractionation determined by the ratio of intercellular/atmospheric CO_2 under different environmental conditions. *Oecologia*, 52, 121–137.

FIELD, C. D. (1984) Movement of ions and water into the xylem sap of tropical mangroves. *Physiology and Management of Mangroves* (ed. H. J. TEAS) W. Junk Publishers, The Hague, pp 49–52.

FLOWERS, T. J., HAJIBAGHERI, M. A. & CLIPSON, N. J. W. (1986) Halophytes. *Quarterly Review of Biology*, 61, 313–337.

GILL, A. M. & TOMLINSON, P. B. (1977) Studies on the growth of red mangrove (*Rhizophora mangle* L.). 4. The adult root system. *Biotropica*, 9, 145–155.

GREENWAY, H. & OSMOND, C. B. (1972) Salt responses of enzymes from species differing in salt tolerance. *Plant Physiology*, 49, 256–259.

HARVEY, H. W. (1966) *The Chemistry and Fertility of Seawater*. Cambridge University Press, Cambridge.

HOWES, B. W., HOWARTH, R. W., TEAL, J. M. & VALIELA, I. (1981) Oxidation-reduction potentials in a salt marsh: Spatial patterns and interactions with primary production. *Limnology and Oceanography*, 26, 350–360.

HUTCHINGS, P. & SAENGER, P. (1987) *Ecology of Mangroves*. Queensland University Press, St. Lucia.

JACKSON, P. C., MEINZER, F. C., GOLDSTEIN, G., HOLBROOK, N. M., CAVELIER, J. & RADA, F. (1993) Environmental and physiological influences on carbon isotope composition of gap and understory plants in a lowland tropical forest. *Stable Isotopes and Plant Carbon-Water Relations* (eds. J. EHLERINGER, A. HALL & G. FARQUHAR) Academic Press, San Diego, pp 131–140.

JESCHKE, W. D. (1984) K$^+$ – Na$^+$ exchange at cellular membranes, intracellular compartmentation of cations, and salt tolerance. *Salinity Tolerance in Plants: Strategies for Crop Improvement* (eds. R. C. STAPLES & G. H. TOENISSEN) John Wiley and Sons, New York, pp 37–65.

JESCHKE, W. D., PATE, J. S. & ATKINS, C. A. (1987) Partitioning of K$^+$, Na^{++}, Mg^{++}, and Ca^{++} through xylem and phloem to component organs of nodulated white lupin under mild salinity. *Journal of Plant Physiology*, **128**, 77–93.

JIMENEZ, J. A. & SAUTER, K. (1991) Structure and dynamics of mangrove forests along a flooding gradient. *Estuaries*, **14**, 49–56.

JONES, H. G. & SUTHERLAND, R. A. (1991) Stomatal control of xylem embolism. *Plant, Cell and Environment*, **14**, 607–612.

KOMIYAMA, A., OGINO, K., AKSORNKOAE, S. & SABHARSI, S. (1987) Root biomass of a mangrove forest in southern Thailand. I. Estimation by the trench method and the zonal structure of root biomass. *Journal of Tropical Ecology*, **3**, 97–108.

KRAUSE, G. H. & BEHREND, U. (1986) pH-dependent chlorophyll fluorescence quenching indicating a mechanism of protection against photoinhibition of chloroplasts. *FEBS Letters*, **200**, 298–302.

KRIEDEMANN, P. E. (1986) Stomatal and photosynthetic limitations to leaf growth. *Australian Journal of Plant Physiology*, **13**, 15–32.

KRIEDEMANN, P. E. & SANDS, R. (1984) Salt resistance and adaptation to root-zone hypoxia in sunflower. *Australian Journal of Plant Physiology*, **11**, 287–301.

KROM, M. D. & BERNER, R. A. (1980) Adsorption of phosphate in anoxic marine sediments. *Limnology and Oceonography*, **25**, 797–806.

LEIGH, R. A., CHATER, M., STOREY, R. & JOHNSTON, A. E. (1986) Accumulation and sub-cellular distribution of cations in relation to growth of potassium-deficient barley. *Plant, Cell and Environment*, **9**, 595–604.

LIN, G. & STERNBERG, L. da S. L. (1992a) Differences in morphology, carbon isotope ratios, and photosynthesis between scrub and fringe mangroves in Florida, USA. *Aquatic Botany*, **42**, 303–313.

LIN, G. & STERNBERG, L. da S. L. (1992b) Comparative study of water uptake and photosynthetic gas exchange between scrub and fringe red mangrove, *Rhizophora mangle* L. *Oecologia*, **90**, 399–403.

LIN, G. & STERNBERG, L. da S. L. (1992c) Effects of growth form, salinity, nutrient, and sulphide on photosynthesis, carbon isotope discrimination, and growth of red mangrove (*Rhizophora mangle* L.). *Australian Journal of Plant Physiology*, **19**, 509–517.

LIN, G. & STERNBERG, L. da S. L. (1993a) Effects of salinity fluctuation on photosynthetic gas exchange and plant growth of the red mangrove (*Rhizophora mangle* L.). *Journal of Experimental Botany*, **44**, 9–16.

LIN, G. & STERNBERG, L. da S. L. (1993b) Hydrogen isotopic fractionation by plant roots during water uptake in coastal wetland plants. *Stable Isotopes and Plant Carbon-Water Relations* (eds. J. EHLERINGER, A. HALL & G. FARQUHAR) Academic Press, San Diego, pp 497–510.

LIN, G. & STERNBERG, L. da S. L. (1994) Utilization of surface water by red mangrove (*Rhizophora mangle* L.): An isotopic study. *Bulletin of Marine Science*, **54**, 94–102.

LONG, S. P. & BAKER, N. R. (1986). Saline terrestrial environments. *Photosynthesis in Contrasting Environments* (eds. N. R. BAKER & S. P. LONG) Elsevier Science Publishers, Amsterdam, pp 63–102.

LOVELOCK, C. E. & CLOUGH, B. F. (1992) Influence of solar radiation and leaf angle on xanthophyll concentrations in mangroves. *Oecologia*, **91**, 518–525.

LOVELOCK, C. E., CLOUGH, B. F. & WOODROW, I. E. (1992) Distribution and accumulation of ultraviolet-radiation-absorbing compounds in leaves of tropical mangroves. *Planta*, **188**, 143–154.

MACNAE, W. (1968) A general account of the fauna and flora of mangrove swamps and forests in the Indo-West-Pacific region. *Advances in Marine Biology*, **6**, 73–270.

MALLERY, C. H. & TEAS, H. J. (1984) The mineral ion relations of mangroves. I. Root cell compartments in a salt excluder and a salt excreter species at low salinities. *Plant and Cell Physiology*, **25**, 1123–1131.

McCREE, K. J. (1986) Whole-plant carbon balance during osmotic adjustment to drought and salinity stress. *Australian Journal of Plant Physiology*, **13**, 33–43.

McKEE, K. L. (1993) Soil physicochemical patterns and mangrove species distribution – Reciprocal effects? *Journal of Ecology*, **81**, 477–487.

McKEE, K. L. & MENDELSSOHN, I. A. (1987) Root metabolism in the black mangrove (*Avicennia germinans* (L.): Response to hypoxia. *Environmental and Experimental Botany*, **27**, 147–156.

MEDINA, E., CUEVAS, E., POPP, M. & LUGO, A. E. (1990) Soil salinity, sun exposure, and growth of *Acrostichum aureum*, the mangrove fern. *Botanical Gazette*, **151**, 41–49.

MEINZER, F., RUNDEL, P. W., SHARAFI, R. & RICHTER, H. (1986) Turgor and osmotic relations of the desert shrub *Larrea tridentata*. *Plant, Cell and Environment*, **3**, 131–140.

MENDELSSOHN, I. A., McKEE, K. L. & PATRICK, W. H. (1981) Oxygen deficiency in *Spartina alterniflora* roots: Metabolic adaptation to anoxia. *Science*, **214**, 439–441.

MILLER, P. C. (1972) Bioclimate, leaf temperature and primary production in red mangrove canopies in South Florida. *Ecology*, **53**, 22–45.

MOON, G. J., CLOUGH, B. F., PETERSON, C. A. & ALLAWAY, W. G. (1986) Apolastic and symplastic pathways in *Avicennia marina* (Forsk.) Vierh. roots revealed by fluorescent tracer dyes. *Australian Journal of Plant Physiology*, **13**, 637–648.

MOORE, R. T., MILLER, P. C., ALBRIGHT, D. & TIESZEN, L. L. (1972) Comparative gas exchange characteristics of three mangrove species in winter. *Photosynthetica*, **6**, 387–193.

MOORE, R. T., MILLER, P. C., EHLERINGER, J. & LAWRENCE, W. (1973) Seasonal trends in gas exchange characteristics of three mangrove species. *Photosynthetica*, **7**, 387–394.

MUNNS, R. (1985) Na$^+$, K$^+$ and CI$^-$ in xylem sap flowing to shoots of NaCl-treated barley. *Journal of Experimental Botany*, **36**, 1032–1042.

MUNNS, R. (1993) Physiological processes limiting plant growth in saline soils: Some dogmas and hypotheses. *Plant, Cell and Environment*, **16**, 15–24.

MUNNS, R. & TERMAAT, A. (1986) Whole plant responses to salinity. *Australian Journal of Plant Physiology*, **13**, 143–160.

NAIDOO, G. (1983) Effects of flooding on leaf water potential and stomatal resistance in *Bruguiera gymnorrhiza* (L.) LAM. *New Phytologist*, **93**, 369–376.

NAIDOO, G. (1985) Effects of waterlogging and salinity on plant water relations and on the accumulation of solutes in three mangrove species. *Aquatic Botany*, **22**, 133–143.

NAIDOO, G. (1989) Seasonal plant water relations in a South African mangrove swamp. *Aquatic Botany*, **33**, 87–100.

NICKERSON, N. H. & THIBODEAU, F. R. (1985) Association between pore water sulfide concentrations and the distribution of mangroves. *Biogeochemistry*, **1**, 183–192.

OSMOND, C. B. (1981) Photorespiration and photoinhibition: Some implications for the energetics of photosynthesis. *Biochimica Biophysica Acta*, **639**, 77–89.

OSMOND, C. B. (1994) What is photoinhibition? Some insights from comparisons of shade and sun plants. *Photoinhibition of Photosynthesis: From Molecular Mechanisms to the Field* (eds. N. R. BAKER & J. R. BOWYER) Bios Scientific Publishers, Oxford, pp 1–24.

PASSIOURA, J. B., BALL, M. C. & KNIGHT, J. H. (1992) Mangroves may salinise the soil and in so doing, limit their transpiration rate. *Functional Ecology*, **6**, 476–481.

PEZESHKI, S. R., DeLAUNE, R. D. & PATRICK, Jr. W. H., (1990) Differential response of selected mangroves to soil flooding and salinity: Gas exchange and biomass partitioning. *Canadian Journal of Forestry*, **20**, 869–874.

POPP, M. (1984a) Chemical composition of Australian mangroves. I. Inorganic ions and organic acids. *Zietschrift Pflanzenphysiology*, **113**, 395–409.

POPP, M. (1984b) Chemical composition of Australian mangroves. II. Low molecular weight carbohydrates. *Zietschrift Pflanzenphysiology*, **113**, 411–421.

POPP, M., LARHER, F. & WEIGEL, P. (1984) Chemical composition of Australian mangroves. III. Free amino acids, total methylated onium compounds and total nitrogen. *Zietschrift Pflanzenphysiology*, **114**, 15–25.

POPP, M., POLANIA, J. & WEIPER, M. (1993) Physiological adaptations to different salinity levels in mangrove. *Towards the Rational Use of High Salinity Tolerant Plants*, Vol 1. (eds. H. LIETH & A. AL MASOOM) Kluwer Academic Publishers, Amsterdam, pp 217–224.

RABINOWITZ, D. (1978) Early growth of mangrove seedlings in Panama, and an hypothesis concerning the relationship of dispersal and zonation. *Journal of Biogeography*, **5**, 113–133.

RADA, F., GOLDSTEIN, G., OROZCO, A., MONTILLA, M., ZABALA, O. & AZOCAR, A. (1989) Osmotic and turgor relations of three mangrove ecosystem species. *Australian Journal of Plant Physiology*, **16**, 477–486.

ROBERTSON, A. I. & DANIEL, P. A. (1989) The influence of crabs on litter processing in high intertidal mangrove forests in tropical Australia. *Oecologia*, **78**, 191–198.

SAENGER, P. & MOVERLEY, J. (1985) Vegetative phenology of mangroves along the Queensland coastline. *Proceedings of the Ecological Society of Australia*, **13**, 257–265.

SAENGER, P., SPECHT, M. M., SPECHT, R. L. & CHAPMAN, V. J. (1977) Mangal and coastal salt-marsh communities in Australasia. *Wet Coastal Ecosystems* (ed. V. J. CHAPMAN) Elsevier, Amsterdam, pp 293–345.

SCHOLANDER, P. F. (1968) How mangroves desalinate seawater. *Physiologia Plantarum*, **21**, 251–261.

SCHOLANDER, P. F., BRADSTREET, E. D., HAMMEL, H. T. & HEMMINGSEN, E. A. (1966) Sap concentrations in halophytes and some other plants. *Plant Physiology*, **41**, 529–532.

SCHOLANDER, P. F., HAMMEL, H. T., HEMMINGSEN, E. A. & BRADSTREET, E. D. (1964) Hydrostatic pressure and osmotic potential in leaves of mangroves and some other plants. *Proceedings of the National Academy of Science USA*, **52**, 119–125.

SCHOLANDER, P. F., HAMMEL, H. T., HEMMINGSEN, E. A. & GAREY, W. (1962) Salt balance in mangroves. *Plant Physiology*, **37**, 722–729.

SCHOLANDER, P. F., VAN DAM, L. & SCHOLANDER, S. I. (1955) Gas exchange in the roots of mangroves. *American Journal of Botany*, **42**, 92–98.

SEMENIUK, V. (1983) Mangrove distribution in northwestern Australia in relationship to regional and local freshwater seepage. *Vegetatio*, **53**, 11–31.

SMITH, J. A. C., POPP, M., LUTTGE, U., CRAM, W. J., DIAZ, M., GRIFFITHS, H., LEE, H. S. J., MEDINA, E., SCHAFER, C., STIMMEL, K-H. & THONKE, B. (1989) Ecophysiology of xerophytic and halophytic vegetation of a coastal alluvial plain in northern Venezuela. VI. Water relations and gas exchange of mangroves. *New Phytologist*, **111**, 293–307.

SMITH, T. J., III (1987) Seed predation in relation to tree dominance and distribution in mangrove forests. *Ecology*, **68**, 266–273.

SMITH, T. J., III (1988) Differential distribution between subspecies of the mangrove *Ceriops tagal*: Competitive interactions along a salinity gradient. *Aquatic Botany*, **32**, 79–89.

SMITH, T. J., III (1992) Forest structure. *Tropical Mangrove Ecosystems* (eds. A. I. ROBERTSON & D. M. ALONGI) Coastal and Estuarine Studies # 41, American Geophysical Union, Washington, D. C., pp 101–136.

SMITH, T. J., III, BOTO, K. G., FRUSHER, S. D. & GIDDINS, R. L. (1991) Keystone species and mangrove forest dynamics: The influence of burrowing by crabs on soil nutrient status and forest productivity. *Estuarine, Coastal and Shelf Science*, **33**, 419–432.

SOMMER, C., THONKE, B. & POPP, M. (1990) The compatibility of D-pinitol and 1D-1-O-methylmucoinositol with malate dehydrogenase activity. *Botanica Acta*, **103**, 270–273.

SOTO, R. (1988) Geometry, biomass allocation, and leaf life-span of *Avicennia germinans* (L.) (Avicenniaceae) along a salinity gradient in Salinas, Puntarenas, Costa Rica. *Revista de Biologia Tropical*, **36**, 309–323.

SPERRY, J. S. (1986) Relationship of xylem pressure potential, stomatal closure, and shoot morphology in the palm *Rhapis excelsa*. *Plant Physiology*, **80**, 110–116.

SPERRY, J. S. & TYREE, M. T. (1988) Mechanism of water stress-induced xylem embolism. *Plant Physiology*, **88**, 581–604.

SPERRY, J. S., TYREE, M. T. & DONNELLY, J. R. (1988) Vulnerability of xylem to embolism in a mangrove vs an inland species of Rhizophoraceae. *Physiologia Plantarum*, **74**, 276–283.

STERNBERG, L. da S. L., ISH-SHALOM, N., ROSS, M. & ÒBREIN, J. (1991) Water relations of coastal plant communities near the ocean/freshwater boundary. *Oecologia*, **88**, 305–310.

TERMAAT, A., PASSIOURA, J. B. & MUNNS, R. E. (1985) Shoot turgor does not limit shoot growth of NaCl-affected wheat and barley. *Plant Physiology*, **77**, 869–872.

THIBODEAU, F. R. & NICKERSON, N. H. (1986) Differential oxidation of mangrove substrate by *Avicennia germinans* and *Rhizophora mangle*. *American Journal of Botany*, **73**, 512–516.

THOM, B. G. (1967). Mangrove ecology and deltaic geomorphology, Tabasco, Mexico. *Journal of Ecology*, **55**, 301–343.

TOMLINSON, P. B. (1986) *The Botany of Mangroves*. Cambridge University Press, Cambridge, pp 62–115.

TYREE, M. T. & DIXON, M. A. (1986) Water stress-induced cavitation and embolism in some woody plants. *Physiologia Plantarum*, **66**, 397–405.

TYREE, M. T. & SPERRY, J. S. (1988) Do woody plants operate near the point of catastrophic xylem dysfunction caused by dynamic water stress? Answers from a model. *Plant Physiology*, **88**, 574–580.

TYREE, M. T., FISCUS, E. L., WULLSCHLEGER, S. D. & DIXON, M. A. (1986) Detection of xylem cavitation in corn under field conditions. *Plant Physiology*, **82**, 597–599.

VALIELA, I. (1984) *Marine Ecological Processes*. Springer-Verlag, New York.

VAN DER MOEZEL, P. G., WATSON, L. E., PEARCE-PINTO, G. V. N. & BELL, D. T. (1988) The response of six *Eucalyptus* species and *Casuarina obesa* to the combined effect of salinity and waterlogging. *Australian Journal of Plant Physiology*, **15**, 465–474.

VITOUSEK, P. M. (1984) Litterfall, nutrient cycling, and nutrient limitation in tropical forests. *Ecology*, **65**, 285–298.

WATSON, J. G. (1928) Mangrove forests of the Malay Peninsula. *Malayan Forest Records*, **6**, 1–275.

WERNER, A. & STELZER, R. (1990) Physiological responses of the mangrove *Rhizophora mangle* grown in the absence and presence of NaCl. *Plant, Cell and Environment*, **13**, 243–255.

WHITMORE, T. C. (1992) *An Introduction to Tropical Rainforests*. Clarendon Press, Oxford.

WOLANSKI, E. & RIDD, P. (1986) Tidal mixing and trapping in mangrove swamps. *Estuarine, Coastal and Shelf Science*, **25**, 43–51.

ZUBERER, D. A. & SILVER, W. S. (1978) Biological dinitrogen fixation (acetylene reduction) associated with Florida mangroves. *Applied Environmental Microbiology*, **35**, 567–575.

17

Ecotypic Variation in the Physiology of Tropical Plants

Kevin P. Hogan

Understanding the factors that determine the distributions of plant species has long been an objective of plant ecology. A basic premise of plant ecophysiology is that plant distributions are strongly influenced by interactions between the plant and its physical environment (Grace, 1987; Osmond et al., 1987; Woodward, 1987). Thus, ecophysiology has a natural interface with the study of plant distributions. There are at least four patterns of interacting physiological and ecological factors that could allow a plant to achieve a wide distribution.

1. *Microenvironment specialization.* The species may occur in different microenvironments throughout its ecological or geographic range (for example, on ridgetops in wetter areas and in depressions in drier areas). The physical environment to which it is exposed is similar at each location.

2. *Tolerance.* The species may be able to survive and reproduce under a range of physical conditions, even in the absence of any plastic or genetically-based responses.

3. *Physiological plasticity.* The species may have the capacity to vary its morphology and physiology to suit a range of environmental conditions such that a given genotype may be widely distributed and genetic differentiation among populations is low.

4. *Ecotypic differentiation.* The species may be genetically differentiated into locally-adapted ecotypes according to varying environmental conditions.

In spite of the abundance of studies on physiological ecotypes, we lack a synthesis that would allow predictive generalizations concerning the role of ecophysiological responses in determining species distributions. Such a synthesis may only be attained when studies explicitly consider alternative hypotheses (e.g., those above), and combine physiological ecology with studies of population genetics, reproductive biology, and demography. A synthesis will require knowing when ecotypic differentiation is *not* observed, so studies with "negative" results may be of particular value.

Understanding when local differentiation will be favored over phenotypic plasticity is important for several practical reasons. First, predicting the consequences of global climate change requires that we have an understanding of the genetic basis for adaptation within species (Holt, 1990). Second, deforestation in the tropics is leaving small patches of forest with presumably little gene flow between them (Lovejoy et al., 1984; Hall, Orrell & Bawa, 1994). Small populations of narrowly adapted ecotypes may be especially susceptible to fluctuations of climate (cf. Leigh et al., 1989). Finally, to select suitable source material for restoration of human-perturbed areas (Janzen, 1986), it will be important to know whether the plants are closely adapted to environmental conditions.

The study of ecotypic differentiation is fundamental to evolutionary biology. Genetic adaptation to microenvironments may maintain within-population genetic variation that can serve as the basis for adaptive evolution (Van Valen, 1965), and speciation may be possible due to selection pressures along environmental gradients (Endler, 1977). Stated most generally, the question is whether physiology matters in an ecological context. Although numerous studies have shown apparently adaptive variation in physiological responses along environmental or successional gradients, a valid hypothesis is that species distributions are determined by ecological factors such as seed dispersal, seed predation, herbivory, competition, etc., while physiological responses may be plastic or rapidly-evolving adaptations rather than determining factors in plant distribution. Furthermore, knowledge of the distribution of genetic variation in physiological adaptations will tell us something about whether and how natural selection is acting. In-particular, studies of adaptive responses of wild tropical plants to variation in mineral nutrients are especially needed given that nutrients have marked effects on tropical forest species composition (Cavalier, Chapter 14), and studies from temperate cultivated and wild plants indicate genetic control of nutrient uptake processes (e.g., Woolhouse, 1969).

In this chapter, I discuss examples of ecophysiological adaptation to environmental conditions. The discussion is organized around environmental factors, and focuses on examples of genetically-based variation in physiological responses of naturally-occurring (i.e. not cultivated) plants. Ecotypic variation in tropical forest plants has received much less attention than in temperate plants (e.g., Abrams, 1994), so some examples are from temperate plants when necessary to indicate potential areas of investigation. Finally, I discuss aspects of plant biology that are likely to influence the evolution of physiological ecotypes, and then suggest directions for future research.

17.1 ALTITUDE

17.1.1 Variation in CO_2 pressure

The most obvious and probably the most studied environmental gradient is altitude, which is also the most complex. Many of the trends discussed here are also discussed by Cavelier (Chapter 14), but the present discussion is restricted to examples where the data are suggestive of ecotypic differentiation over altitudinal gradients. Temperature, precipitation, atmospheric pressure, solar radiation, wind, soils, and CO_2 partial pressure may all vary along an altitudinal transect. Increased diffusivity of CO_2 may compensate for the decrease in partial pressure (Smith & Donahue, 1991), leaving photosynthesis unaffected. If lower CO_2 partial pressure with increased altitude imposes an increased limitation on photosynthesis, we could expect altitudinal trends in physiological or leaf anatomical characteristics affecting gas exchange (Cavelier, Chapter 14). Several studies have noted an increase in stomatal density with increasing altitude in temperate plants (Körner, Bannister & Mark, 1986), but Körner et al. (1983) noted the opposite trend for some tropical plants. However, stomatal density seems to be related to photosynthetic capacity (Körner, Scheel

& Baur, 1979), which is affected by temperature, light, and mineral nutrition. Nevertheless, alpine plants generally have a higher efficiency of carbon dioxide uptake than closely related lowland plants, and maintain this advantage at higher CO_2 pressures (Körner & Diemer, 1994). Reports of higher assimilation rates in high-altitude ecotypes in common-garden studies (e.g., Gurevitch, 1992) should consider the altitudinal decline in CO_2 partial pressure if estimates of photosynthesis under field conditions are desired. An alternative hypothesis is that it is reduction of water loss rather than enhanced carbon gain that is the adaptive function of increased CO_2 uptake efficiency in high-altitude plants (Gurevitch, 1992).

17.1.2   Plant size

A decrease in plant size with increasing altitude is often observed in studies of ecotypic variation (Clausen, Keck & Hiesey, 1940). This may be a direct response to wind (Jaffe, 1980). For example, the shorter, thicker stems of the elfin-forest tree *Didymopanax pittieri* (Araliaceae) near ridgetops compared to plants in more sheltered locations may be a morphogenetic response to wind (Lawton, 1982). However, high altitude *Metrosideros* genotypes in Hawaii were shorter and had smaller leaves in common-garden experiments (Corn & Hiesey, 1973). Reduced wind stress in shorter plants may permit higher leaf temperatures and reduce evaporative demand. Thus, wind stress may be a selective factor in determining plant height. In the temperate herbaceous perrenial *Stellaria longipes* (Caryophyllaceae), an alpine ecotype showed greater height plasticity than a prairie ecotype (Emery, Reid & Chinnappa, 1994). While wind treatment inhibited growth in both ecotypes, ethylene production (which is involved in the control of height growth) was stimulated in the alpine but not the prairie ecotype.

17.1.3   Seed size and germination

Seed size and germination responses vary considerably within tropical species (Vázquez-Yanes & Orozco-Segovia, Chapter 18; Kitajima, Chapter 19). Seed size and mass increased with altitude among populations of *Gliricidia sepium* (Papilionaceae), a neotropical tree (Salazar, 1986). This result was based on seeds collected from natural stands. No reciprocal transplants were done, so the effects could be environmental. However, when more recent data for this species (Jon Llap et al., 1990) are included in the analysis, there is no relationship between seed size or mass and altitude (Figure 17.1a). Nevertheless, several patterns emerge (Figure 17.1). First, seed mass increases towards the north and towards the west (Figure 17.1b,c). Second, seed mass decreases with distance from the population with the highest mean seed mass (Figure 17.1d). Finally, seed mass increases as annual rainfall decreases and number of dry months increases (Figure 17.1e,f). Thus, even if experimentation were to show that variation in seed size or mass among populations has a genetic basis, the data would support several interpretations, e.g., geographic clines, alleles for larger seed size "diffusing" from some center of origin, or response to differences in water availability. Therefore, even with a reasonably large number of sites (22 in the two studies combined), if the effect of a single environmental variable is of interest, study populations should be chosen to avoid potentially confounding covariates.

Similarly, the tree *Erythrina poeppigiana* (Leguminosae) produced larger seeds at higher altitudes in Costa Rican populations (Salazar & Vásquez, 1988). Temperature decreased with altitude, but there was no clear pattern in precipitation. Since this species was introduced to Costa Rica only 80 years ago, the variation could be environmentally induced. Unfortunately, the seeds studied were collected directly from the original populations at different altitudes, so environmental effects cannot be ruled out. In contrast, *Brosimum alicastrum* showed a slight decrease in seed mass with altitude (Lopez-Mata, 1987). Greater seed mass was associated with lower annual rainfall, and higher allocation to roots may enhance the probabilities of establishment in a drier environment (Lopez-Mata, 1987), but because seeds were collected from existing plants, it is not known whether these differences were environmentally or genetically controlled. Trends in seed mass for this species are also consistent with a geographic cline.

17.1.4 Temperature

Janzen (1967) has argued that stability of temperature regimes in the tropics may allow organisms to become more narrowly adapted to the conditions at a given altitude. Indirectly supporting this hypothesis is the observation that taxonomic richness increases more rapidly with the maximum altitude of tropical than temperate archipelagos

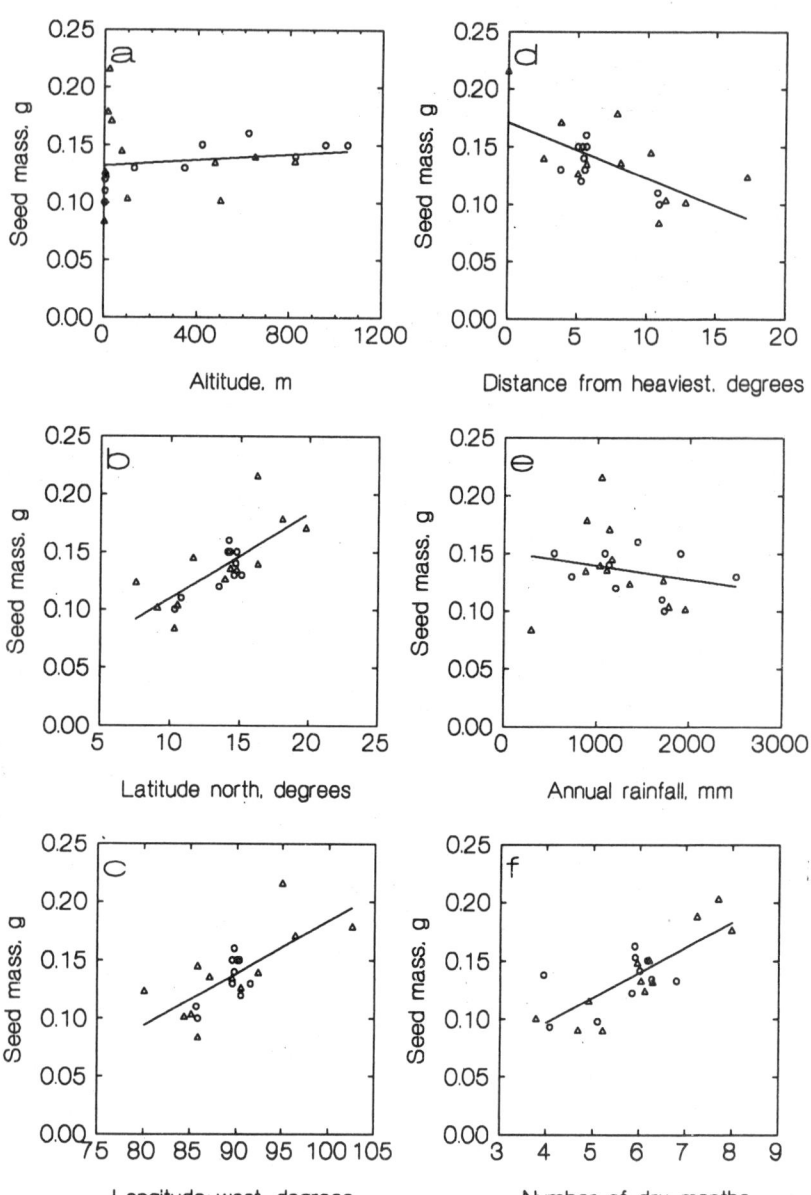

Figure 17.1. *The relationship between environmental variables, location, and seed mass of seeds collected in situ from different populations of the Central American dry forest tree* Gliricidium sepia. *Circles represent data from Salazar (1986), white triangles represent data from Jon Llap et al. (1990).*

(Smith, 1990). In a more direct test of the Janzen hypothesis, Smith (1975) looked at seed germination along an altitudinal gradient using a reciprocal transplant experiment and found that germination percentage declined with decreasing temperature in seeds of plants from lower altitudes, while seeds collected at higher altitudes germinated well at a wide range of temperatures. He noted that the differences could be environmental rather than genetic, since the seeds were collected from naturally-occurring plants in the field. Nevertheless, these results provide an interesting example that counters the generalization that stress-tolerant plants are often unable to take advantage of more favorable environmental conditions (Grime, 1977).

Tropical plants of high altitudes must be adapted to withstand low night temperatures, perhaps below freezing, during most or all of the year. As expected, tropical alpine species show maximal photosynthetic rates at temperatures that are relatively low for tropical plants (8–14°C), and recover photosynthetic activity rapidly after chilling or even freezing temperatures (Schulze et al., 1985). Supercooling capacity and chilling resistance are correlated with solute content (Rada et al., 1985; Earnshaw et al., 1990). However, in Espeletia schultzii, the relationship between leaf water potential and supercooling point is different among populations from different altitudes (Rada et al., 1987), so other factors are probably involved in freezing-point depression. Leaf cells were smaller, and apoplastic water content was lower at higher altitudes (Rada et al., 1987). The relationships among cell solute potential, electrolyte content, and altitude differ among species, each of which was studied at several altitudes (Earnshaw et al., 1990), suggesting that the mechanisms resulting in freezing-point differences among populations vary among species.

The temperature optimum of photosynthesis was highest in plants grown at warmer temperatures, for both higher (4500 m) and lower (3450 m and 3550 m) altitude populations of Espeletia schultzii (Baruch, 1979). The 3450 m population showed higher photosynthetic rates when acclimated to higher (30°C day/23°C night) temperatures, while the 3550 m and 4200 m populations showed highest photosynthetic rates when acclimated to lower (7°C day/2°C night) temperatures.

Leaf pubescence increased with altitude in E. schultzii, permitting higher leaf temperatures by reducing sensible heat flux (Baruch, 1979; also Meinzer, Goldstein & Rundel, 1985), while reducing light interception in E. schultzii by only 10-15% (Meinzer, Goldstein & Rundel, 1985). This difference in pubescence is maintained when plants from different altitudes are grown together under controlled conditions (Baruch, 1979). Other aspects of leaf morphology that affect leaf temperatures are likely to vary with altitude. The temperate herb *Achillea millefolia* showed genetically-based differences in leaf dissection that result in greater leaf warming in alpine ecotypes (Gurevitch, 1988).

The wild tomato species *Lycopersicon hirsutum* (Solanaceae) has a wide altitudinal distribution in the Andes mountains, and showed ecotypic differentiation with respect to chilling tolerance in common garden studies (Patterson, Paull & Smillie, 1978). Chilling-sensitive genotypes showed reduced germination at low temperatures, turgor loss, inhibited chlorophyll development, root death, and whole plant mortality (Patterson, Paull & Smillie, 1978). Although these symptoms showed strong trends with respect to the altitude of origin of the genotype, at some altitudes there was considerable variation among genotypes.

One of the best documented ecophysiological gradients in the tropics is the replacement of C_4 grasses by C_3 grasses with increasing altitude (Chazdon, 1978; Rundel, 1980) due to temperature effects on quantum efficiency (Ehleringer, 1978) or to changes in water relations (Teeri & Stowe, 1976). C_4 photosynthesis is found in plants of cool climates (Long, 1983) and shady habitats (Robichaux & Pearcy, 1980), although the existence of C_4 trees in the understory of Hawaiian forests may be partly a consequence of the depauperate island flora. Plants from a high altitude (3280 m) population of the C_4 grass *Miscanthus floridulus* in Papua New Guinea were more chilling-tolerant than plants from a lower altitude (2600 m) (Earnshaw et al., 1990).

The question of C_3 vs. C_4 photosynthesis becomes relevant in a discussion of intraspecific variation because C_3/C_4 intermediates are known to exist (Brown & Bouton, 1993; Medina, Chapter 2), and expression of different photosynthetic traits could vary among individuals within a species. Photosynthetic responses varied among four populations of *Mullugo verticillata* (Sayre & Kennedy, 1977). A site in Kansas had the highest evaporative demand and nearly the lowest rainfall of four populations studied. At this site, photosynthetic rates were highest, CO_2 compensation point was lowest, and the enhancement of photosynthesis at 2% O_2 relative to 21% O_2 was the lowest of the four sites, indicating more C_4 photosynthesis at the Kansas site. A site in Mexico had higher temperatures, but summer rainfall was much higher (91.3 vs. 34.5 cm), suggesting that water stress rather than temperature was the factor selecting for greater C_4 photosynthetic activity.

Ecological success across a large altitudinal range does not require ecotypic adaptation to temperature. Populations of the grass *Pen-*

nisetum setaceum did not differ significantly in the response of photosynthesis or stomatal conductance to temperature (Williams & Black, 1993). The temperature optimum for photosynthesis differed significantly among "genets" within populations, but the species showed no electrophoretically-detectable variation in the Hawaiian islands, is an obligately apomictic triploid, and may be a single clone (Williams & Black, 1993). Its extensive distribution may be due to tolerance and/or plasticity.

17.1.5 Ultraviolet radiation

The intensity of biologically-damaging ultraviolet radiation increases with altitude (Caldwell, 1968). Plants vary in sensitivity to damage by UV radiation (Caldwell & Robberecht, 1983). Fourteen of 33 Hawaiian species studied showed decreased height with increasing exposure to UV-B, while eight species showed decreased biomass (Sullivan, Teramura & Ziska, 1992). Sensitivity to UV-B was lower in species grown from seed collected at higher elevations. Similarly, in two species grown from seed collected from both low and high altitude Hawaiian sites, photosynthesis and growth decreased with increasing UV exposure in low elevation plants, but were unaffected in high elevation plants (Ziska, Teramura & Sullivan, 1992). UV-absorbing compounds increased with increasing UV exposure in low elevation plants, but were consistently high in high elevation plants.

17.2 MOISTURE GRADIENTS

17.2.1 Water in the soil

Variability in soil moisture undoubtedly plays an important role in determining the ecological and geographic distribution of plant species. Smith (1984) noted that the tropical alpine caulescent rosette plant *Espeletia schultzii* (Asteraceceae) colonized mesic swales in years of moderate rainfall, but the plants died in years of heavy rain. Conversely, plants in somewhat drier areas may be destroyed by fire if a succession of dry years has allowed the accumulation of combustible material. Mortality of the canopy palm *Socratea exorrhiza* (formerly *S. durissima*) was several times higher during an El Niño drought year (Hogan, 1986). These observations underscore the problem in relating climate to plant distributions. Species distributions and abundance may be affected strongly by extremes and rare events, and may not reflect the mean or median climatic conditions (cf. Osmond et al., 1987).

Few studies address intraspecific variation in physiological responses of non-cultivated tropical plants to soil moisture. The genus *Dubautia* in Hawaii shows interspecific differences in water relations (Robichaux, 1984; Robichaux & Canfield, 1985; Canfield, 1990). Of the species that grow in mesic or dry sites, those that grow on drier sites have lower tissue osmotic potential at full hydration (Robichaux, 1984) and lower bulk tissue elastic moduli (Robichaux & Canfield, 1985) such that they are better able to maintain higher turgor pressures as tissue water content decreases. In the one species, *D. menziesii*, that was collected from two different altitudes (2200 and 2900 m), tissue elastic properties did not differ between the two sites (Robichaux & Canfield, 1985). Tissue elastic properties in *D. ciliolata* material collected in the winter (wetter season) were similar to those of summer-collected material, although the elastic modulus near full hydration was slightly higher in the winter (Robichaux & Canfield, 1985). A comparison of a wet-forest species with a bog species showed that, in contrast to the trend for species from mesic to dry sites, daily minimum water potentials, osmotic potential at full hydration, and the water potential at which turgor reaches zero are significantly lower for the bog species (Canfield, 1990).

The differences in tissue elastic properties clearly have a genetic basis. *Dubautia* species of dry sites have 13 pairs of chromosomes, while those from mesic to wet sites have 14 pairs (Robichaux & Canfield, 1985). The correlation of habitat and tissue elastic properties with chromosome number is probably due to the fact that those species with the same number of chromosomes are very closely related (Witter & Carr, 1988). For two species with the same number of chromosomes that differ in tissue elastic properties, their natural hybrid shows intermediate properties, but its ecological distribution is restricted to the more mesic sites (Robichaux, 1984).

In two populations of the temperate herbaceous plant *Cleome serrulata* (Capparaceae), water potential at full turgor and at zero turgor, relative water content at zero turgor, and apoplastic water content differed along a mild gradient in water stress over distances of about 30 m (Farris, 1987). These differences showed heritable variation when plants were grown (from field-collected seed) under well-watered conditions. Partial correlation analysis suggested that cell wall elasticity increased with decreasing water availability to the

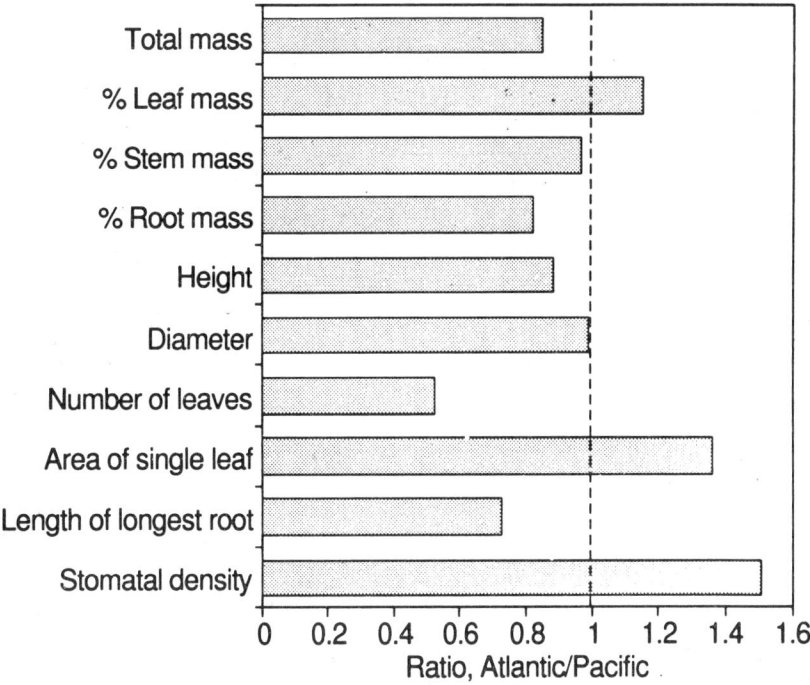

Figure 17.2 Morphological variation among Psychotria horizontalis *plants collected from sites on the Atlantic coast (annual precipitation 3277 mm) and Pacific coast (annual precipitation 1751 mm) of Panama. The plants studied were grown from small cuttings for 13 months in a growing house, and then for an additional 20 months under natural forest understory conditions on Barro Colorado Island, Panama. The bars represent the ratio of the means for Atlantic compared to Pacific coast plants. Data from Hogan et al. (1994).*

maternal plant. These results suggest that variation such as in *Dubautia* noted above could be observed intraspecifically as well.

In the lowland forest understory shrub *Psychotria horizontalis* (Rubiaceae) grown under natural forest conditions from cuttings of plants in wet and dry forest, the dry forest plants had significantly longer roots, and allocation to root mass was higher than in plants grown from wet forest cuttings (Figure 17.2, data from Hogan et al., 1994).

17.2.2 Water in the atmosphere

Carbon uptake by terrestrial plants inevitably results in loss of water vapor to the atmosphere. Water loss can be curtailed by reducing stomatal conductance when the leaf-to-air vapor pressure gradient is high (Schulze, 1982; cf. Mooney et al., 1983). For wet and dry forest plants of *Psychotria horizontalis* grown together from small cuttings for over a year, stomatal conductance was lower in the plants collected from dry forest (Hogan et al., 1994). This difference was attributed to the lower density of stomata in the dry forest plants. *Psychotria marginata* leaves showed morphological and physiological variation, depending on whether the leaves were produced just before the wet or the dry season (Mulkey et al., 1992). Phenotypically different leaves have been shown to be produced in other plants (e.g., Orshansky, 1932; Lewis, 1972). However, in *P. marginata* the dry season pheno-type is produced at the end of the rainy season, more in anticipation rather than in response to dry conditions, and continues to be pro-duced even under conditions of experimental watering (Mulkey et al., 1992). This suggests that there is strong genetic control of leaf ecophysiological phenotypes, and that this control is not proximally responsive to the variations in moisture conditions to which it is ultimately adaptive. Variations in the mechanisms that control the production of these seasonal phenotypes could provide the basis for ecotypic differentiation (cf. West-Eberhard, 1989).

The amount of water lost per unit carbon gained under given environmental conditions can be increased by some combination of higher rates of photosynthesis or lower stomatal conductance. Among populations of the Hawaiian tree *Metrosideros polymorpha* (Myr-taceae), photosynthetic activity was similar, but stomatal conductance varied more than two-fold (Meinzer et al., 1992). This resulted in different carbon stable isotope ratios in leaves of different populations, indicating long-term average differences in water-use efficiency. Un-fortunately, the plants were studied only *in situ* rather than in com-mon gardens or reciprocal transplant experiments, and the sites studied varied in altitude, temperature, rainfall, soil moisture, and exposure to wind, so the adaptive value of the differences in water-use efficiency is not clear. Heritability of carbon isotope ratios has been found in the desert shrub *Gutierrezia microcephala* (Schuster et al., 1992).

17.3 SALINITY GRADIENTS

Tropical mangrove species show readily observable zonation of species along the gradient from more saline conditions close to the sea to less saline conditions further inland (Tomlinson, 1986; Ball, Chapter 16). Growth in saline environments is physiologically costly (e.g., Lin & Sternberg, 1993); the costs and benefits associated with salt tolerance might lead to intra- as well as interspecific differences. Carbon assimilation, stomatal conductance, and leaf intercellular CO_2 concentration differed significantly among 19 mangrove species in 9 estuaries in northeastern Australia and Papua New Guinea, 12 of which were studied at two or more sites (Clough & Sim, 1989). Salinity or relative humidity accounted for a significant proportion of the total variation (among species and sites) for all three gas exchange variables, indicating that the species' responses are affected by environmental conditions. Reciprocal transplant studies would be useful to determine whether the intraspecific variation is due to local differentiation of ecotypes or to the induction of plastic physiological responses.

Borrichia frutescens showed morphological and physiological variation along a salinity gradient within stands in a salt marsh in coastal Georgia, U.S.A. (Antlfinger, 1981). Salinity varied from about 25 ppt to 45 ppt over distances of a few meters. Among stands, heritability was greater in morphological than in physiological variables, while along the salinity gradient within stands, the reverse was true.

The mangrove *Avicennia germinans* (Avicenniaceae) showed differences in branching response to the light environment among seedlings from sites that differed in salinity and canopy openness (Soto, 1988; see also Ball, Chapter 16). Seedlings of plants from a high salinity area (where the plants are naturally shorter) produced more lateral branches when planted into the high salinity area, but produced few lateral branches when planted into a low salinity area with tall trees. In contrast, seedlings of tall plants from the low salinity area produced few lateral branches under both sets of conditions (Soto, 1988). In perhaps the only study of a wild tropical plant reciprocally transplanted into field conditions, the results parallel those of Smith's (1975) study of altitudinal differentiation of germination responses in that the ecotype from the more stressful site (high salinity) shows greater plasticity. This contrasts with the tendency for shade species to be more flexible in allocation patterns than light-demanding species (King, 1991; Kitajima, Chapter 19).

17.4 LIGHT

Responses of tropical forest species to variation in the light environment is a recurrent theme in tropical plant ecophysiology (Björkman,

Ludlow & Morrow, 1972; Chazdon et al., Chapter 1). Irradiance can vary greatly over space (Evans, Whitmore & Wong, 1960) and time (Pearcy, 1990), and even long-term averages vary greatly over space and time (Smith, Hogan & Idol, 1992). Generally, growth is enhanced by light intensities greater than shaded understory conditions (Hartshorn, 1978; Denslow, 1980), and many herbs reproduce sexually primarily in treefall gaps (Smith, 1987), but high irradiance can be damaging for some species (Mulkey & Pearcy, 1992; Kamaluddin & Grace, 1992; Araus & Hogan, 1994).

Many tropical species have been shown to have the capacity to acclimate their photosynthetic responses to the ambient light level during growth (Boardman, 1977; Chazdon et al., Chapter 1; Strauss-Debenedetti & Bazzaz, Chapter 6). Early successional species often have a high capacity to acclimate to different light levels (Bazzaz & Pickett, 1980), but other studies found considerable acclimation capacity in shade-tolerant species (Mulkey, 1986; Hogan, 1988; Newell et al., 1993). Few studies of light acclimation have controlled for variation among plant genotypes. However, clones of the pioneer wet forest shrub species *Piper sancti-felicis* (Piperaceae), derived from cuttings from a single individual, acclimated photosynthetic capacity to ambient natural light conditions along a transect through a treefall gap (Chazdon, 1992). A congeneric species, *P. hispidum*, occurs naturally in a wide range of light conditions in the wet forest at Los Tuxtlas, Mexico. Seedlings transplanted from open sites into more closed sites acquired the pigmentation and leaf anatomy typical of plants of shaded sites within about 300 days (Tinoco Oranguren & Vazquez-Yanes, 1983). Transplanting from shaded to more open sites was not possible because seedlings were rare in the shade and shaded plants produced few seeds, indicating that *P. hispidum* is a light-demanding species that can persist vegetatively for a time under shaded conditions (Fredeen & Field, Chapter 20).

Several studies of temperate species address genotypic differences in photosynthetic characters. Among three populations of the temperate grass *Danthonia spicata* from sites differing in light levels, light-saturated photosynthesis did not differ among populations in spite of heritable variation among genotypes (Scheiner, Gurevitch & Teeri, 1984). Photosynthetic rates did not differ among populations or light levels when grown at 22% or 100% of full sunlight, indicating low phenotypic plasticity in this trait. Photosynthetic rate was the only one of more than twenty traits measured that did not show significant variation among populations. Other studies have also shown a lack of differentiation among populations with respect to the light environment, in spite of heritable variation in photosynthetic rates (references in Scheiner, Gurevitch & Teeri, 1984). The maintainence of within-population genetic variation in photosynthetic rates may be due to spatial and temporal variation in the light environment (Scheiner, Gurevitch & Teeri, 1984; cf. Antonovics, 1971). However, in the temperate herb *Plantago major* ssp. *major,* two genotypes (one each from a shaded and an exposed site) showed differences in photosynthetic capacity, dark respiration, and other physiological variables (Kuiper & Smid, 1985; see references therein for other studies of temperate herbs with similar results). Clones of *Phlox paniculata* cultivars did not differ in photosynthetic capacity at the single leaf level, but did differ in whole-plant photosynthetic capacity and unit leaf rate (dry mass increase per unit leaf area per unit time) (Garbutt, 1986), emphasizing the importance of plant architecture.

17.5 NUTRIENTS

Tropical latitudes contain a high diversity of soil types, and tropical soils vary greatly in their nutrient status (Lathwell & Grove, 1986). Compared to physiological responses to variation in light, tropical plant responses to nutrients have received much less attention, although species composition of tropical vegetation is known to change markedly depending on soil nutrient status (Richards, 1952; Huston, 1980; Whitmore, 1984). Most studies of plant-nutrient relationships in the tropics report on vegetation structure (Grubb & Whitmore, 1967), species composition (Huston, 1980), or community-wide averages of leaf attributes such as thickness, toughness, and nutrient content (Cuevas & Medina, 1988). Those that report on variation among individuals to different levels of soil nutrients often report only growth, which tells us little about how physiological processes might be adapted to different nutrient levels.

Variation among genotypes or ecotypes in their response to soil nutrient levels is known for agricultural (El Bassam, Dambroth & Loughman, 1990) and wild (Goodman, 1969; Woolhouse, 1969) plants, but has received little attention in wild tropical plants. As a consequence of varying levels of soil phosphate, the Central American trees *Gliricidium sepium* and *Leucaena lecocephala* (Leguminosae) showed variation among provenances and isolines in growth, phosphate uptake, and phosphate use efficiency (Arias et al., 1991). Variation among provenances, isolines, and ecotypes under controlled experi-

mental conditions may reflect genetic variation in the control of the
the physiological mechanisms that affect nutrient uptake (Goodman,.
1969; Woolhouse, 1969; El Bassam, Dambroth & Loughman, 1990).
Such variation is likely to be of ecological and evolutionary importance
for tropical forest plants.

Studies on ecotypic responses to nutrients should take symbiotic
relations into consideration. Among five genotypes of the dry forest
canopy tree *Gliricidium sepia* (Leguminosae) and five strains of *Rhizobium*, the bacterial strain that gave the highest rate of nitrogen
fixation was different for each of the tree genotypes (Awonaike, Hardarson & Kumarasinghe, 1992). Mineral nutrition also affects nitrogen fixation: two tropical genotypes of common bean (*Phaseolus vulgaris*) differing in molybdenum accumulation showed differing rates
of N fixation (Brodrick & Giller, 1991). Mycorrhizae are important for
mineral nutrition in tropical plants (Janos, 1980; Sanford & Cuevas,
Chapter 10). Ectomycorrhizal fungi show host specificity, while V-A
mycorrhizae, the most common type for tropical species, are generalists (Alexander, 1989), so variation among plant genotypes may be
more important in plant species infected by ectomycorrhizae.

Whether variation is found among genotypes in response to nutrients may depend on the nutrient levels used in the experiment: nitrogen-based photosynthetic capacity differed among genotypes at low
but not high nitrogen concentrations (Evans, 1991). Responses to
nutrients will vary depending on which nutrient is studied. Studies
such as the interspecific comparison of the response to different nutrients carried out by Denslow, Vitousek and Schultz (1987) should be
conducted on different genotypes of single species.

17.6 BIOTIC INTERACTIONS

Intraspecific genetic variation that affects interactions with herbivores, seed predators, pathogens, and symbionts is amply documented
for temperate cultivated and wild plants. Such variation certainly
occurs in wild tropical plants. Genetic variation in plant secondary
compounds, which could affect herbivores or pathogens, has been
documented (Bazzaz et al., 1975; Langenheim et al., 1978). In the
understory shrub *Psychotria horizontalis*, the effect of canopy openness on tannin production and leaf toughness varied among genotypes
(Sagers, 1992), so that defense against herbivores varies among genotypes and with level of irradiance. In *Piper arieianum*, an understory

shrub of neotropical wet forest, levels of herbivory varied significantly among genotypes, and the range of variation was sufficient to cause variation in growth and reproductive output (Marquis, 1987), indicating that herbivores may be acting as a selective factor in plant evolution. However, the rankings of genotypes according to levels of herbivory may change over time, perhaps because genotypes differ in the relative importance of different herbivore species (Marquis, 1990). Thus, variation in the herbivore community is important, and genetic adaptation to biotic interactions may be more complex than adaptation to the physical environment.

17.7 EVOLUTION OF ECOTYPES

17.7.1 Breeding systems and life histories

Differentiation of ecotypes adapted to local conditions implies that local populations are genetically distinct. Physiological responses will evolve in part in response to local selection pressures, but selection may be overcome by gene flow (Slatkin, 1987). To understand the processes leading to ecotypic differentiation, and to arrive at predictive generalizations, we need a better understanding of gene flow and the genetic structure of tropical plant populations (Loveless & Hamrick, 1984). Pollen and seed dispersal are the mechanisms of gene flow, and it may be possible to predict levels of among-site genetic variation based on the reproductive biology of the species (Figure 17.3). Dioecy is more common in tropical than temperate trees, and most of the hermaphroditic tropical species have self-incompatibility mechanisms that enforce outbreeding (Bawa, 1979), so that gene flow among populations may be high. Nevertheless, apomixis is known to occur in tropical trees (Kaur et al., 1978).

A persistent seed bank could help maintain genetic diversity within a population by allowing some genotypes to persist through unfavorable years. Among tropical forest plants, pioneer species and gap colonizers are most likely to persist in the seed bank, while primary forest species show little or no seed dormancy (Garwood, 1989; Vázquez-Yanes & Orozco-Segovia, 1993; Vásquez-Yanes, Chapter 18) but can live for centuries as established, reproductive plants (Lieberman et al., 1985). Similarly, even understory herbs that depend on treefall gaps for sexual reproduction may persist for years either as supressed plants or in the seed bank (Smith, 1987). Such life-history variation probably affects the susceptibility of a species to selection for local ecotypes.

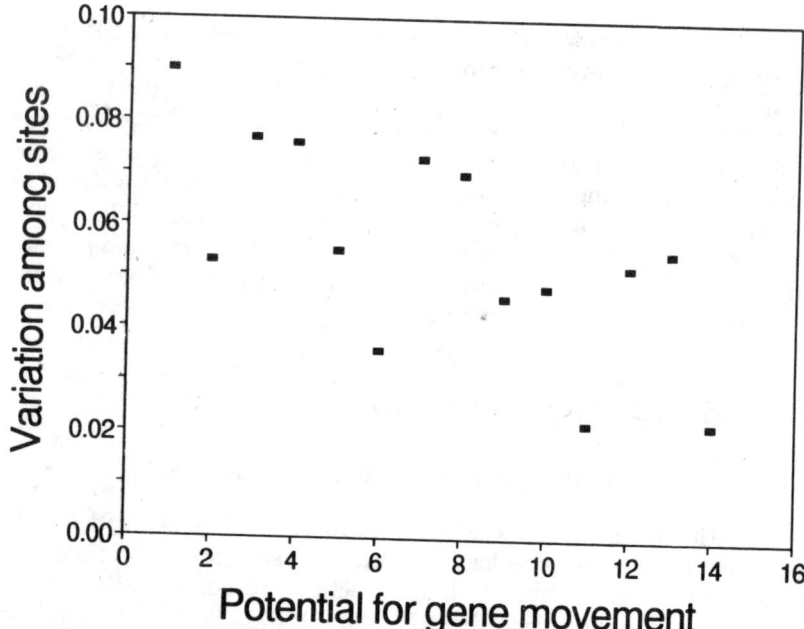

Fig. 17.3 The relationship between genetic variation among sites and potential gene movement for several Panamanian forest species. Higher numbers on the x-axis imply greater potential gene flow, estimated from reproductive biology and seed dispersal mechanisms (data from Hamrick & Loveless, 1989).

Tropical woody species generally have higher levels of genetic variability than temperate woody species, and compared to temperate species, a higher proportion of this variation is within rather than among populations (Hamrick & Loveless, 1986; Hamrick & Murawski, 1991; Eguiarte, Perez-Nasser & Piñero, 1992; Perez-Nasser, Eguiarte & Piñero, 1993). This suggests that gene flow may be high enough to reduce local differentiation and inhibit the formation of ecotypes, but even in large populations with high gene flow rates, adults can show more among-site differentiation than seeds (Eguiarte, Perez-Nasser & Piñero, 1992), suggesting that selection pressures between germination and maturity can be strong and locally variable.

17.7.2 Founder effects and drift

The tree *Acacia mangium* (Leguminosae) of Indonesia and tropical Australia shows low levels of genetic diversity (Moran, Muona & Bell,

1989), perhaps as a consequence of having been restricted to small isolated refuges during past climatic changes (Moran, Muona & Bell, 1989). Other *Acacia* of similar distribution are more tolerant of salt (*A. auriculiformis*) or altitudinal variation (*A. crassicarpa*), which Moran et al. (1989) hypothesized may have allowed them to maintain a wider distribution over geological time and thereby preserve their genetic diversity. *A. auriculiformis* shows high genetic diversity, and differentiation among populations is much higher than in other studies (Hamrick & Murawski, 1991; Wickneswari & Norwati, 1993), but *A. auriculiformis* collection sites were more widely distributed than those for *Acacia* in most other studies (but not those in the *A. mangium* study).

Populations founded by a few individuals may show low levels of genetic diversity. Founder effects and patterns of distribution over geological time may explain currently observed genetic structure of some tropical species, e.g. species in the genus *Dubautia* in Hawaii (Witter & Carr, 1988) and the genus *Coffea* in Africa (Berthaud, 1985). Thus, understanding of the historical changes in the distribution of a species may be helpful in predicting the likelihood of local differentiation. Furthermore, the higher species richness of tropical compared to temperate forests implies that tropical plants will generally occur at lower population densities. In particular, rare species may be of special interest. For example, the understory plant *Lacandonia schismatica* (Lacandoniaceae) is known only from a small population at a single site in Mexico, where it was found to be genetically uniform at all loci studied (Coello, Escalante & Soberón, 1993; see also Kitamura & Rahman, 1992). In general, genetic diversity within populations is positively related to population size (Ellstrand & Elam, 1993), and is greater in widespread species than in those with restricted distributions (Hamrick et al., 1991).

Nevertheless, it is unclear whether low genetic diversity is a cause or consequence of a restricted distribution. *Torreya taxifolia* is confined to cool refugia in northern Florida, from which it failed to migrate northward following the last glacial maximum (Falk, 1992). Its ability to adapt genetically may be limited by the failure of sexual reproduction because the heat-stressed adult plants succumb to pathogenic fungi. The species may be locked into a self-perpetuating interaction between climatic stress, a restricted distribution, and reduced genetic variability. In contrast, the cedar glade endemic herb *Echinacea tennesseensis* (Asteraceae) shows no differences in photosynthetic responses to light or water stress, compared to the more widespread and more genetically variable congener *E. angustifolia* (Baskauf & Eickmeyer, 1994).

Not all genetic variation within or among populations is adaptive (Sultan, 1987). Genetic differentiation among sites could be a result of drift or historical factors rather than selection and adaptation. For example, variation among Hawaiian populations of *Bidens* (Asteraceae) was almost as great as variation among subspecies and species, and there was no correlation between genetic and morphological variation (Helenurm & Ganders, 1985). Similarly, study of isozyme variation in Hawaiian taxa of the tree *Metrosideros spp.* (Myrtaceae) showed a high degree of genetic variation within populations, while genetic similarity among taxa was in some cases as high as among populations within taxa, in spite of obvious morphological variation (Aradhya, Mueller-Dombois & Ranker, 1991). Nine populations of the temperate annual herbaceous plant *Phlox drummondii* in Texas showed strong morphological and electrophoretically-detectable genetic variation among populations, but there was no evidence that differentiation among populations was adaptive (Schwaegerle & Bazzaz, 1987).

17.7.3   Alternatives and limitations to ecotypic differentiation

Phenotypic plasticity may allow a species to attain an ecologically or geographically wide distribution in spite of environmental variability. In this sense, plasticity may be thought of as an alternative to ecotypic differentiation. Furthermore, phenotypic plasticity may shelter the genotype from selection, thus facilitating the maintenance of genetic diversity within populations. In this way, plasticity may allow organisms to invade new ecological space to which they might subsequently become adapted through selection on the genetic control of developmental processes (West-Eberhard, 1989). Thus, phenotypic plasticity may act as a facilitator of ecotypic differentiation.

For these reasons, phenotypic plasticity is closely linked to ecotypic differentiation, and should be considered in both theoretical and empirical studies of ecotypic differentiation. Phenotypic plasticity is reviewed elsewhere (e.g., Schlichting, 1986; Sultan, 1987; Scheiner, 1993). It should be kept in mind that plasticity itself is subject to natural selection and thus is a trait that may differ among ecotypes (e.g., Emery, Reid & Chinnappa, 1994). The studies of Sultan and Bazzaz (1993a, b, c) and Jordan (1991) are of the sort of work needed to reveal the adaptive value of variation among populations and among genotypes. Sultan (1987) notes that studies that use analysis of variance models to partition variance into genotypic and environmental components do not reveal the adaptive value of plastic responses, and recommends the quantification of norms of reaction to environmental factors. Variation among phenotypes should be compared for plants of the same biomass as well as the same age to reveal variation independent of differences in growth rate (Rice & Bazzaz, 1989; Coleman, McConnaughy & Ackerly, 1994). While this point has been made in the context of phenotypic plasticity, it should be considered for ecotypic variation as well, especially where environmental conditions in reciprocal transplant studies may cause variation in growth rates.

Whether selection favors phenotypic plasticity or local adaptation into ecotypes will depend on the underlying genetic variability for the traits under selection, and in particular on genetic correlations among traits (Via & Lande, 1985; Arnold, 1992). Under natural conditions, many physiological and morphological features combine to influence plant fitness. Selection on these features will be complicated by genetic corrrelations among them, and by unavoidable trade-offs such as between carbon gain and water loss (Geber & Dawson, 1990 and references therein). An alternative hypothesis is that correlation among characters, perhaps an effect of common underlying control, may allow major changes in the phenotype with relatively small genetic changes (Chapin, Autumn & Pugnaire, 1993).

17.8. CONCLUSIONS

17.8.1 What is known

Numerous studies discussed above have documented variation within species in their physiological responses to environmental variables. It has been predicted that genetic differentiation of ecotypes is not likely to be an important source of habitat breadth in persistent rainforest species (Walters & Field, 1987), but where it has been looked for, considerable intraspecific variation has been found, and may be as high as interspecific variation (Ceulemans et al., 1984). Recently, microsite genetically-based differentiation has been found for oaks that have high rates of gene flow (Sork, Stowe & Hochwender, 1993). Several studies have documented intraspecific variation in relation to environmental factors under field conditions. Some have shown intraspecific variation in physiology or morphology without linking this variation to environmental factors (Salazar, 1983; Ceulemans et al.,

1984; Lapido et al., 1984; Ottosen, Rosenquist & Ogren, 1989). Such studies are useful for selecting genetic lines for agriculture, plantation forestry, or breeding programs. However, understanding the biology and evolution of non-cultivated plants will require different kinds of research linking ecotypic variation to specific environmental factors.

17.8.2 Why tropical plants are especially interesting

Because of the rapid deforestation currently underway in many tropical areas, and the fact that tropical plants are still relatively poorly studied (despite considerable progress in recent years), several features of their biology merit emphasis here. Compared to most temperate plants, tropical plants often occur at lower population density, are more likely to be animal dispersed and/or pollinated, and are more likely to have outbreeding mechanisms such as dioecy or self-incompatibility. Tropical plants also may be more restricted in altitudinal range, are less likely to persist for a long time in a soil seed bank, and are predominantly woody and perennial. These characteristics and others are reviewed by Janzen (1975), Longman and Jeník (1974), Vázquez-Yanes (Chapter 18), and Bawa (1979). Most of these differences are expected to reduce genetic differentiation among populations while promoting genetic diversity within populations (Loveless & Hamrick, 1984). Available evidence on genetic variation within and among tropical plant populations is consistent with these predictions (Hamrick & Loveless, 1989). Thus, the probability of ecotypic differentiation in tropical plants may be lower than in temperate plants, but small population sizes and aggregated distributions (Hubbell, 1979) could increase the probability of differentiation.

17.8.3 What needs to be done

We need more ecophysiological studies explicitly linked to studies of population genetics to determine whether the genetic variability being found in tropical plants is adaptive. It should be kept in mind that the genetic structure of present-day populations of long-lived plants may not reflect the current patterns of gene flow disrupted by ongoing deforestation (Hall, Orrell & Bawa, 1994). We need reciprocal transplant and common-garden studies of wild tropical plants from different habitats and of known genetic origin to document the ecological consequences of physiological and morphological variation among genotypes under different environmental conditions. Virtually no study of ecotypic adaptation in a wild tropical plant has experimentally separated correlated environmental factors. Given the observed relationship between potential gene flow and genetic differentiation among populations, such studies should contrast species with different reproductive and dispersal modes. Studies should be designed to contribute towards generalizations, contrasting plants of different life forms, longevity, taxonomic groups, population structures, and geographic distributions. Tropical montane plants may be especially vulnerable to global warming because the height of mountain ranges imposes an upper limit to possible shifts in plant distributions (Peters & Darling, 1985). For conservation reasons, the genotypic and plastic components of variation in these plants deserve special attention. For lowland plants, genotypic and plastic variation in response to nutrients is an important area of research that has received little attention.

Studies of ecotypic variation have a long history in plant ecology. However, much progress has been made in evolutionary theory and in technology for the study of population genetics and of ecophysiology under field conditions. We are now at a point where integrated studies of population biology, genetics, and ecophysiology can make important contributions to evolutionary biology and conservation.

ACKNOWLEDGEMENTS

I thank J. Hunt, J. Kallarackal, and the editors for comments on earlier drafts of the manuscript. I thank P. Cushing, N. Garwood, and the librarians of the Smithsonian Tropical Research Institute and The Natural History Museum (London) for bibliographic assistance.

REFERENCES

ABRAMS, M. D. (1994) Genotypic and phenotypic variation as stress adaptations in temperate tree species: A review of several case studies. *Tree Physiology*, **14**, 833–842.

ALEXANDER, I. (1989) Mycorrhizas in tropical forest. *Mineral Nutrients in Tropical Forest and Savanna Ecosystems* (ed. J. PROCTOR) *Special Publications Series of the British Ecological Society*, **9**, Blackwell Scientific Publications, Oxford.

ANTLFINGER, A. E. (1981) The genetic basis of microdifferentiation in natural and experimental populations of *Borrichia frutescens* in relation to salinity. *Evolution*, **35**, 1056–1068.

ANTONOVICS, J. (1971) The effects of a heterogeneous environment on the genetics of natural populations. *American Scientist*, **59**, 593–599.

ARADHYA, K. M., MUELLER-DOMBOIS, D., & RANKER, T. A. (1991) Genetic evidence of recent and incipient speciation in the evolution of *Hawaiian Metrosideros* (Myrtaceae). *Heredity*, **67**, 129–138.

ARAUS, J. L. & HOGAN, K. P. (1994) Leaf structure and patterns of photoinhibition in two neotropical palms in clearings and forest understory during the dry season. *American Journal of Botany*, **81**, 726–738.

ARIAS, I., KOOMEN, I., DODD, J. C., WHITE, R. P., & HAYMAN, D. S. (1991) Growth responses of mycorrhizal and non-mycorrhizal tropical forage species to different levels of soil phosphate. *Plant and Soil*, **132**, 253–260.

ARNOLD, S. J. (1992) Constraints on phenotypic evolution. *American Naturalist*, **140**, S85–S107.

AWONAIKE, K. O., HARDARSON, G., & KUMARASINGHE, K. S. (1992) Biological nitrogen fixation of *Gliricidia sepium / Rhizobium* symbiosis as influenced by plant genotype, bacterial strain and their interactions. *Tropical Agriculture*, **69**, 381–385.

BARUCH, Z. (1979) Elevational differentiation in *Espeletia schultzii* (Compositae), a giant rosette plant of the Venezuelan paramos. *Ecology*, **60**, 85–98.

BASKAUF, C. J. & EICKMEYER, W. G. (1994) Comparative ecophysiology of a rare and a widespread species of *Echinacea* (Asteraceae). *American Journal of Botany*, **81**, 958–964.

BAWA, K. S. (1979) Breeding systems of trees in a tropical wet forest. *New Zealand Journal of Botany*, **17**, 521–524.

BAZZAZ, F. A. & PICKETT, S. T. A. (1980) Physiological ecology of tropical succession: A comparative review. *Annual Review of Ecology and Systematics*, **11**, 287–310.

BAZZAZ, F. A., DUSEK, D., SEIGLER, D. S., & HANEY, A. W. (1975) Photosynthesis and cannabinoid content of temperate and tropical populations of *Cannabis sativa*. *Biochemical Systematics and Ecology*, **3**, 15–18.

BERTHAUD, J. (1985) Gene flow and population structure in *Coffea canephora* coffee populations in Africa. *Genetic Differentiation and Dispersal in Plants* (eds. P. JACQUARD, G. HEIM & J. ANTONOVICS) NATO ASI Series, Vol G5, Springer-Verlag, Berlin.

BJÖRKMAN, O., LUDLOW, M. M., & MORROW, P. A. (1972) Photosynthetic performance of two rainforest species in their native habitat and an analysis of their gas exchange. *Carnegie Institute of Washington Yearbook*, **71**, 94–102.

BOARDMAN, N. K. (1977) Comparative photosynthesis of sun and shade plants. *Annual Review of Plant Physiology*, **28**, 355–377.

BRODRICK, S. J. & GILLER, K. E. (1991) Genotypic differences in molybdenum accumulation affects N_2-fixation in tropical *Phaseolus vulgaris*. *Journal of Experimental Botany*, **42**, 1339–1343.

BROWN, R. H. & BOUTON, J. H. (1993) Physiology and genetics of interspecific hybrids between photosynthetic types. *Annual Review of Plant Physiology and Plant Molecular Biology*, **44**, 435–456.

CALDWELL, M. M. (1968) Solar ultraviolet radiations as an ecological factor for alpine plants. *Ecological Monographs*, **38**, 243–268.

CALDWELL, M. M. & ROBBERECHT, R. (1983) Protective mechanisms and acclimation to solar ultraviolet-B radiation in *Oenothera stricta*. *Plant, Cell and Environment*, **6**, 477–485.

CANFIELD, J. E. (1990) Plant water deficits, osmotic properties, and hydraulic resistances of Hawaiian *Dubautia* species from adjacent bog and wet-forest habitats. *Pacific Science*, **44**, 449–455.

CEULEMANS, R., GABRIELS, R., IMPENS, I., YOON, P. K., LEONG, W., & NG, A. P. (1984) Comparative study of photosynthesis in several *Hevea brasiliensis* clones and *Hevea* species under tropical field conditions. *Tropical Agriculture*, **61**, 273–275.

CHAPIN, F. S., III, AUTUMN, K., & PUGNAIRE, F. (1993) Evolution of suites of traits in response to environmental stress. *American Naturalist*, **142**, S78–S92.

CHAZDON, R. L. (1978) Ecological aspects of the distribution of C_4 carbon pathway grasses on selected habitats of Costa Rica. *Biotropica*, **10**, 265–269.

CHAZDON, R. L. (1992) Photosynthetic plasticity of two rainforest shrubs across natural gap transects. *Oecologia*, **92**, 586–595.

CLAUSEN, J., KECK, D. D., & HIESEY, W. M. (1940) *Experimental Studies on the Nature of Species. I. Effect of Varied Environments on Western North American Plants*. Publication 520 Edition, Carnegie Institute of Washington, Washington, D. C.

CLOUGH, B. F. & SIM, R. G. (1989) Changes in gas exchange characteristics and water use efficiency of mangroves in response to salinity and vapor pressure deficit. *Oecologia*, **79**, 38–44.

COELLO, G., ESCALANTE, A., & SOBERÓN, J. (1993) Lack of genetic variation in *Lacandonia schismatica* (Lacondoniaceae: Triuridales) in its only known locality. *Annals of the Missouri Botanical Garden*, **80**, 898–901.

COLEMAN, J. S., MCCONNAUGHY, K. D. M., & ACKERLY, D. D. (1994) Interpreting phenotypic variation in plants. *Trends in Ecology and Evolution*, **9**, 187–191.

CORN, C. A. & HIESEY, W. M. (1973) Altitudinal variation in Hawaiian *Metrosideros*. *American Journal of Botany*, **60**, 991–1002.

CUEVAS, E. & MEDINA, E. (1988) Nutrient dynamics within Amazonian forests. *Oecologia*, **76**, 222–235.

DENSLOW, J. S. (1980) Gap partitioning among tropical forest trees. *Biotropica*, **12**, 47–55.

DENSLOW, J. S., VITOUSEK, P. M., & SCHULTZ, J. C. (1987) Bioassays of nutrient limitation in a tropical rainforest soil. *Oecologia*, **74**, 370–376.

EARNSHAW, M. J., CARVER, K. A., GUNN, T. C., KERENGA, K., HARVEY, V., GRIFFITHS, H., & BROADMEADOW, M. S. J. (1990) Photosynthetic pathway, chilling tolerance, and cell sap osmotic potential values of grasses along an altitudinal gradient in Papua New Guinea. *Oecologia*, **84**, 280–288.

EGUIARTE, L. E., PEREZ-NASSER, N., & PIÑERO, D. (1992) Genetic structure, outcrossing rate, and heterosis in *Astrocaryum mexicanum* (tropical palm): Implications for evolution and conservation. *Heredity*, **69**, 217–228.

EHLERINGER, J. R. (1978) Implications of quantum yield differences on the distributions of C_3 and C_4 grasses. *Oecologia*, **31**, 255–267.

EL BASSAM, N., DAMBROTH, M., & LOUGHMAN, B. C. (eds.) (1990) *Genetic Aspects of Plant Mineral Nutrition*. Kluwer Academic Publishers, Dordrecht.

ELLSTRAND, N. C. & ELAM, D. R. (1993) Population genetic consequences of small population size: Implications for plant conservation. *Annual Review of Ecology and Systematics*, **24**, 217–242.

EMERY, R. J. N., REID, D. M., & CHINNAPPA, C. C. (1994) Phenotypic plasticity of stem elongation in two ecotypes of *Stellaria longipes*: The role of ethylene and response to wind. *Plant, Cell and Environment*, **17**, 691–700.

ENDLER, J. A. (1977) *Geographic Variation, Speciation, and Clines*. Princeton University Press, Princeton, New Jersey.

EVANS, A. S. (1991) Leaf physiological aspects of nitrogen-use efficiency in *Brassica campestris* L.: Quantitative genetic variation across nutrient treatments. *Theoretical and Applied Genetics*, **81**, 64–70.

EVANS, G. C., WHITMORE, T. C., & WONG, T. K. (1960) The distribution of light reaching the ground vegetation in a tropical rainforest. *Journal of Ecology*, **48**, 193–204.

FALK, D. A. (1992) From conservation biology to conservation practice: Strategies for protecting plant diversity. *Conservation Biology: The Theory and Practice of Nature Conservation, Preservation, and Management* (eds. P. L. FIEDLER & S. K. JAIN) Chapman and Hall, New York and London.

FARRIS, M. A. (1987) Natural selection on the plant-water relations of *Cleome serrulata* growing along natural moisture gradients. *Oecologia*, **72**, 434–439.

GARBUTT, K. (1986) Genetic differentiation in leaf and whole plant photosynthetic capacity and unit leaf rate among clones of *Phlox paniculata*. *American Journal of Botany*, **73**, 1364–1371.

GARWOOD, N. C. (1989) Tropical soil seed banks: A review. *Ecology of Soil Seed Banks* (eds. M. A. LECK, V. T. PARKER & R. L. SIMPSON) Academic Press, San Diego, California.

GEBER, M. A. & DAWSON, T. E. (1990) Genetic variation in and covariation between leaf gas exchange, morphology, and development in *Polygonum arenastrum*, an annual plant. *Oecologia*, **85**, 153–158.

GOODMAN, P. J. (1969) Intraspecific variation in mineral nutrition of plants from different habitats. *Ecological Aspects of the Mineral Nutrition of Plants* (ed. I. H. RORISON), Blackwell Scientific Publishers, Oxford.

GRACE, J. (1987) Climatic tolerance and the distribution of plants. *New Phytologist*, **106**, 113–130.

GRIME, J. P. (1977) Evidence for the existence of three primary strategies in plants and its relevance to ecological and evolutionary theory. *American Naturalist*, **111**, 1169–1194.

GRUBB, P. J. & WHITMORE, T. C. (1967) A comparison of montane and lowland forest in Ecuador. III. The light reaching the ground vegetation. *Journal of Ecology*, **55**, 33–57.

GUREVITCH, J. (1988) Variation in leaf dissection and leaf energy budgets among populations of *Achillea* from an altitudinal gradient. *American Journal of Botany*, **75**, 1298–1306.

GUREVITCH, J. (1992) Differences in photosynthetic rate in populations of *Achillea lanulosa* from two altitudes. *Functional Ecology*, **6**, 568–574.

HALL, P., ORRELL, L. C., & BAWA, K. S. (1994) Genetic diversity and mating system in a tropical tree, *Carapa guianensis* (Meliaceae). *American Journal of Botany*, **81**, 1104–1111.

HAMRICK, J. L. & LOVELESS, M. D. (1986) Isozyme variation in tropical trees: Procedures and preliminary results. *Biotropica*, **18**, 201–207.

HAMRICK, J. L. & LOVELESS, M. D. (1989) The genetic structure of tropical tree populations: Associations with reproductive biology. The Evolutionary Ecology of Plants (eds. J. H. BOCK & Y. B. LINHART) Westview Press, Boulder Colorado.

HAMRICK, J. L. & MURAWSKI, D. A. (1991) Levels of allozyme diversity in populations of uncommon neotropical tree species. *Journal of Tropical Ecology*, **7**, 395–399.

HAMRICK, J. L., GODT, M. J. W., MURAWSKI, D. A., & LOVELESS, M. D. (1991) Correlations between species traits and allozyme diversity: Implications for conservation biology. *Genetics and Conservation of Rare Plants* (eds. D. A. FALK & K. E. HOLSINGER) Oxford University Press, Oxford.

HARTSHORN, G. S. (1978) Treefalls and tropical forest dynamics. *Tropical Trees as Living Systems* (eds. P. B. TOMLINSON & M. H. ZIMMERMANN) Cambridge University Press, Cambridge.

HELENURM, K. & GANDERS, F. R. (1985) Adaptive radiation and genetic differentiation in Hawaiian *Bidens*. *Evolution*, **39**, 753–765.

HOGAN, K. P. (1986) Plant architecture and population ecology in the palms *Socratea durissima* and *Scheelea zonensis* on Barro Colorado Island, Panama. *Principes*, **30**, 105–107.

HOGAN, K. P. (1988) Photosynthesis in two neotropical palm species. *Functional Ecology*, **2**, 371–377.

HOGAN, K. P., SMITH, A. P., ARAUS, J. L., & SAAVEDRA, A. (1994) Ecotypic differentiation of gas exchange responses and leaf anatomy in a tropical forest understory shrub from areas of contrasting rainfall regimes. *Tree Physiology*, **14**, 819–831.

HOLT, R. D. (1990) The microevolutionary consequences of climate change. *Trends in Ecology and Evolution*, **5**, 311–315.

HUBBELL, S. P. (1979) Tree dispersion, abundance, and diversity in a tropical dry forest. *Science*, **203**, 1299–1309.

HUSTON, M. (1980) Soil nutrients and tree species richness in Costa Rican forests. *Journal of Biogeography*, **7**, 147–157.

JAFFE, M. J. (1980) Morphogenetic response of plants to mechanical stress or stimuli. *Bioscience*, **30**, 239–243.

JANOS, D. P. (1980) Mycorrhizae influence tropical succession. *Biotropica*, **12**, 56–64.

JANZEN, D. H. (1967) Why mountain passes are higher in the tropics. *American Naturalist*, **101**, 233–249.

JANZEN, D. H. (1975) *Ecology of Plants in the Tropics*. Edward Arnold Publishers, London.

JANZEN, D. H. (1986) *Guanacaste National Park: Tropical Ecological and Cultural Restoration*. Editorial Universidad Estatal a Distancia, San Jose, Costa Rica.

JON LLAP, R., CAMACHO HERNÁNDEZ, Y., & SÁNCHEZ, G. (1990) Comportamiento juvenil de procedencias de *Gliricidium sepium* de la region de origen. *El Chasqui*, **22**, 7–13.

JORDAN, N. (1991) Multivariate analysis of selection in experimental populations derived from hybridization of two ecotypes of the annual plant *Diodia teres* W. (Rubiaceae). *Evolution*, **45**, 1760–1772.

KAMALUDDIN, M. & GRACE, J. (1992) Acclimation in seedlings of a tropical tree, *Bischofia javanica*, following a stepwise reduction in light. *Annals of Botany*, **69**, 557–562.

KAUR, A., HA, C. A., JONG, K., SANDS, V. E., CHAN, H. T., SOEPADMO, E., & ASHTON, P. S. (1978) Apomixis may be widespread among trees of the climax rainforest. *Nature*, **271**, 440–442.

KING, D. A. (1991) Correlations between biomass allocation, relative growth rate, and light environment in tropical forest saplings. *Functional Ecology*, **5**, 485–492.

KITAMURA, K. & RAHMAN, M. Y. B. A. (1992) Genetic diversity among natural populations of *Agathis borneensis* (Araucariaceae), a tropical rainforest conifer from Brunei Darussalam, Borneo, Southeast Asia. *Canadian Journal of Botany*, **70**, 1945–1949.

KÖRNER, C. & DIEMER, M. (1994) Evidence that plants from high altitudes retain their greater photosynthetic efficiency under elevated CO_2. *Functional Ecology*, **8**, 58–68.

KÖRNER, C., ALLISON, A., & HILSCHER, H. (1983) Altitudinal variation of leaf diffusive conductance and leaf anatomy in heliophytes of montane New Guinea and their interrelation with microclimate. *Flora*, **174**, 91–135.

KÖRNER, C., BANNISTER, P., & MARK, A. F. (1986) Altitudinal variation in stomatal conductance, nitrogen content, and leaf anatomy in different plant life forms in New Zealand. *Oecologia*, **69**, 577–588.

KÖRNER, C., SCHEEL, J. A., & BAUR, H. (1979) Maximum leaf diffusive conductance in vascular plants. *Photosynthetica*, **13**, 45–82.

KUIPER, D. & SMID, A. (1985) Genetic differentiation and phenotypic plasticity in *Plantago major* ssp. major: I. The effect of differences in level of irradiance on growth, photosynthesis, respiration, and chlorophyll content. *Physiologia Plantarum*, **65**, 520–528.

LANGENHEIM, J. H., STUBBLEBINE, W. H., LINCOLN, D. E., & FOSTER, C. E. (1978) Implications of variations in resin composition among organs, tissues, and populations in the tropical legume *Hymenea*. *Biochemical Systematics and Ecology*, **6**, 299–313.

LAPIDO, D. O., GRACE, J., SANDFORD, A. P., & LEAKEY, R. R. B. (1984) Clonal variation in photosynthetic and respiration rates and diffusion resistances in the tropical hardwood *Triplochiton scleroxylon* K. Schum. *Photosynthetica*, **18**, 20–27.

LATHWELL, D. J. & GROVE, T. L. (1986) Soil–plant relationships in the tropics. *Annual Review of Ecology and Systematics*, **17**, 1–16.

LAWTON, R. O. (1982) Wind stress and elfin stature in a montane rainforest tree: An adaptive explanation. *American Journal of Botany*, **69**, 1224–1230.

LEIGH, E. G., JR., WINDSOR, D. M., RAND, A. S., & FOSTER, R. B. (1989) The impact of the "El Niño" drought of 1982-83 on a Panamanian semideciduous forest. *Global Ecological Consequences of the 1982–83 El Niño-Southern Oscillation* (ed. P. W. GLYNN) *Oceanography Series*, **52**, Elsevier Press, Amsterdam.

LEWIS, M. C. (1972) The physiological significance of variation in leaf structure. *Science Progress*, **60**, 25–51.

LIEBERMAN, D., LIEBERMAN, M., HARTSHORN, G., & PERALTA, R. (1985) Growth rates and age-size relationships of tropical wet forest trees in Costa Rica. *Journal of Tropical Ecology*, **1**, 97–109.

LIN, G. H. & STERNBERG, L. D. L. (1993) Effects of salinity fluctuation on photosynthetic gas exchange and plant growth of the red mangrove (*Rhizophora mangle* L). *Journal of Experimental Botany*, **44**, 9–16.

LONG, S. P. (1983) C_4 photosynthesis at low temperatures. *Plant, Cell and Environment*, **6**, 345–363.

LONGMAN, K. A. & JENIK, J. (1974) *Tropical Forest and Its Environment*. Longman, London.

LOPEZ-MATA, L. (1987) Genecological differentiation in provenances of *Brosimum alicastrum* – A tree of moist tropical forests. *Forest Ecology and Management*, **21**, 197–208.

LOVEJOY, T. E., RANKIN, J. M., BIERREGAARD, R. O., BROWN, K. S., JR., EMMONS, L. H., & VAN DER BOORT, M. (1984) Ecosystem decay of Amazon forest remnants. *Extinctions* (ed. M. H. NITECKI) University of Chicago Press, Chicago, Illinois.

LOVELESS, M. D. & HAMRICK, J. L. (1984) Ecological determinants of genetic structure in plant populations. *Annual Review of Ecology and Systematics*, **15**, 65–95.

MARQUIS, R. (1987) Variación en la herbivoría foliar y su importancia selectiva en *Piper arieianum* (Piperaceae). *Revista de Biología Tropical*, **35** (Suppl.), 133–149.

MARQUIS, R. (1990) Genotypic variation in leaf damage in *Piper arieianum* (Piperaceae) by a multispecies assemblage of herbivores. *Evolution*, **44**, 104–120.

MEINZER, F. & GOLDSTEIN, G. (1985) Some consequences of leaf pubescence in the Andean giant rosette plant *Espeletia timotensis*. *Ecology*, **66**, 512–520.

MEINZER, F. C., GOLDSTEIN, G., & RUNDEL, P. W. (1985) Morphological changes along an altitude gradient and their consequences for an Andean giant rosette plant. *Oecologia*, **65**, 278–283.

MEINZER, F. C., RUNDEL, P. W., GOLDSTEIN, G., & SHARIFI, M. R. (1992) Carbon isotope composition in relation to leaf gas exchange and environmental conditions in Hawaiian *Metrosideros polymorpha* populations. *Oecologia*, **91**, 305–311.

MOONEY, H. A., FIELD, C., VASQUEZ YANES, C., & CHU, C. (1983) Environmental controls on stomatal conductance in a shrub of the humid tropics. *Proceedings of the National Academy of Sciences, USA*, **80**, 1295–1297.

MORAN, G. F., MUONA, O., & BELL, J. C. (1989) *Acacia mangium*: A tropical forest tree of the coastal lowlands with low genetic diversity. *Evolution*, **43**, 231–235.

MULKEY, S. S. (1986) Photosynthetic acclimation and water-use efficiency of three species of understory herbaceous bamboo (Gramineae) in Panama. *Oecologia*, **70**, 514–519.

MULKEY, S. S. & PEARCY, R. W. (1992) Interactions between acclimation and photoinhibition of photosynthesis of a tropical forest understory herb, *Alocasia macrorrhiza*, during simulated canopy gap formation. *Functional Ecology*, **6**, 719–729.

MULKEY, S. S., SMITH, A. P., WRIGHT, S. J., MACHADO, J. L., & DUDLEY, R. (1992) Contrasting leaf phenotypes control seasonal variation in water loss in a tropical forest shrub. *Proceedings of the National Academy of Sciences of the United States of America*, **89**, 9084–9088.

NEWELL, E. A., McDONALD, E. P., STRAIN, B. R., & DENSLOW, J. S. (1993) Photosynthetic responses of *Miconia* species to canopy openings in a lowland tropical rainforest. *Oecologia*, **94**, 49–56.

ORSHANSKY, G. (1932) Seasonal leaf dimorphism in *Ononis natrix* L. *Palestine Journal of Botany*, **1**, 233–234.

OSMOND, C. B., AUSTIN, M. P., BERRY, J. A., BILLINGS, W. D., BOYER, J. S., DACEY, J. W. H., NOBEL, P. S., SMITH, S. D., & WINNER, W. E. (1987) Stress physiology and the distribution of plants. *Bioscience*, **37**, 38–49.

OTTOSEN, C. -O., ROSENQUIST, E., & ÖGREN, E. (1989) Clonal variation in oxygen evolution in *Ficus benjamina* L. *Photosynthetica*, **23**, 537–542.

PATTERSON, B. D., PAULL, R., & SMILLIE, R. M. (1978) Chilling resistance in *Lycopersicon hirsutum* Humb. and Bonpl., a wild tomato with a wide altitudinal distribution. *Australian Journal of Plant Physiology*, **5**, 609–617.

PEARCY, R. W. (1990) Sunflecks and photosynthesis in plant canopies. *Annual Review of Plant Physiology and Plant Molecular Biology*, **41**, 421–453.

PEREZ-NASSER, N., EGUIARTE, L. E., & PIÑERO, D. (1993) Mating system and genetic structure of the distylous tropical tree *Psychotria faxlucens* (Rubiaceae). *American Journal of Botany*, **80**, 45–52.

PETERS, R. L. & DARLING, J. D. S. (1985) The greenhouse effect and nature reserves. *Bioscience*, **35**, 707–717.

RADA, F., GOLDSTEIN, G., AZOCAR, A., & MEINZER, F. (1985) Daily and seasonal osmotic changes in a tropical treeline species. *Journal of Experimental Botany*, **36**, 989–1000.

RADA, F., GOLDSTEIN, G., AZOCAR, A., & TORRES, F. (1987) Supercooling along an altitudinal gradient in *Espeletia schultzii*, a caulescent giant rosette species. *Journal of Experimental Botany*, **38**, 491–497.

RICE, S. A. & BAZZAZ, F. A. (1989) Quantification of plasticity of plant traits in response to light intensity: Comparing phenotypes at a common weight. *Oecologia*, **78**, 502–507.

RICHARDS, P. W. (1952) *The Tropical Rainforest, 1957 Reprint Edition*. Cambridge University Press, Cambridge.

ROBICHAUX, R. H. (1984) Variation in tissue water relations of two sympatric Hawaiian *Dubautia* species and natural hybrids. *Oecologia*, **65**, 75–81.

ROBICHAUX, R. H. & CANFIELD, J. E. (1985) Tissue elastic properties of eight Hawaiian *Dubautia* species that differ in habitat and diploid chromosome number. *Oecologia*, **66**, 77–80.

ROBICHAUX, R. H. & PEARCY, R. W. (1980) Photosynthetic responses of C_3 and C_4 species from cool shaded habitats in Hawaii. *Oecologia*, **47**, 106–109.

RUNDEL, P. W. (1980) The ecological distribution of C_4 and C_3 grasses in the Hawaiian islands. *Oecologia*, **45**, 354–359.

SAGERS, C. L. (1992) Plasticity of plant defenses in a neotropical shrub: Effects of light and genotype. *Bulletin of the Ecological Society of America*, **73**, 332.

SALAZAR, R. (1983) Genetic variation in needles of *Pinus caribaea* var. *hondurensis* Barr. et Golf. from natural stands. *Silvae Genetica*, **32**, 52–59.

SALAZAR, R. (1986) Genetic variation in seeds and seedlings of ten provenances of *Gliricidia sepium* (Jacq.) Steud. *Forest Ecology and Management*, **16**, 391–401.

SALAZAR, R. & VÁSQUEZ, M. S. (1988) Variación genética en ocho procedencias de *Erythrina poeppigiana* en Costa Rica. *Turrialba*, **38**, 71–81.

SAYRE, R. T. & KENNEDY, R. A. (1977) Ecotypic differences in the C_3 and C_4 photosynthetic activity in *Mollugo verticillata*, a C_3-C_4 intermediate. *Planta*, **134**, 257–262.

SCHEINER, S. M. (1993) Genetics and evolution of phenotypic plasticity. *Annual Review of Ecology and Systematics*, **24**, 35–68.

SCHEINER, S. M., GUREVITCH, J., & TEERI, J. A. (1984) A genetic analysis of the photosynthetic properties of populations of *Danthonia spicata* that have different growth responses to light level. *Oecologia*, **64**, 74–77.

SCHLICHTING, C. D. (1986) The evolution of phenotypic plasticity in plants. *Annual Review of Ecology and Systematics*, **17**, 667–693.

SCHULZE, E. -D. (1982) Plant life forms and their carbon, water, and nutrient relations. *Encyclopedia of Plant Physiology* (eds. O. L. LANGE, P. S. NOBEL, C. B. OSMOND & H. ZIEGLER) **12B**, Springer-Verlag, Berlin.

SCHULZE, E. D., BECK, E., SCHEIBE, R., & ZIEGLER, P. (1985) Carbon dioxide assimilation and stomatal response of afroapline giant rosette plants. *Oecologia*, **65**, 207–213.

SCHUSTER, W. S. F., PHILIPS, S. L., SANDQUIST, D. R., & EHLERINGER, J. R. (1992) Heritability of carbon isotope discrimination in *Gutierrezia microcephala* (Asteraceae). *American Journal of Botany*, **79**, 216–221.

SCHWAEGERLE, K. E. & BAZZAZ, F. A. (1987) Differentiation among nine populations of *Phlox*: Response to environmental gradients. *Ecology*, **68**, 54–64.

SLATKIN, M. (1987) Gene flow and the population structure of natural populations. *Science*, **236**, 787–792.

SMITH, A. P. (1975) Altitudinal seed ecotypes from the Venezuelan Andes. *American Midland Naturalist*, **94**, 247–250.

SMITH, A. P. (1984) Postdispersal parent-offspring conflict in plants: Antecedent and hypothesis from the Andes. *American Naturalist*, **123**, 354–370.

SMITH, A. P. (1987) Respuestas de hierbas del sotobosque tropical a claros ocasionados por la caída árboles. *Revista de Biologia Tropical*, **35**, 111–118.

SMITH, A. P. (1990) Does the correlation of elevation with plant taxonomic richness vary with latitude? *Biotropica*, **20**, 259–261.

SMITH, W. K. & DONAHUE, R. A. (1991) Simulated influence of altitude on photosynthetic CO_2 uptake potential in plants. *Plant, Cell and Environment*, **14**, 133–136.

SMITH, A. P., HOGAN, K. P., & IDOL, J. R. (1992) Spatial and temporal patterns of light and canopy structure in a lowland tropical moist forest. *Biotropica*, **24**, 503–511.

SORK, V. L., STOWE, K. A., & HOCHWENDER, C. (1993) Evidence for local adaptation in closely adjacent subpopulations of Northern Red Oak (*Quercus rubra* L.) expressed as resistance to leaf herbivores. *American Naturalist*, **142**, 928-936.

SOTO, R. (1988) Geometry, biomass allocation, and leaf life-span of *Avicennia germinans* L. (Avicenniaceae) along a salinity gradient in Salinas, Puntarenas, Costa Rica. *Revista de Biología Tropical*, **36**, 309–323.

SULLIVAN, J. H., TERAMURA, A. H., & ZISKA, L. H. (1992) Variation in UV-B sensitivity in plants from a 3,000-m elevational gradient in Hawaii. *American Journal of Botany*, **79**, 737–743.

SULTAN, S. E. (1987) Evolutionary implications of phenotypic plasticity in plants. *Evolutionary Biology*, **21**, 127–178.

SULTAN, S. E. & BAZZAZ, F. A. (1993a) Phenotypic plasticity in *Polygonum persicaria*: I. Diversity and uniformity of genotypic norms of reaction to light. *Evolution*, **47**, 1009–1031.

SULTAN, S. E. & BAZZAZ, F. A. (1993b) Phenotypic plasticity in *Polygonum persicaria*. II. Norms of reaction to soil moisture and the maintenance of genetic diversity. *Evolution*, **47**, 1032–1049.

SULTAN, S. E. & BAZZAZ, F. A. (1993c) Phenotypic plasticity in *Polygonum persicaria*. III. The evolution of ecological breadth for nutrient environment. *Evolution*, **47**, 1050–1071.

TEERI, J. A. & STOWE, L. G. (1976) Climatic patterns and distributions of C_4 grasses in North America. *Oecologia*, **23**, 1–12.

TINOCO ORANGUREN, C. & VÁZQUEZ-YANES, C. (1983) Diferencias en poblaciones de *Piper hispidum* bajo condiciones de luz contrastante en una selva alta perennifolia. *Biotica*, **8**, 281–294.

TOMLINSON, P. B. (1986) *The Botany of Mangroves*. Cambridge University Press, Cambridge.

VAN VALEN, L. (1965) Morphological variation and the width of the ecological niche. *American Naturalist*, **99**, 377–390.

VÁZQUEZ-YANES, C. & OROZCO-SEGOVIA, A. (1993) Patterns of seed longevity and germination in the tropical rainforest. *Annual Review of Ecology and Systematics*, **24**, 69–87.

VIA, S. & LANDE, R. (1985) Genotype-environment interaction and the evolution of phenotypic plasticity. *Evolution*, **39**, 505–522.

WALTERS, M. B. & FIELD, C. B. (1987) Photosynthetic light acclimation in two rainforest *Piper species* with different ecological amplitudes. *Oecologia*, **72**, 449–456.

WEST-EBERHARD, M. J. (1989) Phenotypic plasticity and the origins of diversity. *Annual Review of Ecology and Systematics*, **20**, 249–278.

WHITMORE, T. C. (1984) *Tropical Rain Forests of the Far East, Second Edition*. Clarendon Press, Oxford, England.

WICKNESWARI, R. & NORWATI, M. (1993) Genetic diversity of natural populations of *Acacia auriculariformis*. *Australian Journal of Botany*, **41**, 65–77.

WILLIAMS, D. G. & BLACK, R. A. (1993) Phenotypic variation in contrasting temperature environments: Growth and photosynthesis in *Pennisetum setaceum* from different altitudes on Hawaii. *Functional Ecology*, **7**, 623–633.

WITTER, M. S. & CARR, G. D. (1988) Adaptive radiation and genetic differentiation in the Hawaiian silversword alliance (Compositae: Madiinae). *Evolution*, **42**, 1278–1287.

WOODWARD, F. I. (1987) *Climate and Plant Distribution*. Cambridge University Press, Cambridge.

WOOLHOUSE, H. W. (1969) Differences in the properties of the acid phosphatases of plant roots and their significance in the evolution of edaphic ecotypes. *Ecological Aspects of the Mineral Nutrition of Plants* (ed. I. H. RORISON) Blackwell Scientific Publishers, Oxford.

ZISKA, L. H., TERAMURA, A. H., & SULLIVAN, J. H. (1992) Physiological sensitivity of plants along an elevational gradient to UV-B radiation. *American Journal of Botany*, **79**, 863–871.

IV

Ecophysiology of Forest Regeneration and Succession

INTRODUCTION

The subjects of forest dynamics and ecophysiology intersect in the analysis of regeneration responses of individual species. Species that exhibit different patterns of abundance throughout phases of forest succession are likely to vary in functional traits affecting seed germination, early seedling establishment, maximum growth potential, and efficiencies of resource acquisition and assimilation. Thus, it is not surprising that much of the research in tropical plant ecophysiology has focused on describing functional traits across broad, successional groupings of species. Each of the four chapters of this section focuses on ecological patterns in functional traits that underlie interspecific variation in regeneration responses.

Regeneration responses encompass all phases of an individual's life cycle, beginning with seed germination. Vázquez-Yanes and Orozco-

Segovia (Chapter 18) review our state of knowledge on seed dormancy and longevity of tropical rainforest species. Most tropical rainforest seeds germinate or die shortly after dispersal. Notable exceptions to this trend are species with innate dormancy due to embryo immaturity, such as palms, and species with enforced dormancy. The latter are poorly represented by tropical forest trees, and are most frequently invasive weeds and shade-intolerant pioneer species. Pioneer species have photoblastic seeds that germinate in response to high red:far red ratios, thereby ensuring that seedlings will establish in microsites with high light availability.

Seeds of tropical rainforests are generally short-lived in the soil compared to those of other plant communities. High metabolic (respiration) rates of seeds are directly linked to longevity; seeds with an uninterrupted respiratory rate that remain moist in the soil rapidly exhaust energetic reserves, reducing longevity. Seed storage banks are generally considered poor means of conserving tropical plant germplasm due to injurious effects of dehydration on seed viability of most species.

Seed germination is only the first step in regeneration; conditions must also be right to ensure successful seedling establishment. Kitajima (Chapter 19) focuses on ecophysiology of tropical tree seedlings. Collectively, tropical tree species have slightly higher mean seed mass than temperate tree species, and seed mass is greater for shade-tolerant species than for shade-intolerant pioneers. Across species, relative growth rate of seedlings and seedling mortality are negatively correlated with seed mass; larger seeds yield slow-growing seedlings with low rates of mortality. High seedling survival in the forest understory requires sturdy construction, slow growth, and allocation of non-structural carbohydrates to storage reserves; these traits are inconsistent with high growth rates. Kitajima (Chapter 19) further emphasizes the importance of a well-developed root system for seedling survival in shade.

Most ecophysiological studies of photosynthesis and light acclimation in tropical trees have been conducted on seedlings because their small size facilitates experimental manipulations. As Kitajima points out, however, photosynthetic behavior is not an adequate predictor of seedling survival potential. Further studies of seedling ecophysiology need to address the integration of growth and survival processes at the whole-plant level.

Comparative studies within a genus can reveal underlying patterns of adaptation and acclimation to different environments. Fredeen and Field (Chapter 20) provide an example of this approach, based on field studies of the diverse shrub genus *Piper*. Intrinsic growth potential is certainly a critical trait affecting the ecological distribution of species. Studies on *Piper* suggest that species differences in growth and photosynthesis are more pronounced when compared at high light levels than at low light levels. Pathways of nitrogen uptake and assimilation also vary between gap and understory species. Fredeen and Field's synthesis suggests that the ecophysiological basis for distributions of shade-tolerant species are far better understood than for shade-intolerant species.

Ackerly (Chapter 21) examines growth of shade-intolerant pioneer species as an integration of leaf- and canopy-level processes. Pioneer species form an unambiguous ecological group with several unifying functional traits. Among these are specialized seed germination requirements favoring disturbance, rapid height growth, early reproductive maturity, and high reproductive allocation. Rapid growth of tropical pioneers is facilitated by the capacity for continuous and unrestricted leaf production. Associated with rapid height growth are strong apical dominance, high rates of leaf turnover, short leaf lifespan, and low-density wood. Ackerly's analysis of the carbon economy of leaf clusters of different pioneer species provides important insights into integration of growth processes that extend far beyond tropical pioneer species.

Collectively, these chapters emphasize that no single trait adequately predicts species regeneration requirements. Rather, it is the complex integration of leaf- and whole-plant growth processes that underlies the physiological basis of regeneration responses.

18

Physiological Ecology of Seed Dormancy and Longevity

C. Vázquez-Yanes and A. Orozco-Segovia

18.1 INTRODUCTION

Studies of seed biology represent a sizable part of existing botanical research due to the importance of seeds for plant propagation, forest regeneration, and conservation. From an ecological perspective, knowledge of plant community structure, dynamics, and succession must include information about plant reproduction, dispersal, and establishment in which seed characteristics are central. It is well known that the vast diversity of plants that typify the tropical rain-forests generate an impressive multiplicity of seed characteristics in aspects as distinct as season of production, volume of the seed crop, number of seeds per fruit, seed size, shape, morphology, anatomy, moisture content, nature of reserves, and presence of secondary compounds. Seed weights alone range seven orders of magnitude between about 0.1 mg in some Orchidaceae, Piperaceae, and Melastomataceae to near 1000 g in some riparian trees and in members of the families Lecythidaceae, Palmae, Fabaceae, and others (Foster, 1986).

Physiological characteristics of seeds, like basal metabolism at dispersal, development of photosynthesis in the embryo, duration of quiescence, type and periodicity of dormancy, speed of germination, germination display, and longevity in natural, moist or dry conditions also diverge greatly. In a single family, the Fabaceae, longevity is very different among species. Some of them produce seeds that begin to germinate inside the pod, before dispersal, while others produce hardcoated seeds that may have a dormant longevity of decades (Vázquez-Yanes & Orozco-Segovia, 1984).

Tropical rainforest woody plants usually show important seed size variability in the same population and even within the same seed crop of a single plant. Seeds of *Shorea trapezifolia* from Sri Lanka show interesting intraspecific germination variability depending on date of seed fall and maturity and type of tree (De Zoysa & Ashton, 1991). Germination extending over a period of time from days to months is also frequent, and populations of the same species covering a wide geographical distribution may show significant variability in seed viability, speed of germination, and germination responses to temperature (Vázquez-Yanes & Orozco-Segovia, 1990a).

Dormancy is the failure of the living seed to germinate when moisture, air and suitable temperature for metabolic activity and radicle emergence prevail in its environment. If water is in short supply, delaying germination, seeds are better described as quiescent. During dormancy, metabolism is often arrested or retarded and the development of the embryo inside the seed is often, but not always, interrupted. Dormancy has at least two components: a means of maintaining metabolic arrest, and a means of perceiving and predicting the state of the environment. Three types of dormancy have been defined: innate, enforced, and induced (*sensu* Harper, 1957 cited in Murdoch & Ellis, 1992). Innate dormancy refers to the inability of seeds to germinate under favorable conditions immediately upon removal from the parent plant, while enforced dormancy refers to delayed germination mediated by an environmental sensor of some kind which detects properties of the surroundings (temperature and/or light) and their changes that might correspond to the arrival of favorable conditions for germination and establishment. Induced dormancy is an acquired inability to germinate caused by some event after ripening. As in many other plant communities, environmentally enforced dormancy by light quality and/or temperature conditions seems to be common among early heliophile pioneers and colonizer plants in all the tropical forests of the world (Vázquez-Yanes & Orozco-Segovia, 1994). Most of the physioecological research on tropical rainforest seeds has been directed to the study of this kind of dormancy among early successional plants.

In this chapter, we review the physiological characteristics of tropical seeds. We consider the controls and characteristics of dormancy, seed longevity as a function of moisture content and availability, and practical aspects of seed storage. Studies of seed storage are timely because of their importance for *ex situ* germplasm conservation schemes. Previous review papers address parts of these and other aspects of seed biology in the tropical forest (Chin & Roberts, 1980; Foster, 1986; Garwood, 1989; Janzen & Vázquez-Yanes, 1990; Vázquez-Yanes & Orozco-Segovia, 1984, 1990a, 1993, 1994, Whitmore, 1983; Willan, 1985).

18.2 DORMANCY

18.2.1 Innate dormancy

Most tropical rainforest woody plant seeds germinate soon after dispersal (Machargue & Hartshorn, 1983), and some may present a period of quiescence in the soil often linked to dry periods or to the presence of a hard watertight seed coat (Acuña & Garwood, 1987; Kandya, 1990), although in most cases, germination takes place within the seed production year. However, retarded germination that can last for months after seed dissemination is common and may be unrelated to water availability or environmentally enforced dormancy. Very little information exists about the physiology of this behavior (Ng & Mat Asri, 1991). There are documented cases of innate dormancy due to embryo immaturity (in palms, Chin, Krishnapillay & Alang, 1988) or the presence of endogenous inhibitory compounds, but very little is known about the nature of this kind of dormancy in the tropics (Vázquez-Yanes & Orozco-Segovia, 1993). Outside the tropics, innate dormancy is often linked with after-ripening processes that take place during the cold season of the year (Murdoch & Ellis, 1992).

One of the mechanisms that might be involved in the germination delay entails the presence of high concentrations of phenols in many tropical forest seeds. Phenolic compounds of different molecular weights have been related to defensive aspects of plant-microorganism and plant-animal interactions. Some of them exhibit an inhibitory effect on the germination of seeds of wild plants, including endogenous seed compounds like the coumarins (Emu-Kwesi & Dumbroff, 1980; Siqueira et al., 1991; Qi et al., 1993). Oxidation and leaching of the phenols through time when the seeds remain in the soil may account for the end of dormancy (McDonough & Chadwick, 1970). Differences in the rate of leaching might help to explain variation in seed germination timing within seed cohorts. Soluble germination inhibitors have been detected in the seeds of *Chlorophora excelsa* (Adesomoju & Fasidi, 1984).

18.2.2   Environmentally enforced dormancy

Seeds showing this type of dormancy have environmental sensors that detect changes in the surroundings (mainly, temperature and/or light) that can correspond to the arrival of favorable conditions for germination and establishment. Such environmental changes may include the initiation of a wider diurnal temperature variation in the soil, a change in the quality of the light arriving at the forest floor as a consequence of canopy gap formation, the disappearance of the litter layer, and the exhumation of the seeds caused by soil disturbance (Pons, 1992; Probert, 1992; Vázquez-Yanes & Orozco-Segovia, 1994).

In nature, enforced dormancy is most frequent in weeds with opportunistic behavior, either ruderals from disturbed places or among crop weeds. It is much less frequent among primary forest trees of any kind. Nevertheless, detailed descriptions of enforced dormancy behavior have been made for tropical forests throughout the world. The plants most often showing enforced seed dormancy are the fast-growing, short-living heliophile pioneers, which are often the most abundant components of the soil seed bank (Valio & Joly, 1979; Whitmore, 1983; Uhl & Clark, 1983; Garwood, 1989; Thompson, 1992; Ellison et al., 1993; Vázquez-Yanes & Orozco-Segovia, 1993; 1994). Enforced seed dormancy has also been found among strangler trees, some understory shrubs, and spores of ferns (Pérez-García, Orozco-Segovia & Riba, 1994). Rainforest seeds with enforced dormancy that require light for germination may remain dormant in darkness for a long time, even in the continuous presence of available water and appropriate temperatures for germination. They germinate quickly as soon as light is provided and often exhibit high viability (Orozco-Segovia & Vázquez-Yanes, 1990).

The phytochrome family of pigments are the sensors of photon flux and light quality changes in photoblastic species. This is the most common environmentally-enforced dormancy type found in the tropical rainforest. Phytochrome operates in nature as a signal-transducing photoreceptor enabling the plant to acquire information on the light environment, and allowing modulation of physiological and developmental processes according to environmental change (Furuya, 1987). Photoblastic seeds may detect a transference from a dark environment to an illuminated environment when buried seeds are exhumed. Such seeds may also detect the change in the red (655–665 nm): far red (725–735 nm) ratio (R:FR) of the light. Canopy gap formation is the cause of this change (see Chazdon et al., Chapter 1 for details). It is well known that the R:FR ratio of light beneath a forest canopy is inhibitory for germination of many photoblastic seeds in diverse kinds of forest ecosystems (Pons, 1992). The sudden establishment of a soil temperature fluctuation regime, characteristic of many gaps, is also a cue for germination. Germination may occur in response to this signal alone, or in interaction with light produced by canopy gap formation (Probert, 1992).

Sometimes ecological research fails to detect enforced dormancy because certain treatments and germination conditions are required to make it apparent. Photoblastic seeds germinate readily soon after dispersal in the presence of sunlight. Sunlight may be direct, but diffuse light from artificial filters or shade cloth can also trigger germination (Ng, 1978). Abnormal temperature conditions may also induce germination in the dark, giving the impression that seeds lack any dormancy. Despite the fact that the relationship between light quality and germination has been well known for many years recent reports from the tropics include experiments on seed germination where the light quality (R:FR ratio) is not measured (Raich & Gong, 1990; Ellison et al., 1993). Shaded conditions generated by shade cloth or so-called "green" filters usually do not transmit R:FR ratios similar to those found in closed-canopy forest. Often black plastic bags are used for storing and transporting soil with seeds. Unfortunately these bags are not as dark as they seem to be; usually they transmit more than enough light to encourage germination of moist seeds, as most seeds require only very dim radiant energy of the appropriate R:FR ratio to initiate germination. For example, _Cecropia obtusifolia_ seeds can be stimulated to 80% germination by exposure to light of only 0.026 μmol m^{-2} s^{-1} while buried in soil at a depth of 4 mm (Bliss & Smith, 1985). For _Piper auritum_, even lower photon flux is needed for inducing germination (Orozco-Segovia, Sánchez-Coronado & Vázquez-Yanes, 1993a).

As environmentally-enforced dormancy has been analyzed in detail in previous review papers (Vázquez-Yanes & Orozco-Segovia, 1993, 1994), here we will concentrate on the latest findings on the subject of light-controlled germination: (1) relationships between light and temperature, (2) the responses of photoblastic seeds to light quality, (3) sunflecks and germination of seeds inside the forest, (4) maternal light environment and seed response to light, (5) light quality beneath the litter and finally, (6) seed dormancy changes through time in artificial and natural storage.

1. Experiments have been performed with a rainforest tree from Southeastern México, *Urera caracasana*, which becomes established in gaps and persists in the understory. The light requirement for germination is reduced under the influence of temperature alternation (Orozco-Segovia et al., 1987). The complexity of the interaction between light and temperature is illustrated in Figure 18.1. In dry and dark-imbibed seeds stored in the laboratory, the photoperiod required for germination increases, and the need for wide variation in temperature decreases. On the other hand, for seeds buried in the forest soil, the light and alternating temperature requirements for germination disappear with time and only a small proportion of seeds remain light sensitive. In the pioneer tree *Cecropia obtusifolia*, temperature alternations may even induce partial germination in darkness, as shown in Figure 18.2 (Vázquez-Yanes & Orozco-Segovia, 1990b).

2. Photoblastic seeds from different species contain different levels of total phytochrome and active phytochrome. Thus, they may vary greatly in response to different R:FR ratios. For example, four *Piper* species have photoblastic seeds that do not germinate in darkness: *P. auritum* and *P. umbellatum* do not germinate in the low R:FR ratios characteristic of the forest understory, but *P. aequale* and *P. hispidum* germinate well in such environments (Orozco-Segovia & Vázquez-Yanes, 1989). Figures 18.3 and 18.4 show the comparative effect on germination of four light conditions in different species of photoblastic and light-indifferent tropical seeds. Thus, species with photoblastic seeds that do not germinate in darkness can have a different response to far red light. This indicates that failure to germinate in dark is not enough to predict the behavior of the seeds in the forest light environment.

3. The frequency of sunflecks is a function of the degree of canopy openness and plays an important role in creating heterogeneous light environments in the understory. The effect of sunflecks on the germination of photoblastic seeds in forest understory is complex because it depends on the mean R:FR ratio of the prevailing diffuse light and not on the frequency of sunflecks reaching the soil. This light heterogeneity affects germination of photoblastic seeds and may determine the

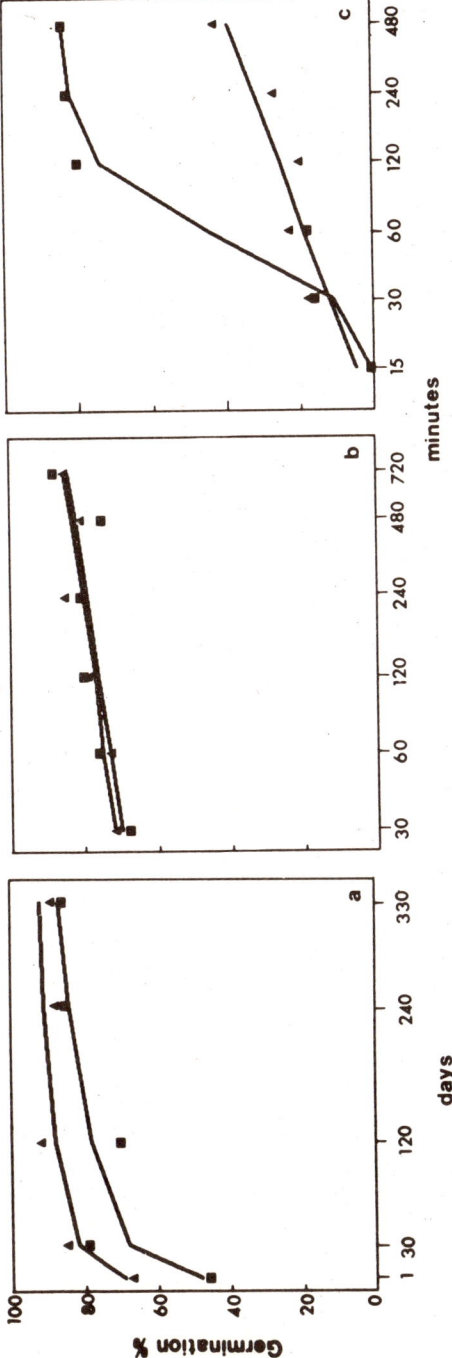

Figure 18.1.   Germination of Urera caracasana: a) through time with (■) 12 h d⁻¹ and (▲) 24 h d⁻¹ of white light; b) germination under 12 h d⁻¹ at 25°C with an interval of temperature increase to 35°C during the periods of time stated in X-axis for (■) seeds four months old, and for (▲) seeds eight months old; c) seeds placed in darkness at 25°C, exposed simultaneously to periods of white light and higher temperature (35°C) for times indicated in X-axis; symbols are the same as in (b).

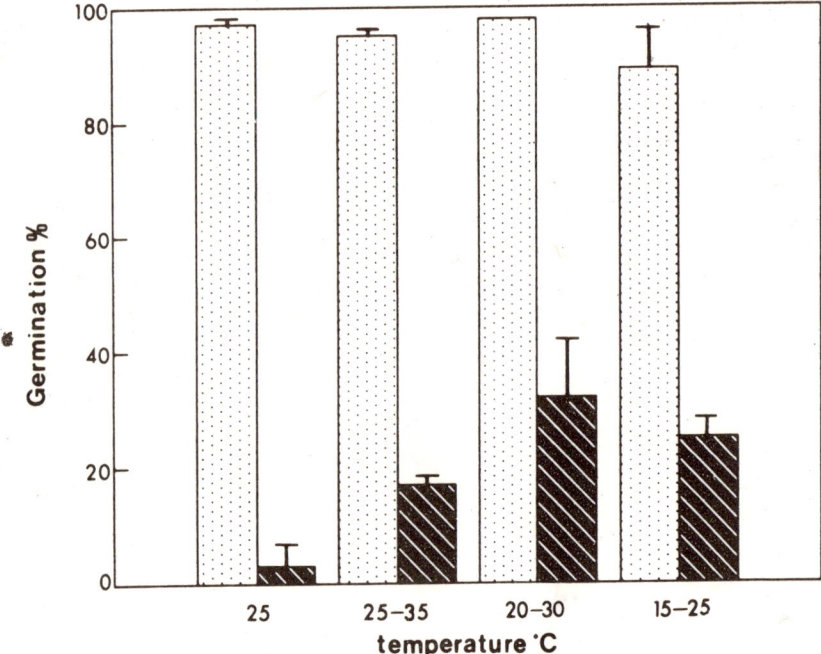

Figure 18.2.   Germination of Cecropia obtusifolia *in (⬚) 12 h d⁻¹ of white light and (■) total darkness, at a constant temperature and three alternating temperatures.*

amount and spatial distribution of seeds that will germinate or remain dormant (Orozco-Segovia & Vázquez-Yanes, 1989; Orozco-Segovia, Sánchez-Coronado & Vázquez-Yanes, 1993a).

   4.   The light quality of the environment where the spikes of *P. auritum* ripen has an effect on the balance of phytochrome in the seeds. In turn, this balance affects the sensitivity of the seeds to light. Therefore, the light heterogeneity experienced by developing seeds creates an intrafamily seed heteromorphism in the germination response to different light environments (Orozco-Segovia, Sánchez-Coronado & Vázquez-Yanes, 1993b).

   5.   Recent research has demonstrated that in rainforest the layer of litter is an important element controlling germination. Litter may either inhibit germination in some species, or enhance proper hydration and germination of the seeds of other species (Guzmán-Grajales & Walker, 1991; Molofsky & Augspurger, 1992). The light quality beneath litter is strongly inhibitory of the germination of

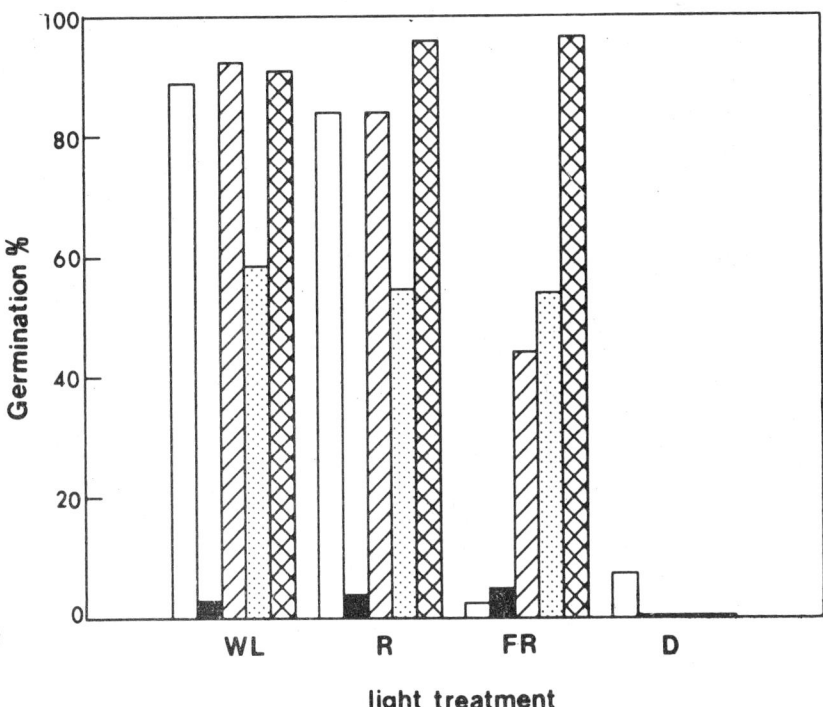

Fig. 18.3. Germination of (□) Cecropia obtusifolia, (■) Ficus aurea, (▨) F. insipida, (▦) F. petenensis and (⊠) F. yopoyensis after 20 days under four different light conditions: WL = white light, R = red light, FR = far red light (12 h photoperiod), and D = darkness.

photoblastic seeds and therefore litter may be an important factor enforcing the dormancy of the seeds of the soil seed bank, or may affect the successful establishment of seedlings coming from small seeds (Vázquez-Yanes et al., 1990; Vázquez-Yanes & Orozco-Segovia, 1992).

6.   After dispersal, seeds buried in the soil, or those stored at room temperature in the laboratory, may change their response to light. More often, aging makes seeds less sensitive to light quality and they may require shorter periods of illumination for germination. The light requirement for germination may facilitate the incorporation of seeds into the permanent forest soil seed bank, but this does not mean that the natural cause of dormancy remains the same through time. Other environmental variables like O_2 and CO_2 levels may replace light as a dormancy-enforcing factor (Orozco-Segovia & Vázquez-Yanes, 1989).

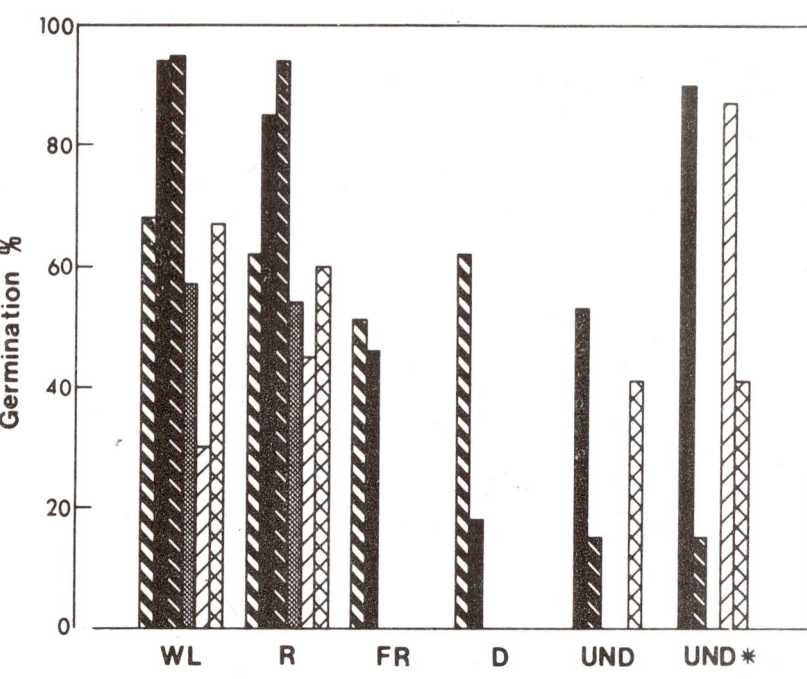

Figure 18.4.   Germination of (▨) Heliocarpus appendiculatus, (▦) Piper hispidum, (▨) P. auritum, (▨) P. umbellatum, (▨) P. aequale and (▨) Urera caracasana after 20 days under six different light conditions: WL = white light, R = red light, FR = far red light (12 h photoperiod), D = darkness, UND = beneath the forest canopy during 15 days, and UND* = same condition as the previous one during four months (in petri dishes).

18.3   LONGEVITY

18.3.1   Seed longevity in the soil

The mean time that seeds remain viable in the soil under the influence of the natural environment is their ecological longevity. The ecological longevities of seeds from tropical species are among the shortest of all species. This is because these seeds tend to germinate fast in the continuously warm and humid forest environment, or they are promptly killed by predators and parasites. Alternatively, they may die of old age after exhausting energetic reserves (Janzen & Vázquez-Yanes, 1990). In many cases, the natural selection has shifted the waiting per' ' for environmental conditions favoring successful establishment from he seed to the seedling stage, (Kitajima, 1992; Chapter 19). Consequently, most seeds are not equipped with anatomical and physiological features needed for retaining a long viability in natural or artificial storage conditions (Priestley, 1986) Objective evaluations of seed longevity in the forest soil exist for only a handful of species. These experiments involve the burial of seeds inside net sacks or clay pots for different periods of time. The enclosure excludes the effect of some seed predators (Juliano, 1940; Holthuijzen & Boerboom, 1982; Pérez-Nasser & Vazquez-Yanes, 1986; Orozco-Segovia & Vázquez-Yanes, 1989, 1990). The largest experiment of this kind examined the survival of 50 rainforest species of seeds buried in the forest floor inside nylon bags (Hopkins & Graham, 1987). In these conditions, seeds that remain viable for a few years are usually hard-coated legumes and small light-sensitive weeds and pioneers. This latter group has enforced dormancy, which prevents germination inside the bags. In contrast, soft-coated seeds of primary forest trees did not show any lasting longevity.

Seed longevity in the soil is also dependent on density factors and can vary in time and space corresponding to changes in population densities of parent plants, seed predators and parasites, and seasonal variation in crop size and quality. Therefore, evaluations of the mean seed longevity restricted to one place and to one season cannot provide a universal evaluation of this parameter for any species (Alvarez-Buylla & Martínez-Ramos, 1990). Factors that independently or in combination may extend seed longevity in the forest soil are: (1) the presence of a dormancy mechanism that prevents fast germination, (2) the degree of interruption of respiratory metabolism in seeds, (3) the presence of a hard coat, (4) the production of abundant seed crops which may allow some of the seeds to survive parasites and predators, and (5) the presence of chemical defenses that prevent parasitism and predation and may also delay germination.

Some of the physiological mechanisms determining the ecological longevity of seeds are linked to their metabolic rate (Murdoch & Ellis, 1992). A seed showing a continuous and uninterrupted respiratory rate will exhaust energetic reserves very fast, preventing germination after some time in the soil (Garwood & Lighton, 1990; Rodríguez-Hernández & Vázquez-Yanes, unpublished data). It is likely that many species of tropical rainforest seeds have uninterrupted respiration between dissemination and germination, and this is reflected by their lack of persistence through time. High metabolic rates in the seeds are directly linked to storability, which is discussed later in this chapter. Unfortunately, comparative studies of respiration among seeds from species with different representation in the seed bank of the forest are yet to be performed. We speculate that the most persistent seeds in the soil must have very low or interrupted respiration, in contrast to the seeds that germinate upon arrival at the ground. In some species, like balsa, a hard coat acts as insulation from deleterious factors in the soil and as a barrier preventing hydration, which may induce metabolic activity (Vázquez-Yanes & Pérez-García, 1976).

18.3.2   Longevity of moist seeds

Before dehiscence, seed maturation is often, but not always, accompanied by seed desiccation. In many cases, seeds retain a relatively high moisture content until dispersal. Such is the case for some fleshy-fruited crop species like the tomato (Berry & Bewley, 1991). However, most seed research begins with the drying of the seeds before experimentation. This practice may alter the way seeds commonly behave in nature, so it is not always an appropriate procedure for physioecological studies. This is particularly important for rainforest species, which generally do not endure deep dehydration between seed maturation and germination. An exception to this are some wind-dispersed seeds produced during the dry season and some seeds with impermeable seed coats (Triviño, De Acosta & Castillo, 1989-90; Vázquez-Yanes & Orozco-Segovia, 1990a). Most rainforest seeds probably remain water-imbibed during the transition through the ripening, dispersal, quiescence, and germination phases.

Longevity of most seeds may be prolonged significantly by deep drying and storage at very low temperatures. At intermediate moisture contents and temperatures, seed longevity usually shortens considerably due to cell damage and microbial development. However, dormant photoblastic seeds, fully imbibed in darkness, may remain viable for a much longer time than those stored in intermediate conditions. This interesting behavior has been studied in lettuce seeds (Rao, Roberts & Ellis, 1987). In contrast to partially dry seeds, the dormant fully imbibed ones may maintain enzyme, membrane, and DNA repair mechanisms that are able to eliminate damage as it occurs. Membranes may remain intact and oxidation reactions may be retarded (Villiers, 1973). This phenomenon may account for the extended longevity shown by photoblastic weed seeds buried in temperate soils, either in seed burial experiments or in ancient soil samples (Priestley, 1986).

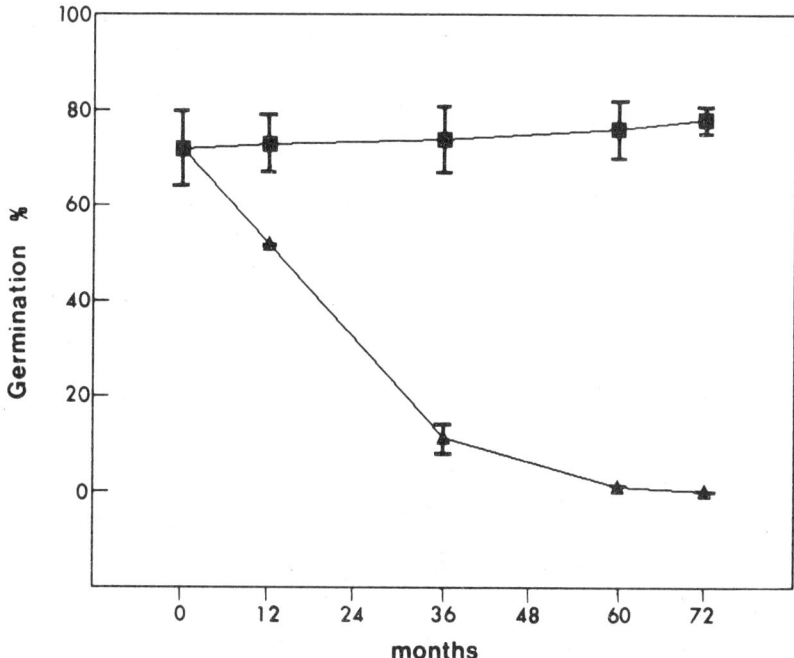

Figure 18.5. Germination of Piper aequale *after the number of months stated in X-axis of storage in two conditions: (▲) air-dried seeds in brown paper bags stored in the laboratory at room temperature (22°C ± 1), (■) water-imbibed seeds seeded in petri dishes wrapped with aluminum foil to insure total darkness, stored in the same conditions as above.*

These facts prompted the design of imbibed seed storage experiments with photoblastic seeds of the rainforest shrub *Piper aequale* placed in petri dishes stored in dark conditions (Orozco-Segovia & Vázquez-Yanes, 1990). Fully hydrated light-sensitive seeds of this species may remain dormant for years in complete darkness. This experiment has been in progress for more than five years and there are still many viable seeds in the stored samples (Figure 18.5). The potential for storage of seeds of most rainforest species has not yet been explored. It may provide an explanation for cases of longevity of several years shown by some photoblastic seeds buried in tropical forest soils (Holthuijzen & Boerboom, 1982; Pérez-Nasser & Vázquez-Yanes, 1986).

18.3.3   Longevity of dry seeds

Potential longevity is the maximum duration of germination capacity (viability) of dormant seeds in optimal storage conditions. Most seeds showing significant ecological longevity also have extended longevity in optimal storage conditions. For example, seeds with hard, watertight coats often live a long time either in the natural substratum or in artificial storage (Rolston, 1978). But even some seeds showing short ecological longevities might remain alive for a very long time if they are stored under stable, low moisture, low temperature conditions. Extended potential longevity is a valuable attribute of many seeds because it simplifies germplasm conservation through seed storage.

Storage longevity depends upon several factors of intrinsic and external nature like heredity, seed structure, composition, physiology, seed provenance, quality and maturity, seed handling, and constancy of the storage environment (Villiers, 1973). Intrinsic factors limit the storage time for some seeds, while others may remain alive for a very long (indefinite?) time when stored in the right environmental conditions. This is due to the fact that the seeds of different plant species have different metabolic rates and dormancy mechanisms which influence the duration of their viability in the soil or in artificial storage (Priestley, 1986). Many seeds may stop respiration while they undergo deep dehydration and then may endure cold storage conditions without suffering damage, while other seed species die when they lose water and/or stop respiring. They lack the capability to rearrange cellular components when free water in the cells is removed during dehydration, so the functional structure cannot be recovered after rehydration (Vertucci, 1989). Alternatively, the presence of free water restricts the protective effect of freezing due to the formation of ice crystals (Roberts & Ellis, 1989).

The expression of genes codifying some proteins (dehydrins) during seed development has been correlated with the acquisition of desiccation tolerance, although a few somewhat desiccation-sensitive temperate tree seeds also produce them (Finch-Savage, Pramanik & Bewley, 1994). Other features that had been linked to desiccation resistance include the accumulation of other peculiar polypeptides and proteins (Dure, 1993), the presence of antioxidants and unsaturated fatty acids in cellular membranes, the replacement of water by polyhydroxyl compounds that stabilize macromolecules as water is withdrawn (Pammenter, Vertucci & Berjak, 1991), and the formation of intracellular glasses (semi-solids) of soluble sugars (Koster, 1991). Abscisic acid (ABA) is the primary hormone produced in the developing embryo (endogenous), or in the parent plant (exogenous), which propels the physiological, cellular and molecular responses that lead to the development of desiccation tolerance (Ferrant, Pammenter & Berjak, 1993). Developing seeds often contain ABA (Pence, 1991). Comprehensive reviews about the development of these processes in maturing seeds have been published (Stanwood & McDonald, 1989; Kermode, 1990; Galau, Jakobsen & Hughes, 1991).

After the classification by Roberts (1973), the three categories of seed storage behavior that are currently widely used are orthodox, recalcitrant, and intermediate (Ellis, Hong & Roberts, 1991; Hong, unpublished manual). Orthodox seeds can be dried to low moisture content without damage at least down to seed moisture content in equilibrium with 10% relative humidity (at this value, starchy seeds have moisture contents close to 5%, whereas oily seeds are 2–3% drier). Over a wide range of storage environments, their longevities increase with a decrease in seed storage moisture content and temperature in a predictable way. The International Plant Genetic Resources Institute (IPGRI, FAO, Rome) preferred condition for long-term storage of these seeds is 5% moisture content at −18°C. Some tropical tree genera showing this seed storage behavior are *Cedrela, Ceiba, Ochroma,* and *Terminalia.*

A subdivision of this category has been proposed to include a greater diversity of possibilities (Bonner, 1990). True orthodox seeds can be stored for long periods at sub-freezing temperatures if their moisture content is reduced below 10%. For short-term storage of 5 years or less, temperatures of 0 to −5°C are satisfactory. For longer storage, less than −15°C is commonly used. Sub-orthodox seeds are those that can be stored under the same conditions as true orthodox seeds, but for shorter periods due to their high lipid content or because of fragile structure or thin seed coats. Some tree seeds once ranked as recalcitrant through careful drying are now known to share the characteristics of this group.

Recalcitrant seeds cannot be dried without damage below a relatively high moisture content. There is great variation in the critical moisture content below which viability is reduced among species. Some seeds begin to die rapidly in equilibrium with a relative humidity of 98–99%, and many seeds are killed when dried to moisture contents in equilibrium with 60–70% relative humidity (about 16–30% moisture content, fresh weight basis). There is still no satisfactory method for maintaining the long-term viability of seeds of these species, particularly for those of tropical origin, which survive generally about 2–9 months. A long list of tropical plants shows this behavior. Probably the best studied cases are in the genera *Avicennia, Hevea, Shorea,* and *Theobroma.* As above, this category has been subdivided in two groups (Bonner, 1990). Temperate recalcitrant seeds cannot be dried but they can be stored for several months at near-freezing temperatures, between 0 and 3 °C. These seeds maintain active metabolism and can sometimes germinate during storage. Tropical recalcitrant seeds cannot be dried and are very sensitive to low temperatures even during short periods of time. Due to the diversity of seed peculiarities among species with recalcitrant seeds, it is unlikely that a common treatment to significantly prolong storage longevity will be found soon. Little research has been directed to this objective. Some seeds previously believed to be recalcitrant can be dehydrated carefully to levels where storage is possible. These are now included in the intermediate category.

The distinguishing feature of intermediate seeds is their sensitivity to desiccation at a relatively low moisture content, below about 7–10% (or below 30–50% relative humidity). In addition, the longevity of dry intermediate seeds is reduced with reduction in temperature at cool (below 5 °C) and sub-zero temperatures. The IPGRI preferred conditions for long-term storage are potentially damaging and should not be used, as most of the intermediate seeds will die under these conditions within a few months. Medium-term (10 years) storage for seeds of these species is nevertheless possible. The known cases are *Carica papaya* and *Azadirachta indica.*

Although predictions about seed storage behavior can be made based on ecological traits, life history patterns, and the seed storage characteristics of taxonomically-related plants, trials are required to classify seeds as orthodox or recalcitrant. A testing protocol includes the following. First, the seed sample is divided into equal parts. Then the viability is tested for one part of fresh seeds, and the other part is desiccated before viability is tested. Another test performed after freezing will indicate if the seeds that can endure desiccation are orthodox or intermediate (Chin, 1988).

There are detailed studies on the effect of seed viability under dehydration for several tropical species of economic importance. These include species of the genera *Azadirachta* (Maithani et al., 1989), *Cedrela* (Corbineau, Defresne & Côme, 1985), *Cedrelinga* (Palmeira & Pereira, 1986), *Dipterocarpus* (Maury-Lechon, Hassan, & Bravo, 1981; Tompsett, 1987), *Hevea* (Chin, Ang & Hamzah, 1981), *Hopea* (Corbineau & Côme, 1986), *Hyeronima* (González, 1992), *Podocarpus*

(Dood, van Staden & Smith, 1989), *Shorea* (Sasaki, 1976; Maury-Lechon, Hassan & Bravo, 1981; Nautiyal & Purohit, 1985a,b,c; Nautiyal, Thapliyal & Purohit, 1985; Tompsett, 1985; Corbineau & Côme, 1986, 1989), *Symphonia* (Maury-Lechon, Corbineau & Côme, 1980; Bras & Maury-Lechon, 1986; Corbineau & Côme, 1989), *Theobroma* (King & Roberts, 1982), *Terminalia* (Tompsett, 1986), and several others.

Institutional and commercial forest seed suppliers in the tropics retail mainly exotic species like pines and eucalyptus, and sometimes a few rainforest plants which produce easily-storable seeds like *Ceiba pentandra*, *Ochroma lagopus*, *Cedrela spp*, *Switenia spp*, *Tectona grandis*, and *Terminalia spp*, and a few others. Most rainforest species having valuable or potentially valuable timber, fruit, or medicinal traits are not available in any seed storage bank (Willan, 1985). Seed storage banks in the tropics should be considered as a last resort *ex situ* conservation strategy because of the difficulties of preserving an authentic representation of the genetic variability of any species (Hamilton, 1994), among other drawbacks.

18.4 FUTURE RESEARCH

There are several research priorities that should be covered with respect to tropical seeds. Extremely little research has been focused on the germination response of seeds in natural conditions over a varied range of environmental conditions. Some of the exceptions are the works of Garwood (1983), Raich and Gong (1990), Augspurger and Kitajima (1992), and Ellison et al. (1993). This kind of research suggests a valuable alternative for the study of seed behavior when scarce technological resources limit the development of more basic physiological research in the laboratory. Much remains to be learned of metabolic factors determining endogenous dormancy and factors that induce heteromorphic germination in the high humidity, high temperature environment of the tropical rainforest. The broad area of seed metabolism must be investigated during stages other than dormancy and germination, including dispersal, quiescence, and seed maturation. The study of fruits and seeds still promises fascinating discoveries in basic biology (e.g., Murray et al., 1994).

Most tropical rainforests of the world are being destroyed at an alarming speed and many of their species are on the verge of extinction. In this context, knowledge about rainforest seeds is essential for forest scientists and technicians working on any aspect of rainforest ecology and management. Additionally, seeds provide a natural vehicle for plant reproduction, germplasm collection, transport and storage of plants, and preservation of genetic variation. From an applied point of view, we need to know more about factors that determine longevity in storage, and mechanisms to prolong it for conservation purposes. A better knowledge of tropical rainforest seeds will help us to find .alternatives to eucalyptus, pine, casuarina, and others that plague the tropical seed storage banks and which, unfortunately, most often come to the minds of foresters when they think of plants for reforestation.

ACKNOWLEDGEMENTS

This paper was written when one of us (CVY) was "Charles Bullard Fellow" (Harvard Forest) at Harvard University. Facilities in which to work, as well as the stimulating academic environment provided by Dr. F. A. Bazzaz and his team at the Department of Organismic and Evolutionary Biology, are deeply appreciated. We are also grateful to Ma. Esther Sánchez Coronado and Mariana Rojas Aréchiga for technical support during several steps of the elaboration of this review.

REFERENCES

ACUÑA, P. & GARWOOD, N. C. (1987) Efecto de la luz y de la escarificación en la germinación de las semillas de cinco especies de árboles tropicales secundarios. *Revista de Biología Tropical*, **35**, 203–207.

ADESOMOJU, A. A. & FASIDI, I. O. (1984) Comparative study of some endogenous phytohormones in dormant and germinated seeds of *Chlorophora excelsa* (Welw.) Benth. *Nigerian Journal of Science*, **18**, 6–8.

ALVAREZ-BUYLLA, E. & MARTÍNEZ-RAMOS, M. (1990) Seed banks versus seed rain in the regeneration of a tropical pioneer tree. *Oecologia*, **84**, 314–325.

AUSPURGER, C. J. & KITAJIMA, K. (1992) Experimental studies of seedling recruitment from contrasting seed distributions. *Ecology*, **73**, 1270–1284.

BERRY, T. & BEWLEY, D. (1991) Seeds of tomato (*Lycopersicon esculentum* Mill.), which develop in a fully hydrated environment in the fruit, switch from a developmental to a germinative mode without a requirement for desiccation. *Planta*, **186**, 27–34.

BLISS, D. & SMITH, H. (1985) Penetration of light into soil and its role in the control of seed germination. *Plant, Cell & Environment*, **8**, 475–483.

BONNER, F. T. (1990) Storage of seeds: Potential and limitations for germplasm conservation. *Forest Ecology and Management*, **35**, 35–43.

BRAS, P. & MAURY-LECHON, G. (1986) Graines forestières tropicales de type fortement hydraté: La conservation et ses effects, exemple du *Symphonia globulifera* L. f. de Guyane Francaise. *Bois et Forêst des Tropiques,* **212,** 35–46.

CHIN, H. F. (1988) *Recalcitrant Seeds: A Status Report.* International Board for Plant Genetic Resources, FAO, Rome.

CHIN, H. F. & E. H. ROBERTS. (1980) *Recalcitrant Crop Seeds.* Tropical Press SDN, Kuala Lumpur, Malaysia.

CHIN, H. F., ANG, B. B. & HAMZAH, S. (1981) The effect of moisture and temperature on the ultrastructure and viability of seeds of *Hevea brasiliensis. Seed, Science and Technology,* **9,** 411–422.

CHIN, H. F., KRISHNAPILLAY, B. & ALANG, Z. C. (1988) Breaking dormancy in kentia palm seeds by infusion technique. *Pertanika,* **11,** 137–141.

CORBINEAU, F. & CÔME, D. (1986) Experiments on germination and storage of the seeds of two dipterocarp: *Shorea roxburghii* and *Hopea odorata. The Malaysian Forester,* **49,** 371–381.

CORBINEAU, F. & CÔME, D. (1989) Germination and storage of recalcitrant seeds of some tropical forest tree species. *Forest Tree Physiology,* **46,** 89–91.

CORBINEAU, F., DEFRESNE, S. & CÔME, D. (1985) Quelques caractéristiques de la germination des graines et de la croissance des plantules de *Cedrela odorata* L. (Méliacées). *Bois et Forêst des Tropiques,* **207,** 17–22.

DE ZOYSA, N. D. & ASHTON, P. M. S. (1991) Germination and survival of *Shorea trapezifolia:* Effects of dewinging, seed maturity, and different light and soil microenvironments. *Journal of Tropical Forest Science,* **4,** 52–63.

DOOD, M. C., VAN STADEN, J. & SMITH, M. T. (1989) Seed development in *Podocarpus henkelii:* An ultrastructural and biochemical study. *Annals of Botany,* **64,** 297–310.

DURE L., III (1993) A repeating 11-mer amino acid motif and plant desiccation. *The Plant Journal,* **3,** 363–369.

ELLIS, R. H., HONG, T. D. & ROBERTS, E. H. (1991) Effect of storage temperature and moisture on the germination of papaya seeds. *Seed Science Research,* **1,** 69–72.

ELLISON, A., DENSLOW, J. S., LOISELLE, B. A. & BRENES, M. (1993) Seed and seedling ecology of neotropical Melastomataceae. *Ecology,* **74,** 1733–1739.

EMU-KWESI, L. & DUMBROFF, E. B. (1980) Changes in phenolic inhibitors in seeds of *Acer saccharum* during stratification. *Journal of Experimental Botany,* **31,** 425–436.

FERRANT, J. M., PAMMENTER N. W. & BERJAK, P. (1993) Seed development in relation to desiccation tolerance: A comparison between desiccation-sensitive (recalcitrant) seeds of *Avicennia marina* and desiccation-tolerant types. *Seed Science Research,* **3,** 1–13.

FINCH-SAVAGE, W. E., PRAMANIK, S. K. & BEWLEY, J. D. (1994) The expression of dehydrin proteins in desiccation-sensitive (recalcitrant) seeds of temperate trees. *Planta,* **193,** 478–485.

FOSTER, S. A. (1986) On the adaptive value of large seeds for tropical moist forest trees: A review and synthesis. *Botanical Review*, **52**, 260–299.

FURUYA, M. (1987) *Phytochrome and Photoregulation in Plants.* Academic Press, Tokyo.

GALAU, G. A., JAKOBSEN, K. S. & HUGHES, D. W. (1991) The controls of late dicot embryogenesis and early germination. *Physiologia Plantarum*, **81**, 280–288.

GARWOOD, N. C. (1983) Seed germination in a seasonal tropical forest in Panama: A community study. *Ecological Monographs*, **53**, 159–181.

GARWOOD, N. C. (1989) Tropical soil seed banks: A review. *Ecology of Soil Seed Banks* (eds. M. A. LECK, V. T. PARKER & R. L. SIMPSON) Academic Press, San Diego, California, pp 149–209.

GARWOOD, N. C. & LIGHTON, J. R. B. (1990) Physiological ecology of seed respiration in some tropical species. *New Phytologist*, **115**, 549–558.

GONZÁLEZ, E. J. (1992) Humedad y germinación de semillas de *Hyeronima alchorneoides* (Euphorbiaceae). *Revista de Biología Tropical*, **40**, 139–141.

GUZMÁN-GRAJALES, S. M. & WALKER, L. R. (1991) Differential seedling response to litter after Hurricane Hugo in the Luquillo Experimental Forest, Puerto Rico. *Biotropica*, **23**, 407–413.

HAMILTON, M. B. (1994) *Ex situ* conservation of wild plant species: Time to reassess the genetic assumptions and implications of seed banks. *Conservation Biology*, **8**, 39–49.

HOLTHUIJZEN, A. M. A. & BOERBOOM J. H. A. (1982) The *Cecropia* seed bank in the Surinam lowland rainforest. *Biotropica*, **14**, 62–68.

HOPKINS, M. & GRAHAM, A. W. (1987) The viability of seeds of rainforest species after experimental soil burials under tropical wet lowland forest in northeastern Australia. *Australian Journal of Ecology*, **12**, 97–108.

JANZEN, D. H. & VÁZQUEZ-YANES, C. (1990) Aspects of tropical seed ecology of relevance to management of tropical forested wildlands. *Rainforest Regeneration and Management* (eds. A. GÓMEZ-POMPA, T. C. WHITMORE & M. HADLEY) MAB Series, UNESCO & Parthenon Publishing, Carnforth, Lancaster, pp 137–157.

JULIANO, J. B. (1940) Viability of some Philippine weed seeds. *Philippine Agriculturalist*, **29**, 313–326.

KANDYA, S. (1990) Mechanism of scarification to remove mechanical and physical dormancy of seeds in some important forest tree species. *Journal of Tropical Forestry*, **6**, 242–247.

KERMODE, A. R. (1990) Regulatory mechanisms involved in the transition from seed development to germination. *Critical Reviews in Plant Science*, **9**, 155–195.

KING, M. W. & ROBERTS, E. H. (1982) The imbibed storage of cocoa (*Theobroma cacao*) seeds. *Seed Science and Technology*, **10**, 535–540.

KITAJIMA, K. (1992) Relationship between photosynthesis and thickness of cotyledons for tropical tree species. *Functional Ecology*, **6**, 582–89.

KOSTER, K. L. (1991) Glass formation and desiccation tolerance in seeds. *Plant Physiology*, **96**, 302–304.

MACHARGUE, L. A. & HARTSHORN, G. S. (1983) Seed and seedling ecology of *Carapa guianensis*. *Turrialba*, **33**, 399–404.

MAITHANI, G. P., BAHUGUNA, V. K., RAWATT, M. M. S. & SOOD, O. P. (1989) Fruit maturity and interrelated effects of temperature and container on longevity of neem (*Azadirachta indica*) seeds. *Indian Forester*, **115**, 89–97.

MAURY-LECHON, G., CORBINEAU, F. & CÔME, D. (1980) Donees préliminaires sur la germination des graines et la conservation des plantules de *Symphonia globulifera* L. f. (Guttifère). *Bois et Forêst des Tropiques*, **193**, 35–40.

MAURY-LECHON, G., HASSAN, A. M. & BRAVO, D. R. (1981) Seed storage of *Shorea parviflora* and *Dipterocarpus humeratus*. *The Malaysian Forester*, **44**, 267–280.

MCDONOUGH, W. T. & CHADWICK, D. L. (1970) Pre-emergence leaching losses from seeds. *Plant & Soil*, **32**, 327–334.

MOLOFSKY, J. & AUGSPURGER, C. K. (1992) The effect of leaf litter on early seedling establishment in a tropical forest. *Ecology*, **73**, 68–77.

MURDOCH, A. J. & ELLIS, R. H. (1992) Longevity, viability, and dormancy. *Seeds: The Ecology of Regeneration in Plant Communities* (ed. M. FENNER) CAB International, Wallingford, United Kingdom, pp 193–229.

MURRAY, K. G., RUSSELL, S., PICONE, C. M., WINNETT-MURRAY, K., SHERWOOD, W. & KUHLMANN, M. L. (1994) Fruit laxatives and seed passage rates in frugivores: Consequences for plant reproductive success. *Ecology*, **75**, 989–994.

NAUTIYAL, A. R. & PUROHIT, A. N. (1985a) Seed viability in soil I. Physiological and biochemical aspects of seed development in *Shorea robusta*. *Seed Science and Technology*, **13**, 59–68.

NAUTIYAL, A. R. & PUROHIT, A. N. (1985b) Seed viability in soil II. Physiological and biochemical aspects of aging in seeds of *Shorea robusta*. *Seed Science and Technology*, **13**, 69–76.

NAUTIYAL, A. R. & PUROHIT, A. N. (1985c) Seed viability in soil III. Membrane disruption in aging seeds of *Shorea robusta*. *Seed Science and Technology*, **13**, 77–82.

NAUTIYAL, A. R., THAPLIYAL, A. P. & PUROHIT, A. N. (1985) Seed viability in sal IV. Protein changes accompanying loss of viability in *Shorea robusta*. *Seed Science and Technology*, **13**, 83–86.

NG, F. S. P. (1978) Strategies of establishment in Malaysian forest trees. *Tropical Trees as Living Systems* (eds. P. B. TOMLINSON & M. H. ZIMMERMANN) Cambridge University Press, Cambridge, pp 129–162.

NG, F. S. P. & MAT ASRI, N. S. (1991) *Germination and Seedling Records.* Forest Research Institute, Malaysia, Research Pamphlet 108.

OROZCO-SEGOVIA, A. & VÁZQUEZ-YANES, C. (1989) Light effect on seed germination in *Piper* L. *Acta Oecologica Oecologia Plantarum*, 10, 123–146.

OROZCO-SEGOVIA, A. & VÁZQUEZ-YANES, C. (1990) Effect of moisture on longevity in seeds of some rainforest species. *Biotropica*, 22, 215–216.

OROZCO-SEGOVIA, A., SÁNCHEZ-CORONADO, M. E. & VÁZQUEZ-YANES C. (1993a) Light environment and phytochrome controlled germination in *Piper auritum*. *Functional Ecology*, 7, 585–590.

OROZCO-SEGOVIA, A., SÁNCHEZ-CORONADO, M. E. & VÁZQUEZ-YANES C. (1993b) Effect of maternal light environment on seed germination in *Piper auritum*. *Functional Ecology*, 7, 395–402.

OROZCO-SEGOVIA, A., VÁZQUEZ-YANES, C., COATES-ESTRADA, R & PEREZ-NASSER, N. (1987) Ecophysiological characteristics of the seed of the tropical forest pioneer *Urera caracasana* (Urticaceae). *Tree Physiology*, 3, 375–386.

PALMEIRA, V. & PEREIRA, A. (1986) Consevaçao de sementes de cedrorana (*Cedrelinga catenaeformis* Ducke)-Leguminosae. *Acta Amazonica*, 16/17, 549–556.

PAMMENTER, N. W., VERTUCCI, C. W. & BERJAK, P. (1991) Homeohydrous (recalcitrant) seeds: Dehydration, the state of water, and viability characteristics in *Landolphia kirkii*. *Plant Physiology*, 96, 1093–1098.

PENCE, V. C. (1991) Abscisic acid in developing zygotic embryos of *Theobroma cacao*. *Plant Physiology*, 95, 1291–1293.

PÉREZ-GARCIA, B., OROZCO-SEGOVIA, A. & RIBA, R. (1994) The effects of white light, far red light, darkness, and moisture on the germination of spores of *Lygodium heterodoxum* Kuntze (Schizaceae). *American Journal of Botany*, 81, 1367–1370.

PÉREZ-NASSER, N. & VÁZQUEZ-YANES, C. (1986) Longevity of buried seeds from some tropical rainforest trees and shrubs of Veracruz, México. *Malaysian Forester*, 94, 352–356.

PONS, T. L. (1992) Seed responses to light. *Seeds: The Ecology of Regeneration in Plant Communities* (ed. M. FENNER). CAB International, Wallingford, United Kingdom, pp 259–284.

PRIESTLEY, D. A. (1986) *Seed Aging: Implications for Seed Storage and Persistence in the Soil.* Comstock, Cornell University Press.

PROBERT, R. J. (1992) The role of temperature in germination ecophysiology. *Seeds: The Ecology of Regeneration in Plant Communities* (ed. M. FENNER) CAB International, Wallingford, United Kingdom, pp 285–395.

QI, M. Q., UPADHYAYA, M. K., FURNESS, N. H. & ELLIS, B. E. (1993) Mechanism of seed dormancy in *Cynoglossun officinale* L. *Journal of Plant Physiology*, 142, 325–330.

RAICH, J. W., & GONG, W. K. (1990) Effects of canopy openings on tree seed germination in a Malaysian dipterocarp forest. *Journal of Tropical Ecology*, 6, 203–217.

RAO, N. K., ROBERTS, E. H. & ELLIS, R. H. (1987). Loss of viability in lettuce seeds and the accumulation of chromosome damage under different storage conditions. *Annals of Botany*, **60**, 85–96.

ROBERTS, E. H. (1973) Predicting the storage life of seeds. *Seed Science & Technology*, **1**, 499–514.

ROBERTS, E. H. & ELLIS, R. H. (1989) Water and seed survival. *Annals of Botany*, **63**, 39–52.

ROLSTON, M. P. (1978) Water impermeable seed dormancy. *The Botanical Review*, **44**, 365–396.

SASAKI, S. (1976) The physiology, storage and germination of timber seeds. *Seed Technology in the Tropics*, **1**, 111–115.

SIQUEIRA, J. O., NAIR, M. G., HAMMERSCHMIDT, R. & SAFIR, G. R. (1991) Significance of phenolic compounds in plant-soil-microbial systems. *Critical Reviews in Plant Sciences*, **10**, 63–121.

STANWOOD, P. C. & MCDONALD, M. B. (eds.) (1989) *Seed Moisture*, Crop Science Society of America, Publication 14, Madison, Wisconsin, USA.

THOMPSON, K. (1992) The functional ecology of seed banks. *Seeds: The Ecology of Regeneration in Plant Communities* (ed. M. FENNER) CAB International, Wallingford, United Kingdom.

TOMPSETT, P. B. (1985) The influence of moisture content and storage temperature on the viability of *Shorea almon, Shorea robusta*, and *Shorea roxburghii* seed. *Canadian Journal of Forestry Research*, **15**, 1074–1079.

TOMPSETT, P. B. (1986) The effect of temperature and moisture content on the longevity of seed of *Ulmus carpinifolia* and *Terminalia brasii. Annals of Botany*, **57**, 875–883.

TOMPSETT, P. B. (1987) Desiccation and storage studies on *Dipterocarpus* seeds. *Annals of Applied Biology*, **110**, 371–379.

TRIVIÑO, D., DE ACOSTA, R. S. & CASTILLO, A. (1989–90) *Técnicas de manejo de semillas para algunas especies forestales neotropicales en Colombia*, Serie de Documentación No. 10, Bogotá, Editorial Gente Nueva.

UHL, C. & CLARK, K. (1983) Seed ecology of selected Amazon basin successional species. *Botanical Gazette*, **141**, 419–425.

VALIO, I. F. M. & JOLY C. A. (1979) Light sensitivity of the seeds on the distribution of *Cecropia glaziovi* Snethlange (Moraceae). *Zeitschrift für Pflanzenphysiologie*, **91**, 371–76.

VÁZQUEZ-YANES, C. & OROZCO-SEGOVIA, A. (1984) Ecophysiology of seed germination in the tropical humid forests of the world: A review. *Physiological Ecology of Plants of the Wet Tropics* (ed. E. MEDINA, H. A. MOONEY & C. VÁZQUEZ-YANES) Dr. W. Junk Publishers, The Hague, pp 37-50.

VÁZQUEZ-YANES, C. & OROZCO-SEGOVIA, A. (1990a) Seed dormancy in the tropical rainforest. *Reproductive Ecology of Tropical Forest Plants: Man and the Biosphere Series 7* (eds. K. S. BAWA & M. HADLEY) UNESCO-Parthenon Publishing, Carnforth, United Kingdom, pp 247–259.

VÁZQUEZ-YANES, C. & OROZCO-SEGOVIA, A. (1990b) Ecological significance of light-controlled seed germination in two contrasting tropical habitats. *Oecologia*, **83**, 171–175.

VÁZQUEZ-YANES, C. & OROZCO-SEGOVIA, A. (1992) Effect of litter from a tropical rainforest on tree seed germination and establishment under controlled conditions. *Tree Physiology*, **11**, 391–400.

VÁZQUEZ-YANES, C. & OROZCO-SEGOVIA, A. (1993) Patterns of seed longevity and germination in the tropical rainforest. *Annual Review of Ecology and Systematics*, **24**, 69–87.

VÁZQUEZ-YANES, C. & OROZCO-SEGOVIA, A. (1994) Signals for seeds to sense and respond to gaps. *Exploitation of Environmental Heterogeneity by Plants: Ecophysiological Processes Above and Below Ground* (eds. M. CALD-WELL & R. PEARCY) Academic Press, New York, pp 261–318.

VÁZQUEZ-YANES, C. & PÉREZ-GARCÍA, B. (1976) Notas sobre la morfología y la anatomía de la testa de las semillas de *Ochroma lagopus* Sw. *Turrialba*, **26**, 310–311.

VÁZQUEZ-YANES, C., OROZCO-SEGOVIA, A., RINCÓN, E., SÁNCHEZ-CORONADO, M. E., HUANTE, P., TOLEDO, J. R. & BARRADAS, V. L. (1990) Light beneath the litter in a tropical forest: Effect on seed germination. *Ecology*, **71**, 1952–1958.

VERTUCCI, C. W. (1989) The effects of low water content on physiological activities of seeds. *Physiologia Plantarum*, **77**, 172–176.

VILLIERS, T. A. (1973) Ageing and the longevity of seeds in field conditions. *Seed Ecology* (ed. W. HEYDECKER) Pennsylvania State University Press, University Park, Pennsylvania, pp 265–289.

WHITMORE, T. C. (1983) Secondary succession from seed in tropical rainforest. *Forestry Abstracts*, **44**, 767–779.

WILLAN, R. L. (1985) *A Guide to Forest Seed Handling with Special Reference to the Tropics.* FAO, Rome, Forestry Paper 20/2.

19

Ecophysiology of Tropical Tree Seedlings

Kaoru Kitajima

19.1 INTRODUCTION

The ecology of reproduction is of key importance for understanding population and community ecology of plants in tropical forests. High morphological diversity of flowers, fruits, and seeds found among tropical tree species is considered to be a component of regeneration niche diversification in these forests. Seedlings of tropical forest trees also exhibit higher morphological diversity than temperate trees, but little is known about the functional significance of this morphological diversity. The seedling phase is uniquely different from the later stages of plant life in terms of dependency on maternally-derived resources and rapid developmental changes in morphology and allocation patterns. Given the small size and high mortality of seedlings, differences in form, function, and development must have great ecological significance.

In this chapter, I focus on ecophysiological problems specific to seedlings from a whole plant perspective, rather than repeating the detailed discussions in other chapters of this volume that are applicable to both seedlings and larger plants. I begin by clarifying the ecophysiological definition of seedlings and discussing the relative importance of environmental variables that affect the survival and growth of tropical tree seedlings. The remaining part of my review is divided into two parts, according to the two overlapping phases of seedling development of tropical trees: the period of active utilization of seed reserves and the period of autotrophic growth. Issues related to seed reserve utilization are discussed with respect to variation in type and amount of seed reserves, morphology and function of cotyledons, patterns of seed reserve utilization, and biosynthetic efficiency of constructing seedling tissue using seed reserves. Finally, I cover the issues relevant after seedlings start deriving the majority of resources autotrophically. I discuss the morphological and physiological basis for seedling shade tolerance, traits that enable high growth rates, below-ground process, and response to temporal environmental heterogeneity. Throughout this review, I attempt to link these issues to the regeneration habits of tropical tree species.

19.1.1 Ecophysiological definition of seedlings

What is a seedling? In a strict physiological sense, a plant developed from a seed is a seedling as long as it depends on seed reserves (Fenner, 1987). Unfortunately, the time when a young plant becomes independent of its seed reserves is not readily quantifiable. In some species, seed reserves remain inside a storage organ (endosperm or storage-type cotyledons) for a long time after germination. However, the time when such storage organs become abscised may not coincide with the end of seed reserve dependency. The developing seedling axis (hypocotyl and epicotyl) may serve as temporary storage organ as well, and the transition of resource dependency from internal to external sources occurs gradually (Fenner, 1987). Using functional growth analysis, it is possible to distinguish several stages during this gradual change of seed reserve utilization (Figure 19.1; Kitajima, 1992a).

Initially after radicle emergence, a developing seedling acquires all necessary resources from seed reserves and its growth rate is independent of external resource availability. This is the stage of complete dependency on seed reserves. Then, following development of organs necessary for autotrophy, such as photosynthetic cotyledons or leaves

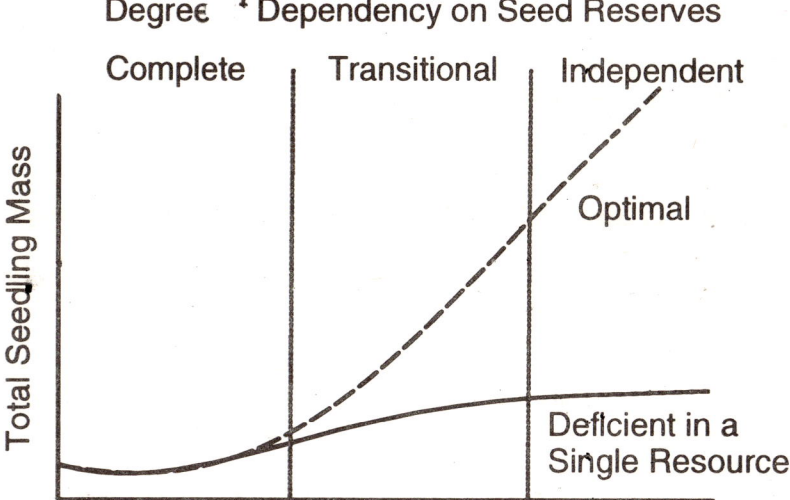

Figure 19.1. Typical growth curves of seedling biomass under the optimal conditions and a condition deficient in supply of a particular resource type (light availability, nitrogen in soil, etc.), illustrating the change in degree of dependency on seed reserves during early seedling stage. The diagram is based on typical relationships between log (seedling mass) and time after germination fitted to third order polynomials in an experimental study for three species (Kitajima, 1992a). While seedlings are completely dependent on seed reserves, there is no difference in seedling mass between optimal and resource deficient conditions. As seedlings start depending on external sources, the growth curves under the two conditions depart from each other. For a while, seedlings depend on both seed-derived and external sources for a given resource supply (transitional phase). Eventually, seedling growth rate becomes completely independent of seed reserves.

for acquisition of energy and roots for acquisition of mineral nutrients, a seedling starts to uptake externally available resources. During this transitional stage, a seedling utilizes both internal (seed-derived) and external sources with increasing dependency on the latter, until the former becomes of negligible importance. Long after all seed reserves are exhausted, seedling size is still a correlate of seed size rather than seedling growth rate (Augspurger, 1984b; Ernst, 1988; Kitajima, 1992a; Osunkoya et al., 1993). Thus, even during the autotrophic phase, ecological performance of a seedling is likely to hinge on contribution from seed reserves for a long time. The actual duration of each stage varies from several days to several months depending on resource type, microenvironmental conditions, and species. In summary, a young plant should be called a seedling as long as a significant proportion of its biomass is constructed from seed reserves, rather than from resources acquired autotrophically.

19.1.2   Tropical forest environments for seedlings

Where and when do seedlings occur? Spatial distributions of treefall gaps, branch holes, and type of overstory trees modify light and other physical factors (Denslow, 1987). Gap distributions affect spatial patterns of seed dispersal (Schupp et al., 1989; Augspurger & Franson, 1988), control of seed dormancy in pioneer species (Vásquez-Yanes & Orozco-Segovia, Chapter 18), and where seedlings survive and grow up to saplings (Augspurger, 1984a; Garwood, 1986; Brokaw & Scheiner, 1989; Welden et al., 1991; Kitajima, unpublished). Many researchers distinguish regeneration guilds of tropical trees based on species response to the treefall gaps in these aspects. Pioneers are those that germinate and survive only in clearings and light gaps, while juveniles of shade-tolerant species survive well in shaded understories. Between these two extremes, interspecific variation of seedling shade tolerance is continuous rather than disjunct (Augspurger, 1984a,b), and it is impossible to put species into clear dichotomous groups by any ecological traits (Brokaw & Scheiner, 1989; Martínez-Ramos, Alvarez-Buylla & Sarukhán, 1989). Nevertheless, it is helpful to contrast the suites of seedling traits typically associated with pioneers and light-gap species vs. shade-tolerant species (Table 19.1), which somewhat parallels the contrast between early vs. late successional species (Bazzaz 1979; Strauss-Debenedetti & Bazzaz, Chapter 6).

Light availability is the primary limiting factor of seedling growth, as demonstrated in several cross-factorial experiments controlling nutrient or water supply with light availability (Kitajima, 1992a; Burslem, 1995). Light environment for seedlings in the forest floor is extremely heterogeneous in small spatial and temporal scales. The majority of the forest floor under closed canopy typically receives 1–2% of photosynthetically active radiation (PAR) above the forest (Chazdon et al., Chapter 1). Treefall gaps where direct sunlight can reach germinating seeds and seedlings occupy only a small fraction of the forest (Brokaw, 1982; Sanford, Braker & Hartshorn, 1986). Between these two extremes, the light environment varies over a very small spatial scale such that spatial autocorrelation disappears above

Table 19.1. *Typical suites of ecological and physiological traits at seed and early seedling stage that are associated with contrasting regeneration habits of tropical tree species. Trends without clear support from published studies are followed by (?). All these traits should be considered as continuous variation between two extremes in the shade tolerant—light demanding continuum of seedling regeneration habits.*

	Shade Tolerant ⟷ Light Demanding	
Seed size	large	small
Lipid or nitrogen contents in seed	variable	variable
Germination in gap microclimate	same or decreased	enhanced
Primary function of cotyledons	storage	photosynthetic
Time until seed reserve exhaustion	slow	fast
Initial seedling size	large	small
Leaf mass per unit area	high	low
Maximum photosynthetic rate sun-acclimated leaves		
per unit mass	low	high
per unit leaf area	low?	high?
Respiration rate for shade-acclimated leaves	(no difference)	
Root:shoot ratio	high	low
LAR	low	high
Wood density	high	low
Allocation to defensive traits	high?	low?
Rate of acclimation	slow	fast
Phenotypic plasticity	low?	high?
Inherent RGR	low	high
Seedling survivorship in shade	high	low

2.5 m distance (Becker & Smith, 1990; Smith, Hogan & Idol, 1992). Thus, the forest light environment is quite heterogeneous for seedlings. In the understory, even a small increase in total daily PAR from 0.5% to 2.0% results in a sharp increase in seedling survivorship (Osunkoya et al., 1992). Within a single treefall gap, light environment varies greatly, such that seedlings at the edge of a very large gap may receive more light than another seedling in the center of a small gap (Chazdon, 1992; Brown, 1993). Spatial variation in light availability leads to variation in other physical and biotic environmental factors, including temperature, soil moisture, herbivore abundance, and activity of pathogenic fungi and bacteria. Thus, successful seedling establishment in the understory or light gaps hinges upon species-specific responses to these multiple factors confounded with light environment, not merely upon the light intensity, spectral quality, or sunfleck frequency.

In seasonal tropical forests, moisture availability drops rapidly in the shallow rooting zone during the dry season (Becker et al., 1988; Wright, 1991). Most species avoid germination during the dry season. The peak of seed dispersal and germination in a seasonal forest community in Panama occurs at the beginning of the rainy season, and many seeds dispersed in late rainy season delay germination until the following rainy season (Garwood, 1983). Thus, only seedlings that have survived through the rainy season experience drought stress. Although evapotranspirational demands are much higher in light gaps, seedling survivorship through the first year is disproportionately higher in light gaps than in the shaded understory for most tree species (Augspurger, 1984a; Garwood, 1986). The advantage of faster growth in light gaps during the rainy season outweighs the disadvantage of greater transpiration during the dry season, even for slow-growing shade-tolerant species. Topography and substrate type also affect the moisture availability in soil within a forest. Within the 50 ha forest dynamics plot in Panama, tree species diversity is higher on slopes than on the flat plateau, possibly because seedling survival is better on slopes that retain higher soil water potential during the dry season (Becker et al., 1988; S. Hubbell, personal communication).

19.2 SEED RESERVE UTILIZATION

19.2.1 Quantity and quality of seed reserves

Tropical tree species as a group appear to have slightly larger mean seed mass than temperate tree species, but the range of interspecific variation may be at the same order of magnitude (typically in the range of 10^{-4} to 10 g; Foster & Janson, 1985; Westoby, Jurado & Leishman, 1992; Kitajima, 1992a). As an overall trend within tropical forest communities, seed mass is greater for shade-tolerant species than for pioneers and light-demanding species (Foster & Janson, 1985; Kitajima, 1992a; Osunkoya et al., 1994), as is found in non-tropical communities (Grime & Jeffrey, 1965; Mazer, 1989; Westoby, Jurado and Leishman, 1992). Among 16 tree species, seed mass (excluding seed coat) is negatively correlated with seedling mortality rates (Kitajima, 1992a). Also, there is a negative interspecific correlation between seed mass and relative growth rates of seedlings such that large-seeded species grow slowly (Kitajima, 1992a; Osunkoya et al., 1993, 1994; Rincón & Huante, 1993; R. Boot, personal communication). Mechanistic understanding of these relationships requires close examination of how seed reserves are utilized.

Several hypotheses attempt to explain ecophysiological advantages of larger seeds for seedling survival in shade. One view is that a large seed supports the seedling's carbon demand for a longer period (Thompson, 1987). A contrasting view is that a large seed creates a large seedling and thus an important initial size advantage (Fenner, 1987; Westoby, Juardo & Leishman, 1992). This hypothesis is favored by the findings that a large initial seedling size promotes emergence though leaf litter (Molofsky & Augspurger, 1992), and probably provides resistance to physical disturbance (Aide, 1987; Núñez-Farfán & Dirzo, 1988; Clark & Clark, 1989). The third view is that the relationship between seed mass and regeneration habitat is indirect via cotyledon functional morphology (Miquel, 1987; Hladik & Miquel, 1991; Kitajima, 1992a). Photosynthetic cotyledons are usually associated with small seeds, while storage cotyledons are common among large-seeded species (Ng, 1978; Hladik & Miquel, 1991; Kitajima, 1992a). Maintaining energy storage in cotyledons may provide an important advantage in the shaded understory, where daily carbon balance may stay below the compensation point for several cloudy days at a time. Stored reserves in cotyledons also help seedlings recover from herbivory and physical damage (K. Harms & J. Darling, personal communication). These three views are interrelated and not mutually exclusive. Further, it is important to consider other ecological constraints as well. For example, selective pressure for increasing seed number and dispensability may be achieved by small seeds. In order to minimize the disadvantage of small initial seedling size, such species should be selected to allocate all seed reserves into the development of photosynthetic cotyledons.

The total amount of seed reserves available for a seedling is not determined by the seed mass alone, but is also influenced by the chemical composition in the seed. The three major types of seed reserves are energy (carbon), protein (nitrogen), and other minerals. Concentration of energy reserves can be approximated by heat of combustion, which mainly reflects the ratio of lipid (ca. 39 KJ g^{-1}) to starch (17.5 KJ g^{-1}), and to a minor extent the protein content (ca. 24 KJ g^{-1}). The lipid contents and their distribution in seeds vary widely among phylogenies (Vaughan, 1970; Barclay & Earl, 1974). Since lipid contents in seeds tend to be higher for woodland species than herbaceous species from open habitats, Levin (1974) hypothesized a possible advantage of high energy concentrations in seeds for seedling establishment in shaded environments. The same selective force that favors heavier seeds as discussed in the previous paragraph may favor higher lipid contents in seeds. However, among 12 tropical species belonging to three families, there was no correlation between seed lipid contents or degree of shade tolerance and seed mass (Kitajima, 1992a). This is perhaps not surprising. Because the metabolic cost of using lipid as energy storage instead of starch is high (see section 19.2.3), increasing the total amount of energy available for seedlings by increasing lipid content is bioenergetically more expensive than increasing seed mass without changing lipid content. Thus, existence of high lipid contents in seeds of some species must indicate a compensating selection when lighter seeds are strongly selected, for example, in order for better dispersal.

Higher contents of nitrogen and other mineral nutrients per unit seed mass may be selected for in small-seeded species. Pate et al. (1986) demonstrated that a standardized nutrient score that accounts for nitrogen, phosphorous, calcium, and magnesium concentration in seeds is negatively correlated with seed mass among 70 species in Proteaceae. Likewise, for 24 temperate herbaceous species in Compositae, there is a negative correlation between seed mass and percentage mineral ash content (Fenner, 1983). These trends might be related to the fact that small-seeded species tend to have more photosynthetic cotyledons. Within-species variation of mineral concentration in seeds is small relative to the variation in seed mass, and the total nutrient content is largely a function of seed mass rather than nutrient concentration (Oladokun, 1989). Yet, phosphorous and calcium concentration in seeds of *Cola acuminata* (Sterculiaceae) and calcium concentration in seeds of *C. nitida* appear to be negatively correlated with their seed mass ranging from 1–25 g and 1–40 g, respectively (Oladokun, 1989).

19.2.2   Cotyledon functional morphology

Ecophysiology of cotyledons deserves special attention because of the high diversity of morphology, degree of exposure, position, and associated functions of cotyledons found among tropical tree species (Ng, 1978; de Vogel, 1980; Garwood, 1983; Miquel, 1987; Hladik & Miquel, 1991; Kitajima, 1992b). Before seed germination, cotyledons absorb resources from the endosperm and the mother plant (Murray, 1985). During and after germination, cotyledons transfer reserve materials (lipids, carbohydrates, mineral nutrients) into developing shoot and roots (Marshall & Kozlowski, 1975, 1976; Olofinboba, 1975; Ashcroft & Murray, 1979; Ernst, 1988). Cotyledons of some species serve strictly as organs to store and transfer seed reserves throughout their lifespan, while cotyledons of other species develop a second function, photosynthetic carbon assimilation (Lovell & Moore, 1971; Marshall & Kozlowski, 1974; Kitajima, 1992b). Completely storage cotyledons are globoid, remaining partially or completely in the seed coat (cryptocotylar), and are typically positioned at or below the ground level (hypogeal). Photosynthetic cotyledons are free of seed coat (phanerocotylar), thin and paper-like (papyraceous), and raised above ground (epigeal). Some species have intermediate cotyledon types that are free of seed coat (phanerocotylar), globoid or in the shape of thick plate (> 1 mm in thickness), green with chloroplasts in cells close to the surface, and raised above ground (epigeal) or at the surface of the ground (semi-hypogeal).

Cotyledon thickness varies continuously across the two groups from 0.2 mm to 5 mm among epigeal phanerocotylar species (Kitajima, 1992a,b). Their photosynthetic capacity per unit mass of cotyledons measured by leaf-disk oxygen electrode is inversely correlated with cotyledon thickness (1/cotyledon thickness), and positively correlated with cotyledon SLA (cotyledon area divided by cotyledon mass) (Kitajima, 1992b). This relationship exists because the ratio of storage cells (inside) to photosynthetic cells increases with cotyledon thickness. Photosynthetic cells are restricted to within 200 μm from the surface due to light extinction through green tissue (Knapp et al., 1988). Thin cotyledons (e.g., 0.2 mm) have very high rates of gross photosynthesis (e.g., 0.6 μmol O_2 g^{-1} s^{-1}), while cotyledons thicker than 1 mm have photosynthetic rates just high enough to balance respiration and achieve a small but positive net photosynthesis. Dark respiration rates per unit mass do not vary with cotyledon thickness, indicating similar respiration rates for storage cells and photosynthetic cells. Per unit area, gross photosynthetic rates are high (e.g., 16 μmol O_2 m^{-2} s^{-1}) and do not differ significantly among species with various cotyledon thicknesses, indicating a similar total amount of photosynthetic machinery per unit area. Stomatal frequencies on the abaxial surface of cotyledons are positively and linearly correlated with maximum photosynthetic rates and dark respiration rates per unit mass, but not with photosynthetic or respiration rates per unit area. The relationships among stomatal distribution, photosynthetic capacity, and leaf thickness in cotyledons differ from those found in true leaves (Mott, Gibson & O'Leary, 1982).

Physiological function of cotyledons is of great importance in determining growth response of seedlings to light environment. When there is sufficient light available, photosynthesis by thick epigeal cotyledons can supply all the energy required for non-photosynthetic function, such as modification and export of stored reserves. As a result, seedlings with cotyledons with dual functions (storage plus photosynthesis) export and use up reserves in cotyledons faster in sun than in shade, while cotyledon retention time is minimally affected for species with cryptocotylar storage cotyledons (Kitajima, 1992a). The species with dual-function cotyledons also show greater difference in RGR between sun and shade environments than species with strictly storage cotyledons. Within a given light environment, cotyledon retention time is shortest for the species with photosynthetic storage cotyledons, followed by those with completely storage cotyledons, and is indefinitely long for species with thin photosynthetic cotyledons. Species with dual-function cotyledons expand true leaves faster than those with completely photosynthetic cotyledons.

19.2.3   Tissue construction using seed reserves

What is the mass of seedling constructed from reserves stored in the seed alone? Do species with higher lipid content create heavier seedlings than species with equivalent seed mass but with lower lipid content? Is lipid or starch the more economical storage form for efficient energy retrieval? In order to assess the ecological and evolutionary significance of parental investment into individual offspring, we must address these questions. Techniques for the analysis of biosynthetic efficiency of tissue construction (Penning de Vries, Brunsting & Van Laar, 1974; Williams et al., 1987) can be applied for theoretical calculation of seedling tissue construction from seed reserves of different chemical composition (Penning de Vries & Van Laar, 1977). Early seedling development is the process by which seed tissues rich in reserves are transformed into seedling tissues. The relative biomass retention during this process depends solely on the seed energy reserves and should reflect the original energy concentration in seed, as well as the efficiency of overall metabolism for seedling development.

Energy contents per unit seed mass range from 15.4 to 29.6 KJ g^{-1} among 12 tropical woody species belonging to 3 families, reflecting their differences in seed lipid contents (Kitajima, 1992a). In a theoretical calculation of construction costs for typical seedling tissue, seed tissues whose carbon reserves consist solely of starch can produce seedling biomass 8% less than the total seed mass, while seed tissues with 50% lipid contents can be modified into seedling tissues with 30% more biomass than initial total seed mass (Kitajima, 1992a). This calculation assumed that the biosynthesis occurs along the most efficient pathways and that there is no maintenance respiration. In reality, the relative increase of biomass with increase in seed lipid contents may be smaller. When seedlings of 12 tropical tree species are compared, the higher the energy content per unit seed mass, the greater the seedling biomass created per unit seed mass. However, this is a less cost-effective way of increasing seedling size than simply increasing mass because a 100% increase in cost of increasing energy content per unit seed mass returns only a 40% increase in seedling biomass (Kitajima, 1992a). Part of this lower energy retrieval efficiency from lipid-rich seed reserves is due to the low efficiency of the catabolic pathway for lipids in plants (Chapin, 1989). High lipid contents in seed reserves and cotyledons of some tropical tree species contrast sharply with the almost complete lack of lipid as carbon storage in leaves, stems and roots of adult plants (Golley, 1961, 1969) and seedling tissues other than cotyledons (Kitajima, 1992a).

Different mineral nutrients in seeds are exhausted at different rates (Marshall & Kozlowski, 1975; Brookes, Wingston & Bourne, 1980). In various temperate herbaceous species, nitrogen is the first among mineral nutrients to become insufficient in supply from seed reserves alone (Fenner, 1986; Fenner & Lee, 1989). Nitrogen, most of which is stored in the form of storage protein in seeds, is used for synthesis of various enzymes necessary during seedling development, including those of early photosynthetic organs, such as photosynthetic cotyledons or the first leaves. Higher nitrogen contents in a seed support the seedling's nitrogen demand for a longer period (Kitajima, 1992a). High nitrogen concentration in seeds may lend an energetic advantage. As long as seedlings' nitrogen demands are met by nitrogen reserves in seeds, seedlings do not need to expend any energy on uptake and reduce soil nitrogen. Up to 20% of nitrogen reserve in seeds may be leached out during germination (Simon & Harun, 1972; Penning de Vries & Van Laar, 1977; Kitajima, 1992a). Although this loss seems very high, the remaining amount of nitrogen is probably enough to support slow seedling development in shade for more than 2 months (Kitajima, 1992a). Little is known about utilization of other mineral nutrients in seeds of tropical tree species. Seedlings of temperate *Quercus* species depend on cotyledon reserves rather than on soil for phosphorus and potassium for most of the first year (Brookes, Wingston & Bourne, 1980).

19.3 AUTOTROPHIC PHASE

19.3.1 Adaptive traits for survival in shade

Once seedlings use up seed reserves, maintenance of a positive net carbon gain is a prerequisite for survival. What is the functional basis for seedling survival in shade beyond the seed-reserve dependency period? What suites of seedling traits make some species survive better than others in shade? Many ecologists define shade tolerance as an ability to survive and grow under low light (Hartshorn, 1978; Brokaw & Scheiner, 1989). However, maintaining positive carbon balance in shade does not necessarily mean growing fast. Seedlings of some species may survive in the shaded understory for a long time with little or no growth (suppression stage) until they are released from shade by increased canopy openness (Whitmore, 1978; Brokaw, 1985; Canham, 1985). Thus, it is inappropriate to use growth rates in shade alone as measure of shade tolerance (e.g., Bongers & Popma, 1990). Rather, there are theoretical and empirical reasons to believe that growth rates are negatively correlated with survivorship in the shaded understory.

Relative growth rates of plants in nature reflect both innate physiological factors (e.g., photosynthesis, respiration, and senescence) as well as external ecological factors (loss of tissue to consumers, including herbivores and pathogens) as summarized in the following simple equation:

Realized RGR = Inherent RGR of the species − Rate of Tissue Loss to Consumers.

The relative balance of the two terms in this equation differs between resource-rich and resource-poor environments (Coley, Bryant & Chapin, 1985; Tilman, 1988). The relative importance of minimizing the second term is greater in the energy-limited understory than in the light gaps. According to the theory of carbon allocation in plants (Mooney, 1972; Coley, Bryant & Chapin, 1985; Chapin, Schulze & Mooney, 1990), inherent growth rates are determined by relative allocation into growth, maintenance, storage, and defense. Defensive traits of plants can be either structural (thick cuticle, bark, tough support tissue with high lignin content) or chemical (tannins, alkaloids). An increase in defense allocation in order to minimize tissue loss to consumers inevitably decreases allocation for construction of new assimilatory organs, such as leaves, and as a result, the first term in the above equation. This is an allocation based trade-off between maximization of inherent growth rate and minimization of tissue loss rates.

For a higher survival in shade, should seedlings be constructed to maximize the efficiency of carbon capture in the shade (first term in the above equation) or to minimize the tissue loss to herbivores and pathogens (the second term in the equation)? If maximization of inherent growth rates is important for survival in shade, shade-tolerant species should have greater efficiency of light harvesting and lower respiration rates than shade-intolerant species in comparable shaded environments. If maintenance of positive carbon balance and survival of seedlings in the shade hinge upon protection from herbivores and pathogens, shade-tolerant species should have greater allocation to defense, and as a result, lower inherent growth rates than shade-intolerant species. The importance of protection from herbivores for seedling survival is clear from the results of caging experiments (Sork, 1987; Osunkoya et al., 1992; Molofsky & Fisher, 1993). I have examined the relationship between survival potential and growth potential by combining a comparative data set of seedling survivorship in shade (Augspurger, 1984b) and inherent growth rates determined under controlled sun and shade environments in a growth house where tissue loss to pathogens and herbivores was minimal (Kitajima, 1994). The result was in support of the second hypothesis. All species perform better in sun than in shade, growing faster and surviving better. However, in a given light environment, either sun or shade, species that survive well grow slowly, and species that grow fast survive relatively poorly.

The above result suggests that seedlings of shade-tolerant species are likely to have greater defense allocation than shade-intolerant species. It has been demonstrated that negative interspecific correlation exists between the degrees of defensive traits and the growth rates for saplings of tropical (Coley, 1983) and temperate species (Shure & Wilson, 1993; Dudt & Shure, 1994). For tropical tree seedlings, stem wood density (dry mass per unit green volume of stem) is negatively correlated with inherent growth rates and mortality rates of the species (Kitajima, 1994). For seedlings and saplings, the actual cost of defense is yet to be determined. This is perhaps because it is not easy to separate defense allocation from allocation for growth, especially for structural defense. For example, construction of secondary cell walls is a process of growth, but construction of sturdier tissue rich in lignin content costs more since construction cost per unit mass of lignin is much higher than cellulose (Williams et al., 1987). Perhaps the best way to determine defense costs is to identify all defensive traits of seedlings (e.g., leaf thickness, toughness, wood density, resins) and calculate the increase in cost of tissue construction per unit mass relative to the cost without these traits.

Another mechanism for a negative interspecific correlation between shade tolerance and inherent growth rates is the difference in storage allocation among species. As discussed earlier, some species set aside a significant proportion of seed reserves in storage organs, such as storage cotyledons, endosperm, and modified hypocotyl. Similarly, there may be interspecific differences in the proportion of current photosynthate allocated for nonstructural carbohydrates stored in stems and roots. In either case, maintenance of storage inevitably lowers the allocation for new growth, and results in lower growth rates than otherwise. However, such storage may be of great importance in the understory where recovery from loss of tissue to herbivores or fallen branches may be extremely slow if seedlings depend only on current photosynthetic income. Storage is also important as a mechanism to maintain the long-term carbon balance since light availability changes due to weather patterns, fallen leaves covering seedlings, and closure of nearby light gaps or branch holes. A comparative study of nonstructural carbohydrates in seedling stems and roots between shade-tolerant and shade-intolerant species would help clarify the role of storage allocation.

19.3.2 Traits that enable high growth rates

While shade-tolerant species appear to be selected for greater defense and storage allocations at the expense of seedling growth rates, shade-intolerant species whose natural regeneration habitats are open sites appear to be selected for high inherent growth rates (Figure 19.2., Appendix 19.1). The competitive advantage of fast growth realizable in light gaps perhaps overrides the disadvantage associated with low allocation to defense and storage allocation. What traits, in addition to lowered defense and storage allocation, make seedlings of some species grow faster than others? Insights in answering this question may be derived from theoretical and experimental studies of growth analysis (Hunt, 1982; Poorter, 1990; Poorter & Remkes, 1990; Ackerly, Chapter 21) in which relative growth rate (RGR) is treated as a product of net dry-mass gain rate per unit leaf mass (ULR, unit leaf rate) and ratio of leaf area to plant mass (LAR, leaf area ratio). ULR

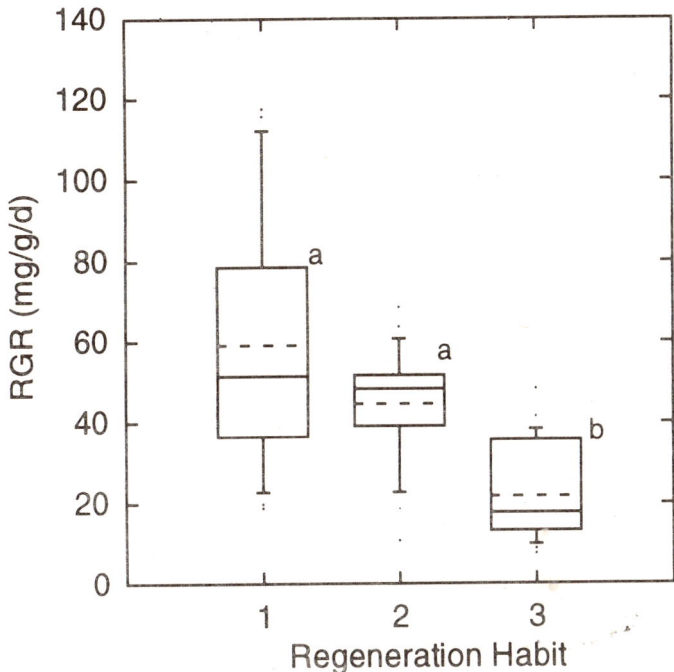

Figure 19.2.   Range of relative growth rate (RGR) of tropical tree seedlings for contrasting regeneration habits from various published studies (Appendix 19.1, excluding two studies conducted in growth chambers). Regeneration habits are grouped roughly as (1) pioneer, (2) light gap dependent, and (3) shade-tolerant species based on information supplied with each study. Box plots display 10, 25, 50, 75, and 90th percentile, actual extreme values (points), and means (broken line) for each group. Although there is a great overlap in the range of RGR among three groups, significant differences exist among groups in one-way analysis of variance ignoring seedling age and harvest intervals. Different letters indicate a significant difference between two groups by Tukey-Kramer HSD test.

is analogous to the net photosynthetic rates of leaves from which whole-plant respiration rates are subtracted. By examining correlations between RGR and its theoretical components, it is possible to determine whether the variation in RGR is attributable to the variation in physiological performance or to the variation in allocation patterns.

When seedlings of a given species raised at different light availabilities are compared, RGR is a positive correlate of ULR but not of LAR (Wadsworth & Lawton, 1968; Okali, 1972; Whitmore & Bowen, 1983;

Popma & Bongers, 1988, 1991; Kitajima, 1992a). In fact, the phenotypic response to increased light availability is to have lower LAR, diminishing the increase of RGR accompanying the increase in ULR. In addition, RGR, LAR and ULR of a given species change through developmental stages and seedling age (Goodall, 1950; Okali, 1971; Fasehun & Audu, 1980; Whitmore & Gong, 1983). They are also affected by the length of harvest intervals (RGR and ULR are lower for longer harvest intervals), and the environments before and during the experiment. When data from different studies in Appendix 19.1 are analyzed together in a general linear model to examine the effects of regeneration habitat, seedling age, and duration of experiment on RGR, LAR, and ULR, duration of experiment always has a significant interaction with regeneration habitat. LAR generally declines with seedling age, while RGR and ULR are lower with longer harvest intervals. Thus, comparisons across different studies require caution, and approaches to studying interspecific variation must be standardized.

For interspecific variation, both ULR and LAR are significant correlates of RGR. What is the relative importance of physiology (photosynthesis and respiration) and allocation patterns (leaf area ratio, leaf mass per unit mass, root:shoot ratio, etc.) as the determinants of interspecific variation in RGR? For temperate herbaceous or tree species (Poorter & Remkes, 1990), LAR, but not ULR, is a significant correlate of RGR. In contrast, RGR of tropical tree species tends to be better correlated with ULR than with LAR (Oberbauer & Donnely, 1986; Popma & Bongers, 1988; Kitajima, 1992a; Osunkoya et al., 1994; Ackerly, Chapter 21). Very high RGR that some pioneers exhibit is attributable to their high ULR. For example, *Ceiba pentandra* and *Ochroma lagopus* have very high ULR, comparable to that of herbaceous *Helianthus annus*, and higher than other temperate or tropical tree species examined (Okali, 1971; Whitmore & Gong, 1983; Oberbauer & Donnely, 1986). The higher correlation between ULR and RGR than between LAR and RGR must be interpreted with caution because ULR and RGR are statistically not independent (equations of both contain change in total plant mass between two harvest times as dividend). When actual leaf gas exchange rates are determined in conjunction with growth analysis, LAR explains a greater portion of interspecific variation in RGR than light-saturated photosynthetic rates (Ramos & Grace, 1990; Kitajima, 1992a, 1994). This is in agreement with the results for temperate tree seedlings (Walters, Kruger & Reich, 1993a,b) and herbaceous species (Poorter & Remke & Lambers, 1990).

Interspecific variation in LAR can be attributed to specific mor-

Table 19.2.   Interspecific correlations between LAR and other allocation traits for seedlings of neotropical trees raised under high light availability at two different developmental stages (Stage 3: with the first set of true leaves fully expanded; Stage 4: cotyledons are abscised or 10 weeks after Stage 3, whichever was earlier). Because of the difficulty of assigning storage cotylendon's mass to stem or root, the analysis at Stage 3 was restricted to species with thin photosynthetic cotyledons (27 phanerocotylar spp). For Stage 4, analysis was also done for a larger number of species, including those with storage cotylendons that had been abscised (49 spp). SLA = specific leaf area; LWR = leaf mass / total seedling mass; root:shoot = root mass / shoot mass; leaf:stem = leaf mass / stem mass (data from Kitajima, 1992a).

	Stage 3 (28 phanerocotylar spp)		Stage 4 (28 phanerocotylar spp)		Stage 4 (49 species)	
	r^2	P	r^2	P	r^2	P
Cotyledon SLA	0.66	0.0001	0.34	0.002	—	—
Leaf SLA	0.57	0.0001	0.77	0.0001	0.68	0.0001
LWR	0.08	NS	0.43	0.0001	0.24	0.0004
root:shoot	0.04	NS	0.10	NS	0.13	0.01
leaf:stem	0.09	NS	0.28	0.004	0.12	0.02

phological traits; a high LAR may be a result of either high specific leaf area (SLA = leaf area per unit leaf mass), high leaf weight ratio (LWR = leaf mass divided by total plant mass), high leaf:stem ratio, and/or low root:shoot ratio. For seedlings of 28 tree species with thin photosynthetic cotyledons studied by Kitajima (1992a), SLA of cotyledons and true leaves is a much better correlate of LAR than LWR, root:shoot ratio, or leaf:stem ratio at two developmental stages (at full expansion of first set of leaves and 10 weeks later, Table 19.2). Although the total cotyledon area is only 3.1% of the total true leaf area at 10 weeks after the expansion of the first true leaves (median of 28 species with thin photosynthetic cotyledons in Table 19.2), cotyledon SLA remains positively correlated with LAR because species with high cotyledon SLA tend to have high SLA for true leaves as well ($r^2 = 0.45$).

Although SLA is generally considered a morphological variable, it is inseparable from physiological traits. Photosynthetic rate per unit mass is equal to photosynthetic rate per unit area times SLA. Photosynthetic capacity (A_{max}) per unit cotyledon mass is a correlate of cotyledon SLA, while A_{max} per unit cotyledon area varies little among species (Kitajima, 1992b). Correlation of leaf A_{max} with RGR depends on whether A_{max} is expressed on an area or mass basis. RGR has a better correlation with A_{max} per unit leaf mass than A_{max} per unit leaf area for 13 tree species ($r^2 = 0.49$ and 0.34, respectively; Kitajima, 1994). Similar results have been obtained for temperate tree seedlings (Walters, Kruger & Reich, 1993b). Thus, efficiency of photosynthetic carbon gain and RGR of seedlings hinge upon SLA, a measure of how a unit mass of photosynthetic machinery is displayed in the horizontal plane. As seedlings grow and accumulate more leaves, self-shading may become a constraining factor for efficiency of photosynthetic carbon gain, especially for pioneer seedlings growing in open space (Coombe & Hatfield, 1962; D. Ackerly, personal communication). Species with high SLA have shorter leaf life-spans and faster leaf turnover rates of seedlings (Bongers & Popma, 1990), perhaps because self-shading is a greater constraint for seedlings with high SLA, LAR, and RGR, or because leaves with high SLA (i.e., low mass per unit area) tend to be less defended against herbivores as well (Coley, Bryant & Chapin, 1985).

19.3.3 Belowground process

Development of photosynthetic activity appears to precede development of belowground resource acquisition, even though radicle emergence and anchorage to the ground are the first steps of seed germination. Seedlings of most species germinate during the rainy season (Garwood, 1983), and the irregular rainfall during this period seems to have little effect on seedling survival even on sandy soil (Blain & Kellman, 1991). Drought stress becomes an issue only for seedlings that survive the rainy season and enter the dry season. In regions without an extensive dry season, seedlings may never encounter drought stress (Steege, 1994). In light gaps, seedling survival through the dry season is correlated with seedling size (Garwood, 1986; Brown & Whitmore, 1992). When seedlings are grown in deep shade, shade-tolerant species that are likely to survive until the beginning of the dry season have greater root mass and root:shoot ratios than less shade-tolerant species at the same age or developmental stage (Kitajima, unpublished). These trends suggest that seedlings entering the dry season with well-developed roots tolerate drought stress better. Vulnerability of seedlings to drought stress may hinge upon several aspects, such as relative amount of root tissue, fine root development, symbiotic relationship with mycorrhizal fungi, anatomy of xylem tissues, conductivity of root and stem, and control of water loss. The degree of fine vs. tap root development varies greatly among tropical tree species (Huante, Rincon & Gavito, 1992; Garwood, 1995;

Sanford & Cuevas, Ch. pter 10), although the functional significance of this variation is little known. Branching patterns of roots, either dichotomous or herring-bone pattern, influence the ratio of root length to root mass (Huante, Rincon & Gavito, 1992). Mycorrhizal seedlings of *Gmelina arborea* have higher xylem P concentrations, maintain a higher water potential gradient via osmoregulation, and keep stomates open at lower soil water content during drought than non-mycorrhizal seedlings (Osonubi, 1989). Conductivities of roots and stems of seedlings are much higher in pioneers than in slow-growing shade-tolerant species, consistent with their higher transpiration rates (M. Tyree, personal communication).

Seedlings respond to light availability at an earlier stage than they respond to soil nutrient availability because mineral nutrient reserves in seeds appear to be sufficient to support seedling growth for at least several weeks (Kitajima, 1992a). Seedlings growing fast under sunny conditions start depending on soil mineral nutrients earlier than seedlings growing in shade. As long as seedlings stay in shade, soil mineral nutrients do not limit growth of seedlings for a very long time. Experimental studies show that the growth of 1–2 year-old seedlings of shade-tolerant species with mycorrhizal association does not respond to the addition of phosphorous or other nutrients (Sundralingham, 1983; Turner, Brown & Newton, 1993; Burslem, Grubb & Turner, 1995). Which soil mineral nutrient becomes most limiting to seedling growth differs among tropical tree species (Burslem, Turner & Grubb, 1994; Burslem, Grubb & Turner, 1995), perhaps reflecting the tissue mineral concentration derived from seeds, degree and type of mycorrhizal association, and soil type. Although nitrogen may be exhausted more rapidly from seed reserves than other mineral nutrients, nitrogen availability in tropical soil limits seedling growth less often than phosphorous availability (Kadir, Hamzah & Sundralingam, 1988; Burslem, Turner & Grubb, 1994).

Establishment of mycorrhizae is of critical importance for uptake of mineral nutrients for the majority of the tree species in many phosphorous-poor lowland tropical soils (Janos, 1980, 1983; Alexander, 1989; Alexander, Ahmad & See, 1992). The majority of mycorrhizal associations are with non-host-specific vescular-arbuscular (VA) fungi, although some dominant tree species (e.g., Dipterocarpaceae spp. and some Fabaceae spp.) form ectomycorrhizal associations. Both VA and ectomycorrhizal associations appear to enhance availability of mineral nutrients and water for seedlings, and enhance growth and survival of tropical tree seedlings in nutrient-poor soil (Janos, 1980; Tomer et al., 1985; Alexander, 1989; Alexander, Ahmad & See, 1992;

Habte & Fox, 1989; Osonubi, 1989; Herrera et al., 1992; Michelson, 1992). However, carbon drain by mycorrhizal fungi may slow down seedling growth of shade-tolerant species with inherently slow growth, even though their phosphorous nutrition is improved (McGee, 1990). Seedlings appear to play a passive role in the establishment of the symbiotic relationship with fungi. When seedling roots become infected is solely a function of the availability of fungal spores (and proximity to the infected adult tree roots in the case of ectomycorrhizae, Alexander, Ahmad & See, 1992), and is independent of seedlings' demand for mineral nutrients (Janos, 1980). Mycorrhizal infection may start while cotyledons are still attached (Janos, 1980; Herrera et al., 1992), or as early as 20 days after germination (Alexander, Ahmad & See, 1992). Cotyledon retention may be longer for infected seedlings, probably because mineral nutrient reserves in cotyledons are withdrawn more slowly in mycorrhizal seedlings (Janos, 1980). The percentage of infected roots gradually increases over the course of several months (Alexander, Ahmad & See, 1992). The degree of mycorrhizal infection as well as its positive effects on seedling growth appear to remain small in the shaded understory (McGee, 1990; Alexander, Ahmad & See, 1992). The carbon cost of mycorrhizae relative to its benefit may be greater in shade than in sun. It is not clear whether seedlings control the development of mycorrhizal infection in response to light availability. A possible mechanism for the low level of development of mycorrhizal infection in shade is the low carbohydrate level in seedling stems and roots in shade (Sasaki & Mori, 1981).

19.3.4   Acclimation of seedlings to gap opening and reshading

Seedling leaves of most tropical tree species, both shade-tolerant and intolerant, exhibit acclimation of morphological and photosynthetic traits to light environment (Morikawa, Inoue & Sasaki, 1980; Sasaki & Mori, 1981; Fetcher, Strain & Oberbauer, 1983; Oberbauer and Strain, 1984, 1986b; Fetcher et al., 1987; Pearcy, 1987; Thompson, Stocker & Kriedmann, 1988; Kitajima & Augspurger, 1989; Ramos and Grace, 1990; Riddoch et al., 1991; Riddoch, Lehto & Grace, 1991; Strauss-Debenedetti and Bazzaz, 1991; Kamaluddin and Grace, 1992a,b, 1993; Turnbull, Doley & Yates, 1993; Kitajima, 1994; Press et al., 1995). For a general discussion of the ecophysiology of acclimation and phenotypic plasticity in response to light environment, readers are referred to other chapters in the book (Chazdon et al., Chapter 1; Straus-Debennedetti & Bazzaz, Chapter 6). Here, I would like to focus on two issues: first, comparison of acclimation response between seedlings and adults, and second and more importantly, the ecological significance of acclimation responses to the seedling regeneration habits.

In the majority of tropical tree species, seedling leaves are smaller and thinner, with low mass per unit area, than adult leaves, until they reach a certain size or developmental stage (Fetcher, Oberbauer & Chazdon, 1994). In shade-tolerant species, light-saturated net photosynthetic rates of sun-phenotype seedling leaves may be similar to those of adults (Oberbauer & Strain, 1986a,b; Pearcy, 1987). However, in pioneers and late secondary successional species, seedling leaves generally do not achieve photosynthetic rates as high as adults. Adult leaves of pioneers typically have maximum photosynthetic rates above 20 μmol m^{-2} s^{-1} (Stephens & Waggoner, 1970; Koyama, 1981; Zotz & Winter, Chapter 3; Kitajima, Mulkey & Wright, unpublished), while seedling leaves of these species may typically have much lower rates, especially on an area basis (8–12 μmol CO_2 m^{-2} s^{-1}) (Fetcher et al., 1987; Kitajima, 1994). An exceptionally high rate of 27 μmol CO_2 m^{-2} s^{-1} was recorded for *Ochroma lagopus* raised in a growth chamber (Oberbauer & Strain, 1984). However, sun-grown seedlings of the same species raised in native habitat (clearing under clear plastic cover) in another study had only 9.2 μmol m^{-2} s^{-1} (Fetcher, et al., 1987). Slight differences in age or nutrient availability may affect seedling leaf photosynthetic capacity of pioneer species with inherently high growth rates. Differences in maximum photosynthetic capacity achievable by seedling leaves between pioneers and shade-tolerant species may be much smaller than previously considered (Bazzaz & Pickett, 1980; Oberbauer & Strain, 1984). In order to address the nature of developmental constraints, future studies should follow the chronological and developmental order of leaves and change in leaf size (especially of leaf mass per unit area) in conjunction with gas exchange measurement.

In order to discuss the ecological relevance of acclimation responses for seedlings, it is useful to distinguish between initial acclimation response (expression of phenotypic plasticity in the first set of leaves developed with seed reserves) and secondary acclimation responses (change in phenotype from the initial phenotype following a change in light environment). This distinction is important in order to compare the results from various studies. The initial acclimation response is purely a function of genotype. In contrast, the second type of responses are influenced not only by the degree and direction of change in light environment, but also by the ontogenetic stage, the size, and the allometry (SLA, LAR) and stored reserve level of plants at the time of light change (Fetcher, Strain & Oberbauer, 1983; Kitajima & Augspurger, 1989; Rice & Bazzaz, 1989; Strauss-Debenedetti & Bazzaz, 1991; Kamaluddin & Grace, 1993; Walters, Kruger & Reich, 1993a, b).

The initial acclimation response is measured by comparing the seedlings grown under contrasting light environments continuously from immediately after germination, while the second type of response is quantified after switching the light environment for older seedlings that have become independent of seed reserves (e.g., reciprocal change between shade and sun or from partially shaded pre-treatment level to shadier and sunnier light levels). The choice of experimental design depends on the ecological question addressed by each study. For example, if the objective is to examine the phenotypic plasticity of the species *per se*, or to assess the importance of acclimation responses for the seedling establishment, the first type of experiment is the best choice. However, for a study of how seedlings of shade-tolerant species that are established and acclimated to the shaded understory respond to the sudden opening of a light gap, the second type of experiment is more appropriate. Strauss-Debenedetti and Bazzaz discuss the differences between these types of studies in Chapter 6 of this volume.

The ecological significance of acclimation responses of seedlings and post-seedling juveniles should be greater for shade-tolerant climax species whose seedlings survive well in the shaded understory and typically experience the opening and closing of light gaps several times before reaching sapling size. Whether the acclimation potential is related to regeneration niches, which in turn is a correlate of the successional status of the species, depends on whether magnitude or rate of acclimation is measured. Most comparative studies cited above found no clear difference in the degree of phenotypic plasticity between early successional and late successional species. However, rates of acclimation appear to be faster in inherently fast-growing species that produce more leaves for a unit duration of time than in slow-growing species (Strauss-Debenedetti & Bazzaz, 1991). As a result, in a typical acclimation study that compares most recently fully expanded leaves after a certain fixed period following light change, fast-growing species should exhibit greater change than slow-growing species. Some shade-tolerant species may exhibit photoinhibition and growth decline when light availability is above 50% of full sun (Nicholson, 1960; Sasaki & Mori, 1981; Langenheim et al., 1984, Turner & Newton, 1990; Kamaluddin & Grace, 1992a). Photoinhibitory light may occur in the center of very large gaps or clearings (Brown & Whitmore, 1992), but it is rare inside natural forests. Photoinjuries and water stress are a more important issue for restoration after heavy disturbance, such as logging, wind storms, hurricanes, etc., than for understanding treefall gap dynamics.

Reshading accompanying gap closures may be stressful since seedlings need to adjust their physiology and allocation back to that of shade-phenotypes while they experience carbon deprivation. Reciprocal light-switching experiments adequately simulate the sudden increase of light levels with light gap opening, but not the closure of gaps which occurs gradually. Typical treefall gaps close within two to three years in the low strata close to the ground (N. Brokaw, C. Horwitz & D. Ackerly, personal communications). In a cohort of shade-tolerant *Tachigalia versicolor* monitored for 9 years since germination, some individuals that had experienced one or more release events achieved heights between 0.5–3 m, while those that remained in shade were still 0.2–0.4 m tall. The mortality during the eighth year was independent of plant size, as a result of high mortality of both reshaded individuals and very small individuals in continuous deep shade. In experimental studies of reciprocal light switching, sun-phenotype leaves of shade-tolerant tree seedlings showed reduced dark respiration similar to the level of shade-phenotype leaves within several days (Kamaluddin & Grace, 1992b; Turnbull, Doley & Yates, 1993). Nevertheless, in natural populations of *Tachigalia*, recently reshaded individuals appeared to lose sun-phenotype leaves while retaining older leaves produced before the light-gap event. Newly expanding leaves of reshaded individuals appeared especially vulnerable to pathogenic fungi, became necrotic and were abscised before expansion was complete. Thus, even for very shade-tolerant species such as *Tachigalia*, reshading may be a period of great stress. There is little information on how seedlings and post-seedling juveniles of slow-growing species acclimate back to shade after enjoying greater light availability inside or near gaps. Experimental studies applying gradual light reduction, or in closing natural gaps, will close a gap in our understanding of carbon economy during reshading.

19.4 FUTURE DIRECTIONS FOR STUDIES OF SEEDLING ECOPHYSIOLOGY

Many ecophysiological studies of tropical trees have used seedlings because their small sizes are suitable for experimental manipulations. Unfortunately, we may reach erroneous conclusions if we fail to pay attention to the ecology of seedlings. For example, physiological ecologists frequently assume that higher photosynthetic income and higher growth rate result in better survival. However, recent comparative studies of seedlings of shade-tolerant vs. intolerant species clearly point to a need for decoupling growth potential from survival potential. Seedlings of pioneer and light-gap dependent species that suffer high mortality in shade have higher growth potential than seedlings of shade-tolerant series when both are raised under the same degree of shade. Despite the expectations based on carbon optimization theory, seedling leaves of shade-tolerant and intolerant species are not different in their light compensation points, dark respiration, or quantum yield when both types of species are raised under the same degree of shade. One key to solve this apparent puzzle may lie in whole-plant perspectives incorporating the role of seed reserves during early seedling establishment phase. The physiological ecology of tropical tree seedlings will benefit from better linkage to field studies, more whole-plant perspectives, clear understanding of the degree and duration of seed reserve dependency, and longer-term study of acclimation responses of seedlings of slow-growing non-pioneer species in dynamic light environments in the forest.

ACKNOWLEDGEMENTS

I dedicate this article to the late Dr. Alan Smith, who supported my research at the Smithsonian Tropical Research Institute as my fellowship advisor during 1986-1990. I am indebted to Dr. Carol Augspurger for her guidance throughout my career. Her comparative work of tropical seedling ecology provided a foundation for my seedling ecophysiology research. The manuscript was greatly improved by comments from the reviewers.

REFERENCES

AIDE, T. M. (1987) Limbfalls: A major cause of sapling mortality for tropical forest. *Biotropica*, **19**, 284–285.

ALEXANDER, I. (1989) Mycorrhizas in tropical forests. *Mineral Nutrients in Tropical Forest and Savanna Ecosystems.* (ed. J. PROCTOR), Blackwell Scientific Publishers, Oxford, pp 169–188.

ALEXANDER, I., AHMAD, N. & SEE, L. S. (1992) The role of mycorrhizas in the regeneration of some Malaysian forest trees. *Philosophical Transaction of the Royal Society of London, Series B,* **335**, 379–388.

ASHCROFT, W. J. & MURRAY, D. R. (1979) The dual functions of the cotyledons of *Acacia iteaphylla* F. Muell. (Mimosoideae). *Australian Journal of Botany,* **27**, 343–352.

AUGSPURGER, C. K. (1984a) Seedling survival of tropical tree species: Interactions of dispersal distance, light gaps, and pathogens. *Ecology,* **65**, 1705–1712.

AUGSPURGER, C. K. (1984b) Light requirements of neotropical tree seedlings: A comparative study of growth and survival. *Journal of Ecology,* **72**, 777–795.

AUGSPURGER, C. K. & FRANSON, S. E. (1988) Input of wind-dispersed seeds into light-gaps and forest sites in a neotropical forest. *Journal of Tropical Ecology,* **4**, 239–252.

BARCLAY, A. S. & EARL, F. R. (1974) Chemical analyses of seeds III. Oil and protein content of 1253 species. *Economic Botany,* **28**, 178–236.

BAZZAZ, F. A. (1979) Physiological ecology of plant succession. *Annual Review of Ecology and Systematics,* **10**, 351–371.

BAZZAZ, F. A. & PICKETT, S. T. A. (1980) Physiological ecology of tropical succession: A comparative review. *Annual Review of Ecology and Systematics,* **11**, 287–310.

BECKER, P. & SMITH, A. P. (1990) Spatial autocorrelation of solar radiation in a tropical moist forest understory. *Agriculture and Forestry Meteorology,* **52**, 373–379.

BECKER, P., RABENOLD, P. E., IDOL, J. R. & SMITH, A. P. (1988) Water potential gradients for gaps and slopes in a Panamanian tropical moist forest's dry season. *Journal of Tropical Ecology,* **4**, 173–184.

BLAIN, D. & KELLMAN, M. (1991) The effects of water supply on tree seed germination and seedling survival in a tropical seasonal forest in Veracruz, Mexico. *Journal of Tropical Ecology,* **7**, 69–83.

BONGERS, F. & POPMA, J. (1990) Leaf dynamics of seedlings of rainforest species in relation to canopy gaps. *Oecologia,* **82**, 122–127.

BROKAW, N. V. L. (1982) Treefalls: Frequency, timing, and consequences. *The Ecology of a Tropical Forest: Seasonal Rhythms and Long-Term Changes* (eds. E. G. LEIGH, JR., A. S. RAND. & D. M. WINDSOR) Smithsonian Institution Press, Washington, D.C., pp 101–108.

BROKAW, N. V. L. (1985) Gap-phase regeneration in a tropical forest. *Ecology,* **66**, 682–687.

BROKAW, N. V. L. & SCHEINER, S. (1989) Species composition in gaps and structure of a tropical forest. *Ecology*, **70**, 538–540.

BROOKES, P. D., WINGSTON, D. L. & BOURNE, W. F. (1980) The dependence of *Quercus robur* and *Q. petraea* seedlings on cotyledon potassium, magnesium calcium, and phosphorus during the first year of growth. *Forestry*, **53**, 167–177.

BROWN, N. D. (1993) The implications of climate and gap microclimate for seedling growth conditions in a Bornean lowland rainforest. *Journal of Tropical Ecology*, **9**, 153–168.

BROWN, N. D. & WHITMORE, T. C. (1992) Do dipterocarp seedlings really partition tropical rainforest gaps? *Philosophical Transactions of the Royal Society of London, Series B*, **335**, 369–378.

BURSLEM, D. F. R. P. (1995) Differential responses to nutrients, shade and drought among tree seedlings of lowland tropical forest in Singapore. *The Ecology of Tropical Forest Seedlings* (ed. M. D. SWAIN) Parthenon, Carnforth, in press.

BURSLEM, D. F. R. P., GRUBB, P. J. & TURNER, I. M. (1995) Responses to nutrient addition among shade-tolerant tree seedlings of lowland tropical rainforest in Singapore. *Journal of Ecology*, **83**, 113–122.

BURSLEM, D. F. R. P., TURNER, I. M. & GRUBB, P. J. (1994) Mineral nutrient status of coastal hill dipterocarp forest and adinandra belukar in Singapore: Bioassays of nutrient limitation. *Journal of Tropical Ecology*, **10**, 579–599.

CANHAM, C. D. (1985) Suppression and release during canopy recruitment in *Acer saccharum*. *Bulletin of the Torrey Botanical Club*, **112**, 134–145.

CHAPIN, F. S. (1989) The cost of tundra plant structures: Evaluation of concepts and currencies. *American Naturalist*, **133**, 1–19.

CHAPIN, F. S., SCHULZE, E.-D. & MOONEY, H. A. (1990) The ecology and economics of storage in plants. *Annual Review of Ecology and Systematics*, **21**, 432–447.

CHAZDON, R. L. (1992) Photosynthetic plasticity of two rainforest shrubs across natural gap transects. *Oecologia*, **92**, 586–595.

CLARK, D. B. & CLARK, D. A. (1989) The role of physical damage in the seedling mortality regime of a neotropical rainforest. *Oikos*, **55**, 225–230.

COLEY, P. D. (1983) Herbivory and defensive characteristics of tree species in a lowland tropical forest. *Ecological Monographs*, **53**, 209-233.

COLEY, P. D., BRYANT, J. P. & CHAPIN, F. S., III. (1985) Resource availability and plant anti-herbivore defense. *Science*, **230**, 895–899.

COOMBE, D. E. (1960) An analysis of the growth of *Trema guineensis*. *Journal of Ecology*, **48**, 219–231.

COOMBE, D. E. & HATFIELD, W. (1962) An analysis of the growth of *Musanga cecropioides*. *Journal of Ecology*, **50**, 221–234.

DE VOGEL, E. F. (1980) *Seedlings of Dicotyledons.* Centre for Agricultural Publishing and Documentation (PUDOC), Wageningen.

DENSLOW, J. (1987) Tropical rainforest gaps and tree species diversity. *Annual Review of Ecology and Systematics*, **18**, 431–451.

DUDT, J. F. & SHURE, D. J. (1994) The influence of light and nutrients on foliar phenolics and insect herbivory. *Ecology*, **75**, 86–98.

ERNST, W. H. O. (1988) Seed and seedling ecology of *Bachystegia spiciformis*, a predominant tree component in miombo woodlands in south central Africa. *Forest Ecology and Management*, **25**, 195-210.

FASEHUN, F. E. & AUDU, M. (1980) Comparative seedling growth and respiration of four tropical hardwood species. *Photosynthetica*, **14**, 193–197.

FENNER, M. (1983) Relationships between seed weight, ash content, and seedling growth in twenty-four species of Compositae. *New Phytologist*, **95**, 697–706.

FENNER, M. (1986) A bioassay to determine the limiting minerals for seeds from nutrient-deprived *Senecio vulgaris* plants. *Journal of Ecology*, **74**, 497–505.

FENNER, M. (1987) Seedlings. *New Phytologist*, **106**, 35–47.

FENNER, M. & LEE, W. G. (1989) Growth of seedlings of pasture grasses and legumes deprived of single mineral nutrients. *Journal of Applied Ecology*, **26**, 223–232.

FETCHER, N., OBERBAUER, S. F. & CHAZDON, R. L. (1994) Physiological ecology of plants. *La Selva, Ecology and Natural History of a Neotropical Rain Forest* (eds. L. A. MACDADE, K. S. BAWA, H. A. HESPENHEIDE & G. S. HARTSHORN) University of Chicago Press, Chicago, pp. 128–141.

FETCHER, N., OBERBAUER, S. F., ROJAS, G. & STRAIN, B. R. (1987) Efectos del régimen de luz sobre la fotosíntesis y el crecimiento en plántulas de árboles de un bosque lluvioso tropical de Costa Rica. *Revista de Biología Tropical*, **35**, 97–110.

FETCHER, N., STRAIN, B. R. & OBERBAUER, S. F. (1983) Effects of light regime on the growth, leaf morphology, and water relations of two species of tropical trees. *Oecologia*, **58**, 314–319.

FOSTER, S. A. & JANSON, C. H. (1985) The relationship between seed size and establishment conditions in tropical woody plants. *Ecology*, **66**, 773–780.

GARWOOD, N. C. (1983) Seed germination in a seasonal tropical forest in Panama: A community study. *Ecological Monographs*, **53**, 159–181.

GARWOOD, N. C. (1986) Constraints on the timing of seed germination in a tropical forest. *Frugivores and Seed Dispersal* (eds. A. ESTRADA & T. H. FLEMING) Dr W. Junk Publishers, Dordrecht, pp 347–355.

GARWOOD, N. C. (1995) Functional morphology of tropical tree seedlings. *The Ecology of Tropical Forest Seedlings* (ed. M. D. SWAIN) Parthenon, Carnforth, in press.

GOLLEY, F. B. (1961) Energy values of ecological materials. *Ecology*, **42**, 581–584.

GOLLEY, F. B. (1969) Caloric value of wet tropical forest vegetation. *Ecology* **50**, 517–519.

GOODALL, D. W. (1950) Growth analysis of cacao seedlings. *Journal of Ecology*, **16**, 292–306.

GRIME, J. P. & JEFFREY, D. W. (1965) Seedling establishment in vertical gradients of sunlight. *Journal of Ecology*, **53**, 621–642.

HABTE, M. & FOX, R. L. (1989) *Leucaena leucocephala* seedling response to vesicular-arbuscular mycorrhizal inoculation in soils with varying levels of inherent mycorrhizal effectiveness. *Biology and Fertility of Soils*, **8**, 111–115.

HARTSHORN, G. S. (1978) Tree falls and tropical forest dynamics. *Tropical Trees as Living Systems* (eds. P. B. THOMLINSON & M. H. ZIMMERMAN) Cambridge University Press, Cambridge, pp 627–638.

HERRERA, R. A., CAPOTE, R. P., MENÈDEZ, L. & RODRÍGUEZ, M. E. (1992) Silvigenesis stages and the role of mycorrhiza in natural regeneration in Sierra del Rosario, Cuba. *Rain Forest Regeneration and Management* (eds. A. GOMEZ-POMPA, T. C. WHITMORE & M. HADLEY) Parthenon, Carnforth, pp 211–221.

HLADIK, A. & MIQUEL, S. (1991) Seedling types and plant establishment in an African rainforest. *Reproductive Ecology of Tropical Forest Plants* (eds. K. S. BAWA & M. HADLEY) Carnforth, Parthenon, pp 317–319.

HUANTE, P., RINCON, E. & GAVITO, M. (1992) Root system analysis of seedlings of seven tree species from a tropical dry forest in Mexico. *Trees*, **6**, 77–82.

HUNT, R. (1982) *Plant Growth Curves: The Functional Approach to Plant Growth Analysis.* Edward Arnold, London.

JANOS, D. P. (1980) Vesicular-arbuscular mycorrhizae affect lowland tropical rainforest plant growth. *Ecology*, **61**, 151–162.

JANOS, D. P. (1983) Tropical mycorrhizas, nutrient cycles, and plant growth. *Tropical Rain Forest: Ecology and Management* (eds. S. L. SUTTON, T. C. WHITMORE & A. C. CHADWICK) Blackwell Scientific Publishers, Oxford, pp 327–345.

KADIR, W. R. A., HAMZAH, A. & SUNDRALINGAM, P. (1988) Effect of nitrogen and phosphorus on the early growth of three exotic plantation species in peninsular Malaysia. *Journal of Tropical Ecology*, **1**, 178–187.

KAMALUDDIN, M. & GRACE, J. (1992a) Acclimation in seedlings of a tropical tree, *Bischofia javanica*, following a stepwise reduction in light. *Annals of Botany*, **69**, 557–562.

KAMALUDDIN, M. & GRACE, J. (1992b) Photoinhibition and light acclimation in seedlings of *Bischofia javanica*, a tropical forest tree from Asia. *Annals of Botany*, **69**, 47–52.

KAMALUDDIN, M. & GRACE, J. (1993) Growth and photosynthesis of tropical forest tree seedlings (*Bischofia javanica* Blume) as influenced by a change in light availability. *Tree Physiology*, **13**, 189–201.

KITAJIMA, K. (1992a) *The Importance of Cotyledon Functional Morphology and Patterns of Seed Reserve Utilization for the Physiological Ecology of Neotropical Tree Seedlings.* Ph. D. thesis, University of Illinois, Urbana.

KITAJIMA, K. (1992b) Relationship between photosynthesis and thickness of cotyledons for tropical tree species. *Functional Ecology,* **6**, 582–584.

KITAJIMA, K. (1994) Relative importance of photosynthetic traits and allocation patterns as correlates of seedling shade tolerance of 13 tropical trees. *Oecologia,* **98**, 419–428.

KITAJIMA, K. & AUGSPURGER, C. K. (1989) Seed and seedling ecology of a monocarpic tropical tree, *Tachigalia versicolor. Ecology,* **70**, 1102–1114.

KNAPP, A. K., VOGELMANN, T. C., MCCLEAN, T. M. & SMITH, W. K. (1988) Light and chlorophyll gradients within *Cucurbita* cotyledons. *Plant, Cell and Environment,* **11**, 257–263.

KOYAMA, H. (1981) Photosynthetic rates in lowland rainforest trees of peninsular Malaysia. *Japanese Journal of Ecology,* **31**, 361–369.

LANGENHEIM, J. H., OSMOND, C. B., BROOKS, A. & FERRA, P. J. (1984) Photosynthetic responses to light in seedlings of selected Amazonian and Australian rainforest tree species. *Oecologia,* **63**, 215–224.

LEVIN, D. A. (1974) The oil content of seeds: An ecological perspective. *American Naturalist,* **108**, 193–206.

LOVELL, P. & MOORE, K. (1971) A comparative study of the role of the cotyledons in seedling development. *Journal of Experimental Botany,* **22**, 153–162.

MARSHALL, P. E. & KOZLOWSKI, T. T. (1974) Photosynthetic activity of cotyledons and foliage leaves of young angiosperm seedlings. *Canadian Journal of Botany,* **52**, 2023–2032.

MARSHALL, P. E. & KOZLOWSKI, T. T. (1975) Changes in mineral contents of cotyledons and young seedlings of woody angiosperms. *Canadian Journal of Botany,* **53**, 2026–2031.

MARSHALL, P. E. & KOZLOWSKI, T. T. (1976) Compositional changes in cotyledons of woody angiosperms. *Canadian Journal of Botany,* **54**, 2473–2477.

MARTÍNEZ-RAMOS, M., ALVAREZ-BUYLLA, E. & SARUKHÁN, J. (1989) Tree demography and gap dynamics in a tropical rainforest. *Ecology,* **70**, 555–558.

MAZER, S. J. (1989) Ecological, taxonomic, and life history correlates of seed mass among Indiana dune angiosperms. *Ecological Monographs,* **59**, 153–175.

MCGEE, P. A. (1990) Survival and growth of seedlings of coachwood (*Ceratopetalum apetalum*): Effects of shade, mycorrhizas and a companion plant. *Australian Journal of Botany,* **38**, 583–592.

MICHELSON, A. (1992) Mycorrhiza and root nodulation in tree seedlings from 5 nurseries in Ethiopia and Somalia. *Forest Ecology and Management,* **48**, 335–344.

MIQUEL, S. (1987) Morphologie fonctionnelle de plantules d'espèces forestières du Gabon. *Bulletin du Muséum National d'Histoire Naturelle, Paris, 4èmme série, Section B, Adansonia,* **9**, 101–102.

MOLOFSKY, J. & AUGSPURGER, C. K. (1992) The effect of leaf litter on early seedling establishment in a tropical forest. *Ecology,* **73**, 68–77.

MOLOFSKY, J. & FISHER, B. L. (1993) Habitat and predation effects on seedling survival and growth in shade-tolerant tropical trees. *Ecology,* **74**, 261–264.

MOONEY, H. A. (1972) The carbon balance of plants. *Annual Review of Ecology and Systematics,* **3**, 315–346.

MORIKAWA, Y., INOUE, T. & SASAKI, S. (1980) Light photosynthesis curves in *Shorea talura* seedlings grown under various light intensities. *Bulletin of the Forestry and Forest Products Research Institute,* **309**, 109–115.

MOTT, K. A., GIBSON, A. C. & O'LEARY, J. W. (1982) The adaptive significance of amphistomatic leaves. *Plant, Cell and Environment,* **5**, 455–460.

MURRAY, D. R. (1985) *Seed Physiology, Vol. 2, Germination and Reserve Mobilization.* Academic Press, Sydney.

NG, F. S. P. (1978) Strategies of establishment in Malayan forest trees. *Tropical Trees as Living Systems* (eds. P. B. TOMLINSON & M. H. ZIMMER-MAN) Cambridge University Press, Cambridge, pp 129–162.

NICHOLSON, D. I. (1960) Light requirements of seedlings of five species of Dipterocarpaceae. *Malaysian Forester,* **23**, 344–356.

NÚÑEZ-FARFÁN, J. & DIRZO, R. (1988) Within-gap spatial heterogeneity and seedling performance in a Mexican tropical forest. *Oikos,* **51**, 274-284.

OBERBAUER, S. F. & DONNELY, M. A. (1986) Growth analysis and successional status of Costa Rican rainforest trees. *New Phytologist,* **104**, 517–521.

OBERBAUER, S. F. & STRAIN, B. R. (1984) Photosynthesis and successional status of Costa Rican rainforest trees. *Photosynthesis Resecrch,* **5**, 227–232.

OBERBAUER, S. F. & STRAIN, B. R. (1986a) Effects of canopy position and irradiance on the leaf physiology and morphology of *Pentaclethra macroloba* (Mimosase). *American Journal of Botany,* **73**, 409–416.

OBERBAUER, S. F. & STRAIN, B. R. (1986b) Effects of light regime on the growth and physiology of *Pentaclethra macroloba* (Mimosaceae). *Journal of Tropical Ecology,* **1**, 303–320.

OKALI, D. U. U. (1971) Rates of dry-matter production in some tropical forest tree seedlings. *Annals of Botany,* **35**, 87–97.

OKALI, D. U. U. (1972) Growth rates of some West African forest tree seedlings in shade. *Annals of Botany,* **36**, 953–959.

OLADOKUN, M. A. O. (1989) Nut weight and nutrient contents of *Cola acuminata* and *C. nitida* (Sterculiaceae). *Economic Botany,* **43**, 17–22.

OLOFINBOBA, M. O. (1975) Studies on seedlings of *Theobroma cacao* L., variety F3 Amazon. I. Role of cotyledons in seedling development. *Turrialba,* **25**, 121–127.

OSONUBI, O. (1989) Osmotic adjustment in mycorrhizal *Gmelina arborea* Roxb. seedlings. *Journal of Tropical Ecology,* **3**, 143–151.

OSUNKOYA, O. O., ASH, J. E., GRAHAM, A. W. & HOPKINS, M. S. (1993) Growth of tree seedlings in tropical rainforests of North Queensland, Australia. *Journal of Tropical Ecology*, **9**, 1–18.

OSUNKOYA, O. O., ASH, J. E., HOPKINS, M. S. & GRAHAM, A. W. (1992) Factors affecting survival of tree seedlings in North Queensland rainforests. *Oecologia*, **91**, 569–578.

OSUNKOYA, O. O., ASH, J. E., HOPKINS, M. S. & GRAHAM, A. W. (1994) Influence of seed size and seedling ecological attributes on shade tolerance of rain forest tree species in northern Queensland. *Journal of Ecology*, **82**, 149–163.

PATE, J. S., RASINS, E., RULLO, J. & KUO, J. (1986) Seed nutrient reserves of Proteaceae with special reference to protein bodies and their inclusions. *Annals of Botany*, **57**, 747–770.

PEARCY, R. W. (1987) Photosynthetic gas exchange responses of Australian tropical forest trees in canopy, gap, and understory micro-environment. *Functional Ecology*, **1**, 169–178.

PENNING DE VRIES, F. W. T., BRUNSTING, A. H. M. & VAN LAAR, H. H. (1974) Products, requirements and efficiency of biosynthesis: A quantitative approach. *Journal of Theoretical Biology*, **45**, 339–377.

PENNING DE VRIES, F. W. T. & VAN LAAR, H. H. (1977) Substrate utilization in germinating seeds. *Environmental Effects on Crop Physiology* (eds. J. J. LANDSBERG & C. V. CUTTING) Academic Press, London, pp 217–228.

POORTER, H. (1990) Interspecific variation in relative growth rate: On ecological causes and physiological consequences. *Causes and Consequences of Variation in Growth Rate and Productivity of Higher Plants* (ed. H. LAMBERS) PSB Academic Publishing, The Hague, pp 45–68.

POORTER, H. & REMKES, C. (1990) Leaf area ratio and net assimilation rate of 24 wild species differing in relative growth rate. *Oecologia*, **83**, 553–559.

POORTER, H., REMKE, C., & LAMBERS, H. (1990) Carbon and nitrogen economy of 24 wild species differing in relative growth rate. *Plant Physiology*, **94**, 621–627.

POPMA, J. & BONGERS, F. (1988) The effect of canopy gaps on growth and morphology of seedlings of rainforest species. *Oecologia*, **75**, 625–632.

POPMA, J. & BONGERS, F. (1991) Acclimation of seedlings of three Mexican tropical rainforest species to a change in light availability. *Journal of Tropical Ecology*, **7**, 85–97.

PRESS, M. C., BROWN, N. D., BARKER, M. G. & ZIPPERLEN, S. W. (1995) Photosynthetic responses to light in tropical rainforest tree seedlings. *The Ecology of Tropical Forest Seedlings* (ed. M. D. SWAIN) Parthenon, Carnforth, in press.

RAMOS, J. & GRACE, J. (1990) The effects of shade on the gas exchange of seedlings of four tropical trees from Mexico. *Functional Ecology*, **4**, 667–677.

RICE, S. A. & BAZZAZ, F. A. (1989) Quantification of plasticity of plant traits in response to light intensity: Comparing phenotypes at a common weight. *Oecologia*, **78**, 502–507.

RIDDOCH, I., GRACE, J., FASEHUN, F. E., RIDDOCH, B. & LADIPO, D. O. (1991) Photosynthesis and successional status of seedlings in a tropical semideciduous rainforest in Nigeria. *Journal of Ecology*, **79**, 491–503.

RIDDOCH, I., LEHTO, T. & GRACE, J. (1991) Photosynthesis of tropical tree seedlings in relation to light and nutrient supply. *New Phytologist*, **119**, 137–147.

RINCÓN, E. & HUANTE, P. (1993) Growth responses of tropical deciduous tree seedlings to contrasting light conditions. *Trees*, **7**, 202–207.

SANFORD, R. L. J., BRAKER, H. E. & HARTSHORN, G. S. (1986) Canopy openings in a primary neotropical lowland forest. *Journal of Tropical Ecology*, **2**, 277–282.

SASAKI, S. & MORI, T. (1981) Growth responses of dipterocarp seedlings to light. *Malaysian Forester*, **44**, 319–345.

SCHUPP, E. W., HOWE, H. F., AUGSPURGER, C. K. & LEVEY, D. J. (1989) Arrival and survival in tropical treefall gaps. *Ecology*, **70**, 562–564.

SHURE, D. R. & WILSON, L. A. (1993) Patch-size effects on plant phenolics in successional openings of the Southern Appalachians. *Ecology*, **74**, 55–67.

SIMON, E. W. & HARUN, R. M. R. (1972) Leakage during seed imbibition. *Journal of Experimental Botany*, **23**, 1076–1085.

SMITH, A. P., HOGAN, K. P. & IDOL, J. R. (1992) Spatial and temporal patterns of light and canopy structure in a lowland tropical moist forest. *Biotropica*, **24**, 503–511.

SORK, V. L. (1987) Effects of predation and light on seedling establishment in *Gustavia superba*. *Ecology*, **68**, 1341-1350.

STEEGE, H. T. (1994) Flooding and drought tolerance in seeds and seedlings of two *Mora* species segregated along a soil hydrological gradient in the tropical rainforest of Guyana. *Oecologia*, **100**, 356–367.

STEPHENS, C. R. & WAGGONER, P. E. (1970) Carbon dioxide exchange of a tropical rainforest. Part II. *Bioscience*, **20**, 1054-1059.

STRAUSS-DEBENEDETTI, S. & BAZZAZ, F. A. (1991) Plasticity and acclimation to light in tropical Moraceae of different successional positions. *Oecologia*, **87**, 377-387.

SUNDRALINGHAM, P. (1983) Responses of potted seedlings of *Dryobalanops aromatica* and *Dryobalanops oblongifolia* to commercial fertilizers. *Malaysian Forester*, **46**, 86-92.

THOMPSON, K. (1987) Seeds and seed banks. *New Phytologist*, **106** (**Suppl.**), 23-34.

THOMPSON, W. A., STOCKER, G. C. & KRIEDMANN, P. E. (1988) Growth and photosynthetic response to light and nutrients of *Flindersia brayleyana* F. Muell., a rainforest tree with broad tolerance to sun and shade. *Australian Journal of Plant Physiology*, **15**, 299-315.

TILMAN, D. (1988) *Dynamics and Structure of Plant Communities.* Princeton University Press, Princeton.

TOMER, G. S., SHRIVASTAVA, S. K., GONTIA, A. S., KHARE, A. K. & SHRIVASTAVA, M. K. (1985) Influence of endomycorrhiza in relation to nutrient application on the growth of *Leucaena leucocephala. Journal of Tropical Forestry*, **1**, 156-159.

TURNBULL, M. H., DOLEY, D. & YATES, D. (1993) The dynamics of photosynthetic acclimation to changes in light quantity and quality in three Australian rainforest tree species. *Oecologia*, **94**, 218-228.

TURNER, I. M. & NEWTON, A. C. (1990) The initial responses of some tropical rainforest tree seedlings to a large gap environment. *Journal of Applied Ecology*, **27**, 605-608.

TURNER, I. M., BROWN, N. D., NEWTON, A. C. (1993) The effect of fertilizer application on dipterocarp seedling growth and mycorrhizal infection. *Forest Ecology and Management*, **57**, 329-337.

VAUGHAN, J. G. (1970) *The Structure and Utilization of Oil Seeds.* Chapman and Hall, London.

WADSWORTH, R. M. & LAWTON, J. R. S. (1968) The effect of light intensity on the growth of seedlings of some tropical tree species. *Journal of West African Scientific Association*, **13**, 212-214.

WALTERS, M. B., KRUGER, E. L. & REICH, P. B. (1993a) Growth, biomass distribution and CO$_2$ exchange of northern hardwood seedlings in high and low light: Relationships with successional status and shade tolerance. *Oecologia*, **94**, 7-16.

WALTERS, M. B., KRUGER, E. L. & REICH, P. B. (1993b) Relative growth rate in relation to physiological and morphological traits for northern hardwood tree seedlings: Species, light environment and ontogenetic considerations. *Oecologia*, **96**, 219-236.

WELDEN, C. W., HEWETT, S. W., HUBBELL, S. P. & FOSTER, R. B. (1991) Sapling survival, growth, and recruitment: Relationship to canopy height in a neotropical forest. *Ecology*, **72**, 35-50.

WESTOBY, M., JURADO, E. & LEISHMAN, M. (1992) Comparative evolutionary ecology of seed size. *Trends in Ecology and Evolution*, **7**, 368-372.

WHITMORE, T. C. (1978) Gaps in the forest canopy. *Tropical Trees as Living Systems* (eds. P. B. TOMLINSON & M. H. ZIMMERMANN) Cambridge University Press, Cambridge, pp 639-655.

WHITMORE, T. C. & BOWEN, M. R. (1983) Growth analyses of some *Agathis* species. *Malaysian Forester*, **46**, 186-196.

WHITMORE, T. C. & GONG, W.-K. (1983) Growth analysis of the seedlings of balsa, *Ochroma lagopus. New Phytologist*, **95**, 301-311.

WILLIAMS, K., PERCIVAL, F., MERINO, J. & MOONEY, H. (1987) Estimation of tissue construction cost from heat of combustion and organic nitrogen content. *Plant, Cell and Environment*, **10**, 725-734.

WRIGHT, S. J. (1991) Seasonal drought and the phenology of understory shrubs in a tropical moist forest. *Ecology*, **72**, 1643-1657.

APPENDIX 19.1

Maximum relative growth rates of tropical tree seedlings under sunny conditions measured in various studies. Approximate regeneration habits and successional status of the species are grouped as (1) pioneer and light demanding, (2) non-pioneer whose seedlings survive only in large light gaps, and (3) non-pioneer whose seedlings may survive in the shaded understory. The environments under which seedlings are grown are indicated as (I) in clearing by the native forest, protected by insect screen and/or clear plastic roof, (II) glass house in temperate region with additional light, and (III) growth chamber (not included in Fig. 19.2 or statistical analysis). If light was reduced by moderate shading of 10-20%, it is indicated as (*). Seedling age is the time from germination at the initiation of the growth experiment. Duration is the time between the initial and the final harvests. LAR is the average for the two harvests.

Species	Family	Reg. Habit	Env.	Age	Duration (d)	RGR (mg/g/d)	LAR (cm²/g)	ULR (g/m²/d)
Goodall (1950)								
Theobroma cacao	Sterculiaceae	3	I*	12 wk	42	18	130	1.1
Coombe (1960)								
Trema guineensis	Ulmaceae	1	II	ca 3 mo	7	53	255	2.1
Coombe and Hatfield (1962)								
Trema guineensis	Ulmaceae	1	II	8 mo	13	118	236	4.5
Musanga cecropioides	Moraceae	1	II	8 mo	13	20	149	1.5
Wadsworth and Lawton (1968)								
Eucalyptus deglupta	Myrtaceae	2?	I*	3 mo	46	50	100	3.2
Nauclea diderrichii	Rubiaceae	1?	I*	3 mo	46	50	130	3.0
Aucoumea klaineana	Burseraceae	2?	I*	3 mo	46	37	165	2.0
Pinus caribea	Pinaceae	2?	I*	3 mo	46	29	55	3.4
Khaya grandifoliola	Meliaceae	3?	I*	3 mo	46	20	110	2.0
Okali (1971)								
Terminalia ivorensis	Combretaceae	1	I	13 wk	28	97	290	5.9
Ceiba pentandra	Bombacaceae	1	I	6 wk	21	80	102	8.6
Ceiba pentandra	Bombacaceae	1	I	14 wk	28	51	88	7.7
Ceiba pentandra	Bombacaceae	1	I	51 wk	21	44	57	8.6
Chlorophora excelsa	Moraceae	1	I	13 wk	28	40	246	2.0
Musanga cecropioides	Moraceae	1	I	40 wk	21	26	93	3.5
Fasehum and Audu (1980) (using their 4th period)								
Chlorophora regia	Moraceae	1	I	ca 2 mo	14	51	101	4.3

Species	Family	Reg. Habit	Env.	Age	Duration (d)	RGR (mg/g/d)	LAR (cm²/g)	ULR (g/m²/d)
Gmelina arborea	Verbenaceae	1	I	ca 2 mo	14	46	68	8.3
Terminalia superba	Combretaceae	1	I	ca 2 mo	14	33	82	5.7
Terminalia ivorensis	Combretaceae	1	I	ca 2 mo	14	33	78	5.2
Whitmore and Bowen (1983)								
Agathis macrophylla	Araucariaceae	3	II	ca 300 d	29	14	67	2.1
Agathis robusta	Araucariaceae	3	II	ca 300 d	29	13	63	2.1
Whitmore and Gong (1983)								
Ochroma lagopus	Bombacaceae	1	II	44 d	7	109	232	4.7
Oberbauer and Donnely (1986)								
Ochroma lagopus	Bombacaceae	1	I	4 wk	7	116	158	7.4
Heliocarpus appendiculatus	Tiliaceae	1	I	14 wk	7	77	127	6.0
Cordia alliodora	Boraginaceae	2	I	10 wk	7	69	139	5.9
Terminalia oblonga	Combretaceae	2	I	14 wk	7	19	130	1.4
Pentaclethra macroloba	Leguminosae	3	I	10 wk	7	21	64	3.1
Brosimum alicastrum	Moraceae	3	I	14 wk	7	21	58	3.7
Popma and Bongers (1988)								
Cecropia obtusifolia	Moraceae	1	I	1-2 mo	99	19	325	1.0
Amphitecta tuxtlensis	Bignoniaceae	3	I	1-2 mo	210	15	208	2.6
Pseudolmedia oxyphyllaria	Moraceae	3	I	1-2 mo	264	15	169	2.6
Psychotria simiarum	Rubiaceae	3	I	1-2 mo	270	13	250	1.2
Brosimum alicastrum	Moraceae	3	I	1-2 mo	210	11	152	1.8

Species	Family	Reg. Habit	Env.	Age	Duration (d)	RGR (mg/g/d)	LAR (cm²/g)	ULR (g/m²/d)
Myriocarpa longipes	Urticaceae	2	I	1-2 mo	210	11	118	1.0
Cordia megalantha	Boraginaceae	3	I	1-2 mo	237	11	260	1.1
Lonchocarpus guatemalensis	Leguminosae	3	I	1-2 mo	237	10	167	1.5
Omphalea oleifera	Euphorbiaceae	3	I	1-2 mo	237	9	79	1.0
Poulsenia armata	Moraceae	3	I	1-2 mo	273	8	83	0.9

Kitajima (1992a (for species with photosynthetic cotyledons only, using harvests at stage 3 and 4, see Table 19.2)

Species	Family	Reg. Habit	Env.	Age	Duration (d)	RGR (mg/g/d)	LAR (cm²/g)	ULR (g/m²/d)
Guazuma ulmifolia	Sterculiaceae	1	I	14 d	68	67	268	3.8
Ceiba pentandra	Bombacaceae	2	I	22 d	54	64	188	4.9
Zuelania guidonia	Flacourtiaceae	2	I	43 d	53	57	181	3.5
Ochroma pyramidale	Bombacaceae	1	I	18 d	60	55	267	2.3
Apeiba membranacea	Tiliaceae	2	I	18 d	73	53	252	2.4
Hasseltia floribunda	Flacourtiaceae	2	I	33 d	68	52	216	2.5
Tabebuia rosea	Bignoniaceae	2	I	20 d	67	51	231	2.7
Luehea seemannii	Tiliaceae	2	I	16 d	76	49	248	1.9
Pseudobombax septenatum	Bombacaceae	1	I	17 d	50	49	168	2.9
Hyeronima laxiflora	Euphorbiaceae	2	I	29 d	78	48	192	2.4
Psychotoria marginata	Rubiaceae	3	I	54 d	63	48	233	2.3
Cordia alliodora	Boraginaceae	2	I	24 d	76	47	154	3.2
Cordia bicolor	Boraginaceae	2	I	21 d	66	46	162	3.0
Tabebuia guayacan	Bignoniaceae	2	I	18 d	70	46	171	2.9
Lafoensia punicifolia	Lythraceae	3	I	12 d	75	42	217	2.2
Bombacopsis quinata	Bombacaceae	2	I	18 d	83	40	202	2.4

Species	Family	Reg. Habit	Env.	Age	Duration (d)	RGR (mg/g/d)	LAR (cm²/g)	ULR (g/m²/d)
Adelia triloba	Euphorbiaceae	2	I	34 d	61	39	206	1.9
Psychotoria limonensis	Rubiaceae	3	I	58 d	70	38	219	1.9
Randia armata	Rubiaceae	3	I	78 d	56	37	132	3.4
Genipa americana	Rubiaceae	3	I	43 d	69	36	148	2.7
Hybanthus prunifolius	Violaceae	3	I	21 d	79	36	237	1.8
Cavanillesia platanifolia	Bombacaceae	2	I	18 d	70	29	167	1.9
Tocoyena pittieri	Rubiaceae	3	I	71 d	77	22	132	1.8
Heisteria concinna	Olacaceae	3	I	83 d	71	20	104	2.0
Guapira standleyanum	Nictagiaceae	3	I	35 d	70	15	98	1.7
Anaxagorea panamensis	Annonaceae	3	I	77 d	72	14	132	1.1
Kamaluddin and Grace (1993)								
Bischofia javanica	Euphorbiaceae	2	III					
Rincón and Huante (1993) (tropical dry forest species)								
Heliocarpus pallidus	Tiliaceae	1	III	36 d	16	146	227	7.4
Apoplanesia paniculata	Leguminosae	1	III	36 d	16	131	198	6.7
Amphipterigium adstringens	Julianaceae	3	III	36 d	16	111	132	9.3
Caesalpinia platyloba	Leguminosae	3	III	36 d	16	66	108	7.8
Caesalpinia eriostachys	Leguminosae	3	III	36 d	16	57	124	8.6

20

Ecophysiological Constraints on the Distribution of *Piper* Species

Arthur L. Fredeen and Christopher B. Field

In the shade, light strongly limits to plant growth and successful species typically possess features that increase both the capture and conservation of light energy. In the sun, light is often non- or co-limiting for growth and successful species often possess traits that enable the dissipation of light energy and the partitioning of resources to enhance the capture of other limiting resources. At least in concept, it is unclear why phenotypic plasticity cannot allow all plants to acclimate to any resource level(s). At a fundamental level, the answer must be that phenotypic plasticity has limits. Such limits could affect individual traits. Alternatively, the real limitation to phenotypic plasticity may reside in the difficulty of successfully orchestrating the large number of phenological, morphological, and physiological changes necessary for success in environments representing extremes in resource availability.

This chapter is intended to help describe what makes a species successful in a limited range of rainforest environments by providing

an overview of what we and others have learned about possible plant
characteristics affecting the distribution of species of the genus *Piper*
across gradients of light availability in tropical rainforest. The
greatest emphasis will be given to those *Piper* species that occur in
lowland tropical rainforest at the Estación Biológia Tropical, Los Tuxtlas in Veracruz, Mexico (18°36′N 95°07′W), which is operated by the
Universidad Nacional Autónoma de Mexico (UNAM).

20.1   *PIPER*: A MODEL SYSTEM

Lowland tropical rainforest provides an unparalleled system for the
study of adaptation and acclimation of plants to gradients in light
availability across time and space. Rainforest microsites often span
three orders of magnitude in photon receipt (e.g., Yoda, 1974; see also
Chazdon et al., Chapter 1). In addition, the euphotic or upper layer of
the forest canopy selectively removes photons in the red and blue
regions of the visible spectrum, corresponding with chlorophyll absorption, thus specifically depleting the light of those wavelengths
most useful for photosynthesis in the understory. Light is also highly
variable in the horizontal and temporal dimensions. Gaps or clearings,
often caused by treefall events, juxtapose high-light and understory
as well as early and late succession environments. In the temporal
domain, plants in the understory may receive over half of their PFD
in sunflecks, i.e., in brief periods of direct solar irradiance (Chazdon
et al., Chapter 1; Pearcy, 1983).

One of the more common understory genera in both the old and new
world tropics is *Piper*. Taxonomy in the genus *Piper* (Piperaceae) is
very confusing, but there may be as many as 1200 species in the
neotropics, 150 species in Central America and Mexico alone (Burger,
1972) with as many as 40 species occurring in a single forest (Opler,
Frankie & Baker, 1980). *Piper* species span the entire breadth of
rainforest light habitats, including species restricted to primary forest, secondary forest, and recent gaps and clearings, as well as species
that occur across the entire successional sequence (Gómez-Pompa,
1971). Interspecific comparisons are facilitated by similarities in form.
Although they differ greatly in their horizontal distribution, they are
more uniformly distributed in the vertical direction, i.e., almost all
species are non-epiphytic small shrubs and less commonly, trees
(mostly 1–2 meters in height).

Another reason for studying *Piper* is that the genus has been
examined from a wide range of perspectives in the neotropics, i.e.,

controls on dispersal (Fleming & Heithaus, 1981; Fleming, 1985), seed germination (Vázquez-Yanes, 1976; Orozco-Segovia & Vázquez-Yanes, 1989; Vázquez-Yanes et al., 1990; Orozco-Segovia, Sanchez-Coronado & Vázquez-Yanes, 1993b), maternal effects on seed germination (Orozco-Segovia, Sanchez-Coronado & Vázquez-Yanes , 1993a), floral biology (Gómez-Pompa & Vázquez-Yanes, 1974), photosynthesis (Chiariello, Field & Mooney, 1987; Walters & Field, 1987; Field, 1988; Chazdon, 1992; Tinoco-Ojanguren & Pearcy, 1992, 1993a, 1993b), respiration (Fredeen & Field, 1991; Chazdon & Kaufmann, 1993), water relations (Mooney, Field & Vázquez-Yanes, 1984; Chiariello, Field & Mooney, 1987), mineral nutrition (Denslow, Vitousek & Schultz, 1987; Fredeen, Griffin & Field, 1991; Fredeen & Field, 1992), canopy architecture (Chazdon, Williams & Field, 1988), leaf longevity and cost (Williams, Field & Mooney, 1989), leaf anatomy (Chazdon & Kaufmann, 1993), herbivory (Marquis, 1984;1992), seed predation (Greig, 1993a), insect associations (Risch & Rickson, 1981), phenology (Opler, Frankie & Baker, 1980; Marquis, 1988), growth (Denslow et al., 1990; Sanchez-Coronado, Rincòn & Vázquez-Yanes, 1990; Fredeen et al., 1995), and vegetative reproduction (Greig, 1993b). With this broad base of ecological and physiological understanding, it has been possible to focus our search for constraints on the distribution of *Piper* species in tropical forest, and indeed, help to clarify the general ecological and physiological processes that may govern the distribution of species across resource gradients.

20.2 GROWTH AND CARBON METABOLISM

20.2.1 Growth rates, leaf longevity, and leaf construction cost

It has been suggested that one of the most general features of plants occurring in resource-limited environments is low intrinsic growth rates (Grime, 1977; Chapin et al., 1987). For example, fast-growing species typically predominate in disturbed and productive habitats, while slow-growing species predominate in stable and climax habitats (Grime & Hunt, 1975). Chronic resource limitation(s) presumably puts strong evolutionary pressure on plants to down-regulate growth-related processes and physiology, since unchecked growth under low resource conditions is potentially deleterious to a plant's vegetative or reproductive success, although experimental examples of this are scarce.

We compared seedling growth in four *Piper* species found in rainforest at Los Tuxtlas; *P. auritum* (gap specialist), *P. hispidum* (generalist; occupying both gap and understory microsites), and *P. lapathifolium* and *P. aequale* (understory specialists). Plants were grown for six to seven months in four-liter containers at high (50% of full sun) or low (15% full sun) light in a factorial arrangement with either ambient or low red to far-red light (R/FR) under constant high humidity greenhouses at Stanford, CA. Light quality had no significant effect on either growth or net assimilation rate in any of the species, whereas all species had reduced growth rates at low light relative to high light (Fredeen et al., 1995). Growth rates were higher in the gap and generalist species than in the understory species at both light levels, but less so at the lower light level (Figure 20.1). Hence, our results support the hypothesis that species from low resource environments, in this case low light, do have intrinsically lower growth rates. Conversely, higher growth rates in the gap and generalist species at low light, relative to the understory species, provide no explanation for the paucity of these species in low-light environments in the rainforest. Two studies to date have examined relative growth rates of gap and understory *Piper* species *in situ* and neither provides a clear demonstration of superior relative growth in high light-requiring species at high light nor in shade-tolerant species in understory light (Denslow et al., 1990; Sanchez-Coronado, Rincòn & Vázquez-Yanes, 1990). Therefore, relative growth rates do not appear to be essential determinants of rainforest habitat preference.

A more commonly observed effect of low light on plant growth is a shift to higher shoot production relative to root (see Björkman, 1981), while a major effect of shade light quality, i.e., reduced R/FR, appears to be increased stem elongation (e.g., Corré, 1983). Neither shoot versus root partitioning nor stem elongation was significantly affected by R/FR in the *Piper* species we examined (Fredeen et al., 1995). However, we did observe a significant difference among species with respect to shoot versus root partitioning at high and low light. With gap (*P. auritum*) and generalist (*P. hispidum*) species, shoot/root ratios were significantly reduced in response to increased light intensity, i.e., shoot/root ratios ranged from 4 to 6 at low light (15% of full sun) and from 2 to 4 at high light (50% of full sun) (Fredeen et al., 1995). In comparison, the understory species *P. aequale* and *P. lapathifolium* exhibited shoot/root ratios of 5 at both low and high light intensities. We are curious as to whether inflexibility in shoot versus root partitioning is a common feature of understory species, or whether this example is unique to these species or set of growing conditions.

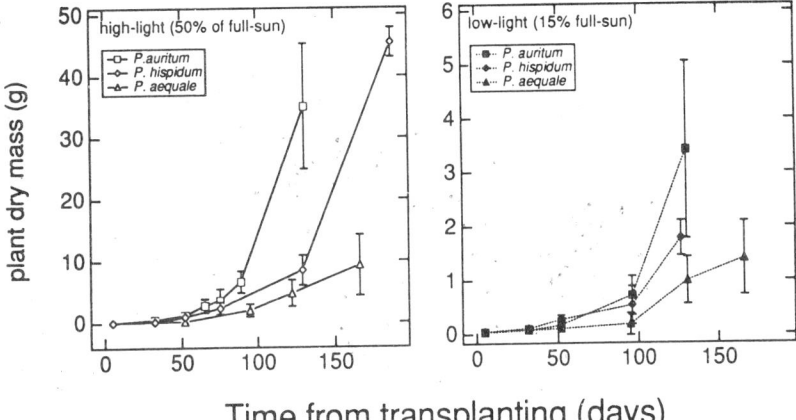

Figure 20.1.   Plant dry mass accumulation in seedlings of three species of Piper: P. auritum *(gap specialist),* P. hispidum *(generalist with respect to light habitat), and* P. aequale *(understory shade specialist). Plants were grown in 50% full-sun (left plot) or 15% full-sun (right plot). Means for three or four replicates are shown ± SD.*

Lower leaf and root turnover rates are characteristic of slower-growing species (Chapin, 1980). From a theoretical perspective, production of plant parts must provide a return on investment for growth to occur (Bloom, Chapin & Mooney, 1985). Hence, a leaf produced in the understory will need to have greatly increased longevity or increased light-use efficiency over its gap counterpart to provide a similar return. In fact, understory *Piper* species do exhibit enhanced utilization of light energy in the form of lightflecks (Tinoco-Ojanguren & Pearcy, 1992) and increased leaf longevity (Williams, Field & Mooney, 1989) over generalists and gap species. Williams, Field & Mooney (1989) originally hypothesized that a positive relationship should exist between construction cost and leaf longevity since the potential for herbivore attack increases with leaf age, potentially increasing the need for costly protective structures or compounds in long-lived leaves (but see Gulmon and Mooney, 1986). Among the *Piper* species occurring at Los Tuxtlas, the reverse was true. However, leaf longevity scaled positively with the ratio of carbon cost:daily carbon income, consistent with the notion that leaves need to be retained longer in the understory to achieve a positive carbon balance.

20.2.2 Photosynthesis and light

Acclimation responses to shade, relative to high light, involve numerous components, such as thinner leaves, larger chloroplasts, and enhanced thylakoid to stromal volume and pigment content. Acclimation and adaptation of photosynthesis to sun and shade have been intensively studied for more than 30 years (Björkman, 1981; Field, 1988; Chazdon et al., Chapter 1). For instance, shade species typically have higher photosynthetic light-use efficiencies and lower light compensation points and respiration rates. In the now classic studies of sun and shade ecotypes of *Solidago virgaurea* (Björkman & Holmgren, 1963) and *Solanum dulcamara* (Gauhl, 1969; 1976), it was demonstrated that the shade ecotypes were incapable of developing sufficient photosynthetic capacity to acclimate to high light. However, later studies have suggested that obligate shade behavior, at least in the case of *Solanum dulcamara*, may have been related to viral infection (Osmond, 1983). Additional studies of photosynthetic light acclimation in rainforest tree and shrub species (e.g., Turnbull, 1991; Thompson, Huang & Kriedemann, 1992) suggest that neither photosynthetic capacity nor ability to acclimate to different light levels is distinctly correlated with successional status or sun versus shade preference (see Strauss-Debenedetti & Bazzaz, Chapter 6).

Among *Piper* species, a common consensus has not been reached. Several recent studies show an apparent inability of understory species (*P. arieianum* and *P. aequale*) to acclimate photosynthesis to high light levels, especially when compared to gap species (*P. sancti-felicis* and *P. auritum*) (Chazdon, 1992, and Tinoco-Ojanguren & Pearcy, 1992, respectively). In contrast, Walters and Field (1987) and Field (1988) found no significant differences in photosynthesis and acclimation potential between the gap specialist, *P. auritum* and the generalist *P. hispidum*. In more recent laboratory experiments, rates of photosynthesis at saturating light and either saturating or ambient CO_2, as well as the initial slopes of the photosynthesis:intercellular $[CO_2]$ relationship, were found to be similar for high- and low-light grown gap and understory *Piper* species (Fredeen et al., 1995). Furthermore, ability to acclimate photosynthesis to high and low light, defined as the increase in maximum photosynthesis in high-light plants (grown at 50% full sun) versus low-light plants (grown at 15% full sun), was not different among the *Piper* species (Figure 20.2). What is clear is that ability to acclimate photosynthesis to high light is not uniformly lacking in understory species, and gap species typically have a greater ability to acclimate to low light than understory species to high light.

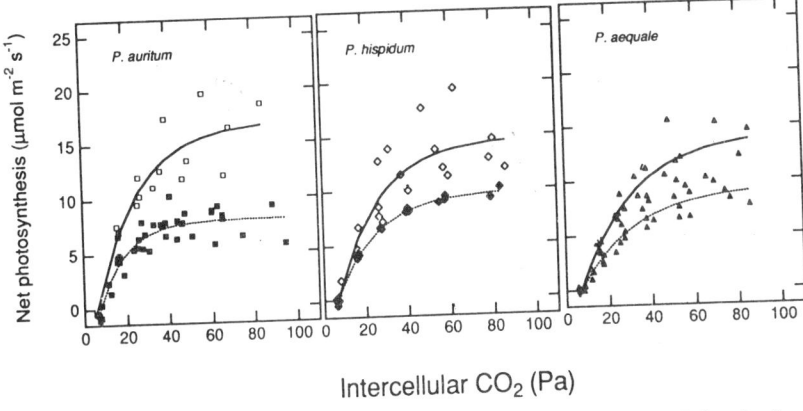

Figure 20.2. Light-saturated net photosynthesis versus intercellular leaf [CO₂] in P. auritum, P. hispidum, and P. aequale in plants grown at either 50% full sun (open symbols) or 15% full sun (closed symbols). Three to five replicates at each light level are shown with a best fit gaussian curve.

There are several reasons why the ability to acclimate photosynthesis to high light conditions may not always be lacking in low light-adapted species. First, many late successional or understory species germinate in gaps or margins and have seedlings that presumably benefit from these high light conditions while in early successional habitats. An example of this is *Piper aequale*, which requires high light for germination but otherwise is a permanent resident of the understory (Vázquez-Yanes, 1976; Orozco-Segovia & Vázquez-Yanes, 1989). Second, a large fraction of an understory plant's photosynthesis can occur in bursts during and subsequent to high intensity sunflecks (Chazdon et al., Chapter 1; Pearcy, 1990; Woodward, 1990), thus maintaining the benefit of a higher photosynthetic capacity than would be predicted from the otherwise low light intensity in the understory. For example, the understory species *P. aequale* appears to be better adapted to utilize sunfleck energy for photosynthesis than the gap species *P. auritum* (Tinoco-Ojanguren & Pearcy, 1992).

20.2.3 Respiration and the alternative pathway

Reduced leaf respiration is a fairly general response to reduced irradiance (Bazzaz & Carlson, 1982; Björkman, 1981; Fitter & Hay, 1987), and low rates of respiration appear to be critical for the persistence of plants in shade environments (Hutchinson, 1967; Loach, 1967; Cross, 1975). More recent experiments involving *Piper* species have corroborated these earlier conclusions. For example, Walters and Field (1987) and Field (1988) demonstrated an inverse relationship between *in situ* light environment and leaf respiration in a range of *Piper* species growing in rainforest. The interpretations from these studies were limited by the sensitivity of the infrared gas analyzer and effects of recent history on dark respiration (Sharp, Matthews & Boyer, 1984). We have since studied dark respiration with an oxygen electrode (Delieu & Walker, 1981), which besides having increased sensitivity, has allowed us to utilize selective inhibitors of the oxidases of the two major respiratory pathways and to control leaf temperature. We studied respiration at Los Tuxtlas on leaf punches from individuals of six *Piper* species growing in rainforest at the high- and low-light extremes of their natural distributions (Fredeen & Field, 1991). Species normally found in large gaps and clearings (*P. auritum and P. umbellatum*) had approximately twice the dark respiration per unit of leaf area or dry mass as species found predominantly in shaded understory sites (*P. aequale, P. lapathifolium*, and *P. amalago*). Within a species, dark respiration was lower in individuals from low-light than high-light sites (Figure 20.3). Over all species, leaf respiration was positively correlated with the average daily photosynthetically active photon flux density at each site (estimated from hemispherical photographs (Chazdon & Field, 1987), and negatively correlated with mean leaf longevity.

It has been asserted that the alternative oxidase pathway of respiration in plants is energetically wasteful (Siedow & Berthold, 1986), in part because it has no known function in plants apart from thermogenic flowering in some plants, e.g., *Sauromatum guttatum* in the Araceae (McIntosh, 1994). As a result, we expected that the alternative pathway would be absent or reduced in understory species. In greenhouse experiments, leaf respiration was monitored in three *Piper* species grown under a common light level (full sun). Under these conditions, total leaf respiration was 2–3–fold higher in the gap species (*P. auritum*) than in the generalist (*P. hispidum*) and shade species (*P. aequale*) (Figure 20.3). Potassium cyanide and salicylhydroxamic acid (SHAM) were used to inhibit the cytochrome and alternative oxidases, respectively. Contrary to our hypothesis, all species exhibited alternative pathway respiration. Furthermore, both the capacity and the *in vivo* engagement of the alternative pathway were similar as a percent of total respiration in shade species as in the gap species (Fredeen & Field, 1991). As a result, we are tempted to

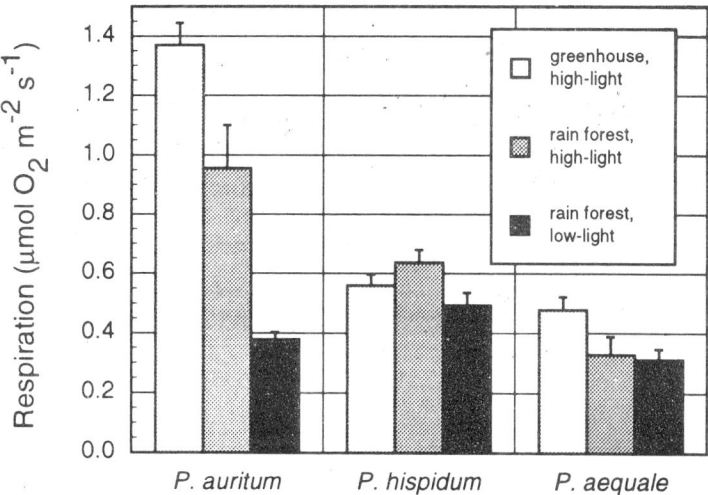

Figure 20.3.   Leaf respiration in seedlings of greenhouse-grown P. auritum, P. hispidum, *and* P. aequale *at full-sun or in plants from natural high and low light extremes in rainforest (Los Tuxtlas, Mexico). Means for three or four replicates are shown* $\pm$ *SD.*

speculate that the alternative pathway is not simply an energetically wasteful process, but has some function in plants.

20.3   NITROGEN ACQUISITION AND ASSIMILATION

Nitrogen is probably the most commonly limiting nutrient in terrestrial systems (Vitousek & Howarth, 1991). Nitrogen limitation results in part because it is required at such high concentrations (Haynes, 1986), but also because the plant-available forms are particularly labile and susceptible to leaching and gaseous loss. Next to photosynthetic carbon reduction, nitrogen acquisition and assimilation is the primary energy-requiring process in plants (Chapin et al., 1987). Under light- or energy-limiting conditions, nitrate assimilation requires approximately 15% of the energy processed within a plant (Penning De Vries, Brunsting & Van Laar, 1974). In contrast, when light is not limiting, leaf assimilation of nitrate may be driven largely by photosynthetic electron transport (Wallace, 1987) with only minor carbon costs to the plant (Bloom et al., 1989; McDermitt & Loomis,

1981). Given the importance of nitrogen for plant growth and physiology as well as its large potential energetic cost under light-limiting conditions, we hypothesized that nitrogen acquisition would show strong acclimation and adaptation to light level.

20.3.1 Nitrate reductase as an *in situ* indicator of nitrate use

Given the inherent difficulties in assessing nitrogen acquisition *in situ*, nitrate reductase activity provides an index of potential nitrate use and of the location of nitrate assimilation. There is strong evidence at the biochemical level that nitrate reductase activity should reflect plant uptake or level of nitrate in plant tissue. Nitrate reductase is the rate-limiting enzyme in the assimilation of nitrate (Campbell, 1988), and the enzyme is strongly induced by its substrate, nitrate (Remmler & Campbell, 1986). Although nitrate reductase activity rarely corresponds with nitrate uptake at the whole plant level (Doddema, Hofstra & Feenstra, 1978; MacKown, Jackson & Volk, 1982; Rao & Rains, 1976), the activity assay remains an important tool for the determination of nitrate use in natural systems, especially in forest ecosystems where root systems can be extensive.

In *Piper* species native to Los Tuxtlas, pioneer/gap species had higher levels of nitrate reductase activity located predominantly in the leaf, relative to generalist and understory species, which had lower nitrate reductase activities in both shoot (Figure 20.4) and root (Fredeen, Griffin & Field, 1991). The pattern is similar for Australian rainforest species. (Stewart, Hegarty & Specht, 1988). In the gap species *P. auritum*, foliar *in vitro* nitrate reductase activity was positively related to nitrate net uptake (Fredeen & Field, 1992). In contrast, these same parameters were uncorrelated in other *Piper* species (generalist and understory) (Fredeen & Field, unpublished results), perhaps because they have less of their total nitrate reductase activity in the leaf and because foliar nitrate reductase activity in these species appears to be less responsive to nitrate additions to the root system than in gap species (Fredeen, Griffin & Field, 1991). Another possibility is that foliar nitrate reductase plays a secondary role in iron assimilation (Campbell & Redinbaugh, 1984), a property that might explain the requirement for constitutive activities in understory and generalist species.

To address the possibility that higher nitrate reductase activities in gap species were due to higher soil nitrate levels (or lower ammonium

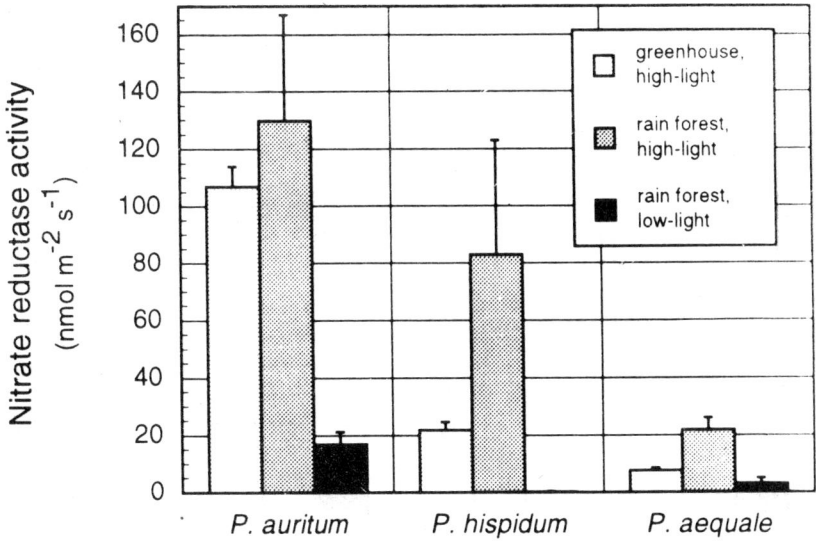

Figure 20.4.   In vivo leaf nitrate reductase activities in seedlings of greenhouse-grown P. auritum, P. hispidum, and P. aequale at full sun or in plants from natural high and low light extremes in rainforest (Los Tuxtlas, Mexico). Means for three or four replicates are shown ± SD.

levels) in gaps relative to understory, we analyzed soil nitrogen dynamics in primary and secondary forest (~ 25 years of regrowth) at Los Tuxtlas, including some samples from forest gaps. Mineral nitrogen (nitrate and ammonium) levels and net nitrogen mineralization rates were remarkably similar between primary and secondary forest and across understory light environments (Fredeen, Griffin & Field, 1991), comparable to the observations of Vitousek and Denslow (1986) and Matson et al. (1987). The lack of variation in nitrate and ammonium levels argues against the possibility that soil nitrate availability restricted nitrate reductase activity in the understory. However, since growth rates of shade species are low (Fredeen et al., 1995), the demands on soil nitrogen must also be low, preserving the possibility that nitrogen demands are met primarily by ammonium. In addition, differences in rooting depth or root activity between gap and understory species could result in very different mineral nitrogen environments. At Los Tuxtlas, nitrate was the predominant form of soil mineral nitrogen from 0–10 cm, while ammonium was the predominant form below 50 cm (Fredeen, Griffin & Field, 1991).

20.3.2 Preference for nitrate versus ammonium

Ammonium and nitrate are the two primary sources of mineral nitrogen utilized by plants. Nitrate assimilation typically carries a greater energetic cost than ammonium (Chapin et al., 1987; Pate, 1986); the conversion of nitrate to ammonium in plants requires 347 kJ mole^{-1}. In addition, the uptake of nitrate is always energy requiring (ATP) while ammonium has a dual uptake system, one energy dependent and the other not (Lewis, 1986). Despite the obvious difference between the nitrogen forms with respect to energy cost, especially under light-limiting conditions, few assessments of the influence of light level on nitrate versus ammonium uptake in plants have been made.

To test the hypothesis that understory species preferentially take up ammonium, we examined the effects of light level and nitrogen form (ammonium and/or nitrate) on ammonium and nitrate uptake in a range of *Piper* species We hypothesized that all *Piper* species would exhibit an increased preference for ammonium relative to nitrate under low light (a general acclimation response), but that understory species would exhibit a greater preference for ammonium than gap species at all light levels (an adaptation response). Three species of *Piper, P. auritum* (gap), *P. hispidum* (generalist), and *P. aequale* (understory), were grown hydroponically at ~50% of full sun, and then transferred to root cuvettes for open-system net uptake determinations at 25°C. Nutrient uptake was examined in solutions containing 200 μM nitrate and/or 200 μM ammonium at two light levels: low (50 μmol m^{-2} s^{-1}) and high (500 μmol m^{-2} s^{-1}). At both light levels, uptake rates were similar for nitrate and ammonium when either form was supplied alone (Fredeen & Field, 1992). With both forms provided simultaneously, preference for ammonium relative to nitrate was higher in the understory species than in gap species, and intermediate in the generalist species (Figure 20.5). The preferential uptake of ammonium was accentuated in all species at low light, but especially in the understory species, confirming our initial hypothesis.

To our knowledge, the precise nature by which ambient light level affects the ratio of ammonium:nitrate uptake by roots has not been explored. Since changes in extent of ammonium repression of nitrate uptake, after increasing or decreasing the light level, required more than 24 hours to stabilize (Fredeen & Field, 1992), adjustments at the molecular, biochemical, and physiological levels were possible. The most obvious adjustment to a 10-fold reduction in light level is a reduction in photosynthesis and hence, carbohydrate availability. The

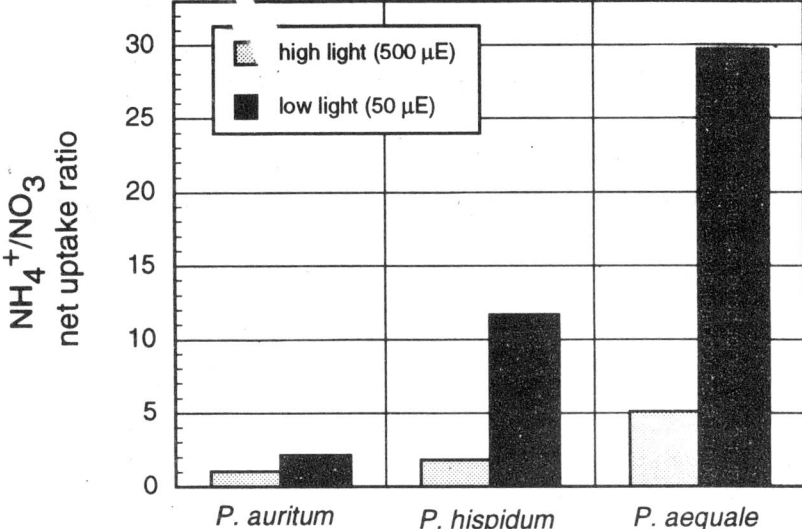

Figure 20.5.   The ratio of net ammonium uptake to net nitrate uptake in P. auritum, P. hispidum, *and* P. aequale *grown in hydroponics at 50% full sun; equilibrated to solutions containing 200 μM of both ammonium and nitrate at either high light (500 μE m^{-2} s^{-1}) or low light (50 μE m^{-2} s^{-1}) (n = 4).*

latter is known to be important for nitrate (Sasakawa & Yamamoto, 1978) and ammonium (Reisenauer, 1978) uptake.

20.4   PLANT ARCHITECTURE AND LEAF MORPHOLOGY

A broad survey comparing the architecture of *Piper* gap and understory specialists has not been made. Nevertheless, several studies have examined aspects of the architecture of the *Piper* species at Los Tuxtlas. For example, *P. auritum* exhibits an architecture typical of species of rainforest clearings and gaps (Chiariello, Field & Mooney, 1987), namely, the arrangement of relatively large leaves in a hemispherical monolayer (Ackerly, Chapter 21; Whitmore, 1975). Leaf morphology in *Piper auritum*, i.e. a stiff midrib and large flexible lamina, allows for increased leaf-level water use efficiency in plants receiving direct sunlight (Chiariello, Field & Mooney, 1987), primarily through the relatively rapid wilting, and subsequent recovery, of the leaf lamina. In contrast, understory *Piper* species tend to be more woody with small, rigid, long-lived and deeply pigmented leaves arranged more randomly in space (Chazdon, Williams & Field, 1988).

20.5   SEEDLING SURVIVORSHIP, HERBIBORY, AND PLANT ASSOCIATIONS

Seedling mortality may be the greatest single factor determining natural selection (Cook, 1979). Factors that may affect seedling mortality that have not already been discussed above include debris fall, resistance to pathogens, invertebrate or vertebrate attack or herbivory, and establishment of mutualistic relationships (e.g., mycorrhizal or insect). Several of these factors are probably not important determinants of the spatial distribution of *Piper* species. Seed dispersal agents, primarily frugivorous bats, are probably similar for all species of *Piper* (Fleming, 1985) and hence, are probably not a factor. However, reliance on seed for regeneration does appear to differ greatly between gap and understory Piper species (Greig, 1993b). At La Selva, *Piper* gap species regenerate both vegetatively and from seed, while understory species, especially those occurring late in succession, rely almost exclusively on vegetative propagation. High levels of vegetative propagation by understory species may be a consequence of (1) greatly reduced fecundities in shade-tolerant *Piper* species resulting from increased seed predation and infructescence damage (Greig, 1993a), (2) higher rates of seedling mortality in the understory (e.g., Osunkjoya et al., 1992), attributable in some cases primarily to debris fall (Clark & Clark, 1987), (3) a preponderance of severed and pinned branch regrowth in the understory where the incidence of branch injury is higher (Gartner, 1989), and (4) stored carbohydrate reserves. Maternal light environment, viz. the R/FR ratio, may predispose germination of seed from gap species to similar light environments (Orozco-Segovia, Sanchez-Coronado & Vázquez-Yanes, 1993a), minimizing germination in habitats unfavorable for growth.

Herbivory has not been studied systematically on *Piper* species at Los Tuxtlas. In *P. arieianum*, an understory species common in Costa Rica, single defoliation events negatively impact both seed production and growth rate in subsequent years (Marquis, 1984). Both the pattern of folivory and time of year appear to be important variables in determining the severity of the impact in this species (Marquis, 1992). The Piperaceae produce an abundance of secondary plant products, and there are a number of clues from the phytochemical literature suggesting that many of these products are important in protecting *Piper* from herbivores and pathogens, e.g., the kavalactones in *Piper methysticum*, widely disseminated and cultivated by Pacific islanders for its medicinal and narcotic/hypnotic properties (Lebot & Lévesque, 1989; Hänsel, 1968). The diuretic, soporific, antiepileptic, spasmolytic, and analgesic properties would presumably restrict herbivory, while the bactericidal and antimycotic properties could reduce pathogen-related injury. We are not aware of studies that have looked at the protective role of these compounds in *Piper* seedling establishment and survival and plant growth. Foliar phenolic concentrations were similar in a gap and understory *Piper* species at La Selva (Denslow et al., 1990), but this pattern may not be typical for other species and other classes of compounds.

Insect associations with *Piper* species may also play a role in determining success of *Piper* species in understory and gap. Removal of ants from *Acacia* species in gap environments, for example, can result in complete defoliation of the plant (Janzen, 1975). Burger (1972) alluded to the observation of Janzen that the stems of certain *Piper* species are hollowed to support colonization by ants. Many *Piper* species, including most of those occurring at Los Tuxtlas, also produce a multitude of minute foliar sacules, presumably food bodies, which may facilitate the maintenance of protective ant associations. At least in one case, the presence of a single species of ant (*Pheidole bicornis*) results in the stimulation of food body production in a *Piper* species (*P. cenocladum*) (Risch & Rickson, 1981). In this case, the ant/plant association is apparently mutualistic, the plant providing food (and shelter inside hollowed stems) and the ant providing protection from herbivores and vines. The authors speculated that constitutive food body production, as observed in gap species such as *Cecropia*, *Acacia*, and *Macaranga*, is not observed in understory species such as *P. cenocladum* because the carbon expenditures would be untenable in light depauperate microsites.

20.6 CONCLUSIONS AND FUTURE DIRECTIONS

A suite of physiological and morphological adaptations confer tolerance to either high or low light extremes in rainforest. No single adaptation in the genus *Piper* seems to predominate, although some generalizations emerge. First, understory plants often have reduced capacities for a number of physiological processes, in comparison to high light species, when both are grown in high light. The evidence for photosynthesis is mixed, but the capacity for growth, respiration, nitrate uptake and reduction are constrained in understory species and may play a role in restricting them to understory environments. The ability of understory species to selectively utilize ammonium and sunflecks in shade environments may be a significant adaptive feature in energy-restricted environments, but it provides little, if any, benefit in high-light environments.

In the reverse situation, gap plants fully acclimate photosynthesis to low light levels relative to understory species, but consistently have higher rates of growth, respiration, nitrate reductase activity, and nitrate uptake at any given light level. Gap species appear to be unable to achieve states of depressed physiological activity required in deep shade (see Kitajima, Chapter 19). Gap species also appear to have limited ability to extend leaf longevity (unpublished data), a feature that appears critical for plant survival in low light environments. For other factors that are functionally important and vary consistently between gap and understory species at Los Tuxtlas, (for example, leaf morphology), we do not yet have a quantitative understanding of the implications.

Our evidence is consistent with the hypothesis that genetically determined growth potential is a central feature of adaptation to habitats of contrasting resource availability (Chapin, 1980, 1991). However, at least with *Piper* species, intrinsic growth rates per se do not appear to be the critical determinants of habitat occurrence. Detailed studies that quantify the sensitivity of physiological and architectural processes to precisely known differences in the environment will address some of the outstanding issues. Others will require new approaches, including studies with transgenic plants (e.g. Stitt & Quick, 1989) and with integrated models (e.g., Reynolds, Hilbert & Kemp, 1992).

ACKNOWLEDGEMENTS

We gratefully acknowledge support from the National Science Foundation grants BSR8415875 and BSR8717422 to C.B.F. that supported much of this research. The authors are indebted to C. Vázquez-Yanes for assistance and guidance in the field and comments and suggestions on earlier versions of this chapter. Thoughtful reviews by R. L. Chazdon, S. S. Mulkey, and E. Newell were very useful in the writing of this chapter. This publication is CIWDPB # 1185.

REFERENCES

BAZZAZ, F. A. & CARLSON, R. W. (1982) Photosynthetic acclimation to variability in the light environment of early and late successional plants. *Oecologia*, **54**, 313–316.

BJÖRKMAN, O. (1981) Responses to different quantum flux densities. *Physiological Plant Ecology. Volume 1. Responses to the Physiological Environment*. Springer-Verlag, New York.

BJÖRKMAN, O. & HOLMGREN, P. (1963) Adaptability of the photosynthetic apparatus to light intensity in ecotypes from exposed and shaded habitats. *Physiologia Plantarum*, **16**, 889–914.

BLOOM, A. J., CALDWELL, R. M., FINAZZO, J., WARNER, R. L. & WEISSBART, J. (1989) Oxygen and carbon dioxide fluxes from barley shoots depend on nitrate assimilation. *Plant Physiology*, **91**, 352–356.

BLOOM, A. J., CHAPIN, F. S., III & MOONEY, H. A. (1985) Resource limitation in plants – An economic analogy. *Annual Review of Ecology and Systematics*, **16**, 363–392.

BURGER, W. C. (1972) Evolutionary trends in the Central American species of *Piper* (Piperaceae). *Brittonia*, **24**, 356–362.

CAMPBELL, W. H. (1988) Nitrate reductase and its role in nitrate assimilation in plants. *Physiologia Plantarum*, **74**, 214–219.

CAMPBELL, W. H. & REDINBAUGH, M. G. (1984) Ferric-citrate reductase activity of nitrate reductase and its role in iron assimilation by plants. *Journal of Plant Nutrition*, **7**, 799–806.

CHAPIN, F. S., III (1980) The mineral nutrition of wild plants. *Annual Review of Ecology and Systematics*, **11**, 233–260.

CHAPIN, F. S., III (1991) Integrated responses of plants to stress. *Bioscience*, **41**, 29–36.

CHAPIN, F. S., III, BLOOM, A. J., FIELD, C. B. & WARING, R. H. (1987) Interactions of environmental factors in controlling plant growth. *Bioscience*, **37**, 49–57.

CHAZDON, R. L. (1992) Photosynthetic plasticity of two rainforest shrubs across natural gap transects. *Oecologia*, **92**, 586–595.

CHAZDON, R. L. & FIELD, C. B. (1987) Photographic estimation of photosynthetically active radiation: Evaluation of a computerized technique. *Oecologia*, **73**, 586–595.

CHAZDON, R. L. & KAUFMANN, S. (1993) Plasticity of leaf anatomy in relation to photosynthetic light acclimation. *Functional Ecology*, **7**, 385–394.

CHAZDON, R. L., WILLIAMS, K. & FIELD, C. B. (1988) Interactions between crown structure and light environment in five rainforest *Piper* species. *American Journal of Botany*, **75**, 1459–1471.

CHIARIELLO, N. R., FIELD, C. B. & MOONEY, H. A. (1987) Midday wilting in a tropical pioneer tree. *Functional Ecology*, **1**, 3–11.

CLARK, D. B. & CLARK, D. A. (1987) An experimental method for community-level assessment of components of seedling mortality, with data from a tropical rainforest. *Bulletin of the Ecological Society of America*, **68**, 280.

COOK, R. E. (1979) Patterns of juvenile mortality and recruitment in plants. *Topics in Plant Population Biology* (eds. O. T. SOLBRIG, S. JAIN, G. B. JOHNSON, & P. H. RAVEN) Columbia University Press, New York.

CORRÉ, W. J. (1983) Growth and morphogenesis of sun and shade plants. II. The influence of light quality. *Acta Botanica Neerlandica*, **32**, 185–202.

CROSS, J. R. (1975) Biological flora of the British Isles: *Rhododendron ponticum*. *Journal of Ecology*, **63**, 345–359.

DELIEU, T. & WALKER, D. A. (1981) Polarographic measurement of photosynthetic O_2 evolution by leaf discs. *New Phytologist*, **89**, 165–175.

DENSLOW, J. S., SCHULTZ, J. C., VITOUSEK, P. M. & STRAIN, B. R. (1990) Growth responses of tropical shrubs to treefall gap environments. *Ecology*, **71**, 165–170.

DENSLOW, J. S., VITOUSEK, P. M. & SCHULTZ, J. C. (1987) Bioassays of nutrient limitation in a tropical rainforest soil. *Oecologia*, **74**, 370–376.

DODDEMA, H., HOFSTRA, J. J. & FEENSTRA, W. J. (1978) Uptake of nitrate by mutants of *Arabidopsis thaliana*, disturbed in uptake or reduction of nitrate. I. Effect of nitrogen source during growth on uptake of nitrate and chlorate. *Physiologia Plantarum*, **43**, 343–350.

FIELD, C. B. (1988) On the role of photosynthetic responses in constraining the habitat distribution of rainforest plants. *Ecology of Photosynthesis in Sun and Shade* (eds. J. R. EVANS, S. VON CAEMMERER, & W. W. ADAMS III) CSIRO, Australia.

FITTER, A. H. & HAY, R. K. M. (1987) *Environmental Physiology of Plants*. Academic Press, London.

FLEMING, T. H. (1985) Coexistence of five sympatric *Piper* (Piperaceae) species in a dry tropical forest. *Ecology*, **66**, 688–700.

FLEMING, T. H. & HEITHAUS, E.R. (1981). Frugivorous bats, seed shadows, and the structure of tropical forest. *Biotropica Reproductive Botany*, **45**, 33.

FREDEEN, A. L. & FIELD, C. B. (1991) Leaf respiration in *Piper* species native to a Mexican rainforest. *Physiologia Plantarum*, **82**, 85–92.

FREDEEN, A. L. & FIELD, C. B. (1992) Ammonium and nitrate uptake in gap, generalist, and understory species of the genus *Piper*. *Oecologia*, **92**, 207–214.

FREDEEN, A. L., GRIFFIN, K. & FIELD, C. B. (1991) Effects of light quantity and quality and soil nitrogen status on nitrate reductase activity in rainforest species of the genus *Piper*. *Oecologia*, **86**, 441–446.

FREDEEN, A. L., GRIFFIN, K., HENNESSEY, T. L. & FIELD, C. B. (1995) Intrinsic growth rates and photosynthetic properties of gap, generalist, and understory *Piper* species (submitted)

GARTNER, B. (1989) Breakage and regrowth of *Piper* species in rainforest understory. *Biotropica*, **21**, 303–307.

GAUHL, E. (1969) Differential photosynthetic performance of *Solanum dulcamara* ecotypes from shaded and exposed habitats. *Carnegie Insitution of Washington Yearbook*, **67**, 482–487.

GAUHL, E. (1976) Photosynthetic response to varying light intensity in ecotypes of *Solanum dulcamara* L. from shaded and exposed habitats. *Oecologia*, **22**, 274–286.

GÓMEZ-POMPA, A. (1971) Posible papel de la vegetacion secundaria en la evolución de la flora tropical. *Biotropica*, **3**, 125–135.

GÓMEZ-POMPA, A. & VÁZQUEZ-YANES, C. (1974) Studies on the secondary succession of tropical lowlands: The life cycle of secondary species. *First International Congress of Ecology*.

GREIG, N. (1993a) Predispersal seed predation of five *Piper* species in tropical rainforest. *Oecologia*, **92**, 412–420.

GREIG, N. (1993b) Regeneration mode in neotropical *Piper*: Habitat and species comparisons. *Ecology*, **74**, 2125–2135.

GRIME, J. P. (1977) Evidence for the existence of three primary strategies in plants and its relevance to ecological and evolutionary theory. *American Naturalist*, **111**, 1169–1174.

GRIME, J. P. & HUNT, R. (1975) Relative growth rate: Its range and adaptive significance in a local flora. *Journal of Ecology*, **63**, 393–422.

GULMON, S. L. & MOONEY, H.A. (ed.) (1986) Costs of defense on plant productivity. *On the Economy of Plant Form and Function* (ed. T. J. GIVNISH) Cambridge University Press, Cambridge, United Kingdom pp 681–698.

HÄNSEL, R. (1968) Characterization and physiological activity of some Kava constituents. *Pacific Science*, **22**, 369–373.

HAYNES, R. J. (1986) *Uptake and Assimilation of Mineral Nitrogen by Plants*, Academic Press, Orlando.

HUTCHINSON, G. E. (1967) Comparative studies of the ability of species to withstand prolonged periods of darkness. *Journal of Ecology*, **55**, 291–299.

JANZEN, D. H. (1975) *Pseudomyrimex nigropilosa*: A parasite of a mutualism. *Science*, **188**, 936–937.

LEBOT, V. & LÉVESQUE, J. (1989) The origin and distribution of kava (*Piper methysticum* Forts. F., Piperaceae): A phytochemical approach. *Allertonia*, **5**, 223–278.

LEWIS, O. A. M. (1986) *The Processing of Inorganic Nitrogen by the Plant*. Camelot Press Ltd., Southampton.

LOACH, K. (1967) Shade tolerance in tree seedlings. I. Leaf photosynthesis and respiration in plants raised under artificial shade. *New Phytologist*, **66**, 607–621.

MACKOWN, C. T., JACKSON, W. A. & VOLK, P. J. (1982) Restricted nitrate influx and reduction in corn seedlings exposed to ammonium. *Plant Physiology*, **69**, 353–359.

MARQUIS, R. J. (1984) Leaf herbivores decrease fitness of a tropical plant. *Science*, **226**, 537–539.

MARQUIS, R. J. (1988) Phenological variation in the neotropical understory shrub *Piper arieianum*: Causes and consequences. *Ecology*, **69**, 1552–1565.

MARQUIS, R, J. (1992) A bite is a bite is a bite? Constraints on response to folivory in *Piper arieianum* (Piperaceae). *Ecology*, **73**, 143–152.

MATSON, P. A., VITOUSEK, P.M., EWEL, J.J., MAZZARINO, M.J. & ROBERTSON, G.P. (1987) Nitrogen transformations following tropical forest felling and burning on a volcanic soil. *Ecology*, **68**, 491–502.

MCDERMITT, D. K. & LOOMIS, R. S. (1981) Elemental composition of biomass and its relation to energy content, growth efficiency, and growth yield. *Annals of Botany*, **48**, 275–290.

MCINTOSH, L. (1994) Molecular biology of the alternative oxidase. *Plant Physiology*, **105**, 781–786.

MOONEY, H. A., FIELD, C.B. & VÁZQUEZ-YANES, C. (eds.) (1984) Photosynthetic characteristics of wet tropical forest plants. *Physiological Ecology of Plants of the Wet Tropics*. Dr. Junk Publishers, The Hague, 254 p.

OPLER, P. A., FRANKIE, G.W. & BAKER, H.G. (1980) Comparative phenological studies of treelet and shrub species in tropical wet and dry forests in the lowlands of Costa Rica. *Journal of Ecology*, **68**, 167–188.

OROZCO-SEGOVIA, A., VÁZQUEZ-YANES, C. (1989) Light effect on seed germination in *Piper* L. *Acta Oecologia / Oecologia Planta*, **10**, 123–146.

OROZCO-SEGOVIA, A., SANCHEZ-CORONADO, M. E. & VÁZQUEZ-YANES, C. (1993a) Effect of maternal light environment on seed germination in *Piper auritum*. *Functional Ecology*, **7**, 395–402.

OROZCO-SEGOVIA, A., SANCHEZ-CORONADO, M.E. & VÁZQUEZ-YANES, C. (1993b) Light environment and phytochrome controlled germination in *Piper auritum*. *Functional Ecology*, **7**, 585–590.

OSMOND, C. B. (1983) Interactions between irradiance, nitrogen nutrition, and water stress in the sun-shade response of *Solanum dulcamara*. *Oecologia*, **57**, 316–321.

OSUNKJOYA, O. O., ASH, J. E., HOPKINS, M. S. & GRAHAM, A. W. (1992) Factors affecting survival of tree seedlings in North Queensland rainforests. *Oecologia*, **91**, 569–578.

PATE, J. S. (1986) Economy of symbiotic nitrogen fixation. *On the Economy of Plant Form and Function* (ed. T. J. GIVNISH) Cambridge University Press, Cambridge, United Kingdom.

PEARCY, R. W. (1983) The light environment and growth of C_3 and C_4 tree species in the understory of a Hawaiian forest. *Oecologia*, **58**, 19–25.

PEARCY, R. W. (1990) Sunflecks and photosynthesis in plant canopies. *Annual Reviews of Plant Physiology and Plant Molecular Biology*, **41**, 421–453.

PENNING DE VRIES, F. W. T., BRUNSTING, A. H. M. & VAN LAAR H. H. (1974) Products, requirements and efficiency of biosynthesis: A qualitative approach. *Journal of Theoretical Biology*, **45**, 339–377.

RAO, K. P. & RAINS, D. W. (1976) Nitrate absorption by barley. I. Kinetics. *Plant Physiology*, **57**, 55–58.

REISENAUER, H. M. (ed.) (1978) Absorption and utilization of ammonium nitrogen by plants. *Nitrogen in the Environment*. Academic Press, New York.

REMMLER, J. L. & CAMPBELL, W. H. (1986) Regulation of corn leaf nitrate reductase. II. Synthesis and turnover of the enzyme's activity and protein. *Plant Physiology*, **80**, 442–447.

REYNOLDS, J. F., HILBERT, D. W. & KEMP, P. R. (1992) Scaling ecophysiology from the plant to the ecosystem: A conceptual framework. *Scaling Physiological Processes: Leaf to Globe* (eds. J. R. EHLERINGER & C. B. FIELD) Academic Press, San Diego.

RISCH, S. J. & RICKSON, F. R. (1981) Mutualism in which ants must be present before plants produce food bodies. *Nature*, **291**, 149–150.

SANCHEZ-CORONADO, M. E., RINCÓN, E. & VÁZQUEZ-YANES, C. (1990) Growth responses of three contrasting Piper species growing under different light conditions. *Canadian Journal of Botany*, **68**, 1182–1186.

SASAKAWA, H. & YAMAMOTO, Y. (1978) Comparison of the uptake of nitrate and ammonium by rice seedlings. Influences of light, temperature, oxygen concentration, exogenous sucrose, and metabolic inhibitors. *Plant Physiology*, **62**, 665-669.

SIEDOW, J. N. & BERTHOLD, D. A. (1986) The alternative oxidase: A cyanide-resistant respiratory pathway in higher plants. *Physiologia Plantarum*, **66**, 569–573.

SHARP, R. E., MATTHEWS, M. A. & BOYER, J. S. (1984) Kok effect and the quantum yield of photosynthesis. *Plant Physiology*, **75**, 95–101.

STEWART, G. H., HEGARTY, E. E. & SPECHT, R. L. (1988) Inorganic nitrogen assimilation in plants of Australian rainforest communities. *Physiologia Plantarum*, **74**, 26–33.

STITT, M. & QUICK, W. P. (1989) Photosynthetic carbon partitioning, its regulation, and possibilities for manipulation. *Physiologia Plantarum*, **77**, 633–641.

THOMPSON, W. A., HUANG, L. –K. & KRIEDEMANN, P. E. (1992). Photosynthetic response to light and nutrients in sun-tolerant and shade-tolerant rainforest trees. II. Leaf gas exchange and component processes of photosynthesis. *Australian Journal of Plant Physiology*, **19**, 19–42.

TINOCO-OJANGUREN, C. & PEARCY, R. W. (1992) Dynamic stomatal behavior and its role in carbon gain during lightflecks of a gap phase and an understory Piper species acclimated to high and low light. *Oecologia*, **92**, 222–228.

TINOCO-OJANGUREN, C. & PEARCY, R. W. (1993a) Stomatal dynamics and its importance to carbon gain in two rainforest Piper species. I. VPD effects on the transient stomatal response to lightflecks. *Oecologia*, **94**, 288–294.

TINOCO-OJANGUREN, C. & PEARCY, R. W. (1993b) Stomatal dynamics and its importance to carbon gain in two rainforest *Piper* species. II. Stomatal versus biochemical limitations during photosynthetic induction. *Oecologia*, **92**, 222–228.

TURNBULL, M. H. (1991) The effect of light quantity and quality during development on the photosynthetic characteristics of six Australian rainforest tree species. *Oecologia*, **87**, 110–117.

VÁZQUEZ-YANES, C. (1976) *Estudios sobre ecophysiologia de la germinación en una zona cálido-húmeda de México*. CECSA, Mexico City.

VÁZQUEZ-YANES, C., OROZCO-SEGOVIA, A., RINCÓN, E., SÁNCHEZ-CORONADO, M.E., HUANTE, P., TOLEDO, J.R. & VARRADAS, V.L. (1990) Light beneath the litter in a tropical forest: Effect on seed germination. *Ecology*, **71**, 1952–1958.

VITOUSEK, P. M. & DENSLOW, J. S. (1986) Nitrogen and phosphorus availability in treefall gaps of a lowland tropical rainforest. *Journal of Ecology*, **74**, 1167–1178.

VITOUSEK, P. M. & HOWARTH, R. W. (1991) Nitrogen limitation on land and in the sea: How can it occur? *Biogeochemistry*, **13**, 87–115.

WALLACE, W. (1987) Regulation of nitrate utilization in higher plants. *Inorganic Nitrogen Metabolism*. Springer-Verlag, Berlin, pp 223–230.

WALTERS, M. B. & FIELD, C. B. (1987) Photosynthetic light acclimation in two rainforest *Piper* species with different ecological amplitudes. *Oecologia*, **72**, 449–456.

WHITMORE, T. C. (1975) *Tropical Rain Forests of the Far East*. Clarendon, Oxford.

WILLIAMS, K., FIELD, C. B. & MOONEY, H. A. (1989) Relationships among leaf construction cost, leaf longevity, and light environment in rainforest plants of the genus *Piper*. *The American Naturalist*, **133**, 198–211.

WOODWARD, F. I. (1990) From ecosystems to genes: The importance of shade tolerance. *Trends in Ecology and Evolution*, **5**, 111–115.

YODA, K. (1974) Three-dimensional distribution of light intensity in a tropical rainforest of West Malaysia. *Japanese Journal of Ecology*, **24**, 247–254.

21

Canopy Structure and Dynamics: Integration of Growth Processes in Tropical Pioneer Trees

David D. Ackerly

21.1 INTRODUCTION

When a tree falls in a tropical forest, or a landslide exposes an expanse of bare soil, a distinct group of plant species—variously known as pioneers, early successionals, shade-intolerant, or secondary species—appears and dominates the early phase of the successional process. The success of these species following disturbance is attributable to a suite of demographic and physiological traits that comprise a distinctive life history strategy. Although much research remains to be done, a great deal is known about pioneer plant species in tropical forests and other ecosystems. This knowledge reflects the amenability of pioneer plants as subjects of experimental research due to their rapid growth and adaptability to a broad range of environmenta' conditions, and their critical ecological role in the early stage of suc cession. In this chapter, I first examine the life cycle of pioneer trees, highlighting important physiological processes at each stage, and then address in detail the structure and dynamics of the canopy, with an emphasis on leaf lifespan and the interaction between patterns and processes at the level of the individual leaf and of the plant canopy.

21.1.1   What is a pioneer tree species?

Ecologists have struggled for many years with the question of how to recognize and structure the diversity of life history strategies among wet tropical forest tree species. Are there discrete strategies, and if so, how many should or can be recognized? Or does the variation in life history characteristics form a continuum, and if so, how many axes are needed to discriminate variation among species? There are many and varied answers to these questions, but all of them share a few characteristics. First, the responses of species to variation in light levels, and the role of treefall gaps, are critical components of the definition of life history strategies. Secondly, the pioneer strategy is invariably identified as a discrete life history category, while classification of the 'non-pioneers' presents more problems. Acknowledging this situation, Swaine and Whitmore (1988) suggested that only these two groups, pioneer and non-pioneer, may be unambiguously recognized, and proposed that the most broadly applicable diagnostic trait of pioneer species is the light requirement for seed germination (see Vázquez-Yanes & Orozco-Segovia, Chapter 18). This characteristic is accompanied by a suite of physiological and demographic traits characteristic of pioneer species, including long seed dormancy and large seed banks, high photosynthetic and growth rates, rapid and continuous leaf production, early and long flowering time, and rapid acclimation to environmental variation (see thorough discussions in Bazzaz, 1979, 1984, 1991; Bazzaz & Pickett, 1980; Martínez-Ramos, 1985; Swaine & Whitmore, 1988; Chazdon et al., Chapter 1; Strauss-Debenedetti & Bazzaz, Chapter 6).

Pioneer tree species may be defined as those species that germinate and establish in recently disturbed sites and that complete their life cycle without being overtopped by neighboring trees. Within this rubric, two types of specific variations are particularly important. First, species vary in the minimum size of disturbance required for successful establishment, leading to so-called small-gap and large-gap pioneers (e.g., Brokaw, 1987). Secondly, there is considerable variation in longevity and the maximum stature of adults. At one extreme are very short-lived, small trees, such as the neotropical *Carica papaya*, which may reach reproductive maturity in 8 months and complete its life cycle in under 10 years with a maximum height of 5–6 m (Sarukhán, Piñero & Martínez-Ramos, 1985). In the intermediate range are short-lived, canopy height trees, such as the well-known neotropical *Cecropia* sp. and some species of the paleotropical *Macaranga*, which reach heights of 30 + m and may persist for 30–50 years (e.g., Alvarez-Buylla & Martínez-Ramos, 1992). Finally, there are so-called long-lived pioneers, trees that establish and grow rapidly to canopy height in treefall gaps and then persist as canopy dominants, possibly for several centuries (e.g., *Ceiba pentandra*). Most research on the ecophysiology of pioneers has focused on the short-lived species, the primary focus of this review.

21.2 THE CARBON ECONOMY OF PIONEER TREES

The life cycle of pioneer trees may be considered in several critical stages: 1) seed germination and early seedling establishment immediately following disturbance (Vázquez-Yanes & Orozco-Segovia, Chapter 18; Kitajima, Chapter 19); 2) rapid growth of seedlings, accompanied by high mortality of individuals that are shaded by neighboring vegetation; 3) predominant investment in height growth to maintain canopy position through the sapling phase; 4) early attainment of reproductive age, and high reproductive allocation; and 5) senescence and mortality of adults. In Figure 21.1, these stages are illustrated in relation to the life cycle of *Cecropia obtusifolia* (Alvarez-Buylla & Martínez-Ramos, 1992), a dominant pioneer of the forest of Los Tuxtlas, Veracruz, Mexico. In this section, I review the carbon economy of seedling and sapling growth, focusing on the interaction between physiological processes and variation in ambient light environments associated with differences in the size of gaps in which pioneers establish and complete their life cycle.

21.2.1 Growth and biomass allocation in relation to light environments

Growth studies of tropical pioneer tree seedlings were initially conducted to determine the physiological basis of the apparently exceptional growth of tropical vegetation (Coombe, 1960). Relative growth rates (RGR) of pioneer seedlings at high light levels generally range from 0.015 to 0.13 $g\ g^{-1}\ d^{-1}$, and net assimilation rates (NAR) range

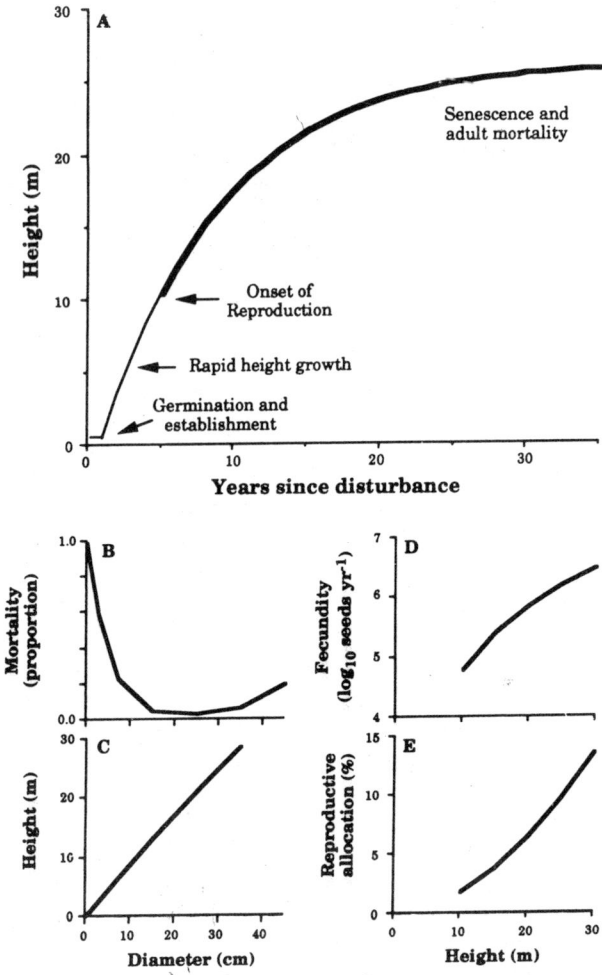

Figure 21.1. *Allometric relationships during the life cycle of a pioneer tree,* Cecropia obtusifolia *(based on allometric equations in Alvarez-Buylla & Martínez-Ramos, 1992). a) Following germination, usually in a recent disturbance, height growth is rapid and reproduction is initiated early. b) Mortality is very high in seedlings and lowest for young, reproductively mature trees, and then increases again in the largest size classes. c) Height increases monotonically with diameter. d) Female fecundity increases exponentially with plant size, and may exceed 4 million seeds per year in large trees. Production of flower buds in males is also size-dependent and is approximately 3.5 times higher than female flower initiation at a given size. e) Reproductive allocation for females, as a proportion of annual productivity, increases with plant size (see text for method of estimating allocation curve).*

from 1 to over 8 g m^{-2} d^{-1} (Coombe, 1960; Coombe and Hadfield, 1962; Okali, 1971, 1972; Whitmore & Gong, 1983; Oberbauer & Donnelly, 1986; Popma & Bongers, 1988; Ackerly, 1993). These rates are similar to those measured on fast-growing temperate woody seedlings, and approach those of the fast-growing herb *Helianthus annuus*. Coombe and Hadfield (1962) concluded that the rapid growth of tropical pioneers is not due to higher efficiency of energy conversion, but rather to the capacity for continuous and unrestricted leaf production (discussed in detail below), and the continuously favorable humid tropical environment.

Assimilation rates determined from whole-plant growth analysis are consistent with lower-level measurements of leaf photosynthetic rates and system level estimates of energy conversion in secondary forests. Light saturated photosynthetic rates of tropical pioneers range from 8–14 μmol m^{-2} s^{-1}, with compensation points of 20–50 μmol m^{-2} d^{-1} and an initial slope of the light response curve of approximately 0.03 (e.g., Fetcher et al., 1987). In an open site with daily PFD of 20 mol m^{-2} d^{-1}, integration of light response parameters of this magnitude with the diurnal time course of PFD predicts diurnal carbon gain of roughly 0.3–0.5 mol m^{-2} d^{-1} (Ackerly, 1993). Assuming that biomass consists of 55% carbon by weight, and construction and maintenance cost of 1.5 g glucose g^{-1} biomass (Griffin, 1994), these rates correspond to net assimilation rates of 4 to 7 g m^{-2} d^{-1}. At the other extreme, Saldarriaga and Luxmoore (1991) measured PAR interception and productivity of secondary forests in the upper Rio Negro valley of Venezuela and estimated that solar energy conversion efficiencies peaked in 1 to 10-year-old stands at about 0.5 g MJ^{-1}. Incoming PAR was 2.75 GJ m^{-2} y^{-1}, giving a primary productivity of 3.76 g m^{-2} d^{-1}; with a leaf area index of 3 to 4, this corresponds to leaf level assimilation rates of 0.95 to 1.25 g m^{-2} d^{-1}. As expected, due to the additional losses of energy that must be considered at each level of organization, net assimilation rates decline from leaf to plant to stand levels.

The relationship between area-based assimilation rates and whole-plant growth rates depends on leaf area ratio (LAR), the ratio of leaf area to total plant biomass (RGR = LAR * NAR). LAR tends to be higher in pioneer than non-pioneer species, at a given light level, with typical values between 0.01 and 0.03 m^2 g^{-1}. Leaf area ratio almost always increases in response to lower light levels (e.g., King, 1994), and due to this compensatory response, relative growth rate is less sensitive than net assimilation rates to reductions in light levels (e.g., Popma & Bongers, 1988). LAR can be decomposed into two components; leaf weight ratio (LWR), the proportion of total biomass invested in leaves, and specific leaf area (SLA), the ratio of leaf area to leaf biomass (a measure of leaf 'thinness') (LAR = LWR * SLA). Changes in SLA at different light levels reflect significant alterations of leaf structure associated with sun/shade acclimation of photosynthetic responses (see Chazdon et al., Chapter 1). Mesic forest plants invariably exhibit increases in SLA at low light, which are often, but not always, paralleled by increases in LWR, leading to observed responses of LAR. Care must be taken in interpreting experimental data on allocational responses because most allocation parameters exhibit marked ontogenetic shifts with increasing plant size (Evans, 1972; Chazdon, 1986; Coleman, McConnaughay & Ackerly, 1994), and plants in different light treatments usually attain very different final biomass at the time of harvest. In one experiment using *Heliocarpus appendiculatus*, seedlings grown at 35 and 100% of full sun attained identical final biomass, allowing comparisons of allocation at similar ontogenetic stage. In this case, LWR was identical at the two light levels while SLA, and consequently LAR as well, was about 60% higher in the lower light level (Ackerly & Bazzaz, 1995a; cf. King, 1994). In general, SLA and LAR exhibit greater and more consistent responses to light gradients than does LWR.

21.2.2   Trade-offs between growth at high and low light levels

Relative growth rates and net assimilation rates of pioneer (and non-pioneer) tree seedlings decline markedly at lower light levels, and several experimental studies provide direct estimates of whole-plant light compensation points (Givnish, 1988) (indirect approaches to compensation point, e.g. King, 1994, are discussed below). In two experimental studies of relative growth rate across light gradients, whole-plant compensation points (i.e., RGR = 0) of five pioneer species ranged from 1 to 5% of full sun (Okali, 1972; Popma & Bongers, 1988). In the most thorough study of this type, Moad (1992) studied growth and survival of 550 naturally established saplings representing 18 species of dipterocarps in the Pasoh Forest Reserve, Malaysia, encompassing a range in life history from 'almost pioneers' (e.g., *Shorea leprosula, S. macroptera, S. acuminata*) to highly shade-tolerant species (e.g., *S. glaucescens, Vatica oblongifolia*). Within each species, relative growth rates for above ground biomass were significantly correlated with light availability (measured as minutes of sunflecks per day, estimated from hemispherical photographs), and the slope of this relationship was strongly and positively correlated with the x-intercept. In other words, species that were most responsive to increased light availability, in terms of enhanced growth rates, also had the highest light compensation points. Species with high light compensation points also had higher sapling mortality rates and shorter leaf life span. Similarly, in a study of 10 species of pioneers and non-pioneers from Los Tuxtlas, species with a higher RGR in full shade exhibited less enhancement of growth in large gap conditions (Popma & Bongers, 1988). These studies provide strong quantitative evidence of a trade-off between rapid growth rates at high light and the ability to establish and survive in low light sites, although there is little evidence for a similar trade-off between growth rates at high vs. low light.

For pioneers, this trade-off implies that species with the highest growth rates in large disturbances will also have larger minimum gap sizes. Brokaw (1987) documented this pattern in his comparative study of three pioneer species, *Trema micrantha, Cecropia insignis,* and *Miconia argentea,* on Barro Colorado Island, Panama. *Trema* has the most rapid height growth rates, reaching 12 m in two years, and after 5 years was the tallest tree in all gaps in which it successfully established; however, it was not observed in gaps less than 350 m². *Cecropia* grew to 4 m in two years, and was the dominant species primarily in gaps between 200 and 350 m², while *Miconia* was the slowest grower (2.5 m tall at 2 years old), but predominated in most of the small gaps between 75 and 200 m². These relationships will result in niche partitioning of the gradient of light availability, but maintenance of high species diversity by such a mechanism requires extremely tight niche packing along this environmental axis. It is not clear at this point whether pioneer and non-pioneer species differ qualitatively in their light response characteristics in terms of whole plant growth rates, or whether there is a continuum of response. In the latter case, these characteristics may provide critical information for quantitative discrimination of life history strategies, but they would not support the delimitation of tropical trees into discrete life history classes.

21.2.3 Allometry and height-diameter growth of saplings

As these studies emphasize, rapid height growth and efficient display of leaf area is extremely important for successful establishment and attainment of reproductive maturity in pioneer trees (see King, 1990). In *Cecropia obtusifolia*, for example, seedling survivorship is markedly higher for individuals that are taller and/or have more leaves, and hence greater access to light. For juvenile plants between 0.3 and 10 cm dbh, almost 70% of mortality observed over a one-year period was attributed to shading (Alvarez-Buylla & Martínez-Ramos, 1992, cf. Augspurger, 1984). Height-growth rates are commonly 2–3 m per year among tropical pioneers, with maximum records as high as 7 and 9 m yr^{-1} for *Trema micrantha* (Ewel, 1980; Brokaw, 1985) and 10 m yr^{-1} for *Alibizzia molucanna* (UNESCO, 1978; see Bazzaz, 1991). First-year height-growth rates are most rapid for the first cohort colonizing a gap, and decline for individuals emerging in subsequent years, again emphasizing the importance of early establishment, rapid growth, and canopy dominance for successful regeneration in gaps (Brokaw, 1985).

The relationship between the amount of biomass invested in support tissue and the height and crown diameter of a sapling depends on the proportion of biomass in trunk, vs. branches and petioles, the diameter and taper of the stem, which determine stem volume, and wood density. Some of the large-leaved pioneers, such as *Cecropia* species, do not branch until heights of 5–10 m, maximizing investment in the trunk. *Trema*, in contrast, which has one of the highest measured height growth rates, has small leaves and starts branching very early (see section 21.3.1 on canopy architecture). Shukla and Rama-krishnan (1984a) compared allocation strategies of two early and two late successional tree species in India. Over the first seven years of growth, the early successionals invested less in root biomass and more in the bole (primary shoot), but there were not consistent differences in proportional branch biomass. Allometry of crown width vs. sapling height has been investigated for understory species vs. saplings of canopy trees (Kohyama, 1987; King, 1990), but pioneers have not been included in these studies. For the trunk itself, *Cecropia* has a larger diameter relative to its height than the saplings of understory and non-pioneer species (cf. allometric equations in Alvarez-Buylla & Martínez-Ramos, 1992 and King, 1990), and so has a larger stem volume. Whether this is true for pioneers in general is not clear.

King (1994) has presented novel analysis of height growth utilizing several new parameters related to growth analysis. The instantaneous rate of height growth (dh/dt) is considered as the product of growth rate, G (above ground biomass gain per unit leaf area per unit time, closely analogous to net assimilation rate) and height growth efficiency (HE, the amount of height extension achieved per unit of new biomass). HE is calculated as LA(dh/dM), the product of sapling leaf area and the rate of height growth per unit change in biomass, and has units of $m^3 \, g^{-1}$. This analysis was applied to sapling height growth rates of 10 tree species on Barro Colorado Island, including three pioneers, *Cecropia insignis*, *Miconia argentea*, and *Palicourea guianensis*. Within species, increasing height growth at higher light levels was due primarily to increases in growth rate, G, and to a lesser extent to increases in height growth efficiency, HE. *Cecropia* had the highest height growth rate of the species studied, which also was due primarily to more rapid biomass gain; its height growth efficiency was high, but not outside the range observed for other species.

The most important factor facilitating rapid height growth per unit biomass invested in the stem is the very low stem and wood density of tropical pioneers (Wiemann & Williamson, 1989a; King, 1991). Several of these species, such as *Ochroma pyramidale* (balsa), are sources of the world's lightest woods (Ewel, 1980). Tropical pioneer trees are also characterized by extreme radial gradients in specific gravity, with much lighter wood at the pith than immediately under the cambium at the outside of the trunk (Wiemann & Williamson, 1988, 1989a). In the case of one *Ceiba pentandra* tree with a diameter of 102 cm, specific gravity changed from $0.09 \, g \, cm^{-3}$ at the pith to $0.33 \, g \, cm^{-3}$ in the outer 3 cm of wood. Stems of *Cecropia* and some *Macaranga* species have an even more extreme gradient, as their internodes are hollow. These gradients are biomechanically sensible, as a hollow cylinder is stronger than a solid structure made with the same amount of material (Niklas, 1992). The gradients are much less pronounced in non-pioneer tropical trees, tropical montane and dry forest trees, and temperate trees (Wiemann & Williamson, 1989a,b). These authors suggested that these gradients reflect more intense selection in rainforest pioneers, compared to temperature species, for rapid height growth in seedlings and young saplings (when the 'old' wood of the future tree is being produced), followed by the necessity for increased biomechanical stability later in life. This suggestion is supported by the observation in *Ochroma pyramidale* that the wood produced at a given time in the life of the tree, at all heights along the trunk, is of similar specific gravity (Rueda & Williamson, 1992). It is also conceivable that the intensity of competition is similar in temperate and tropical successional environments, but that the physiological requirements of overwintering preclude the very low specific gravity observed in the tropics.

From a developmental and morphological perspective, height growth can be viewed as the product of internode length and leaf (= node) production rates on the central axis of the plant. In his study of height growth, King (1994) found that in pioneer and non-pioneer species both internode length and leaf production rate increased with increasing light levels (see below for additional discussion on leaf dynamics). *Cecropia insignis* had the most rapid rate of leaf production and longer than average internodes (among species), resulting in the most rapid height-growth rates. The other two pioneers, *Miconia argentea* and *Palicourea guianensis*, had long internodes, but did not have more rapid leaf production rates than non-pioneers.

21.2.4 Reproductive allocation

Pioneer trees, like other colonizing organisms, initiate reproduction at an early age and most species produce large numbers of very small seeds (Bazzaz, 1979, 1984). In general, direct estimates of reproductive allocation are extremely difficult to obtain for trees. It is possible to obtain an indirect estimate for *Cecropia obtusifolia*, a dioecious species, based on the allometric and demographic study of Alvarez-Buylla and Martínez-Ramos (1992). The total investment in seeds for a female tree can be estimated from the allometric relationships between height, diameter, and fecundity (Figure 21.1) and the mean seed weight of 0.0009 g (Ibarra-Manriquez & Oyama, 1992). For a 20 m tall tree, seed production is estimated at 654 g per year, and total leaf area, which also scales with height, is approximately 28 m^2. Assuming a net assimilation rate of 1 g m^{-2} d^{-1} (measured on seedlings [Popma & Bongers, 1988]), annual carbon gain for this plant would be 10233 g, resulting in a reproductive allocation of 6.4%. Similar calculations generate the curve shown in Fig. 21.1e, predicting that reproductive allocation increases with plant size (and hence age), reaching almost 20% in the largest trees. Male trees produce more than three times as many flower buds as females, at an equivalent size, but the biomass investment in male function, and the investment in pollinator and disperser rewards in both males and females, have not been estimated.

21.3 CANOPY STRUCTURE AND DYNAMICS: INTEGRATION OF PHYSIOLOGICAL PROCESSES

I now turn from this overview of the life cycle to a detailed consideration of canopy structure and the dynamics of leaf populations, with particular focus on leaf life span. The objective of this portion of the chapter is to address the coupling between morphological structure and physiological processes of the canopy. The interaction between structure and function is complex. When we look at a plant, we see a static snapshot, a moment in the ontogeny; the structure of the canopy reflects the cumulative outcome of growth processes up to that moment and is also the context for the continued unfolding of those processes. The distribution of leaves within the canopy reflects variation in rates of leaf production, coupled with patterns of leaf abscission; these processes in turn are strongly influenced by patterns of light availability, while light availability is in part a product of the pattern of leaf distribution, and so on. These interrelated components have been addressed from distinct viewpoints, ranging from the strongly morphological and developmental studies of plant architecture by Hallé and colleagues (Hallé, Oldeman & Tomlinson, 1978; Oldeman & van Dijk, 1991) and the economic models of plant form and function (Givnish, 1986) to the empirical and quantitative studies of leaf physiology and life span (e.g., Williams, Field & Mooney, 1989; Reich, Walters & Ellsworth, 1992). Here I review a range of such research, with an emphasis on pioneer trees, focusing on the interactions between structure and morphology on the one hand, and leaf physiology and dynamics on the other. Herbert (Chapter 5) provides a detailed theoretical discussion of plant geometry in relation to photosynthesis. A broad synthesis of pattern and process in plant canopies remains one of the outstanding challenges of plant ecophysiology.

21.3.1 Canopy architecture, leaf display and light interception

Horn (1971) proposed that a tree should add, or maintain, additional layers of leaves below the upper canopy as long as those leaves are above their light compensation point. In high-light environments, more light will penetrate through upper leaf layers, leading to the prediction that early successional trees should be multilayered and late successional and understory species monolayered, predictions that are generally supported in temperate successional environments. Ashton (1978) and others have subsequently noted that canopy structure of many tropical pioneers does not follow these predictions. The 'classic' pioneer species of all three continents, *Cecropia*, *Musanga*, and *Macaranga*, have sparse branching systems with small clusters of large leaves at the tips, forming an umbrella-shaped, monolayered canopy. Orians (1982) suggested that the combination of intense competition for light and rapid height growth in tropical forest gaps favors the vertically-oriented monolayer because only the upper leaves will be energetically profitable. Ashton (1978) argued that these large-leaved species with minimal leaf overlap have extremely high growth rates, but are also more drought sensitive, and observed that they are much less common in windy locations, possibly explaining their restriction to the lowland tropics. Another common generalization is that plants of high-light environments tend to have orthotropic, vertically-oriented shoots with leaves arranged in spiral phyllotaxy, while shade plants have plagiotropic shoots with distichous phyllotaxy (Leigh, 1975; Hallé, Oldeman & Tomlinson, 1978). Orthotropic shoots promote rapid height growth and minimize the support costs per unit leaf mass, and some degree of self-shading among leaves may minimize heat load without reducing light levels below the photosynthetic saturation point (Givnish, 1984).

Examination of the pioneer tree flora reveals an additional pattern that has received little attention in these discussions. The 'classic' tropical pioneer architecture described above is not observed in all pioneer species. A number of taxa, such as the genus *Trema* which is also found throughout the tropics, have a strong, orthotropic central axis that produces a large number of plagiotropic lateral branches bearing many small leaves. The 'architectural spectrum' of secondary forests, based on the distribution of architectural models, illustrates the importance of these two contrasting types of canopy structure, the former (e.g., *Cecropia*) belonging to Rauh's model and the latter to Roux's model (Hallé, Oldeman & Tomlinson, 1978). For example, in secondary vegetation of French Guiana, Rauh's model accounted for 35% of the trees (belonging to 2 *Cecropia* sp. and *Loreya mespiloides*), while Roux's model was observed in 28% of the trees (e.g., *Goupia glabra* and *Xylopia nitida*) (de Foresta, 1983). Scaronne's, Leeuwenberg's, Massart's, and Attim's models also recur in species of secondary vegetation in various locations (de Foresta, 1983).

This spectrum of architectural types is apparent in Ashton's (1978) leaf size data from vegetation plots at the Pasoh Forest Reserve, Malaysia, which indicate that the pioneer tree community is characterized by greater variation in leaf size than the primary forest species (CV = 102 vs. 66%, respectively) (calculated from Figure 25.5 in Ashton, 1978). In addition, there is a negative correlation among species between average leaf size and the proportion of nodes on the trunk that produced branches, and a positive correlation between leaf size and twig diameter, illustrating 'Corner's rules' (Hallé, Oldeman & Tomlinson, 1978, p. 83). These studies indicate that no one architectural pattern can be considered typical or optimal for pioneer species. Pioneer species span the entire range of these associations, from large-leaved, sparsely branched taxa with thick twigs, to those with small leaves, densely-branched canopies and thin twigs.

Ackerly and Bazzaz (1995b; Ackerly, 1993) examined the efficiency of light interception in four species representing these contrasting architectural patterns; *Trema micrantha, Heliocarpus appendiculatus, Piper auritum*, and *Cecropia obtusifolia*. A total of 94 seedlings were transplanted into natural gaps at Los Tuxtlas (Veracruz, Mexico) and after a period of adjustment to the gap environment, detailed quantitative measures of crown architecture were obtained in order to reconstruct leaf display using a computer model (YPLANT, R. Pearcy, unpublished data). Based on these reconstructions, combined with a hemispherical photograph above each seedling, whole-plant light interception efficiency was estimated, integrating over all sky directions. This integrated three-dimensional estimate was tightly correlated with the vertical projection efficiency of leaf area (the ratio of projected to total leaf area) (Figure 21.2a illustrates results for the two 'extremes,' *Cecropia* and *Trema*), although Niklas (1988) has demonstrated that these two efficiency measures can be decoupled in plants that receive a great deal of lateral light.

The efficiency of vertical projection can be considered as the product of two components, angular efficiency (the total vertical projection of individual leaves divided by total leaf area, which is reduced by non-horizontal leaf angles) and exposure efficiency (the vertical projection of the seedling divided by the summed projection of individual leaves, inversely related to the extent of self-shading) (Chazdon, 1985). The relative contribution of reductions in the two components varied among species (Figure 21.2b). In *Trema*, with many small leaves displayed on a central axis and a few branches, leaves were held close to horizontal but there was considerable leaf overlap and self-shading. *Cecropia*, in contrast, maintained few leaves on its single main axis, and had little self-shading, but the leaves were held at lower angles, on average, reducing displayed leaf area. Despite these differences, the two species exhibited comparable reductions in overall efficiency of light interception with increasing total leaf area (Figure 21.2c), emphasizing the potential for functional convergence among plants that are architecturally quite distinct. Is the persistence of these species the result of ecological convergence and competitive equivalence despite their divergent morphology, or do these architectural

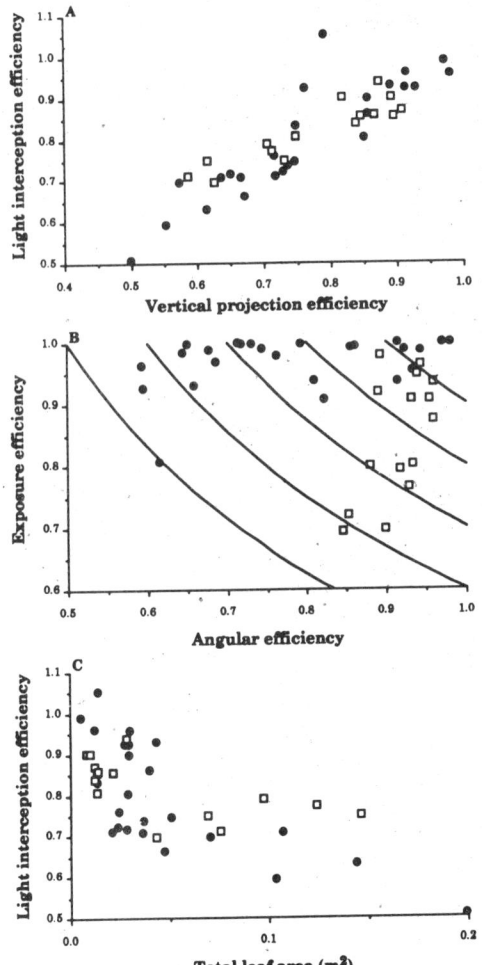

Figure 21.2. a) Whole-plant light interception, integrated over all sky directions, was estimated for seedlings of Cecropia obtusifolia and Trema micrantha from detailed three-dimensional computer reconstructions of leaf display coupled with analysis of hemispherical photographs. Light interception efficiency was strongly correlated with the vertical projection efficiency of the crown (the ratio of vertically projected area to total leaf area). b) Vertical projection efficiency can be broken into two components of leaf display: angular efficiency, which is reduced by non-horizontal leaf angles, and exposure efficiency, which is reduced by leaf overlap (see Chazdon, 1985). The reduction in light interception efficiency in larger plants of Trema was due primarily to leaf overlap and reduced exposure efficiency, while the reduction in Cecropia was due to non-horizontal leaf angles. c) Estimated light interception efficiency (the ratio of total light interception to light interception by a flat surface of the same total leaf area) declined with plant size in both species.

differences reflect the outcome of selection operating on subtle variations in life history among species of pioneers? The answer to this question awaits a broader understanding of evolutionary and ecological processes in tropical forests.

21.3.2 Leaf production and life span in pioneer trees

In temperate regions, leaf production is necessarily constrained to the favorable growing season. Consequently, leaf habit (deciduous vs. evergreen) and leaf life span correspond fairly closely and may be treated synonymously: if individual leaf life span is under one year, the tree is deciduous, and if life span exceeds one year the tree is evergreen (Chabot & Hicks, 1982). In the tropics, leaf habit and leaf life span are decoupled, as leaves can be produced throughout the year and an evergreen canopy may be maintained with either short or long-lived leaves (Kikuzawa, 1989). To recognize these patterns, it is useful to replace the conventional classification of deciduous/evergreen with a tripartite distinction between 1) evergreen and evergrowing, 2) evergreen with periodic growth, and 3) deciduous (Shukla & Ramakrishnan, 1982; cf. Koriba, 1958). For evergrowing species, evergreenness is maintained by continuous leaf production, regardless of leaf life span. In species with periodic growth, the canopy will be deciduous if leaf life span is shorter than the interval between leaf flushes and evergreen if leaf life span is longer than this interval. This distinction is critical in the study of tropical plant leafing phenology because theoretical studies of leaf habit in relation to seasonality do not address the variation in life span among evergreen species of continuously favorable environments.

Continuous leaf production throughout the year is observed in wet forests where the dry season is short and not too severe, and among canopy trees is most commonly observed in pioneer species (Wright, Chapter 15). For example, in La Selva (Costa Rica, 10° N), continuous leafing at the population level was observed in 26 of a sample of 93 overstory species (Frankie, Baker & Opler, 1974), and many of the 26 can be considered pioneers (D. A. Clark, personal communication). In a survey of 46 tree species on Barro Colorado Island (Panama, 9° N), the CV of leaf production rates throughout the year (a measure of periodicity) was significantly lower for pioneer than non-pioneer species (Coley, 1983). Leaf life span was under 9 months for 19 of 20 pioneer species, and over 10 months for 18 of 21 non-pioneers (Coley, 1988). In Los Tuxtlas (Veracruz, Mexico, 18° N), monitoring of individual shoots of *Heliocarpus appendiculatus* confirmed that individual apices produce leaves throughout the year, and there is no evidence of changes in leaf production rates or life span during the two month dry season (Figure 21.3a; D. Ackerly, unpublished observation). Thus, continuous leafing in pioneer trees is observed at all hierarchical levels (Newstrom et al., 1994), from the individual shoot and the tree to the population and community. In seasonal tropical forests, most tree species are deciduous, but among overstory species, those with continuous leafing are invariably pioneers (Frankie, Baker & Opler, 1974; Shukla & Ramakrishnan, 1982), and when they are not ever-growing, they are still characterized by more extended periods of leaf production and shorter leaf life span than co-occurring late successional species (e.g., Boojh & Ramakrishnan, 1982; Shukla & Rama-krishnan, 1984b; Lowman, 1992).

Cambial activity and stem growth tend to exhibit the same pattern (i.e., periodic or evergrowing) as leaf production. In a study of 12 tree species at La Selva, three exhibited continuous cambial activity (*Hampea appendiculata, Laetia procera,* and *Virola sebifera*), all of which have continuous leaf production. Species with intermittent leaf production (including *Goethalsia meiantha*) also had periodic stem growth, and the periods of stem dormancy usually coincide with periods of leaf flush (Breitsprecher & Bethel, 1990, cf. with data in Frankie, Baker & Opler, 1974).

21.3.3 Leaf life span and leaf numbers on shoots of evergrowing plants

In order to understand the mechanisms underlying the structure and dynamics of leaf populations, it is useful to consider two components (Ackerly, 1993): 1) the number and dynamics of actively growing shoots in the tree crown, and 2) the dynamics of each leaf cluster, the leaves on an individual shoot (Fisher, 1986). The dynamics of branching and of the numbers of active apical meristems are not well studied in tropical plants (cf. Maillette, 1982). In *Cecropia obtusifolia*, branching starts relatively late in life, when plants reach about 10 m in height, and the number of active apices increases exponentially to as many as 100 at maximum height (Alvarez-Buylla & Martínez-Ramos, 1992). However, the number of branches that have died during this period, and therefore the total number produced, has not been measured. Branches may be quite transient; for example, the small pioneer

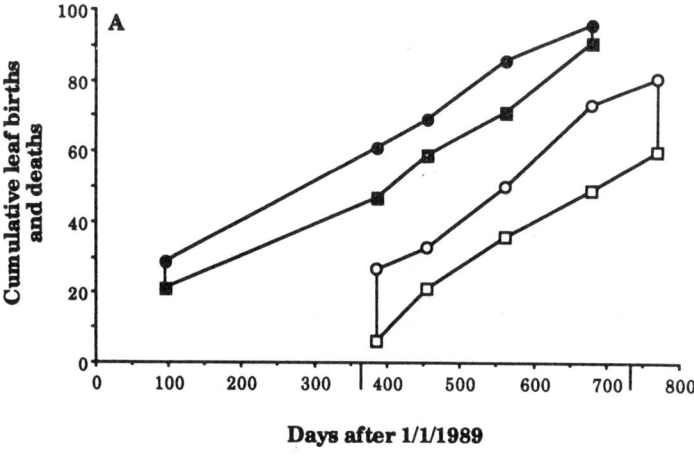

Days after 1/1/1989

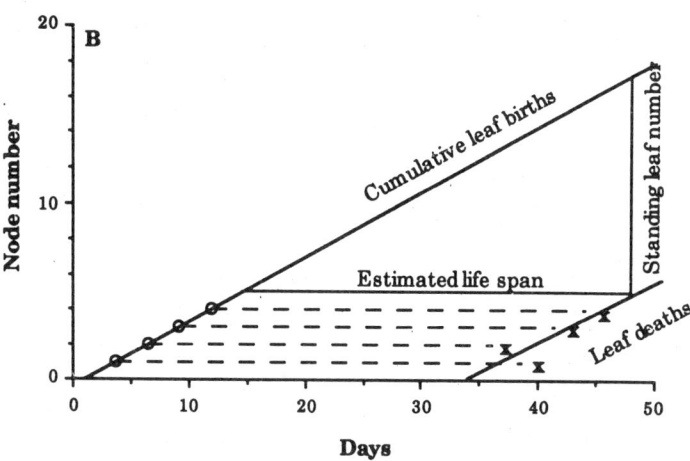

Days

Figure 21.3.   a) In evergrowing, evergreen plants, such as most tropical pioneers, leaf birth and death take place throughout the year, as illustrated for these two canopy level shoots of Heliocarpus appendiculatus, *monitored for over a year each (D. Ackerly, unpublished data). Leaf life span on these shoots is about three months, meaning that the entire leaf population of this tree is replaced four times per year. b) When the number of leaves on an individual shoot is more or less constant (i.e., leaf birth and death rates are equal), the ratio of leaf number to life span will equal leaf birth rate. This relationship provides a simple coupling between canopy structure and dynamics and is the basis for models of the carbon economy of individual shoots discussed in the text.*

tree *Piper auritum* produces plagiotropic lateral branches that form an umbrella-shaped canopy. Each branch bears up to 10 or 12 leaves, and then leaf initiation stops, older leaves senesce and abscise, and soon the entire branch is dropped. The interval between production of successive branches (which is the same as the interval between production of main stem leaves) is approximately one month, and a mature plant may have 8 to 12 branches, suggesting that individual branches survive about one year (D. Ackerly, unpublished observation). However, in plants with higher-order branching, the life span of individual apical meristems and of the branch segments and branches they produce (i.e., modules, *sensu* Tomlinson, 1978) will be highly variable as some are shed quickly and others persist to form the skeleton of the mature canopy. These processes occur over long time scales and have received little attention.

Within each leaf cluster, the number of leaves is regulated by the relative rates and seasonal patterns of leaf birth, at the apex, and death (Kikuzawa, 1983, 1984). In tropical, evergrowing plants, including pioneer trees and understory rosette plants such as palms, cycads, and tree ferns, leaf birth and death rates are often comparable such that the number of leaves in a cluster is more or less constant. For example, in *Cecropia obtusifolia*, leaf number increases only slightly before branching begins, from about 5 to 10 in plants from 0.3 to 10 m tall. Subsequently, there is a very strong and linear correlation between the total number of branches and the total number of leaves. The intercept of this relationship is near zero, and the slope is 8.35, which is therefore the mean leaf number per branch tip (Alvarez-Buylla & Martínez-Ramos, 1992; cf. Koike, 1986).

For individual shoots with continuous leaf production and loss, there is a very simple mathematical relationship between leaf number, leaf birth rate, and leaf lifespan, analogous to the relationship between standing biomass, productivity, and residence time, respectively, in ecosystems. This relationship is clearly illustrated by plotting the cumulative number of leaf births and deaths against time for an individual shoot (Figure 21.3). The vertical distance between these lines is the number of leaves present at a particular moment in time, while the horizontal distance is the life span of a particular leaf. Consequently, assuming that leaf birth and death rates are equal, the leaf birth rate (r) can be calculated as the ratio of leaf number (N) to lifespan (L):

$$r = N/L$$

$$\text{or} \qquad\qquad (1)$$

$$L = N * P$$

where P is the plastochron interval, the time interval between initiation of successive leaves ($= 1/r$) (Corner, 1966; Chazdon, 1986; Ackerly & Bazzaz, 1995a; King, 1994; also see Gill & Tomlinson, 1971). Due to the simple arithmetic relationship among these parameters, the values of any two will determine the third; consequently, variation in all three parameters is constrained to only two independent dimensions. This relationship can be graphically expressed by plotting values of standing leaf number on the vertical axis in relation to leaf life span on the horizontal axis, as shown in Figure 21.4 for a variety of evergrowing tropical species (Figure 21.4, see Ackerly, 1993 for data sources). Each point on this plot represents a particular value of leaf birth rate as well, depending on the ratio of N to L. As a result, there is a series of diagonal lines along which leaf birth rates are constant; these lines diverge from the origin if the two axes are linear and are parallel if the axes are log-transformed. In this sample, comprised primarily of palms and pioneer trees, leaf life span varies over a 160-fold range, from 32 to 5200 days, leaf production rates vary by 87-fold, from 0.7 to 122 leaves yr^{-1}, while leaf number only varies by 15-fold, from 3 to 45. The short life span and rapid leaf production of pioneer trees are immediately apparent, compared to the longer life spans of the palms and understory plants, but the ranges of leaf numbers (per shoot) are similar. How can these patterns of variation, within and between groups, be explained in relation to canopy structure and leaf physiology? The next three sections explore the answers to this question.

21.3.4 Leaf life span and physiological processes

The relationships between leaf life span and physiological processes were first addressed in studies of carbon gain (Mooney & Gulmon, 1982) and leaf habit (Chabot & Hicks, 1982). In recent years, these studies have been extended by comparative studies of tropical trees (Coley, 1983, 1988; Williams, Field & Mooney, 1989; Reich et al., 1991, 1994; Reich, Alsworth & Uhl, 1994) and synthetic review of patterns across numerous vegetation types (Reich, Walters & Ellsworth, 1992; Reich, 1993). Leaf life span is associated with a broad spectrum of physiological parameters, in particular those associated with rates of carbon gain (Table 21.1). I will first address some theoretical considerations underlying these relationships among life span, carbon assimilation rates, and in section 21.3.5, self-shading patterns. An understanding of these relationships will then provide the context for a

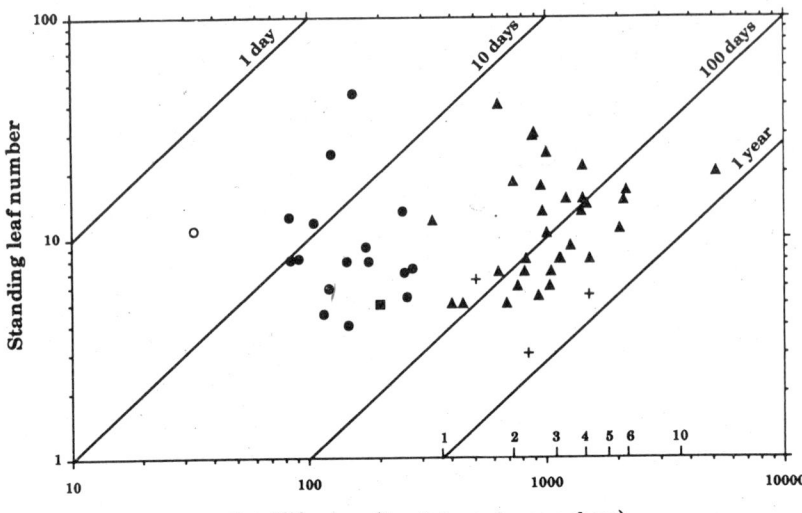

Leaf life span (days below axis, years above)

Figure 21.4.   The evergrowing, evergreen leaf habit is observed in a number of tropical plants, including pioneer trees (circles), rosette plants such as palms (triangles), tree ferns, and cycads (crosses), and mangroves (square). (The open circle, with the shortest leaf life span, was measured in a greenhouse.) Given the simple relationship between leaf number, life span, and birth rate (Figure 21.3), variation in these three parameters can be plotted in two dimensions (Ackerly, 1993). The diagonal lines on this plot represent isoclines of constant plastochron interval (=1/birth rate), the time interval between production of successive leaves. Pioneer trees have short leaf life span and rapid leaf initiation rates, while palms have longer life span and slower initiation rates. The number of leaves maintained on individual shoots (not on the entire tree) is similar in the two species. Details ard sources of data are available in Ackerly (1993) or directly from the author.

review of economic models of optimal leaf life span and canopy structure (section 21.3.6).

The relationship between assimilation rates and leaf life span is illuminated by an examination of the carbon economy of the individual leaf cluster (Ackerly, 1993). Above, leaf birth rate was defined as the ratio of leaf number to life span for a leaf cluster with equal birth and death rates. An alternative formulation of leaf birth rate can be derived from rates of carbon assimilation and allocation. The total daily assimilation of carbon by a leaf cluster (A_T, g carbohydrate d^{-1}) is given by:

$$A_T = NmA_m \qquad (2)$$

where N = leaf number, m = leaf mass (g), and A_m = mean assimila-

Table 21.1. Physiological and structural traits correlated with short leaf life span among temperate and tropical trees (from Chabot & Hicks, 1982; Mooney & Gulmon, 1982; Coley, 1983, 1988; Williams et al., 1989; Reich et al., 1991, 1992, 1994; Moad, 1992).

Traits that exhibit higher values in species with short-lived leaves

Light saturated photosynthetic rates per unit leaf mass
Light saturated photosynthetic rates per unit leaf area
Conductance
Leaf nitrogen content per unit leaf mass
Nitrogen resorption efficiency
Leaf tissue construction cost (in rainforest plants only?)
Density of leaf hairs
Rates of herbivory
Specific leaf area (i.e., thinner leaves)
Seedling relative growth rates
Seedling leaf area ratio
Sapling height growth rates
Whole plant light compensation point
Sensitivity of relative growth rate to increased light levels

Traits that exhibit lower values in short-lived leaves

Leaf tannin, lignin, phenol, and fiber content (per unit mass)
Leaf toughness

Variables not correlated with leaf life span

Leaf nitrogen concentration per unit leaf area

tion rate on a mass basis, including maintenance respiration (g carbohydrate g^{-1} leaf d^{-1}). If a proportion F of available carbon is allocated to new leaf production at the apex of the shoot, and $(1-F)$ is exported from the shoot to the rest of the plant, then $F \cdot A_T$ represents the amount of carbon available for leaf production, assuming physiological autonomy of the shoot. Given a leaf construction cost (c, g carbohydrate g^{-1} leaf) that incorporates the material and respiration costs of leaf tissue as well as marginal costs of support (petiole construction and stem thickening, including respiration costs), the rate of leaf production will be the ratio of available carbon to the product of leaf mass and unit cost. Combining these terms, along with Equations (1) and (2) above gives:

$$r = \frac{N}{L} = \frac{NmA_mF}{mc} \tag{3}$$

which simplifies to the following expression for leaf life span:

$$L = \frac{c}{FA_m} \tag{4}$$

It is important to note that this is not a prediction of optimal behavior; rather, it is a descriptive equation for the behavior of a single leaf cluster that follows from the assumption of equal leaf birth and death rates. This assumption does not imply that the plant is not growing, as the assimilate exported from the cluster can be used to produce new branches, and as argued above, increased plant size depends more on increasing the number of leaf clusters than on increases in the number of leaves per cluster (in plants with only a single apical meristem, e.g., palms and early growth stages of species such as *Cecropia*, increasing individual leaf size is also very important). How does this equation relate to observed relationships with leaf life span (Table 21.1)? In an empirical study of rainforest plants in the genus *Piper*, Williams, Field and Mooney (1989) found a strong positive correlation between life span and the ratio of c to A_m, which they considered as a measure of the time required for payback of leaf construction costs. Which of the three parameters is most important in generating these correlations? In the *Piper* species, leaf construction cost (c) was higher in species with short-lived leaves, opposite to the direction predicted above, but the variation among species and environments was very small (Fredeen & Field, Chapter 20). In contrast, construction costs tend to be higher for long-lived leaves of evergreen species in more seasonal environments (Mooney & Gulmon, 1982; Sobrado, 1991), but recall that the equation above has been developed for evergrowing, evergreen plants. Allocation to leaf production (F), at the whole plant level and presumably within the individual cluster, increases at low light, and is usually higher in non-pioneer than pioneer species (King, 1994), both of which would predict shorter leaf life span in these cases, whereas the opposite is the case. Consequently, the basis for the relationship above is the very strong negative correlation between leaf life span (L) and assimilation per unit leaf mass (A_m), and the empirical pattern across a wide range of species approximates an inverse relationship, as predicted (Reich, Walters & Ellsworth, 1992).

Closer analysis of Equation (2) also emphasizes the relationship between leaf life span and relative growth rate. Assume first that the proportional allocation of standing biomass to leaves, LWR, is similar to the dynamic allocation fraction, F, an approximation that will be best for young plants before onset of leaf mortality. Second, replace A_m, the mass-based assimilation rates, by the product of specific leaf area (m^2 g^{-1}) and area-based assimilation rates (A_a, g carbohydrate m^{2} d^{-1}), and then invert the relationship, giving:

$$\frac{1}{L} = (LWR \cdot SLA) \frac{A_a}{c} \qquad (5)$$

The first two terms here comprise leaf area ratio, while the last term, area-based assimilation divided by the cost of constructing new biomass (recalling that c was defined as leaf construction cost, *including* costs of support biomass), approximates net assimilation rate, the rate of biomass gain per unit leaf area. Thus, the inverse of leaf life span closely approximates the formula for relative growth rate (RGR = LAR * NAR). An analogous relationship is apparent from the inverse of Equation (1): $1/L = r/N$, i.e., the inverse of leaf life span equals the rate of new leaf production per leaf currently on the plant, a measure of relative production rate. Empirically, leaf life span exhibits a negative correlation with seedling relative growth rate, approximating an inverse relationship (Reich, Walters & Ellsworth, 1992).

King (1994) independently arrived at the same relationship with respect to the problem of whole-plant light compensation points. A plant at its growth compensation point will gain just enough carbon to maintain its leaf area, so that the rates of production and loss of leaf area will be equal, which is equivalent to the assumption of constant leaf numbers above. Based on separate equations for the rate of leaf area production and leaf area loss, the compensation growth rate (G_a, plant biomass gain per unit leaf area) at which production and loss are equal is:

$$G_a = \frac{LMA}{F \cdot L} \tag{6}$$

where LMA is leaf mass per area and other symbols are as above. Converting G to mass-based growth rate (G_m) and rearranging gives:

$$L = \frac{1}{F \cdot G_m} \tag{7}$$

Thus, the growth rate necessary to maintain plant leaf area is inversely related to leaf life span, and since G is strongly correlated with light environment, the light compensation point will also be negatively correlated with life span. From this analysis, and a year-long study of sapling growth, King determined that the compensation point for three pioneer species ranged from 3.7% to 5.2% of maximum diffuse radiation, while seven non-pioneers ranged from 0.9 to 3.4%, values that are comparable to those measured directly by growth analysis (see section 21.2.2).

The relationships between carbon gain, growth, and leaf life span suggest some of the mechanisms that may underlie patterns of intraspecific variation and plasticity in leaf dynamics. These patterns can be examined by plotting the responses to environmental conditions as vectors on the plot of leaf number, birth rate, and life span (Figure 21.5). There are six qualitatively distinct axes of variation on this plot (and 12 different directions of change) along which at least two of the three parameters vary (Figure 21.5a). Leaf life span in pioneer trees generally declines in response to increased resource availability and growth rates. However, this decline may be associated with increased or decreased leaf numbers and leaf birth rates, depending on the proportional changes in each factor (Figure 21.5b). For example, in seedlings of *Heliocarpus appendiculatus*, leaf number and birth rate increased, and life span declined, at high nutrients (Ackerly & Bazzaz, 1995a), and similar patterns have been reported in response to high light in *Cecropia obtusifolia* (Strauss-Debenedetti, 1989) and *Psychotria simiarum* (Bongers & Popma, 1990). In other cases, increased light availability causes a decline in L and an increase in r, with no change in leaf numbers (*Omphalaea oleifera* and *Myriocarpa longipes*, Bongers & Popma, 1990; *Astrocaryum mexicanum* in gap vs. understory sites, Piñero, Sarukhán & González, 1977; Piñero et al., 1986; Martínez-Ramos, Sarukhán & Piñero, 1988). And finally, in seedlings of *H. appendiculatus* grown at two light levels, L and N declined in parallel in high light, such that r remained constant (Figure 21.5b). Removal of leaves by herbivores would also cause simultaneous decreases in N and L, as demonstrated in an artificial defoliation experiment with *H. appendiculatus* (Núñez-Farfán & Dirzo, 1991 and unpublished data), and may depress leaf production rates (Marquis, 1992), leading to simultaneous decline in N, L, and r. The decline in leaf life span in response to high resource availability is consistent with the relationships with leaf physiology and growth, as outlined above. But what explains the variation, within and among species, in the number of leaves maintained on the shoot? In general, this question has attracted much less attention than the factors regulating leaf life span, and is the principal focus of the next two sections.

21.3.5   Canopy structure, self-shading, and leaf physiology

The discussion of carbon gain and leaf dynamics above is based on mean assimilation rates for the leaves in a cluster, or on a plant, as if all leaves were identical. It is critical to an understanding of canopy

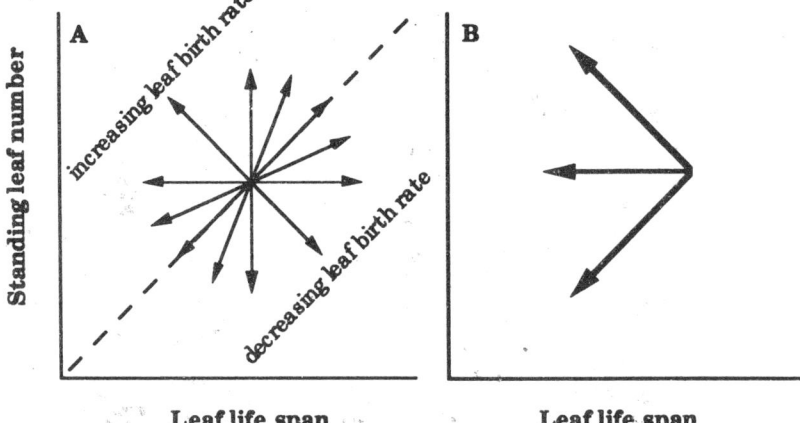

Figure 21.5. Vectors of change in leaf number, life span, and leaf initiation rates in response to environmental conditions. a) There are six possible axes of change along which two or three of these parameters change. Shifts to the left reflect a decline in leaf life span and downward indicates a reduction in leaf numbers. The dotted line represents a constant leaf birth rate, such that vectors crossing this line to the left and up indicate an increase in birth rates. b) In many species, leaf life span declines in response to increasing resource availability and higher plant growth rates; leaf numbers may increase, decrease, or remain constant, and leaf birth rate may increase or remain constant (see details in text).

structure and dynamics to examine the variation among leaves in microenvironmental conditions and physiological activity. The complex, three-dimensional structure of tree canopies generates considerable heterogeneity in environmental conditions among leaves on a single plant, due to self-shading and shading by neighbors. These shading patterns are an important factor regulating canopy structure, as the pattern of light availability influences carbon gain, leaf production, and abscission, which in turn influence the dynamics of the canopy and subsequent patterns of light availability among leaves. The morphological basis of canopy architecture and leaf display was discussed above; in this section, I examine canopy structure from the perspective of shading patterns among leaves and the consequences for leaf physiology and canopy function.

The distribution of light availability among leaves depends on plant growth form (Herbert, Chapter 5). On orthotropic shoots, considerable self-shading is expected among leaves on the shoot, while on plagiotropic branches, leaves on a single branch may not shade each other, but will cast shade on branches below. In a field study of 1–2 m-tall saplings of five pioneer species in Los Tuxtlas, the average ratio of highest to lowest daily PFD measured on individual leaves within a single plant was 3.67, with a range from 1.4 to 9 (Traw & Ackerly, 1995; cf. Oberbauer et al., 1988). In three of the pioneers, *Cecropia obtusifolia, Heliocarpus appendiculatus* and *Ficus insipida*, there was a significant exponential decline in daily PFD with leaf position (counting from the apex), while in the other two, *Urera caracasana* and *Piper auritum*, the variation was not significantly correlated with leaf position. In *Heliocarpus*, light levels on leaves illuminated by a direct light source above the plant peaked at the fourth leaf from the apex, where leaf angles were closest to horizontal, but did not decline markedly until leaves 8 and 9, where the 3/8 phyllotactic spiral of this species completes its first full cycle (Ackerly & Bazzaz, 1995a); this effect of phyllotaxy was not apparent under diffuse light or under field conditions (Traw & Ackerly, 1995).

In a study of five species of *Piper*, Chazdon, Williams and Field (1988) distinguished the effect of self-shading from effects of spatial heterogeneity in the ambient light environment by comparing variability in PFD among leaves with variability in spatial arrays of sensors placed above the crown. These data demonstrated that the overall level of variability among leaves is higher in understory plants, but that self-shading may be more important as a source of heterogeneity in open sites where the ambient environment is less variable on small spatial scales. Two species, *Piper auritum*, an open site pioneer, and *P. lapathifolium*, of the understory, exhibited little variation due to canopy structure and self-shading; both of these have horizontal leaves and little leaf overlap, illustrating convergence on monolayer leaf display in plants of open and closed conditions (cf. Orians, 1982).

The process of leaf senescence, with declining leaf nitrogen concentrations and photosynthetic capacity, parallels patterns of declining light availability with increasing leaf age and overtopping by younger leaves. Field (1983) argued that the gradients of nitrogen availability from young to old leaves enhance carbon gain at the canopy level by maximizing nitrogen use efficiency (NUE) averaged over low and high light positions. NUE will always be highest if all leaves are exposed to high light, but when this is not the case, a positive association between leaf light environment and nitrogen concentration will enhance canopy-level photosynthetic NUE. This hypothesis is supported by numerous studies relating gradients of light, nitrogen, and photosynthetic capacity in plants of different growth forms and habitats (e.g., Mooney et al., 1981; Hirose & Werger, 1987; Field, 1988; Hirose, Werger & van Rheenan, 1989; Ackerly, 1992).

Theoretical studies suggest that the gradients of nitrogen concentrations and assimilation rates should be steeper when the *absolute* range of light levels (the difference between the highest and lowest PFD) increases, even if the *relative* range (the ratio of highest to lowest) is constant (Ackerly, 1993). This is important because shading reduces available light by a certain proportion, and as ambient light levels increase, that proportion will represent a greater absolute reduction. Thus, even for similar morphologies, nitrogen allocation gradients should be steeper in high light and hence may be steeper in pioneers. This prediction was supported in an experimental study of intraspecific variation in *Heliocarpus appendiculatus* in which gradients of leaf-level PFD, nitrogen concentration (per unit mass) and light saturated assimilation rate (per unit area) were steeper in high than low light conditions (Ackerly & Bazzaz, 1995a). Pioneer species also exhibit significantly higher resorption of nitrogen than non-pioneers (Reich, Ellsworth & Uhl, 1994), which may reflect a greater importance of non-uniform nitrogen allocation in the canopies of high light species.

The theory of optimal nitrogen allocation addresses the static problem of how to maximize daily carbon gain, given a particular distribution of PFD availability among leaves in a canopy. In the short-term, NUE will be maximized by removing nitrogen from shaded leaves and dropping them, such that all leaves are in high light positions. However, long-term nitrogen use efficiency (NUE) depends critically on leaf life span, as total carbon assimilation per unit nitrogen in a leaf will increase if the leaf is maintained on the plant longer. These two conditions introduce a trade-off between short-term and long-term NUE, as plants with longer leaf life span will also have more leaves in shaded positions (Field, 1988; Fredeen & Field, Chapter 20). Two species, *Piper auritum* and *P. hispidum*, illustrate contrasting solutions to this trade-off. *P. auritum* is restricted to high light sites, and there is little overlap among leaves, so instantaneous NUE is very high. Avoidance of self-shading depends on short leaf life span, however, so the high photosynthetic NUE of each leaf is maintained for only a short time. *P. hispidum*, in contrast, occupies a broader range of light environments and there is greater self-shading among leaves (Chazdon, Williams & Field, 1988), so instantaneous daily NUE of the canopy is reduced. But this is compensated by longer leaf life span such that lifetime NUE of each leaf is enhanced (Field, 1988; cf. Fredeen & Field, Chapter 20). This comparison illustrates the importance of the coupling between leaf-level physiology and the structure and dynamics of the canopy in order to understand the mechanisms of whole-plant carbon balance.

Empirical evidence provides qualitative support for the predictions of nitrogen allocation theory, but the physiological mechanisms underlying these patterns are not clear. Three different factors may be involved in nitrogen allocation among leaves, and resorption during the leaf life cycle: 1) the chronological age of a leaf, 2) leaf position on the shoot, relative to the apex and other leaves, and 3) the ambient light environment of the leaf. Each of these factors changes progressively during the life cycle of a leaf, but the relationships among them vary under different circumstances. For example, the series of leaves along a shoot always bear the same relative order in both age and position, but the quantitative relationship will depend on the phenology of leaf production. In evergrowing plants, the age of a leaf will equal the product of its position (counting from the apex) and the plastochron interval, so in shoots or plants with different leaf birth rates, the age of leaves in any particular position will differ. In terms of self-shading patterns, leaf position impacts directly on light environment, while leaf age has no direct effect *per se*, but again the exact relationship depends on shoot orientation, leaf size and shape, internode length and the spatio-temporal structure of the ambient light environment (Niklas, 1988, 1989; Ackerly, 1993; Ackerly & Bazzaz, 1995b).

Several approaches have been taken to distinguish the relative role of these factors. In the vine *Syngonium podophyllum*, which exhibits alternating periods of upward and downward growth, leaf nitrogen concentration was correlated with light level, and not with leaf position, supporting a direct effect of relative light availability (Ackerly, 1992). A similar result was obtained for a species of *Carex* in which, due to the basal meristem of the leaves, the older leaf tissue is in a higher light environment (Hirose, Werger & van Rheenen 1989). In *Heliocarpus*, experimental alteration of the light gradient such that light levels of lower leaves were reduced in one treatment caused a reduction in leaf life span, suggesting a direct effect of light on senescence, but this was not accompanied by a change in the nitrogen gradient (Ackerly & Bazzaz, 1995a). In the study of pioneer species introduced above, Traw and Ackerly (1995) took a statistical approach, conducting a path analysis to test the relative contributions of leaf position and light level. In four of the five species, leaf position had a strong direct effect on nitrogen concentration, while direct effects of light environment were weak; variation in specific leaf area also explained a significant portion of variation in leaf nitrogen concentration in all species. These results suggest that the decline in leaf nitrogen concentration at lower leaf positions is an ontogenetic pattern that may have evolved due to the association between leaf position and light environment, but it does not appear to be a direct response to light gradients. Thus, there is no one answer regarding the relative importance of direct responses to the light environment vs. ontogenetic patterns associated with leaf position along the shoot. It is likely that the role of these two factors varies among species and depends on the amount of variation in the light environments of different leaves within an individual.

21.3.6 Economic and optimality models of leaf life span and canopy structure

Economic models of leaf life span have focused on the relative costs and benefits of maintaining a leaf through an unfavorable period (e.g., Chabot & Hicks, 1982). As mentioned above, these models fail to distinguish leaf life span from leaf habit of the canopy and do not apply to the question of optimal leaf life span in evergrowing evergreens, such as tropical pioneers, growing in a continuously favorable environment.

Williams, Field and Money (1989), in their study of seven rainforest species of *Piper*, started with the proposition that leaves that are more expensive should be longer lived in order to 'pay back' their own cost. However, they observed the opposite: species occupying high light environments had shorter leaf life span, higher construction costs, and also higher assimilation rates. The ratio of construction cost to assimilation rate, a measure of the 'payback time' required to assimilate as much carbon as was invested in the leaf, was positively correlated with leaf life span. Interestingly, the payback time represented only about 10% of leaf life span in short-lived leaves, but increased to over 100% in longer-lived leaves, indicating that these leaves were either operating below their lifetime compensation point or that costs were overestimated or assimilation rates underestimated. A full cost accounting of leaf construction and maintenance must also consider the cost of support structures, root and transport tissue to supply water, losses to herbivory and pathogens, and related factors (Givnish, 1984, 1988) which would increase the payback time even further.

Kikuzawa (1991) presented a formal economic model for optimal leaf life span in a continuously favorable environment as part of his analysis of why evergreen plants have two peaks of abundance, in the tropics and at high latitudes. His model is based on maximization of net carbon gain of a leaf in relation to its initial net assimilation rate at full expansion on a mass basis (a_0), construction cost (c), and the leaf age at which assimilation reaches zero (b), assuming a linear decline following full expansion. Based on these parameters, optimal life span in a continuously favorable environment is:

$$L_{opt} = \left(\frac{2bc}{a_0} \right)^{1/2} \tag{8}$$

Note that this expression resembles those derived above (e.g., Equation 2 as it includes the ratio of construction cost to assimilation rates and is qualitatively consistent with available evidence in that respect. However, this model has two important implications. First, the model provides no prediction of optimal life span if the decline in assimilation rate is not incorporated (i.e., if b = ∞). This is intuitively reasonable because if a leaf never loses its assimilatory potential, why should it ever be discarded? Thus, for continuously favorable environments, some parameter describing declining assimilation over time may be necessary to construct predictive models of leaf life span.

Secondly, recall that b is the leaf age at which assimilation reaches zero due to physiological senescence. However, the predicted value for optimal leaf life span may be considerably shorter than this measure of physiological death. For example, substituting several reasonable values for the leaf of a pioneer tree into Equation (8) (a_0 = 0.15 g glucose g^{-1} leaf d^{-1}, c = 1.5 g glucose g^{-1} leaf, b = 100 days) generates the prediction that optimal leaf life span is 45 days. The implication of this prediction is that the leaf would be dropped when its assimilation rate is still 55% of maximum (based on the assumption of a linear decline in assimilation following full explansion). Empirical evidence, however, strongly suggests that leaves are maintained until assimilation has dropped much further and is approaching zero (e.g., Ackerly & Bazzaz, 1995a), and I have elaborated this carbon gain model further to address this discrepancy (Ackerly, 1993).

In place of a single leaf, let us consider the carbon economy of a leaf cluster, as discussed above, in relation to height growth of a shoot. In addition, as argued above, suppose that the decline in leaf assimilation rate is not dependent on leaf age, but on the position of a leaf on a shoot, and define a parameter z, analogous to b, which is the leaf position at which assimilation reaches 0. The magnitude of z will depend on the morphology of the leaf cluster and resulting patterns of self-shading, coupled with the species photosynthetic physiology and nitrogen allocation gradients. Finally, consider an alternative optimality criterion based on maximization of the rate of leaf production at the shoot apex, an important component of rapid height growth. The relationship between carbon gain and leaf production requires consideration of F, the proportion of assimilate allocated to new leaf production. This height-growth model generates predictions for both optimal leaf number and, indirectly, leaf life span on a shoot. Optimal leaf number is equal to z, the leaf position at which assimilation reaches zero, and hence the model predicts that leaves are abscised when they reach null carbon balance. Optimal leaf life span is:

$$L_{opt} = \frac{2c}{F_{max} a_0} \tag{9}$$

In other words, leaf production rates are maximized by investing as much as possible in new leaf production, consistent with requirements for new root and support tissue; this allocation level, along with assimilation rates and tissue construction costs, will determine leaf life span (cf. Equation (4) (see Ackerly, 1993 for full presentation of this model, including other combinations of senescence patterns and optimality criteria).

This model was tested in a field study at Los Tuxtlas in March of 1993 by examining leaf carbon balance and dynamics in 16 individuals, 1–2 m tall, of 4 species of pioneers, *Carica papaya, Cecropia obtusifolia, Hampea nutricia* and *Heliocarpus appendiculatus* (D. Ackerly, unpublished data). On each plant, daily carbon balance on five leaves, from upper to lower positions, was determined by measurement of photosynthesis at 3 or 4 times during the day, and 1 to 4 times at night, for a total of over 500 individual measurements. Leaf production and life span were determined from a second census of the plants a month after the photosynthetic measurements were obtained. The highest daily assimilation rate measured on each plant was taken as an estimate of a_0, regressions of assimilation rate on leaf age and position were used to estimate b and z, respectively, and construction cost was estimated from the mass of leaves, petioles, and internodes, with a respiratory cost of 0.5 g glucose g^{-1} leaf and 0.2 g glucose g^{-1} of petiole and stem (cf. Griffin, 1994). Daily assimilation rate at leaf death was calculated from the regression of assimilation rate on leaf position, based on the observed leaf number on each plant. Optimal leaf life span and assimilation at leaf death were predicted under the carbon gain model of Kikuzawa, calculated from Equations (3) and (4), and compared with predictions for optimal leaf number

(N = z) and assimilation at leaf death ($a_x = 0$) under the height growth model just presented.

For the carbon gain model, there was a positive correlation between observed and predicted values, for both leaf life span and assimilation at leaf death, but for life span observed values were significantly greater than predicted, and for assimilation they were significantly lower than predicted. For the height growth model, in contrast, observed values of leaf number corresponded closely with model predictions, though they were slightly lower on average. Predicted values of assimilation at leaf death ranged from -0.06 to 0.1, measured as a proportion of maximum assimilation rates, with a mode just above 0. Thus, these observations support the idea that leaves are maintained until they reach a null carbon balance. This result provides support for a model of leaf dynamics that incorporates self-shading and effects of leaf position rather than chronological age as the primary factors influencing leaf senescence.

21.4   QUESTIONS FOR FUTURE RESEARCH

Despite advances in our understanding of growth processes in pioneer trees, many questions remain. Below, I suggest a variety of questions for future research.

1. The physiological ecology of large saplings and mature trees in tropical forests has received little attention, and is a difficult topic due to their physical size and the long time-scale of many processes. What are the dynamic patterns of biomass allocation in later stages of the life cycle? Is there a significant trade-off between reproductive output and continued rapid height growth? What are the primary mortality factors, and would alternative allocation strategies reduce the influence of these factors? If so, at what cost to growth and reproduction? The demographic consequences of altering patterns of allocation and growth can be explored using matrix models of population dynamics, requiring a synthesis of ecophysiology and population biology.

2. Study of the dynamics of plant canopies has focused on turnover of leaf populations. What are the dynamics of the larger components of the crown? How long do branches live? What is the distribution of branch life span during the life cycle of a tree? What determines which new shoots will persist and become large branches, and which will die and fall off? What are the physiological mechanisms regulating

branch production, growth, and abscission, how do such mechanisms evolve, and what are the evolutionary consequences of variation in branching processes?

3. As discussed in section 21.3.1, there is considerable architectural variation within the guild of pioneer trees in the tropical forests throughout the world. This diversity is often overshadowed by the dramatic convergence on the large-leaved, umbrella morphology in species of *Cecropia, Musanga,* and *Macaranga.* What are the functional consequences of this variation for growth, competition, and reproductive success? Is the variation in architectural types associated with differences in life history, or in species' nutrient or water requirements, and if so, how are morphology and life history related? If there is not a consistent association, how can the diversity of morphology within communities be explained (cf. Givnish, 1987)?

4. Finally, the synthesis of dynamic and static models in canopy ecophysiology remains an outstanding challenge. For example, the theory of nitrogen allocation predicts the optimal, instantaneous distribution of nitrogen to maximize daily carbon gain. How are these gradients of nitrogen related to the temporal processes of nitrogen reallocation among leaves? The model presented in section 21.3.6 generated predictions of optimal leaf dynamics based on the pattern of declining assimilation with leaf position. Can this assimilation pattern be predicted from knowledge of shoot morphology and self-shading patterns, coupled with optimal nitrogen allocation theory? What are the trade-offs between instantaneous and long-term nitrogen use efficiency (Field, 1988), and what determines the outcome of these trade-offs in different species? These are just a few of the outstanding questions that emerge at the interface of morphological and physiological approaches to the study of plant canopies. They are by no means unique to the study of pioneer trees. For these and other questions, the rapid and continuous growth of tropical pioneer species provide an important model for the study of plant function, a foundation for understanding the additional complexity introduced by climatic seasonality, extreme environments, and more complex selective regimes of long-lived species.

ACKNOWLEDGEMENTS

This chapter is dedicated to Fakhri A. Bazzaz, in acknowledgment of his unflagging inspiration and support. I am grateful to the many ecologists whose research has made this review possible, in particular E. Alvarez-Buylla, C. B. Field, K. Kikuzawa, D. A. King, M. Martínez-Ramos, and P. B. Reich. I thank the editors for the invitation to contribute this chapter, and P. S. Ashton, F. A. Bazzaz, R. L. Chazdon, R. Gonzalez, S. S. Mulkey, S. Strauss-Debenedetti, P. Wayne, and an anonymous reviewer for critical comments and answers to queries that greatly improved the final product. Research leading to this review was supported by Harvard University and by a Doctoral Dissertation Improvement Grant from the National Science Foundation.

REFERENCES

ACKERLY, D. D. (1992) Light, leaf age, and leaf nitrogen concentration in a tropical vine. *Oecologia*, **89**, 596-600.

ACKERLY, D. D. (1993) *Phenotypic plasticity and the scale of environmental heterogeneity: Studies of tropical pioneer trees in variable light environments.* Ph. D. thesis, Harvard University, Cambridge, Massachusetts.

ACKERLY, D. D. & BAZZAZ, F. A. (1995a) Leaf dynamics, self-shading, and carbon gain in seedlings of a tropical pioneer tree. *Oecologia*, **101**, 289-296.

ACKERLY, D. D. & BAZZAZ, F. A. (1995b) Seedling crown orientation and interception of diffuse radiation in tropical forest gaps. *Ecology*, **76**, 1134-1146.

ALVAREZ-BUYLLA, E. R. & MARTÍNEZ-RAMOS, M. (1992) Demography and allometry of *Cecropia obtusifolia*, a neotropical pioneer tree – An evaluation of the climax-pioneer paradigm for tropical rainforests. *Journal of Ecology*, **80**, 275-290.

ASHTON, P. S. (1978) Crown characteristics of tropical trees. *Tropical Trees as Living Systems* (eds. P. B. TOMLINSON & M. H. ZIMMERMAN) Cambridge University Press, Cambridge, pp 591-615.

AUGSPURGER, C. K. (1984) Light requirements of neotropical tree seedlings: A comparative study of growth and survival. *Journal of Ecology*, **72**, 777-795.

BAZZAZ, F. A. (1979) The physiological ecology of plant succession. *Annual Review of Ecology and Systematics*, **10**, 351-371.

BAZZAZ, F. A. (1984) Dynamics of wet tropical forests and their species strategies. *Physiological Ecology of Plants of the Wet Tropics* (eds. E. MEDINA, H. A. MOONEY & C. VÁZQUEZ-YANES) Dr W. Junk Publishers, The Hague, pp 233-243.

BAZZAZ, F. A. (1991) Regeneration of tropical forests: Physiological responses of pioneer and secondary species. *Rainforest Regeneration and Management* (eds. A. GÓMEZ-POMPA, T. C. WHITMORE & M. HADLEY) Parthenon Publishing Group, Parkridge, New Jersey, pp 91-117.

BAZZAZ, F. A. & PICKETT S. T. A. (1980) Physiological ecology of tropical succession: A comparative review. *Annual Review of Ecology and Systematics*, **11**, 287-310.

BONGERS, F. & POPMA, J. (1990) Leaf dynamics of seedlings of rain-forest species in relation to canopy gaps. *Oecologia*, **82**, 122-127.

BOOJH, R. & RAMAKRISHNAN, P. S. (1982) Growth strategy of trees related to successional status II. Leaf dynamics. *Forest Ecology and Management*, **4**, 375-386.

BREITSPRECHER, A. & BETHEL, J. S. (1990) Stem-growth periodicity of trees in a tropical wet forest of Costa Rica. *Ecology*, **71**, 1156-1164.

BROKAW, N. V. L. (1985) Treefalls, regrowth, and community structure in tropical forests. *The Ecology of Natural Disturbance and Patch Dynamics* (eds. S. T. A. PICKETT & P. S. WHITE) Academic Press, New York, pp 53-69.

BROKAW, N. V. L. (1987) Gap-phase regeneration of three pioneer species in a tropical forest. *Journal of Ecology*, **75**, 9-19.

CHABOT, B. F. & HICKS, D. J. (1982) The ecology of leaf life spans. *Annual Review of Ecology and Systematics*, **13**, 229-259.

CHAZDON, R. L. (1985) Leaf display, canopy structure, and light interception of two understory palm species. *American Journal of Botany*, **72**, 1493-1502.

CHAZDON, R. L. (1986) The costs of leaf support in understory palms: Economy versus safety. *American Naturalist*, **127**, 9-30.

CHAZDON, R. L., WILLIAMS, K. & FIELD, C. B. (1988) Interactions between crown structure and light environment in five rainforest *Piper* species. *American Journal of Botany*, **75**, 1459-1471.

COLEMAN, J. S., MCCONNAUGHAY, K. D. M. & ACKERLY, D. D. (1994) Interpreting phenotypic variation in plants. *Trends in Ecology and Evolution*, **9**, 187-191.

COLEY, P. D. (1983) Herbivory and defensive characteristics of tree species in a lowland tropical forest. *Ecological Monographs*, **53**, 209-233.

COLEY, P. D. (1988) Effects of plant growth rate and leaf lifetime on the amount and type of anti-herbivore defense. *Oecologia*, **74**, 531-536.

COOMBE, D. E. (1960) An analysis of the growth of *Trema guineensis*. *Journal of Ecology*, **48**, 219-231.

COOMBE, D. E. & HADFIELD, W. (1962) An analysis of the growth of *Musanga cecropioides*. *Journal of Ecology*, **50**, 221-234.

CORNER, E. J. H. (1966) The Natural History of Palms. University of California Press, Berkeley, California.

DE FORESTA, H. (1983) Le spectre architectural: Application à l'étude des relations entre architecture des arbres et écologie forestière. *Bulletin du Museum National D'histoire Naturelle. 4E Ser, Section B, Adansonia*, **5**, 295-302.

EVANS, G. C. (1972) *The Quantitative Analysis of Plant Growth*. University of California Press, Berkeley, California.

EWEL, J. (1980) Tropical succession: Manifold routes to maturity. *Biotropica*, **12S**, 2-7.

FETCHER, N., OBERBAUER, S. F., ROJAS, G. & STRAIN, B. R. (1987) Efectos del régimen de luz sobre la fotosíntesis y el crecimiento en plántulas de árboles de un bosque lluvioso tropical de Costa Rica. *Revista de Biologia Tropical*, **35**, (**Suppl. 1**), 97-110.

FIELD, C. B. (1983) Allocating leaf nitrogen for the maximization of carbon gain: Leaf age as a control on the allocation program. *Oecologia*, **56**, 341-347.

FIELD, C. B. (1988) On the role of photosynthetic responses in constraining the habitat distribution of rainforest plants. *Australian Journal of Plant Physiology*, **15**, 343-358.

FISHER, J. B. (1986) Branching patterns and angles in trees. *On the Economy of Plant Form and Function* (eds. T. J. GIVNISH) Cambridge University Press, Cambridge, pp 493-524.

FRANKIE, G. W., BAKER, H. G. & OPLER, P. A. (1974) Comparative phenological studies of trees in tropical wet and dry forests in the lowlands of Costa Rica. *Journal of Ecology*, **62**, 881-919.

GILL, A. M. & TOMLINSON, P. B. (1971) Studies on the growth of red mangrove (*Rhizophora mangle* L.) 3. Phenology of the shoot. *Biotropica*, **3**, 109-124.

GIVNISH, T. J. (1984) Leaf and canopy adaptations in tropical forests. *Physiological Ecology of Plants of the Wet Tropics* (eds. E. MEDINA, H. A. MOONEY & C. VÁZQUEZ-YANES) Dr W. Junk Publishers, The Hague, pp 51-84.

GIVNISH, T. J. (ed) (1986) *On the Economy of Plant Form and Function* Cambridge University Press, Cambridge.

GIVNISH, T. J. (1987) Comparative studies of leaf form: Assessing the relative roles of selective pressures and phylogenetic constraints. *New Phytologist*, **106S**, 131-160.

GIVNISH, T. J. (1988) Adaptation to sun and shade: A whole-plant perspective. *Australian Journal of Plant Physiology*, **15**, 63-92.

GRIFFIN, K. L. (1994) Calorimetric estimates of construction cost and their use in ecological studies. *Functional Ecology*, **8**, 551-562.

HALLÉ, F., OLDEMAN, R. A. A. & TOMLINSON, P. B. (1978) *Tropical Trees and Forests: An Architectural Analysis*. Springer-Verlag, Berlin.

HIROSE & T., WERGER, M. J. A. (1987) Nitrogen use efficiency in instantaneous and daily photosynthesis of leaves in the canopy of a *Solidago altissima* stand. *Physiologia Plantarum*, **70**, 215-222.

HIROSE, T., WERGER, M. J. A. & VAN RHEENEN, J. W. A. (1989) Canopy development and leaf nitrogen distribution in a stand of *Carex acutiformis*. *Ecology*, **70**, 1610-1618.

HORN, H. (1971) *The Adaptive Geometry of Trees.* Princeton University Press, Princeton, New Jersey.

IBARRA-MANRIQUEZ, G. & OYAMA, K. (1992) Ecological correlates of reproductive traits of Mexican rain forest trees. *American Journal of Botany,* **79,** 383-394.

KIKUZĀWA, K. (1983) Leaf survival of woody plants in deciduous broad-leaved forests. 1. Tall trees. *Canadian Journal of Botany,* **61,** 2133-2139.

KIKUZAWA, K. (1984) Leaf survival of woody plants in deciduous broad-leaved forests. 2. Small trees and shrubs. *Canadian Journal of Botany,* **62,** 2551-2556.

KIKUZAWA, K. (1989) Ecology and evolution of phenological pattern, leaf longevity, and leaf habit. *Evolutionary Trends in Plants,* **3,** 105-110.

KIKUZAWA, K. (1991) A cost-benefit analysis of leaf habit and leaf longevity of trees and their geographical pattern. *American Naturalist,* **138,** 1250-1263.

KING, D. A. (1990) The adaptive significance of tree height. *American Naturalist,* **135,** 809-828.

KING, D. A. (1991) Correlations between biomass allocation, relative growth rate, and light environment in tropical forest saplings. *Functional Ecology,* **5,** 485-492.

KING, D. A. (1994) Influence of light level on the growth and morphology of saplings in a Panamanian forest. *American Journal of Botany,* **81,** 948-957.

KOHYAMA, T. (1987) Significance of architecture and allometry in saplings. *Functional Ecology,* **1,** 399-404.

KOIKE, F. (1986) Canopy dynamics estimated from shoot morphology in an evergreen broad-leaved forest. *Oecologia,* **70,** 348-350.

KORIBA, K. (1958) On the periodicity of tree-growth in the tropics, with reference to the mode of branching, the leaf-fall, and the formation of the resting bud. *Gardens Bulletin, Singapore,* **17,** 11-81.

LEIGH, E. G. (1975) Structure and climate in tropical rainforests. *Annual Review of Ecology and Systematics,* **6,** 67-86.

LOWMAN, M. D. (1992) Leaf growth dynamics and herbivory in five species of Australian rainforest canopy trees. *Journal of Ecology,* **80,** 433-447.

MAILLETTE, L. (1982) Structural dynamics of silver birch II. A matrix model of the bud populations. *Journal of Applied Ecology,* **19,** 219-238.

MARQUIS, R. J. (1992) A bite is a bite is a bite? Constraints on response to folivory in *Piper arieianum* (Piperaceae). *Ecology,* **73,** 143-152.

MARTÍNEZ-RAMOS, M. (1985) Claros, ciclos vitales de los árboles tropicales y regeneración natural de las selvas altas perennifolias. *Investigaciones Sobre la Regeneración de Selvas Altas en Veracruz, México. II* (eds. A. GÓMEZ-POMPA & R. S. DEL AMO) Edit. Alhambra Mexicana, Mexico City, Mexico, pp 191-240.

MARTÍNEZ-RAMOS, M., SARUKHÁN, J. & PIÑERO, D. (1988) The demography of tropical trees in the context of forest gap dynamics. *Plant Population Ecology* (eds. A. J. DAVY, M. J. HUTCHINGS & A. R. WATKINSON) Blackwell Scientific Publications, Oxford, pp 293-313.

MOAD, A. (1992) *Dipterocarp Juvenile Growth and Understory Light in Malaysian Tropical Forest*. Ph.D. thesis, Harvard University.

MOONEY, H. A. & GULMON, S. L. (1982) Constraints on leaf structure and function in reference to herbivory. *Bioscience*, **32**, 198-206.

MOONEY, H. A., FIELD, C., GULMON, S. L. & BAZZAZ, F. A. (1981) Photosynthetic capacity in relation to leaf position in desert versus old-field annuals. *Oecologia*, **50**, 109-112.

NEWSTROM, L. E., FRANKIE, G. W. & BAKER, H. G. (1994) A new classification for plant phenology based on flowering patterns in lowland tropical rainforest trees at La Selva, Costa Rica. *Biotropica*, **26**, 141-159.

NIKLAS K. J. (1988) The role of phyllotactic pattern as a "developmental constraint" on the interception of light by leaf surfaces. *Evolution*, **42**, 1-16.

NIKLAS, K. J. (1989) The effect of leaf-lobing on the interception of direct solar radiation. *Oecologia*, **80**, 59-64.

Niklas, K. J. (1992) *Plant Biomechanics*. Chicago University Press, Chicago.

NUÑEZ-FARFÁN, J. & DIRZO, R. (1991) Effects of defoliation on growth and survival of the saplings of a gap-colonizing neotropical tree. *Journal of Vegetation Science*, **2**, 459-464.

OBERBAUER, S. F. & DONNELLY, M. A. (1986) Growth analysis and successional status of Costa Rican rainforest trees. *New Phytologist*, **104**, 517-521.

OBERBAUER, S. F., CLARK, D. B., CLARK, D. A. & QUESADA, M. (1988) Crown light environments of saplings of two species of rainforest emergent trees. *Oecologia*, **75**, 207-212.

OKALI, D. U. U. (1971) Rates of dry-matter production in some tropical forest-tree species. *Annals of Botany*, **35**, 87-97.

OKALI, D. U. U. (1972) Growth-rates of some West African forest-tree seedlings in shade. *Annals of Botany*, **36**, 953-959.

OLDEMAN, R. A. A. & VAN DIJK, J. (1991) Diagnosis of the temperament of tropical rainforest trees. *Rainforest Regeneration and Management* (eds. A. GÓMEZ-POMPA, T. C. WHITMORE & M. HADLEY) Parthenon Publishing Group, Parkridge, New Jersey, pp 21-65.

ORIANS, G. H. (1982) The influence of tree-falls in tropical forests on tree species richness. *Tropical Ecology*, **23**, 255-279.

PIÑERO, D., MARTÍNEZ-RAMOS, M., MENDOZA, A., ALVAREZ-BUYLLA, E. & SARUKHAN, J. (1986) Demographic studies in *Astrocaryum mexicanum* and their use in understanding community dynamics. *Principes*, **30**, 108-116.

PIÑERO, D., SARUKHAN, J. & GONZÁLEZ, E. (1977) Estudios demograficos en plantas. *Astrocaryum mexicanum* Liebm. I. Estructura de las poblaciones. *Boletín de la Sociedad Botánica de México*, **37**, 69-118.

POPMA, J. & BONGERS, F. (1988) The effect of canopy gaps on growth and morphology of seedlings of rainforest species. *Oecologia*, **75**, 625-632.

REICH, P. B. (1993) Reconciling apparent discrepancies among studies relating lifespan, structure, and function of leaves in contrasting plant life forms and climates: "The blind men and the elephant retold". *Functional Ecology*, **7**, 721-725.

REICH, P. B., ELLSWORTH, D. S. & UHL, C. (1994) Leaf carbon and nutrient assimilation and conservation in species of differing successional status in an oligotrophic Amazonian forest. *Functional Ecology* in press.

REICH, P. B., UHL, C., WALTERS, M. B. & ELLSWORTH, D. S. (1991) Leaf lifespan as a determinant of leaf structure and function among 23 Amazonian tree species. *Oecologia*, **86**, 16-24.

REICH, P. B., WALTERS, M. B. & ELLSWORTH, D. S. (1992) Leaf lifespan in relation to leaf, plant, and stand characteristics among diverse ecosystems. *Ecological Monographs*, **62**, 365-392.

REICH, P. B., WALTERS, M. B., ELLSWORTH, D. S. & UHL, C. (1994) Photosynthesis-nitrogen relations in Amazonian tree species I. Patterns among species and communities. *Oecologia*, **97**, 62-72.

RUEDA, R. & WILLIAMSON, G. B. (1992) Radial and vertical wood specific gravity in *Ochroma pyramidale* (Cav. ex Lam.) Urb. (Bombacaceae). *Biotropica*, **24**, 512-518.

SALDARRIAGA, J. G. & LUXMOORE, R. J. (1991) Solar energy conversion efficiencies during succession of a tropical forest in Amazonia. *Journal of Tropical Ecology*, **7**, 233-242.

SARUKHÁN, J., PIÑERO, D. & MARTÍNEZ-RAMOS, M. (1985) Plant demography: A community-level interpretation. *Studies on Plant Demography: A Festschrift for John L. Harper* (ed. J. WHITE), Academic Press, London, pp 17-31.

SHUKLA, R. P. & RAMAKRISHNAN, P. S. (1982) Phenology of trees in a subtropical humid forest in northeastern India. *Vegetatio*, **49**, 103-109.

SHUKLA, R. P. & RAMAKRISHNAN, P. S. (1984a) Biomass allocation strategies and productivity of tropical trees related to successional status. *Forest Ecology and Management*, **9**, 315-324.

SHUKLA, R. P. & RAMAKRISHNAN, P. S. (1984b) Leaf dynamics of tropical trees related to successional status. *New Phytologist*, **97**, 697-706.

SOBRADO, M. A. (1991) Cost-benefit relationships in deciduous and evergreen leaves of tropical dry forest species. *Functional Ecology*, **5**, 608-616.

STRAUSS-DEBENEDETTI, S. I. (1989) *Responses to Light in Tropical Moraceae of Different Successional Stages*. Ph. D. thesis, Yale University, New Haven, Connecticut.

SWAINE, M. D. & WHITMORE, T. C. (1988) On the definition of ecological species groups in tropical rainforests. *Vegetatio*, **75**, 81-86.

TOMLINSON, P. (1978) Branching and axis differentiation in tropical trees. *Tropical Trees as Living Systems* (eds. P. TOMLINSON & M. H. ZIMMERMAN), Cambridge University Press, Cambridge, pp 187-207.

TRAW, M. B. & ACKERLY, D. D. (1995) Leaf age, light levels and nitrogen allocation in five species of rainforest pioneer trees. *American Journal of Botany*, in press.

UNESCO (1978) *Tropical Forest Ecosystems*, UNESCO, Paris.

WHITMORE, T. C. & GONG, W.-K. (1983) Growth analysis of the seedlings of Balsa, *Ochroma lagopus*. *New Phytologist*, **95**, 305-312.

WIEMANN, M. C. & WILLIAMSON, G. B. (1988) Extreme radial changes in wood specific gravity in some tropical pioneers. *Wood and Fiber Science*, **20**, 344-349.

WIEMANN, M. C. & WILLIAMSON, G. B. (1989a) Radial gradients in the specific gravity of wood in some tropical and temperate trees. *Forest Science*, **35**, 197-210.

WIEMANN, M. C. & WILLIAMSON, G. B. (1989b) Wood specific gravity gradients in tropical dry and montane rainforest trees. *American Journal of Botany*, **76**, 924-928.

WILLIAMS, K., FIELD, C. B. & MOONEY, H. A. (1989) Relationships among leaf construction cost, leaf longevity, and light environments in rainforest plants of the genus *Piper*. *American Naturalist*, **133**, 198-211.

Index